Understanding Space
An Introduction to Astronautics

SPACE TECHNOLOGY SERIES

*This book is published as part of the Space Technology Series,
a cooperative activity of the United States Department of Defense and the National
Aeronautics and Space Administration.*

Wiley J. Larson
Managing Editor

From Kluwer and Microcosm Publishers:

Space Mission Analysis and Design Third Edition by Larson and Wertz.

Spacecraft Structures and Mechanisms: From Concept to Launch by Sarafin.

Reducing Space Mission Cost by Wertz and Larson.

From McGraw-Hill:

Understanding Space: An Introduction to Astronautics Second Edition by Sellers.

Space Propulsion Analysis and Design by Humble, Henry, and Larson.

Cost-Effective Space Mission Operations by Boden and Larson.

Fundamentals of Astrodynamics and Applications by Vallado.

Applied Modeling and Simulation: An Integrated Approach to Development and Operation by Cloud and Rainey.

Human Spaceflight: Mission Analysis and Design by Larson and Pranke.

The Lunar Base Handbook by Eckart.

Future Books in the Series:

Space Transportation Systems Design and Operations by Larson.

Visit the website:

www.understandingspace.com

Revised Second Edition
UNDERSTANDING SPACE
An Introduction to Astronautics

Jerry Jon Sellers

With Contributions by:
Williams J. Astore
Robert B. Giffen
Wiley J. Larson

all from
United States Air Force Academy

Editor
Douglas H. Kirkpatrick
Illustrated by:
Dale Gay
Text Design by:
Anita Shute

Space Technology Series

Boston   Burr Ridge, IL   Dubuque, IA   Madison, WI   New York   San Francisco   St. Louis
Bangkok   Bogotá   Caracas   Lisbon   London   Madrid
Mexico City   Milan   New Delhi   Seoul   Singapore   Sydney   Taipei   Toronto

UNDERSTANDING SPACE
An Introduction to Astronautics
Revised Second Edition

McGraw-Hill's Custom Publishing consists of products that are produced from camera-ready copy. Peer review, class testing, and accuracy are primarily the responsibility of the author(s).

1 2 3 4 5 6 7 8 9 0 WCK WCK 0 9 8 7 6 5 4 3

ISBN 0-07-294364-5

Editor: Judith Wetherington
Production Editor: Carrie Braun
Cover Design: Dale Gay
Text Design: Anita Shute
Printer/Binder: Quebecor World

This book is dedicated to the thousands of people who've devoted their lives to exploring space.

Understanding Space
An Introduction to Astronautics

Chapter 1   **Space in Our Lives**   **1**

Why Space?   3
    The Space Imperative   3
    Using Space   5

Elements of a Space Mission   13
    The Mission   13
    The Spacecraft   14
    Trajectories and Orbits   15
    Launch Vehicles   16
    Mission Operations Systems   18
    Mission Management and Operations   18
    The Space Mission Architecture in Action   21

Chapter 2   **Exploring Space**   **29**

Early Space Explorers   33
    Astronomy Begins   33
    Reordering the Universe   35

Entering Space   44
    The Age of Rockets   44
    Sputnik: The Russian Moon   46
    Armstrong's Small Step   49
    Satellites and Interplanetary Probes   51

Space Comes of Age   53
    Space International   53
    Space Science Big and Small   54
    The New High Ground   61
    The Future   62

Chapter 3   **The Space Environment**   **71**

Cosmic Perspective   73
    Where is Space?   73
    The Solar System   74
    The Cosmos   76

The Space Environment and Spacecraft   79
    Gravity   79
    Atmosphere   81
    Vacuum   82

	Micrometeoroids and Space Junk	84
	The Radiation Environment	85
	Charged Particles	86
	Living and Working in Space	**91**
	Free fall	91
	Radiation and Charged Particles	93
	Psychological Effects	95
Chapter 4	**Understanding Orbits**	**103**
	Orbital Motion	**105**
	Baseballs in Orbit	105
	Analyzing Motion	107
	Newton's Laws	**109**
	Weight, Mass, and Inertia	109
	Momentum	111
	Changing Momentum	114
	Action and Reaction	115
	Gravity	116
	Laws of Conservation	**123**
	Momentum	123
	Energy	124
	The Restricted Two-body Problem	**130**
	Coordinate Systems	130
	Equation of Motion	132
	Simplifying Assumptions	133
	Orbital Geometry	135
	Constants of Orbital Motion	**140**
	Specific Mechanical Energy	140
	Specific Angular Momentum	143
Chapter 5	**Describing Orbits**	**153**
	Orbital Elements	**155**
	Defining the Classic Orbital Elements (COEs)	156
	Alternate Orbital Elements	164
	Computing Orbital Elements	**167**
	Finding Semimajor Axis, a	167
	Finding Eccentricity, e	168
	Finding Inclination, i	168
	Finding Right Ascension of the Ascending Node, Ω	169
	Finding Argument of Perigee, ω	171
	Finding True Anomaly, ν	172
	Spacecraft Ground Tracks	**179**

Chapter 6   **Maneuvering In Space**   **191**

Hohmann Transfers   193

Plane Changes   203
    Simple Plane Changes   203
    Combined Plane Changes   205

Rendezvous   208
    Coplanar Rendezvous   209
    Co-orbital Rendezvous   213

Chapter 7   **Interplanetary Travel**   **221**

Planning for Interplanetary Travel   223
    Coordinate Systems   223
    Equation of Motion   224
    Simplifying Assumptions   224

The Patched-conic Approximation   227
    Elliptical Hohmann Transfer between Planets—
        Problem 1   230
    Hyperbolic Earth Departure—Problem 2   238
    Hyperbolic Planetary Arrival—Problem 3   243
    Transfer Time of Flight   248
    Phasing of Planets for Rendezvous   248

Gravity-assist Trajectories   252

Chapter 8   **Predicting Orbits**   **259**

Predicting an Orbit (Kepler's Problem)   261
    Kepler's Equation and Time of Flight   262

Orbital Perturbations   272
    Atmospheric Drag   273
    Earth's Oblateness   273
    Other Perturbations   277

Predicting Orbits in the Real World   280

Chapter 9   **Getting To Orbit**   **289**

Launch Windows and Times   291
    Launch Windows   291
    Launch Time   292

When and Where to Launch   298

Launch Velocity   308

Chapter 10   **Returning from Space: Re-entry**   **323**

Analyzing Re-entry Motion   325
    Trade-offs for Re-entry Design   325
    The Motion Analysis Process   327
    Re-entry Motion Analysis in Action   332

Options for Trajectory Design   335
    Trajectory and Deceleration   335
    Trajectory and Heating   337
    Trajectory and Accuracy   340
    Trajectory and the Re-entry Corridor   340

Options for Vehicle Design   342
    Vehicle Shape   342
    Thermal-protection Systems   345

Lifting Re-entry   350

Chapter 11   **Space Systems Engineering**   **359**

Space Mission Design   361
    The Systems Engineering Process   361
    Designing Payloads and Subsystems   368
    The Design Process   374

Remote-sensing Payloads   382
    The Electromagnetic Spectrum   383
    Seeing through the Atmosphere   385
    What We See   386
    Payload Sensors   388
    Payload Design   393

Chapter 12   **Space Vehicle Control Systems**   **401**

Control Systems   403

Attitude Control   407
    Having the Right Attitude   407
    Attitude Dynamics   410
    Disturbance Torques   413
    Spacecraft Attitude Sensors   417
    Spacecraft Attitude Actuators   422
    The Controller   429

Orbit Control   434
    Space Vehicle Dynamics   434
    Navigation—The Sensor   435
    Rockets—The Actuators   437
    Guidance—The Controller   438

Chapter 13   Spacecraft Subsystems   447

Communication and Data-handling Subsystem
(CDHS)   449
    System Overview   449
    Basic Principles   450
    Systems Engineering   457

Electrical Power Subsystem (EPS)   461
    Basic Principles   462
    Systems Engineering   474

Environmental Control and Life-support
   Subsystem (ECLSS)   480
    System Overview   480
    Basic Principles of Thermal Control   481
    Basic Principles of Life Support   488
    Systems Engineering   493

Structures and Mechanisms   498
    System Overview   498
    Basic Principles   501
    Systems Engineering   509

Chapter 14   Rockets and Launch Vehicles   531

Rocket Science   533
    Thrust   533
    The Rocket Equation   535
    Rockets   539

Propulsion Systems   560
    Propellant Management   561
    Thermodynamic Rockets   564
    Electrodynamic Rockets   574
    System Selection and Testing   576
    Exotic Propulsion Methods   578

Launch Vehicles   586
    Launch-vehicle Subsystems   586
    Staging   591

Chapter 15   Space Operations   607

Mission Operations Systems   609
    Spacecraft Manufacturing   610
    Operations   614
    Communication   617
    Satellite Control Networks   623

Mission Management and Operations 630
    Mission Teams 631
    Mission Management 638
    Spacecraft Autonomy 645

Chapter 16   **Using Space** **653**

The Space Industry 655
    Globalization 656
    Commercialization 657
    Capital Market Acceptance 660
    Emergence of New Industry Leaders 661

Space Politics 662
    Political Motives 662
    Laws, Regulations, and Policies 664

Space Economics 668
    Lifecycle Costs 668
    Cost Estimating 675
    Return on Investment 676
    The FireSat Mission 677

Appendix A   **Math Review** **685**

Trigonometry 685
    Trigonometric Functions 685
    Angle Measurements 687
    Spherical Trigonometry 687

Vector Math 689
    Definitions 689
    Vector Components 689
    Vector Operations 689
    Transforming Vector Coordinates 694

Calculus 695
    Definitions 695

Appendix B   **Units and Constants** **699**

Canonical Units 699
    Canonical Units for Earth 699
    Solar Canonical Units 700

Unit Conversions 701

Constants 706

Greek Alphabet 708

Appendix C   Derivations   **709**

Restricted Two-body Equation of Motion   709

Constants of Motion   710
    Proving Specific Mechanical Energy is Constant   710
    Proving Specific Angular Momentum is Constant   712

Solving the Two-body
Equation of Motion   714

Relating the Energy Equation
to the Semimajor Axis   718

The Eccentricity Vector   720

Deriving the Period Equation
for an Elliptical Orbit   723

Finding Position and Velocity
Vectors from COEs   724

$V_{burnout}$ in SEZ Coordinates   728

Deriving the Rocket Equation   729

Deriving the Potential Energy Equation and
Discovering the Potential Energy Well   731

Appendix D   Solar and Planetary Data   **733**

Physical Properties of the Sun   733

Physical Properties of the Earth   734

Physical Properties of the Moon   735

Planetary Data   737

Spheres of Influence for the Planets   738

Appendix E   Motion of Ballistic Vehicles   **741**

Equation of Motion   741
    Ground-track Geometry   742
    Trajectory Geometry   744
    Maximum Range   745
    Time of Flight   745
    Rotating-Earth Correction   747
    Error Analysis   747

**Appendix F   Answers to Numerical
Mission Problems**   **749**

Preface

This 2nd edition of *Understanding Space* gives us the opportunity to update and expand the discussions on the elements of space missions. Our goal is to give the reader a more comprehensive overview of space systems engineering and how we apply it to spacecraft subsystems, rockets, and operations systems. In this second edition we've updated everything, added about 20% new material and developed a full-color format, all to better help you understand (and enjoy!) space.

Space travel and exploration are exciting topics; yet, many people shy away from them because they seem complex. The study of astronautics and space missions can be difficult at times, but our goal in this book is to *bring space down to Earth*. If we're successful—and you'll be the judge—after studying this book you should understand the concepts and principles of spaceflight, space vehicles, launch systems, and space operations.

We want to help you understand space missions while developing enthusiasm and curiosity about this very exciting topic. We've been inspired by the thousands of people who've explored space—from the people who've studied and documented the heavens, to the people who've given their lives flying there. We hope to inspire you! Whether you're interested in engineering, business, politics, or teaching—you can make a difference. We need talented people to lead the way in exploring space, the stars, and galaxies, and you are our hope for the future.

This book is intended for use in a first course in astronautics, as well as a guide for people needing to understand the "big picture" of space. Practicing engineers and managers of space-related projects will benefit from the brief explanations of concepts. Even if you're a junior or senior in high school and have a strong background in physics and math, come on in—you'll do fine!

If you don't like equations—don't worry! The book is laid out so you can learn the necessary concepts from the text without having to read or manipulate the equations. The equations are for those of you who want to be more fully grounded in the basics of astronautics.

We've included helpful features in this book to make it easier for you to use. The first page of each chapter contains

- An outline of the chapter, so you know what's coming

- An "In This Chapter You'll Learn To. . ." box that tells you what you should learn in the chapter

- A "You Should Already Know. . ." box, so you can review material that you'll need to understand the chapter

Each section of the chapter contains

- An "In This Section You'll Learn To. . ." box that gives you learning outcomes

- A detailed section review which summarizes key concepts and lists key terms and equations

Within each chapter you'll find

- Key terms italicized and defined

- Full-color diagrams and pictures "worth a thousand words"

- Tables summarizing important information and concepts

- Key equations boxed

- Detailed, step-by-step solutions to real-world example problems

- "Astro Fun Facts" to provide interesting insights and space trivia

At the end of each chapter you'll find

- A list of references for further study

- Problems and discussion questions, so you can practice what you've read. Astronautics is not a spectator sport—the real learning happens when you actually do what you've studied.

- Mission profiles designed to give you insight on specific programs and a starting point for discussion

We hope these features help you learn how exciting space can be!

Acknowledgments

Books, like space missions, are a team effort. This book is the result of several years of effort by an international team of government, industry, and academic professionals. The Department of Astronautics, United States Air Force Academy, provided unwavering support for the project. Robert B. Giffen and Michael DeLorenzo, past and present Department Heads, respectively, furnished the time, encouragement, and resources necessary to complete this edition. The entire Department of Astronautics, most notably Dave Cloud and Jack Ferguson, along with many cadets reviewed numerous drafts and provided very useful comments and suggestions. Michael Caylor and his Understanding Space review team did an incredible job of reviewing and commenting on the contents, finding and correcting errors, and enhancing the presentation of the material. The review team included Dr. Werner Balogh, United Nations Office of Outer Space Affairs (and currently the Austrian Space Agency); Dr. Tarik Kaya, NASA Goddard Space Flight Center (and the International Space University); Ms. Elizabeth Bloomer, NASA Johnson Space Center; Vadim Zakirov, University of Surrey; and Dr. Gabrielle Belle, U.S. Air Force Academy. Thanks also goes to additional reviews and helpful comments provided by Ron Humble and Tim Lawrence. Connie Bryant did a great deal of picture scanning for us.

We'd especially like to thank our illustrator, Dale Gay, for his creative ideas and patience in creating several hundred full-color illustrations. These truly make the book come alive. The contributing authors—Bill Astore, Julie Chesley, and Bob Giffen—provided key expertise on important topics and helped make the book complete. Their names are on the chapters they contributed. McGraw-Hill was exceptionally helpful during the development and we'd thank our publisher, Margaret Hollander, for her patience and guidance.

For the new material on space systems engineering and subsystems, we're grateful for the help and support of the Surrey Space Centre and Surrey Satellite Technology, Ltd., U.K. We'd especially like to thank Craig Underwood for his advise on the FireSat nanosatellite concept, Maarten "Max" Meerman for all his help with the FireSat mechanical design, engineering drawings, and microsatellite photographs, and Martin Sweeting for making their contributions possible.

We'd also like to thank the NASA Public Affairs Offices at Johnson Space Center in Houston, Texas, Kennedy Space Center in Florida, Ames Research Center at Moffett Field, California, Marshall Space Flight Center in Huntsville, Alabama, and Jet Propulsion Laboratory in Pasadena, California, for their help with photographs.

Leadership, funding, and support essential to developing this book were also provided by the following organizations

- Space and Missile Center, Los Angeles Air Force Base, California, including most program offices

- U.S. Air Force Space Command and Air Force Research Laboratory

- Naval Research Laboratory, Office of Naval Research

- Army Laboratory Command

- National Aeronautics and Space Administration including Headquarters, Goddard Space Flight Center, Johnson Space Center, Kennedy Space Center, Glen Research Center, and Jet Propulsion Laboratory

- U.S. Departments of Commerce, Transportation, and Energy

- Industry sponsors including The Boeing Company and Lockheed Martin

- European Space Agency

Getting time and money to develop much-needed reference material is exceptionally difficult in the aerospace community. We are deeply indebted to the sponsoring organizations for their support and their recognition of the importance of projects such as this one.

The OAO Corporation and the National Northern Education Foundation (NNEF), in Colorado Springs, Colorado, also provided exceptional contract support for the project. Richard Affeld and Jerry Worden of OAO and David Nelson of NNEF were particularly helpful throughout the development period.

Again we owe special thanks to Anita Shute for literally making this book happen. She took our crude, often illegible drafts and sketches and created the product you'll be reading. Her creative ideas and talent are surpassed only by her hard work and patience!

We sincerely hope this book will be useful to you in your study of astronautics. We've made every effort to eliminate mathematical and factual errors, but some may have slipped by us. Please send any errors, omissions, corrections, or comments to us, so we can incorporate them in the next edition of the book. Good luck and aim for the stars!

April, 2000

Jerry Jon Sellers
Author and Editor

Douglas Kirkpatrick
Editor

Wiley J. Larson
Managing Editor

Department of Astronautics
United States Air Force Academy
USAF Academy, Colorado 80840
Voice: 719-333-4110   FAX: 719-333-3723
Email: Wiley.Larson@usafa.af.mil

From a cosmic perspective, Earth's a very small place. This view of the spiral galaxy M4414 shows what the Milky Way may look like, if we could get away and look back at it. *(Courtesy of the Association of Universities for Research in Astronomy, Inc./Space Telescope Science Institute)*

Space in Our Lives

1

In This Chapter You'll Learn to...

☛ List and describe the unique advantages of space and some of the missions that capitalize on them

☛ Identify the elements that make up a space mission

You Should Already Know...

❑ Nothing about space yet. That's why we're here!

Outline

1.1 Why Space?
The Space Imperative
Using Space

1.2 Elements of a Space Mission
The Mission
The Spacecraft
Trajectories and Orbits
Launch Vehicles
Mission Operations Systems
Mission Management and
    Operations
The Space Mission Architecture
    in Action

Space. The Final Frontier. These are the voyages of the Starship Enterprise. Its continuing mission—to explore strange new worlds, to seek out new life and new civilizations, to boldly go where no one has gone before!

Star Trek—The Next Generation

Why study space? Why should you invest the considerable time and effort needed to understand the basics of planet and satellite motion, rocket propulsion, and spacecraft design—this vast area of knowledge we call astronautics? The reasons are both poetic and practical.

The poetic reasons are embodied in the quotation at the beginning of this chapter. Trying to understand the mysterious beauty of the universe, "to boldly go where no one has gone before," has always been a fundamental human urge. Gazing into the sky on a starry night, you can share an experience common to the entire history of humankind. As you ponder the fuzzy expanse of the Milky Way and the brighter shine and odd motion of the planets, you can almost feel a bond with ancient shepherds who looked at the same sky and pondered the same questions thousands of years ago.

The changing yet predictable face of the night sky has always inspired our imagination and caused us to ask questions about something greater than ourselves. This quest for an understanding of space has ultimately given us greater control over our destiny on Earth. Early star gazers contemplated the heavens with their eyes alone and learned to construct calendars enabling them to predict spring flooding and decide when to plant and harvest. Modern-day star gazers study the heavens with sophisticated ground and space instruments, enabling them to push our understanding of the universe far beyond what the unaided eye can see, such as in Figure 1.1.

To study space, then, is to grapple with questions as old as humanity. Understanding how the complex mechanisms of the universe work gives us a greater appreciation for its graceful and poetic beauty.

While the practical reasons for studying space are much more down to Earth, you can easily see them, as well, when you gaze at the night sky on a clear night. The intent sky watcher can witness a sight that only the current generation of humans has been able to see—tiny points of light streaking across the background of stars. They move too fast to be stars or planets. They don't brighten and die out like meteors or "falling stars." This now common sight would have startled and terrified ancient star gazers, for they're not the work of gods but the work of people. They are spacecraft. We see sunlight glinting off their shiny surfaces as they patiently circle Earth. These reliable fellow travellers with Earth enable us to manage resources and communicate on a global scale.

Since the dawn of the Space Age only a few decades ago, we have come to rely more and more on spacecraft for a variety of needs. Daily weather forecasts, instantaneous worldwide communication, and a constant ability to record high-resolution images of vital regions are all examples of space technology that we've come to take for granted. Studying space offers us a chance to understand and appreciate the complex requirements of this technology. See Figure 1-2.

Throughout this book, we'll focus primarily on the practical aspects of space—What's it like? And how do we get there? How do we use space for our benefit? In doing so, we hope to inspire a keen appreciation and sense of poetic wonder about the mystery of space—the final frontier.

Figure 1-1.   Hubble Image. This photo of a star forming region is one of thousands sent to Earth by the Hubble Space Telescope. *(Courtesy of the Association of Universities for Research in Astronomy, Inc./Space Telescope Science Institute)*

Figure 1-2.   Milstar Communication Satellite. The state-of-the-art Milstar satellite system provides worldwide communications to thousands of users simultaneously. *(Courtesy of the U.S. Air Force)*

1.1 Why Space?

▦ In This Section You'll Learn To...

- ☛ List and describe the advantages offered by space and the unique space environment
- ☛ Describe current space missions

The Space Imperative

Getting into space is dangerous and expensive. So why bother? Space offers several compelling advantages for modern society

- A global perspective—the ultimate high ground
- A clear view of the heavens—unobscured by the atmosphere
- A free-fall environment—enabling us to develop advanced materials impossible to make on Earth
- Abundant resources—such as solar energy and extraterrestrial materials
- A unique challenge as the final frontier

Let's explore each of these advantages in turn to see their potential benefit to Earth.

Space offers a global perspective. As Figure 1-3 shows, the higher you are, the more of Earth's surface you can see. For thousands of years, kings and rulers took advantage of this fact by putting lookout posts atop the tallest mountains to survey more of their realm and warn of would-be attackers. Throughout history, many battles have been fought to "take the high ground." Space takes this quest for greater perspective to its ultimate end. From the vantage point of space, we can view large areas of Earth's surface. Orbiting spacecraft can thus serve as "eyes and ears in the sky" to provide a variety of useful services.

Space offers a clear view of the heavens. When we look at stars in the night sky, we see their characteristic twinkle. This twinkle, caused by the blurring of "starlight" as it passes through the atmosphere, we know as *scintillation.* The atmosphere blurs some light, but it blocks other light altogether, which frustrates astronomers who need access to all the regions of the electromagnetic spectrum to fully explore the universe. By placing observatories in space, we can get instruments above the atmosphere and gain an unobscured view of the universe, as depicted in Figure 1-4. The Hubble Space Telescope, the Gamma Ray Observatory, and the Chandra X-ray Observatory are all armed with sensors operating far beyond the range of human senses. Results using these instruments from the unique vantage point of space are revolutionizing our understanding of the cosmos.

Figure 1-3.   A Global Perspective. Space is the ultimate high ground; it allows us to view large parts of Earth at once for various applications. *(Courtesy of Analytical Graphics, Inc.)*

Figure 1-4.   Space Astronomy. Earth's atmosphere obscures our view of space, so we put satellites, like the Hubble Space Telescope, above the atmosphere to see better. *(Courtesy of NASA/Johnson Space Center)*

Figure 1-5. Free-fall Environment. Astronauts Duque (right-side up) and Lindsey (upside down) enjoy the free fall experience on STS-95. *(Courtesy of NASA/Johnson Space Center)*

Space offers a free-fall environment enabling manufacturing processes not possible on Earth's surface. To form certain new metal alloys, for example, we must blend two or more metals in just the right proportion. Unfortunately, gravity tends to pull heavier metals to the bottom of their container, making a uniform mixture difficult to obtain. But space offers the solution. A manufacturing plant in orbit (and everything in it) is literally falling toward Earth, but never hitting it. This is a condition known as free fall (NOT zero gravity, as we'll see later). In *free fall* there are no contact forces on an object, so, we say it is weightless, making uniform mixtures of dissimilar materials possible. We'll explore this concept in greater detail in Chapter 3. Unencumbered by the weight felt on Earth's surface, factories in orbit have the potential to create exotic new materials for computer components or other applications, as well as promising new pharmaceutical products to battle disease on Earth. Studying the effects of weightlessness on plant, animal, and human physiology also gives us greater insight into how disease and aging affect us (Figure 1-5).

Space offers abundant resources. While some people argue about how to carve the pie of Earth's finite resources into ever smaller pieces, others contend that we need only bake a bigger pie. The bounty of the solar system offers an untapped reserve of minerals and energy to sustain the spread of mankind beyond the cradle of Earth. Spacecraft now use only one of these abundant resources—limitless solar energy. But scientists have speculated that we could use lunar resources, or even those from the asteroids, to fuel a growing space-based economy. Lunar soil, for example, is known to be rich in oxygen and aluminum. We could use this oxygen in rocket engines and for humans to breathe. Aluminum is an important

Astro Fun Fact
Shot Towers

In the mid sixteenth century, Italian weapon makers developed a secret method to manufacture lead shot for use in muskets. Finding that gravity tended to misshape the shots when traditionally cast, the Italians devised a system that employed principles of free fall. In this process, molten lead was dropped through a tiny opening at a height of about 100 m (300 ft.) from a "shot tower." As the molten lead plummeted, it cooled into a near perfect sphere. At journey's end, the lead fell into a pool of cold water where it quickly hardened. As time passed, shot towers became common throughout Europe and the United States. More cost-effective and advanced methods have now replaced them.

Burrard, Sir Major Gerald. The Modern Shotgun Volume II: The Cartridge. London: Herbert Jenkins Ltd., 1955.

Deane. Deane's Manual of Fire Arms. London: Longman, Brown, Green, Longmans and Robers, 1858.

Contributed by Troy Kitch, the U.S. Air Force Academy

metal for various industrial uses. It is also possible that water ice may be trapped in eternally-dark craters at the Lunar poles. These resources, coupled with the human drive to explore, mean the sky is truly the limit!

Finally, space offers an advantage simply as a frontier. The human condition has always improved as new frontiers were challenged. As a stimulus for increased technological advances, and a crucible for creating greater economic expansion, space offers a limitless challenge that compels our attention. Many people have compared the challenges of space to those faced by the first explorers to the New World. European settlers explored the apparently limitless resources, struggling at first, then slowly creating a productive society out of the wilderness.

We're still a long way from placing colonies on the Moon or Mars. But already the lure of this final frontier has affected us. Audiences spend millions of dollars each year on inspiring movies such as *Star Wars*, *Star Trek*, *Independence Day*, and *Contact*. The Apollo Moon landings and scores of Space Shuttle flights have captured the wonder and imagination of people across the planet. NASA records thousands of hits per day on their Mars Mission websites. Future missions promise to be even more captivating as a greater number of humans join in the quest for space. For each of us "space" means something different, as illustrated in Figure 1-6.

Figure 1-6.   Space. Space is many things to many people. It's the wonder of the stars, rockets, spacecraft, and all the other aspects of the final frontier.

Using Space

Although we have not yet realized the full potential of space, over the years we've learned to take advantage of several of its unique attributes in ways that affect all of us. The most common space missions fall into four general areas

- Communications
- Remote sensing
- Navigation
- Science and exploration

Let's briefly look at each of these missions to see how they are changing the way we live in and understand our world.

Space-based Communications

In October 1945, scientist and science-fiction writer Arthur C. Clarke (author of classics, such as *2001: A Space Odyssey*) proposed an idea that would change the course of civilization.

One orbit, with a radius of 42,000 km, has a period (the time it takes to go once around the Earth) of exactly 24 hours. A body in such an orbit, if its plane coincided with that of the Earth's equator, would revolve with the Earth and would thus be stationary above the same spot ... [a satellite] in this orbit could be provided with receiving and transmitting equipment and could act as a repeater to relay transmissions between any two points on the hemisphere beneath .... A transmission received from any point on the hemisphere could be broadcast to the whole visible face of the globe. (From Wireless World [Canuto and Chagas, 1978].)

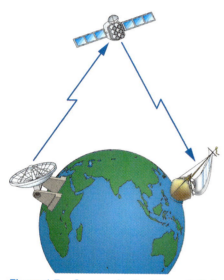

Figure 1-7. Communication Through Satellites. A satellite's global perspective allows users in remote parts of the world to talk to each other.

Figure 1-8. Satellite Communications. The explosion in satellite communication technology has shrunk the world and linked us more tightly together in a global community. Here we see a soldier sending a message through a portable ground station. *(Courtesy of Rockwell Collins)*

The information age was born. Clarke proposed a unique application of the global perspective space offers. Although two people on Earth may be too far apart to see each other directly, they can both "see" the same spacecraft in high orbit, as shown in Figure 1-7, and that spacecraft can relay messages from one point to another.

Few ideas have had a greater impact in shrinking the apparent size of the world. With the launch of the first experimental communications satellite, Echo I, into Earth orbit in 1960, Clarke's fanciful idea showed promise of becoming reality. Although Echo I was little more than a reflective balloon in low-Earth orbit, radio signals were bounced off it, demonstrating that space could be used to broaden our horizons of communication. An explosion of technology to exploit this idea quickly followed.

Without spacecraft, global communications as we know it would not be possible. We now use spacecraft for most commercial and governmental communications, as well as domestic cable television. Live television broadcasts by satellite from remote regions of the globe are now common on the nightly news. Relief workers in remote areas can stay in continuous contact with their home offices, enabling them to better distribute aid to desperate refugees. Figure 1-8 shows a soldier in the field sending a message via satellite. Military commanders now rely almost totally on spacecraft, such as the Defense Satellite Communication System and the Milstar system, to communicate with forces deployed worldwide.

Communication satellites have also been a boon to world development. Canuto and Chagas [1978] described how the launch of the Palapa A and

B satellites, for example, allowed the tropical island country of Indonesia to expand telephone service from a mere 625 phones in 1969 (limited by isolated population centers), to more than 233,000 only five years later (Figure 1-9). This veritable explosion in the ability to communicate has been credited with greatly improving the nation's economy and expanding its gross national product, thus benefiting all citizens. All this from only two satellites! Other developing nations have also realized similar benefits. Many credit the worldwide marketplace of ideas ushered in by satellites with the former Soviet Union's collapse and the rejection of closed, authoritarian regimes.

Today, a large collection of spacecraft in low-Earth orbit form a global cellular telephone network. With this network in place, anyone with one of the small portable phones can call any other telephone on the planet. Now, no matter where you go on Earth, you are always able to phone home. We can only imagine how this expanded ability to communicate will further shrink the global village.

Remote-sensing Satellites

Remote-sensing satellites use modern instruments to gather information about the nature and condition of Earth's land, sea, and atmosphere. Located in the "high ground" of space, these satellites use sensors that can "see" a broad area and report very fine details about the weather, the terrain, and the environment. The sensors receive electromagnetic emissions in various spectral bands that show what objects are visible, such as clouds, hills, lakes, and many other phenomena below. These instruments can detect an object's temperature and composition (concrete, metal, dirt, etc.), the wind's direction and speed, and environmental conditions, such as erosion, fires, and pollution. With these sophisticated satellites, we can learn much about the world we live in (Figure 1-10).

For decades, military "spy satellites" have kept tabs on the activities of potential adversaries using remote-sensing technology. These data have been essential in determining troop movements and violations of international treaties. During the Gulf War, for example, remote-sensing satellites gave the United Nations alliance a decisive edge. The United Nations' forces knew nearly all Iraqi troop deployments, whereas the Iraqis, lacking these sensors, didn't know where allied troops were. Furthermore, early-warning satellites, originally orbited to detect strategic missile launches against the United States, proved equally effective in detecting the launch of the smaller, Scud, missiles against allied targets. This early warning gave the Patriot antimissile batteries time to prepare for the Scuds.

Military remote-sensing technology has also had valuable civilian applications. The United States' Landsat and France's SPOT (Satellite Pour l'Observation de la Terre) systems are good examples. Landsat and SPOT satellites produce detailed images of urban and agricultural regions, as demonstrated in Figure 1-11 of Washington, D.C., and Figure 1-12 of Kansas. These satellites "spy" on crops, ocean currents, and natural resources to aid farmers, resource managers, and demographic planners.

Figure 1-9. Palapa A Coverage. The coverage for the Palapa A satellite means telephone service is available from one end of Indonesia to the other. *(Courtesy of Hughes Space and Communications Company)*

Figure 1-10. Earth Observation Satellite (EOS). This satellite takes high-resolution images of sites on Earth's surface. Government agencies and commercial firms use the images for many purposes, such as city planning and market growth analysis. *(Courtesy of NASA/Goddard Space Flight Center)*

Figure 1-11. City Planning from Space. Government officials can use remote-sensing images, such as this one from the Landsat spacecraft, for urban planning. In this image of Washington, D.C., specialists merged a Landsat image and a Mir space station photograph to get 2-m resolution. *(Courtesy of NASA/Goddard Space Flight Center)*

In countries where the failure of a harvest may mean the difference between bounty and starvation, spacecraft have helped planners manage scarce resources and head off potential disasters before insects or other blights could wipe out an entire crop. For example, in agricultural regions near the fringes of the Sahara desert in Africa, scientists used Landsat images to predict where locust swarms were breeding. Then, they were able to prevent the locusts from swarming, saving large areas of crop land.

Remote-sensing data can also help us manage other scarce resources by showing us the best places to drill for water or oil. From space, astronauts can easily see fires burning in the rain forests of South America as trees are cleared for farms and roads. Remote-sensing spacecraft have become a formidable weapon against the destruction of the environment because they can systematically monitor large areas to assess the spread of pollution and other damage.

Figure 1-12.   Two Views of Kansas. A remote-sensing image from the French SPOT satellite shows irrigated fields (the circular areas) in Kansas. Red means crops are growing. Light blue means the fields lay fallow. On the right, astronaut Joe Engle, a native of Kansas, gives his opinion of being in space. *(Courtesy of NASA/Goddard Space Flight Center and Johnson Space Center)*

Figure 1-13. Viewing Hurricanes From Space. It's hard to imagine a world without weather satellites. Images of hurricanes and other severe storms provide timely warning to those in their path and save countless lives every year. *(Courtesy of NASA/Johnson Space Center)*

Remote-sensing technology has also helped map makers. With satellite imagery, they can make maps in a fraction of the time it would take using a laborious ground survey. This enables city planners to better keep up with urban sprawl and gives deployed troops the latest maps of unfamiliar terrain.

National weather forecasts usually begin with a current satellite view of Earth. At a glance, any of us can tell which parts of the country are clear or cloudy. When they put the satellite map in motion, we easily see the direction of clouds and storms. An untold number of lives are saved every year by this simple ability to track the paths of hurricanes and other deadly storms, such as the one shown in Figure 1-13. By providing farmers valuable climatic data and agricultural planners information about potential floods and other weather-related disasters, this technology has markedly improved food production and crop management worldwide.

Overall, we've come to rely more and more on the ability to monitor and map our entire planet. As the pressure builds to better manage scarce resources and assess environmental damage, we'll call upon remote-sensing spacecraft to do even more.

Navigation Satellites

Satellites have revolutionized navigation—determining where you are and where you're going. The Global Positioning System (GPS), developed by the U.S. Department of Defense, and the GLONASS system, developed by the Russian Federation, use a small armada of satellites to help people, airplanes, ships, and ground vehicles navigate around the globe.

Besides supporting military operations, this system also offers incredible civilian applications. Surveyors, pilots, boaters, hikers, and many others who have a simple, low-cost receiver, can have instant information on where they are—with mind-boggling accuracy. With four satellites in view, as shown in Figure 1-14, it can "fix" a position to within a hundred meters. In fact, the biggest problem some users face is that the fix from GPS is more accurate than many maps!

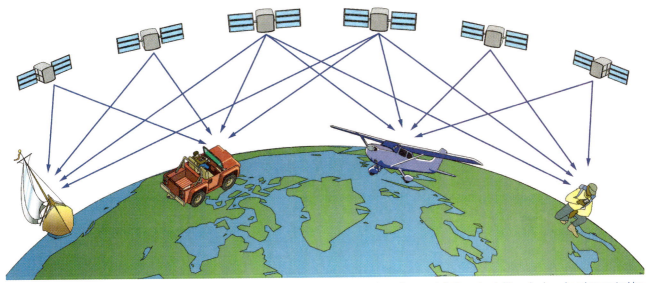

Figure 1-14.   Global Positioning System (GPS). The GPS space segment consists of a constellation of satellites deployed and operated by the U.S. Air Force. GPS has literally revolutionized navigation by providing highly accurate position, velocity, and time information to users on Earth.

Car manufacturers now offer GPS receivers as a standard feature on some models. Now you can easily find your way across a strange city without ever consulting a map. You simply put in the location you're trying to reach, and the system tells you how to get there. No more stops at gas stations to ask directions!

Science and Exploration Satellites

Since the dawn of the space age, scientists have launched dozens of satellites for purely scientific purposes. These mechanical explorers have helped to answer (and raise) basic questions about the nature of Earth, the solar system, and the universe. In the 1960s and 1970s, the United States launched the Pioneer series of spacecraft, to explore Venus, Mercury, and the Sun. The Mariner spacecraft flew by Mars to give us the first close-up view of the Red Planet. In 1976, two Viking spacecraft landed on Mars to do experiments designed to search for life on the one planet in our solar system whose environment most closely resembles Earth's. In the 1970s and 1980s, the Voyager spacecraft took us on a grand tour of the outer planets, beginning with Jupiter and followed by Saturn, Uranus, and Neptune. The Magellan spacecraft, launched in 1989, has mapped the surface of Venus beneath its dense layer of clouds, as shown in Figure 1-15. The first mobile Martian—the Sojourner rover—part of the Mars Pathfinder mission, fascinated Earthlings in 1997, as it made the first, tentative exploration of the Martian surface. The Hubble Space Telescope orbits Earth every 90 minutes and returns glorious images of our solar-system neighbors, as well as deep-space phenomena that greatly expand our knowledge, as shown in Figures 1-1 and 1-16. Although all of these missions have answered many questions about space, they have also raised many other questions which await future generations of robotic and human explorers.

Since the launch of cosmonaut Yuri Gagarin on April 12, 1961, space has been home to humans as well as machines. In the space of a generation, humans have gone from minute-long missions in cramped capsules to a year-long mission in a space station. The motivation for sending humans into space was at first purely political, as we'll see in Chapter 2. But scientific advances in exploration, physiology, material processing, and environmental observation have proved that, for widely varying missions, humans' unique ability to adapt under stress to changing conditions make them essential to mission success.

Future Space Missions

What does the future hold? In these times of changing world order and continuous budget fluctuation, it's impossible to predict. The International Space Station is under construction. The debate continues about sending humans back to the Moon, this time to stay, and then on to Mars, as shown in Figure 1-17.

As we become more concerned about damage to Earth's environment, we look to space for solutions. We continue to use our remote sensing satellites to monitor the health of the planet. Data from these satellites help us assess the extent of environmental damage and prepare better programs for cleaning up the environment and preventing future damage. Figure 1-18 shows one example of monitoring the environment from space. We use these images, taken by the Total Ozone Mapping Spectrometer, to track the

Figure 1-15.   Images of Venus. Magellan, another interplanetary spacecraft, has provided a wealth of scientific data. Using its powerful synthetic aperture radar to pierce the dense clouds of Venus, it has mapped the Venusian surface in detail. *(Courtesy of NASA/Jet Propulsion Laboratory)*

Figure 1-16.   Hubble Image. This photo of Spiral Galaxy M100 is one of thousands sent to Earth by the Hubble Space Telescope. *(Courtesy of the Association of Universities for Research in Astronomy, Inc./Space Telescope Science Institute)*

Figure 1-17.   Going to Mars. Future human missions may explore the canyons of Mars for signs that life may have once flourished there, only to be extinguished as the planet's atmosphere diminished. *(Courtesy of NASA/Ames Research Center)*

ozone concentration which protects us from harmful ultraviolet radiation. Concerns about its depletion have mobilized scientists to monitor and study it in greater detail using a variety of space-based sensors.

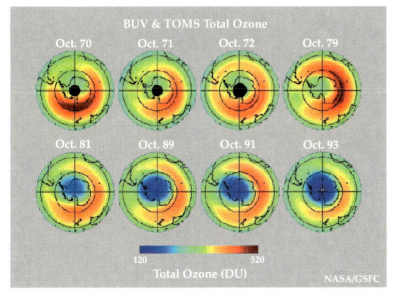

Figure 1-18.   Monitoring Ozone. Images from the Nimbus 4 Backscatter Ultraviolet (BUV) instrument for 1970–1973, and Total Ozone Mapping Spectrometer (TOMS) for 1979–1993, show variations in the ozone amounts over Antarctica. A DU is a Dobson Unit. 300 DUs is equivalent to a 3 mm thick layer of ozone at standard sea level atmospheric pressure. Black dots indicate no data. Note that the amount of dark blue (low total ozone) grows over the years. *(Courtesy of NASA/Goddard Space Flight Center)*

The International Space Station had its first module launched in November 1998, and will continue adding modules through 2004 (Figure 1-19). The research onboard this modern vessel will advance our understanding of life and help us improve our quality of life worldwide.

The manned mission to Mars continues to gain momentum, but overcoming the obstacles for this mission will take experts in many fields and several governments to commit resources. The rewards for exploring Mars are many and varied, including medical research, Martian resource evaluation, and scientific innovation.

The eventual course of the space program is very much up to you. Whether we continue to expand and test the boundaries of human experience or retreat from it depends on the level of interest and technical competence of the general public. By reading this book, you've already accepted the challenge to learn about space. In this study of astronautics you too can explore the final frontier.

Figure 1-19.   International Space Station (ISS). The ISS will provide a free fall laboratory for studying many aspects of spaceflight. *(Courtesy of NASA/Ames Research Center)*

▤ Section Review

Key Terms	Key Concepts

Key Terms

free fall
remote-sensing
scintillation

Key Concepts

➤ Space offers several unique advantages which make its exploration essential for modern society
  - Global perspective
  - A clear view of the universe without the adverse effects of the atmosphere
  - A free-fall environment
  - Abundant resources
  - A final frontier

➤ Since the beginning of the space age, missions have evolved to take advantage of space
  - Communications satellites tie together remote regions of the globe
  - Remote-sensing satellites observe the Earth from space, providing weather forecasts, essential military information, and valuable data to help us better manage Earth's resources
  - Navigation satellites revolutionize how we travel on Earth
  - Scientific spacecraft explore the Earth and the outer reaches of the solar system and peer to the edge of the universe
  - Manned spacecraft provide valuable information about living and working in space and experiment with processing important materials

1.2 Elements of a Space Mission

In This Section You'll Learn to...

☞ Identify the elements common to all space missions and how they work together for success

Now that you understand a little more about why we go to space, let's begin exploring how. In this section we introduce the basic building blocks, or elements, of space missions. These elements form the basis for our exploration of *astronautics* (the science and technology of spaceflight) in the rest of the book.

When you see a weather map on the nightly news or use the phone to make an overseas call, you may not think about the complex network of facilities that make these communications possible. If you think about space missions at all, you may picture an ungainly electronic box with solar panels and antennas somewhere out in space—a spacecraft. However, while a spacecraft represents the result of years of planning, designing, building, and testing by a veritable army of engineers, managers, operators, and technicians, it is only one small piece of a vast array of technology needed to do a job in space.

We define the *space mission architecture*, shown in Figure 1-20, as the collection of spacecraft, orbits, launch vehicles, operations networks, and all other things that make a space mission possible. Let's briefly look at each of these elements to see how they fit together.

The Mission

At the heart of the space mission architecture is the mission. Simply stated, the *mission* is why we're going to space. All space missions begin with a need, such as the need to communicate between different parts of the world (Figure 1-21) or to monitor pollution in the upper atmosphere. This need creates the mission. Understanding this need is central to understanding the entire space mission architecture. For any mission, no matter how complex, we must understand the need well enough to write a succinct *mission statement* that tells us three things

- The mission *objective*—*why* do the mission

- The mission *users or customers*—*who* will benefit

- The mission *operations concept*—*how* the mission elements will work together

For example, the mission objective for a hypothetical mission to warn us about forest fires might look like this

> *Mission objective—Detect and locate forest fires
> worldwide and provide timely notification to users.*

Figure 1-21.   Iridium Phone. The need for a *global* cellular telephone service triggered the iridium commercial enterprise. With these hand-held phones, you can phone anyone on Earth from anywhere on Earth. *(Courtesy of Personal Satellite Network, Inc.)*

Figure 1-20. Space Mission Architecture. A space mission requires a variety of interrelated elements, collectively known as the space mission architecture. Central to the architecture is the Space Mission. Surrounding the architecture is the Mission Management and Operations.

This mission objective tells us the "why" of the mission: we'll explore the "who" and "how" of this space scenario in much greater detail starting in Chapter 11. For now, simply realize that we must answer each of these important questions before we can develop a cohesive mission architecture.

We'll begin investigating the elements of a space mission architecture by looking at the most obvious element—the spacecraft.

The Spacecraft

The word "spacecraft" may lead you to conjure up images of the starship *Enterprise* or sleek flying saucers from all those 1950s Sci-Fi movies. In reality, spacecraft tend to be more squat and ungainly than sleek and streamlined. The reasons for this are purely practical—we build spacecraft to perform a specific mission in an efficient, cost-effective manner. In the vacuum of space, there's no need to be streamlined. When it comes to spacecraft, form must follow function. In Chapter 11, we'll learn more about spacecraft functions, and resulting forms. For now, it's sufficient to understand that we can conceptually divide any spacecraft into two basic parts—the payload and the spacecraft bus.

The *payload* is the part of the spacecraft that actually performs the mission. Naturally, the type of payload a spacecraft has depends directly on the type of mission it's performing. For example, the payload for a mission to monitor Earth's ozone layer could be an array of scientific sensors, each designed to measure some aspect of this life-protecting chemical compound (Figure 1-22). As this example illustrates, we design payloads to interact with the primary focus for the mission, called the *subject*. In this example, the subject would be the ozone. If our mission objective were to monitor forest fires, the subject would be the fire and we would design spacecraft payloads that could detect the unique characteristics or "signature" of forest fires, such as their light, heat, or smoke. As we'll see in Chapter 11, understanding the subject, and its unique properties, are critical to designing space payloads to detect or interact with them.

Figure 1-22.   Upper Atmospheric Research Satellite (UARS). The payloads for the UARS are sensitive instruments, which take images of various chemicals in Earth's atmosphere. *(Courtesy of NASA/Goddard Space Flight Center)*

The spacecraft bus does not arrive every morning at 7:16 to deliver the payload to school. But the functions performed by a spacecraft bus aren't that different from those a common school bus does. Without the spacecraft bus, the payload couldn't do its job. The *spacecraft bus* provides all the "housekeeping" functions necessary to make the payload work. The bus includes various subsystems that produce and distribute electrical power, maintain the correct temperature, process and store data, communicate with other spacecraft and Earth-bound operators, control the spacecraft's orientation, and hold everything together (Figure 1-23). We'll learn more about spacecraft bus design in Chapter 11 and explore the fundamentals of all bus subsystems in Chapter 12–14. It's the spacecraft's job to carry out the mission, but it can't do that unless it's in the right place at the right time. The next important element of the space mission architecture is concerned with making sure the spacecraft gets to where it needs to go.

Figure 1-23.   Defense Satellite Communication System (DSCS). The spacecraft bus for this DSCS III spacecraft provides power, attitude control, thermal control, and communication with mission operators. *(Courtesy of the U.S. Air Force)*

Trajectories and Orbits

A *trajectory* is the path an object follows through space. In getting a spacecraft from the launch pad into space, a launch vehicle follows a carefully-chosen ascent trajectory designed to lift it efficiently out of Earth's atmosphere. Once in space, the spacecraft resides in an *orbit*. We'll look at orbits in great detail in later chapters, but for now it's useful to think of an orbit as a fixed "racetrack" on which the spacecraft travels around a planet or other celestial body. Similar to car racetracks, orbits usually have an oval shape, as shown in Figure 1-24. Just as planets orbit the Sun, we can place spacecraft into orbit around Earth.

When selecting an orbit for a particular satellite mission, we need to know where the spacecraft needs to point its instruments and antennas.

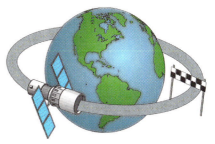

Figure 1-24.   The Orbit. We can think of an orbit as a fixed racetrack in space that the spacecraft drives on. Depending on the mission, this racetrack's size, shape, and orientation will vary.

We can put a spacecraft into one of a limitless number of orbits, but we must choose the orbit which best fulfills the mission. For instance, suppose our mission is to provide continuous communication between New York and Los Angeles. Our subject—the primary focus for the mission—is the communication equipment located in these two cities, so we want to position our spacecraft in an orbit that allows it to always see both cities. The orbit's size, shape, and orientation determine whether the payload can observe these subjects and carry out the mission.

Just as climbing ten flights of stairs takes more energy than climbing only one, putting a spacecraft into a higher (larger) orbit requires more energy, meaning a bigger launch vehicle and greater expense. The orbit's size (height) also determines how much of Earth's surface the spacecraft instruments can see. Naturally, the higher the orbit, the more total area they can see at once. But just as our eyes are limited in how much of a scene we can see without moving them or turning our head, a spacecraft payload has similar limitations. We define the payload's *field-of-view (FOV)*, as shown in Figure 1-25, to be the cone of visibility for a particular sensor. Our eyes, for example, have a useful field of view of about 204 degrees, meaning without moving our eyes or turning our head, we can see about 204 degrees of the scene around us. Depending on the sensor's field of view and the height of its orbit, a specific total area on Earth's surface is visible at any one time. We call the linear width or diameter of this area the *swath width*, as shown in Figure 1-25.

Some missions require continuous coverage of a point on Earth or the ability to communicate simultaneously with every point on Earth. When this happens, a single spacecraft can't satisfy the mission need. Instead, we build a fleet of identical spacecraft and place them in different orbits to provide the necessary coverage. We call this collection of cooperating spacecraft a *constellation*.

The Global Positioning System (GPS) mission requirement is a good example of one that requires a constellation of satellites to do the job. The mission statement called for every point on Earth be in view of at least four GPS satellites at any one time. This was impossible to do with just four satellites at any altitude. Instead, mission planners designed the GPS constellation to contain 24 satellites working together to continuously cover the world (Figure 1-26).

Another constellation of spacecraft called the Iridium System, provides global coverage for personal communications. This constellation of 66 satellites operates in low orbits. This new mobile telephone service is revolutionizing the industry with person-to-person phone links, meaning we can have our own, individual phone number and call any other telephone on Earth from virtually anywhere, at any time.

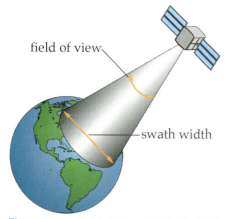

Figure 1-25. Field-of-View (FOV). The FOV of a spacecraft defines the area of coverage on Earth's surface, called the swath width.

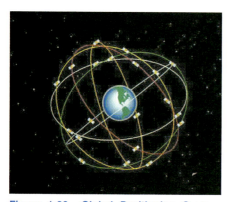

Figure 1-26. Global Positioning System (GPS) Constellation. The GPS constellation guarantees that every point on the globe receives at least four satellite signals simultaneously, for accurate position, velocity, and time computations. *(Courtesy of the National Air and Space Museum)*

Launch Vehicles

Now that we know where the spacecraft's going, we can determine *how* to get it there. As we said, it takes energy to get into orbit—the higher the orbit, the greater the energy. Because the size of a spacecraft's orbit determines its energy, we need something to deliver the spacecraft to the right mission orbit—a rocket. The thunderous energy released in a rocket's fiery blast-off provides the velocity for our spacecraft to "slip the surly bonds of Earth" (as John Gillespie Magee wrote in his poem, "High Flight") and enter the realm of space, as the Shuttle demonstrates in Figure 1-27.

A *launch vehicle* is the rocket we see sitting on the launch pad during countdown. It provides the necessary velocity change to get a spacecraft into space. At lift-off, the launch vehicle blasts almost straight up to gain altitude rapidly and get out of the dense atmosphere which slows it down due to drag. When it gets high enough, it slowly pitches over to gain horizontal velocity. As we'll see later, this horizontal velocity keeps a spacecraft in orbit.

As we'll see in Chapter 14, current technology limits make it very difficult to build a single rocket that can deliver a spacecraft efficiently into orbit. Instead, a launch vehicle consists of a series of smaller rockets that ignite, provide thrust, and then burn out in succession, each one handing off to the next one like runners in a relay race. These smaller rockets are *stages*. In most cases, a launch vehicle uses at least three stages to reach the mission orbit. For example, the Ariane V launch vehicle, shown in Figure 1-28, is a three-stage booster used by the European Space Agency (ESA).

For certain missions, the launch vehicle can't deliver a spacecraft to its final orbit by itself. Instead, when the launch vehicle finishes its job, it leaves the spacecraft in a parking orbit. A *parking orbit* is a temporary orbit where the spacecraft stays until transferring to its final mission orbit. After the spacecraft is in its parking orbit, a final "kick" sends it into a transfer orbit. A *transfer orbit* is an intermediate orbit that takes the spacecraft from its parking orbit to its final, mission orbit. With one more kick, the spacecraft accelerates to stay in its mission orbit and can get started with business, as shown in Figure 1-29.

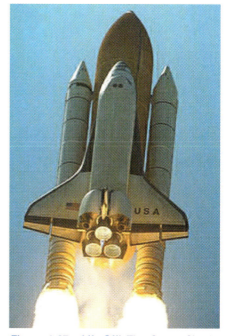

Figure 1-27. Lift Off! The Space Shuttle acts as a booster to lift satellites into low-Earth orbit. From there, an upperstage moves the satellite into a higher orbit. *(Courtesy of NASA/Johnson Space Center)*

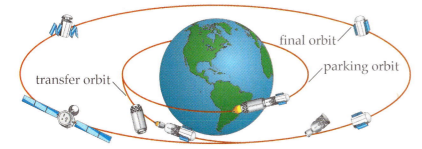

Figure 1-29. Space Mission Orbits. We use the booster primarily to deliver a spacecraft into a low-altitude parking orbit. From this point an upperstage moves the spacecraft into a transfer orbit, and then to the mission orbit.

The extra kicks of energy needed to transfer the spacecraft from its parking orbit to its mission orbit comes from an *upperstage*. In some cases, the upperstage is actually part of the spacecraft, sharing the plumbing and propellant which the spacecraft will use later to orient itself and maintain its orbit. In other cases, the upperstage is an autonomous

Figure 1-28. Ariane. The European Space Agency's Ariane V booster lifts commercial satellites into orbit. Here we see it lifting off from its pad in Kourou, French Guyana, South America. *(Courtesy of Service Optique CSG; Copyright by Arianespace/European Space Agency/CNES)*

Figure 1-30. The Inertial Upperstage (IUS). The IUS, attached to the Magellan spacecraft, boosted it to Venus. *(Courtesy of NASA/Johnson Space Center)*

spacecraft with the one-shot mission of delivering the spacecraft to its mission orbit. In the latter case, the upperstage releases the spacecraft once it completes its job, then moves out of the way by de-orbiting to burn up in the atmosphere or by raising its orbit a bit (and becoming another piece of space junk). Regardless of how it is configured, the upperstage consists mainly of a rocket engine (or engines) and the propellent needed to change the spacecraft's energy enough to enter the desired mission orbit. Figure 1-30 shows the upperstage used to send the Magellan spacecraft to Venus.

After a spacecraft reaches its mission orbit, it may still need rocket engines to keep it in place or maneuver to another orbit. These relatively small rocket engines are *thrusters* and they adjust the spacecraft's orientation and maintain the orbit's size and shape, both of which can change over time due to external forces. We'll learn more about rockets of all shapes and sizes in Chapter 14.

Mission Operations Systems

As you can imagine, designing, building, and launching space missions requires a number of large, expensive facilities. Communicating with and controlling fleets of spacecraft once they're in orbit requires even more expensive facilities. The *mission operations system* include the ground and space-based infrastructure needed to coordinate all other elements of the space mission architecture. It is the "glue" that holds the mission together.

As we'll see in Chapter 15, operations systems include manufacturing and testing facilities to build the spacecraft, launch facilities to prepare the launch vehicle and get it safely off the ground, and communication networks and operations centers used by the *flight-control team* to coordinate activities once it's in space.

One of the critical aspects of linking all these far-flung elements together is the communication process. Figure 1-31 shows the components of a typical communication network. Whether we're talking to our friend across a noisy room or to a spacecraft on the edge of the solar system, the basic problems are the same. We'll see how to deal with these problems in greater detail in Chapter 15.

Mission Management and Operations

So far, most of our discussion of space missions has focused on hardware—spacecraft, launch vehicles, and operations facilities. But while the mission statement may be the heart of the mission, and the hardware the tools, the mission still needs a brain. No matter how much we spend on advanced technology and complex systems there is still the need for people. People are the most important element of any space mission. Without people handling various jobs and services, all the expensive hardware is useless.

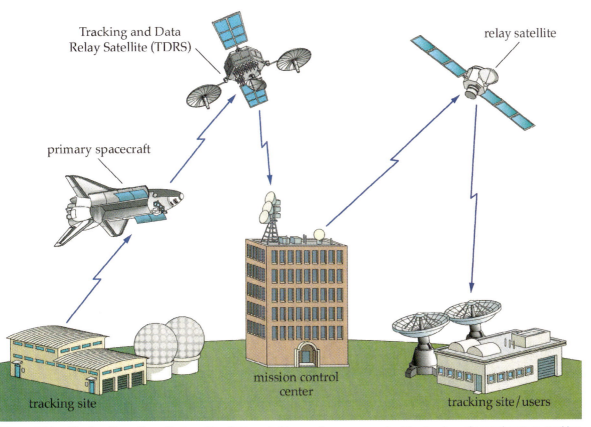

Tracking and Data
Relay Satellite (TDRS)

relay satellite

primary spacecraft

tracking site

mission control
center

tracking site/users

Figure 1-31.   Mission Operations System. The flight-control team relies on a complex infrastructure of control centers, tracking sites, satellites, and relay satellites to keep them in contact with spacecraft and users. In this example, data goes to the Space Shuttle from a tracking site, which relays it through another satellite, such as the Tracking and Data Relay Satellite (TDRS), back to the control center. The network then passes the data to users through a third relay satellite.

Hollywood tends to show us only the most "glamorous" space jobs—astronauts doing tasks during a space walk or diligent engineers hunched over computers in the Mission Control Center (Figure 1-32). But you don't have to be an astronaut or even a rocket scientist to work with space. Thousands of jobs in the aerospace industry require only a desire to work hard and get the job done. Many of these jobs are in space mission management and operations. *Mission management and operations* encompasses all of the "cradle to grave" activities needed to take a mission from a blank sheet of paper to on-orbit reality, to the time when they turn out the lights and everyone moves on to a new mission.

Mission managers lead the program from the beginning. The mission management team must define the mission statement and lay out a workable mission architecture to make it happen. Team members are involved with every element of the mission architecture, including

- Designing, building, and testing the spacecraft
- Performing complex analysis to determine the necessary mission orbit

Figure 1-32.   Mission Control Center. After several tense days, the mission control team at the Johnson Space Center watch the Apollo 13 crew arrive on the recovery ship after splashdown. *(Courtesy of NASA/Johnson Space Center)*

- Identifying a launch vehicle (or designing, building and testing a new one!) and launching the spacecraft to its mission orbit

- Bringing together the far-flung components of the mission operations systems to allow the flight-control team to run the entire mission

But mission management is far more than just technical support. From food services to legal services, a diverse and dedicated team is needed to get any space mission off the ground. It can take a vast army of people to manage thousands of separate tasks, perform accounting services, receive raw materials, ship products, and do all the other work associated with any space mission. Sure, an astronaut turning a bolt to fix a satellite gets his or her picture on the evening news, but someone had to make the wrench, and someone else had to place it in the toolbox before launch.

As soon as the spacecraft gets to orbit, mission operations begin. The first word spoken by humans from the surface of the Moon was "Houston." Neil Armstrong was calling back to the Mission Control Center at Johnson Space Center in Houston, Texas, to let them know the Eagle had successfully landed. To the anxious Flight Director and his operations team, that first transmission from the lunar surface was important "mission data." In the design of an operations concept to support our mission statement, we have to consider how we will collect, store, and deliver the mission data to users or customers on Earth. Furthermore, we have to factor in how the flight-control team will receive and monitor data on the spacecraft's health and to build in ground control for commanding the spacecraft's functions from the complex, minute to minute, activities on the Space Shuttle, to the far more relaxed activities for less complex, small satellites, as shown in Figure 1-33.

It would be nice if, once we deploy a spacecraft to its final orbit, it would work day after day on its own. Then users on Earth could go about their business without concern for the spacecraft's "care and feeding." Unfortunately, this automatic mode is not yet possible. Modern spacecraft, despite their sophistication, require a lot of attention from a team of flight controllers on the ground.

Figure 1-33. Small Satellite Ground Station. The size and complexity of the control center and flight-control team depends on the mission. Here a single operator controls over a dozen small satellites. *(Courtesy of Surrey Satellite Technology, Ltd., U.K.)*

The *mission operations team* monitors the spacecraft's health and status to ensure it operates properly. Should trouble arise, flight controllers have an arsenal of procedures they can use to nurse the spacecraft back to health. The flight-control team usually operates from a Mission (or Operations) Control Center (MCC or OCC) such as the one in Houston, Texas, used for United States' manned missions and shown in Figure 1-34. U.S. military operators and their contractor support teams control Department of Defense robotic satellites at similar MCCs (or OCCs) at Schriever Air Force Base, in Colorado Springs, Colorado, and Onizuka Air Station, Sunnyvale, California. A new OCC is under construction at Schriever AFB—the old Falcon AFB.

Within the mission's operation center, team members hold positions that follow the spacecraft's functional lines. For example, one person may monitor the spacecraft's path through space while another keeps an eye

Figure 1-34.   The Space Shuttle Mission Control Center in Houston, Texas. Space operations involves monitoring and controlling spacecraft from the ground. Here, flight controllers attend to their Guidance/Navigation, Propulsion, and Flight Dynamics consoles. *(Courtesy of NASA/Johnson Space Center)*

on the electrical-power system. The lead mission operator, called the *flight director* (*operations director* or *mission director*), orchestrates the inputs from each of the flight-control disciplines. Flight directors make decisions about the spacecraft's condition and the important mission data, based on recommendations and their own experience and judgment. We'll examine the specific day-to-day responsibilities of mission operators in greater detail in Chapter 15.

The Space Mission Architecture in Action

Now that we've defined all these separate mission elements, let's look at an actual space mission to see how it works in practice. NASA launched Space Shuttle mission STS-95 from the Kennedy Space Center (KSC) in Florida on October 29, 1998. The primary objectives of this mission were to deploy three science and engineering satellites, run experiments on human physiology, and operate microgravity tests. The three satellites were the Spartan 201 Solar Observer, the International Extreme Ultraviolet Hitchhiker Experiment, and the HST Optical Systems Test platform. In Figure 1-35, we show how all the elements for this mission tie together.

Throughout the rest of this book, we'll focus our attention on the individual elements that make up a space mission. We'll begin putting missions into perspective by reviewing the history of spaceflight in Chapter 2. Next, we'll set the stage for our understanding of space by exploring the unique demands of this hostile environment in Chapter 3. In Chapters 4–10, we'll consider orbits and trajectories to see how their behavior affects mission planning. In Chapters 11–13, we turn our attention to the spacecraft to learn how all payloads and their supporting subsystems tie together to make an effective mission. In Chapter 14 we'll focus on rockets to see how they provide the transportation to get

Communication Network

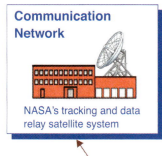

NASA's tracking and data relay satellite system

STS-95 Crew Members

Clockwise from top: Scott Parazynski, John Glenn, Curtis Brown, Steven Lindsey, Stephen Robinson, Pedro Duque, and Chiaki Mukai.

Trajectory and Orbits

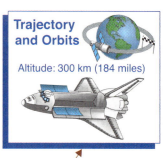

Altitude: 300 km (184 miles)

Launch Vehicle

The Space Shuttle delivers the crew and cargo to low-Earth orbit.

Spacecraft

Deploying the Spartan satellite.

Mission Operations Systems

Operation Concept: Ground controllers in Houston, Texas, monitor and support the Shuttle crew around the clock for this 10-day mission.

Figure 1-35.   STS-95 Space Mission Architecture. *(All photos courtesy of NASA/Johnson Space Center)*

spacecraft into space and move them around as necessary. Chapter 15 looks at the remaining two elements of a space mission—operations systems and mission management and operations. There we explore complex communication networks and see how to manage and operate successful missions. Finally, in Chapter 16, we look at trends in space missions, describe how space policy affects missions and how the bottom line, cost, affects everything we do in space.

▬ Section Review

Key Terms

astronautics
constellation
customers
field-of-view (FOV)
flight-control team
flight director
launch vehicle
mission
mission director
mission management and operations
mission operations system
mission operations team
mission statement
objective
operations concept
operations director
orbit
parking orbit
payload
space mission architecture
spacecraft bus
stages
subject
swath width
thrusters
trajectory
transfer orbit
upperstage
users

Key Concepts

➤ Central to understanding any space mission is the mission itself

- The mission statement clearly identifies the major objectives of the mission (why we do it), the users (who will benefit), and the operations concept (how all the pieces fit together)

➤ A space mission architecture includes the following elements

- The spacecraft—composed of the bus, which does essential housekeeping, and the payload, that performs the mission

- The trajectories and orbits—the path the spacecraft follows through space. This includes the orbit (or racetrack) the spacecraft follows around the Earth.

- Launch vehicles—the rockets which propel the spacecraft into space and maneuver it along its mission orbit

- The mission operations systems—the "glue" that holds the mission together. It consists of all the infrastructure needed to get the mission off the ground, and keep it there, such as manufacturing facilities, launch sites, communications networks, and mission operations centers.

- Mission management and operations—the brains of a space mission. An army of people make a mission successful. From the initial idea to the end of the mission, individuals doing their jobs well ensure the mission products meet the users' needs.

References

Canuto, Vittorio and Carlos Chagas. *The Impact of Space Exploration on Mankind*. Pontificaia Academia Scientiarum, proceedings of a study week held October 1–5, 1984, Ex Aedibus Academicis In Civitate Vaticana, 1986.

Wertz, James R. and Wiley J. Larson. *Space Mission Analysis and Design*. Third edition. Dordrecht, Netherlands: Kluwer Academic Publishers, 1999.

Wilson, Andrew (ed.), *Space Directory 1990–91*. Jane's information group. Alexandria, VA, 1990.

Mission Problems

1.1   Why Space?

1   What five unique advantages of space make its exploitation imperative for modern society?

2   What are the four primary space missions in use today? Give an example of how each has affected, or could affect, your life.

1.2   Elements of a Space Mission

3   The mission statement tells us what three things?

4   What are the elements of a space mission?

5   List the two basic parts of a spacecraft and discuss what they do for the mission.

6   What is an orbit? How does changing its size affect the energy required to get into it and the swath width available to any payload in this orbit?

7   What is a parking orbit? A transfer orbit?

8   Describe what an upperstage does.

9   Why do we say that the operations network is the "glue" that holds the other elements together?

10   What is the mission management and operations element?

11   Use this mission scenario to answer the following questions. NASA launches the Space Shuttle from Kennedy Space Center on a mission to deploy a spacecraft that will monitor Earth's upper atmosphere. Once deployed from the low Shuttle orbit, an inertial upperstage (IUS) will boost the spacecraft into its transfer orbit and then to its mission orbit. Once in place, it will monitor Earth's atmosphere and relay the data to scientists on Earth through the Tracking and Data Relay Satellite (TDRS).

a)   What is the mission of this Shuttle launch?

b)   What is the mission of the spacecraft?

c)   Discuss the spacecraft, trajectory, upperstage, and mission management and operations.

d)   Who are the mission users?

e)   What is the subject of the mission?

f)   What part does TDRS play in this mission?

g)   Briefly discuss ideas for the operations concept.

For Discussion

12 What future missions could exploit the free-fall environment of space?

13 What future space missions could exploit lunar-based resources?

14 You hear a television commentator say the Space Shuttle's missions are a waste of money. How would you respond to this charge?

Projects

15 Moderate a debate between sides for and against space exploration. Outline what points you'd expect each side to make.

16 Given the following mission statement, select appropriate elements to accomplish the task.

    Mission Statement: To monitor iceflows in the Arctic Ocean and warn ships in the area.

17 Obtain information from NASA on an upcoming space mission and prepare a short briefing on it to present to your class.

18 Write a justification for a manned mission to Mars. List and explain each element of the mission. Compile a list of skills needed by each member of the astronaut crew and the mission team.

Notes

Mission Profile—Voyager

The Voyager program consisted of two spacecraft launched by NASA in late 1977 to tour the outer planets, taking pictures and sensor measurements along the way. Voyager 2 actually launched a month prior to Voyager 1, which flew on a shorter, faster path. This shorter trajectory enabled Voyager 1 to arrive at the first planet, Jupiter, four months before Voyager 2. The timing of the operation was critical. Jupiter, Saturn, Uranus, and Neptune align themselves for such a mission only once every 175 years. The results from the Voyager program have answered and raised many basic questions about the origin of our solar system.

Mission Overview

NASA engineers designed the Voyager spacecraft with two objectives in mind. First, they built two identical spacecraft for redundancy. They feared that the available technology meant at least one of the spacecraft would fail. Second, they planned to visit only Jupiter and Saturn, with a possibility of visiting Neptune and Uranus, if the spacecraft lasted long enough. It was generally agreed that five years was the limit on spacecraft lifetimes. In the end, both spacecraft performed far better than anyone wildly imagined. Today they continue their voyage through empty space beyond our solar system, their mission complete.

Mission Data

The Voyager spacecraft used the gravity of the planets they visited to slingshot themselves to their next target. This gravity assist (described in Chap. 7) shortened each spacecraft's voyage by many years.

Voyager 1 headed into deep space after probing Saturn's rings. Voyager 2, however, successfully probed Neptune and Uranus, as well.

Voyager 1 discovered that one of Jupiter's moons, Io, has an active volcano spouting lava 160 km (100 mi.) into space.

Miranda, one of Uranus' 15 known moons, has been called the "strangest body in the solar system." Discovered by Voyager 2, it's only 480 km (300 miles) across and constantly churning itself inside-out. Scientists believe this is caused by the strong gravity from Uranus reacting with a process called differentiation (where the densest material on the moon migrates to the core). The result is a moon which looks like "scoops of marble-fudge ice cream"—the dense and light materials mixed randomly in jigsaw fashion.

Mission Impact

The overwhelming success of the Voyager mission has prompted a new surge of planetary exploration by NASA. Two of these are the Cassini mission to explore Saturn and the Galileo mission to study Jupiter. These two new missions by NASA will help to answer the new questions the Voyager missions have uncovered.

Voyager Mission. The Voyager spacecraft points its sensitive instruments toward Saturn and keeps its high-gain antenna directed at Earth. *(Courtesy of NASA/Jet Propulsion Laboratory)*

For Discussion

• The major problem with space exploration is exorbitant cost. Do you think the United States should spend more money on future exploratory missions? What about teaming up with other advanced countries?

• What is the benefit for humans to uncover the mysteries and perplexities of our solar system? Do you think there will be pay back in natural resources?

Contributor

Troy Kitch, the U.S. Air Force Academy

References

Davis, Joel. *FLYBY: The Interplanetary Odyssey of Voyager 2.* New York: Atheneum, 1987.

Evans, Barry. *The Wrong Way Comet and Other Mysteries of Our Solar System.* Blue Ridge Summit: Tab Books, 1992.

Vogt, Gregory. *Voyager.* Brookfield: The Millbrook Press, 1991.

Buzz Aldrin poses against the stark lunar landscape. Neil Armstrong can be seen reflected in his helmet. *(Courtesy of NASA/Johnson Space Center)*

Exploring Space

2

William J. Astore
the U.S. Air Force Academy

▦ In This Chapter You'll Learn to...

- ☛ Describe how early space explorers used their eyes and minds to explore space and contribute to our understanding of it
- ☛ Explain the beginnings of the Space Age and the significant events that have led to our current capabilities in space
- ☛ Describe emerging space trends, to include the growing commercialization of space

▦ You Should Already Know...

- ❑ Nothing yet; we'll explore space together

▦ Outline

2.1 Early Space Explorers
Astronomy Begins
Reordering the Universe

2.2 Entering Space
The Age of Rockets
Sputnik: The Russian Moon
Armstrong's Small Step
Satellites and Interplanetary
Probes

2.3 Space Comes of Age
Space International
Space Science Big and Small
The New High Ground
The Future

It is difficult to say what is impossible, for the dream of yesterday is the hope of today and the reality of tomorrow.

Robert H. Goddard

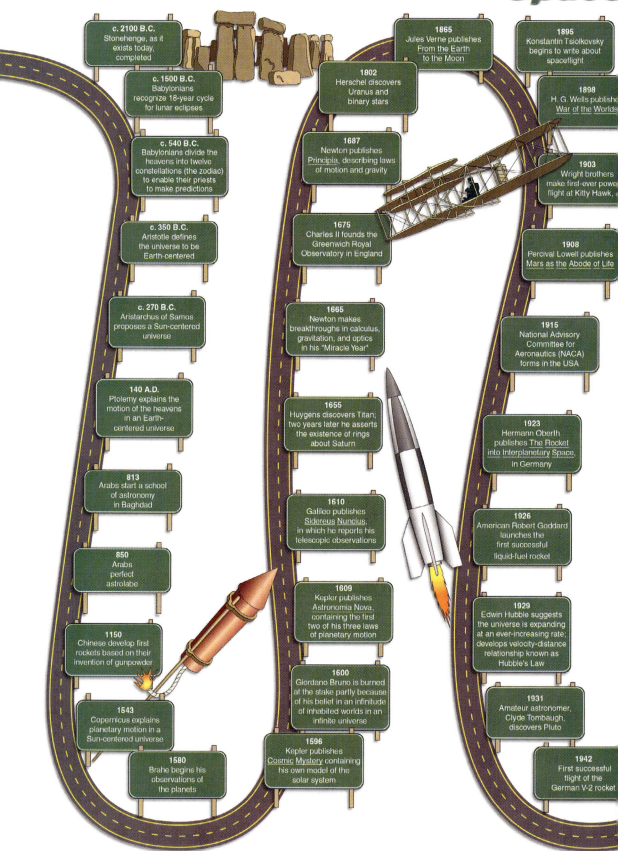

c. 2100 B.C.
Stonehenge, as it
exists today,
completed

1865
Jules Verne publishes
From the Earth
to the Moon

1895
Konstantin Tsiolkovsky
begins to write about
spaceflight

c. 1500 B.C.
Babylonians
recognize 18-year cycle
for lunar eclipses

1802
Herschel discovers
Uranus and
binary stars

1898
H. G. Wells publishes
War of the Worlds

c. 540 B.C.
Babylonians divide the
heavens into twelve
constellations (the zodiac)
to enable their priests
to make predictions

1687
Newton publishes
Principia, describing laws
of motion and gravity

1903
Wright brothers
make first-ever powered
flight at Kitty Hawk,

c. 350 B.C.
Aristotle defines
the universe to be
Earth-centered

1675
Charles II founds the
Greenwich Royal
Observatory in England

1908
Percival Lowell publishes
Mars as the Abode of Life

c. 270 B.C.
Aristarchus of Samos
proposes a Sun-centered
universe

1665
Newton makes
breakthroughs in calculus,
gravitation, and optics
in his "Miracle Year"

1915
National Advisory
Committee for
Aeronautics (NACA)
forms in the USA

140 A.D.
Ptolemy explains the
motion of the heavens
in an Earth-
centered universe

1655
Huygens discovers Titan;
two years later he asserts
the existence of rings
about Saturn

1923
Hermann Oberth
publishes The Rocket
into Interplanetary Space,
in Germany

813
Arabs start a school
of astronomy
in Baghdad

1610
Galileo publishes
Sidereus Nuncius,
in which he reports his
telescopic observations

1926
American Robert Goddard
launches the
first successful
liquid-fuel rocket

850
Arabs
perfect
astrolabe

1609
Kepler publishes
Astronomia Nova,
containing the first
two of his three laws
of planetary motion

1929
Edwin Hubble suggests
the universe is expanding
at an ever-increasing rate;
develops velocity-distance
relationship known as
Hubble's Law

1150
Chinese develop first
rockets based on their
invention of gunpowder

1600
Giordano Bruno is burned
at the stake partly because
of his belief in an infinitude
of inhabited worlds in an
infinite universe

1931
Amateur astronomer,
Clyde Tombaugh,
discovers Pluto

1543
Copernicus explains
planetary motion in a
Sun-centered universe

1596
Kepler publishes
Cosmic Mystery containing
his own model of the
solar system

1942
First successful
flight of the
German V-2 rocket

1580
Brahe begins his
observations of
the planets

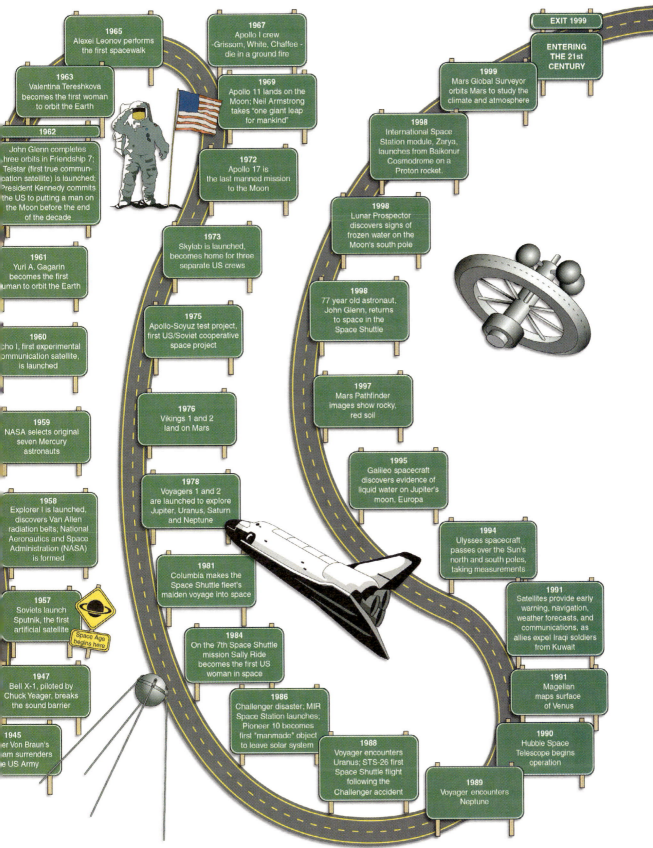

meline

1965
Alexei Leonov performs the first spacewalk

1967
Apollo I crew -Grissom, White, Chaffee - die in a ground fire

1963
Valentina Tereshkova becomes the first woman to orbit the Earth

1969
Apollo 11 lands on the Moon; Neil Armstrong takes "one giant leap for mankind"

1962
John Glenn completes three orbits in Friendship 7; Telstar (first true communication satellite) is launched; President Kennedy commits the US to putting a man on the Moon before the end of the decade

1972
Apollo 17 is the last manned mission to the Moon

1961
Yuri A. Gagarin becomes the first human to orbit the Earth

1973
Skylab is launched, becomes home for three separate US crews

1960
Echo I, first experimental communication satellite, is launched

1975
Apollo-Soyuz test project, first US/Soviet cooperative space project

1959
NASA selects original seven Mercury astronauts

1976
Vikings 1 and 2 land on Mars

1958
Explorer I is launched, discovers Van Allen radiation belts; National Aeronautics and Space Administration (NASA) is formed

1978
Voyagers 1 and 2 are launched to explore Jupiter, Uranus, Saturn and Neptune

1957
Soviets launch Sputnik, the first artificial satellite

Space Age begins here

1981
Columbia makes the Space Shuttle fleet's maiden voyage into space

1947
Bell X-1, piloted by Chuck Yeager, breaks the sound barrier

1984
On the 7th Space Shuttle mission Sally Ride becomes the first US woman in space

1945
her Von Braun's am surrenders he US Army

1986
Challenger disaster; MIR Space Station launches; Pioneer 10 becomes first "manmade" object to leave solar system

1988
Voyager encounters Uranus; STS-26 first Space Shuttle flight following the Challenger accident

1989
Voyager encounters Neptune

EXIT 1999

ENTERING THE 21st CENTURY

1999
Mars Global Surveyor orbits Mars to study the climate and atmosphere

1998
International Space Station module, Zarya, launches from Baikonur Cosmodrome on a Proton rocket.

1998
Lunar Prospector discovers signs of frozen water on the Moon's south pole

1998
77 year old astronaut, John Glenn, returns to space in the Space Shuttle

1997
Mars Pathfinder images show rocky, red soil

1995
Galileo spacecraft discovers evidence of liquid water on Jupiter's moon, Europa

1994
Ulysses spacecraft passes over the Sun's north and south poles, taking measurements

1991
Satellites provide early warning, navigation, weather forecasts, and communications, as allies expel Iraqi soldiers from Kuwait

1991
Magellan maps surface of Venus

1990
Hubble Space Telescope begins operation

You don't ever have to leave Earth to explore space. Long before rockets and interplanetary probes escaped Earth's atmosphere, people explored the heavens with their eyes and imagination. Later, with the aid of telescopes and other instruments, humans continued their quest to understand and bring order to the heavens. With order came a deeper understanding of humanity's place in the universe.

Thousands of years ago, the priestly classes of ancient Egypt and Babylon carefully observed the heavens to plan religious festivals, to control the planting and harvesting of various crops, and to understand at least partially the realm in which they believed many of their gods lived. Later, philosophers such as Aristotle and Ptolemy developed complex theories to explain and predict the motions of the Sun, Moon, planets, and stars.

The theories of Aristotle and Ptolemy dominated astronomy and our understanding of the heavens well into the 1600s. Combining ancient traditions with new observations and insights, natural philosophers such as Nicolaus Copernicus, Johannes Kepler, and Galileo Galilei offered rival explanations from the 1500s onward. Using their ideas and Isaac Newton's new tools of physics, astronomers in the 1700s and 1800s made several startling discoveries, including two new planets—Uranus (Figure 2-1) and Neptune (Figure 2-2). As we moved into the 20th century, physical exploration of space became possible. Advances in technology, accelerated by World War II, made missiles and eventually large rockets available, allowing us to escape Earth entirely. In this chapter, we'll follow the trailblazers who have led us from our earliest attempts to explore space to our explorations of the Moon and beyond.

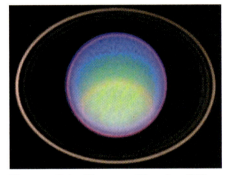

Figure 2-1. Uranus. Though William Herschel discovered Uranus in 1781, we didn't see it this well until the Hubble Space Telescope took this image in 1996. *(Courtesy of the Association of Universities for Research in Astronomy, Inc./Space Telescope Science Institute)*

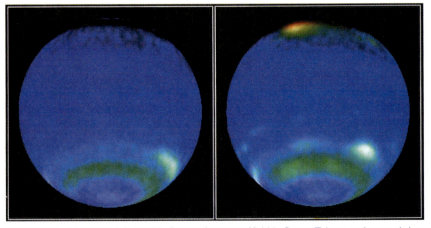

Figure 2-2. Neptune. It is a cold, distant planet, yet Hubble Space Telescope images bring it to life and tell us much about its make up and atmospheric activity. *(Courtesy of the Association of Universities for Research in Astronomy, Inc./Space Telescope Science Institute)*

2.1 Early Space Explorers

In This Section You'll Learn To...

- Explain the two traditions of thought established by Aristotle and Ptolemy that dominated astronomy into the 1600s
- Discuss the contributions to astronomy made by prominent philosophers and scientists in the modern age

Astronomy Begins

More than 4000 years ago, the Egyptians and Babylonians were, for the most part, content with practical and religious applications of their heavenly observations. They developed calendars to control agriculture and star charts both to predict eclipses and to show how the movements of the Sun and planets influenced human lives (astrology). But the ancient Greeks took a more contemplative approach to studying space. They held that *astronomy*—the science of the heavens—was a divine practice best understood through physical theories. Based on observations, aesthetic arguments, and common sense, the Greek philosopher Aristotle (384–322 B.C.) developed a complex, mechanical model of the universe. He also developed comprehensive rules to explain changes such as the motion of objects.

Explaining how and why objects change their position can be difficult, and Aristotle made mistakes. For example, he reasoned that if you dropped two balls, one heavy and one light at precisely the same time, the heavier ball would fall faster to hit the ground first, as illustrated in Figure 2-3. Galileo would later prove Aristotle wrong. But his rigorous logic set an example for future natural philosophers to follow.

Looking to the heavens, Greek philosophers, such as Aristotle, saw perfection. Because the circle was perfectly symmetric, the Greeks surmised that the paths of the planets and stars must be circular. Furthermore, because the gods must consider Earth to be of central importance in the universe, it must occupy the center of creation with everything else revolving around it.

In this *geostatic* (Earth not moving) and *geocentric* (Earth-centered) universe, Aristotle believed solid crystalline spheres carried the five known planets, as well as the Moon and Sun, in circular paths around the Earth. An outermost crystalline sphere held the stars and bounded the universe. In Aristotle's model, an "unmoved mover," or god, inspired these spheres to circle Earth.

Aristotle further divided his universe into two sections—a *sublunar realm* (everything beneath the Moon's sphere) and a *superlunar realm* (everything from the Moon up to the sphere of the fixed stars), as seen in

Figure 2-3. Aristotle's Rules of Motion. Aristotle predicted that heavy objects fall faster than light objects.

Figure 2-4.   Aristotle's Model. The universe divided into two sections—a sublunar and a superlunar realm—each having its own distinct elements and physical laws. *(Courtesy of Sigloch Edition)*

Figure 2-5.   Astrolabe. Arabic scholars used an astrolabe to determine latitude, tell time, and make astronomical calculations. It revolutionized astronomy and navigation. *(Courtesy of Sigloch Edition)*

Figure 2-4. Humans lived in the imperfect sublunar realm, consisting of four elements—earth, water, air, and fire. Earth and water naturally moved *down*—air and fire tended to move *up*. The perfect superlunar realm, in contrast, was made up of a fifth element (aether) whose natural motion was circular. In separating Earth from the heavens and using different laws of physics for each, Aristotle complicated the efforts of future astronomers.

Although this model of the universe may seem strange to us, Aristotle developed it from extensive observations combined with a strong dose of common sense. What should concern us most is not the accuracy but the audacity of Aristotle's vision of the universe. With the power of his mind alone, Aristotle explored and ordered the heavens. His geocentric model of the universe dominated astronomy for 2000 years.

Astronomy in the ancient world reached its peak of refinement in about 140 A.D. with Ptolemy's *Almagest*. Following Greek tradition, Ptolemy calculated orbits for the Sun, Moon, and planets using complex combinations of circles. These combinations, known as eccentrics, epicycles, and equants, were not meant to represent physical reality— they were merely devices for calculating and predicting motion. Like Aristotle, Ptolemy held that heavenly bodies—suspended in solid crystalline spheres, composed of aether—followed their natural tendency to circle Earth. In the eyes of the ancients, describing motion (kinematics) and explaining the causes of motion (dynamics) were two separate problems. It would take almost 1500 years before Kepler healed this split.

Astronomers made further strides during the Middle Ages, with Arabic contributions being especially noteworthy. While Europe struggled through the Dark Ages, Arabic astronomers translated the *Almagest* and other ancient texts. They developed a learned tradition of commentary about these texts, which Copernicus later found invaluable in his reform of Ptolemy. Arabs also perfected the astrolabe, a sophisticated observational instrument, shown in Figure 2-5, used to chart the courses of the stars and aid travellers in navigation. Their observations, collected in the Toledan tables, formed the basis of the Alfonsine Tables used for astronomical calculations in the west from the 13th century to the mid-16th century. Moreover, Arabic numerals, combined with the Hindu concept of zero, replaced the far clumsier Roman numerals. Together with Arabic advances in trigonometry, this new numbering system greatly enhanced computational astronomy. Our language today bears continuing witness to Arabic contributions—we adopted algebra, nadir, zenith, and other words and concepts from them.

With the fall of Toledo, Spain, in 1085, Arabic translations of and commentaries about ancient Greek and Roman works became available to the west, touching off a renaissance in 12th-century Europe. Once again, Europeans turned their attention to the heavens. But because medieval scholasticism had made Aristotle's principles into dogma, centuries passed before fundamental breakthroughs occurred in astronomy.

Reordering the Universe

With the Renaissance and humanism came a renewed emphasis on the accessibility of the heavens to human thought. Nicolaus Copernicus (1473–1543), a Renaissance humanist and Catholic cleric (Figure 2-6), reordered the universe and enlarged humanity's horizons within it. He placed the Sun near the center of the solar system, as shown in Figure 2-7, and had Earth rotate on its axis once a day while revolving about the Sun once a year. Copernicus promoted his *heliocentric* (sun-centered) vision of the universe in his *On the Revolutions of the Celestial Spheres*, which he dedicated to Pope Paul III in 1543.

A heliocentric universe, he explained, is more symmetric, simpler, and matches observations better than Aristotle's and Ptolemy's geocentric model. For example, Copernicus explained it was simpler to attribute the observed rotation of the sphere of the fixed stars (he didn't abandon Aristotle's notion of solid crystalline spheres) to Earth's own daily rotation than to imagine the immense sphere of the fixed stars rotating at near infinite speed about a fixed Earth.

Copernicus further observed that, with respect to a viewer located on Earth, the planets occasionally appear to back up in their orbits as they move against the background of the fixed stars. Ptolemy had resorted to complex combinations of circles to explain this retrograde or backward motion of the planets. But Copernicus cleverly explained that this motion was simply the effect of Earth overtaking, and being overtaken by, the planets as they all revolve about the Sun.

Copernicus' heliocentric system had its drawbacks, however. Copernicus couldn't prove Earth moved, and he couldn't explain why Earth rotated on its axis while revolving about the Sun. He also adhered to the Greek tradition that orbits follow uniform circles, so his geometry was nearly as complex and physically erroneous as Ptolemy's. In addition, Copernicus wrestled with the problem of *parallax*—the apparent shift in the position of bodies when viewed from different locations. If Earth truly revolved about the Sun, critics noted, a viewer stationed on Earth should see an apparent shift in position of a closer star with respect to its more distant neighbors. Because no one saw this shift, Copernicus' sun-centered system was suspect. In response, Copernicus speculated that the stars must be at vast distances from Earth, but such distances were far too great for most people to contemplate at the time, so this idea was also widely rejected.

Copernicus saw himself more as a reformer than as a revolutionary. Nevertheless, he did revolutionize astronomy and challenge humanity's view of itself and the world. The reality of his system was quickly denied by Catholics and Protestants alike, with Martin Luther bluntly asserting: "This fool [Copernicus] wishes us to reverse the entire science of astronomy...sacred Scripture tells us that Joshua commanded the Sun to stand still [Joshua 10:12–13], and not the Earth." Still, Catholics and Protestants could accept the Copernican hypothesis as a useful tool for astronomical calculations and calendar reform as long as it wasn't used to represent reality.

Figure 2-6.   Nicolaus Copernicus. He reordered the universe and enlarged humanity's horizons. *(Courtesy of Western Civilization Collection, the U.S. Air Force Academy)*

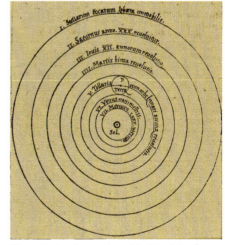

Figure 2-7.   Copernican Model of the Solar System. Copernicus placed the Sun near the center of the universe with the planets moving around it in circular orbits. *(Courtesy of Sigloch Edition)*

Figure 2-8. Tycho Brahe. He made valuable, astronomical observations, overturning Aristotle's theories and paving the way for later theoricians. *(Courtesy of Western Civilization Collection, the U.S. Air Force Academy)*

Figure 2-9. Brahe's Quadrant. He used this 90° arc to precisely measure the angles of celestial bodies above the horizon. *(Courtesy of Western Civilization Collection, the U.S. Air Force Academy).*

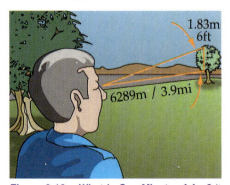

Figure 2-10.   What is One Minute of Arc? It is the angle that a 1.83 m (6 ft.) tree makes when we see it 6289 m (3.9 mi.) away.

Because of these physical and religious problems, only a few scholars dared to embrace Copernicanism. Those who did were staggered by its implications. If Earth were just another planet, and the heavens were far more vast than previously believed, then perhaps an infinite number of inhabited planets were orbiting an infinite number of suns, and perhaps the heavens themselves were infinite. Giordano Bruno (1548–1600) promoted these views, but because of their radical nature and his unorthodox religious views, he was burned at the stake in 1600.

Ironically, Bruno's vision of an infinite number of inhabited worlds occupying an infinite universe derived from his belief that an omnipotent God could create nothing less. Eventually, other intrepid explorers seeking to plumb the depths of space would share his vision. But his imaginative insights were ultimately less productive than more traditional observational astronomy, especially as practiced at this time by Tycho Brahe (1546–1601).

Brahe (Figure 2-8) rebelled against his parents, who wanted him to study law and serve the Danish King at court in typical Renaissance fashion. Instead, he studied astronomy and built, on the island of Hven in the Danish Sound, a castle-observatory known as Uraniborg, or "heavenly castle." Brahe's castle preserved his status as a knight, and a knight he was, both by position and temperament. Never one to duck a challenge, Brahe once dueled with another Danish nobleman and lost part of his nose, which he ingeniously reconstructed out of gold, silver, and wax.

Brahe brought the same ingenuity and tenacity to observational astronomy. He obtained the best observing instruments of his time and pushed them to the limits of their accuracy to achieve observations precise to approximately one minute of arc (Figure 2-9). If you were to draw a circle and divide it into 360 equal parts, the angle described would be a *degree*. If you then divide each degree into 60 equal parts, you would get one *minute of arc*. Figure 2-10 gives an idea of how small one minute of arc is.

Brahe observed the supernova of 1572 and the comet of 1577. He calculated that the nova was far beyond the sphere of the Moon and that the comet's orbit intersected those of the planets. Thus, he concluded that change does occur in the superlunar realm and that solid crystalline spheres don't exist in space. In a sense, he shattered Aristotle's solid spheres theory, concluding that space was imperfect and empty except for the Sun, Moon, planets, and stars.

Although Brahe's findings were revolutionary, he couldn't bring himself to embrace the Copernican model of the solar system. Instead, he kept the Earth at the center of everything in a complex, geo-heliocentric model of the universe. In this model, the Moon and Sun revolved about Earth, with everything else revolving about the Sun. This alternative model preserved many of the merits of the Copernican system while keeping Christians safely at the center of everything. Many scholars who could not accept Copernicanism, such as the Jesuits, adopted Brahe's system.

Brahe didn't take full advantage of his new, more precise observations, partly because he wasn't a skilled mathematician. But Johannes Kepler (1571–1630), shown in Figure 2-11, was. Astronomers, Kepler held, were priests of nature who God called to interpret His creation. Because God plainly chose to manifest Himself in nature, the study of the heavens would undoubtedly be pleasing to God and as holy as the study of Scripture.

Inspired by this perceived holy decree, Kepler explored the universe, trying to redraw in his own mind God's harmonious blueprint for it. By the age of twenty-five, Kepler published the *Cosmic Mystery*, revealing God's model of the universe. Although his model attracted few supporters in 1596 and seems bizarre to students today, Kepler insisted throughout his life that this model was his monumental achievement. He even tried to sell his duke on the idea of creating, out of gold and silver, a mechanical miniature of this model which would double as an elaborate alcoholic drink dispenser! Having failed with this clever appeal for support, Kepler sought out and eventually began working with Brahe in 1600.

The Brahe-Kepler collaboration would be short-lived, for Brahe died in 1601. Before his death, Brahe challenged Kepler to calculate the orbit of Mars. Brahe's choice of planets was fortunate, for of the six planets then known, Mars had the second most eccentric orbit (Mercury was the most eccentric). Eccentric means "off center," and *eccentricity* describes the deviation of a shape from a perfect circle. A circle has an eccentricity of zero, and an ellipse has an eccentricity between zero and one. As Kepler began to pore through Brahe's observations of Mars, he found a disturbing discrepancy. Mars' orbit wasn't circular. He consistently calculated a difference of eight minutes of arc between what he expected for a circular orbit and Brahe's observations.

Figure 2-11. Johannes Kepler. He struggled to find harmony in the motion of the planets. *(Courtesy of Western Civilization Collection, the U.S. Air Force Academy)*

Astro Fun Fact
"There is Nothing New Under the Sun"

Was Copernicus the first scientist to place the Sun, not Earth, in the center of the solar system? No! In the 5th century B.C., Philolaus, a Pythagorean, suggested that the Earth rotated on its axis once a day while revolving about a central fire (not the Sun). Observers on Earth couldn't see this central fire, Philolaus explained, because a "counter-earth" blocked the view, shielding Earth from direct exposure to heat. In approximately 320 B.C., Aristarchus was born in Samos of the Ancient Greek Empire. The profession he grew into was astronomy, so he moved to Alexandria, then the cultural hub for natural philosophers. His work centered on determining the distance from Earth to the Sun and the Moon. He did this through geometric measurements of the Moon's phases and the size of Earth's shadow during lunar eclipses. He eventually showed that the Sun was enormously larger than Earth. Therefore, he believed that the Sun and not Earth occupied the center of the known universe. More radically, he correctly surmised that Earth spun daily on its axis, while revolving yearly around the Sun. Their findings were too revolutionary for their times, and if not for the fact that Archimedes mentioned Aristarchus' work in some of his writings and Copernicus mentioned Philolaus, their ideas would be lost in obscurity.

Asimov, Isaac. Asimov's Biographical Encyclopedia of Science and Technology. Garden City, NJ: Doubleday & Co. Inc., 1972.

Contributed by Thomas L. Yoder, the U.S. Air Force Academy

Some astronomers would have ignored this discrepancy. Not Kepler, however. He simply couldn't disregard Brahe's data. Instead, he began to look for other shapes that would match the observations. After wrestling with ovals for a short time he arrived at the idea that the planets moved around the Sun in elliptical orbits, with the Sun not at the center but at a *focus*. (Focus comes from a Latin term meaning hearth or fireplace.) Confident in his own mathematical abilities and in Brahe's data, Kepler codified this discovery into a Law of Motion. Although this was actually the second law he discovered, we call this Kepler's First Law, shown in Figure 2-12.

Kepler's First Law. The orbits of the planets are ellipses with the Sun at one focus.

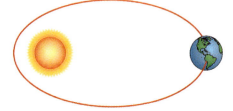

Figure 2-12. Kepler's First Law. Kepler's First Law states that the orbits of the planets are ellipses with the Sun at one of the foci, as shown here in this greatly exaggerated view of Earth's orbit around the Sun.

With his Second Law, Kepler began to hint at the Law of Universal Gravitation that Newton would discover decades later. By studying individual "slices" of the orbit of Mars versus the time between observations, Kepler noticed that a line between the Sun and Mars swept out equal areas in equal times. For instance, if a planet moved through two separate arcs of its orbit, each in 30 days, both arcs would define the same area. To account for this, he reasoned that as a planet draws closer to the Sun it must move faster (to sweep out the same area), and when it is farther from the Sun, it must slow down. Figure 2-13 shows this varying motion.

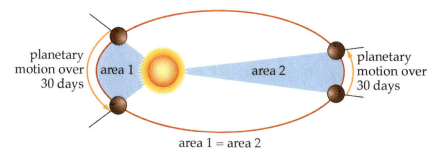

planetary motion over 30 days   area 1   area 2   planetary motion over 30 days

area 1 = area 2

Figure 2-13. Kepler's Second Law. Kepler's Second Law states that planets (or anything else in orbit) sweep out equal areas in equal times.

Kepler's Second Law. The line joining a planet to the Sun sweeps out equal areas in equal times.

Kepler developed his first two laws between 1600 and 1606 and published them in 1609. Ten years later, Kepler discovered his Third Law while searching for the notes he believed the planets sang as they orbited the Sun! Again, after much trial and error, Kepler formulated a relationship, known today as his Third Law.

Kepler's Third Law. The square of the orbital period—the time it takes to complete one orbit—is directly proportional to the cube of the mean or average distance between the Sun and the planet.

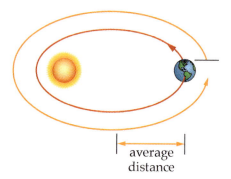

average distance

Figure 2-14. Kepler's Third Law. Kepler's Third Law states that square of an orbit's period is proportional to the cube of the average distance between the planet and the Sun.

Figure 2-14 illustrates these parameters. As we'll see in Chapter 4, when we look at orbital motion in earnest, Kepler's Third Law allows us to predict orbits not only of planets but also of moons, satellites, and space shuttles.

Together with his three laws for describing planetary motion, Kepler's astronomy brought a new emphasis on finding and quantifying the physical causes of motion. Kepler fervently believed that God had drawn His plan of the universe along mathematical lines and implemented it using only physical causes. Thus, he united the geometrical or kinematic description of orbits with their physical or dynamic cause.

Kepler also used his imagination to explore space, "traveling" to the constellation of Orion in an attempt to prove the universe was finite. (Harmony and proportion were everything to Kepler, and an infinite universe seemed to lack both.) In 1608 he wrote a fictional account of a Moon voyage (the *Somnium*) which was published posthumously in 1634. In the *Somnium*, Kepler "mind-trips" to the Moon with the help of magic, where he discovers the Moon is an inhospitable place inhabited by specially-adapted Moon creatures. Kepler's *Somnium* would eventually inspire other authors to explore space through imaginative fiction, including Jules Verne in *From the Earth to the Moon* (1865) and H.G. Wells in *The First Men in the Moon* (1900).

Figure 2-15.   Galileo Galilei. Galileo used the telescope to revolutionize our understanding of the universe. *(Courtesy of Western Civilization Collection, the U.S. Air Force Academy)*

Up to Kepler's time, humanity's efforts to explore the universe had been remarkably successful but constrained by the limits of human eyesight. But this was to change. In 1609 an innovative mathematician, Galileo Galilei (1564–1642), shown in Figure 2-15, heard of a new optical device which could magnify objects so they would appear to be closer and brighter than when seen with the unaided eye. Building a telescope that could magnify an image 20 times, Galileo ushered in a new era of space exploration. He made startling telescopic observations of the Moon, the planets, and the stars, thereby attaining stardom in the eyes of his peers and potential patrons.

Observing the Moon, Galileo noticed it looked remarkably like the Earth's surface, with mountains, valleys, and even seas. Looking at the Sun, Galileo saw blemishes or sunspots. These observations disproved Aristotle's claim that the Moon and Sun were perfect and wholly different from Earth. Observing the planets, Galileo noticed that Jupiter had four moons or satellites (a word Kepler coined in 1611) that moved about it. These Jovian moons disproved Aristotle's claim that everything revolved about Earth. Meanwhile, the fact that Venus exhibited phases like the Moon implied that Venus orbited the Sun, not Earth.

Observing the stars, Galileo solved the mystery of the "milkiness" of our Milky Way galaxy. He explained it was due to the radiance of countless faint stars which the unaided eye couldn't resolve. Galileo further noticed that his telescope didn't magnify the stars, which seemed to confirm Copernicus' guess about their vast distance from Earth (Figure 2-16).

Galileo quickly published his telescopic discoveries in the *Starry Messenger* in 1610. This book, written in a popular, non-technical style, presented a formidable array of observational evidence against Aristotle's

Figure 2-16.   Galileo's Telescope. Through this crude device, Galileo discovered many amazing truths that disproved earlier theories. *(Courtesy of Siglich Edition)*

and Ptolemy's geocentric universe. Galileo at first had to overcome people's suspicions, especially as to the trustworthiness of telescopes, which used glass of uncertain quality. As he did so, he used his telescope to subvert Aristotle's distinction between the sub- and superlunar realms and to unify terrestrial and heavenly phenomena.

Almost immediately after Galileo published the *Starry Messenger*, people began to argue by analogy that, if the Moon looks like Earth, perhaps it too is inhabited. The search for extraterrestrial life encouraged experts and laymen alike to explore the heavens. As early as 1638, in his book *The Discovery of a World in the Moone*, John Wilkins encouraged people to colonize the Moon by venturing out into space in "flying chariots."

Galileo also reformed Aristotle's physics. He rolled a sphere down a grooved ramp and used a water clock to measure the time it took to reach bottom. He repeated the experiment with heavier and lighter spheres, as well as steeper and shallower ramps, and cleverly extended his results to objects in free fall. Through these experiments, Galileo discovered, contrary to Aristotle, that all objects fall at the same rate regardless of their weight, as illustrated in Figure 2-17.

Galileo further contradicted Aristotle as to why objects, once in motion, tend to keep going. Aristotle held that objects in "violent" motion, such as arrows shot from bows, keep going only as long as something is physically in touch with them, pushing them onward. Once this push died out, they resumed their natural motion and dropped straight to Earth. Galileo showed that objects in uniform motion keep going unless disturbed by some outside influence. He wrongly believed that this uniform motion was circular, and he never used the term "inertia." Nevertheless, we recognize him today for refining the concept of inertia, which we'll explore in Chapter 4.

Another concept Galileo refined was *relativity* (often termed Galilean relativity to distinguish it from Albert Einstein's theory of relativity). Galileo wrote,

> *Imagine two observers, one standing on a moving ship's deck at sea, the other standing still on shore. A sailor near the top of the ship's mast drops an object to the observer on deck. To this observer, the object falls straight down. To the observer on shore, however, who does not share the horizontal motion of the ship, the object follows a parabolic course as it falls. Both observers are correct!* [Galileo, 1632]

In other words, motion depends on the perspective or frame of reference of the observer.

To complete the astronomical revolution, which Copernicus had started and which Brahe, Kepler, and Galileo had advanced, the terrestrial and heavenly realms had to be united under one set of natural laws. Isaac Newton (1642–1727), shown in Figure 2-18, answered this challenge. Newton was a mercurial person, a brilliant natural philosopher, and mathematician, who provided a majestic vision of

Figure 2-17. Everything Falls at the Same Rate. Galileo was the first to demonstrate through experiment that all masses, regardless of size, fall at the same rate when dropped from the same height (neglecting air resistance).

Figure 2-18.   Isaac Newton. Newton was perhaps the greatest physicist who ever lived. He developed calculus (independent of Gottfried Leibniz), worked out laws of motion and gravity, and experimented with optics. Yet, he spent more time studying alchemy and biblical chronology! *(Courtesy of Western Civilization Collection, the U.S. Air Force Academy)*

nature's unity and simplicity. During his "miracle year" in 1665, Newton invented calculus, developed his law of gravitation, and performed critical experiments in optics. Newton later developed the first "Newtonian reflector," as shown in Figure 2-19. Extending Galileo's groundbreaking work in dynamics, Newton published his three laws of motion and the law of universal gravitation in the *Mathematical Principles of Natural Philosophy*, in 1687. With these laws one could explain and predict motion not only on Earth but also in tides, comets, moons, planets—in other words, motion everywhere.

Newton's crowning achievement helped inspire the Enlightenment of the 18th century, an age when philosophers believed the universe was thoroughly rational and understandable. Motivated by this belief, and Newton's shining example, astronomers in the 18th century confidently explored the night sky. Some worked in state-supported observatories to determine longitudinal position at sea by using celestial observations. Others, like William Herschel (1738–1822), tried to find evidence of extraterrestrial life. Herschel never found his moon-dwellers, but with help from his sister Caroline, he shocked the world in 1781 when he accidently discovered Uranus (Figure 2-20). As astronomers studied Uranus, they noticed its orbit wobbled slightly. John Couch Adams (1819–1892) and Urbain Leverrier (1811–1877) used this wobble, known as an orbital *perturbation*, to calculate the location of a new planet which, obeying Newton's Law of Gravity, would cause the wobble. Observing the specified coordinates, astronomers at the Berlin observatory located Neptune in 1846.

Other startling discoveries in the 19th century stemmed from developments in *spectroscopy* (the study of radiated energy in visible bands) and photography. By analyzing star spectra, William Huggins (1824-1910) showed that stars are composed of the same elements as those found here on Earth. His work overthrew once and for all the ancient belief that the heavens consisted of a unique element—aether. He also proved conclusively that some nebulas were gaseous. (Many astronomers like William Herschel had suggested that, as more powerful telescopes became available, all nebulas would eventually be resolved into stars.) Using spectroscopes, astronomers could determine star distances, whether stars were single or double, their approximate surface temperature, and whether they were approaching or receding from Earth (as measured by their Doppler shift). They could even discover new elements, as Joseph Norman Lockyer did in identifying helium in 1868 through spectroscopic analysis of the Sun.

Huggins' and Lockyer's work marked the beginning of astrophysics and brought to fruition Tycho Brahe's quest to unify terrestrial chemistry with astronomy. Meanwhile, photography proved equally revelatory. A camera's ability to collect light through prolonged exposures, for example, provided clear evidence in 1889 that Andromeda was a galaxy, not an incipient solar system. Similar to telescopes, spectroscopes and cameras helped us to extend our explorations of the heavens.

Figure 2-19. Newton's Telescope. While Galileo had used refracting telescopes, Newton developed the first reflecting telescope. *(Courtesy of Sigloch Edition)*

Figure 2-20. Herschel's Telescope. This huge instrument helped Herschel make many planetary observations. *(Courtesy of Sigloch Edition)*

Astro Fun Fact
Measuring the Gravity Constant

In 1798, a physicist named Henry Cavendish performed an exciting experiment that helped him determine the gravitational constant, G. Cavendish took a rod with a tiny, solid sphere at each end and hung it from a torsion balance, using a thin metal thread. Then he placed two big, solid spheres close to and level with the small ones, but not attached. Slowly, the small spheres moved toward the big ones, due to the gravitational pull. By choosing an appropriate material for the thread, he simulated a free-fall condition between the big and small spheres.

Measuring the time of the "free fall" and the distance s with $s = (a/2)t^2$, he calculated the gravitational acceleration. Using the force $F = ma = GmM/r^2$ with r being the distance between the two centers of mass, he was able to compute G. Although he never intended to determine the gravitational constant, because his goal was to find the mass of Earth, we recognize him as the first to compute a value for G.

Shaefer, Bergmann. Lehrbuch der Experimentalphysik, Vol. 1.

Contributed by Gabriele Belle, the U.S. Air Force Academy

The 20th century witnessed equally remarkable discoveries by astronomers. Until 1918, astronomers believed that our solar system was near the center of the Milky Way, that the Milky Way was the only galaxy in the universe, and that its approximate size was a few thousand light-years across. (A *light year* is the distance light travels in one year at a speed of 300,000 km/s [186,000 mi./s], which is about 9460 billion km [5.88×10^{12} mi.].) By 1930 astronomers realized that our solar system was closer to the edge than the center of our galaxy, that other galaxies existed beyond the Milky Way, that the universe was expanding, and that previous estimates of our galaxy's size had been ten times too small.

Two American astronomers, Harlow Shapley (1885–1972) and Edwin Hubble (1889–1953), were most responsible for these radical shifts. Shapley determined in 1918 that our solar system was near the fringes of the Milky Way. Using the 250-cm reflecting telescope at Mount Wilson observatory, Hubble roughly determined the size and structure of the universe through a velocity-distance relationship now known as Hubble's Law. By examining the *Doppler* or *red-shift* of stars, he also determined the universe was expanding at an increasing rate. The red shift comes from the apparent lengthening of electromagnetic waves as a source and observer move apart. At this point, astronomers began to speculate that a huge explosion, or "Big Bang," marked the beginning of the universe.

While Hubble and others contemplated an expanding universe, Albert Einstein (1879–1955), shown in Figure 2-21, revolutionized physics with his concepts of relativity, the space-time continuum, and his famous equation, $E = mc^2$. This equation showed the equivalence of mass and energy related by a constant, the speed of light. Combined with the discovery of radioactivity in 1896 by Henri Becquerel (1852–1908), Einstein's equation explained in broad terms the inner workings of the Sun.

Figure 2-21.   Albert Einstein. Einstein revolutionized physics with his concepts of relativity, space-time, and the now famous equation, $E = mc^2$. *(Courtesy of Western Civilization Collection, the U.S. Air Force Academy)*

Neptune's discovery in the previous century had been a triumph for Newton's laws. But those laws didn't explain why, among other things, Mercury's orbit changes slightly over time. Einstein's general theory of relativity accurately predicted this motion. Einstein explained that any amount of mass curves the space surrounding it, and that gravity is a manifestation of this "warped" space. Furthermore, Einstein showed that the passage of time is not constant, but relative to the observer. This means two objects traveling at different speeds will observe time passing at different rates. This concept has profound implications for satellites such as the Global Positioning System, which rely on highly accurate atomic clocks.

By the dawn of the Space Age, astronomers had constructed a view of the universe radically different from earlier concepts. We continue to explore the universe with our minds and Earth-based instruments. But since 1957, we've also been able to launch probes into space to explore the universe directly. Thus, advances in our understanding of the universe increasingly depend on efforts to send these probes, and people, into space.

Section Review

Key Terms

astronomy
degree
Doppler
eccentricity
focus
geocentric
geostatic
heliocentric
light year
minute of arc
parallax
perturbation
red-shift
relativity
spectroscopy
sublunar realm
superlunar realm

Key Concepts

➤ Two distinct traditions existed in astronomy through the early 1600s
  - Aristotle's geocentric universe of concentric spheres
  - Ptolemy's complex combinations of circles used to calculate orbits for the Sun, Moon, and planets

➤ Several natural philosophers and scientists reformed our concept of space from 1500 to the 20th century
  - Copernicus defined a heliocentric (sun-centered) universe
  - Brahe vastly improved the precision of astronomical observations
  - Kepler developed his three laws of motion
    - The orbits of the planets are ellipses with the Sun at one focus
    - Orbits sweep out equal areas in equal times
    - The square of the orbital period is proportional to the cube of the mean distance from the Sun
  - Galileo developed dynamics and made key telescopic discoveries
  - Newton developed his three laws of motion and the law of universal gravitation
  - Shapley proved our solar system was near the fringe, not the center, of our galaxy
  - Hubble helped show that our galaxy was only one of billions of galaxies, and that the universe was expanding at an ever-increasing rate, perhaps due to a "Big Bang" at the beginning of time
  - Einstein developed the theory of relativity and the relationship between mass and energy described by $E = mc^2$

2.2 Entering Space

▬ In This Section You'll Learn To...

☞ Describe the rapid changes in space exploration in the 20th century from the first crude rockets to space shuttles

We've shown that we don't need to leave the Earth's atmosphere to explore space. With our senses, imagination, and instruments such as the telescope, we can discover new features of and raise new questions about the universe. But somehow, that's not enough. People long to go there. Even before the myth of Daedalus and Icarus, we dreamed of flying. On December 17, 1903, Orville and Wilbur Wright made this dream come true at Kitty Hawk, North Carolina. From then on, advances in flight and rocketry have led us to the edge of space and beyond.

The Age of Rockets

Kepler journeyed to the Moon by magic, and Jules Verne's hero in *From the Earth to the Moon* was fired out of an immensely powerful cannon. But rockets made spaceflight possible. Military requirements initially drove rocket development. The first recorded military use of rockets came in 1232 A.D., when the Chin Tartars defended Kai-feng-fu in China by firing rockets at the attacking Mongols.

In the early 19th century William Congreve (1772–1828), a British colonel and artillery expert, developed incendiary rockets based on models captured in India (Figure 2-22). Congreve's rockets, powered by black powder, ranged in size from 3 to 23 kg (6.6 to 50 lbs.). During the Napoleonic wars, the British fired two hundred of these rockets in thirty minutes against the French at Boulogne, setting the town on fire. The British also fired rockets against Fort McHenry in Baltimore, Maryland, during the War of 1812, with the "rocket's red glare" inspiring Francis Scott Key to pen the United States National Anthem. After 1815 conventional artillery rapidly improved, however, and the British Army lost interest in rockets.

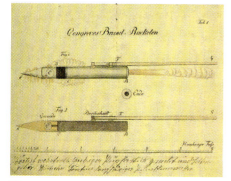

Figure 2-22. Congreve's Rockets. Small rockets, such as these, helped British soldiers turn the tide of battle in the early 19th century. *(Courtesy of Sigloch Edition)*

Waning military interest in rockets didn't deter theoretical studies, however. One of the first people to research rocket-powered spaceflight was Konstantin E. Tsiolkovsky (1857–1935), the father of Russian cosmonautics. In the 1880s he calculated the velocity (known as "escape velocity") required for a journey beyond the Earth's atmosphere. He also suggested that burning a combination of liquid hydrogen and liquid oxygen could improve rocket efficiency. (The Space Shuttle's main engines run on these propellants.) Inspired by Tsiolkovsky's brilliance, the former Soviet Union became the first country to endorse and support the goal of spaceflight, creating in 1924 the Bureau for the Study of the Problems of Rockets.

Astro Fun Fact
The Father of Cosmonautics

Konstantin Tsiolkovsky (kon-stan-teen see-ol-koff-skee) (1857–1935) grew up deaf in Russia, and without the benefit of the usual schools, he educated himself. Surprisingly, with little financial support or engineering background, he applied Newton's third law (action and reaction) to describe the possibility of rocket flight into outer space. Although he did no experiments, his theory, written in 1903, on overcoming Earth's gravity using rockets, was the basis for later tests by Russian engineers. His "reaction vehicles," as he called them, could use liquid fuels to gain enough velocity to rise above Earth. Other imaginative ideas he presented were reaction vehicles for interplanetary flight, multistage rockets, and artificial Earth satellites, including manned space platforms. He was a forward-thinking theorist, who we recognize today as the Father of Cosmonautics.

The Columbia Encyclopedia, Fifth Edition Copyright ©1993, Columbia University Press.

(Courtesy of Vladimir Lytkin)

The United States, in contrast, lagged far behind, except for a single visionary, Robert H. Goddard (1882–1945), shown with one of his first rockets in Figure 2-23. He experimented with liquid-fuel rockets, successfully launching the first in history on March 16, 1926. A skilled engineer and brilliant theorist, Goddard believed that a powerful-enough rocket could reach the Moon or Mars, but he couldn't garner support from the United States government for his ideas.

A far different state of affairs existed in Germany. Hermann J. Oberth's (1894–1989) work on the mathematical theory of spaceflight and his book *The Rocket into Planetary Space* (1923) fostered the growth of rocketry in Germany and led to the founding of the Society for Space Travel in July 1927. Several German rocket societies flourished in the 1920s and 1930s, composed mainly of students and their professors. German government support of these organizations began in the mid-1930s and resulted directly from the Treaty of Versailles that ended World War I. The treaty severely limited Germany's development and production of heavy artillery. After Adolf Hitler assumed power in 1933, the German military saw rockets as a means to deliver warheads over long distances without violating the treaty. The Nazi regime thus supported several rocket societies. Wernher von Braun (1912–1977), a young member of one of these societies, progressed rapidly and finally led the development of the V-2 rocket, the world's first *ballistic missile*. But Von Braun's life-long goal was to develop launch vehicles for interplanetary flight. His rocket research would culminate in the Saturn V moon rocket used in the U.S. Apollo program.

Figure 2-23. Robert Goddard. Goddard pioneered the field of liquid-fueled rocketry. He's shown here with one of his early models. *(Courtesy of NASA/Goddard Space Flight Center)*

Figure 2-24. V-2 at White Sands. At the White Sands Missile Range in New Mexico, U.S. engineers used captured V-2 rockets to advance knowledge of rocketry and the upper atmosphere. *(Courtesy of NASA/White Sands Missile Range)*

Figure 2-25. Sputnik. Weighing less than 100 kg and looking like a basketball trailing four long antennas, Sputnik shook the world when it became the first man-made satellite in 1957. *(Courtesy of Siglich Edition)*

The Germans launched more than two thousand V-2s, armed with one-ton warheads, against allied targets during the last years of World War II. Towards the end of the war, the Allies frantically sought to recruit German rocket scientists. The United States hit paydirt with Project Paperclip, when Von Braun and his research team, carrying their records with them, surrendered to the Americans in 1945. The Russians also recruited heavily, signing German scientists as the technical nucleus of the effort that eventually produced Sputnik.

Besides recruiting German rocket scientists, the United States captured the enormous Mittelwerke underground rocket factory in the Harz Mountains of central Germany, as well as enough components to assemble 68 V-2 rockets. Using these captured V-2s as sounding rockets, with scientific payloads in place of warheads, the V-2 Upper Atmosphere Research Panel studied Earth's atmosphere between 1946 and 1952 and inaugurated the science of X-ray astronomy. The V-2s launched at White Sands, New Mexico, yielded new information about Earth's atmosphere and magnetic field, as well as solar radiation and cosmic rays (Figure 2-24). Until 1957, American astronomers studied space at a leisurely pace. However, Cold War tensions between the U.S. and U.S.S.R. made national security, not astronomy, a priority. Scientists developed rockets such as the Thor Intermediate Range Ballistic Missile (IRBM) and Atlas Intercontinental Ballistic Missile (ICBM), not to explore space, but to deliver nuclear warheads.

Sputnik: The Russian Moon

In the 1950s, the distinction between airplane and rocket research began to blur. Pilots assigned to the National Advisory Committee for Aeronautics (NACA) in the United States flew experimental aircraft, such as the Bell X-1A and X-2, to the edge of Earth's atmosphere. In this era, many aerospace experts believed that pilots flying "spaceplanes" would be the first to explore space. A strong candidate was North American Aviation's X-15, shown in Figure 2-26, a rocket-propelled spaceplane able to exceed Mach 8 (eight times the speed of sound) and climb more than 112 km (70 mi.) above Earth. Confident in its technological supremacy, the United States was shocked when the Russians launched the unmanned satellite Sputnik, shown in Figure 2-25, into orbit on October 4, 1957. Sputnik changed everything. Space became the new high ground in the Cold War, and, in the crisis atmosphere following Sputnik, Americans believed they had to occupy it first.

At least initially, however, the Russians surged ahead in what the media quickly labeled the Space Race. One month after Sputnik, the Russians launched Sputnik II. It carried a dog named Laika, the first living creature to orbit Earth. In the meantime, the United States desperately sought to orbit its own satellite. The first attempt, a Navy Vanguard rocket, exploded on the launch pad on December 6, 1957, in front of a national television audience. Fortunately, Von Braun and his team had been working since 1950 for the Army Ballistic Missile Agency

Figure 2-26. **X-15 Rocket Plane.** The X-15 was piloted to the edge of space and helped develop modern aeronautics and astronautics. *(Courtesy of NASA/Dryden Flight Research Center)*

on the Redstone rocket, a further development of the V-2. Using a modified Redstone, the United States successfully launched its first satellite, Explorer 1, on January 31, 1958.

After the launch of Explorer 1, shown in Figure 2-27, Americans clamored for more space feats to match and surpass the Russians. Most shocking to United States scientists was the ability of their Russian counterparts to orbit heavy payloads. Explorer 1 (Figure 2-28) weighed a feather-light 14 kg (30 lbs.), whereas the first Sputnik weighed 84 kg (185 lbs.). Sputnik III, a geophysical laboratory orbited on May 15, 1958, weighed a whopping 1350 kg (2970 lbs.). The Russians also enthralled the world in October, 1959, when their space probe Luna 3 took the first photographs of the dark side of the Moon.

Without question the Soviet Premier, Nikita Khrushchev (1894–1971), exploited the space feats of Sergei P. Korolev (1907–1966), the "Grand Designer" of Soviet Rocketry in a deliberate propaganda campaign to prove that Communism was superior to Democracy. Meanwhile, calls went out across the Western world to improve science education in schools and to mobilize the best and brightest scientists to counter the presumed Russian technical superiority. The capstone for this effort was the creation of the National Aeronautics and Space Administration (NASA) on October 1, 1958.

NASA was created to consolidate U.S. space efforts under a single civilian agency and to put a man in space before the Russians. It also managed the rapidly increasing budget devoted to the United States space effort, which rocketed from $90 million for the *entire* effort in 1958 to $3.7 billion for NASA *alone* in 1963. These were heady days for NASA, a time of great successes and equally spectacular failures. In this scientific dimension of the Cold War, United States astronauts became more like swashbucklers defending the ship of state than sage scientists seeking

Figure 2-27. **Explorer 1.** Explorer 1 was the first United States satellite launched after many frustrating failures. *(Courtesy of NASA/ Jet Propulsion Laboratory)*

Figure 2-28. **Explorer 1 Satellite.** This small satellite measured the charged particles that surround Earth and comprise the Van Allen belts. *(Courtesy of NASA/Jet Propulsion Laboratory)*

Figure 2-29.   Mercury Program Astronauts. These men were the first U.S. astronauts. They endured seemingly endless tests of endurance and extremes to establish the envelope for human survival in space. *(Courtesy of NASA/ Johnson Space Center)*

Figure 2-30.   Yuri Gagarin. The Soviet Union scored another first in the space race when Yuri Gagarin became the first human to orbit Earth. *(Courtesy of NASA/Goddard Space Flight Center)*

Figure 2-31.   Echo I. From meager hardware, such as this "balloon" satellite, long-distance communication via satellite got its start. *(Courtesy of Siglich Edition)*

knowledge. President Dwight D. Eisenhower (1890–1969) reinforced this nationalistic mood when he recruited the original seven Mercury astronauts exclusively from the ranks of military test pilots, as shown in Figure 2-29. The Mercury program suffered serious initial teething pains. Mercury-Atlas 1, launched on July 29, 1960, exploded one minute after lift-off. On November 21, 1960, Mercury-Redstone 1 ignited momentarily, climbed about 10 cm (4 in.) off the launch pad before its engine cutoff, then settled back down on the launch pad as the escape tower blew. Despite these failures, NASA persisted. On January 21, 1961, Mercury-Redstone 2 successfully launched a chimpanzee called Ham on a suborbital flight.

Once again, however, the Russians caused the world to hold its collective breath when on April 12, 1961, Major Yuri A. Gagarin, shown in Figure 2-30, completed an orbit around Earth in Vostok I. Here was proof positive, in Khrushchev's eyes, of the superiority of Communism. After all, the United States could only launch astronauts Alan B. Shepard, Jr., and Virgil I. "Gus" Grissom on suborbital flights in May and July, 1961. The Russians astounded the world once more on August 7, 1961, by launching cosmonaut Gherman S. Titov on a day-long, seventeen-orbit flight about the Earth. After what some Americans no doubt thought was an eternity, the United States finally orbited its first astronaut, John H. Glenn, Jr., on February 20, 1962. Glenn's achievement revived America's sagging morale, although he completed only three orbits due to a problem with his capsule's heat shield.

From 1961–1965, the Russians orbited more cosmonauts (including one woman, Valentina Tereshkova, on June 16, 1963), for longer periods, and always sooner than their American rivals. The Russians always seemed to be a few weeks or months ahead. For example, cosmonaut Alexei Leonov beat astronaut Edward H. White II by eleven weeks for the honor of the first spacewalk, which took place on March 18, 1965.

The United States recorded some space firsts in communication, however. Echo I (Figure 2-31), launched on August 12, 1960, was an aluminum-coated, 30.4 meter-diameter plastic sphere that passively reflected voice and picture signals. It demonstrated the feasibility of satellite communications. Echo's success paved the way for Telstar, the first active communications satellite. Launched on July 10, 1962, it relayed communication between far-flung points on the Earth. Echo and Telstar were the forerunners of the incredible variety of *communication satellites* now circling the Earth, upon which our modern information society depends.

Echo, Telstar, and interplanetary probes such as the Mariner series helped reinflate sagging American prestige in the early 1960s. But the United States needed something more dramatic, and President John F. Kennedy (1917–1963) supplied it. In a speech before a joint session of Congress in 1961, he launched the nation on a bold, new course. "I believe this nation should commit itself to achieving the goal, before this decade is out, of landing a man on the Moon and returning him safely to the Earth." For Kennedy, the clearest path to United States pre-eminence in space led directly to the Moon.

Astro Fun Fact
Disney and the Russian Rocket Program

The Russian rocket program was unwittingly boosted by the Walt Disney movie, <u>Man In Space</u>, shown in August, 1955, at an international space conference in Copenhagen which two Russian rocket scientists attended. The Russian newspaper, Pravda, reported on the conference and increased the Russian feeling that the United States was serious about space. The Pravda article also emphasized the need for more Russian public awareness of spaceflight. Professor Leonid I. Sedov, Chairman of the U.S.S.R. Academy of Science's Interdepartmental Commission on Interplanetary Communications, addressed this issue in his Pravda article of September 26, 1955: "A popular-science artistic cinema film entitled <u>Man In Space</u>, released by the American director Walt Disney and the German rocket-missile expert Von Braun, chief designer of the V-2, was shown at the Congress... it would be desirable that new popular-science films devoted to the problems of interplanetary travels be shown in our country in the near future. It is also extremely important to increase the interest of the general public in the problem of astronautics. Here is a worthwhile field of activity for scientists, writers, artists, and for many works of Russian culture." As a result, the Russians became more concerned with the United States' space effort and focused more on their own efforts to launch satellites.

Krieger, F.J. <u>Behind the Sputniks—A Survey of Soviet Space Science</u>. Washington, D.C.: Public Affairs Press, 1958.

Contributed by Dr. Jackson R. Ferguson, Jr., the U.S. Air Force Academy

Armstrong's Small Step

The Apollo program to put an American on the Moon by 1970 was meant to accept the challenge of the space race with the Russians and prove which country was technically superior. By 1963, Apollo already consumed two-thirds of the United States' total space budget. Support for Apollo was strong among Americans, but the program was not without critics. Former President Eisenhower remarked that "Anybody who would spend 40 billion dollars in a race to the Moon for national prestige is nuts."

Critics notwithstanding, Neil Armstrong's and Buzz Aldrin's first footprints in the lunar soil on July 20, 1969, as shown in Figure 2-32, marked an astonishing technical achievement. The world watched the dramatic exploits of Apollo 11 with great excitement and a profound sense of awe. President Richard M. Nixon was perhaps only slightly exaggerating when he described Apollo 11's mission as the "greatest week in the history of the world since the Creation."

Begun as a nationalistic exercise in technological chest-thumping, Apollo became a spiritual adventure, a wonderous experience that made people stop and think about life and its deeper meaning. The plaque left behind on the Apollo 11 lunar module perhaps best summed up the mission. **"Here men from the planet Earth first set foot upon the Moon July 1969, A.D. We came in peace for all mankind."** A gold olive branch remains today at the Sea of Tranquility, left behind by the Apollo 11 astronauts as a symbol of peace.

Figure 2-32.   First on the Moon. *In 1969, Neil Armstrong and Buzz Aldrin left their footprints on the surface of the Moon. (Courtesy of NASA/Johnson Space Center)*

Figure 2-33. Lunar Goal Achieved. Astronaut Harrison Schmitt (geologist) takes samples of lunar dust and rock. *(Courtesy of NASA/Johnson Space Center)*

Figure 2-34. Skylab. Launched on May 14, 1973, Skylab was America's first space station. During nine months, three different crews called it home. Pulled back to Earth because of atmospheric drag, Skylab burned up in 1979. *(Courtesy of NASA/Johnson Space Center)*

Figure 2-36. Mir Space Station. Astronauts in the Russian space station set many records, including Valeri Poliakov's longest stay of 438 days. *(Courtesy of NASA/Johnson Space Center)*

Apollo was perhaps NASA's greatest triumph. Even when disaster struck 330,000 km (205,000 mi.) from the Earth during the Apollo 13 mission, NASA quickly rallied and brought the astronauts home safely. After Apollo 11, there would be five additional Moon landings between November, 1969, and December, 1972. The six Apollo Moon missions brought back 382 kg (840 lbs.) of Moon rocks and soil. Scientists had hoped these rocks would be a "Rosetta stone," helping them date the beginning of the solar system. Regrettably, the rocks turned out to be millions of years younger than expected. Still, the Apollo missions revealed much about the nature of the Moon, although some scientists remained critical of Apollo's scientific worth. This criticism came partly because only on the last mission (Apollo 17) did NASA send a scientist, Dr. Harrison H. (Jack) Schmitt, shown in Figure 2-33, to explore the Moon.

After the triumph of Apollo, NASA's achievements were considerable if somewhat anticlimactic. For example, in a display of brilliant improvisation, NASA converted a Saturn V third-stage into a laboratory module called Skylab, shown in Figure 2-34. After suffering external damage during launch on May 14, 1973, Skylab was successfully repaired by space-walking astronauts and served three separate crews as a laboratory in space for 28, 59, and 84 days, respectively.

In July, 1975, space became a forum for international cooperation between strong and sometimes bitter rivals when Apollo 18 docked with a Soviet Soyuz spacecraft. A triumph for detente, Apollo-Soyuz was perhaps a harbinger of the future of space exploration—progress through cooperation and friendly competition.

With the end of the Apollo program, NASA began work on a new challenge—the Space Shuttle. On April 12, 1981, twenty years to the day after Gagarin's historic flight, astronauts John Young and Robert Crippen piloted the reusable spacecraft Columbia on its maiden flight. Columbia, joined by her sister ships Challenger, Discovery, Atlantis, and later Endeavour, proved the versatility of the shuttle fleet. Shuttle missions have been used to deploy satellites, launch interplanetary probes, rescue and repair satellites, conduct experiments, as shown in Figure 2-35, and monitor the Earth's environment.

While the U.S. focused on the Shuttle, the Russian space program took a different path. Relying on expendable Soyuz boosters, the Russians concentrated their efforts on a series of space stations designed to extend the length of human presence in space. From 1971–1982, the Russians launched seven increasingly capable variations of Salyut space stations, home to dozens of crews. In 1986, the expandable Mir (or "peace") Space Station marked a new generation of space habitat (Figure 2-36). In 1988, cosmonauts Vladimir Titov and Musa Manarov became the first humans to spend more than one year in space aboard Mir.

Partly in response to these efforts, and to exploit the capabilities of the Space Shuttle, in 1984 President Reagan committed NASA to building its own space station—Freedom. Ironically, the collapse of the former Soviet Union opened the way for these space rivals to join forces, so the Space Shuttle docked with Mir and the Russians became partners on the re-designed International Space Station.

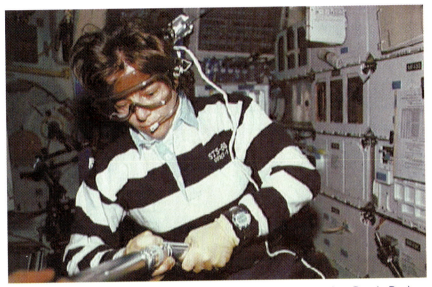

Figure 2-35.   Experiments on the Space Shuttle. Shuttle crewmember, Bonnie Dunbar, conducts a variety of experiments in biology, life science, and material processing. *(Courtesy of NASA/Johnson Space Center)*

Satellites and Interplanetary Probes

While manned space programs drew the lion's share of media attention during the Cold War, satellites and interplanetary probes were quietly revolutionizing our lives and our knowledge of the solar system. Satellites serve as navigation beacons, relay stations for radio and television signals, and other forms of communication. They watch our weather and help us verify international treaties. Satellites are able to discover phenomena that cannot be observed or measured from Earth's surface because of atmospheric interference. While the race to the Moon was on in the 1960s, scientists used satellites to explore the atmosphere. They studied gravitational, electrical, and magnetic fields, energetic particles, and other space phenomena. In addition, projects such as the United States Land Remote Sensing Program (Landsat) started in the early 70's and proved invaluable for tracking and managing Earth's resources.

We've also gained invaluable insights into the nature of our solar system with interplanetary probes. On February 12, 1961, the former Soviet Union launched Venera 1, which passed within 60,000 km (37,200 mi.) of Venus and analyzed its atmosphere. In the 1960s Rangers 7-9 took more than 17,000 detailed photographs of the Moon's surface in preparation for the Apollo landings. The Surveyor probes extended this investigation by testing the composition of lunar soil to ensure it would support the Apollo lunar lander. Between 1962 and 1973, seven Mariner probes studied Mercury, Venus, and Mars. Many practical lessons for Earth emerged from these studies. Large dust storms that Mariner 9 detected on Mars in 1971 led scientists to conclude that these storms cooled Mars' surface. Scientists, such as Carl Sagan, have since speculated about potentially disastrous cooling of Earth's surface from large dust clouds that would be raised by a nuclear war. Similarly, studies of Venus's atmosphere have shed light on the greenhouse effect and global warming from carbon dioxide emissions on Earth.

In the 1970s the Viking and Voyager series of probes led to fundamental breakthroughs in planetary science. Although Vikings 1 and 2 did not discover life on Mars in 1976, they did analyze Martian soil, weather, and atmospheric composition as well as capture more than 4500 spectacular images of the red-planet's surface. Images of Jupiter, Saturn, and their moons, taken by Voyagers 1 and 2 from 1979–1981, were even more spectacular. Among many notable discoveries, Voyager pictures revealed that Jupiter's Great Red Spot was an enormous storm and that Saturn's rings were amazingly complex (and beautiful) with unanticipated kinks and spokes. Originally intended to explore Jupiter and Saturn only, Voyager 2 went on to take revelatory photographs of Uranus in 1986 and Neptune in 1989.

We've also moved some of our most complex and capable scientific instruments into space to escape the interference of Earth's atmosphere. Perhaps the most famous of these space instruments is the Hubble Space Telescope, named in honor of astronomer Edwin Hubble and launched in 1990. The Shuttle has also carried astronomical observatories with specialized instruments to observe ultraviolet and X-ray phenomena, bands of the electromagnetic spectrum that we simply can't perceive with our own eyes.

Today, satellites, interplanetary probes, and space-based instruments continue to revolutionize our lives and understanding of the universe. In Section 2.3 we'll examine some of the more important discoveries and accomplishments made in the 1990s by these technologies.

▤ Section Review

Key Terms

ballistic missile
communication satellites

Key Concepts

➤ Rockets evolved from military weapons in the 1800s to launch vehicles for exploring space after World War II

➤ Sputnik, launched by the former Soviet Union on October 4, 1957, was the first artificial satellite to orbit Earth

➤ Yuri Gagarin was the first human to orbit Earth on April 12, 1961

➤ The space race between the United States and former Soviet Union culminated in Apollo 11, when Neil Armstrong and Buzz Aldrin became the first humans to walk on the Moon

➤ Satellites revolutionized communication and military intelligence and surveillance

➤ Interplanetary probes, such as Viking and Voyager, greatly extended our knowledge of the solar system in the 1970s and 1980s

2.3 Space Comes of Age

▬ In This Section You'll Learn To...

- ☛ Describe some of the major trends in space during the 1990s
- ☛ Discuss recent scientific and commercial space achievements

In the 1990s space exploration and exploitation entered a new stage of maturity. Space exploration witnessed remarkable discoveries by the repaired Hubble Space Telescope and by a series of low-cost, interplanetary probes, such as the Mars Pathfinder. As a result of the Gulf War and other commitments, the U.S. military gained a keener appreciation for the importance of space assets in conducting operations and maintaining world-wide command and control. Finally, commercial space missions surged and, for the first time, civilian spending on space surpassed government spending.

We can identify four major trends in space that characterized the last decade of the 20th century, and set the stage for the beginning of the new millennium:

1) *Space International*—Greater international cooperation in manned spaceflight

2) *Space Science, Big and Small*—Big high-cost missions made remarkable advances, but smaller and less expensive missions began to show their worth

3) *Space Incorporated*—Commercial missions began to dominate space activities

4) *The New High Ground*—Increased importance of space in military planning and operations

In this section we'll examine these trends to understand the never-ending wonder of space exploration and the increasingly routine nature of commercial space exploitation. We'll look at the increasing maturity of space efforts in the 1990s, starting with exploration and science and ending with new commercial and military applications in space.

Space International

Throughout the 90's, the Space Shuttle remained NASA's workhorse for putting people and important payloads into space. One of the most highly publicized shuttle flights, perhaps of all time, took place in 1998 with John Glenn's nostalgic return to space (Figure 2-37). During this and dozens of other missions, Shuttle crews continued to launch important spacecraft, such as Galileo to study Jupiter and its moons, retrieve and

Figure 2-37. John Glenn in the Shuttle. Here, Astronaut John Glenn reviews documents before performing more experiments onboard the Shuttle. *(Courtesy of NASA/Johnson Space Center)*

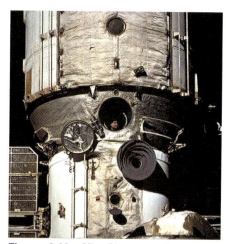

Figure 2-38. Mir Close-up. Over many years, this space station housed astronauts from a variety of nations. Lessons learned will help the International Space Station succeed. *(Courtesy of NASA/Johnson Space Center)*

Figure 2-39. International Space Station (ISS). This artist's concept shows how complex the space station will be. With international cooperation, the cost, technical aspects, and daily operations will be shared by many nations. *(Courtesy of NASA/Johnson Space Center)*

Figure 2-40. Zarya Control Module. Zarya (meaning "sunrise") launched on a Russian Proton rocket from the Baikonur Cosmodrome in November, 1998. The International Space Station was off the ground. *(Courtesy of NASA/Johnson Space Center)*

repair scientific platforms, such as Hubble, conduct valuable experiments in microgravity and other areas of space science, and lay the foundations for the International Space Station.

A major feature of manned space programs in the 1990s was increased cooperation between the United States and the former Soviet Union. Bitter rivals since the dawn of the Space Age, the two superpowers had only cooperated in space once before, during the Apollo/Soyuz mission. But the break-up of the former Soviet Union in the late 80's made cooperation between the two world space powers not only possible, but necessary. With the Shuttle, NASA astronauts had nearly routine access to space. With their Mir space station, the Russians had an outpost in low-Earth orbit (Figure 2-38). By combining the two assets, both countries could more easily achieve important scientific, technical, and operational goals. The Shuttle docked with Mir nine times between 1995 and 1998, with seven American astronauts joining cosmonauts for extended stays on Mir. Some of these stays were punctuated by moments of grave concern, as by the mid-1990s Mir approached the end of its useful life. Its age contributed to a host of major problems with computers, loss of cabin pressure, onboard fires and even a collision with a relief craft. If anything, Mir's problems proved that humans can resolve life-threatening problems while living and working in space. Using Mir, a Russian cosmonaut, Valeri Poliakov, set a space endurance record by spending 438 days in orbit.

When President Reagan began the American space station "Freedom" in 1984, it was in response to Mir's capabilities. But in the 1990's, the newly-named International Space Station (ISS) brought the Russians onboard as major partners (Figure 2-39). Not only are the U.S. and Russia involved: the ISS represents unprecedented international collaboration among 16 nations. The Russians are building the service module that will provide living quarters and a control center, the U.S. is building a laboratory module, the Japanese are providing an experiment module, and the Canadians are building the main robotic arm for the station.

The first element of the space station, the Russian Zarya Control Module, launched in November 1998, from Baikonur Cosmodrome, atop a Proton rocket, was followed closely by the first Shuttle assembly flight in December 1998 (Figure 2-40). On future missions, the growing station will add the main components and astronauts will outfit it with essential equipment and provisions. When completed, ISS will cover the area of two football fields consisting of over 460 tons of structures and equipment.

Space Science Big and Small

As the decade began, a few big projects dominated space science, such as Magellan and Hubble. We'll start by reviewing the major successes of these large space programs, then turn our attention to the smaller, cheaper missions that characterized space science missions at the end of the '90s and into the 21st century.

Big Missions

Magellan. From 1990–1994 the Magellan spacecraft (Figure 2-41) used its powerful synthetic aperture radar to "peel back" the dense cloud cover of Earth's sister planet, mapping 98% of Venus's surface, at 100-meter resolution. These amazing images revealed a planet whose surface is young and changing due to volcanic eruptions. Because its thick atmosphere traps heat in a run-away greenhouse effect, Venus is the solar system's hottest planet, with surface temperatures reaching 482° C.

Galileo. Another expensive and complex mission, the ambitious $1.35 billion Galileo spacecraft (Figure 2-42) launched in 1989 to explore Jupiter, initially seemed crippled by a jammed main antenna and a malfunctioning data-storage tape recorder. Fortunately, NASA engineers improvised and employed a much slower, but, nevertheless effective, secondary antenna to preserve 70% of the mission's objectives. On its way to Jupiter, Galileo took the first close-up photographs of asteroids and discovered Dactyl, a mile-wide moonlet orbiting the asteroid Ida. In July 1994, scientists positioned Galileo to study the spectacular impact of Jupiter by comet Shoemaker-Levy 9. The largest fragment of this comet, approximately 3 km across, sent hot gas plumes 3500 km into space, producing shock waves equivalent to several million megatons of TNT. The potential life-ending implications of a similar impact on Earth's surface were not lost on scientists or the general public and contributed to Hollywood's penchant for disaster movies such as *Deep Impact* and *Armageddon*.

Galileo fulfilled another mission objective in December 1995 when its probe plunged through Jupiter's atmosphere, transmitting data for an hour before being crushed by atmospheric pressure. Scientists quickly learned that many of their predictions about Jupiter's atmosphere were wrong. The probe revealed Jupiter's atmosphere was drier, more turbulent, and windier than predicted, with wind speeds exceeding 560 km/hr (350 m.p.h.) in Jupiter's upper atmosphere. It also proved thicker and hotter than predicted, with temperatures exceeding 700° C. Perhaps most remarkably, Galileo later uncovered intriguing evidence that suggested that liquid water exists beneath the frozen crust of one of Jupiter's moons, Europa, and possibly Callisto, as well. Some scientists now suggest that the best chance of finding life in our solar system may reside in Europa's liquid oceans, which may receive heat from the radioactive decay of Europa's core and immense gravitational squeezing applied to Europa by Jupiter. The combination of liquid water and heat suggests that Europa could support life in the form of simple microorganisms. Galileo continues to study Europa, as well as volcanic activity on the moon, Io, with Galileo's last objective being a close look at Io's rapidly changing surface from a height of 300 km.

Ulysses. NASA has also studied our source of energy and life: the Sun. Working with the European Space Agency (ESA), NASA developed Ulysses (Figure 2-43), which flew over the south and north poles of the Sun in 1994 and 1995. Ulysses measured the solar corona, solar wind, and other properties of the heliosphere. In 1999 and 2000, Ulysses will re-examine these properties under the conditions of solar maximum (the

Figure 2-41.   Magellan. Named for the 16th-century Portuguese explorer, this spacecraft used radar to map Venus from an elliptical, polar orbit. *(Courtesy of NASA)/Johnson Space Center*

Figure 2-42.   Galileo. This artist's conception of the Galileo mission shows the spacecraft with Jupiter behind it and a close up of the moon, Io, in front of it. *(Courtesy of NASA/Jet Propulsion Laboratory)*

Figure 2-43. Ulysses. NASA and the European Space Agency collaborated on this mission to explore the Sun's polar atmosphere. It passed the Sun in 1994 and 1995, and will visit the same path in 2000 and 2001. *(Courtesy of NASA/Jet Propulsion Laboratory)*

Figure 2-44.   Cassini. This science mission must fly-by Venus and Earth to get a gravity-assist to Saturn. *(Courtesy of NASA/Jet Propulsion Laboratory)*

Figure 2-45. Shuttle Astronaut Repairs Hubble. While travelling 28,440 km/hr (17,000 m.p.h.), astronauts Gregory Harbaugh and Joseph Tanner ignore their "head-down" position to replace a vital part. *(Courtesy of NASA/ Johnson Space Center)*

Figure   2-46.   Galaxies.   Hubble   Space Telescope images reveal many new galaxies, each composed of millions of stars. *(Courtesy of the Association of Universities for Research in Astronomy, Inc./Space Telescope Science Institute)*

eleven-year cycle of solar activity). Another NASA and ESA project, the Solar and Heliospheric Observatory (SOHO), is currently studying the Sun's internal structure, its outer atmosphere, and the origins and acceleration of the solar wind. Finally, as part of NASA's Small Explorer Program, the Transition Region and Coronal Explorer (TRACE) probe is studying the Sun's magnetic fields and their relationship to heating within the Sun's corona.

Cassini. The first decade of the new millennium promises equally remarkable discoveries. Already on its way to explore Saturn is Cassini (Figure 2-44). Scheduled to reach Saturn in July 2004, Cassini will deploy a probe (Huygens) built by ESA that will parachute to the surface of Titan, Saturn's Earth-sized moon, to search for life. Cassini will remain in orbit for four years to study Saturn and its rings and moons. Cassini may be the last of the multi-billion-dollar probes.

Hubble. When it comes to space astronomy, we could call the '90s the "Hubble Decade." Dogged by problems when the Shuttle first deployed it in 1988, Shuttle astronauts later repaired and upgraded it during subsequent missions (Figure 2-45). The upgraded Hubble Space Telescope and its suite of improved cameras and instruments, have made a long series of remarkable observations of the Universe. Within our solar system, Hubble has revealed vast storms on Saturn, the presence of dense hydrocarbons in Titan's atmosphere, and drastic changes over several days in Neptune's cloud features. Peering beyond our solar system, Hubble has recorded stars being born, stars in their death throes, and galaxies colliding, and provided deeper understanding of black holes, quasars, and other structural elements of the universe.

In 1998 Hubble detected an enormously powerful black hole at the center of galaxy Centaurus A, whose mass is the equivalent of nearly one billion Sun-like stars, compressed into an area roughly the size of our solar system. Most startling of all were the results of the Deep Field photograph, produced when Hubble focused for ten days on a small, seemingly unimportant slice of the sky. Where Earth-bound telescopes could resolve little of interest, Hubble revealed at least 1500 galaxies (Figure 2-46), some of which may be 12 billion light years distant. This startling photograph suggests there are perhaps 50 billion galaxies in our universe, five times as many as scientists had predicted. Finally, in May 1999, astronomers completed an eight-year study to measure the size and age of the universe. With data from Hubble, scientists narrowed the age of the universe to between 12 and 13.5 billion years, a vast improvement over earlier estimates of between 10 and 20 billion years.

As is the nature of basic research, missions such as Hubble often raise as many questions as they answer. To answer these perplexing questions, NASA and ESA are planning to launch and operate new instruments into space in the early 21st century. The first of these is the Chandra X-Ray Observatory (Figure 2-47). Launched in 1999, this powerful instrument has already begun to answer questions about dark matter and the source of extremely powerful energy and radiation emissions, emanating from the centers of many distant galaxies. NASA has also begun planning the Next

Generation Space Telescope (NGST), a successor to the Hubble that could launch as early as 2007. To add to the data collection, the ESA is working on two space observatories to launch in 2007. The first, Planck, will study the background radiation in space, the faint echoes of the Big Bang of 12–13.5 billion years ago, to enhance our understanding of the universe's structure and rate of expansion. Complementing Planck will be Far Infrared and Submillimetre Telescope (FIRST), which will study planetary systems and the evolution of galaxies in a previously neglected wavelength band (80–670 microns).

Small Missions

Billion-dollar-plus programs such as Cassini and Hubble represent the culmination of more than ten years of effort by thousands of scientists and engineers world-wide. But the constrained budgets that followed the end of the Cold War, as well as the well-publicized problems with Hubble and the complete loss of the Mars Observer in 1993, brought an end to the era of Big Science. NASA Administrator Dan Goldin laid down the gauntlet when he challenged NASA, and the entire space industry, to pursue a new era of "faster, better, cheaper" missions. Missions such as Pathfinder, Lunar Prospector and a host of other government, academic and commercial small satellite missions rose to the challenge and vindicated the wisdom of this new strategy.

An example of NASA's new strategy is Stardust (Figure 2-48). Launched in 1999, Stardust will rendezvous with comet Wild-2 in January 2004, approaching to within 150 km of its nucleus. Stardust's mission is to collect cometary fragments as well as interstellar dust and return them to Earth for analysis in January 2006 (the first return of extraterrestrial material since Apollo). Scientists believe that these cometary fragments will provide information on the evolution of our solar system, including its early composition and dynamics. Stardust may even provide clues as to how life evolved here on Earth. Some scientists have suggested that comets "seeded" Earth with carbon-based molecules, the building blocks of life, in the first billion years of Earth's existence. NASA also has begun planning a mission to Pluto, the only planet yet to be visited by a probe. After examining Pluto and its moon Charon in 2010 or later, the probe may continue on to explore the Edgeworth-Kuiper Disk of "ice dwarfs" or minor planets located at the edge of our solar system.

Back to Mars. Following the loss of Mars Observer, NASA redoubled its efforts to explore the Red Planet, this time through a series of small, less ambitious missions. Pathfinder and Global Surveyor were the first of these. In July 1997 Pathfinder and its rover, Sojourner, explored the Ares Vallis region, an ancient flood plain. Pathfinder sent back spectacular photographs of the Red Planet's surface (Figure 2-49) and showed that it could make a safe and inexpensive landing (cushioned by low-tech airbags) on Mars and that small robots could move across its surface to conduct rock and soil analysis. Pathfinder's pictures captivated a world audience, with NASA's Internet web site experiencing an unprecedented 100 million "hits" in a single day.

Figure 2-47. Chandra X-Ray Observatory. This advanced satellite, launched in July, 1999, will greatly advance our understanding of the universe by detecting high energy, short-wavelength radiation from distant stars and galaxies. *(Courtesy of NASA/Chandra X-ray Center/ Smithsonian Astrophysical Observatory)*

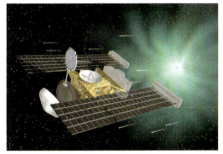

Figure 2-48. Stardust. This Discovery-program mission is the first robotic mission to return samples of our Solar System to Earth. *(Courtesy of NASA/Jet Propulsion Laboratory)*

Figure 2-49. Mars' Twin Peaks. This Mars Pathfinder image of the Martian landscape makes the red planet look habitable. *(Courtesy of NASA/Jet Propulsion Laboratory)*

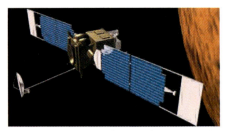

Figure 2-50. Mars Global Surveyor. This successful mission has mapped the entire Martian surface, so planners can locate future landing sites with potentially valuable resources. *(Courtesy of NASA/Jet Propulsion Laboratory)*

Figure 2-51. Clementine. Launched in 1994, this low-cost mission validated several new space technologies and took images of the Moon's south pole that indicate water ice is present. *(Courtesy of Lawrence Livermore National Laboratory)*

Figure 2-52. Lunar Prospector. This Discovery-program mission took images of water ice on the Moon, then performed a controlled crash landing in a small crater on the south pole. *(Courtesy of NASA/Ames Research Center)*

Building on Pathfinder's success was Global Surveyor, which, in 1998, observed the progress of a massive Martian dust storm, whose effects could be detected as high as eighty miles above the surface. In keeping with NASA's new low-cost approach, Global Surveyor employed aerobraking (using the drag created by Mars' atmosphere) rather than expending fuel to slow its speed. (This concept is explained in more detail in Chapter 10.) In March 1999, Global Surveyor began to map Mars' surface and study the role of dust and water in Mars' climate (Figure 2-50). Global Surveyor quickly uncovered evidence of dynamic changes in Mars' crust that are similar to plate tectonic shifts in Earth's geological history. The mission also produced the first three-dimensional, global map of Mars, which measured the Hellas impact crater in Mars' southern hemisphere as being 1300 miles across and nearly six miles deep. NASA concluded that the asteroid that hit Mars threw debris nearly 2500 miles across the planet's surface, enough to cover the continental U.S. in a two-mile thick layer of Martian dust. Despite the back-to-back losses of the Mars Climate Orbiter and Mars Polar Lander missions in late 1999, NASA will continue to focus on understanding the Red Planet.

Back to the Moon and Beyond. The last manned mission to the Moon was Apollo 17 in 1972. For nearly 25 years, the Moon was neglected until a new series of "faster, better, cheaper" missions again visited Earth's closest neighbor. Clementine was a U.S. DoD mission testing technologies developed for the Strategic Defense Initiative (Figure 2-51). In 1994 this small spacecraft sent images of the Moon's poles, suggesting that ice from ancient comet impacts may lie in craters forever in shadow.

Following this success, NASA mounted the first mission in its small, low-cost Discovery series: Lunar Prospector (Figure 2-52). Produced for less than $63 million, and launched in 1998, true to its name it discovered further signs of ice at the Lunar poles. The presence of perhaps 6 billion metric tons of frozen water would be a major boost to future plans for human exploration of the Moon. Having a ready supply of drinking water would be important enough, but by separating the hydrogen and oxygen that make up water, future Lunar settlers could produce rocket propellant and other vital products that would help them "live off the land," and depend less on costly supplies from Earth. In addition to this important find, Lunar Prospector uncovered evidence that supports the theory that the Moon was torn from Earth, billions of years ago, in a collision with an object roughly the size of Mars.

Future space settlers will need other sources of raw materials as well. Examining the practicality of mining operations on asteroids is just one of the objectives of another low-cost mission: the Near-Earth Asteroid Rendezvous (NEAR). By examining the near-Earth asteroid Eros, NEAR promises to reveal much richer knowledge of asteroids, including their structure and composition, possibly identifying the presence of iron, iridium and other important minerals. NEAR rendezvoused with Eros in February 2000.

Space Incorporated

One of the more significant trends of the 1990's was increased globalization and *commercialization of space*. Before then, space was the domain of the U.S., U.S.S.R., China, Japan, and European countries, independently and as part of the European Space Agency (ESA). In the 1990's, the ability to build and launch missions was no longer confined to these countries. Other countries joined the "space club," including Israel, Brazil, and India, with their own launch vehicles and satellites. Korea, Thailand, and Chile started their own national programs by building satellites and launching them on foreign launch vehicles. Even more significant than this globalization of space has been the increasing commercial value of the high frontier. For the first time since the dawn of the Space Age, commercial investment in space surpassed government spending. This commercial growth is a significant milestone on the road to the stars.

Although Hollywood has tended to portray space as a hostile and largely empty realm, to commercial companies, space is a place rich with potential. Inexhaustible energy from the Sun, mineral wealth on the Moon and asteroids, unique microgravity conditions for manufacturing exotic materials and medicines of unprecedented purity—all of these opportunities, and more, await intrepid entrepreneurs. Visionaries speak of hypersonic space planes transporting packages and people from New York to Tokyo in a matter of minutes, of tourism and vacations in space, and of colonization of our solar system in the next few decades. Ultimately, however, corporate visionaries will have to budget carefully. Unlike the government, corporations risk their own capital, and shareholders expect a return on their investment. Commercial activities that prosper in space, therefore, must produce profits for shareholders.

Figure 2-53. GPS Surveying. Precision surveying is one of many industries that benefit from the GPS navigation signal. *(Courtesy of Leica Geosystems, Inc.)*

Currently, the most profitable sector of civilian space activity involves information and communications. The Global Positioning System (GPS) was developed by the U.S. military to provide pinpoint navigation information for airplanes, ships, and troops world-wide. However, the civilian applications of this now essential system have created an $8 billion industry (Figure 2-53), involving everything from hand-held receivers for camping to satellite-guided navigation systems for your car.

Communication services may offer the biggest bonanza for space commercialization. Several companies are currently building global communication networks to support cellular phones and high-speed, digital, data transmission. The first of these to come to market is the 66-satellite constellation, called the Iridium System. Built by Motorola, at a cost of $5 billion, as part of a world-wide consortium of companies, Iridium LLC now has launched its global cellular phone service (Figure 2-54). Anywhere on Earth, from the top of Mount Everest to the South Pole, is a phone call away.

Several other companies plan to build low-Earth-orbit communication constellations to provide world-wide telephone and internet services. By some industry projections, perhaps 1800 new satellites will begin operating between 1999 and 2008 (to join the 500 or so currently in orbit), with U.S. government launches accounting for only ten percent of these.

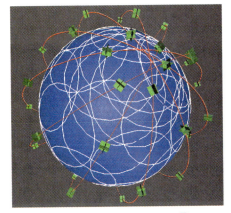

Figure 2-54. Iridium Coverage. To guarantee that global phone calls have high reliability, the Iridium constellation uses 66 satellites that cover the globe. *(Courtesy of Personal Satellite Network)*

Figure 2-55.   Pegasus. It's small but offers a flexible launch site and azimuth for small spacecraft. Here, a Pegasus XL begins its trajectory into space. *(Courtesy of Orbital Sciences Corporation)*

Figure 2-56.   Proton. This launch vehicle carries the Inmarsat III communication satellite to its geostationary orbit from Baikonur Cosmodrome, in Khazakstan. *(Courtesy of International Launch Services)*

One of the biggest drivers of increased commercialization has been the availability, and world-wide marketplace for launch services. Arianespace, a consortium of European countries headed by France, led the way in capitalizing on industry demands for reliable, affordable, launch opportunities. The Ariane IV launcher was the commercial workhorse of the 1990's, launching dozens of communication and remote sensing payloads. The newest launcher in the series, Ariane V, offers increased capability to boost larger and more complex geostationary platforms.

Small launchers, aimed at the increasing market for small satellites, also came of age during the '90s. The Pegasus launch vehicle, built by Orbital Sciences Corporation in the U.S., resembles a small winged airplane when a converted L1011 jet carries it aloft (Figure 2-55). With a 1000 kg (2205 lbs.) spacecraft tucked inside the nosecone, it drops off the mother plane at an altitude of 10,700 m (35,000 ft.), then ignites a solid rocket to propel it into low-Earth orbit.

The end of the Cold War also opened Russian launch vehicles to the world market. The Russians have used their traditional, dedicated launchers such as the Proton (Figure 2-56), Zenit, and Tsyklon to deliver commercial payloads to orbit. They also used converted ICBMs, such as the SS-18 Satan, now called the Dnepr launch vehicle. The U.S. has also converted some of its ICBM stockpile into launch vehicles, such as the Titan II and Minuteman missiles. Issues about stifling the U.S. launch market have made this a slower process than in the former Soviet Union.

The reduced cost of access to space, especially for small spacecraft that can "hitch hike" a low-cost ride, has been a boon to the small satellite industry. The Ariane launcher pioneered this capability with its Ariane Structure for Auxiliary Payloads (ASAP). The Ariane IV ASAP can carry up to six 50-kg "microsatellites," as secondary payloads, each about the size of a small filing cabinet. Its bigger brother, the Ariane V, can carry small satellites up to 100-kg in mass. These opportunities, together with low-cost surplus Russian launchers, enabled universities and small companies to build inexpensive spacecraft, using off-the-shelf terrestrial technology, rather than the specialized components developed especially for space. Surrey Satellite Technology, Ltd., a university-owned company, has focused exclusively on this market, building over a dozen microsatellites for a variety of commercial, scientific, and government applications. Other universities and organizations around the world have followed suit, turning microsatellites from interesting toys into serious business.

But space launches are still expensive, at least $20,000 per kg delivered to orbit, making commercial exploitation of space risky. So, commercial ventures need innovative launch ideas. Several companies are competing to produce reliable, and in some cases reusable launch vehicles at much lower cost and with a higher launch rate than is currently possible using the Space Shuttle or expendable rockets. Within the U.S., the Air Force-sponsored Evolved Expendable Launch Vehicle (EELV) project aims at building a reduced-cost launcher by focusing on simplified design and operations, using conventional technology. Other programs are trying to reduce launch costs by applying unconventional technology. The K-1 launcher proposed by Kistler Aerospace, for example, will use rocket engines developed for the Soviet Lunar program, with airbags to recover its two-stage reusable vehicle. The X-33 is a single-stage-to-orbit proof-of-concept being developed by a NASA-Industry team. This program is pushing the state-of-the-art in propulsion, automatic controls, and structural materials.

Modeled after the aviation prizes of the early 20th century that spurred pioneers such as Charles Lindbergh to push the envelope of technical capability, the X-Prize hopes to encourage launch vehicle innovation. Its sponsors offer up to $10M to the first group that can launch a person to an altitude of 160 km (100 mi.), return him or her safely to Earth, and repeat it within a few weeks. Many organizations around the world have thrown their hat in the ring, but so far, no one has claimed the prize. The group that does may well point the way to a new era of space travel for ordinary citizens and open the way for a new industry of space tourism.

The New High Ground

The increasing importance of space as the new high ground for military operations became obvious during the 1990's. During the Gulf War to expel Iraqi soldiers from Kuwait in 1991, space assets provided navigation, communications, intelligence, and imagery that were essential to increasing the combat effectiveness of the allied coalition, while limiting civilian casualties in Iraq. Space was so crucial to nearly every aspect of the operation that some military leaders called it the first true "space war." They began to speak of space as a strategic center of gravity essential for national security that we had to defend and control.

To the U.S. military, space does not represent an entirely new mission, but rather, a new environment that can enhance traditional missions. In defining the importance of space to national security, the Air Force identifies ways in which effective exploitation of space can enhance global awareness, global reach, and global power. Historically, satellites have played crucial roles in enhancing global awareness. These roles include intelligence, surveillance, and reconnaissance (ISR), as well as weather prediction (Figure 2-57) and early warning of attacks (Figure 2-58). ISR includes high-resolution imagery, electronic eavesdropping, treaty monitoring and verification, determining enemy capabilities and movements, and related functions. Global reach—the ability to deploy troops or weapons anywhere in the world, rapidly and effectively—depends crucially on satellite communications. Likewise, global power depends on communications for command and control, and precise coordinates for better weapon accuracy. Global Positioning System (GPS) satellites (Figure 2-59) are especially important to ensuring that weapons hit their intended military targets.

In attacking targets, the military needs to find, lock on, track, target, and engage the enemy. Engaging a target then leads to "kill assessment"—determining whether the weapon hit the target, and if so, how badly it damaged the target (known as battle damage assessment, or BDA). Sensors located in space are essential to all of these attack phases,

Figure 2-57. Defense Meteorological Satellite Program. A small constellation of military weather satellites provide global coverage for military operations. *(Courtesy of the U.S. Air Force)*

Figure 2-58. Defense Support Program (DSP). DSP satellites sit in geostationary orbits, "watching" constantly for ballistic missile launches. *(Courtesy of the U.S. Air Force)*

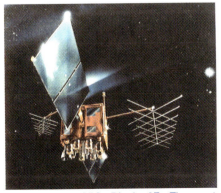

Figure 2-59. GPS Block 2F. The next generation of GPS satellites will make improvements in signal security and system accuracy. *(Courtesy of the U.S. Air Force)*

except engagement—narrowly defined here as placing bombs on target. For example, spacecraft track targets from space with sophisticated infrared, optical, radar, or other sensors. After an attack, these same sensors gather data for the BDA phase. If the military ever places automated weapons, such as lasers, in space, it is conceivable that all phases of an attack would take place from orbit. Conceptually, military theorists suggest that space-based offensive weapons would constitute a military-technical revolution analogous to the invention of gunpowder weapons or airplanes.

If space promises to revolutionize offensive warfare, it is equally likely that space may revolutionize defensive warfare. Space defense today focuses on intercepting ballistic missiles that may carry nuclear, biological, or chemical warheads (known as weapons of mass destruction). The U.S. military is currently considering deploying a limited defensive system against ballistic missile attacks by rogue states.

In the next fifty years, space promises to become a new arena of intense economic competition. As space becomes increasingly important to communication, manufacturing, mining, and other commercial activities, space assets become richer targets. Space assets are important enough today, such that many countries treat attacks upon them as acts of war. The United States considered the threat of such attacks to be serious enough to warrant the creation in 1993 of a Space Warfare Center at Schriever Air Force Base, Colorado.

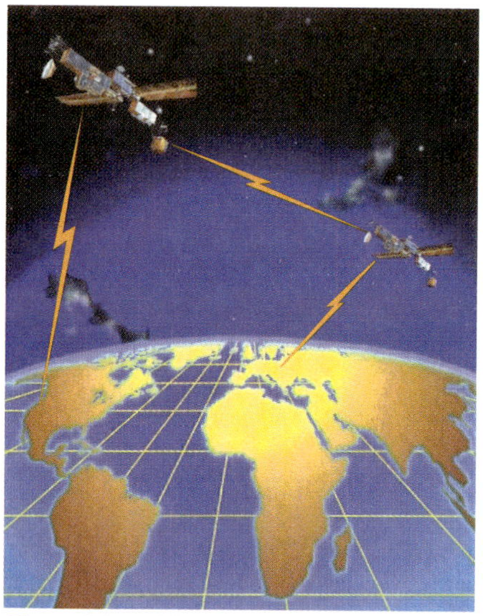

Ironically, this realization of the critical role space plays in national security comes as overall military influence in space began to wane. For most of the Space Age, government and military needs dominated. But as space became increasingly commercialized, this influence began to diminish. Dr. Daniel Hastings, Chief Scientist of the Air Force, noted that, if commercial companies can build cheaper launch vehicles and communication networks, the government would be foolish to duplicate these efforts. Instead, the government should focus on missions and technologies vital to national security, such as ballistic-missile early warning and defense.

Despite potential technical and political hurdles, it is likely that military reliance on space will continue to increase. Not only does space provide global perspective as the new high ground, it also provides an arena for global communication (Figure 2-60) and information gathering. Already, U.S. military leadership has suggested the U.S. Air Force evolve to become the Air & Space Force, and eventually the Space & Air Force. Starships operated by a "Star Fleet" as depicted in *Star Trek* may eventually become science fact instead of fiction.

Figure 2-60.   Milstar Crosslinks. The Milstar constellation provides worldwide communication using crosslinks between spacecraft. *(Courtesy of the U.S. Air Force)*

The Future

Clearly, space remains an environment of untold, unimaginable, and perhaps unfathomable possibilities and wonders. Yet space also remains an unforgiving environment, and therefore space exploration is inherently risky. The *Challenger* explosion in 1986 sobered Americans to the risks of space exploration. And a string of six U.S. rocket-launch failures in 1998–1999 reminds us that spaceflight is not yet routine. Goals worth achieving, however, usually carry with them a measure of risk, and men and women have shown themselves willing to take great risks for even greater rewards.

The most ambitious goal for the first part of the 21st century would be a manned mission to Mars (Figure 2-61). Although a mission to Mars remains uncertain, other manned missions in space hold considerable promise. Men and women will continue to fly on the Shuttle and its successor, and to live and work in orbit onboard the International Space Station. Where humanity goes from here will be exciting to see. The astrophysicist Freeman Dyson has suggested that the next fifty years may witness an era of cheap unmanned missions, whereas the following fifty years may inaugurate an era of cheap manned spaceflight throughout our solar system. Manned interstellar missions over immense distances, however, will most likely remain impractical for the foreseeable future.

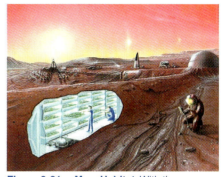

Figure 2-61.   Mars Habitat. With the success of the recent Mars robotic explorers, many people envision a day when humans will live on Mars. *(Courtesy of NASA/Ames Research Center)*

New interplanetary probes will continue to amaze us with unexpected information about our solar system. Many practical lessons have emerged from studies of Earth's neighbors, and this will doubtless continue. Further studies of Venus's atmosphere, for example, may shed more light on the greenhouse effect and global warming from carbon dioxide emissions on our own planet. Many, many scientific questions about our universe remain unanswered. Many also remain unasked. Black holes, naked singularities, pulsars, quasars, the Big Bang theory, the expanding universe—scientists, with the help of space-based instruments and probes, have much left to explore.

This chapter has provided historical perspective on the evolution of humanity's knowledge and exploration of space. During the rest of the book, we'll focus on scientific and technical aspects of exploring space. We'll look at orbits, spacecraft, rockets, and operational concepts that are critical to space missions. But in Chapter 16, we'll return to these emerging space trends to explore how technical, political, and economic issues form the context for future missions and are shaping our destiny in the High Frontier. Yet ultimately it's our human qualities that will shape our destiny in the universe. Men and women shall continue to quest for knowledge about space, and to explore realms beyond the warm and reassuring glow of Earth's atmosphere, as long as we remain creative, imaginative, and strong-minded.

▦ Section Review

Key Terms	Key Concepts

Key Terms

commercialization of space

Key Concepts

➤ Space exploration and science made great strides during the 1990s

- Magellan's synthetic aperture radar mapped more than 98% of Venus's surface

- Mars Pathfinder successfully landed and explored a small part of the Martian surface. The Mars Global Surveyor orbited Mars and mapped most of its surface, watching for large dust storms.

- Lunar Prospector orbited the Moon and discovered vast amounts of water ice that makes a lunar base more feasible

- The Galileo spacecraft took unique photos of the comet Shoemaker-Levy 9 as it smashed into Jupiter

- Ulysses orbited the Sun in a polar orbit and gathered data on the solar corona, solar wind and other properties of the heliosphere

- The Hubble Space Telescope expanded our understanding of our solar system and the universe with spectacular photos of the outer planets and their moons, giant black holes, and previously unseen galaxies

➤ Manned spaceflight continued to be productive in low-Earth orbit

- The Space Shuttle launched space probes, deployed Earth satellites, docked with the Mir space station, and conducted numerous experiments in space science

- The Mir housed several international astronaut teams, which conducted experiments, learned to live for long periods in free-fall, and solved major equipment problems with limited resources

- The first components of the International Space Station arrived in orbit using the U.S. Space Shuttle and a Russian Proton booster

➤ Military use of space leaped forward in many areas

- Intelligence gathering, surveillance, and reconnaissance continue to be important

- Military satellites provide secure communication capability; routine military calls use commercial satellite services

- The Global Positioning System (GPS) revolutionized the way planes, ships, and ground vehicles navigate and deliver weapons

- Fielding an Antiballistic Missile system remains a high U.S. military priority, with plans to use spaceborne sensors for locating and tracking enemy missiles

Continued on next page

▦ Section Review (Continued)

Key Concepts (Continued)

➤ Commercial space activities experienced tremendous growth in the 90s

- Communication constellations took shape, enabling global cellular telephone services

- Commercial uses of GPS blossomed into a vast industry. These uses include surveying; land, sea, and air navigation; accurate crop fertilizing and watering; delivery fleet location and optimal control; and recreational travel.

- Ventures to design and build single-stage-to-orbit launch vehicles pushed the state of the art in propulsion, hypersonic control, and high-temperature/high pressure materials

➤ A manned mission to Mars may become reality with a strong design team's effort to hold down costs, while planning a safe, productive journey to the Red Planet

▆ For Further Reading

Baker, David. *The History of Manned Space Flight*. New York: Crown Publishers, 1981.

Baucom, Donald R. *The Origins of SDI, 1944–1983*. Lawrence, KS: University Press of Kansas, 1992.

Buedeler, Werner. *Geschichte der Raumfahrt*. Germany: Sigloch Edition, 1982.

Burrows, William E. *This New Ocean: The Story of the First Space Age*. New York: Random House, 1998.

Chaikin, Andrew. *A Man on the Moon: The Voyage of the Apollo Astronauts*. New York: Viking, 1994.

Compton, William D. *Where No Man Has Gone Before: A History of Apollo Lunar Exploration Missions*. Washington, D.C.: National Aeronautics and Space Administration, 1989.

Crowe, Michael J. *Extraterrestrial Life Debate, 1750 - 1900: The Idea of a Plurality of Worlds*. Cambridge and New York: Cambridge University Press, 1986.

Dick, Steven J. *Plurality of Worlds: The Origins of the Extraterrestrial Life Debate from Democritus to Kant*. Cambridge and New York: Cambridge University Press, 1982.

Dyson, Freeman J. "The Future of Space Exploration: Warm-Blooded Plants and Freeze-Dried Fish." *Atlantic Monthly*, November 1997, 71-80.

Galilei, Galileo. *Sidereus Nuncius*. 1610. Translated with introduction, conclusion, and notes by Albert Van Helden. Chicago, IL: University of Chicago Press, 1989.

Galilei, Galileo. *Dialogue Concerning the Two Chief World Systems*. (orig. 1632) Translated by Stillman Drake. Berkeley, California: University of California Press, 1967.

Gingerich, Owen. "Islamic Astronomy." *Scientific American*, April 1986, pp. 74-83.

Gray, Colin S. *American Military Space Policy*. Cambridge, MA: Abt Books, 1983.

Hays, Peter L. et al., eds. *Spacepower for a New Millennium: Space and U.S. National Security*. New York: McGraw-Hill, 1999.

Hearnshaw, J.B. *The Analysis of Starlight: One Hundred and Fifty Years of Astronomical Spectroscopy*. Cambridge, MA: Cambridge University Press, 1986.

Hetherington, Norriss S. *Science and Objectivity: Episodes in the History of Astronomy*. Ames, IA: Iowa University Press, 1988.

Johnson, Dana J., Scott Pace and C. Bryan Gabbard. *Space: Emerging Options for National Power*. Santa Monica, CA: RAND, 1998.

Kepler, Johannes. *IOH. Keppleri Mathematitici Olim Imperatorii Somnium, Sev Opus posthumum De Astronomia Lunari. Divulgatum M. Ludovico Kepplero Filio, Medicinae Candidato*. 1634.

Krieger, Firmin J. *Behind the Sputniks: A Survey of Soviet Space Science*. The Rand Corporation: Washington, D.C.: Public Affairs Press, 1958.

Launius, Roger D. "Toward an Understanding of the Space Shuttle: A Historiographical Essay." *Air Power History*, Winter 1992, 3-18.

Lewis, John S. *Mining the Sky: Untold Riches from the Asteroids, Comets, and Planets*. Reading, MA: Helix Books, 1997.

Lewis, Richard. *Space in the 21st Century*. New York: Columbia University Press, 1990.

Logsdon, John M. *The Decision to Go to the Moon: Project Apollo and the National Interest*. Cambridge, MA: MIT Press, 1970.

McCurdy, Howard E. *Space and the American Imagination*. Washington, D.C.: Smithsonian Institution Press, 1997.

McCurdy, Howard E. *Inside NASA: High Technology and Organizational Change in the U.S. Space Program*. Baltimore, MD: Johns Hopkins University Press, 1993.

McDougall, Walter A. *...the Heavens and the Earth: A Political History of the Space Age*. New York: Basic Books, 1985.

Michener, James. *Space*. New York: Ballantine Books, 1982.

Newell, Homer E. *Beyond the Atmosphere: Early Years of Space Science*. Washington, D.C.: Scientific and Technical Information Branch, NASA, 1980.

Peebles, Curtis. *High Frontier: The United States Air Force and the Military Space Program*. Washington, D.C.: Air Force History and Museums Program, 1997.

Preston, Bob. *Plowshares and Power: The Military Use of Civil Space*. Washington: National Defense University Press, 1995.

Rycroft, Michael (ed.), *The Cambridge Encyclopedia of Space*. New York, NY: Press Syndicate of the University of Cambridge, 1990.

Sheehan, William. *Planets and Perception: Telescopic Views and Interpretations, 1609–1909*. Tucson, AZ: University of Arizona Press, 1988.

Smith, Robert W. *The Expanding Universe: Astronomy's 'Great Debate' 1900–1931*. Cambridge: Cambridge University Press, 1982.

Spires, David N. *Beyond Horizons: A Half Century of Space Leadership*. Colorado Springs, CO: Air Force Space Command, 1997.

Van Helden, Albert. *Measuring the Universe: Cosmic Dimensions from Aristarchus to Halley*. Chicago, IL: University of Chicago Press, 1985.

Vaughan, Diane. *The Challenger Launch Decision: Risky Technology, Culture, and Deviance at NASA*. Chicago, IL: University of Chicago Press, 1996.

Wilson, Andrew. *Space Directory*. Alexandria, VA: Jane's Information Group Inc., 1990.

Winter, Frank H. *Rockets into Space*. Cambridge, MA: Harvard University Press, 1989.

Winter, Frank H. *The First Golden Age of Rocketry; Congreve and Hale Rockets of the Nineteenth Century*. Washington, D.C.: Smithsonian Institution Press, 1989.

Wolfe, Tom. *The Right Stuff*. New York: Farrar, Straus & Giroux, 1979.

Zubrin, Robert. *The Case for Mars: The Plan to Settle the Red Planet and Why We Must*. New York: The Free Press, 1996.

Mission Problems

2.1 Early Space Explorers

1   Why did astronomers continue to believe the planets followed circular orbits from Aristotle's time until Kepler's discovery?

2   Why were people reluctant to adopt Copernicus's heliocentric system of the universe?

3   Why might we call Brahe and Kepler a perfect team?

4   List four theories from Aristotle that Galileo disproved.

5   How did Newton complete the Astronomical Revolution of the 17th century?

6   Why are instruments such as telescopes such powerful tools for learning and discovery?

7   How did our concept of the universe change in the first few decades of the 20th century?

2.2 Entering Space

8 What are the various ways that humans can explore space?

9 Why might we call 1957–1965 the years of Russian supremacy in space?

10 What was the chief legacy of the Apollo manned missions to the Moon?

2.3 Space Comes of Age

11 What are some of the key discoveries made during the 1990s with the Hubble Space Telescope?

12 Why should we build an International Space Station? What are its benefits and drawbacks?

13 How might we reduce costs of a manned mission to Mars? If humans venture to Mars, which planet should we visit next?

14 Why are commercial companies becoming interested in space?

15 What missions do militaries hope to accomplish in space? Should weapons be deployed in space?

16 Besides Earth, where are we most likely to find life in our solar system?

For Discussion

17 Do we need to launch men and women into space, or should we rely exclusively on probes and Earth-based instruments to explore space?

18 In exploring space, can we learn anything that can help us solve problems here on Earth?

19 How much do movies and television shows such as *2001: A Space Odyssey*, *Alien*, *Star Wars*, *Star Trek*, and *Contact* point the way to humanity's future in space?

20 Has the search for extraterrestrials been important to the development of astronomy? Should we continue searching for other life forms and inhabited planets?

The German Army first developed the V-2 rocket in [19]42. V-2 stood for "Vengeance Weapon 2," indicative of [Hi]tler's wish for a weapon which could conquer the [w]orld. Yet, the scientists who developed the missile [w]orked independently of Hitler's influence until the [mi]ssile was ready. They called their experimental rocket [th]e A-4. Hitler actually had little interest in the rocket's [de]velopment and failed to adequately fund the project [un]til it was too late to decisively employ it in the Second [W]orld War. The rocket had no single inventor. Rather, it [re]sulted from a team effort of individual Germans, who [en]visioned the good and ill of applied modern rocketry.

Mission Overview

[Th]e V-2 (or A-4) program had obscure beginnings in [th]e late 1920s. Many of the members of the initial V-2 [te]am wished to create a long range rocket, which [m]ight serve as a stepping stone for future spaceflight [ap]plications. Yet the mission remained open-ended: [th]e small team of scientists at Peenemuende (located [on] the Baltic coast) did not focus on the future. Because [th]e Treaty of Versailles (ending World War I) forbade [G]ermans from producing mass artillery, the project's [in]itial drive was to devise a new powerful weapon not [ou]tlawed by the treaty. The Germans officially [co]nstructed and funded the missile as a tactical [w]eapon with improved capabilities and a longer range [th]an the existing long-range artillery.

Mission Data

The V-2 was a single-stage rocket which burned a liquid oxygen and kerosene mixture for a thrust of 244,600 N (55,000 lbs.)

The maximum design range of the missile was 275 km (171 mi.), enabling Hitler to bomb London, as well as continental allied countries.

The maximum altitude reached by the V-2 was 83 km (52 mi), for a total trajectory distance of 190 km (118 mi). These distances were extraordinarily better than any previous missile achieved.

The overall missile length was 14 m (46 ft.), with a maximum width of 3.57 m (11 ft. 8.5 in.)

They never mass-produced the V-2 at Peenemuende. They moved production to mainland Germany in 1944, and built 3,745 and launched most of those for the Axis war effort.

✓ The V-2 design team produced 60,000 necessary changes after they considered it ready for mass production.

✓ The primary advantage of the V-2 was its low cost: $38,000. This compared favorably to the $1,250,000 cost of a manned German bomber.

Mission Impact

While many of the original Peenemuende team dreamed of possible implications for space travel, the V-2 was first and foremost a machine of war. Yet, it was the first supersonic rocket and we generally regard it as a monumental step in the history of modern rocket technology. After the war ended, the V-2 technology and many German scientists came to the United States, forming the foundation of the future U.S. space program.

(Courtesy of Sigloch Edition)

For Discussion

• Can you think of any other technological advances that began and grew through military channels?

• Was the V-2 truly the beginning of the drive towards space travel? What important developments led to the U.S. drive to place a man on the moon?

• What happened to many of the top scientists from Peenemuende after World War II? Do you think that scientific interest should supersede political agendas?

Contributor

Troy Kitch, the U.S. Air Force Academy

References

Emme, Eugene M., (ed.) *The History of Rocket Technology,* "The German V-2" by Walter R. Dornberger. Detroit, MI: Wayne State University Press, 1964.

Gatland, Kenneth. *Missiles and Rockets.* New York: Macmillan Publishing Co., 1975.

An astronaut's eye view of our blue planet. (*Courtesy of NASA/Johnson Space Center*)

The Space Environment

3

In This Chapter You'll Learn to...

- ☛ Explain where space begins and describe our place in the universe
- ☛ List the major hazards of the space environment and describe their effects on spacecraft
- ☛ List and describe the major hazards of the space environment that pose a problem for humans living and working in space

You Should Already Know...

- ❑ The elements of a space mission (Chapter 1)

Outline

3.1 Cosmic Perspective
Where is Space?
The Solar System
The Cosmos

3.2 The Space Environment and Spacecraft
Gravity
Atmosphere
Vacuum
Micrometeoroids and Space
    Junk
The Radiation Environment
Charged Particles

3.3 Living and Working in Space
Free fall
Radiation and Charged Particles
Psychological Effects

In space, no one can hear you whine.

Anonymous

Space is a place. Some people think of space as a nebulous region far above their heads—extending out to infinity. But for us, space is a place where things happen: spacecraft orbit Earth, planets orbit the Sun, and the Sun revolves around the center of our galaxy.

In this chapter we'll look at this place we call space, exploring where it begins and how far it extends. We'll see that space is actually very close (Figure 3-1). Then, starting with our "local neighborhood," we'll take a mind-expanding tour beyond the galaxy to see what's in space. Next we'll see what space is like. Before taking any trip, we usually check the weather, so we'll know whether to pack a swim suit or a parka. In the same way, we'll look at the space environment to see how we must prepare ourselves and our machines to handle this hostile environment.

Figure 3-1.   Earth and Moon. Although in the night sky the Moon looks really far away, Earth's atmosphere is relatively shallow, so space is close. *(Courtesy of NASA/Ames Research Center)*

3.1 Cosmic Perspective

In This Section You'll Learn to...

☞ Explain where space is and how it's defined

☞ Describe the primary outputs from the Sun that dominate the space environment

☞ Provide some perspective on the size of space

Where is Space?

If space is a place, where is it? Safe within the cocoon of Earth's atmosphere, we can stare into the night sky at thousands of stars spanning millions of light years. We know space begins somewhere above our heads, but how far? If we "push the envelope" of a powerful jet fighter plane, we can barely make it to a height where the sky takes on a purplish color and stars become visible in daylight. But even then, we're not quite in space. Only by climbing aboard a rocket can we escape Earth's atmosphere into the realm we normally think of as space.

But the line between where the atmosphere ends and space begins is, by no means, clear. In fact, there is no universally accepted definition of precisely where space begins. If you ask NASA or the U.S. Air Force, you'll find their definition of space is somewhat arbitrary. To earn astronaut wings, for example, you must reach an altitude of more than 92.6 km (57.5 mi.) but don't actually have to go into orbit, as illustrated in Figure 3-2. (That's why X-15 pilots and the first United States' astronauts to fly suborbital flights in the Mercury program were able to wear these much-coveted wings.) Although this definition works, it's not very meaningful.

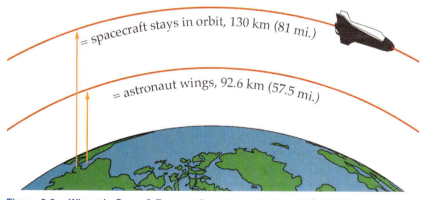

= spacecraft stays in orbit, 130 km (81 mi.)

= astronaut wings, 92.6 km (57.5 mi.)

Figure 3-2. Where is Space? For awarding astronaut wings, NASA defines space at an altitude of 92.6 km (57.5 mi.). For our purposes, space begins where satellites can maintain orbit—about 130 km (81 mi.).

For our purposes, space begins at the altitude where an object in orbit will remain in orbit briefly (only a day or two in some cases) before the

Figure 3-3. Shuttle Orbit Drawn Closer to Scale. (If drawn exactly to scale, you wouldn't be able to see it!) As you can see, space is very close. Space Shuttle orbits are just barely above the atmosphere.

Figure 3-4.   The Sun. It's our source of light and heat, but with the beneficial emissions, come some pretty nasty radiation. This Solar and Heliospheric Observatory (SOHO) satellite using the extreme ultraviolet imaging telescope shows how active our Sun is. *(Courtesy of SOHO/Extreme-ultraviolet Imaging Telescope consortium. SOHO is a project of international cooperation between ESA and NASA)*

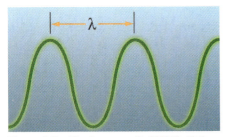

Figure 3-5.   Electromagnetic (EM) Radiation. We classify EM radiation in terms of the wavelength, λ, (or frequency) of the energy.

wispy air molecules in the upper atmosphere drag it back to Earth. This occurs above an altitude of about 130 km (81 mi.). That's about the distance you can drive in your car in just over an hour! So the next time someone asks you, "how do I get to space?" just tell them to "turn straight up and go about 130 km (81 mi.) until the stars come out."

As you can see, space is very close. Normally, when you see drawings of orbits around Earth (as you'll see in later chapters), they look far, far away. But these diagrams are seldom drawn to scale. To put low-Earth orbits (LEO), like the ones flown by the Space Shuttle, into perspective, imagine Earth were the size of a peach—then a typical Shuttle orbit would be just above the fuzz. A diagram closer to scale (but not exactly) is shown in Figure 3-3.

Now that we have some idea of where space is, let's take a grand tour of our "local neighborhood" to see what's out there. We'll begin by looking at the solar system, then expand our view to cover the galaxy.

The Solar System

At the center of the solar system is the star closest to Earth—the Sun (Figure 3-4). As we'll see, the Sun has the biggest effect on the space environment. As stars go, our Sun is quite ordinary. It's just one small, yellow star out of billions in the galaxy. Fueled by nuclear fusion, it combines or "fuses" 600 million tons of hydrogen each second. (Don't worry, at that rate it won't run out of hydrogen for about 5,000,000,000 years!). We're most interested in two by-products of the fusion process

- Electromagnetic radiation

- Charged particles

The energy released by nuclear fusion is governed by Einstein's famous $E = m c^2$ formula. This energy, of course, makes life on Earth possible. And the Sun produces lots of energy, enough each second to supply all the energy the United States needs for about 624 million years! This energy is primarily in the form of electromagnetic radiation. In a clear, blue sky, the Sun appears as an intensely bright circle of light. With your eyes closed on a summer day, you can feel the Sun's heat beating on you. But light and heat are only part of it's *electromagnetic (EM) radiation*. The term "radiation" often conjures up visions of nuclear wars and mutant space creatures, but EM radiation is something we live with every day. EM radiation is a way for energy to get from one place to another. We can think of the Sun's intense energy as radiating from its surface in all directions in waves. We classify these waves of radiant energy in terms of the distance between wave crests, or *wavelength, λ*, as in Figure 3-5.

What difference does changing the wavelength make? If you've ever seen a rainbow on a sunny spring day, you've seen the awesome beauty of changing the wavelength of EM radiation by only 0.0000003 meters $(9.8 \times 10^{-7}$ ft.)! The colors of the rainbow, from violet to red, represent only a very small fraction of the entire electromagnetic spectrum. This spectrum spans from high energy X-rays (like you get in the dentist's office) at one end, to long-wavelength radio waves (like your favorite FM station) at the other. Light and all radiation move at the speed of light—300,000 km/s or more than 671 million m.p.h.! As we'll see, solar radiation can be both helpful and harmful to spacecraft and humans in space. We'll learn more about the uses for EM radiation in Chapter 11.

The other fusion by-product we're concerned with is charged particles. Scientists model atoms with three building-block particles—protons, electrons, and neutrons, as illustrated in Figure 3-6. Protons and electrons are *charged particles*. Protons have a positive charge, and electrons have a negative charge. The neutron, because it doesn't have a charge, is neutral. Protons and neutrons make up the nucleus or center of an atom. Electrons swirl around this dense nucleus.

During fusion, the Sun's interior generates intense heat (more than 1,000,000° C). At these temperatures, a fourth state of matter exists. We're all familiar with the other three states of matter—solid, liquid, and gas. If we take a block of ice (a solid) and heat it, we get water (a liquid). If we continue to heat the water, it begins to boil, and turns into steam (a gas). However, if we continue to heat the steam, we'd eventually get to a point where the water molecules begin to break down. Eventually, the atoms will break into their basic particles and form a hot *plasma*. Thus, inside the Sun, we have a swirling hot soup of charged particles—free electrons and protons. (A neutron quickly decays into a proton plus an electron.)

These charged particles in the Sun don't stay put. All charged particles respond to electric and magnetic fields. Your television set, for example, takes advantage of this by using a magnet to focus a beam of electrons at the screen to make it glow. Similarly, the Sun has an intense magnetic field, so electrons and protons shoot away from the Sun at speeds of 300 to 700 km/s (about 671,000 to 1,566,000 m.p.h.). This stream of charged particles flying off the Sun is called the *solar wind*.

Occasionally, areas of the Sun's surface erupt in gigantic bursts of charged particles called *solar particle events* or *solar flares*, shown in Figure 3-7, that make all of the nuclear weapons on Earth look like pop guns. Lasting only a few days or less, these flares are sometimes so violent they extend out to Earth's orbit (150 million km or 93 million mi.)! Fortunately, such large flares are infrequent (every few years or so) and concentrated in specific regions of space, so they usually miss Earth. Later, we'll see what kinds of problems these charged particles from the solar wind and solar flares pose to machines and humans in space.

Besides the star of the show, the Sun, nine planets, dozens of moons, and thousands of asteroids are in our solar system (Figure 3-8). The planets range from the small terrestrial-class ones—Mercury, Venus, Earth, and Mars—to the mighty gas giants—Jupiter, Saturn, Uranus, and Neptune. Tiny Pluto is all alone at the edge of the solar system and may be a lost moon of Neptune. Figure 3-9 tries to give some perspective on the size of the solar system, and Appendix D.4 gives some physical data on the planets. However, because we tend to spend most of our time near Earth, we'll focus our discussion of the space environment on spacecraft and astronauts in Earth orbits.

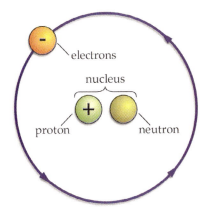

Figure 3-6.   The Atom. The nucleus of an atom contains positively charged protons and neutral neutrons. Around the nucleus are negatively charged electrons.

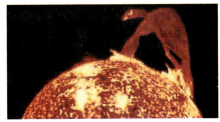

Figure 3-7.   Solar Flares. They fly out from the Sun long distances, at high speeds, and can disrupt radio signals on Earth, and disturb spacecraft orbits near Earth. *(Courtesy of NASA/ Johnson Space Center)*

Figure 3-8.   Solar System. Nine planets and many other objects orbit the Sun, which holds the solar system together with its gravity. *(Courtesy of NASA/Jet Propulsion Laboratory)*

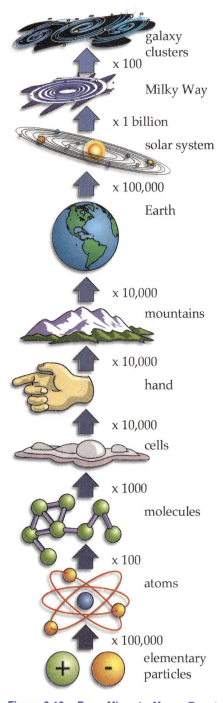

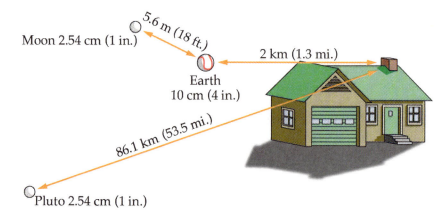

Figure 3-9.   The Solar System in Perspective. If the Earth were the size of a baseball, about 10 cm (~4 in.) in diameter, the Moon would be only 2.54 cm (1 in.) in diameter and about 5.6 m (18 ft.) away. At the same scale the Sun would be a ball 10 m (33 ft.) in diameter (about the size and volume of a small two-bedroom house); it would be more than 2 km (nearly 1.3 mi.) away. Again, keeping the same scale, the smallest planet Pluto would be about the same size as Earth's Moon, 2.54 cm (1 in.), and 86.1 km (53.5 mi.) away from the house-sized Sun.

The Cosmos

Space is big. Really BIG. Besides our Sun, more than 300 billion other stars are in our neighborhood—the Milky Way galaxy. Because the distances involved are so vast, normal human reckoning (kilometers or miles) loses meaning. When trying to understand the importance of charged particles in the grand scheme of the universe, for example, the mind boggles. Figure 3-10 tries to put human references on a scale with the other micro and macro dimensions of the universe.

One convenient yardstick we use to discuss stellar distances is the light year. One *light year* is the distance light can travel in one year. At 300,000 km/s, this is about 9.46×10^{12} km (about 5.88 trillion mi.). Using this measure, we can begin to describe our location with respect to everything else in the universe. The Milky Way galaxy is spiral shaped and is about 100,000 light years across. Our Sun and its solar system is about half way out from the center (about 25,000 light years) on one of the spiral arms. The Milky Way (and we along with it) slowly revolves around the galactic center, completing one revolution every 240 million years or so. The time it takes to revolve once around the center of the galaxy is sometimes called a *cosmic year*. In these terms, astronomers think our solar system is about 20 cosmic years old (4.8 billion Earth years).

Stars in our galaxy are very spread out. The closest star to our solar system is Proxima Centauri at 4.22 light years or 4.0×10^{13} km away. The Voyager spacecraft, currently moving at 56,400 km/hr. (35,000 m.p.h.), would take more than 80,000 years to get there! Trying to imagine these kinds of distances gives most of us a headache. The nearest galaxy to our own is Andromeda, which is about 2 million light years away. Beyond Andromeda are billions and billions of other galaxies, arranged in strange configurations which astronomers are only now beginning to catalog.

Figure 3-10.   From Micro to Macro. To get an idea about the relative size of things in the universe, start with elementary particles—protons and electrons. You can magnify them 100,000 times to reach the size of an atom, etc.

Figure 3-11 puts the distance between us and our next closest star into understandable terms. Figure 3-12 tries to do the same thing with the size of our galaxy. In the next section we'll beam back closer to home to understand the practical effects of sending machines and humans to explore the vast reaches of the cosmos.

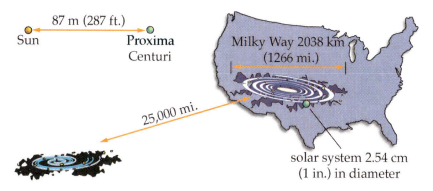

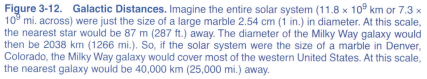

Figure 3-12. Galactic Distances. Imagine the entire solar system (11.8×10^9 km or 7.3×10^9 mi. across) were just the size of a large marble 2.54 cm (1 in.) in diameter. At this scale, the nearest star would be 87 m (287 ft.) away. The diameter of the Milky Way galaxy would then be 2038 km (1266 mi.). So, if the solar system were the size of a marble in Denver, Colorado, the Milky Way galaxy would cover most of the western United States. At this scale, the nearest galaxy would be 40,000 km (25,000 mi.) away.

Figure 3-11. Stellar Distances. Let our Sun (1.4×10^6 km or 8.6×10^6 mi. in diameter) be the size of a large marble, roughly 2.54 cm (1 in.) in diameter. At this scale, the nearest star to our solar system, Proxima Centauri, would be more than 1500 km (932 mi.) away. So, if the Sun were the size of a large marble (2.54 cm or 1 in. in diameter) in Denver, Colorado, the nearest star would be in Chicago, Illinois. At this stellar scale, the diameter of the Milky Way galaxy would then be 33.8 million km (21 million mi.) across! Still too big for us to visualize!

Astro Fun Fact
Message In A Bottle

If you were going to send a "message in a bottle" to another planet, what would you say? Dr. Carl Sagan and a committee of scientists, artists, and musicians tried to answer this question before the Voyager launched in 1977. They developed a multi-media program containing two hours of pictures, greetings, sounds, and music they felt represented Earth's variety of culture. The collection contains such items as Chuck Berry's "Johnny B. Goode," people laughing, a Pakistan street scene, and a map to find your way to the Earth from other galaxies. A record company manufactured a 12-inch copper disc to hold the information, and the record was sealed in a container along with a specially designed phonograph. This package, which scientists estimate could last more than a billion years, was placed onboard each of the Voyager spacecraft. We hope another life form will encounter one of the Voyager probes, construct the phonograph, and play back the sounds of Earth. But don't count on hearing from any space beings—the Voyagers won't visit another star until 80,000 years from now!

(Courtesy of NASA/Jet Propulsion Laboratory)

(Courtesy of NASA/Jet Propulsion Laboratory)

Eberhard, Jonathan. <u>The World on a Record</u>. Science News. Aug. 20, 1977, p. 124–125.

Wilford, John Noble. <u>Some Beings Out There Just May Be Listening</u>. New York Times Magazine. Sept. 4, 1977, p. 12–13.

Contributed by Donald Bridges, the U.S. Air Force Academy

▆ Section Review

Key Terms

charged particles
cosmic year
electromagnetic (EM) radiation
light year
plasma
solar flares
solar particle events
solar wind
wavelength, λ

Key Concepts

➤ For our purposes, space begins at an altitude where a satellite can briefly maintain an orbit. Thus, space is close. It's only about 130 km (81 mi.) straight up.

➤ The Sun is a fairly average yellow star which burns by the heat of nuclear fusion. Its surface temperature is more than 6000 K and its output includes

  • Electromagnetic radiation that we see and feel here on Earth as light and heat

  • Streams of charged particles that sweep out from the Sun as part of the solar wind

  • Solar particle events or solar flares, which are brief but intense periods of charged-particle emissions

➤ Our solar system is about half way out on one of the Milky Way galaxy's spiral arms. Our galaxy is just one of billions and billions of galaxies in the universe.

3.2   The Space Environment and Spacecraft

In This Section You'll Learn to...

- ☞ List and describe major hazards of the space environment and their effect on spacecraft

To build spacecraft that will survive the harsh space environment, we must first understand what hazards they may face. Earth, the Sun, and the cosmos combined offer unique challenges to spacecraft designers, as shown in Figure 3-13.

- The gravitational environment causes some physiological and fluid containment problems but also provides opportunities for manufacturing
- Earth's atmosphere affects a spacecraft, even in orbit
- The vacuum in space above the atmosphere gives spacecraft another challenge
- Natural and man-made objects in space pose collision hazards
- Radiation and charged particles from the Sun and the rest of the universe can severely damage unprotected spacecraft

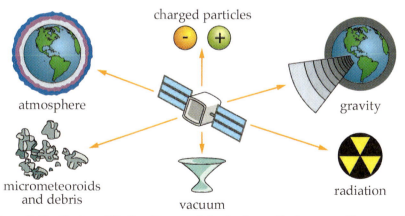

Figure 3-13.   Factors Affecting Spacecraft in the Space Environment. There are six challenges unique to the space environment we deal with—gravity, the atmosphere, vacuum, micrometeoroids and debris, radiation, and charged particles.

Gravity

Whenever we see astronauts on television floating around the Space Shuttle, as in Figure 3-14, we often hear they are in "zero gravity." But this is not true! As we'll see in Chapter 4, all objects attract each other with a gravitational force that depends on their mass (how much "stuff" they

Figure 3-14.   Astronauts in Free Fall. In the free-fall environment, astronauts Julie Payette (left) and Ellen Ochoa (STS-96) easily move supplies from the Shuttle Discovery to the Zarya module of the International Space Station. With no contact forces to slow them down, the supplies need only a gentle push to float smoothly to their new home. *(Courtesy of NASA/Johnson Space Center)*

have). This force decreases as objects get farther away from each other, so gravity doesn't just disappear once we get into space. In a low-Earth orbit, for example, say at an altitude of 300 km, the pull of gravity is still 91% of what it is on Earth's surface.

So why do astronauts float around in their spacecraft? A spacecraft and everything in it are in *free fall*. As the term implies, an object in free fall is falling under the influence of gravity, free from any other forces. Free fall is that momentary feeling you get when you jump off a diving board. It's what skydivers feel before their parachutes open. In free fall you don't feel the force of gravity even though gravity is present. As you sit there in your chair, you don't feel gravity on your behind. You feel the chair pushing up at you with a force equal to the force of gravity. Forces that act only on the surface of an object are *contact forces*. Astronauts in orbit experience no contact forces because they and their spacecraft are in free fall, not in contact with Earth's surface. But if everything in orbit is falling, why doesn't it hit Earth? As we'll see in more detail in Chapter 4, an object in orbit has enough horizontal velocity so that, as it falls, it keeps missing Earth.

Earth's gravitational pull dominates objects close to it. But as spacecraft move into higher orbits, the gravitational pull of the Moon and Sun begin to exert their influence. As we'll see in Chapter 4, for Earth-orbiting applications, we can assume the Moon and Sun have no effect. However, as we'll see in Chapter 7, for interplanetary spacecraft, this assumption isn't true—"the Sun's gravitational pull dominates" for most of an interplanetary trajectory (the Moon has little effect on IP trajectories).

Gravity dictates the size and shape of a spacecraft's orbit. Launch vehicles must first overcome gravity to fling spacecraft into space. Once a spacecraft is in orbit, gravity determines the amount of propellant its engines must use to move between orbits or link up with other spacecraft. Beyond Earth, the gravitational pull of the Moon, the Sun, and other planets similarly shape the spacecraft's path. Gravity is so important to the space environment that an entire branch of astronautics, called *astrodynamics*, deals with quantifying its effects on spacecraft and planetary motion. Chapters 4 through 9 will focus on understanding spacecraft trajectories and the exciting field of astrodynamics.

As we mentioned in Chapter 1, the free-fall environment of space offers many potential opportunities for space manufacturing. On Earth, if we mix two materials, such as rocks and water, the heavier rocks sink to the bottom of the container. In free fall, we can mix materials that won't mix on Earth. Thus, we can make exotic and useful metal alloys for electronics and other applications, or new types of medicines.

However, free fall does have its drawbacks. One area of frustration for engineers is handling fluids in space. Think about the gas gauge in your car. By measuring the height of a floating bulb, you can constantly track the amount of fuel in the tank. But in orbit nothing "floats" in the tank because the liquid and everything else is sloshing around in free fall (Figure 3-15). Thus, fluids are much harder to measure (and pump) in free fall. But these problems are relatively minor compared to the profound physiological

Figure 3-15. Waterball. Astronaut Joseph Kerwin forms a perfect sphere with a large drop of water, which floats freely in the Skylab cabin. Left alone, the water ball may float to a solid surface and coat the surface, making a mess that doesn't run to the floor. *(Courtesy of NASA/Johnson Space Center)*

problems humans experience when exposed to free fall for long periods. We'll look at these problems separately in the next section.

Atmosphere

Earth's atmosphere affects a spacecraft in low-Earth orbit (below about 600 km [375 mi.] altitude), in two ways

- Drag—shortens orbital lifetimes
- Atomic oxygen—degrades spacecraft surfaces

Take a deep breath. The air you breathe makes up Earth's atmosphere. Without it, of course, we'd all die in a few minutes. While this atmosphere forms only a thin layer around Earth, spacecraft in low-Earth orbit can still feel its effects. Over time, it can work to drag a spacecraft back to Earth, and the oxygen in the atmosphere can wreak havoc on many spacecraft materials.

Two terms are important to understanding the atmosphere—pressure and density. *Atmospheric pressure* represents the amount of force per unit area exerted by the weight of the atmosphere pushing on us. *Atmospheric density* tells us how much air is packed into a given volume. As we go higher into the atmosphere, the pressure and density begin to decrease at an ever-increasing rate, as shown in Figure 3-16. Visualize a column of air extending above us into space. As we go higher, there is less volume of air above us, so the pressure (and thus, the density) goes down. If we were to go up in an airplane with a pressure and density meter, we would see that as we go higher, the pressure and density begins to drop off more rapidly.

Earth's atmosphere doesn't just end abruptly. Even at fairly high altitudes, up to 600 km (375 mi.), the atmosphere continues to create drag on orbiting spacecraft. *Drag* is the force you feel pushing your hand backward when you stick it out the window of a car rushing along the freeway. The amount of drag you feel on your hand depends on the air's density, your speed, the shape and size of your hand, and the orientation of your hand with respect to the airflow. Similarly, the drag on spacecraft in orbit depends on these same variables: the air's density plus the spacecraft's speed, shape, size, and orientation to the airflow.

Drag immediately affects spacecraft returning to Earth. For example, as the Space Shuttle re-enters the atmosphere enroute to a landing at Edwards AFB in California, the astronauts use the force of drag to slow the Shuttle (Figure 3-17) from an orbital velocity of over 25 times the speed of sound (27,900 km/hr or 17,300 m.p.h.) to a runway landing at about 360 km/hr. (225 m.p.h.). Similarly, drag quickly affects any spacecraft in a very low orbit (less than 130 km or 81 mi. altitude), pulling them back to a fiery encounter with the atmosphere in a few days or weeks.

The effect of drag on spacecraft in higher orbits is much more variable. Between 130 km and 600 km (81 mi. and 375 mi.), it will vary greatly depending on how the atmosphere changes (expands or contracts) due to variations in solar activity. Acting over months or years, drag can cause spacecraft in these orbits to gradually lose altitude until they re-enter the

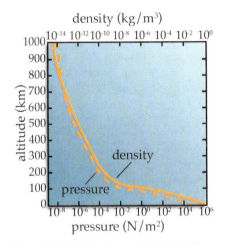

Figure 3-16.   Structure of Earth's Atmosphere. The density of Earth's atmosphere decreases exponentially as we go higher. Even in low-Earth orbit, however, spacecraft can still receive the effects of the atmosphere in the form of drag.

Figure 3-17.   Shuttle Re-entry. Atmospheric drag slows the Shuttle to landing speed, but the air friction heats the protective tiles to extremely high temperatures. *(Courtesy of NASA/Ames Research Center)*

atmosphere and burn up. In 1979, the Skylab space station succumbed to the long-term effects of drag and plunged back to Earth. Above 600 km (375 mi.), the atmosphere is so thin the drag effect is almost insignificant. Thus, spacecraft in orbits above 600 km are fairly safe from drag.

Besides drag, we must also consider the nature of air. At sea level, air is about 21% oxygen, 78% nitrogen, and 1% miscellaneous other gasses, such as argon and carbon dioxide. Normally, oxygen atoms like to hang out in groups of two--molecules, abbreviated O_2. Under normal conditions, when an oxygen molecule splits apart for any reason, the atoms quickly reform into a new molecule. In the upper parts of the atmosphere, oxygen molecules are few and far between. When radiation and charged particles cause them to split apart, they're sometimes left by themselves as *atomic oxygen*, abbreviated O.

So what's the problem with O? We've all seen the results of exposing a piece of steel outside for a few months or years—it starts to rust. Chemically speaking, rust is *oxidation*. It occurs when oxygen molecules in the air combine with the metal creating an oxide-rust. This oxidation problem is bad enough with O_2, but when O by itself is present, the reaction is much, much worse. Spacecraft materials exposed to atomic oxygen experience breakdown or "rusting" of their surfaces, which can eventually weaken components, change their thermal characteristics, and degrade sensor performance. One of the goals of NASA's Long Duration Exposure Facility (LDEF), shown in Figure 3-18, was to determine the extent of atomic oxygen damage over time, which it did very well. In many cases, depending on the material, the results were as dramatic as we just described.

On the good side, most atomic oxygen floating around in the upper atmosphere combines with oxygen molecules to form a special molecule, O_3, called *ozone*. Ozone acts like a window shade to block harmful radiation, especially the ultraviolet radiation that causes sunburn and skin cancer. In Chapter 11 we'll learn more about how the atmosphere blocks various types of radiation.

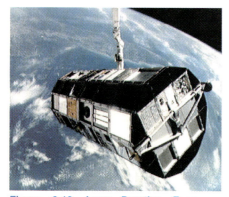

Figure 3-18. Long Duration Exposure Facility (LDEF). The mission of LDEF, deployed by the Space Shuttle (STS-41-C) in April, 1984, was to determine the extent of space environment hazards such as atomic oxygen and micrometeoroids. *(Courtesy of NASA/Johnson Space Center)*

Vacuum

Beyond the thin skin of Earth's atmosphere, we enter the vacuum of space. This vacuum environment creates three potential problems for spacecraft

- Out-gassing—release of gasses from spacecraft materials
- Cold welding—fusing together of metal components
- Heat transfer—limited to radiation

As we've seen, atmospheric density decreases dramatically with altitude. At a height of about 80 km (50 mi.), particle density is 10,000 times less than what it is at sea level. If we go to 960 km (596 mi.), we would find a given volume of space to contain one trillion times less air than at the surface. A pure vacuum, by the strictest definition of the word, is a volume of space completely devoid of all material. In practice, however, a pure vacuum is nearly unattainable. Even at an altitude of 960 km (596 mi.), we still find about 1,000,000 particles per cubic centimeter. So when we talk about the vacuum of space, we're talking about a "near" or "hard" vacuum.

Under standard atmospheric pressure at sea level, air exerts more than 101,325 N/m^2 (14.7 lb./in.^{2}) of force on everything it touches. The soda inside a soda can is under slightly higher pressure, forcing carbon dioxide (CO_2) into the solution. When you open the can, you release the pressure, causing some of the CO_2 to come out of the solution, making it foam. Spacecraft face a similar, but less tasty, problem. Some materials used in their construction, especially composites, such as graphite/epoxy, can trap tiny bubbles of gas while under atmospheric pressure. When this pressure is released in the vacuum of space, the gasses begin to escape. This release of trapped gasses in a vacuum is called *out-gassing*. Usually, out-gassing is not a big problem; however, in some cases, the gasses can coat delicate sensors, such as lenses or cause electronic components to arc, damaging them. When this happens, out-gassing can be destructive. For this reason, we must carefully select and test materials used on spacecraft. We often "bake" a spacecraft in a thermal-vacuum chamber prior to flight, as shown in Figure 3-19, to ensure it won't outgas in space.

Another problem created by vacuum is cold welding. *Cold welding* occurs between mechanical parts that have very little separation between them. When we test the moving part on Earth, a tiny air space may allow the parts to move freely. After launch, the hard vacuum in space eliminates this tiny air space, causing the two parts to effectively "weld" together. When this happens, ground controllers must try various techniques to "unstick" the two parts. For example, they may expose one part to the Sun and the other to shade so that differential heating causes the parts to expand and contract, respectively, allowing them to separate.

Due to cold welding, as well as practical concerns about mechanical failure, spacecraft designers carefully try to avoid the use of moving parts. However, in some cases, such as with spinning wheels used to control spacecraft attitude, there is no choice. On Earth, moving parts, like you find in your car engine, are protected by lubricants such as oil. Similarly, spacecraft components sometimes need lubrication. However, because of the surrounding vacuum, we must select these lubricants carefully, so they don't evaporate or outgas. Dry graphite (the "lead" in your pencil) is an effective lubricant because it lubricates well and won't evaporate into the vacuum as a common oil would.

Finally, the vacuum environment creates a problem with heat transfer. As we'll see in greater detail in Chapter 13, heat gets from one place to another in three ways. *Conduction* is heat flow directly from one point to another through a medium. If you hold a piece of metal in a fire long enough, you'll quickly discover how conduction works when it burns your fingers (Figure 3-20). The second method of heat transfer is convection. *Convection* takes place when gravity, wind, or some other force moves a liquid or gas over a hot surface (Figure 3-21). Heat transfers from the surface to the fluid. Convection takes place whenever we feel

Figure 3-19. Spacecraft in a Vacuum Chamber. Prior to flight, spacecraft undergo rigorous tests, including exposure to a hard vacuum in vacuum chambers. In this way we can test for problems with out-gassing, cold welding, or heat transfer. *(Courtesy of Surrey Satellite Technologies, Ltd., U.K.)*

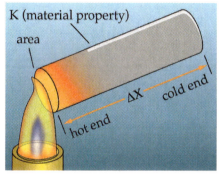

Figure 3-20. Conduction. Heat flows by conduction through an object from the hot end to the cool end. Spacecraft use conduction to remove heat from hot components.

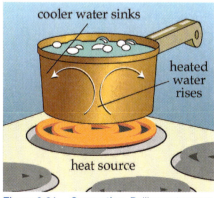

Figure 3-21. Convection. Boiling water on a stove shows how convection moves heat through a fluid from the fluid near a hot surface to the cooler fluid on top. Special devices on spacecraft use convection to remove heat from a hot components.

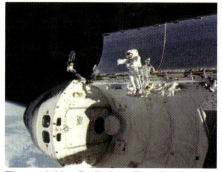

Figure 3-22. Radiation. The Shuttle Bay doors contain radiators that collect heat from the equipment bay and dump it into space. Because objects emit radiation, the bay door radiators efficiently remove heat from the Shuttle. *(Courtesy of NASA/Johnson Space Center)*

Figure 3-23. CERISE. The CERISE spacecraft lost its long boom when a piece of an Ariane rocket struck it at orbital speed. Without its boom, the spacecraft could not hold its attitude and perform its mission. *(Courtesy of Surrey Satellite Technologies, Ltd., U.K.)*

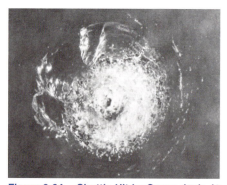

Figure 3-24. Shuttle Hit by Space Junk. At orbital speeds, even a paint flake can cause significant damage. The Space Shuttle was hit by a tiny paint flake, causing this crater in the front windshield. *(Courtesy of NASA/Johnson Space Center)*

chilled by a breeze or boil water on the stove. We can use both of these methods to move heat around inside a spacecraft but not to remove heat from a spacecraft in the free fall, vacuum environment of space. So we're left with the third method—radiation. We've already discussed electromagnetic radiation. *Radiation* is a way to transfer energy from one point to another. The heat you feel coming from the glowing coils of a space heater is radiated heat (Figure 3-22). Because radiation doesn't need a solid or fluid medium, it's the primary method of moving heat into and out of a spacecraft. We'll explore ways to do this in Chapter 13.

Micrometeoroids and Space Junk

The space around Earth is not empty. In fact, it contains lots of debris or space junk most of which we're used to. If you've seen a falling star, you've witnessed just one piece of the more than 20,000 tons of natural materials—dust, meteoroids, asteroids, and comets—that hit Earth every year. For spacecraft or astronauts in orbit, the risk of getting hit by a meteoroid or micrometeoroid, our name for these naturally occurring objects, is remote. However, since the beginning of the space age, debris has begun to accumulate from another source—human beings.

With nearly every space mission, broken spacecraft, pieces of old booster segments or spacecraft, and even an astronaut's glove have been left in space. The environment near Earth is getting full of this space debris (about 2200 tons of it). The problem is posing an increasing risk to spacecraft and astronauts in orbit. A spacecraft in low orbit is now more likely to hit a piece of junk than a piece of natural material. In 1996, the CERISE spacecraft, shown in Figure 3-23, became the first certified victim of space junk when its 6 m gravity-gradient boom was clipped off during a collision with a left-over piece of an Ariane launch vehicle.

Keeping track of all this junk is the job of the North American Aerospace Defense Command (NORAD) in Colorado Springs, Colorado. NORAD uses radar and optical telescopes to track more than 8000 objects, baseball sized and larger, in Earth orbit. Some estimates say at least 40,000 golf-ball-sized pieces (too small for NORAD to track) are also in orbit [Wertz and Larson, 1999]. To make matters worse, there also may be billions of much smaller pieces—paint flakes, slivers of metal, etc.

If you get hit by a paint flake no big deal, right? Wrong! In low-Earth orbit, this tiny chunk is moving at fantastic speeds—7000 m/s or greater when it hits. This gives it a great amount of energy—much more than a rifle bullet! The potential danger of all this space junk was brought home during a Space Shuttle mission in 1983. During the mission, a paint flake only 0.2 mm (0.008 in.) in diameter hit the Challenger window, making a crater 4 mm (0.16 in.) wide. Luckily, it didn't go all the way through. The crater, shown in Figure 3-24, cost more than $50,000 to repair. Analysis of other spacecraft shows collisions with very small objects are common. Russian engineers believe a piece of space debris may have incapacitated one of their spacecraft in a transfer orbit.

Because there are billions of very small objects and only thousands of very large objects, spacecraft have a greater chance of getting hit by a very small object. For a spacecraft with a cross-sectional area of 50–200 m² at an altitude of 300 km (186 mi.) (typical for Space Shuttle missions), the chance of getting hit by an object larger than a baseball during one year in orbit is about one in 100,000 or less [Wertz and Larson, 1999]. The chance of getting hit by something only 1 mm or less in diameter, however, is about one hundred times more likely, or about one in a thousand during one year in orbit.

One frightening debris hazard is the collision of two spacecraft at orbital velocity. A collision between two medium-sized spacecraft would result in an enormous amount of high velocity debris. The resulting cloud would expand as it orbited and greatly increase the likelihood of impacting another spacecraft. The domino effect could ruin a band of space for decades. Thus, there is a growing interest in the level of debris at various altitudes.

Right now, there are no plans to clean up this space junk. Some international agreements aim at decreasing the rate at which the junk accumulates—for instance, by requiring operators to boost worn-out spacecraft into "graveyard" orbits. Who knows? Maybe a lucrative 21st century job will be "removing trash from orbit."

The Radiation Environment

As we saw in the previous section, one of the Sun's main outputs is electromagnetic (EM) radiation. Most of this radiation is in the visible and near-infrared parts of the EM spectrum. Of course, we see the light and feel the heat of the Sun every day. However, a smaller but significant part of the Sun's output is at other wavelengths of radiation, such as X-rays and gamma rays.

Spacecraft and astronauts are well above the atmosphere, so they bear the full brunt of the Sun's output. The effect on a spacecraft depends on the wavelength of the radiation. In many cases, visible light hitting the spacecraft solar panels generates electric power through *solar cells* (also called photovoltaic cells). This is a cheap, abundant, and reliable source of electricity for a spacecraft (Figure 3-25); we'll explore it in greater detail in Chapter 13. This radiation can also lead to several problems for spacecraft

- Heating on exposed surfaces
- Degradation or damage to surfaces and electronic components
- Solar pressure

The infrared or thermal radiation a spacecraft endures leads to heating on exposed surfaces that can be either helpful or harmful to the spacecraft, depending on the overall thermal characteristics of its surfaces. Electronics in a spacecraft need to operate at about normal room temperature (20° C or 68° F). In some cases, the Sun's thermal energy can help to warm electronic components. In other cases, this solar input—in addition to the heat generated onboard from the operation of electronic components—can

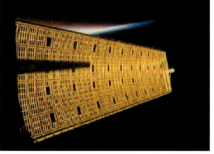

Figure 3-25. Solar Cells. Solar radiation provides electricity to spacecraft through solar cells mounted on solar panels, but it also degrades the solar cells over time, reducing their efficiency. Here the gold colored solar array experiment extends from the Space Shuttle Discovery. *(Courtesy of NASA/Johnson Space Center)*

make the spacecraft too hot. As we'll see in Chapter 13, we must design the spacecraft's thermal control system to moderate its temperature.

Normally, the EM radiation in the other regions of the spectrum have little effect on a spacecraft. However, prolonged exposure to ultraviolet radiation can begin to degrade spacecraft coatings. This radiation is especially harmful to solar cells, but it can also harm electronic components, requiring them to be shielded, or *hardened*, to handle the environment. In addition, during intense solar flares, bursts of radiation in the radio region of the spectrum can interfere with communications equipment onboard.

When you hold your hand up to the Sun, all you feel is heat. However, all that light hitting your hand is also exerting a very small amount of pressure. Earlier, we said EM radiation could be thought of as waves, like ripples on a pond. Another way to look at it is as tiny bundles of energy called photons. *Photons* are massless bundles of energy that move at the speed of light. These photons strike your hand, exerting pressure similar in effect to atmospheric drag (Figure 3-26). But this *solar pressure* is much, much smaller than drag. In fact, it's only about 5 N of force (about one pound) for a square kilometer of surface (one-third square mile). While that may not sound like much, over time this solar pressure can disturb the orientation of spacecraft, causing them to point in the wrong direction. We'll learn more about solar pressure effects in Chapter 12. In Chapter 14, we'll see how we may use this effect to sail around the solar system.

Figure 3-26.   Solar Max Spacecraft. Spacecraft with large surface areas, such as solar panels, must correct for the pressure from solar radiation that may change their attitude. *(Courtesy of NASA/Johnson Space Center)*

Charged Particles

Perhaps the most dangerous aspect of the space environment is the pervasive influence of charged particles. Three primary sources for these particles are

- The solar wind and flares
- Galactic cosmic rays (GCRs)
- The Van Allen radiation belts

As we saw in Section 3.1, the Sun puts out a stream of charged particles (protons and electrons) as part of the solar wind—at a rate of 1×10^9 kg/s (2.2×10^9 lb/s). During intense solar flares (Figure 3-27), the number of particles ejected can increase dramatically.

As if this source of charged particles wasn't enough, we must also consider high-energy particles from *galactic cosmic rays (GCRs)*. GCRs are particles similar to those found in the solar wind or in solar flares, but they originate outside of the solar system. GCRs represent the solar wind from distant stars, the remnants of exploded stars, or, perhaps, shrapnel from the "Big Bang" explosion that created the Universe. In many cases, however, GCRs are much more massive and energetic than particles of solar origin. Ironically, the very thing that protects us on Earth from these charged particles creates a third hazard, potentially harmful to orbiting spacecraft and astronauts—the Van Allen radiation belts.

Figure 3-27.   Solar Flares. Solar flares send many more charged particles into space than usual, so spacecraft orbiting Earth receive many times their normal dose, causing electronic problems. *(Courtesy of NASA/Jet Propulsion Laboratory)*

To understand the Van Allen belts, we must remember that Earth has a strong magnetic field as a result of its liquid iron core. This magnetic field behaves in much the same way as those toy magnets you used to play with as a kid, but it's vastly more powerful. Although you can't feel this field around you, it's always there. Pick up a compass and you'll see how the field moves the needle to point north. Magnets always come with a North Pole at one end and a South Pole at the other. If you've ever played with magnets, you've discovered that the north pole attracts the south pole (and vice versa), whereas two north poles (or south poles) repel each other. These magnetic field lines wrap around Earth to form the *magnetosphere*, as shown in Figure 3-28.

Remember, magnetic fields affect charged particles. This basic principle allows us to "steer" electron beams with magnets inside television sets. Similarly, the solar wind's charged particles and the GCRs form streams which hit Earth's magnetic field like a hard rain hitting an umbrella. Just as the umbrella deflects the raindrops over its curved surface, Earth's magnetic field wards off the charged particles, keeping us safe. (For Sci-fi buffs, perhaps a more appropriate analogy is the fictional force field or "shields" from Star Trek, used to divert Romulan disrupter beams, protecting the ship.)

The point of contact between the solar wind and Earth's magnetic field is the *shock front* or *bow shock*. As the solar wind bends around Earth's magnetic field, it stretches out the field lines along with it, as you can see in Figure 3-29. In the electromagnetic spectrum, Earth resembles a boat traveling through the water with a wake behind it. Inside the shock front, the point of contact between the charged particles of the solar wind and the magnetic field lines is the *magnetopause*, and the area directly behind the Earth is the *magnetotail*. As we'll see, charged particles can affect spacecraft orbiting well within Earth's protective magnetosphere.

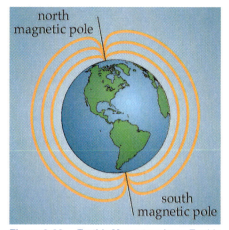

Figure 3-28.   Earth's Magnetosphere. Earth's liquid iron core creates a strong magnetic field. This field is represented by field lines extending from the south magnetic pole to the north magnetic pole. The volume this field encloses is the magnetosphere.

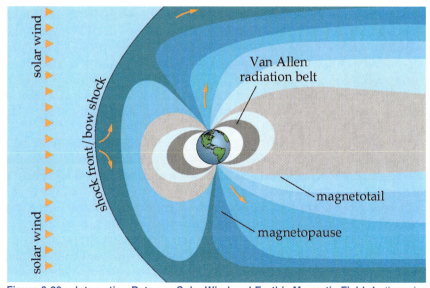

Figure 3-29.   Interaction Between Solar Wind and Earth's Magnetic Field. As the solar wind and GCRs hit Earth's magnetosphere, they are deflected, keeping us safe.

As the solar wind interacts with Earth's magnetic field, some high-energy particles get trapped and concentrated between field lines. These areas of concentration are the *Van Allen radiation belts*, named after Professor James Van Allen of the University of Iowa. Professor Van Allen discovered them based on data collected by Explorer 1, America's first satellite, launched in 1958.

Although we call them "radiation belts," space is not really radioactive. Scientists often lump charged particles with EM radiation and call them radiation because their effects are similar. Realize, however, that we're really dealing with charged particles in this case. (Perhaps we should call the radiation belts, "charged-particle suspenders," because they're really full of charged particles and occupy a region from pole to pole around Earth!)

Whether charged particles come directly from the solar wind, indirectly from the Van Allen belts, or from the other side of the galaxy, they can harm spacecraft in three ways

- Charging
- Sputtering
- Single-event phenomenon

Spacecraft charging isn't something the government does to buy a spacecraft! The effect of charged particles on spacecraft is similar to us walking across a carpeted floor wearing socks. We build up a static charge that discharges when we touch something metallic—resulting in a nasty shock. *Spacecraft charging* results when charges build up on different parts of a spacecraft as it moves through concentrated areas of charged particles. Once this charge builds up, discharge can occur with disastrous effects—damage to surface coatings, degrading of solar panels, loss of power, or switching off or permanently damaging electronics.

Sometimes, these charged particles trapped by the magnetosphere interact with Earth's atmosphere in a dazzling display called the Northern Lights or Aurora Borealis, as shown in Figure 3-30. This light show comes from charged particles streaming toward Earth along magnetic field lines converging at the poles. As the particles interact with the atmosphere, the result is similar to what happens in a neon light—charged particles interact with a gas, exciting it, and making it glow. On Earth we see an eerie curtain of light in the sky.

These particles can also damage a spacecraft's surface because of their high speed. It's as if they were "sand blasting" the spacecraft. We refer to this as *sputtering*. Over a long time, sputtering can damage a spacecraft's thermal coatings and sensors.

Finally, a single charged particle can penetrate deep into the guts of the spacecraft to disrupt electronics. Each disruption is known as a *single event phenomenon (SEP)*. Solar flares and GCR can cause a SEP. One type of SEP is a *single event upset (SEU)* or "bitflip." This occurs when the impact of a high-energy particle resets one part of a computer's memory from 1 to 0, or vice versa. This can cause subtle but significant changes to spacecraft functions. For example, setting a bit from 1 to 0 may cause the spacecraft to

Figure 3-30.   Lights in the Sky. As charged particles from the solar wind interact with Earth's upper atmosphere, they create a spectacular sight known as the Northern (or Southern) Lights. People living in high latitudes can see this light show. Shuttle astronauts took this picture while in orbit. *(Courtesy of NASA/ Johnson Space Center)*

turn off or forget which direction to point its antenna. Some scientists believe an SEU was the cause of problems with the Magellan spacecraft when it first went into orbit around Venus and acted erratically.

It's difficult for us to prevent these random impacts. Spacecraft shielding offers some protection, but spacecraft operators must be aware of the possibility of these events and know how to recover the spacecraft should they occur.

Astro Fun Fact

1977 XF$_{11}$—Is the End Near?

In December, 1997, a new asteroid, designated 1977 XF$_{11}$, about one mile across, was discovered orbiting the Sun. Astronomers discover many new asteroids each year, but this asteroid was special in that its orbit was predicted to pass as close as 28,000 km from Earth on October 26, 2028. Speculation abounded as to whether it might actually hit Earth. In order to get better predictions, old astronomical pictures were searched and the asteroid was found in some of them. Fortunately, using that data showed that 1977 XF$_{11}$ should pass well beyond the Moon's orbit, with very little chance of hitting Earth.

Marsden, Brian G. "One-Mile-Wide Asteroid to Pass Close to the Earth in 2028." Harvard-Smithsonian Center for Astrophysics. Press Release, March 12, 1998.

Contributed by Scott R. Dahlke, the U.S. Air Force Academy

▮ Section Review

Key Terms

astrodynamics
atmospheric density
atmospheric pressure
atomic oxygen
bow shock
cold welding
conduction
contact forces
convection
drag
free fall
galactic cosmic rays (GCRs)
hardened
magnetopause
magnetosphere
magnetotail
out-gassing
oxidation
ozone
photons
radiation
shock front
single event phenomena (SEP)
single event upset (SEU)
solar cells
solar pressure
spacecraft charging
sputtering
Van Allen radiation belts

Key Concepts

➤ Six major environmental factors affect spacecraft in Earth orbit.
  - Gravity
  - Atmosphere
  - Vacuum
  - Micrometeoroids and space junk
  - Radiation
  - Charged particles

➤ Earth exerts a gravitational pull which keeps spacecraft in orbit. We best describe the condition of spacecraft and astronauts in orbit as free fall, because they're falling around Earth.

➤ Earth's atmosphere isn't completely absent in low-Earth orbit. It can cause
  - Drag—which shortens orbit lifetimes
  - Atomic oxygen—which can damage exposed surfaces

➤ In the vacuum of space, spacecraft can experience
  - Out-gassing—a condition in which a material releases trapped gas particles when the atmospheric pressure drops to near zero
  - Cold welding—a condition that can cause metal parts to fuse together
  - Heat transfer problems—a spacecraft can rid itself of heat only through radiation

➤ Micrometeoroids and space junk can damage spacecraft during a high speed impact

➤ Radiation, primarily from the Sun, can cause
  - Heating on exposed surfaces
  - Damage to electronic components and disruption in communication
  - Solar pressure, which can change a spacecraft's orientation

➤ Charged particles come from three sources
  - Solar wind and flares
  - Galactic cosmic rays (GCRs)
  - Van Allen radiation belts

➤ Earth's magnetic field (magnetosphere) protects it from charged particles. The Van Allen radiation belts contain charged particles, trapped and concentrated by this magnetosphere.

➤ Charged particles from all sources can cause
  - Charging
  - Sputtering
  - Single event phenomena (SEP)

3.3 Living and Working in Space

In This Section You'll Learn to...

- Describe the free-fall environment's three effects on the human body
- Discuss the hazards posed to humans from radiation and charged particles
- Discuss the potential psychological challenges of spaceflight

Humans and other living things on Earth have evolved to deal with Earth's unique environment. We have a strong backbone, along with muscle and connective tissue, to support ourselves against the pull of gravity. On Earth, the ozone layer and the magnetosphere protect us from radiation and charged particles. We don't have any natural, biological defenses against them. When we leave Earth to travel into space, however, we must learn to adapt in an entirely different environment. In this section, we'll discover how free fall, radiation, and charged particles can harm humans in space. Then we'll see some of the psychological challenges for astronauts venturing into the final frontier.

Free fall

Earlier, we learned that in space there is no such thing as "zero gravity"; orbiting objects are actually in a free-fall environment. While free fall can benefit engineering and materials processing, it poses a significant hazard to humans. Free fall causes three potentially harmful physiological changes to the human body, as summarized in Figure 3-31.

- Decreased hydrostatic gradient—fluid shift
- Altered vestibular functions—motion sickness
- Reduced load on weight-bearing tissues

Hydrostatic gradient refers to the distribution of fluids in our body. On Earth's surface, gravity acts on this fluid and pulls it into our legs. So, blood pressure is normally higher in our feet than in our heads. Under free fall conditions, the fluid no longer pools in our legs but distributes equally. As a result, fluid pressure in the lower part of the body decreases while pressure in the upper parts of the body increases. The shift of fluid from our legs to our upper body is called a *decreased hydrostatic gradient* or *fluid shift* (Figure 3-32). Each leg can lose as much as 1 liter of fluid and about 10% of its volume. This effect leads to several changes.

To begin with, the kidneys start working overtime to eliminate what they see as "extra" fluid in the upper part of the body. Urination increases, and total body plasma volume can decrease by as much as 20%. One effect of this is a decrease in red blood cell production.

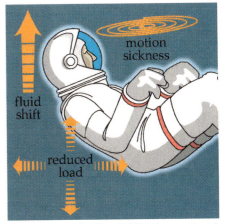

Figure 3-31. The Free-fall Environment and Humans. The free-fall environment offers many hazards to humans living and working in space. These include fluid shift, motion sickness, and reduced load on weight-bearing tissue.

Figure 3-32. Lower Body Negative Pressure Device. To reverse the effects of fluid shift while on orbit, astronauts "soak" in the Lower Body Negative Pressure device, which draws fluid back to their legs and feet. *(Courtesy of NASA/Johnson Space Center)*

The fluid shift also causes *edema* of the face (a red "puffiness"), so astronauts in space appear to be blushing. In addition, the heart begins to beat faster with greater irregularity and it loses mass because it doesn't have to work as hard in free fall. Finally, astronauts experience a minor "head rush" on return to Earth. We call this condition *orthostatic intolerance*—that feeling we sometimes get when we stand up too fast after sitting or lying down for a long time. For astronauts returning from space, this condition is sometimes very pronounced and can cause blackouts.

Vestibular functions have to do with a human's built-in ability to sense movement. If we close our eyes and move our head around, tiny sensors in our inner ear detect this movement. Together, our eyes and inner ears determine our body's orientation and sense acceleration. Our vestibular system allows us to walk without falling down. Sometimes, what we feel with our inner ear and what we see with our eyes gets out of synch (such as on a high-speed roller coaster). When this happens, we can get disoriented or even sick. That also explains why we tend to experience more motion sickness riding in the back seat of a car than while driving—we can feel the motion, but our eyes don't see it.

Because our vestibular system is calibrated to work under a constant gravitational pull on Earth's surface (or 1 "g"), this calibration is thrown off when we go into orbit and enter a free-fall environment. As a result, nearly all astronauts experience some type of motion sickness during the first few days in space until they can re-calibrate. Veteran astronauts report that over repeated spaceflights this calibration time decreases.

Astro Fun Fact
The "Vomit Comet"

How do astronauts train for the free-fall environment of space? They take a ride in a modified Air Force KC-135 aircraft owned and operated by NASA. Affectionately called the "Vomit Comet" by those who've experienced the fun, as well as the not-so fun, aspects of free fall, this plane flies a series of parabolas, alternately climbing and diving to achieve almost a minute of free fall in each parabola. This experience is similar to the momentary lightness we feel when we go over a hill at a high speed in a car. During these precious few seconds of free fall, astronauts can practice getting into space suits or experiment with other equipment specifically designed to function in space.

(Courtesy of NASA/Johnson Space Center)

(Courtesy of NASA/Johnson Space Center)

Free fall results in a loss of cardiovascular conditioning and body fluid volume, skeletal muscle atrophy, loss of lean body mass, and bone degeneration accompanied by calcium loss from the body. These changes may not be detrimental as long as an individual remains in free fall or microgravity. However, they can be debilitating upon return to a higher-gravity environment. Calcium loss and related bone weakening, in particular, seem progressive, and we don't know what level of gravity or exercise (providing stress on the weight-bearing bones) we need to counter the degenerative effects of free fall. However, if unchecked, unacceptable fragility of the bones could develop in a person living in microgravity for 1–2 years [Churchill, 1997]

If you're bedridden for a long time, your muscles will grow weak from lack of use and begin to atrophy. Astronauts in free fall experience a similar reduced load on weight bearing tissue such as on muscles (including the heart) and bones. Muscles lose mass and weaken. Bones lose calcium and weaken. Bone marrow, which produces blood, is also affected, reducing the number of red blood cells.

Scientists are still working on ways to alleviate all these problems of free fall. Vigorous exercise offers some promise in preventing long-term atrophy of muscles (Figure 3-33), but no one has found a way to prevent changes within the bones. Some scientists suggest astronauts should have "artificial gravity" for very long missions, such as missions to Mars. Spinning the spacecraft would produce this force, which would feel like gravity pinning them to the wall. This is the same force we feel when we take a corner very fast in a car and we're pushed to the outside of the curve. This artificial gravity could maintain the load on all weight-bearing tissue and alleviate some of the other detrimental effects of free fall. However, building and operating such a system is an engineering challenge.

Figure 3-33.   Shuttle Exercise. To maintain fitness and control the negative effects of free fall, astronauts workout everyday on one of several aerobic devices on the Shuttle. Here, astronaut Steven Hawley runs on the Shuttle's treadmill. *(Courtesy of NASA/Johnson Space Center)*

Radiation and Charged Particles

As we've seen, the ozone layer and magnetosphere protect us from charged particles and electromagnetic (EM) radiation down here on Earth. In space, however, we're well above the ozone layer and may enter the Van Allen radiation belts or even leave Earth's vicinity altogether, thus exposing ourselves to the full force of galactic cosmic rays (GCRs).

Until now, we've been careful to delineate the differences between the effects of EM radiation and charged particles. However, from the standpoint of biological damage, we can treat exposure to EM radiation and charged particles in much the same way. The overall severity of this damage depends on the total dosage. Dosage is a measure of accumulated radiation or charged particle exposure.

Quantifying the dosage depends on the energy contained in the radiation or particles and the *relative biological effectiveness (RBE)*, rating of the exposure. We measure dosage energy in terms of *RADs*, with one RAD representing 100 erg (10^{-5} J) of energy per gram of target material (1.08×10^{-3} cal/lb.). (This is about as much energy as it takes to lift a paper clip 1 mm [3.9×10^{-2} in.] off a desk). The RBE represents the destructive power of the dosage on living tissue. This depends on whether the exposure is EM radiation (photons) with an RBE of one, or charged particles with an RBE of as much as ten, or more. An RBE of ten is ten times more destructive to tissue than an RBE of one. The total dosage is then quantified as the product of RAD and RBE to get a dosage measurement in *roentgen equivalent man (REM)*. The REM dosage is cumulative over a person's entire lifetime.

The potential effects on humans exposed to radiation and charged particles depend to some extent on the time over which a dosage occurs. For example, a 50-REM dosage accumulated in one day will be much more harmful than the same dosage spread over one year. Such short-term dosages are called *acute dosages*. They tend to be more damaging, primarily because of their effect on fast reproducing cells within our bodies, specifically in the gastrointestinal tract, bone marrow, and testes. Table 3-1 gives the effects of acute dosages on humans, including blood count changes, vomiting, diarrhea, and death. The cumulative effects of dosage spread over much longer periods include cataracts, and various cancers, such as leukemia.

Table 3-1.   Effects of Acute Radiation and Charged Particle Dosages on Humans. (From Nicogossian, et al.) The higher the dosage and the faster it comes, the worse the effects on humans.

Effect	Dosage (REM)
Blood count changes	15–50
Vomiting "effective threshold"[*]	100
Mortality "effective threshold"[*]	150
LD_{50}[†] with minimal supportive care	320–360
LD_{50}[†] with full supportive medical treatment required	480–540

[*] Effective threshold is the lowest dosage causing these effects in at least one member of the exposed population

[†] LD_{50} is the lethal dosage in 50% of the exposed population

Just living on Earth, we all accumulate dosage. For example, living one year in Houston, Texas, (at sea level) gives us a dosage of 0.1 REM. As we get closer to space there is less atmosphere protecting us, so living in Denver, Colorado, (the "Mile-high City") gives us a dosage of twice that amount. Certain medical procedures also contribute to our lifetime dosage. One chest X-ray, for example, gives you 0.01 REM exposure. Table 3-2 shows some typical dosages for various events.

Except for solar flares, which can cause very high short-term dosages with the associated effects, astronauts concern themselves with dosage spread over an entire mission or career. NASA sets dosage limits for astronauts at 50 REM per year. Few astronauts will be in space for a full year, so their dosages will be much less than 50 REM. By comparison, the nuclear industry limits workers to one tenth that, or five REM per year.

Table 3-2. Dosages for Some Common Events (from SICSA Outreach and Nicogossian, et al.).

Event	Dosage (REM)
Transcontinental round trip in a jet	0.004
Chest X-ray (lung dose)	0.01
Living one year in Houston, Texas (sea level)	0.1
Living one year in Denver, Colorado (elev. 1600 m)	0.2
Skylab 3 for 84 days (skin dose)	17.85
Space Shuttle Mission (STS-41D)	0.65

A typical Shuttle mission exposes the crew to a dosage of less than one REM. The main concern is for very long missions, such as in the space station or on a trip to Mars.

For the most part, it is relatively easy to build shielding made of aluminum or other light metals to protect astronauts from the solar EM radiation and the protons from solar wind. In the case of solar flares, long missions may require "storm shelters"—small areas deep within the ship that would protect astronauts for a few days until the flare subsides. However, GCRs cause our greatest concern. Because these particles are so massive, it's impractical to provide enough shielding. To make matters worse, as the GCRs interact with the shield material, they produce secondary radiation (sometimes called "bremsstrahlung" radiation after a German word for braking), which is also harmful.

Space-mission planners try to avoid areas of concentrated charged particles such as those in the Van Allen belts. For example, because space suits provide very little shielding, NASA plans extra vehicular activities (EVA—or space walks) for when astronauts won't pass through the "South Atlantic Anomaly." In this area between South America and Africa, shown in Figure 3-34, the Van Allen belts "dip" toward Earth. Long missions, however, such as those to Mars, will require special safety measures, such as "storm shelters" and a radiation warning device when solar flares erupt. As for GCRs, we need to do more research to better quantify this hazard and to minimize trip times.

Psychological Effects

Because sending humans to space costs so much, we typically try to get our money's worth by scheduling grueling days of activities for the crew. This excessive workload can begin to exhaust even the best crews, seriously degrading their performance, and even endangering the mission. It can also lead to morale problems. For instance, during one United States Skylab mission, the crew actually went on strike for a day to protest the excessive demands on their time. Similar problems have been reported aboard the Russian Mir space station.

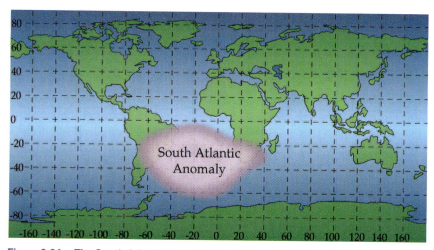

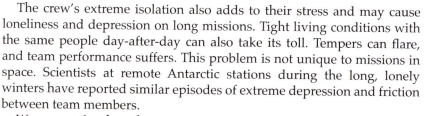

Figure 3-34. The South Atlantic Anomaly. The South Atlantic Anomaly is an area over the Earth where the Van Allen belts "dip" closer to the surface. Astronauts should avoid space walks in this region because of the high concentration of charged particles.

Figure 3-35. Shuttle Close Quarters. Living with seven crew members for ten days on the Shuttle can put a strain on relationships. Careful screening and busy schedules help prevent friction. Here, the crew of STS 96 pose for their traditional inflight portrait. *(Courtesy of NASA/ Johnson Space Center)*

The crew's extreme isolation also adds to their stress and may cause loneliness and depression on long missions. Tight living conditions with the same people day-after-day can also take its toll. Tempers can flare, and team performance suffers. This problem is not unique to missions in space. Scientists at remote Antarctic stations during the long, lonely winters have reported similar episodes of extreme depression and friction between team members.

We must take these human factors into account when planning and designing missions. Crew schedules must include regular breaks or "mini-vacations." On long missions, crews will need frequent contact with loved ones at home to alleviate their isolation. Planners also must select crew members who can work closely, in tight confines, for long periods (Figure 3-35). Psychological diversions such as music, video games, and movies will help on very long missions to relieve boredom.

Section Review

Key Terms

acute dosages
decreased hydrostatic gradient
edema
fluid shift
hydrostatic gradient
orthostatic intolerance
RADs
relative biological effectiveness (RBE)
roentgen equivalent man (REM)
vestibular functions

Key Concepts

➤ Effects of the space environment on humans come from
- Free fall
- Radiation and charged particles
- Psychological effects

➤ The free-fall environment can cause
- Decreased hydrostatic gradient—a condition where fluid in the body shifts to the head
- Altered vestibular functions—motion sickness
- Decreased load on weight bearing tissue—causing weakness in bones and muscles

➤ Depending on the dosage, the radiation and charged particle environment can cause short-term and long-term damage to the human body, or even death

➤ Psychological stresses on astronauts include
- Excessive workload
- Isolation, loneliness, and depression

References

Air University Press. *Space Handbook*. AV-18. Maxwell AFB, AL: 1985.

"Astronomy." August 1987.

Bate, Roger R., Donald D. Mueller, and Jerry E. White. *Fundamentals of Astrodynamics*. New York, NY: Dover Publications, Inc., 1971.

Buedeler, Werner. *Geschichte der Raumfahrt*. Germany: Sigloch Edition, 1982.

Bueche, Frederick J. *Introduction to Physics for Scientists and Engineers*. New York, NY: McGraw-Hill, Inc., 1980.

Chang, Prof. I. Dee (Stanford University), Dr. John Billingham (NASA Ames), and Dr. Alan Hargen (NASA Ames), Spring, 1990. "Colloquium on Life in Space."

Churchill, S.E. ed. 1997. Fundamentals of Space Life Sciences. Vol 1. Melbourne, FL: Krieger Publishing Company.

Concepts in Physics. Del Mar, CA: Communications Research Machines, Inc., 1973.

Concise Science Dictionary. Oxford, UK: Oxford University Press, 1984.

Glover, Thomas J. *Pocket REF*. Morrison, CO: Sequoia Publishing, Inc., 1989.

Goldsmith, Donald. *The Astronomers*. New York, NY: Community Television of Southern California, Inc., 1991.

Gonick, Larry and Art Huffman. *The Cartoon Guide to Physics*. New York, NY: Harpee Perennial, 1990.

Hartman, William K. *Moon and Planets*. Belmont, CA: Wadsworth, Inc., 1983.

Hewitt, Paul G. *Conceptual Physics. A New Introduction to Your Environment*. Boston, MA: Little, Brown and Company, 1981.

Jursa, Adolph S. (ed.). *Handbook of Geophysics and the Space Environment*. Air Force Geophysics Laboratory, Air Force Systems Command USAF, 1985.

King-Hele, Desmond. *Observing Earth Satellites*. New York, NY: Van Nostrand Reinhold Company, Inc., 1983.

NASA. 1994. Designing for Human Presence in Space: An Introduction to Environmental Control and Life Support Systems, NASA RP-1324. Prepared by Paul O. Wieland, National Aeronautics and Space Administration, Marshall Space Flight Center, AL.

Nicogossian, Arnauld E., Carolyn Leach Huntoon, Sam L. Pool. *Space Physiology and Medicine*. 2nd Ed. Philadelphia, PA: Lea & Febiger, 1989.

Rycroft, Michael (ed.), *The Cambridge Encyclopedia of Space*. New York, NY: Press Syndicate of the University of Cambridge, 1990.

Sasakawa International Center for Space Architecture (SICSA) Outreach. July–September 1989. Special Information Topic Issue, "Space Radiation Health Hazards: Assessing and Mitigating the Risks." Vol. 2, No. 3.

Suzlman, F.M. and A.M. Genin, eds. 1994. *Space Biology and Medicine*. Vol. II, Life Support and Habitability, a joint U.S./Russian publication. Washington, D.C. and Moscow, Russia. American Institute of Aeronautics and Astronautics and Nauka Press.

Tascione, Maj. T.F., Maj. R.H. Bloomer, Jr., and Lt. Col. D.J. Evans. *SRII, Introduction to Space Science: Short Course*. USAF Academy, Department of Physics.

Wertz, James R. and Wiley J. Larson. *Space Mission Analysis and Design*. Third edition. Dordrecht, Netherlands: Kluwer Academic Publishers, 1999.

Woodcock, Gordon, *Space Stations and Platforms*. Malabar, FL: Orbit Book Company, 1986.

The World Almanac and Book of Facts. 1991. New York, NY: Pharos Books, 1990.

"Weightlessness and the Human Body," *Scientific American*, Sept. 1998. pp. 58–63, Ronald J. White.

Mission Problems

3.1   Cosmic Perspective

1   Where does Space begin?

2   What object most strongly affects the space environment?

3   What is the star closest to Earth? The second-closest star?

4   List and describe the Sun's two forms of energy output.

5   What are solar flares? How do they differ from the solar wind?

3.2   The Space Environment and Spacecraft

6   List the six major hazards to spacecraft in the space environment.

7   Why are astronauts in space not in a "zero gravity" environment? Why is free fall a better description of the gravity environment?

8   How does the density and pressure of Earth's atmosphere change with altitude?

9   What is atmospheric drag?

10   What is atomic oxygen? What effects can it have on spacecraft?

11   What are the major problems in the vacuum environment of space?

12   Describe the potential hazards to spacecraft from micrometeoroids and space junk.

13   Describe the mechanism that protects the Earth from the effects of solar and cosmic charged particles.

14   What are Galactic Cosmic Rays?

15   What are the Van Allen radiation belts and what do they contain?

16   Describe the potential harmful effects on spacecraft from charged particles.

3.3   Living and Working in Space

17 List and describe the three physiological changes to the human body during free fall.

18 How are dosages of radiation and charged particles quantified?

19 What are the potential short-term and long-term effects to humans of exposure to radiation and charged particles?

20 How do long spaceflights affect astronauts psychologically?

For Discussion

21 Using a basketball to represent the size of the Sun, lay out a scale model of the solar system. How far away would the nearest star have to be?

22 As a spacecraft designer for a human mission to Mars, you must protect the crew from the space environment. Compile a list of all the potential hazards they may face during this multi-year mission and discuss how you plan to deal with them.

Mission Profile—SETI

In the fall of 1992, 500 years after Columbus discovered America, NASA officially began an exciting ten-year mission to search for extraterrestrial intelligence (SETI). In 1993, NASA passed operations to a private group, which calls it, Project Phoenix. Because they can't send a spacecraft over interstellar distances, the mission focuses on radio astronomy as the most probable way to contact extraterrestrial life. As Seth Shostak of the SETI Institute says, "our generation is the first with the capability to address one of mankind's most fundamental questions."—is there other intelligent life in the universe?

Mission Overview

The mission tries to intercept radio signals from other intelligent beings. Through a computer-linked network, operators combine the efforts of two radio telescopes. These telescopic dishes survey the stars and select certain areas for more sensitive searches.

Mission Data

- The radio telescopes scan 57 million frequencies every second with 300 times the sensitivity of any previous system.

- In the first minutes of the mission, the system automatically searched more comprehensively for extraterrestrial intelligence than in all previous attempts combined.

- The system listens for frequencies between 1 and 10 GHz. Below 1 GHz, natural radiation in our galaxy makes communication difficult to discern. Above 10 GHz, radio noise from the atmosphere makes it impossible to hear a transmission. The area between these frequencies offers the strongest possible radio transmission.

- Using stationary dishes at Arecibo, Puerto Rico and Lovell, England, they target only stars similar to our own Sun (about 10% of all known stars).

Mission Impact

With the onset of the ten year NASA mission, SETI was given a substantial credibility boost. Whether or not the new operators will find intelligent life, they will undoubtedly accrue new knowledge of our universe,

as well as upgrade our technological ability. This alone will prove the mission worthy. Yet, if Project Phoenix (SETI) succeeds to find other life in the universe, it will truly be a turning point in human history.

The SETI project employs the 305-m (1000 ft.) Arecibo dish in Puerto Rico, shown here, as well as the 76.2 m (250 ft.) Lovell Telescope at the Jodrell Bank Observatory in England. *(Courtesy of National Astronomy and Ionosphere Center—Arecibo Observatory, a facility of the National Science Foundation. Photo by David Parker.)*

For Discussion

- What is the next step once we find intelligent life?

- Given that radio astronomy is only a best guess for contacting extraterrestrial intelligence, how can operators be sure it is worth the effort and money? Should they continue this mission?

- Is it simply an Earth bias to specifically look at stars similar to our Sun? Could life exist in other scenarios?

Contributor

Troy E. Kitch, the U.S. Air Force Academy

References

Blum, Howard. *Out There*. New York: Simon and Schuster, 1990.

White, Frank. *The SETI Factor*. New York: Walker and Company, 1990.

The lunar module orbits above the Moon's stark landscape with the blue Earth rising above the horizon. *(Courtesy of NASA/ Johnson Space Center)*

Understanding Orbits

4

Outline

4.1 Orbital Motion
Baseballs in Orbit
Analyzing Motion

4.2 Newton's Laws
Weight, Mass, and Inertia
Momentum
Changing Momentum
Action and Reaction
Gravity

4.3 Laws of Conservation
Momentum
Energy

4.4 The Restricted Two-body Problem
Coordinate Systems
Equation of Motion
Simplifying Assumptions
Orbital Geometry

4.5 Constants of Orbital Motion
Specific Mechanical Energy
Specific Angular Momentum

In This Chapter You'll Learn to...

☛ Explain the basic concepts of orbital motion and describe how to analyze them

☛ Explain and use the basic laws of motion Isaac Newton developed

☛ Use Newton's laws of motion to develop a mathematical and geometric representation of orbits

☛ Use two constants of orbital motion—specific mechanical energy and specific angular momentum—to determine important orbital variables

You Should Already Know...

❏ Elements of a space mission (Chapter 1)

❏ Orbital concepts (Chapter 1)

❏ Concepts of vector mathematics (Appendix A.2)

❏ Calculus concepts (Appendix A.3)

❏ Kepler's Laws of Planetary Motion (Chapter 2)

Space is for everybody. It's not just for a few people in science or math, or a select group of astronauts. That's our new frontier out there and it's everybody's business to know about space.

Christa McAuliffe
teacher and astronaut on the
ill-fated Challenger Space Shuttle

Space Mission Architecture. This chapter deals with the Trajectories and Orbits segment of the Space Mission Architecture, introduced in Figure 1-20.

pacecraft work in orbits. In Chapter 1, we described an orbit as a "racetrack" that a spacecraft drives around, as seen in Figure 4-1. Orbits and trajectories are two of the basic elements of any space mission. Understanding this motion may at first seem rather intimidating. After all, to fully describe orbital motion we need some basic physics along with a healthy dose of calculus and geometry. However, as we'll see, spacecraft orbits aren't all that different from the paths of baseballs pitched across home plate. In fact, in most cases, both can be described in terms of the single force pinning you to your chair right now—gravity.

Armed only with an understanding of this single pervasive force, we can predict, explain, and understand the motion of nearly all objects in space, from baseballs to spacecraft, to planets and even entire galaxies. Chapter 4 is just the beginning. Here we'll explore the basic tools for analyzing orbits. In the next several chapters we'll see that, in a way, understanding orbits gives us a crystal ball to see into the future. Once we know an object's position and velocity, as well as the nature of the local gravitational field, we can gaze into this crystal ball to predict where the object will be minutes, hours, or even years from now.

We'll begin by taking a conceptual approach to understanding orbits. Once we have a basic feel for how they work, we'll take a more rigorous approach to describing spacecraft motion. We'll use tools provided by Isaac Newton, who developed some fundamental laws more than 200 years ago that we can use to explain orbits today. Finally, we'll look at some interesting implications of orbital motion that allow us to describe their shape and determine which aspects remain constant when left undisturbed by outside non-gravitational forces.

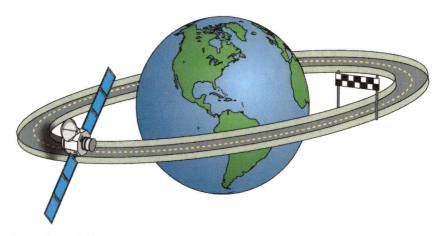

Figure 4-1.   Orbits as Racetracks. Orbits are like giant racetracks on which spacecraft "drive" around Earth.

4.1 Orbital Motion

▬ In This Section You'll Learn to...

- ☞ Explain, conceptually, how an object is put into orbit
- ☞ Describe how to analyze the motion of any object

Baseballs in Orbit

What is an orbit? Sure, we said it was a type of "racetrack" in space that an object drives around, but what makes these racetracks? Throughout the rest of this chapter we'll explore the physical principles that allow orbits to exist, as well as our mathematical representations of them. But before diving into a complicated explanation, let's begin with a simple experiment that illustrates, conceptually, how orbits work. To do this, imagine that we gather a bunch of baseballs and travel to the top of a tall mountain.

Visualize that we are standing on top of this mountain prepared to pitch baseballs to the east. As the balls sail off the summit, what do we see? Besides seeing unsuspecting hikers panting up the trail and running for cover, we should see that the balls follow a curved path. Why is this? The force of our throw is causing them to go outward, but the force of gravity is pulling them down. Therefore, the "compromise" shape of the baseball's path is a curve.

The faster we throw the balls, the farther they go before hitting the ground, as you can see in Figure 4-2. This could lead you to conclude that the faster we throw them the longer it takes before they hit the ground. But is this really the case? Let's try another experiment to see.

As we watch, two baseball players, standing on flat ground, will release baseballs. The first one simply drops a ball from a fixed height. At exactly the same time, the second player throws an identical ball horizontally at the same height as hard as possible. What will we see? If the second player throws a fast ball, it'll travel out about 20 m (60 ft.) or so before it hits the ground. But, the ball dropped by the first player will hit the ground at exactly the same time as the pitched ball, as Figure 4-3 shows!

How can this be? To understand this seeming paradox, we must recognize that, in this case, the motion in one direction is *independent* of motion in another. Thus, while the second player's ball is moving horizontally at 30 km/hr (20 m.p.h.) or so, it's still falling at the same rate as the first ball. This rate is the constant gravitational acceleration of all objects near Earth's surface, 9.798 m/s^2. Thus, they hit the ground at the same time. The only difference is that the pitched ball, because it also has horizontal velocity, will travel some horizontal distance before intercepting the ground.

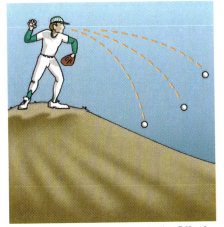

Figure 4-2. Throwing Baseballs Off of a Mountain. When we throw the balls faster, they travel farther before hitting the ground.

Figure 4-3.   Both Balls Hit at the Same Time. A dropped ball and a ball thrown horizontally from the same height will hit the ground at the same time. This is because horizontal and vertical motion are independent. Gravity is acting on both balls equally, pulling them to the ground with exactly the same acceleration of 9.798 m/s^2.

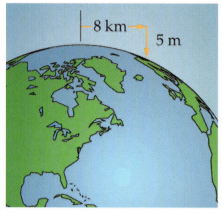

Figure 4-4. Earth's Curvature. Earth's curvature means the surface curves down about 5 m for every 8 km. On the surface of a sphere with that curvature, an object moving at 7.9 km/s is in orbit (ignoring air drag).

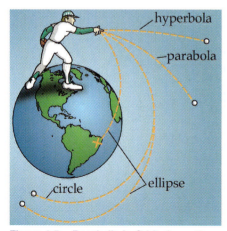

Figure 4-5.   Baseballs in Orbit. As we throw baseballs faster and faster, eventually we can reach a speed at which Earth curves away as fast as the baseball falls, placing the ball in orbit. At exactly the right speed it will be in a circular orbit. A little faster and it's in an elliptical orbit. Even faster and it can escape Earth altogether on a parabolic or hyperbolic trajectory.

Now let's return to the top of our mountain and start throwing our baseballs faster and faster to see what happens. No matter how fast we throw them, the balls still fall at the same rate. However, as we increase their horizontal velocity, they're able to travel farther and farther before they hit the ground. Because Earth is basically spherical in shape, something interesting happens. Earth's spherical shape causes the surface to drop approximately five meters vertically for every eight kilometers horizontally, as shown in Figure 4-4. So, if we were able to throw a baseball at 7.9 km/s (assuming no air resistance), its path would exactly match Earth's curvature. That is, gravity would pull it down about five meters for every eight kilometers it travels, and it would continue around Earth at a constant height. If we forget to duck, it may hit us in the back of the head about 85 minutes later. (Actually, because Earth rotates, it would miss us.) A ball thrown at a speed slower than 7.9 km/s falls faster than Earth curves away beneath it. Thus, it eventually hits the surface. The results of our baseball throwing experiment are shown in Figure 4-5.

If we analyze our various baseball trajectories, we see a whole range of different shapes. Only one velocity produces a perfectly circular trajectory. Slower velocities cause the trajectory to hit the Earth at some point. If we were to project this shape through the Earth, we'd find the trajectory is really a piece of an ellipse (it looks parabolic, but it's actually elliptical). Throwing a ball with a speed slightly faster than the circular velocity, also results in an ellipse. If we throw the ball too hard, it leaves Earth altogether on a parabolic or hyperbolic trajectory, never to return. No matter how hard we throw, our trajectory resembles either a circle, ellipse, parabola, or hyperbola. As we'll see in Section 4.4, these four shapes are *conic sections.*

So an object in orbit is literally falling around Earth, but because of its horizontal velocity it never quite hits the ground. Throughout this book we'll see how important having the right velocity at the right place is in determining the type of orbit we have.

Analyzing Motion

Now that we've looked at orbits conceptually, let's see how we can analyze this motion more rigorously. Chances are, when you first learned to play catch with a baseball, you had problems. Your poor partner had to chase after your first tentative throws, which never seemed to go where you wanted. But gradually, after a little bit of practice (and several exhausted partners), you got better. Eventually, you could place the ball right into your partner's glove, almost without conscious thought.

In fact, expert pitchers don't think about how to throw; they simply concentrate on where to throw. Somehow, their brain calculates the precise path needed to deliver the ball to the desired location. Then it commands the arm to a predetermined release point and time with exactly the right amount of force. All this happens in a matter of seconds, without a thought given to the likes of Isaac Newton and the equations that describe the baseball's motion. "So what?" you may wonder. Why bother with all the equations that describe *why* it travels the way it does?

Unfortunately, to build a pitching machine for a batting cage or to launch a spacecraft into orbit, we can't simply tell the machine or rocket to "take aim and throw." In the case of the rocket especially, we must carefully study its motion between the launch pad and space.

Now, we'll define a system for analyzing all types of motion. It's called the Motion Analysis Process (MAP) checklist and is shown in Figure 4-6. To put the MAP into action, imagine that you must describe the motion of a baseball thrown by our two baseball players in Figure 4-7. How will you go about it?

Figure 4-6. Motion Analysis Process (MAP) Checklist. Apply these steps to learn about moving objects and describe how they will move in the future.

Figure 4-7. Baseball Motion. To analyze the motion of a baseball, or a spacecraft, we must step through the Motion Analysis Process (MAP) checklist.

First of all, you need to define some frame of reference or *coordinate system*. For example, do you want to describe the motion with respect to a nearby building or to the center of Earth? In either case, you must define a reference point and a coordinate frame for the motion you're describing, as shown in Figure 4-8.

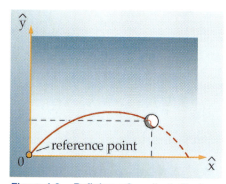

Figure 4-8.   Defining a Coordinate System.
To analyze a baseball's motion, we can define a simple, two-dimensional coordinate system.

Next you need some short-hand way of describing this motion and its relation to the forces involved—a short-hand way we'll call an *equation of motion*. Once you've determined what equation best describes the baseball's motion, you need to simplify it so you can use it. After all, you don't want to try to deal with how the motion of the baseball changes due to the gravitational pull of Venus or every little gust of wind in the park. So you must make some reasonable *simplifying assumptions*. For instance, you could easily assume that the gravitational attraction on the baseball from Venus, for example, is too small to worry about and the drag on the baseball due to air resistance is insignificant. And, in fact, as a good approximation, you could assume that the only force on the baseball comes from Earth's gravity.

With these assumptions made, you can then turn your attention to the finer details of the baseball problem. For example, you want to carefully define where and how the motion of the baseball begins. We call these the *initial conditions* of the problem. If you vary these initial conditions somehow (e.g., you throw the baseball a little harder or in a slightly different direction), the motion of the baseball will change. By assessing how these variations in initial conditions affect where the baseball goes, you can find out how sensitive the trajectory is to small changes or errors in them.

Finally, once you've completed all of these steps, you should verify the entire process by *testing the model* of baseball motion you've developed. Actually throw some baseballs, measure their trajectory deviations, and analyze differences (*error analysis*) between the motion you predict for the baseball and what you find from your tests. If you find significant differences, you may have to change your coordinate system, equation of motion, assumptions, initial conditions, or all of these. With the MAP in mind, we'll begin our investigation of orbital motion in the next section by considering some fundamental laws of motion Isaac Newton developed.

Section Review

Key Terms

conic sections
coordinate system
equation of motion
error analysis
initial conditions
simplifying assumptions
testing the model

Key Concepts

➤ From a conceptual standpoint, orbital motion involves giving something enough horizontal velocity so that, by the time gravity pulls it down, it has traveled far enough to have Earth's surface curve away from it. As a result, it stays above the surface. An object in orbit is essentially falling around the Earth but going so fast it never hits it.

➤ The Motion Analysis Process is a general approach for understanding the motion of any object through space. It consists of

  • A coordinate system

  • An equation of motion

  • Simplifying assumptions

  • Initial conditions

  • Error analysis

  • Testing the model

4.2 Newton's Laws

▬ In This Section You'll Learn to...

- ☛ Explain the concepts of weight, mass, and inertia
- ☛ Explain Newton's laws of motion
- ☛ Use Newton's laws to analyze the simple motion of objects

Since the first caveman threw a rock at a sabre-toothed tiger, we've been intrigued by the study of motion. In our quest to understand nature, we've looked for simple, fundamental laws that all objects obey. These Laws of Motion would apply universally for everything from gumdrops to galaxies (Figure 4-9). They would be unbreakable and empower us to explain the motion of the heavens, understand the paths of the stars, and predict the future position of our Earth. In Chapter 2, we saw how the Greek philosopher Aristotle defined concepts of orbital motion that held favor until challenged by such critical thinkers as Galileo and Kepler. Recall that Kepler gave us three laws to describe planetary motion, but didn't explain their causes. That's where Isaac Newton comes in.

Reflecting on his lifetime of scientific accomplishments, Newton rightly observed that he was able to do so much because he "stood on the shoulders of giants." Armed with Galileo's two basic principles of motion—inertia and relativity—and Kepler's laws of planetary motion, Isaac Newton was poised to determine the basic laws of motion that revolutionized our understanding of the world.

No single person has had as great an impact on science as Isaac Newton. His numerous discoveries and fundamental breakthroughs easily fill a volume the size of this book. Inventing calculus (math students still haven't forgiven him for that!), inventing the reflecting telescope, and defining gravity are just some of his many accomplishments. For our purposes, we'll see that the study of orbits (astrodynamics) builds on four of Newton's laws: three of motion and one describing gravity.

Figure 4-9. Cartwheel Galaxy. Our laws of motion apply universally, including the stars and planets of the Cartwheel Galaxy. *(Courtesy of the Association of Universities for Research in Astronomy, Inc./Space Telescope Science Institute)*

Weight, Mass, and Inertia

Before plunging into a discussion of Newton's many laws, let's take a moment to complicate a topic that, until now, you probably thought you understood very well—*weight*. When we order a "Quarter Pounder with Cheese™" (Figure 4-10), we're describing the weight of the hamburger (before cooking). To measure this weight (say, to determine what it weighs after cooking), we slap the burger on a scale and read the results. If our scale gave weight in metric units, we'd see our quarter-pounder weighs about one newton. This property we call weight is really the result of another, more basic property of the hamburger called "mass" plus the influence of gravity. A hamburger that weighs one newton (1/4 pound)

0.25 lb. = 0.10 kg

Figure 4-10. Quarter Pounder with Cheese™. When we order a Quarter Pounder with Cheese™, we get about 0.1 kg mass of meat.

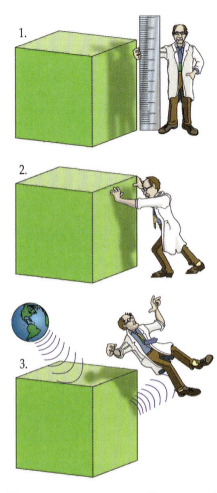

Figure 4-11.   What is Mass? The amount of mass an object has tells us three things about it: (1) how much "stuff" it contains, (2) how much it resists changes in motion—its inertia, and (3) how much gravitational force it exerts and is exerted on it by other masses in the universe.

has a mass of 1/9.798 kg or about 0.1 kg. Knowing the mass of our hamburger, we automatically know three useful things about it, as illustrated in Figure 4-11.

First, *mass* is a measure of how much matter or "stuff" the hamburger contains. The more mass, the more stuff. If we have to haul 200 Quarter Pounders™ to a family picnic, we can add the masses of individual burgers to determine how much total mass we need to carry. Carrying these hamburgers, which have a total mass of 22.5 kg (50 lbs.), will take some planning. Thus, knowing how much stuff one object has is important whenever we must combine it with others (as we do for space missions).

But that's not all. Knowing the mass of an object also tells us how much inertia it has. Galileo first put forth the principle of *inertia* in terms of an object's tendency to stay at rest or in motion unless acted on by an outside influence. To visualize inertia, assume you're in "couch potato" mode in front of the TV, with your work sitting on the desk, calling for your attention. Somehow, you just can't motivate yourself to get up from the couch and start working. You have too much "inertia," so it takes an outside influence (another person or a deep-rooted fear of failure) to overcome that "inertia."

For a given quantity of mass, inertia works in much the same way. An object at rest has a certain amount of inertia, represented by its mass, that must be overcome to get it in motion. Thus, to get the Quarter Pounder™ from its package and into your mouth, you must overcome its inherent inertia. You do that when you pick it up, if you can!

An object already in motion also has inertia by virtue of its mass. To change its direction or speed, we must apply a force. For instance a car skidding on ice slides in a straight line indefinitely (assuming no friction force), or at least until it hits something.

Finally, knowing an object's mass reveals how it affects other objects merely by its presence. There's an old, corny riddle which asks "Which weighs more—a pound of feathers or a pound of lead?" Of course, they weigh the same—one pound. Why is that? Weight is a result of two things—the amount of mass, or "stuff," and gravity. So, assuming we measure the weight of feathers and lead at the same place, their masses are the same. *Gravity* is the tendency for two (or more) chunks of stuff to attract each other. The more stuff (or mass) they have, the more they attract. This natural attraction between chunks of stuff is always there. Thus, our Quarter Pounder™ lying in its package causes a very slight gravitational pull on our fries, milk shake, and all other mass in the universe. (You'd better eat fast!)

Now that you'll never be able to look at a Quarter Pounder™ the same way again, let's see how Isaac Newton used these concepts of mass to develop some basic laws of motion and gravity.

Momentum

Newton's First Law of Motion was actually a variation on Galileo's concept of inertia. He discovered it and other principles of gravity and motion in 1655, when a great plague ravaged England and caused universities to close. At the time, he was a 23-year-old student at Cambridge. Instead of hitting the beach for an extended "spring break," the more scholarly Newton hit the apple orchard for meditation (or so legend has it). But his findings weren't published until 1687—in *The Mathematical Principles of Natural Philosophy*. In this monumental work he stated

Newton's First Law of Motion. A body continues in its state of rest, or of uniform motion in a straight line, unless compelled to change that state by forces impressed upon it.

Newton's First Law says that any object (or chunk of mass) that is at rest will stay at rest forever, unless some force makes it move. Similarly, any object in motion will stay in motion forever, with a constant speed in the same straight-line direction, until some force makes it change either its speed or direction of motion. Try to stop a speeding bullet like the one in Figure 4-12 and you get a good idea how profound Newton's first law is.

One very important aspect of the first law to keep in mind, especially when you study spacecraft motion, is that motion tends to stay in a straight line. Therefore, if you ever see something not moving in a straight line, such as a spacecraft in orbit, some force must be acting on it.

We know that an object at rest is lazy; it doesn't want to start moving and will resist movement to the fullest extent of its mass. We've also discovered that, once it's in motion, it resists any change in its speed or direction. But the amount of resistance for an object at rest and one in motion are not the same! This seeming paradox is due to the concept of momentum. *Momentum* is the amount of resistance an object in motion has to changes in its speed or direction of motion. This momentum is the result of combining an object's mass with its velocity. Because an object's velocity can be either linear or angular, there are two types of momentum: linear and angular.

Let's start with linear momentum. To see how it works, we consider the difference between a bulldozer and a baby carriage moving along a street, as shown in Figure 4-13. Bulldozers are massive machines designed to savagely rip tons of dirt from Earth. Baby carriages are delicate, four-wheeled carts designed to carry cute little babies around the neighborhood. Obviously, a bulldozer has much more mass than a baby carriage, but how does their momentum compare? Unlike inertia, which is a function only of an object's mass, *linear momentum*, $\vec{p}$, is the product of an object's mass, m, and its velocity, $\vec{V}$. [*Note:* because we describe velocity and momentum in terms of magnitude and direction,

Figure 4-12.   Newton's First Law. Any object in motion, such as a speeding bullet, will tend to stay in motion, in a straight line, unless acted on by some outside force (such as gravity or hitting a brick wall.)

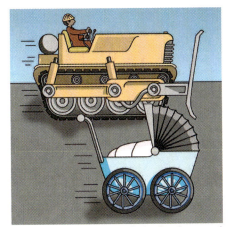

Figure 4-13.   Bulldozer, Baby Carriage, and Momentum. The momentum of any object is the product of its mass and velocity. So, a bulldozer moving at the same speed as a baby carriage has much more momentum, due to its large mass.

111

we treat them and other important concepts as *vector* quantities. Appendix A.2 reviews vector notation and concepts.]

$$\vec{p} = m\vec{V} \tag{4-1}$$

where

$\vec{p}$ = linear momentum vector (kg · m/s)

m = mass (kg)

$\vec{V}$ = velocity vector (m/s)

To compare the linear momentum of the bulldozer and the baby carriage, we'd have to know how fast they were moving. For the two to have the same linear momentum, the baby carriage, being much less massive, would have to be going much, much faster! Example 4-1, at the end of this section, shows this relationship.

Linear momentum is fairly basic because it involves motion in a straight line. Angular momentum, on the other hand, is slightly harder to understand because it deals with angular motion. Let's consider a simple toy top. If we set the top upright on a table, it will fall over, but if we spin it fast enough, the top will seem to defy gravity. A spinning object tends to resist changes in the direction and rate of spin, like the toy top shown in Figure 4-14, just as an object moving in a straight line resists changes to its speed and direction of motion. *Angular momentum, $\vec{H}$*, is the amount of resistance of a spinning object to change in spin rate or direction of spin. Linear momentum is the product of the object's mass, m, (which represents its inertia, or tendency to resist a change in speed and direction), and its velocity, $\vec{V}$. Similarly, angular momentum is the product of an object's resistance to change in spin rate or direction, and its rate of spin. An object's resistance to spin is its *moment of inertia, I*. We represent the *angular velocity*, which is a vector, by $\vec{\Omega}$. So we find the angular momentum vector, $\vec{H}$, using Equation (4-2).

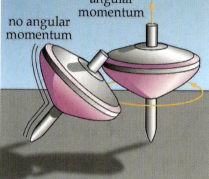

Figure 4-14. Angular Momentum. A non-spinning top (left) falls right over, but a spinning top, because of its angular momentum, resists the force applied by gravity and stays upright.

$$\vec{H} = I\,\vec{\Omega} \tag{4-2}$$

where

$\vec{H}$ = angular momentum vector (kg · m²/s)

I = moment of inertia (kg · m²)

$\vec{\Omega}$ = angular velocity vector (rad/s)

To characterize the direction of angular momentum, we need to examine the angular velocity, $\vec{\Omega}$. Look at the spinning wheel in Figure 4-15 and apply the right-hand rule. With our fingers curled in the direction it's spinning, our thumb points in the direction of the angular velocity vector, $\vec{\Omega}$, and the angular momentum vector, $\vec{H}$.

As Equation (4-2) implies, $\vec{H}$ is always in the same direction as the angular velocity vector, $\vec{\Omega}$. In the next section we'll see that, because of angular momentum, a spinning object resists change to its spin direction and spin rate.

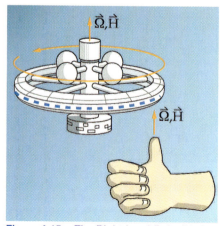

Figure 4-15. The Right-hand Rule. We find the direction of the angular velocity vector, $\vec{\Omega}$, and the angular momentum vector, $\vec{H}$, using the right-hand rule.

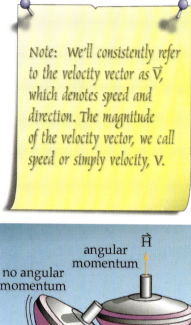

Note: We'll consistently refer to the velocity vector as $\vec{V}$, which denotes speed and direction. The magnitude of the velocity vector, we call speed or simply velocity, V.

We can describe angular momentum in another way. A mass spinning on the end of a string also has angular momentum. In this case, we find it by using the instantaneous tangential velocity of the spinning mass, $\vec{V}$, and the length of the string, $\vec{R}$, also called the *moment arm*. We combine these two with the mass, m, using a cross product (see Appendix A.2) relationship to get $\vec{H}$.

$$\vec{H} = \vec{R} \times m\vec{V} \qquad (4\text{-}3)$$

where

$\vec{H}$ = angular momentum vector $(kg \cdot m^2/s)$
$\vec{R}$ = position (m)
m = mass (kg)
$\vec{V}$ = velocity vector (m/s)

By the nature of the cross product operation, we can tell that $\vec{H}$ must be perpendicular to both $\vec{R}$ and $\vec{V}$. Once again, we can use the right-hand rule to find $\vec{H}$, as shown in Figure 4-16. Example 4-2 analyzes the mass on the end of the string in more detail.

In Section 4.5, we'll see that angular momentum is a very important property of spacecraft orbits. In Chapter 12, we'll find angular momentum is also a useful property for gyroscopes and spacecraft in determining and maintaining their attitude.

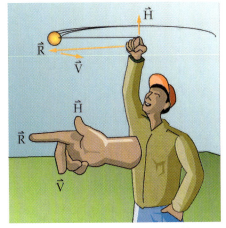

Figure 4-16. Describing Angular Momentum. The direction of the angular momentum vector, $\vec{H}$, is perpendicular to $\vec{R}$ and $\vec{V}$, and follows the right-hand rule.

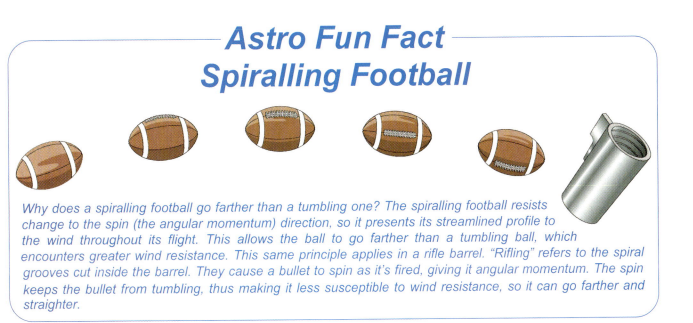

Astro Fun Fact
Spiralling Football

Why does a spiralling football go farther than a tumbling one? The spiralling football resists change to the spin (the angular momentum) direction, so it presents its streamlined profile to the wind throughout its flight. This allows the ball to go farther than a tumbling ball, which encounters greater wind resistance. This same principle applies in a rifle barrel. "Rifling" refers to the spiral grooves cut inside the barrel. They cause a bullet to spin as it's fired, giving it angular momentum. The spin keeps the bullet from tumbling, thus making it less susceptible to wind resistance, so it can go farther and straighter.

Changing Momentum

Now that we've looked at momentum, let's go back to Newton's laws of motion. As we saw, whether we're dealing with linear or angular momentum, both represent the amount a moving object resists change in its direction or speed. Now we can determine what it will take to overcome this resistance using Newton's Second Law.

Newton's Second Law of Motion. The time rate of change of an object's momentum equals the applied force.

In other words, to change an object's momentum very quickly, such as when we hit a fast ball with a bat, the force applied must be relatively high. On the other hand, if we're in no hurry to change the momentum, we can apply a much lower force over much more time.

Let's imagine we see a bulldozer creeping down the street at 1 m/s (3.28 ft./s), as in Figure 4-17. To stop the bulldozer dead in its tracks, we must apply some force, usually by pressing on the brakes. How much force depends on how fast we want to stop the bulldozer. If, for instance, we want to stop it in one second, we'd have to overcome all of its momentum quickly by applying a tremendous force. On the other hand, if we want to bring the bulldozer to a halt over one hour, we could apply a much smaller force. Thus, the larger the force applied to an object, the faster its momentum changes.

Now let's summarize the relationship implied by Newton's Second Law. The shorthand symbol we'll use to represent a force is $\vec{F}$. The symbol $\vec{p}$ represents linear momentum. To represent how fast a quantity is changing, we must introduce some notation from calculus. (See Appendix A.3 for a complete review of these concepts.) We use the Greek symbol "delta," Δ, to represent a very small change in any quantity. Thus, we represent the rate of change of a quantity, such as momentum, $\vec{p}$, over some short length of time, t, as

$$\frac{\Delta \vec{p}}{\Delta t} = \frac{\text{change in momentum}}{\text{change in time}} \qquad (4\text{-}4)$$

This equation shows how fast momentum is changing. We now express Newton's Second Law in symbolic shorthand as

$$\vec{F} = \frac{\Delta \vec{p}}{\Delta t} = \frac{\Delta (m\vec{V})}{\Delta t} \qquad (4\text{-}5)$$

which is true only if Δt is very small.

We can expand this equation by applying the Δ to each term in the parentheses (another concept from calculus), to get

$$\vec{F} = m\frac{\Delta \vec{V}}{\Delta t} + \frac{\Delta m}{\Delta t}\vec{V} \qquad (4\text{-}6)$$

So what can we do with this relationship? Let's begin with $\Delta m / \Delta t$ in the second term. This ratio represents how fast the mass of the object is

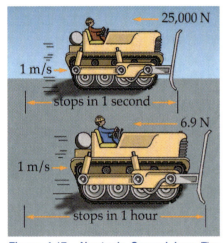

25,000 N

1 m/s

stops in 1 second

6.9 N

1 m/s

stops in 1 hour

Figure 4-17. Newton's Second Law. The force we must apply to stop a moving object depends on how fast we want to change its momentum. If two bulldozers are moving at 1 m/s (about the speed of a brisk walk), we must apply a much, much larger force to stop a bulldozer in one second than to stop it in one hour.

changing. For many cases, the mass of the object won't change, so this term is zero for those cases. (In Chapter 14, we'll see this isn't the case for rockets.) Now, for constant mass problems, we have only the first term in the relationship $\Delta \vec{V}/\Delta t$, which represents how fast velocity is changing. But this is just the definition of acceleration, $\vec{a}$. If we substitute $\vec{a}$ for $\Delta \vec{V}/\Delta t$ into Equation (4-6), we get the more familiar version

$$\boxed{\vec{F} = m\vec{a}}$$ (4-7)

where
$\vec{F}$ = force vector (kg m/s^2 = N)
m = mass (kg)
$\vec{a}$ = acceleration (m/s^2)

Equation (4-7) is arguably one of the most useful equations in all of physics and engineering. It allows us to understand how forces affect the motion of objects. Armed with this simple relationship, we can determine everything from how much force we need to stop a bulldozer, to the amount of acceleration Earth's gravity causes on the Moon. Example 4-3 shows this equation in action.

Action and Reaction

Newton's first two laws alone would have made him famous, but he went on to discover a third law, which describes a very important relationship between action and reaction.

A simple example of Newton's Third Law in action applies to ice skating. Imagine two ice skaters, standing in the middle of the rink, as shown in Figure 4-18. If one gives the other a push, what happens? They both move backward! The first skater exerted a force on the second, but in turn an equal but opposite force is exerted on him, thus sending him backward! In fact, Newton found that the reaction is exactly equal in magnitude but opposite in direction to the original action.

Newton's Third Law of Motion. When body A exerts a force on body B, body B will exert an equal, but opposite, force on body A.

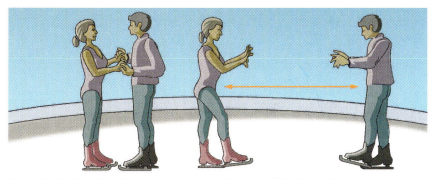

Figure 4-18. Two Ice Skaters Demonstrate Newton's Third Law. If they initially start at rest and the first one pushes against the second, they'll both go backward. The first skater applied an action—pushing—and received an equal but opposite reaction.

In the free-fall environment of space an astronaut must be very conscious of this fact. Suppose an astronaut tries to use a power wrench to turn a simple bolt without the force of gravity to anchor her in place. Unless she braces herself somehow, *she'll* start to spin instead of the bolt!

Gravity

The image most people have of Newton is of a curly-haired man clad in the tights and lace common to the 17th century, seated under an apple tree with an apple about to land on his head. After being hit by one too many apples, he suddenly jumped up and shouted "Eureka! (borrowing a phrase from Archimedes) I've invented gravity!" While this image is more the stuff of Hollywood than historical fact, it contains some truth. Newton did observe falling objects, such as apples, and read extensively Galileo's work on falling objects.

The breakthrough came when Newton reasoned that the force due to gravity must decrease with the square of the distance from the attracting body (Earth). In other words, an object twice as far away from Earth is attracted only one fourth as much. Newton excitedly took observations of the Moon to verify this model of gravity. Unfortunately, his measurements consistently disagreed with his model by one-sixth. Finally, in frustration, Newton abandoned his work on gravity. Years later, however, he found that the value for Earth's mass he had been using in his calculations was off by exactly one-sixth. Thus, his model of gravity had been correct all along! We call it Newton's Law of Universal Gravitation. "Universal" because we believe the same principle must apply everywhere in the universe. In fact, much of modern cosmology— all we know about the structure of the universe—depends on applying this simple law. We can see it applied most simply in Figure 4-19.

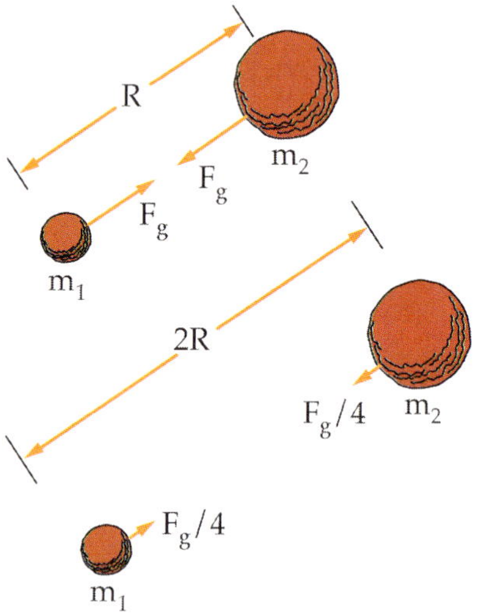

Newton's Law of Universal Gravitation. The force of gravity between two bodies is directly proportional to the product of their two masses and inversely proportional to the square of the distance between them.

Figure 4-19. Newton's Law of Universal Gravitation. The force of attraction between any two masses is directly proportional to the product of their masses and inversely proportional to the square of the distance between them. Thus, if we double the distance between two objects, the gravitational force decreases to 1/4 the original amount.

We can express this in symbolic shorthand as

$$F_g = \frac{G\,m_1 m_2}{R^2} \qquad (4\text{-}8)$$

where

F_g = force due to gravity (N)

G = universal gravitational constant = $6.67 \times 10^{-11}\ \text{N m}^2/\text{kg}^2$

m_1, m_2 = masses of two bodies (kg)

R = distance between the two bodies (m)

So what does this tell us? If we have two bodies, say Earth and the Moon, the force of attraction equals the product of their two masses, times a constant divided by the square of the distance between them. Let's look at some real numbers to see just how hard Earth tugs on the Moon and vice versa, as shown in Figure 4-20. Earth's mass, m_{Earth}, is 5.98×10^{24} kg (give or take a couple of mountains!), and the Moon's mass, m_{Moon}, is 7.35 $\times 10^{22}$ kg. The average distance between the Earth and Moon is about 3.84 $\times 10^8$ m. We already know the gravitational constant, G. Using the relationship for gravitational force we just described, we find

$$F_g = \frac{G m_{Earth} m_{Moon}}{R^2}$$

$$F_g = \frac{\left(6.67 \times 10^{-11} \frac{Nm^2}{kg^2}\right)(5.98 \times 10^{24}\ kg)(7.35 \times 10^{22}\ kg)}{(3.84 \times 10^8\ m)^2}$$

$$F_g = 1.98 \times 10^{20}\ N\ (\text{or about } 4.46 \times 10^{19}\ lb_f)$$

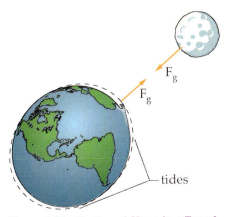

Figure 4-20. Earth and Moon in a Tug-of-War. Because of gravity, the Earth and Moon pull on each other with incredible force, which causes tides on Earth.

In other words, there's a huge force pulling the Earth and Moon together. But do we experience the result of this age-old tug-of-war? You bet we do! The biggest result we see is in ocean tides. The side of Earth closest to the Moon is attracted more than the side away from the Moon (gravity decreases as the square of the distance). Thus, all the ocean water on the side closest to the Moon swells toward the Moon; on the other side, the water swells away from the Moon due to the conservation of angular momentum as Earth rotates. Depending on the height and shape of the ocean floor, tides can raise and lower the sea level in some places more than 5 m (16 ft.). If you think about how much force it would take you to lift half the ocean this much, the incredibly large force we computed above begins to make sense.

It's important to remember that the force of gravity decreases as the square of the distance between masses increases. This means that if you want to weigh less, you should take a trip to the mountains! If you normally live in Houston, Texas, (elevation ~0 ft.) and you take a trip to Leadville, Colorado, (elevation 3048 m or 10,000 ft.), you won't weigh as much. That's because you're a bit farther away from the attracting body (Earth's center). But before you start packing your bags, look closely at what is happening. Your *weight* will change because the force of gravity is slightly less, but your *mass* won't change. Remember, weight measures how much gravity is pulling you down. Mass measures how much stuff you have. So even though the force pulling down on the scale will be slightly less, you'll still have those unwanted bulges.

Because the gravitational force changes, the acceleration due to gravity also changes. We can compute the acceleration due to gravity by combining the relationships expressed in Newton's Second Law of Motion and Newton's Law of Universal Gravitation. We know from Newton's Second Law (dropping vector notation because we're interested only in magnitudes) that

$$F = ma \qquad (4\text{-}9)$$

117

We can substitute this expression into Newton's relationship for gravity (Equation (4-8)) to get an expression for the acceleration of any mass due to Earth's gravity.

$$ma_g = \frac{mG\ m_{Earth}}{R^2}$$

which simplifies to

$$a_g = \frac{G\ m_{Earth}}{R^2}$$

For convenience, we typically combine G and the mass of the central body (Earth in this case) to get a new value we call the *gravitational parameter, μ* (Greek, small mu), where $μ \equiv G\,m$. For Earth, we denote this with a subscript, $μ_{Earth}$.

$$a_g = \frac{μ_{Earth}}{R^2} \qquad (4\text{-}10)$$

where

a_g = acceleration due to gravity (m/s^2)
$μ_{Earth} \equiv G\ m_{Earth} = 3.986 \times 10^{14}\ m^3/s^2$
R = distance to Earth's center (m)

If we substitute the values for $μ_{Earth}$ and use Earth's mean radius (6,378,137.0 m) we get $a_g = 9.798$ m/s^2 at Earth's surface, obviously pulling toward Earth's center. *Note:* we usually use kilometers instead of meters in this equation, because Earth's radius is so large.

Astro Fun Fact
Galileo Was Correct

Nearly 400 years later and more than 400,000 km away, one of Galileo's ideas was finally put to the test. On the Moon during the Apollo 15 mission, in the summer of 1971, astronaut Dave Scott performed a simple experiment: "In my left hand I have a feather. In my right hand, a hammer. I guess one of the reasons we got here today was because of the gentleman named Galileo a long time ago who made a rather significant discovery about falling objects in gravity fields, and we thought, 'where would be a better place to confirm his findings than on the Moon?' And so we'll try it here for you. The feather happens to be appropriately a falcon feather for our Falcon [the name of the lunar lander] and I'll drop the two of them here and, hopefully, they'll hit the ground at the same time." With that, Scott dropped the two objects which impacted the lunar surface simultaneously in the absence of any air resistance. "How about that," Scott exclaimed, "this proves that Mr. Galileo was correct!"

David Baker, PhD, The History of Manned Space Flight. New York, NY: Crown Publishers Inc., 1981.

▬ Section Review

Key Terms

angular momentum, $\vec{H}$
angular velocity, $\vec{\Omega}$
gravitational parameter, μ
gravity
inertia
linear momentum, $\vec{p}$
mass
moment arm
moment of inertia, I
momentum
vector
weight

Key Equations

$$\vec{p} = m\vec{V}$$

$$\vec{H} = I\,\vec{\Omega}$$

$$\vec{H} = \vec{R} \times m\vec{V}$$

$$\vec{F} = m\vec{a}$$

$$F_g = \frac{G\,m_1 m_2}{R^2}$$

$$a_g = \frac{\mu_{Earth}}{R^2}$$

Key Concepts

➤ The mass of an object denotes three things about it
  • How much "stuff" it has
  • How much it resists motion—its inertia
  • How much gravitational attraction it has

➤ Newton's three laws of motion are
  • **First Law.** A body continues in its state of rest, or in uniform motion in a straight line, unless compelled to change that state by forces impressed upon it.
    - The first law says that linear and angular momentum remain unchanged unless acted upon by an external force or torque, respectively
    - Linear momentum, $\vec{p}$, equals an object's mass, m, times its velocity, $\vec{V}$
    - Angular momentum, $\vec{H}$, is the product of an object's moment of inertia, I, (the amount it resists angular motion) and its angular velocity, $\vec{\Omega}$
    - We express angular momentum as a vector cross product of an object's position from the center of rotation, $\vec{R}$ (called its moment arm), and the product of its mass, m, and its instantaneous tangential velocity, $\vec{V}$
  • **Second Law.** The time rate of change of an object's momentum equals the applied force.
  • **Third Law.** When body A exerts a force on body B, body B exerts an equal but opposite force on body A.

➤ Newton's Law of Universal Gravitation. The force of gravity between two bodies (m_1 and m_2) is directly proportional to the product of the two masses and inversely proportional to the square of the distance between them (R).
  • G = universal gravitational constant = 6.67×10^{-11} Nm^2/kg^2
  • We often use the gravitational parameter, μ, to replace G and m. $\mu \equiv G\,m$
    - The gravitational parameter of Earth, μ_{Earth}, is
      $$\mu_{Earth} \equiv G\,m_{Earth} = 3.986 \times 10^{14}\ m^3/s^2,\ \text{or, using kilometer}$$
      instead of meters, $\mu_{Earth} = 3.986 \times 10^5\ km^3/s^2$

Example 4-1

Problem Statement

How fast would a 25 kg baby carriage have to be going to have the same linear momentum as a 25,000 kg bulldozer moving at 1 m/s?

Problem Summary

Given: $m_{bulldozer} = 25,000$ kg
$V_{bulldozer} = 1$ m/s
$m_{baby\ carriage} = 25$ kg

Find: $V_{baby\ carriage}$ to equal momentum of the bulldozer

Problem Diagram

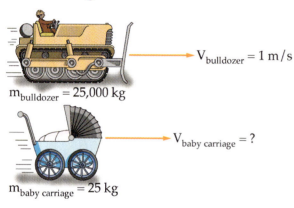

$V_{bulldozer} = 1$ m/s

$m_{bulldozer} = 25,000$ kg

$V_{baby\ carriage} = ?$

$m_{baby\ carriage} = 25$ kg

Conceptual Solution

1) Determine the magnitude of the linear momentum of the bulldozer

$$P_{bulldozer} = m_{bulldozer}\ V_{bulldozer}$$

2) Using the momentum of the bulldozer and the mass of the baby carriage, solve for the required velocity of the baby carriage

$$V_{baby\ carriage} = \frac{P_{bulldozer}}{m_{baby\ carriage}}$$

Analytical Solution

1) Determine linear momentum of bulldozer

$$P_{bulldozer} = m_{bulldozer}\ V_{bulldozer}$$
$$= (25,000\ kg)\ (1\ m/s)$$
$$= 25,000\ kg \cdot m/s$$

2) Solve for required baby carriage velocity

$$V_{baby\ carriage} = \frac{P_{bulldozer}}{m_{baby\ carriage}}$$
$$= \frac{25,000\ kg \cdot m/s}{25\ kg}$$
$$= 1000\ m/s$$

Interpreting the Results

For a baby carriage to have the same linear momentum as a massive bulldozer moving at only m/s (about the speed of a brisk walk), it would have to go 1000 m/s—almost three times the speed of sound!

Example 4-2

Problem Statement

Imagine someone spinning a 0.1 kg ball at the end of a 1.0 m string. The angular momentum of the spinning system is known to be 10 kg m^2/s. If they let go of the string, how fast and in what direction will the ball go?

Problem Summary

Given: $m_{ball} = 0.1$ kg

$H = 10$ kg m^2/s

$R = 1.0$ m

Find: V_{ball} direction when released

Problem Diagram

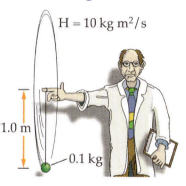

$H = 10$ kg m^2/s

1.0 m

0.1 kg

Conceptual Solution

1) Solve the angular momentum equation for the tangential velocity of the ball

$$\vec{H}_{ball} = \vec{R}_{ball} \times m_{ball}\vec{V}_{ball}$$

$$H_{ball} = R_{ball}m_{ball}V_{ball}$$

$$V_{ball} = \frac{H_{ball}}{R_{ball}m_{ball}}$$

2) By inspection—determine which direction the ball will travel when released.

Analytical Solution

1) Solve for tangential velocity of the ball

$$V_{ball} = \frac{H_{ball}}{R_{ball}m_{ball}}$$

$$= \frac{10 \text{ kg} \cdot \text{m}^2/\text{s}}{(1.0 \text{ m})(0.1 \text{ kg})}$$

$$= 100 \text{ m/s}$$

2) By inspection—when the ball is released, the force of the string is no longer forcing it to go in a circular path, so it will move off on a straight line tangent to the initial path at the point of release.

Interpreting the Results

A 0.1 kg ball on a circular path with a radius of 1.0 m and an angular momentum of 10 kg · m^2/s must be moving at a tangential velocity of 100 m/s. When released, it will fly tangent to the point of release.

Example 4-3

Problem Statement

A placekicker is able to apply a 100 N force to a 1 kg football for a total of 0.1 seconds. Ignoring gravity, how fast will the football be going?

Analytical Solution

1) Solve for ΔV

$$\Delta V = \frac{F \, \Delta t}{m}$$

$$\Delta V = \frac{(100 \text{ N})(0.1 \text{s})}{1 \text{kg}} = 10 \text{ m/s}$$

Problem Summary

Given: $m_{football} = 1 \text{ kg}$
$F_{kicker} = 100 \text{ N}$
$\Delta t = 0.1 \text{ s}$

Find: $\Delta V_{football}$

Interpreting the Results

You can use Newton's Second Law of Motion t[o] analyze the results of applying a given force to a[n] object for some length of time. In this case, a kicke[r] applying 100 N of force will kick a football to a spee[d] of 10 m/s (22 m.p.h.).

Problem Diagram

m = 1 kg

F = 100 N

Conceptual Solution

1) Use Newton's Second Law of Motion to solve for the change in velocity of an object in terms of a force applied over some length of time

$$\vec{F} = m\vec{a}$$

$$F = ma$$

$$F = m\frac{\Delta V}{\Delta t}$$

$$\Delta V = \frac{F\Delta t}{m}$$

4.3 Laws of Conservation

▰ In This Section You'll Learn to...

☛ Describe the basic laws of conservation of momentum and energy and apply them to simple problems

For any mechanical system, basic properties, such as momentum and energy, remain constant. In physics we say that if a certain property or quantity remains unchanged for a given system, that property or quantity is *conserved*. So let's take a look at two basic properties—momentum and energy—to see how they're conserved.

Momentum

One very important implication of Newton's Third Law has to do with the amount of momentum in a system. Newton's Third Law implies the total momentum in a system remains unchanged, or is conserved. We call this *conservation of momentum*.

To understand this concept let's go back to our ice skating example. When the two skaters faced each other, neither of them was moving, so the total momentum of the system was zero. Then the first one pushed on the second, and he moved in one direction with some speed, while she moved in the other. Their speeds won't be the same unless their masses are equal. The first skater moves in one direction with a speed that depends on his mass, while the other moves in the opposite direction with a speed depending on her mass. Now, the second skater's momentum (the product of her mass and velocity) is equal in magnitude, but opposite in direction, to his.

Depending on how we define our frame of reference, the first skater's momentum could be negative while the other's is positive. Adding the momentums, gives us zero, so, the original momentum of the system (the two skaters) hasn't changed. Thus, as Figure 4-21 shows, we say that the system's total momentum is conserved. Example 4-4 also shows this principle in action.

This conservation principle works equally well for angular momentum. You've probably seen a good example of this with figure skaters, who always include a spin in their routines. Remember, once an object (or skater) begins to spin, it has angular momentum.

By watching these skaters closely, you may see them move their arms outward or inward to vary their spin rate. How does this change their spin rate? We know from Equation (4-2) that angular momentum, $\vec{H}$, equals the product of the moment of inertia, I, and the spin rate, $\vec{\Omega}$. The moment of inertia of an object is proportional to its distance from the axis of rotation. To change their moment of inertia, skaters move their arms outward or inward, which increases or decreases the radius, thereby

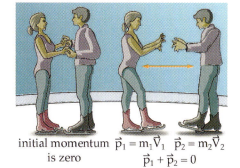

initial momentum is zero $\vec{p}_1 = m_1\vec{V}_1$ $\vec{p}_2 = m_2\vec{V}_2$
$\vec{p}_1 + \vec{p}_2 = 0$

Figure 4-21. Conservation of Momentum. Two people on ice skates demonstrate the concept of conservation of linear momentum. Initially the two are at rest; thus, the momentum of the system is zero. But as one skater pushes on the other, they both start moving in opposite directions. Adding their two momentum vectors together still gives us zero; thus, momentum of the system is conserved.

Figure 4-22. Spinning Slowly. Skaters extend their arms to increase moment of inertia—spinning more slowly.

Figure 4-23. Spinning Quickly. Skaters bring in their arms to decrease moment of inertia—spinning faster. Total angular momentum is the same in both cases.

changing I. Because momentum is conserved, it must stay constant as moment of inertia changes. But the only way this can happen is for the angular velocity, $\vec{\Omega}$, to change. Thus, if skaters put their arms out, as in Figure 4-22, they increase their moment of inertia and spin slower to maintain the same angular momentum. If they bring their arms in, as in Figure 4-23, they decrease their moment of inertia and increase the spin rate to maintain the same angular momentum.

Energy

We've all had those days when somehow we just don't seem to have any energy. But what exactly is energy? Energy can take many forms including electrical, chemical, nuclear, and mechanical. For now, let's deal only with mechanical energy because it's the most important for understanding motion. If you've jumped off a platform, climbed a ladder, or played with a spring, you've experienced mechanical energy. *Total mechanical energy, E,* comes from an object's position and motion. It's composed of *potential energy, PE,* which is due entirely to an object's position and *kinetic energy, KE,* which is due entirely to the object's motion. Total mechanical energy can be only potential, kinetic, or some combination of both

$$E = KE + PE$$ (4-11)

where

E   = total mechanical energy (kg m^2/s^2)

PE = potential energy (kg m^2/s^2)

KE = kinetic energy (kg m^2/s^2)

To better understand what the trade-off between potential and kinetic energy means, we need to understand where it takes place. We say that gravity is a *conservative field*—a field in which total energy is *conserved.* Thus, the sum of PE and KE, or the total E, in a conservative field is constant.

Potential energy is the energy an object in a conservative field has entirely because of its position. We call it "potential" energy because we don't really notice it until something changes. For example, if you pick up a 1 kg (2.2 lb.) mass and raise it above your head, it's higher position gives it more "potential" energy. This potential is realized when you drop the mass and it lands on your foot! To quantify this form of energy, we must derive an expression for the amount of work done by raising the object above a reference point (usually Earth's surface) against the force of gravity (see Appendix C.10). If we raise the object a small distance (a few hundred meters or less), we can assume gravity is constant and we get

$$PE = m \, a_g h$$ (4-12)

where

m = mass (kg)

a_g = acceleration due to gravity (m/s^2)

h = height above a reference point (m)

Thus, to compute an object's potential energy after raising it a small distance, we need to know three things: the amount of mass, m; its position above a reference point, h; and the acceleration due to gravity, a_g, at that reference point. But, if we want to find a spacecraft's potential energy in orbit high above Earth, we can't assume gravity is constant, and we can't use Earth's surface as a convenient reference point anymore. Let's see how we find potential energy in an orbit.

As we know from the last section, the gravitational acceleration varies depending on an object's distance from Earth's center, R. To derive the potential energy equation for this gravitational field (see Appendix C.10), we must determine the amount of work it would take to move the spacecraft from Earth's center to its orbital position, a distance of R. That derivation yields

$$PE = -\frac{m\mu}{R}$$
(4-13)

where

PE = spacecraft's potential energy (kg km^2/s^2)

m = spacecraft's mass (kg)

μ = gravitational parameter (km^3/s^2) = 3.986 × 10^5 km^3/s^2

R = spacecraft's distance from Earth's center (km)

Notice the negative sign in Equation (4-13). This sign is due to the convention we're using, which defines R to be positive outward from Earth's center. We know potential energy should increase as we raise a spacecraft to a higher orbit, so is this still consistent? Yes! As we raise our spacecraft's orbit, R gets bigger, and PE gets less negative—which means it gets bigger too. Remember, for potential energy, –3 is a bigger quantity than –4 because it's less negative. (This approach is analogous to heat: an ice cube at –3 degrees Celsius is "hotter" than one at –4 degrees Celsius.) At the extreme, when R reaches infinity (or close enough), PE approaches zero.

One way to visualize this strange situation is to think about Earth's center being at the bottom of a deep, deep well (Figure 4-24). At the bottom of the well, R is zero, so PE is at a minimum (its largest negative value, PE = –∞). As we begin to climb out of the well, our PE begins to increase (gets less negative) until we reach the lip of the well at R near infinity. At this point, our PE is effectively zero, and for all practical purposes, we have left Earth's gravitational influence. Of course, we never really reach an "infinite" distance from Earth, but as we'll see when we discuss interplanetary travel in Chapter 7, we essentially leave Earth's "gravity well" at a distance of about one million km (621,400 mi.).

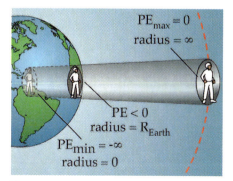

Figure 4-24. Potential Energy (PE). PE increases as we get farther from Earth's center by becoming less negative. It's as if we're climbing out of a deep well.

If you have a 1 kg mass suspended above your head, how do you realize the "potential" of its energy? You let go! Gravity will then cause the mass to accelerate downward, so when it hits the ground (and hopefully not your head or your foot, enroute), it's moving at considerable speed and thus has energy of a different kind—energy of motion, which we call kinetic energy. Similar to linear momentum, kinetic energy is solely a function of an object's mass and its velocity.

$$KE = \frac{1}{2}mV^2 \qquad (4\text{-}14)$$

where
KE = kinetic energy (kg km^2/s^2)
m   = mass (kg)
V   = velocity (km/s)

As we said, total mechanical energy in a conservative field (such as a gravitational field) stays constant. But, a spacecraft in orbit may get close to Earth during part of its orbit and be far away in another part. So, how does it maintain a constant mechanical energy? It must trade the potential energy it loses as it moves closer, for kinetic energy (increased velocity). Then, as it goes farther away, it trades back—kinetic energy goes down, as the potential energy goes up.

The endless trade-off between PE and KE to make this happen goes on all around us—but we often don't notice it. We've all played on a simple playground swing like the one in Figure 4-25. As we swing back and forth, we constantly trade between KE and PE. At the bottom of the arc, we are moving the fastest, so our KE is at a maximum and PE is at a minimum. As we swing up, our speed diminishes until, at the top of the arc, we actually stop briefly. At this point, our KE is zero because we're not moving (velocity is zero), but our PE is at a maximum. The reverse happens as we swing back, this time turning our PE back into energy of motion. If it weren't for friction in the frame attachments and our own wind resistance, once we started on a swing, we'd swing forever even without "pumping." Another way to experience this trade-off between KE and PE is to ride a roller coaster, such as the one illustrated in Example 4-5 at the end of this section.

We can now combine KE and PE to get a new expression for the total mechanical energy of our orbiting spacecraft

$$E = \frac{1}{2}m\,V^2 - \frac{m\mu}{R} \qquad (4\text{-}15)$$

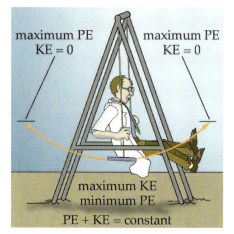

Figure 4-25. Mechanical Energy is Conserved. The total mechanical energy, the sum of kinetic and potential energy, is constant in a conservative field. We can show this with a simple swing. At the bottom of the arc, speed is greatest and height is lowest; hence, KE is at the maximum and PE is at a minimum. As the swing rises to the top of the arc, KE trades for PE until it stops momentarily at the top where PE is maximum and KE is zero.

where

E   = total mechanical energy $(kg\ km^2/s^2)$

m   = mass (kg)

V   = velocity (km/s)

μ   = gravitational parameter (km^3/s^2)

R   = position (km)

Later we'll use this expression to develop some useful tools for analyzing orbital motion.

Section Review

Key Terms

conservation of momentum
conservative field
kinetic energy, KE
potential energy, PE
total mechanical energy, E

Key Equations

$$E = KE + PE$$

$$PE = -\frac{m\mu}{R}$$

$$KE = \frac{1}{2}mV^2$$

$$E = \frac{1}{2}m\ V^2 - \frac{m\mu}{R}$$

Key Concepts

➤ A property is conserved if it stays constant in a system

➤ In the absence of outside forces, linear and angular momentum are conserved

➤ A conservative field, such as gravity, is one in which total mechanical energy is conserved

➤ Total mechanical energy, E, is the sum of potential and kinetic energies

  • Kinetic energy, KE, is energy of motion
  • Potential energy, PE, is energy of position

Example 4-4

Problem Statement

A 50 kg roller skater is motionless holding a 0.5 kg ball. If the skater throws the ball eastward at a velocity of 10 m/s what happens to the skater?

2) Solve for the unknown, $\vec{V}_{skater\,final}$. The direction is found from Newton's Third Law. If the ball is thrown eastward (action) the skater must go westward (reaction).

Problem Summary

Given: $m_{skater} = 50$ kg

$V_{skater\,initial} = 0$ m/s

$V_{ball\,initial} = 0$ m/s

$m_{ball} = 0.5$ kg

$V_{ball\,final} = 10$ m/s

Find: $V_{skater\,final}$ and direction of motion

Problem Diagram

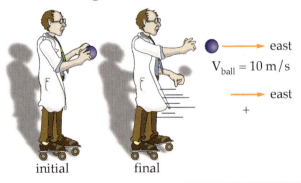

$V_{ball} = 10$ m/s

east

+

east

initial   final

Conceptual Solution

1) Apply the concept of conservation of momentum. The total momentum of the roller skater plus the ball must be the same before and after the ball is thrown.

$$\vec{P}_{initial} = \vec{P}_{final}$$

$$m_{skater}\vec{V}_{skater\,initial} + m_{ball}\vec{V}_{ball\,initial}$$

$$= m_{skater}\vec{V}_{skater\,final} + m_{ball}\vec{V}_{ball\,final}$$

Analytical Solution

1) $(m_{skater}\vec{V}_{skater\,initial}) + (m_{ball}\vec{V}_{ball\,initial})$

$= (m_{skater}\vec{V}_{skater\,final}) + (m_{ball}\vec{V}_{ball\,final})$

$[(50\text{ kg})\,(0\text{ m/s})] + [0.5\text{ kg})\,(0\text{ m/s})]$

$= [(50\text{ kg})\,(\vec{V}_{skater\,final})] + [(0.5\text{ kg})\,(10\text{ m/s})]$

$0 = (50\text{ kg})\,\vec{V}_{skater\,final} + 5\dfrac{\text{kg} \cdot \text{m}}{\text{s}}$

2) Solve for $\vec{V}_{skater\,final}$

$$\vec{V}_{skater\,final} = \dfrac{-5\text{kg} \cdot \text{m}}{50\text{kg} \cdot \text{s}}$$

$$= -0.1\text{ m/s}$$

Negative sign indicates westward travel because we choose eastward for positive.

Interpreting the Results

When a skater throws a ball in one direction, he or she will go in the opposite direction according to Newton's Third Law. We find the velocity in the opposite direction by using the principle of conservation of momentum. This same basic idea is used to propel rockets. They eject mass at some high velocity in one direction and therefore move in the opposite direction.

Example 4-5

roller coaster car (on a frictionless track) begins from
dead stop at the top of the first hill at a height of 50
. How fast will it be going at the top of the second hill
a height of 40 m?

roblem Summary

ven: $V_{initial} = 0 \text{ m/s}$
$h_{final} = 40 \text{ m}$
$h_{initial} = 50 \text{ m}$

nd: V_{final}

roblem Diagram

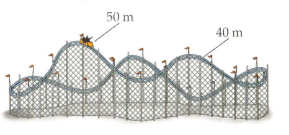

50 m

40 m

onceptual Solution

Find total mechanical energy at the beginning of the
problem [*Hint:* Use PE convention from Equation
(4-12).]

$E_{initial} = KE + PE$

$PE_{initial} = m \, a_g \, h_{initial}$

$KE_{initial} = 1/2 \, mV^2$

By conservation of energy, set this total equal to
the total mechanical energy at the end of the
problem and solve for V_{final}

$E_{initial} = E_{final}$

$m \, a_g \, h_{initial} + 1/2 \, m \, V^2_{initial}$

$= m \, a_g \, h_{final} + 1/2 \, m \, V^2_{final}$

[Assume a_g = constant]

Analytical Solution

1) Find $E_{initial}$

$$E_{initial} = m \, a_g \, h_{initial} + 1/2 \, m \, V^2_{initial}$$

$$\frac{E_{initial}}{m} = (9.798 \text{ m/s}^2)(50 \text{ m}) + 1/2(0 \text{ m/s})^2$$

$$= 489.9 \text{ m}^2/\text{s}^2$$

2) Set $E_{initial} = E_{final}$

$$\frac{E_{initial}}{m} = \frac{E_{final}}{m}$$

$$489.9 \text{ m}^2/\text{s}^2 = a_g \, h_{final} + 1/2 \, V^2_{final}$$

$$= (9.798 \text{ m/s}^2)(40 \text{ m}) + 1/2 \, V^2_{final}$$

$$489.9 \text{ m}^2/\text{s}^2 - 391.9 \text{ m}^2/\text{s}^2 = 1/2 \, V^2_{final}$$

$$V^2_{final} \cong 196 \text{ m}^2/\text{s}^2$$

$$V_{final} \cong 14 \text{ m/s}$$

Interpreting the Results

Starting from the top of the 50 m hill on the roller
coaster, the car is at the point of maximum potential
energy. As it goes down that first hill, it trades potential
for kinetic energy and gains speed. As it starts up the
second hill, the trade-off turns around and it loses
speed, but it's still going 14 m/s at the top of that
second hill (over 30 m.p.h.). Notice we weren't given
the mass of the car in this problem and didn't need it
to find the velocity. This implies the car would reach
the same velocity no matter what its mass is. We'll use
this same concept when we analyze spacecraft motion
and introduce *specific* mechanical energy.

4.4 The Restricted Two-body Problem

▰ In This Section You'll Learn to...

☞ Explain the approach used to develop the restricted two-body equation of motion, including coordinate systems and assumptions

☞ Explain how the solution to the two-body equation of motion dictates orbital geometry

☞ Define and use the terms that describe orbital geometry

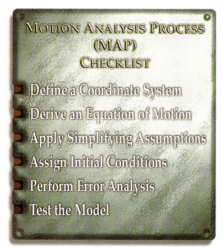

Figure 4-26. Motion Analysis Process (MAP) Checklist. Originally described in Section 4.1 (Figure 4-6), this process applies to balls in flight or spacecraft in orbit.

Earlier, we outlined a general approach to analyzing the motion of an object called the MAP, shown again in Figure 4-26. There we described the motion of a baseball. Now we can use the first three steps of this same method to understand the motion of any object in orbit. A special application of the MAP is the *restricted two-body problem*. Why restricted? As we'll see later in this section, we must restrict our analysis with assumptions we need to make our lives easier. Why two bodies? That's one of the assumptions. Why a problem? Finding an equation to represent this motion has been a classic problem solved and refined by students and mathematicians since Isaac Newton. In this section, we'll rely on the work of the mathematicians who have come before us. So at the end of this section you'll say, "The motion of two bodies? Hey, no problem!"

Coordinate Systems

To be valid, Newton's laws must be expressed in an inertial reference frame, meaning a frame that is not accelerating. To illustrate this, let's suppose we want to describe the flight of a baseball we toss and catch while we're driving in our car. We see the ball go up and down with respect to us. But that's not the whole story. Our car may be accelerating with respect to a police car behind us. Our car and the police car may be accelerating with respect to Earth's surface. And of course we must consider Earth's motion spinning on its axis, Earth's motion around the Sun, the Sun's motion in the Galaxy, the Galaxy's motion through the universe, and the expansion of the universe! These are all accelerating frames of reference for the ball's motion, which complicate our attempt to describe this motion using Newton's laws.

So we can see how this reference frame stuff can get complicated very quickly. Indeed, from astronomical observations, it looks like everything in the universe is accelerating. So how can we find any purely non-accelerating reference? We can't. To apply Newton's laws to our ball, we must select a reference frame that's close enough to, or "sufficiently," inertial for our problem.

Any reference frame is just a collection of unit vectors at right angles to each other that allows us to specify the magnitude and direction of other vectors, such as a spacecraft's position and velocity. This collection of unit vectors allows us to establish the components of vectors in 3-D space. By rigidly defining these unit vectors, we define a coordinate system.

To create a coordinate system we need to specify four pieces of information—an origin, a fundamental plane, a principal direction, and a third axis, as shown in Figure 4-27. The *origin* defines a physically identifiable starting point for the coordinate system. The other three parameters fix the orientation of the frame. The *fundamental plane* contains two axes of the system. Once we know the plane, we can establish the first axis by defining a unit vector that starts at the origin and is perpendicular to this plane. The unit vector in this direction at the origin is one axis. Next, we need a *principal direction* within the plane, which we define by pointing a unit vector toward some visible, distant object, such as a star. Now that we have two directions (the principal direction and an axis perpendicular to the fundamental plane), we can find the third axis using the right-hand rule.

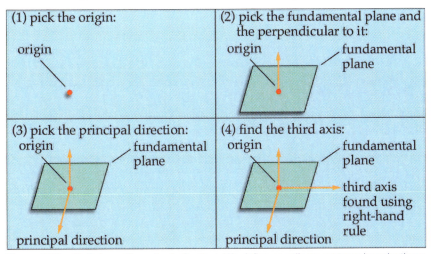

Figure 4-27.   Defining a Coordinate System. We define coordinate systems by selecting a convenient (1) origin; (2) fundamental plane containing the origin and an axis perpendicular to the plane; (3) principal direction within the plane; and (4) third axis using the right-hand rule.

Remember—coordinate systems should make our lives easier. If we choose the correct coordinate system, developing the equations of motion can be simple. If we choose the wrong system, it can be nearly impossible.

For Earth-orbiting spacecraft, we'll choose a tried-and-true system that we know makes solving the equations of motion relatively easy. This *geocentric-equatorial coordinate system* has these characteristics

- Origin—Earth's center (hence the name *geo*centric)
- Fundamental plane—Earth's equator (hence geocentric-*equatorial*). Perpendicular to the plane—North Pole direction
- Principal direction—vernal equinox direction found by drawing a line from Earth to the Sun on the first day of Spring, as shown in Figure 4-28. While this direction may not seem "convenient" to you, it's

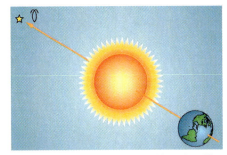

Figure 4-28.   Vernal Equinox Direction. The vernal equinox direction is the principal direction for the geocentric-equatorial coordinate system. It's found by drawing a line from Earth through the Sun on the first day of Spring, usually March 21.

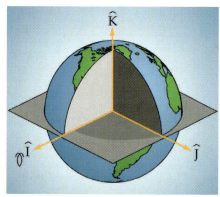

Figure 4-29. Geocentric-equatorial Coordinate System. We define this system by

- Origin—Earth's center
- Fundamental plane—equatorial plane
- Perpendicular to plane—North Pole
- Principal direction—vernal equinox (♈)

We use this coordinate system for analyzing the orbits of Earth-orbiting spacecraft.

significant to astronomers who originally defined the system. Plus it beats any alternatives by a long way, mostly because they move.

- Third axis found using the right-hand rule

Figure 4-29 shows the entire coordinate system.

Equation of Motion

Using the geocentric-equatorial coordinate system, we can safely apply Newton's Second Law to examine the external forces affecting the system, or in this case, a spacecraft. So let's place ourselves on an imaginary spaceship in orbit around Earth and see if we can list the forces on our ship.

- Earth's gravity (Newton wouldn't let us forget this one)
- Drag—if we're a little too close to the atmosphere
- Thrust—if we fire rockets
- 3rd body—gravity from the Sun, Moon, or planets
- Other—just in case we miss something

Astro Fun Fact
Hold that Equinox!

Star gazers, several thousand years ago, first determined the vernal equinox direction. Later, when they noted that it pointed at the first star in the Aries constellation, they called it the First Point of Aries, and used the Aries astrological symbol, ♈ (a rams head), to identify it.

Because Earth's spin axis wobbles a little, the equatorial plane also wobbles, and so, the line of intersection between the equatorial and ecliptic planes (the vernal equinox direction) shifts slightly (westward), about 9 milliarcseconds a day. Over several thousand years, the vernal equinox direction has shifted out of the Aries constellation, through the Pisces constellation, and will soon enter the Aquarius constellation. During this movement, astronomers and astronautical engineers continued to use the name, "First Point of Aries" for the vernal equinox direction, even after it left the Aries constellation.

On January 1, 1998, to avoid the drifting reference problem, astronomers with the International Astronomical Union adopted the International Celestial Reference Frame (ICRF), which pinpoints its principal direction by referencing 608 extragalactic radio stars (mostly pulsars) that are so far away, no one will notice any movement. The principal direction is as close as possible to the vernal equinox direction at noon on January 1, 2000, at the Greenwich Meridian, so the ICRF matches the traditional J2000.0 reference frame. Unfortunately, the vernal equinox will continue to wander, but exacting astronomers and engineers have a new, stationary (nearly inertial) reference.

U.S. Naval Observatory (http://aa.usno.navy.mil)

Wertz, James R. and Wiley J. Larson. Space Mission Analysis and Design. Third edition. Dordrecht, Netherlands: Kluwer Academic Publishers, 1999.

Contributed by Doug Kirkpatrick, the U.S. Air Force Academy

Summing all these forces, shown in Figure 4-30, we get with the following equation of motion

$$\sum \vec{F}_{external} = \vec{F}_{gravity} + \vec{F}_{drag} + \vec{F}_{thrust} + \vec{F}_{3rd\ body} + \vec{F}_{other} = m\vec{a} \quad (4\text{-}16)$$

If we substituted mathematical expressions for the various forces and tried to devise a solution to the equation, we would create a difficult problem—not to mention an enormous headache. So let's examine some reasonable assumptions we can make to simplify the problem.

Simplifying Assumptions

Luckily, we can assume some things about orbital motion that will simplify the problem, but they will "restrict" our solution to cases in which these assumptions apply. Fortunately, this includes most of the situations we'll use. Let's consider the forces on a spacecraft in orbit and assume

- The spacecraft travels high enough above Earth's atmosphere that the drag force is small, $\vec{F}_{drag} \cong 0$

- The spacecraft won't maneuver or change its path, so we ignore the thrust force, $\vec{F}_{thrust} \cong 0$

- We are considering the motion of the spacecraft close to Earth, so we ignore the gravitational attraction of the Sun, the Moon, or any other third body, $\vec{F}_{3rd\ body} \cong 0$. (That's why we call this the two-body problem.)

- Compared to Earth's gravity, other forces such as those due to solar radiation, electromagnetic fields, etc., are negligible, $\vec{F}_{other} \cong 0$

- Earth's mass is much, much larger than the mass of any spacecraft, $m_{Earth} >> m_{spacecraft}$

- Earth is spherically symmetrical with uniform density, so we treat it as a point mass. Thus, we mathematically describe Earth's gravity as acting from its center.

- The spacecraft's mass is constant, $\Delta m = 0$, so Equation 4-7, applies

- The geocentric-equatorial coordinate system is sufficiently inertial, so that Newton's laws apply

After all these assumptions, we're left with gravity as the only force, so our equation of motion becomes $\sum \vec{F}_{external} = \vec{F}_{gravity} = m\vec{a}$, as shown in Figure 4-31. Now we can apply Newton's Law of Universal Gravitation in vector form

$$\vec{F}_{gravity} = -\frac{\mu m}{R^2}\frac{\vec{R}}{R} \quad (4\text{-}17)$$

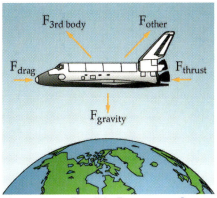

Figure 4-30.   Possible Forces on a Spacecraft. We can brainstorm all the possible forces on a spacecraft to include Earth's gravity, drag, thrust, third-body gravity, and other forces.

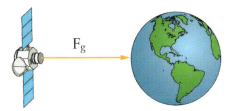

Figure 4-31.   The Force of Gravity. In the restricted two-body problem, we reduce the forces acting on a spacecraft to a single force—Earth's gravity.

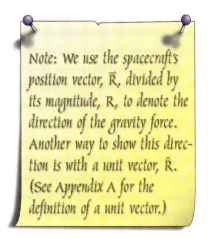

Note: We use the spacecraft's position vector, $\vec{R}$, divided by its magnitude, R, to denote the direction of the gravity force. Another way to show this direction is with a unit vector, $\hat{R}$. (See Appendix A for the definition of a unit vector.)

Substituting the force of gravity equation into the equation of motion, we get

$$\vec{F}_{gravity} = -\frac{\mu m}{R^2}\frac{\vec{R}}{R} = m\vec{a} = m\ddot{\vec{R}}$$

and dividing both sides by m, we arrive at the *restricted two-body equation of motion*

$$\ddot{\vec{R}} + \frac{\mu}{R^2}\frac{\vec{R}}{R} = 0 \qquad (4\text{-}18)$$

where

$\ddot{\vec{R}}$ = spacecraft's acceleration (km/s²)

μ = gravitational parameter (km³/s²) = 3.986×10^5 km³/s² for Earth

$\vec{R}$ = spacecraft's position vector (km)

R = magnitude of the spacecraft's position vector (km)

[*Note:* we use the engineering convention for the second derivative of $\vec{R}$ with respect to time, which is $\ddot{\vec{R}}$, better known as acceleration, $\vec{a}$.]

What can the two-body equation of motion tell us about the movement of a spacecraft around Earth? Unfortunately, in its present form—a second-order, non-linear, vector differential equation—it doesn't help us visualize anything about this movement. So what good is it? To understand the significance of the two-body equation of motion, we must first "solve" it, using a rather complex mathematical derivation (see Appendix C.3). When the smoke clears, we're left with an expression for the magnitude of the position vector (not the velocity) of an object in space in terms of some odd, new variables.

$$R = \frac{k_1}{1 + k_2 \cos v}$$

where

R = magnitude of the spacecraft's position vector, $\vec{R}$

k_1 = constant that depends on μ, $\vec{R}$, and $\vec{V}$

k_2 = constant that depends on μ, $\vec{R}$, and $\vec{V}$

v = (Greek letter "nu") polar angle measured from an orbit's principal axis to $\vec{R}$

This equation is the solution to the restricted, two-body equation of motion and describes the spacecraft's location, R, in terms of two constants and a polar angle, v. You may recognize that this equation also represents a general relationship for any *circle, ellipse, parabola,* or *hyperbola*—commonly known as *conic sections*, shown in Figure 4-32. Now, here's the really significant part of all this—we just proved Kepler's Laws of Planetary Motion! Based on Brahe's data, Kepler showed that the

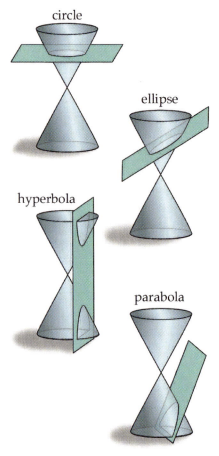

Figure 4-32.   Conic Sections. The solution to the restricted, two-body equation of motion gives the polar equation for a conic section. Conic sections are found by slicing right cones at various angles.

planets' orbits were ellipses but couldn't say why. We've just shown why: any object moving in a gravitational field must follow one of the conic sections. In the case of planets or spacecraft in orbit, this path is an ellipse or a circle (which is just a special case of an ellipse).

Now that we know orbits must follow conic section paths, we can look at some ways to describe the size and shape of an orbit.

Orbital Geometry

Because we're mainly interested in spacecraft orbits, which we know are elliptical, let's look closer at elliptical geometry. Using Figure 4-33 as a reference, let's define some important *geometrical parameters* for an ellipse.

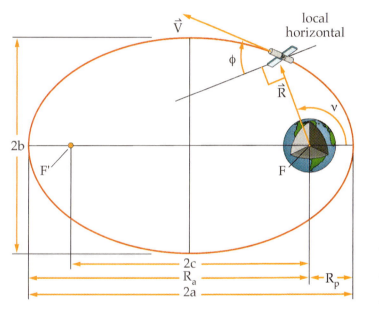

$\vec{R}$ = spacecraft's position vector, measured from Earth's center

$\vec{V}$ = spacecraft's velocity vector

F and F' = primary and vacant foci of the ellipse

R_p = radius of perigee (closest approach)

R_a = radius of apogee (farthest approach)

2a = major axis

2b = minor axis

2c = distance between the foci

a = semimajor axis

b = semiminor axis

ν = true anomaly

ϕ = flight-path angle

Figure 4-33.   Geometry of an Elliptical Orbit. With these parameters, we completely define the size and shape of the orbit.

- R is the radius from the focus of the ellipse (in this case, Earth's center) to the spacecraft

- F and F' are the *primary* (occupied) and *vacant* (unoccupied) *foci*. Earth's center is at the occupied focus.

- R_p is the *radius of periapsis* (radius of the closest approach of the spacecraft to the occupied focus); it's called the radius of *perigee* when the orbit is around Earth

- R_a is the *radius of apoapsis* (radius of the farthest approach of the spacecraft to the occupied focus); it's called the radius of *apogee* when the orbit is around Earth

- 2a is the major axis or the length of the ellipse. One-half of this is "a," or the *semimajor axis* (semi means one half).

$$a = \frac{R_a + R_p}{2} \qquad (4\text{-}19)$$

- 2b is the minor axis or width of the ellipse. One-half of this is "b," or the *semiminor axis*.

- 2c is the distance between the foci, $R_a - R_p$

- v is the *true anomaly* or polar angle measured from perigee to the spacecraft's position vector, $\vec{R}$, in the direction of the spacecraft's motion. It locates the spacecraft in the orbit. For example, if v = 180° the spacecraft is 180° from perigee, putting it at apogee. The range for true anomaly is 0° to 360°.

- φ is the *flight-path angle*, measured from the local horizontal to the velocity vector, $\vec{V}$. At the spacecraft the local horizontal is a line perpendicular to the position vector, $\vec{R}$. When the spacecraft travels from perigee to apogee (outbound), its velocity vector is always above the local horizon (gaining altitude), so φ > 0°. When it travels from apogee to perigee (inbound), its velocity vector is always below the local horizon (losing altitude), so φ < 0°. At exactly perigee and apogee of an elliptical orbit, the velocity vector is parallel to the local horizon, so φ = 0. The maximum value of the flight-path angle is 90°.

- e is the *eccentricity*, which is the ratio of the distance between the foci (2c) to the length of the ellipse (2a)

$$e = \frac{2c}{2a}$$

- Eccentricity defines the shape or type of conic section. Eccentricity is a medieval term representing a conic's degree of noncircularity (meaning "out of center"). Because circular motion was once considered perfect, any deviation was abnormal, or eccentric (maybe you know someone like that). Because the distance between the foci in an ellipse is always less than the length of the ellipse, its eccentricity is between 0 and 1. A circle has e = 0. A very long, narrow ellipse has e approaching 1. A parabola has e = 1 and a hyperbola has e > 1.

With all these geometrical parameters defined, let's look at our polar equation of a conic and substitute for the constants $k_1 = a\,(1 - e^2)$ and $k_2 = e$ (see Appendix C.3). Thus, we have

$$R = \frac{a(1 - e^2)}{1 + e\cos v} \qquad (4\text{-}20)$$

where

R = magnitude of the spacecraft's position vector (km)
a = semimajor axis (km)
e = eccentricity (unitless)
v = true anomaly (deg or rad)

136

To determine the distances at closest approach, R_p, and farthest approach, R_a, we can use this equation.

$$\text{At } v = 0°, R = R_p = \frac{a(1-e^2)}{(1+e\cos(0°))} = a(1-e)$$

$$\text{At } v = 180°, R = R_a = \frac{a(1-e^2)}{(1+e\cos(180°))} = a(1+e)$$

Looking at the geometry of an ellipse, we can see that the length of the ellipse, 2a, equals $(R_a + R_p)$, and the distance between the foci, 2c, is $(R_a - R_p)$. Now, if we want to compute the orbit's eccentricity based on the radii of perigee and apogee, we can use the second part of Equation (4-21). See Example 4-6.

$$e = \frac{2c}{2a} = \frac{R_a - R_p}{R_a + R_p} \qquad (4\text{-}21)$$

Parameters for the ellipse also apply to circular orbits, parabolic trajectories, and hyperbolic trajectories. Figure 4-34 shows a circular orbit, where the radius from Earth's center is constant and equal to the semimajor axis. Therefore, this orbit has no apogee or perigee, and its eccentricity is zero. The flight path angle is always zero.

The parabola in Figure 4-35 represents a minimum escape trajectory or a path that just barely takes a spacecraft away from Earth, never to return. So there is no apogee and no empty focus. Thus, the semimajor axis and the distance between the foci are infinite. We say the eccentricity, e = 1. The true anomaly ranges from 0° to less than 180° on the outbound path. The flight path angle is greater than zero. Of course, if a spacecraft is inbound on a parabolic trajectory, its true anomaly is greater than 180° until it passes perigee, then it resets to 0° and grows to almost 180°. And its flight path angle is less than zero until it passes perigee.

The hyperbola in Figure 4-36 also represents an escape trajectory, so it also has no apogee. It's an unusual shape with a different sign convention. Because the length of the hyperbola (distance between the "ends") bends back on itself, or is measured outside the conic, we define this distance, 2a, as negative. The same convention also applies for the distance between the foci, 2c, so 2c is also negative. But the magnitude of 2c is always larger than the magnitude of 2a, so the eccentricity is greater than 1.0. The true anomaly ranges from 0° to less than 180° on the outbound path and greater than 180° to 0° on the inbound path. The flight path angle is greater than 0° on the outbound path and less than 0° on the inbound path. Table 4-1 summarizes these parameters.

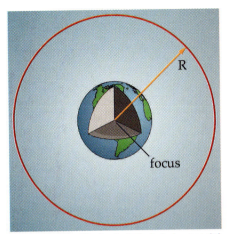

Figure 4-34. Circle. A circle is just a special case of an ellipse.

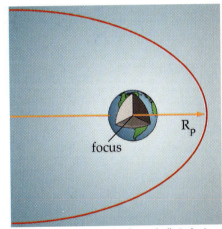

Figure 4-35. Parabola. A parabolic trajectory is a special case which leaves Earth altogether.

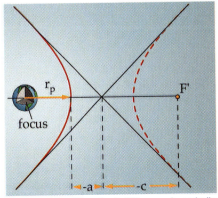

Figure 4-36. Hyperbola. We use a hyperbolic trajectory for interplanetary missions. Notice a real trajectory is around the occupied focus and an imaginary, mirror-image trajectory is around the vacant focus.

Table 4-1.   A Summary of Parameters for Conic Sections.

Conic Section	a = Semimajor Axis	c = One-half the Distance between Foci	e = Eccentricity
circle	$a > 0$	$c = 0$	$e = 0$
ellipse	$a > 0$	$0 < c < a$	$0 < e < 1$
parabola	$a = \infty$	$c = \infty$	$e = 1$
hyperbola	$a < 0$	$|a| < |c| > 0$	$e > 1$

▬ Section Review

Key Terms

apogee
circle
conic sections
eccentricity, e
ellipse
flight-path angle, ϕ
foci
fundamental plane
geocentric-equatorial coordinate system
geometrical parameters
hyperbola
origin
parabola
perigee
primary focus, F
principal direction, $\hat{I}$
radius of apoapsis, R_a
radius of periapsis, R_p
restricted two-body equation of motion
restricted two-body problem
semimajor axis, 2a
semiminor axis, 2b
true anomaly, ν
vacant focus, F'

Key Equations

$$\ddot{\vec{R}} + \frac{\mu}{R^2}\frac{\vec{R}}{R} = 0$$

$$R = \frac{a(1 - e^2)}{1 + e\,\cos\nu}$$

Key Concepts

➤ Combining Newton's Second Law and his Law of Universal Gravitation, we form the restricted two-body equation of motion

- The coordinate system used to derive the two-body equation of motion is the geocentric-equatorial system
  - Origin—Earth's center
  - Fundamental plane—equatorial plane
  - Direction perpendicular to the plane—North Pole direction
  - Principal direction—vernal equinox direction

- In deriving this equation, we assume
  - Drag force is negligible
  - Spacecraft is not thrusting
  - Gravitational pull of third bodies and all other forces are negligible
  - $m_{Earth} \gg m_{spacecraft}$
  - Earth is spherically symmetrical and of uniform density and we can treat it mathematically as a point mass
  - Spacecraft mass is constant, so $\Delta m = 0$
  - The geocentric-equatorial coordinate system is sufficiently inertial for Newton's laws to apply

➤ Solving the restricted two-body equation of motion results in the polar equation for a conic section

➤ Figure 4-33 shows parameters for orbital geometry, and Table 4-1 summarizes parameters for conic sections

Example 4-6

Problem Statement

Suppose a new class of remote-sensing satellite is in an orbit with a perigee radius of 7000 km and an apogee radius of 10,000 km. What is its altitude above Earth when the true anomaly is 90°?

Problem Summary

Given: $R_p = 7000$ km

$R_a = 10,000$ km

Find: Altitude, when $v = 90°$

Problem Diagram

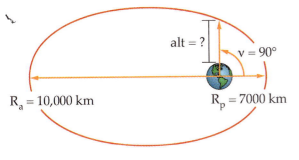

$alt = ?$

$v = 90°$

$R_a = 10,000$ km

$R_p = 7000$ km

Conceptual Solution

1) Find the eccentricity for the orbit using

$$e = \frac{R_a - R_p}{R_a + R_p}$$

2) Find the semimajor axis for the orbit using

$$a = \frac{R_a + R_p}{2}$$

3) Solve for the radius when $v = 90°$ using the polar equation of a conic

$$R = \frac{a(1 - e^2)}{1 + e \cos v}$$

4) The altitude when $v = 90°$ is the radius minus the radius of the Earth, R_{Earth}

$$Alt_{v = 90°} = R_{v = 90°} - R_{Earth}$$

Analytical Solution

1) Find e

$$e = \frac{R_a - R_p}{R_a + R_p} = \frac{10,000 \text{ km} - 7000 \text{ km}}{10,000 \text{ km} + 7000 \text{ km}} = 0.1765$$

2) Find a

$$a = \frac{R_a + R_p}{2} = \frac{10,000 \text{ km} + 7000 \text{ km}}{2} = 8500 \text{ km}$$

3) Find R

$$R = \frac{a(1 - e^2)}{1 + e \cos v} = \frac{(8500 \text{ km})(1 - (0.1765)^2)}{1 + 0.1765 \cos 90°}$$

$$R = 8235.2 \text{ km}$$

4) Find $Alt_{v = 90°}$

$$Alt_{v = 90°} = 8235.2 \text{ km} - 6378.14 \text{ km}$$

$$= 1857.1 \text{ km}$$

Interpreting the Results

When this new remote-sensing spacecraft has reached a point 90° past perigee, it's at an altitude of 1857.1 km.

4.5 Constants of Orbital Motion

▰ In This Section You'll Learn to...

☞ Define the two constants of orbital motion—specific mechanical energy and specific angular momentum

☞ Apply specific mechanical energy to determine orbital velocity and period

☞ Apply the concept of conservation of specific angular momentum to show an orbital plane remains fixed in space

By now you're probably convinced that, with all these flight-path angles, true anomalies, and ellipses flying around, there is nothing consistent about orbits. Well, take heart because we do have constants in astrodynamics. We saw in our discussion of motion in a conservative field that mechanical energy and momentum are conserved. Because orbital motion occurs in a conservative gravitational field, spacecraft conserve mechanical energy and angular momentum. So, now let's see how these principles provide valuable tools for studying orbital motion.

Specific Mechanical Energy

In an earlier section, we referred to equations of motion being like crystal balls, in that they allow us to gaze into the future to predict where an object will be. Mechanical energy provides us with one such crystal ball. Recall in defining mechanical energy, we add potential energy, PE, to kinetic energy, KE. Together, they form a relationship between a spacecraft's mass, m, its position, R, its velocity, V, and the local gravitational parameter, μ (3.986×10^5 km^3/s^2 for Earth).

$$E = \frac{1}{2}mV^2 - \frac{\mu m}{R} \tag{4-22}$$

To generalize this equation, so we don't have to worry about mass, let's divide both sides of the equation by m. Doing so defines a new flavor of mechanical energy called *specific mechanical energy, ε,* which doesn't depend on mass. Thus, we can talk about the energy in a particular orbit, whether the orbiting object is a golf ball or the International Space Station. Specific mechanical energy, ε, is simply the total mechanical energy divided by a spacecraft's mass

$$\varepsilon \equiv \frac{E}{m} \tag{4-23}$$

where ≡ means "defined as," or

$$\boxed{\varepsilon = \frac{V^2}{2} - \frac{\mu}{R}} \tag{4-24}$$

140

where

ε = spacecraft's specific mechanical energy (km^2/s^2)

V = spacecraft's velocity (km/s)

μ = gravitational parameter (km^3/s^2) = 3.986×10^5 km^3/s^2 for Earth

R = spacecraft's distance from Earth's center (km)

Because the specific mechanical energy is conserved (see Appendix C.2), it must be the same at any point along an orbit! As a spacecraft approaches apogee, it is gaining altitude, meaning its R, or distance from Earth's center, increases. This increase in R means it gains potential energy—which actually means the potential energy (PE) gets less negative (because of the way we define it). At the same time, the spacecraft's speed is decreasing and hence it is losing kinetic energy (KE). When it reaches the highest point, its PE is at a maximum. However, because its speed is the slowest at apogee, KE is at a minimum. But the sum of PE and KE—specific mechanical energy—remains constant.

As the spacecraft passes apogee and starts toward perigee, it begins to trade its PE for KE. So, its speed steadily increases until it reaches perigee, where its speed is fastest and its KE is maximum. Again, the sum of potential and kinetic energy—specific mechanical energy—remains constant. Figure 4-37 illustrates these relationships.

The fact that the specific mechanical energy is constant gives us a tremendously powerful tool for analyzing orbits. Look again at the relationship for specific mechanical energy, ε. Notice that ε depends only on position, R, velocity, V, and the local gravitational parameter, μ. This means if we know a spacecraft's position and velocity at any point along its orbit, we know its specific mechanical energy for every point on its orbit.

Another important concept to glean out of the constancy of orbital energy is the relationship between R and V. Assume we know the energy for an orbit. Then, at any given position, R, on that orbit, there is only one possible velocity, V! Thus, if we know the orbital energy and R, we can easily find the velocity at that point. Simply rearranging the relationship for energy gives us an extremely useful expression for velocity.

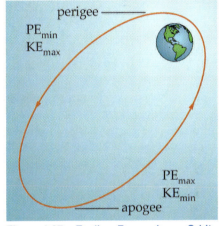

Figure 4-37. **Trading Energy in an Orbit.** An orbit is similar to a swing. PE and KE trade-off throughout the orbit, so their sum is constant.

$$V = \sqrt{2\left(\frac{\mu}{R} + \varepsilon\right)} \qquad (4\text{-}25)$$

where

V = spacecraft's velocity (km/s)

μ = gravitational parameter (km^3/s^2) = 3.986×10^5 km^3/s^2 for Earth

R = spacecraft's distance from Earth's center (km)

ε = spacecraft's specific mechanical energy (km^2/s^2)

We often use this equation to determine velocities while analyzing orbits. For example, during space missions we often have to move space-craft from one orbit to another. We can use this relationship to determine how much we must change the velocity to "drive" over to the new orbit.

Note: We define R from Earth's center, so when you're using orbital altitude remember to add Earth's radius.

Recall from our discussion of conic-section geometry, one parameter represents a spacecraft's mean, or average, distance from the primary focus. This parameter is the semimajor axis, a. We can develop a new relationship for specific mechanical energy which depends only on a and μ. (See Appendix C.4)

$$\varepsilon = -\frac{\mu}{2a}$$ (4-26)

where

ε = spacecraft's specific mechanical energy (km^2/s^2)

μ = gravitational parameter $(km^3/s^2) = 3.986 \times 10^5\ km^3/s^2$ for Earth

a = semimajor axis (km)

This means simply knowing the semimajor axis of a spacecraft's orbit tells us its specific mechanical energy. We can also learn the type of trajectory from the sign of the specific mechanical energy, ε. For a circular or elliptical orbit, ε is *negative* (because a is positive). For a parabola, $\varepsilon = 0$ (because a = ∞). For a hyperbola, ε is *positive* (because a is negative). These are important points to keep in mind as we work orbital problems. If the sign for ε is wrong, the answer probably will be wrong.

Another benefit to knowing a value for energy is that we can determine orbital period. The *orbital period, P,* is the time it takes for a spacecraft to revolve once around its orbit. From Kepler's Third Law of Planetary Motion, which we showed in Chapter 2, P^2 is proportional to a^3, where "a," is the semimajor axis. Using this relationship, we can derive an expression for the orbital period (see Appendix C.6)

$$P = 2\pi\sqrt{\frac{a^3}{\mu}}$$ (4-27)

where

P = period (seconds)

π = 3.14159...(unitless)

a = semimajor axis (km)

μ = gravitational parameter $(km^3/s^2) = 3.986 \times 10^5\ km^3/s^2$ for Earth

Notice that period only has meaning for "closed" conics (circles or ellipses). Period is infinite for a parabola, whose semimajor axis is infinite, and it's an imaginary number for a hyperbola, whose semimajor axis is negative.

Specific mechanical energy, ε, is a very valuable constant of spacecraft motion. With a single observation of position and velocity, we learn much about a spacecraft's orbit. Example 4-7 shows one application of specific mechanical energy.

But ε gives us only part of the story. It tells us the orbit's *size* but doesn't tell us anything about where the orbit is in space. For insight into that important bit of information we need to look at the angular momentum.

Specific Angular Momentum

Recall from our discussion in Section 4.2 that we can find angular momentum from Equation (4-3).

$$\vec{H} = \vec{R} \times m\vec{V}$$

Once again, to uncomplicate our life, we divide both sides of the equation by the mass, m, of the object we're investigating. Doing this, we define the *specific angular momentum, $\vec{h}$*, as

$$\vec{h} \equiv \frac{\vec{H}}{m}$$

where

≡ means "defined as," or

$$\boxed{\vec{h} = \vec{R} \times \vec{V}}$$ (4-28)

where
$\vec{h}$ = spacecraft's specific angular momentum vector (km^2/s)

$\vec{R}$ = spacecraft's position vector (km)

$\vec{V}$ = spacecraft's velocity vector (km/s)

Notice that specific angular momentum is the result of the cross product between two vectors: position and velocity. Recall from geometry that any two lines define a plane. So in this case, $\vec{R}$ and $\vec{V}$ are two lines (vectors having magnitude and direction) that define a plane. We call this plane containing $\vec{R}$ and $\vec{V}$, the *orbital plane*. Because the cross product of any two vectors results in a third vector that is perpendicular to the first two, the angular momentum vector $\vec{h}$ must be perpendicular to $\vec{R}$ and $\vec{V}$. Figure 4-38 shows $\vec{R}$, $\vec{V}$, and $\vec{h}$.

Here's where we need to apply a little deductive reasoning and consider the logical consequence of the facts we know to this point. First of all, as we saw in Section 4.3, angular momentum and, hence, specific angular momentum are constant in magnitude and direction (also see Appendix C.2). Second, $\vec{R}$ and $\vec{V}$ define the orbital plane. Next, $\vec{h}$ is perpendicular to the orbital plane. Therefore, if $\vec{h}$ is always perpendicular to the orbital plane, and $\vec{h}$ is constant, the orbital plane must also be constant. This means that in our restricted, two-body problem the orbital plane is forever frozen in inertial space! However, in reality, as we'll see in Chapter 8, slight disturbances cause the orbital plane to change gradually over time.

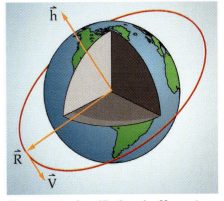

Figure 4-38. Specific Angular Momentum. The specific angular momentum vector, $\vec{h}$, is perpendicular to the orbital plane defined by $\vec{R}$ and $\vec{V}$.

■ Section Review

Key Terms

orbital period, P
orbital plane
specific angular momentum, $\vec{h}$
specific mechanical energy, ε

Key Equations

$$\varepsilon = \frac{V^2}{2} - \frac{\mu}{R}$$

$$V = \sqrt{2\left(\frac{\mu}{R} + \varepsilon\right)}$$

$$\varepsilon = -\frac{\mu}{2a}$$

$$P = 2\pi\sqrt{\frac{a^3}{\mu}}$$

$$\vec{h} = \vec{R} \times \vec{V}$$

Key Concepts

➤ In the absence of any force other than gravity, two quantities remain constant for an orbit

- Specific mechanical energy, ε
- Specific angular momentum, $\vec{h}$

➤ Specific mechanical energy, ε, is defined as $\varepsilon \equiv E/m$

- $\varepsilon < 0$ for circular and elliptical orbits
- $\varepsilon = 0$ for parabolic trajectories
- $\varepsilon > 0$ for hyperbolic trajectories

➤ Specific angular momentum, $\vec{h}$ is defined as $\vec{h} \equiv \vec{H}/m$

- It is constant for an orbit
- Because $\vec{h}$ is constant, orbital planes are fixed in space (neglecting orbital perturbations—see Chapter 8)

Example 4-7

Problem Statement

What is the velocity of the remote-sensing spacecraft discussed in Example 4-6 when the true anomaly is 90°? How long before the spacecraft returns to this point in the orbit?

Problem Summary

Given: $a = 8500$ km
$R_{v\,=\,90°} = 8235.2$ km

Find: $V_{v\,=\,90°}$, P

Problem Diagram

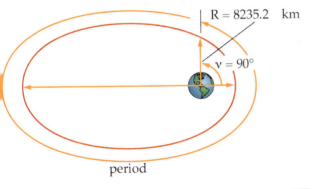

period

Conceptual Solution

1) Find the specific mechanical energy for the orbit from

$$\varepsilon = -\frac{\mu}{2a}$$

2) Knowing R at $v = 90°$, find the velocity there, using

$$V = \sqrt{2\left(\frac{\mu}{R} + \varepsilon\right)}$$

3) Find the orbital period from

$$P = 2\pi\sqrt{\frac{a^3}{\mu}}$$

Analytical Solution

1) Find ε

$$\varepsilon = -\frac{\mu}{2a} = \frac{-3.986 \times 10^5 \dfrac{km^3}{s^2}}{2(8500\ km)} = -23.45\frac{km^2}{s^2}$$

2) Find V

$$V = \sqrt{2\left(\frac{\mu}{R} + \varepsilon\right)}$$

$$= \sqrt{2\left(\frac{3.986 \times 10^5 \dfrac{km^3}{s^2}}{8235.2\ km} - 23.45\frac{km^2}{s^2}\right)}$$

$$= 7.065\ \frac{km}{s}$$

3) Find P

$$P = 2\pi\sqrt{\frac{a^3}{\mu}} = 2\pi\sqrt{\frac{(8500\ km)^3}{3.986 \times 10^5 \dfrac{km^3}{s^2}}}$$

$$= 7799\ seconds$$

$$\cong 130\ minutes$$

Interpreting the Results

At a point in the orbit 90° from perigee, this spacecraft's velocity is 7.065 km/s (15,807 m.p.h.). For this orbit the period is about 130 minutes, or two hours and ten minutes.

References

Bate, Roger R., Donald D. Mueller and Jerry E. White. *Fundamentals of Astrodynamics*. New York, NY: Dover Publications, Inc., 1971.

Boorstin, Daniel J. *The Discoverers*. Random House, 1983.

Concepts in Physics. Del Mar, CA: Communications Research Machines, Inc., 1973.

Feynman, Richard P., Robert B. Leighton, and Matthew Sands. *The Feynman Lectures on Physics*. Reading, MA: Addison-Wesley Publishing Co., 1963.

Gonick, Larry and Art Huffman. *The Cartoon Guide to Physics*. New York, NY: HarperCollins Publishers, 1990.

Hewitt, Paul G. *Conceptual Physics... A New Introduction to Your Environment*. Boston, MA: Little, Brown and Company, 1981.

King-Hele, Desmond. *Observing Earth Satellites*. New York, NY: Van Nostrand Reinhold Company, Inc., 1983.

Szebehely, Victor G. *Adventures in Celestial Mechanics*. Austin, TX: University of Texas Press, 1989.

Thiel, Rudolf. *And There Was Light*. New York: Alfred A Knopf, 1957.

Young, Louise B., Ed., *Exploring the Universe*. Oxford, MA: Oxford University Press, 1971.

Mission Problems

4.1   Orbital Motion

1   Explain how an object's horizontal velocity allows it to achieve orbit.

2   An object in a circular orbit is given a bit of extra velocity. What type of orbit will it now be in?

3   What two types of trajectories completely escape Earth?

4   Explain how you could use the steps in the Motion Analysis Process checklist to analyze the motion of a volleyball being served.

4.2   Newton's Laws

5   What three things does an object's mass tell you about the object?

6   An astronaut on the Moon drops a hammer and a feather from the same height at the same time. Describe what happens and why. Explain the difference if this experiment occurs on Earth.

7   Describe how the recoil you feel when firing a rifle is the result of Newton's Third Law of Motion.

8   You are spinning a 0.25 kg weight over your head at the end of a 0.5 m string. If you let go of the string, the weight will sail off on a tangent at 2 m/s. What is the angular momentum of the spinning weight before release? Because angular momentum is always conserved, where does the angular momentum go after release?

9   The new quarterback throws a football in a perfect spiral with a moment of inertia of 0.001 kg · m². If the football is spinning at 60 r.p.m., what is its angular momentum?

10 Two asteroids pass by each other in the void of interstellar space. Asteroid Zulu has a mass of 1×10^6 kg. Asteroid Echo has a mass of 8×10^6 kg. If the two are separated by 100 m, what is the force of gravitational attraction between them?

11 While flying straight and level in your SR-71 airplane 25,000 m above the surface of Earth, you drop your pencil. What acceleration will the pencil have as it falls toward the cockpit floor?

12 Neglecting air resistance, how fast will a baseball (m = 0.1 kg) be travelling when it hits the ground, if it's dropped from the Empire State Building (about 300 m high)? How long will it take to hit the ground?

13 Match the physical laws on the left with the best term or description on the right. [*Hint:* You may need to review Kepler's Laws from Chapter 2.]

1) Newton's First Law

2) Newton's Second Law

3) Newton's Third Law

4) Newton's Law of Universal Gravitation

5) Kepler's First Law

6) Kepler's Second Law

7) Kepler's Third Law

a) Planetary orbits are ellipses

b) Gravity is the only force on a spacecraft

c) Inertia

d) Action/reaction

e) Relates orbital period to orbital size

f) A net force causes an acceleration

g) Equal areas in equal times

h) Spacecraft may be treated as point masses

i) Force is inversely proportional to the square of the distance

14 For what mass does the gravitational parameter equal 3.986×10^5 km^3/s^2?

15 A rocket engine moves forward by expelling high-velocity gas from the exhaust nozzle. Discuss the physical law which explains why this occurrence produces a force on the rocket.

4.3 Laws of Conservation

16 For an isolated system (one which has no interaction with its surroundings), what quantities are constant according to the laws of physics?

17 You are floating in the cockpit of the Space Shuttle Atlantis when you decide to do somersaults, thus increasing your angular momentum. Why are you floating? If you push off and do forward somersaults, then tuck into as tight a ball as you can, what happens to your angular momentum? Why?

18 Describe the potential, kinetic, and total energy of a baseball that's thrown into the air, reaches its highest point, then falls to the ground.

4.4 The Restricted Two-Body Problem

19 What are the origin, principal direction, and fundamental plane for the geocentric-equatorial coordinate system?

20 What simplifying assumptions do we use to "restrict" the two-body equation of motion?

21 In solving the restricted two-body equation of motion, we obtain the polar equation of a conic section. Why is this significant?

22 Match the following terms with their definitions.

1) $\vec{R}$

2) $\vec{V}$

3) F and F'

4) Radius at perigee (R_p)

5) Radius at apogee (R_a)

6) Major axis (2a)

7) True anomaly (v)

8) Flight-path angle (ϕ)

9) Eccentricity (e)

a) Closest point in an orbit

b) Primary and vacant foci of a conic section

c) Position vector

d) Angle between perigee and the position vector

e) "Out of roundness" of a conic section

f) Distance across the long axis of an ellipse

g) Angle between local horizontal and the velocity vector

h) Velocity vector

i) Furthest point in an orbit

23 A Russian satellite is in Earth orbit with an altitude at perigee of 375 km and an altitude at apogee of 2000 km.

a) What is the semimajor axis of the orbit?

b) What is the eccentricity?

c) If the true anomaly is 175°, what is the satellite's altitude?

d) If the true anomaly is 290°, is the flight-path angle positive or negative? Why?

4.5   Constants of Orbital Motion

24 Where is the potential energy of a spacecraft greater, at perigee or apogee? Why?

25 While co-piloting a futuristic spacecraft, you receive a report of your position and velocity in the geocentric-equatorial frame

$$\vec{R} = 7000\ \hat{I} + 0\ \hat{J} + 0\ \hat{K}\ \text{km}$$

$$\vec{V} = 0\ \hat{I} - 7.063\ \hat{J} + 0\ \hat{K}\ \text{km/s}$$

a) Sketch the spacecraft position and velocity vectors relative to Earth. (*Hint:* Draw the geocentric-equatorial coordinate system first.)

b) What is the specific angular momentum? Draw this vector on the sketch.

c) What does this angular momentum vector tell you about the orientation of your orbit?

d) What is your specific mechanical energy?

e) What is the shape of your trajectory? How can you tell?

26 You are the engineer in charge of a top-secret spy satellite. The satellite will be placed in a circular, sun-synchronous orbit (see Chapter 8), with an altitude of 759 km and a mass of 10,000 kg. A politician who dislikes the project says the satellite poses a danger to the public because of its large kinetic energy so close to Earth. What is the kinetic energy of the satellite? Compare this to the kinetic energy of a 2000 kg truck travelling down the interstate at 65 m.p.h. Is the comparison realistic? Why or why not? (*Hint:* Are the two objects in the same reference frame?)

27 Calculate the altitude needed for a circular, geosynchronous orbit (an orbit whose orbital period matches Earth's rotation rate).

28 Complete the following table with the possible range of values of semimajor axis, eccentricity, and specific mechanical energy.

Conic Section	a = semimajor axis	e = eccentricity	ε = specific mechanical energy
circle			
ellipse			
parabola			
hyperbola			

29 We know that the velocity of a spacecraft in an elliptical orbit is greatest at perigee due to conservation of specific mechanical energy. Relate this fact to Kepler's Second Law.

30 A Mars probe is in a circular orbit around Earth with a radius of 25,000 km. The next step on the way to Mars is to thrust so the probe can enter an escape orbit. (a) Determine the probe's velocity in this circular orbit. (b) Determine the minimum velocity required to enter a parabolic trajectory at that radius. (c) Determine the difference in the specific kinetic energies of the two orbits. (d) Now compare this result to the specific mechanical energy of the original circular orbit. Which one has more specific kinetic energy? Are you surprised? Why or why not?

For Discussion

31 Demonstrate that two-body motion is confined to a plane fixed in space (*Hint:* What quantities are conserved?)

32 For the equations of motion to be correct the coordinate reference frames must be inertial. Is the geocentric-equatorial frame, commonly used for spacecraft, a truly inertial reference frame? Why or why not? If not, why can we use it?

33 We based the equations of motion for spacecraft on the two-body problem. If we wish to design a trajectory for an interplanetary probe to Mars, which two bodies would we consider at the beginning of the flight, when the probe leaves Earth? At the middle, when it's between Earth and Mars? At the end, as it arrives at Mars? What problems, if any, do you think this might produce?

149

Notes

Mission Profile—Apollo

"...I believe that this nation should commit itself to achieving the goal, before this decade is out, of landing a man on the Moon and returning him safely to Earth."

President John F. Kennedy
May 25, 1961

Less than three weeks after Alan Shepard's first suborbital flight, President Kennedy's address to Congress boldly established a Moon landing as a national goal. Over the next eleven years, project Apollo grew from a statement of national intent to a project that successfully launched 11 spacecraft and allowed 12 men to walk on the surface of the Moon.

Mission Overview

Apollo's mission was as simple as President Kennedy's quote—get a man to the Moon and back safely. After this initial goal was accomplished, Apollo astronauts were responsible for collecting scientific data about the Moon and Earth.

Mission Data

✓ Apollo 7, October 11, 1968 (Crew: Cunningham, Eisele, Schirra): First manned Apollo flight. System checkout of command module. Earth orbit only.

✓ Apollo 8, December 21, 1968 (Crew: Anders, Borman, Lovell): First manned launch of Saturn V. First lunar orbit.

✓ Apollo 9, March 3, 1969 (Crew: McDivitt, Schweickart, Scott): First flight, test, and docking. Earth orbit only.

✓ Apollo 10, May 18, 1969 (Crew: Cernan, Stafford, Young): "Dress rehearsal" for lunar landing: included descent of lunar module to 50,000 ft. above lunar surface.

✓ Apollo 11, July 16, 1969 (Crew: Aldrin, Armstrong, Collins): First lunar landing. Armstrong first man to walk on Moon.

✓ Apollo 12, November 14, 1969 (Crew: Bean, Conrad, Gordon): Recovered parts from Surveyor 3. Conducted scientific experiments.

✓ Apollo 13, April 11, 1970 (Crew: Haise, Lovell, Swigert): Mission aborted due to an explosion on the service module on the way to the Moon.

✓ Apollo 14, January 31, 1971 (Crew: Mitchell, Roosa, Shepard): First golf ball hit on the Moon. Conducted scientific experiments.

✓ Apollo 15, July 26, 1971 (Crew: Irwin, Scott, Worden): First use of Lunar Roving Vehicle.

✓ Apollo 16, April 16, 1972 (Crew: Duke, Mattingly, Young): Over 20 hours of extra-vehicular activity on the Moon.

✓ Apollo 17, December 7, 1972 (Crew: Cernan, Evans, Schmidt): Last Apollo Moon landing.

Mission Impact

Apollo met President Kennedy's goal and captured the imagination of the entire world. In addition, the Apollo program provided scientists with invaluable information about the Moon. Unfortunately, the public and Congress soon lost interest in the Moon. NASA shifted its focus to developing the Space Shuttle, and the technological infrastructure to take humans to the Moon and back was laid to rest in museums.

Jim Irwin salutes the flag on Apollo 15. *(Courtesy of NASA/Johnson Space Center)*

For Discussion

* Why did we go to the Moon?
* Some say we spent all that money on Apollo and all we got was "a bunch of rocks." Is this really true?
* What plans do we have to return to the Moon?

Contributor

Todd Lovell, the U.S. Air Force Academy

References

Baker, David. *The History of Manned Spaceflight.* New York: Crown, 1981.

Yenne, Bill. *The Encyclopedia of US Spacecraft*. New York: Exeter, 1985.

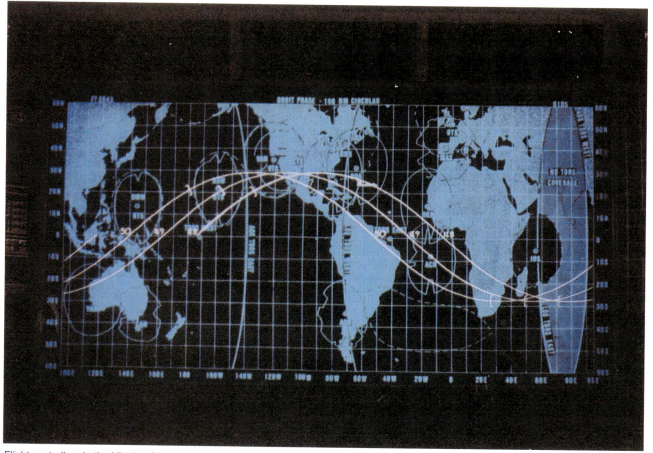

Flight controllers in the Mission Control Center use this large ground track display to diligently monitor the Space Shuttle's path throughout a mission. *(Courtesy of NASA/Johnson Space Center)*

Describing Orbits

5

In This Chapter You'll Learn to...

☛ Define the classic orbital elements (COEs) used to describe the size, shape, and orientation of an orbit and the location of a spacecraft in that orbit

☛ Determine the COEs given the position, $\vec{R}$, and velocity, $\vec{V}$, of a spacecraft at one point in its orbit

☛ Explain and use orbital ground tracks

You Should Already Know...

❏ The restricted two-body equation of motion and its assumptions (Chapter 4)

❏ Orbit specific mechanical energy, ε (Chapter 4)

❏ Orbit specific angular momentum, $\vec{h}$ (Chapter 4)

❏ Definition of vectors and vector operations including dot and cross products (Appendix A.2)

❏ Inverse trigonometric functions $\cos^{-1}$ and $\sin^{-1}$ (Appendix A.1)

Outline

5.1 Orbital Elements
Defining the Classic Orbital Elements (COEs)
Alternate Orbital Elements

5.2 Computing Orbital Elements
Finding Semimajor Axis, a
Finding Eccentricity, e
Finding Inclination, i
Finding Right Ascension of the Ascending Node, Ω
Finding Argument of Perigee, ω
Finding True Anomaly, ν

5.3 Spacecraft Ground Tracks

Space isn't remote at all. It's only an hour's drive away if your car could go straight upwards.

Sir Fred Hoyle, London Observer

Space Mission Architecture. This chapter deals with the Trajectories and Orbits segment of the Space Mission Architecture introduced in Figure 1-20.

In the last chapter we looked at the restricted two-body problem and developed an equation of motion to describe in strictly mathematical terms, how spacecraft move through space. But many times it's not enough to generate a list of numbers that give a spacecraft's position and velocity in inertial space. Often, we want to visualize its orbit with respect to points on Earth. For example, we may want to know when a remote-sensing spacecraft will be over a flood-damaged area (Figure 5-1).

Figure 5-1.   Mississippi River Flooding. Here we show an Earth Observation System view of the river flooding at St. Louis, Missouri, in 1993. *(Courtesy of NASA/Goddard Space Flight Center)*

In this chapter, we'll explore two important tools that help us "see" spacecraft motion—the classic orbital elements (COEs) and ground tracks. Once you get the hang of it, you'll be able to use these COEs to visualize how the orbit looks in space. Ground tracks will allow you to determine when certain parts of the Earth pass into a spacecraft's field of view, and when an observer on Earth can see the spacecraft.

5.1   Orbital Elements

▬ In This Section You'll Learn to...

- ☞ Define the classic orbital elements (COEs)
- ☞ Use the COEs to describe the size, shape, and orientation of an orbit and the location of a spacecraft in that orbit
- ☞ Explain when particular COEs are undefined and which alternate elements we must use in their place

If you're flying an airplane and the ground controllers call you on the radio to ask where you are and where you're going, you must tell them six things: your airplane's

- Latitude
- Longitude
- Altitude
- Horizontal velocity
- Heading (i.e. north, south, etc.)
- Vertical velocity (ascending or descending)

Knowing these things, controllers can then predict your future position.

Space operators do something similar, except they don't ask where the spacecraft is; instead, they use radar at tracking sites to measure it's current position, $\vec{R}$, and velocity, $\vec{V}$. As we'll see in Chapter 8, this information helps them predict the spacecraft's future position and velocity. Notice that position, $\vec{R}$, and velocity, $\vec{V}$, are vectors with three components each. Unfortunately, unlike latitude and longitude used for aircraft, $\vec{R}$ and $\vec{V}$ aren't very useful in visualizing a spacecraft's orbit.

For example, suppose you're given this current position and velocity for a spacecraft

$$\vec{R} = 10{,}000\ \hat{I} + 8000\ \hat{J} - 7000\ \hat{K}\ \ km$$

$$\vec{V} = 4.4\ \hat{I} + 3.1\ \hat{J} - 2.7\ \hat{K}\ \ km/s$$

What could you tell about the orbit's size and shape or the spacecraft's position?

With the tools you've learned, about the only thing you could do is plot $\vec{R}$ and $\vec{V}$ in a 3-dimensional coordinate system and try to visualize the orbit that way. Fortunately, there's an easier way. Hundreds of years ago, Johannes Kepler developed a method for describing orbits that allows us to visualize their size, shape, and orientation, as well as the spacecraft's position within them. Because we still need six quantities to describe an orbit and a spacecraft's place in it, Kepler defined six orbital elements. We call these the *classic orbital elements (COEs)*, and we'll use

Recall that $\vec{V}$ is the velocity vector that describes speed and direction. We use V to denote speed without regard to direction.

155

Classic Orbital Elements (COEs) Checklist

◉ **Orbit's size**

◉ **Orbit's shape**

◉ **Orbit's orientation**
  • Orbital plane in space
  • Orbit within the plane

◉ **Spacecraft's location**

them to tell us the four things we want to know, as summarized in the COEs checklist on the left. In the rest of this section, we'll go through each of the four things on the checklist and learn which COE describe the given properly. As a preview, we'll learn

• Orbital size, uses the semimajor axis, a

• Orbital shape, is defined by eccentricity, e

• Orientation of the orbital plane in space, uses

  - inclination, i

  - right ascension of the ascending node, Ω

• Orientation of the orbit within the plane is defined by argument of perigee, ω and finally

• Spacecraft's location in the orbit is represented by true anomaly, ν

Let's go through these elements to see what each one contributes to our understanding of orbits and check them off one at a time on our COE checklist.

Classic Orbital Elements (COEs) Checklist

✔ **Orbit's size**

◉ **Orbit's shape**

◉ **Orbit's orientation**
  • Orbital plane in space
  • Orbit within the plane

◉ **Spacecraft's location**

Defining the Classic Orbital Elements (COEs)

Let's start with orbital *size*. In Chapter 4 we related the size of an orbit to its specific mechanical energy using the relationship

$$\varepsilon = -\frac{\mu}{2a}$$ (5-1)

where

ε = specific mechanical energy (km^2/s^2)

μ = gravitational parameter of the central body (km^3/s^2)

a = semimajor axis (km)

The semimajor axis, a, describes half the distance across the orbit's major (long) axis, as shown in Figure 5-2, and we use it as our first COE.

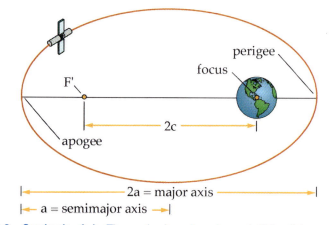

|← ——————— 2a = major axis ——————— →|
|← a = semimajor axis →|

Figure 5-2.   Semimajor Axis. The semimajor axis, a, is one half the distance across the long axis of an ellipse. The distance between the foci (F and F') of the ellipse is 2c.

With the orbit's size accounted for, the next thing we want to know is its shape. In Chapter 4, we described the "out of roundness" of a conic section in terms of its eccentricity, e. *Eccentricity* specifies the shape of an orbit by looking at the ratio of the distance between the two foci and the length of the major axis.

$$e = \frac{2c}{2a} \qquad (5\text{-}2)$$

Table 5-1 summarizes the relationship between an orbit's shape and its eccentricity and Figure 5-3 illustrates this relationship.

Table 5-1.   Relationship Between Conic Section and Eccentricity.

Conic Section	Eccentricity
Circle	e = 0
Ellipse	0 < e < 1
Parabola	e = 1
Hyperbola	e > 1

Now we have two pieces of our orbital puzzle: its size, a, and its shape, e. Next we tackle its orientation in space. In Chapter 4 we learned that because specific angular momentum is constant, an orbital plane is stationary in inertial space. To describe its orientation, we refer to an inertial coordinate system used in Chapter 4—the geocentric-equatorial coordinate system, shown in Figure 5-4. (In the following discussion, we describe angles between key vectors, so make sure you know how to perform dot products [see Appendix A.2 for a review] and how to change from degrees to radians [Appendix A.1].)

The first angle we use to describe the orientation of an orbit with respect to our coordinate system is inclination, i. *Inclination* describes the tilt of the orbital plane with respect to the fundamental plane (the equatorial plane in this case). We could describe this tilt as the angle between the two planes, but this is harder to do mathematically. Instead, we define inclination as the angle between two vectors: one perpendicular to the orbital plane, $\hat{h}$ (the specific angular momentum vector), and one perpendicular to the fundamental plane, $\hat{K}$, as shown in Figure 5-5. Inclination has a range of values from 0° to 180°.

We use inclination to define several different kinds of orbits. For example, an Earth orbit with an inclination of 0° or 180° is an *equatorial orbit*, because it always stays over the equator. If the orbit has i = 90°, we call it a *polar orbit* because it travels over the North and South Poles. We also use the value of inclination to distinguish between two major classes of orbits. If $0° \le i < 90°$, the spacecraft is moving with Earth's rotation (in an easterly direction), and the spacecraft is in a *direct orbit* or *prograde orbit*. If $90° < i \le 180°$, the spacecraft is moving opposite from Earth's rotation (in a westerly direction), so it's in an *indirect orbit* or *retrograde orbit*. Table 5-2 summarizes these orbits.

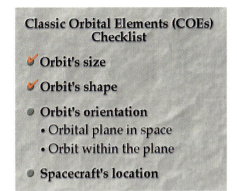

Classic Orbital Elements (COEs) Checklist

✔ **Orbit's size**

✔ **Orbit's shape**

● **Orbit's orientation**
  • Orbital plane in space
  • Orbit within the plane

● **Spacecraft's location**

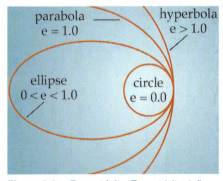

Figure 5-3.   Eccentricity. Eccentricity defines an orbit's shape.

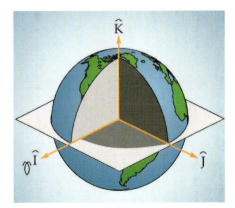

Figure 5-4. The Geocentric-equatorial Coordinate System. We use the geocentric-equatorial coordinate system to reference all orbital elements. The fundamental plane is Earth's equatorial plane, the principal direction ($\hat{I}$) points in the vernal equinox direction, ♈, the $\hat{K}$ unit vector points to the North Pole, and $\hat{J}$ completes the right-hand rule.

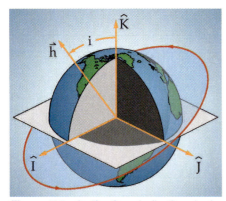

Figure 5-5. Inclination. Inclination, i, describes the tilt of the orbital plane with respect to the equator. The angle between the two planes is the same as the angle between $\hat{K}$ (which is perpendicular to the equator) and $\vec{h}$ (which is perpendicular to the orbital plane).

Thus, inclination is the third COE. It specifies the *tilt* of the orbital plane with respect to the fundamental plane and helps us understand an orbit's orientation with respect to the equator.

The fourth COE is another angle, *right ascension of the ascending node, Ω,* used to describe orbital orientation with respect to the principal direction, $\hat{I}$. Before you give up on this complex-sounding term, let's look at each of its pieces. First of all, what is "right ascension?" It's similar to longitude except its reference point is the vernal equinox and it doesn't rotate with Earth. So, right ascension of the ascending node is an angle we measure along the equator, starting at the $\hat{I}$ direction.

Now let's look at the other part of this new angle's name, "ascending node" (or a node of any kind)? As we just described, the orbital plane normally tilts (is inclined) with respect to the fundamental plane (unless i = 0° or 180°). From plane geometry, you may remember that the intersection of two planes forms a line. In our case, the intersection of the orbital plane and the fundamental plane is the *line of nodes*. The two points at which the orbit crosses the equatorial plane are the nodes. The node where the spacecraft goes from below the equator (Southern Hemisphere) to above the equator (Northern Hemisphere) is the *ascending node*. Similarly, when the spacecraft crosses the equator heading south, it passes through the *descending node*. See Table 5-2.

Table 5-2.   Types of Orbits and Their Inclination.

Inclination	Orbital Type	Diagram
0° or 180°	Equatorial	
90°	Polar	i = 90°
0° ≤ i < 90°	Direct or Prograde (moves in the direction of Earth's rotation)	ascending node
90° < i ≤ 180°	Indirect or Retrograde (moves against the direction of Earth's rotation)	ascending node

Now let's put "right ascension" and "ascending node" together. The right ascension of the ascending node describes the orbital plane's orientation with respect to the principal direction. That is, how is the orbital plane rotated in space? We use the vernal equinox direction or $\hat{I}$ (an inertial reference) as the starting point and measure eastward along the equator to the ascending node. Thus, the right ascension of the ascending node, Ω, is the angle from the principal direction, $\hat{I}$, to the ascending node. It acts like a celestial map reference to give us the *swivel* of the orbit, helping us to better understand its orientation in space. Figure 5-6 illustrates the right ascension of the ascending node. Its range of values is $0° \leq \Omega < 360°$. That's now 4 out of 6 on our COE checklist.

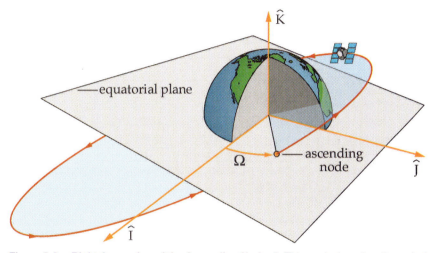

Figure 5-6.   Right Ascension of the Ascending Node, Ω. This angle describes the swivel of the orbital plane with respect to the principal direction. It is the angle along the equator between the principal direction, $\hat{I}$, and the point where the orbital plane crosses the equator from south to north (ascending node), measured eastward.

Let's recap where we are. We now know the orbit's *size*, a, its *shape*, e, its *tilt*, i, and its *swivel*, Ω. But we don't know how the orbit is oriented within the plane. For example, for an elliptical orbit, we may want to know whether perigee (point closest to Earth) is in the Northern or Southern Hemisphere. This is important if we want to take high-resolution pictures of a particular point. So, for this fifth orbital element, we measure the angle along the orbital path between the ascending node and perigee and call it *argument of perigee, ω*. To remove any ambiguities, we always measure this angle in the direction of spacecraft motion.

Where does this unusual sounding term "argument of perigee" come from? To begin with, perigee is an easily identifiable point on the orbit to reference. But why "argument"? Because we're "making clear" (from Latin) where perigee is. So our fifth COE, argument of perigee, ω, is the angle measured in the direction of the spacecraft's motion from the ascending node to perigee. It gives us the orientation of the orbit within the orbital plane, as shown in Figure 5-7. The range on argument of perigee is $0° \leq \omega < 360°$. That's 5 down and 1 to go on our COE checklist.

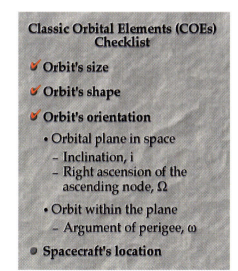

Classic Orbital Elements (COEs) Checklist

✔ **Orbit's size**

✔ **Orbit's shape**

✔ **Orbit's orientation**

  • Orbital plane in space
    – Inclination, i
    – Right ascension of the ascending node, Ω

  • Orbit within the plane
    – Argument of perigee, ω

● **Spacecraft's location**

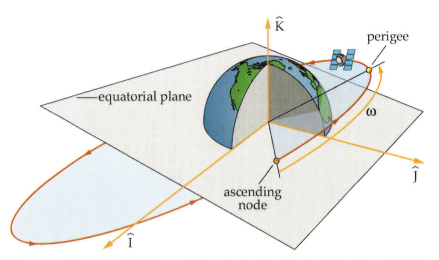

Figure 5-7.   Argument of Perigee, ω. This angle describes the orientation of an orbit within its orbital plane. It is the angle between the ascending node and perigee, measured in the direction of the spacecraft's motion.

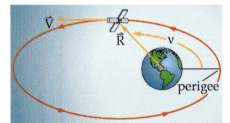

Figure 5-8.   True Anomaly. True anomaly, ν, specifies the location of a spacecraft within the orbit. It is the angle between perigee and the spacecraft's position vector measured in the direction of the spacecraft's motion. Of all the COEs, only true anomaly changes with time (as long as our two-body assumptions hold).

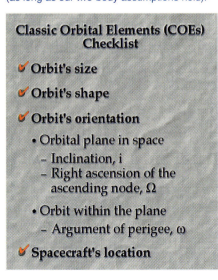

Classic Orbital Elements (COEs) Checklist

✔ **Orbit's size**

✔ **Orbit's shape**

✔ **Orbit's orientation**

• Orbital plane in space

– Inclination, i
– Right ascension of the ascending node, Ω

• Orbit within the plane

– Argument of perigee, ω

✔ **Spacecraft's location**

After specifying the size and shape of the orbit, along with its orientation (tilt and swivel), we still need to find a spacecraft's location within the orbit. As we've already seen in Chapter 4, we can find this using the true anomaly. *True anomaly, ν*, is the angle along the orbital path from perigee to the spacecraft's position vector, $\vec{R}$. Similar to the argument of perigee, we measure true anomaly in the direction of the spacecraft's motion. Figure 5-8 shows true anomaly. Its range of values is $0° \le ν < 360°$.

True anomaly, ν, tells us the location of the spacecraft in its orbit. Of all the COEs, only true anomaly changes with time (while our two-body assumptions hold) as the spacecraft moves in its orbit.

Now that you've seen all six of the COEs, we can show four of them together in Figure 5-9 (we can show size and shape only indirectly in the way we draw the orbit). Table 5-3 summarizes all six. That completes our COE checklist. We've shown all you need to know about describing an orbit and locating a spacecraft within it.

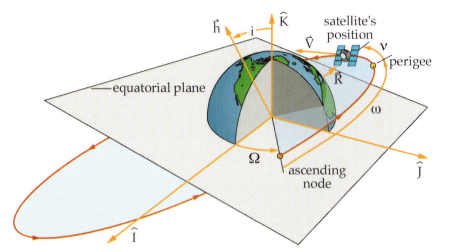

Figure 5-9.   Classic Orbital Elements (COEs). Here we show four of the six COEs. We use the COEs to visualize an orbit and locate a spacecraft in it. The other two COEs, semimajor axis, a, and eccentricity, e, specify the size and shape of an orbit.

Table 5-3.   Summary of Classic Orbital Elements.

Element	Name	Description	Range of Values	Undefined
a	Semimajor axis	Size	Depends on the conic section	Never
e	Eccentricity	Shape	e = 0: circle 0 < e < 1: ellipse	Never
i	Inclination	Tilt, angle from $\hat{K}$ unit vector to specific angular momentum vector $\vec{h}$	$0 \leq i \leq 180°$	Never
Ω	Right ascension of the ascending node	Swivel, angle from vernal equinox to ascending node	$0 \leq \Omega < 360°$	When i = 0 or 180° (equatorial orbit)
ω	Argument of perigee	Angle from ascending node to perigee	$0 \leq \omega < 360°$	When i = 0 or 180° (equatorial orbit) or e = 0 (circular orbit)
ν	True anomaly	Angle from perigee to the spacecraft's position	$0 \leq \nu < 360°$	When e = 0 (circular orbit)

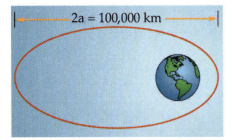

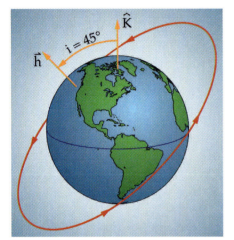

Figure 5-10.   Orbital Size and Shape. Here we show the approximate size and shape of an orbit with a semimajor axis of 50,000 km and an eccentricity of 0.4.

Figure 5-11.   Inclination. This orbit has an inclination of 45°.

By now you may wonder what all these COEs are good for! Let's look at an example to see how they can help us visualize an orbit. Suppose a communication satellite has the following COEs

- Semimajor axis, a = 50,000 km

- Eccentricity, e = 0.4

- Inclination, i = 45°

- Right ascension of the ascending node, Ω = 50°

- Argument of perigee, ω = 110°

- True anomaly, ν = 170°

To begin with, as in Figure 5-10, we can sketch the size and shape of the orbit given the semimajor axis and the eccentricity. The eccentricity of 0.4 indicates an elliptical orbit (it's between 0 and 1). The semimajor axis of 50,000 km tells us how large to draw the orbit.

Now that we see the orbit in two dimensions, we can use the other COEs to visualize how it's oriented in three dimensions. Because the inclination angle is 45°, we know the orbital plane tilts 45° from the equator. We can also describe inclination as the angle between the specific angular momentum vector, $\hat{h}$, and $\vec{K}$ in the geocentric-equatorial coordinate system. So we can sketch the crossing of the two planes in three dimensions as you see in Figure 5-11.

Next, to find the swivel of the orbital plane with respect to the principal direction, we use the right ascension of the ascending node, Ω. After locating the principal direction in the equatorial plane, $\hat{I}$, we swivel the orbital plane by positioning the ascending node 50° east of the $\hat{I}$ vector. What we know so far gives us the picture of the orbit in Figure 5-12.

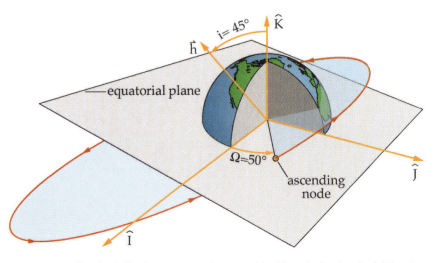

Figure 5-12.   Our Orbit So Far. Here we show an orbit with an inclination, i, of 45° and a right ascension of the ascending node, Ω, of 50°.

So, we've completely specified the orbit's size and shape, as well as the orientation of the orbital plane in space. But we still don't know how the orbit is oriented within the plane. Argument of perigee, ω, comes next. To locate perigee within the orbital plane, we rotate perigee 110° from the ascending node, in the direction of spacecraft motion. Figure 5-13 shows how to orient the orbit in the orbital plane.

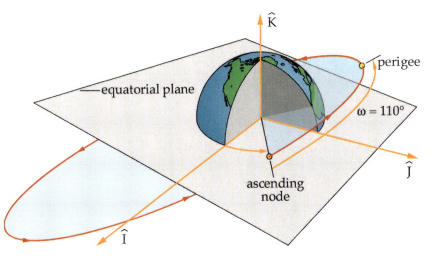

Figure 5-13. Argument of Perigee for the Example. We rotate perigee 110° from the ascending node to determine the argument of perigee, ω, is 110°.

Finally, we locate our communication satellite within the orbit. Using the value of true anomaly, ν, we measure 170° in the direction of spacecraft motion from perigee to the spacecraft's position. And there it is in Figure 5-14!

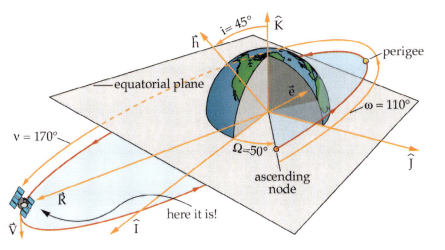

Figure 5-14. Finding the Satellite. Here we show the position of a satellite with the following COEs: a = 50,000 km; i = 45°; e = 0.4; Ω = 50°; ω = 110°; ν = 170°.

As we already know, various missions require different orbits, as described by their COEs. Table 5-4 shows various types of missions and their typical orbits. A *geostationary orbit* is a circular orbit with a period of about 24 hours and inclination of 0°. Geostationary orbits are particularly useful for communication satellites because a spacecraft in this orbit appears motionless to an Earth-based observer, such as a fixed ground station for a cable TV company. *Geosynchronous orbits* are inclined orbits with a period of about 24 hours. A *semi-synchronous orbit* has a period of 12 hours. *Sun-synchronous orbits* are retrograde, low-Earth orbits (LEO) typically inclined 95° to 105° and often used for remote-sensing missions because they pass over nearly every point on Earth's surface. A *Molniya orbit* is a semi-synchronous, eccentric orbit used for some specific communication missions.

Table 5-4.   Orbital Elements for Various Missions.

Mission	Orbital Type	Semimajor Axis (Altitude)	Period	Inclination	Other
• Communication • Early warning • Nuclear detection	Geostationary	42,158 km (35,780 km)	~24 hr	~0°	$e \cong 0$
• Remote sensing	Sun-synchronous	~6500 – 7300 km (~150 – 900 km)	~90 min	~95°	$e \cong 0$
• Navigation – GPS	Semi-synchronous	26,610 km (20,232 km)	12 hr	55°	$e \cong 0$
• Space Shuttle	Low-Earth orbit	~6700 km (~300 km)	~90 min	28.5°, 39°, 51°, or 57°	$e \cong 0$
• Communication/ intelligence	Molniya	26,571 km ($R_p = 7971$ km; $R_a = 45,170$ km)	12 hr	63.4°	$\omega = 270°$ $e = 0.7$

Alternate Orbital Elements

Now that we've shown how to find all the classic orbital elements (COEs), we're ready to share some bad news—they're not always defined! For example, a *circular orbit* has no perigee. In this case, we have no argument of perigee, ω, or true anomaly, ν, because both use perigee as a reference. To correct this deficiency, we bring in an alternate orbital element to replace these two missing angles. In general, whenever we face a peculiar orbit with one or more of the COEs undefined, we work backward from the spacecraft's position vector (the one thing that's always defined) to the next quantity that is defined. For our circular-orbit example, instead of using true anomaly to define position, we use the first alternate element—the argument of latitude, u. We measure *argument of latitude, u*, along the orbital path from the ascending node to the spacecraft's position in the direction of the spacecraft's motion.

Another special situation that requires an alternate element is an equatorial orbit (i = 0° or 180°). In this case, the line of intersection between the equator and the orbital plane is missing (the line of nodes), so the ascending node doesn't exist. This time the right ascension of the ascending node, Ω, and the argument of perigee, ω, are undefined. We replace them with another alternate element, the *longitude of perigee, Π*—the angle measured from the principal direction, Î, to perigee in the direction of the spacecraft's motion.

Finally, a *circular equatorial orbit* has neither perigee *nor* ascending node, so the right ascension of the ascending node, Ω, the argument of perigee, ω, and true anomaly, ν, are all undefined! Instead, we use a final alternate element to replace all of them—the *true longitude, l*. We measure this angle from the principal direction, Î, to the spacecraft's position vector, R̄, in the direction of the spacecraft's motion. Figure 5-15 and Table 5-5 summarize these alternate orbital elements.

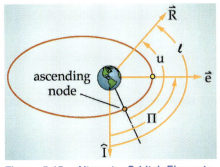

Figure 5-15.   Alternate Orbital Elements. We use the alternate orbital elements when one or more of the classic orbital elements are undefined. u is the argument of latitude. Π is the longitude of perigee. l is the true longitude.

Table 5-5.   Alternate Orbital Elements.

Element	Name	Description	Range of Values	When to Use
u	Argument of latitude	Angle from ascending node to the spacecraft's position	$0° \leq u < 360°$	Use when there is no perigee (e = 0)
Π	Longitude of perigee	Angle from the principal direction to perigee	$0° \leq Π < 360°$	Use when equatorial (i = 0 or 180°) because there is no ascending node
l	True longitude	Angle from the principal direction to the spacecraft's position	$0° \leq l < 360°$	Use when there is no perigee and ascending node (e = 0 and i = 0 or 180°)

Astro Fun Fact
The Number 2

The number 2 plays an exceedingly critical role in our universe. As you have learned, the force of gravity is inversely proportional to the square of the distance between two bodies. But what if the distance were not squared? The answer is disturbing. If the exponent were larger than 2, the orbits of the Earth and Moon would spiral into the Sun. Yet, if the exponent were any less than 2, the orbits would expand away from the Sun into infinity. This holds true for all bodies in the known universe. Geometrically, the number 2 dictates that all orbits must be shaped as closed curves or ellipses. But don't worry about the number 2 suddenly changing! The inverse square law, as applied to the law of universal gravitation, is simply our mathematical way of describing what universally exists in nature.

$$\ddot{\vec{R}} + \frac{\mu}{R^2}\vec{R} = 0$$

Contributed by Dr. Jackson R. Ferguson and Michael Banks, the U.S. Air Force Academy

▰ **Section Review**

Key Terms

argument of latitude, u
argument of perigee, ω
ascending node
circular equatorial orbit
circular orbit
classic orbital elements (COEs)
descending node
direct orbit
eccentricity
equatorial orbit
geostationary orbit
geosynchronous orbit
inclination
indirect orbit
line of nodes
longitude of perigee, Π
Molniya orbit
polar orbit
prograde orbit
retrograde orbit
right ascension of the ascending node, Ω
semi-synchronous orbit
sun-synchronous orbits
true anomaly, ν
true longitude, l

Key Equations

$$\varepsilon = -\frac{\mu}{2a}$$

Key Concepts

➤ To specify a spacecraft's orbit in space, you need to know four things about it
  - Orbit's Size
  - Orbit's Shape
  - Orbit's Orientation
  - Spacecraft's Location

➤ The six classic orbital elements (COEs) specify these four pieces of information
  - Semimajor axis, a—one-half the distance across the long axis of an ellipse. It specifies the orbit's size and relates to an orbit's energy.
  - Eccentricity, e—specifies the shape of an orbit by telling what type of conic section it is
  - Inclination, i—specifies the orientation or tilt of an orbital plane with respect to a fundamental plane, such as the equator
  - Right ascension of the ascending node, Ω—specifies the *orientation* or *swivel* of an orbital plane with respect to the principal direction, $\hat{I}$
  - Argument of perigee, ω—specifies the orientation of an orbit within the plane
  - True anomaly, ν—specifies a spacecraft's location within its orbital plane

➤ Whenever one or more COEs are undefined, you must use the alternate orbital elements

5.2 Computing Orbital Elements

▰ In This Section You'll Learn to...

☞ Determine all six orbital elements, given only the position, $\vec{R}$, and velocity, $\vec{V}$, of a spacecraft at one particular time

Now let's put these classic orbital elements (COEs) to work for us. In real life, we can't measure COEs directly, but we can determine a spacecraft's inertial position and velocity, $\vec{R}$ and $\vec{V}$, using ground-tracking sites. Still, we need some way to convert the $\vec{R}$ and $\vec{V}$ vectors to COEs, so we can make sense of an orbit. As we'll see, armed with just a position vector, $\vec{R}$, and a velocity vector, $\vec{V}$, at a single point in time, we can find all of the orbital elements. This shouldn't be too surprising. We already know we need six pieces of information to define an orbit: the three components of $\vec{R}$; and the three components of $\vec{V}$. If we know $\vec{R}$ and $\vec{V}$, we can compute six different quantities—the COEs—to better visualize the orbit. So, let's go through computing all of the COEs, given just $\vec{R}$ and $\vec{V}$.

Finding Semimajor Axis, a

Recall the semimajor axis, a, tells us the orbit's size and depends on the orbit's specific mechanical energy, ε. Thus, if we know the energy of the orbit, we can determine the semimajor axis. In Chapter 4, we showed specific mechanical energy depended only on the magnitudes of $\vec{R}$ and $\vec{V}$, and the gravitational constant

$$\varepsilon = \frac{V^2}{2} - \frac{\mu}{R} \qquad (5\text{-}3)$$

where
V = magnitude of the spacecraft's velocity vector (km/s)
μ = gravitational parameter (km^3/s^2) = 3.986×10^5 km^3/s^2 for Earth
R = magnitude of the spacecraft's position vector (km)

But we also know that ε relates to semimajor axis through Equation (5-1). So if we know the magnitude of $\vec{R}$ and of $\vec{V}$, we can solve for the energy and thus the semimajor axis.

$$a = -\frac{\mu}{2\varepsilon} \qquad (5\text{-}4)$$

Example 5-1 (Part 1) shows this.

Whenever we solve for semimajor axis (or any other parameter for that matter), it's a good idea to do a "reality check" on our result. For example, an orbiting spacecraft should have a semimajor axis greater than the radius of the planet it's orbiting; otherwise, it would hit the planet! Also be careful with parabolic trajectories where $a = \infty$ and $\varepsilon = 0$, because ε is in the denominator of Equation (5-4).

Finding Eccentricity, e

To determine the eccentricity, we need to define an *eccentricity vector, $\vec{e}$,* that points from Earth's center to perigee and whose magnitude equals the eccentricity, e. $\vec{e}$ relates to position, $\vec{R}$, and velocity, $\vec{V}$, by

$$\vec{e} = \frac{1}{\mu}\left[\left(V^2 - \frac{\mu}{R}\right)\vec{R} - (\vec{R} \cdot \vec{V})\vec{V}\right] \qquad (5\text{-}5)$$

where

$\vec{e}$ = eccentricity vector (unitless, points at perigee)
μ = gravitational parameter $(km^3/s^2) = 3.986 \times 10^5\ km^3/s^2$ for Earth
V = magnitude of $\vec{V}$ (km/s)
R = magnitude of $\vec{R}$ (km)
$\vec{R}$ = position vector (km)
$\vec{V}$ = velocity vector (km/s)

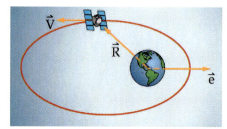

Figure 5-16.   Eccentricity Vector, $\vec{e}$. We find the eccentricity vector, $\vec{e}$, using μ and the position and velocity vectors ($\vec{R}$ and $\vec{V}$). It points at perigee and its magnitude equals the orbit's eccentricity.

Figure 5-16 shows the eccentricity vector for an orbit. Thus, all we need is μ, $\vec{R}$, and $\vec{V}$ to solve for eccentricity, and we find where perigee is as a bonus. We find the value for e by computing the magnitude of $\vec{e}$. This vector will be useful later for computing other COE's that relate to perigee. Because $\vec{e}$ points through perigee, we sometimes call it the perigee vector. Note that it does not exist for circular orbits.

Finding Inclination, i

The other four orbital elements are all angles. To find them, we need to use the definition of the dot product, which allows us to find an angle if we know two appropriate reference vectors. Let's briefly review how this works. For any two vectors $\vec{A}$ and $\vec{B}$, as shown in Figure 5-17, we can say

$$\vec{A} \cdot \vec{B} = AB\cos\theta \qquad (5\text{-}6)$$

where A and B are the magnitudes of the vectors, and θ is the angle between them. Solving for θ gives us

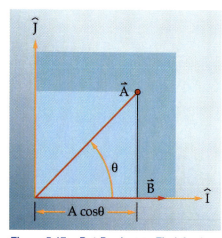

Figure 5-17.   Dot Product to Find Angles. The dot product gives us the angle between two vectors.

$$\theta = \cos^{-1}\left(\frac{\vec{A} \cdot \vec{B}}{AB}\right) \qquad (5\text{-}7)$$

But be careful! When you take an inverse cosine to find θ, as shown in Figure 5-18, two angles are possible: θ and $(360 - \theta)$.

Let's see how we can use this dot product idea to find the inclination angle, i. Recall we defined i as the angle between $\hat{K}$ and $\vec{h}$, as shown in Figure 5-19. We can then apply our dot product relationship to these two vectors to arrive at the following expression for inclination

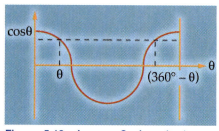

Figure 5-18.   Inverse Cosine. An inverse cosine gives two possible answers: θ and $(360 - \theta)$.

$$i = \cos^{-1}\left(\frac{\hat{K} \cdot \vec{h}}{Kh}\right) \qquad (5\text{-}8)$$

where

i   = inclination (deg or rad)

$\hat{K}$ = unit vector through the North Pole

$\vec{h}$ = specific angular momentum vector (km^2/s)

K = magnitude of $\hat{K}$ = 1

h = magnitude of $\vec{h}$ (km^2/s)

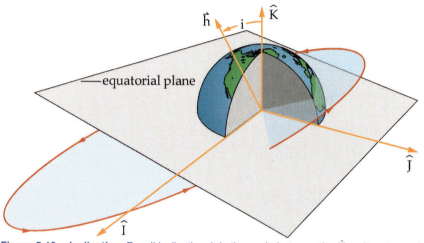

Figure 5-19.   Inclination. Recall inclination, i, is the angle between the $\hat{K}$ unit vector and the specific angular momentum vector, $\vec{h}$.

Because K, the magnitude of $\hat{K}$, is one (it's a unit vector), the denominator reduces to h (the magnitude of $\vec{h}$). Recall from Equation (4-28) that $\vec{h}$ is the cross product of $\vec{R}$ and $\vec{V}$. The quantity $\hat{K} \cdot \vec{h}$ is simply the $\hat{K}$ component of $\vec{h}$ because $\hat{K}$ is a unit vector. Do we have to worry about a quadrant check in this case? No, because the value of inclination is always less than or equal to 180°, so the smaller number will always be the right one.

Finding Right Ascension of the Ascending Node, Ω

We can find the right ascension of the ascending node, Ω, using the same basic approach we used to find inclination. From the definition of Ω, we know it's the angle between the principal direction, $\hat{I}$, and the ascending node. Now we need some vectors. Can we define a vector that points at the ascending node? You bet! If we draw a vector from Earth's center, pointing at the ascending node, we'll notice it lies along the intersection of two planes—the orbital plane and the equatorial plane. Thus, this new vector, which we call the *ascending node vector, $\vec{n}$*, must be perpendicular to $\hat{K}$ and $\vec{h}$, as shown in Figure 5-20. (That's because $\hat{K}$ is perpendicular to the equatorial plane, and $\vec{h}$ is perpendicular to the orbital plane.) Using the definition of cross product (and the right hand rule), we get

$$\boxed{\vec{n} = \hat{K} \times \vec{h}} \qquad (5\text{-}9)$$

where

$\vec{n}$ = ascending node vector (km²/s, points at the ascending node)

$\hat{K}$ = unit vector through the North Pole

$\vec{h}$ = specific angular momentum vector (km²/s)

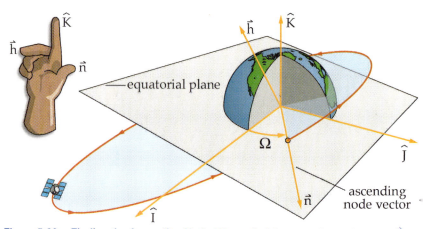

Figure 5-20.   Finding the Ascending Node. We can find the ascending node vector, $\vec{n}$, by using the right-hand rule. Point your index finger at $\hat{K}$ and your middle finger at $\vec{h}$. Your thumb will point in the direction of $\vec{n}$.

While $\vec{n}$ inherits the units of $\vec{h}$, km²/s, these units are physically irrelevant to the problem.

Because Ω is the angle between $\hat{I}$ and $\vec{n}$, we can use the dot product relationship again to find the right ascension of the ascending node

$$\Omega = \cos^{-1}\left(\frac{\hat{I} \cdot \vec{n}}{In}\right) \qquad (5\text{-}10)$$

where

Ω = right ascension of the ascending node (deg or rad)

$\hat{I}$ = unit vector in the principal direction

$\vec{n}$ = ascending node vector (km²/s, points at the ascending node)

I = magnitude of $\hat{I}$ = 1

n = magnitude of $\vec{n}$ (km²/s)

The right ascension of the ascending node can range between 0° and 360°, so a quadrant check is necessary. How do we decide which quadrant Ω belongs in? Looking at Figure 5-21, we see the equatorial plane and the location of the ascending node vector, $\vec{n}$. Notice the $\hat{J}$ component of $\vec{n}$, n_J, tells us which side of the $\hat{I}$ axis $\vec{n}$ is on. If $\vec{n}$ is on the positive $\hat{J}$ side, then n_J is positive and Ω lies between 0° and 180°. If $\vec{n}$ is on the negative $\hat{J}$ side, then n_J is negative and Ω lies between 180° and 360°. Note that if $\vec{n}$ aligns with the positive or negative $\hat{I}$ axis, then Ω is either 0° or 180°, respectively. Thus, we can write a logic statement for this quadrant check

If $n_J \geq 0$ then $0 \leq \Omega \leq 180°$

If $n_J < 0$ then $180° < \Omega < 360°$

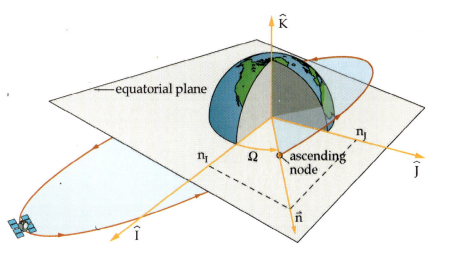

Figure 5-21.   Quadrant Check for Ω. We can find the quadrant for the right ascension of the ascending node, Ω, by looking at the sign of the $\hat{J}$ component of $\vec{n}$, n_J. If n_J is greater than zero, Ω is between 0 and 180°. If n_J is less than zero, Ω is between 180° and 360°.

Finding Argument of Perigee, ω

The argument of perigee, ω, locates perigee in the orbital plane. Remember, we defined it as the angle between the ascending node and perigee, as shown in Figure 5-22. We already know where the ascending node is from the ascending node vector, $\hat{n}$. We also know the eccentricity vector, $\vec{e}$, points at perigee. Using our dot product relationship once again, we can solve for the argument of perigee, ω.

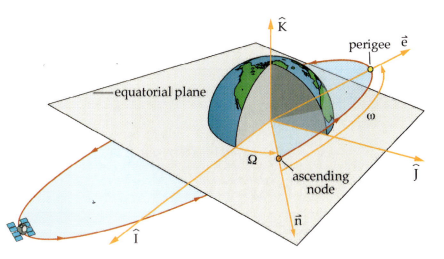

Figure 5-22.   Finding the Argument of Perigee, ω. We can find the argument of perigee, ω, as the angle between the ascending node vector, $\hat{n}$, and the eccentricity vector, $\vec{e}$.

$$\omega = \cos^{-1}\left(\frac{\vec{n} \cdot \vec{e}}{ne}\right) \qquad (5\text{-}11)$$

where

ω = argument of perigee (deg or rad)

$\vec{n}$ = ascending node vector (km^2/s, points at the ascending node)

$\vec{e}$ = eccentricity vector (unitless, points at perigee)

n = magnitude of $\vec{n}$ (km^2/s)

e = magnitude of $\vec{e}$ (unitless)

Once more, we have two possible answers that satisfy the equation, so we have another quadrant-identity crisis. How do we know which quadrant ω belongs in? In Figure 5-23, we can see that if ω is between 0° and 180°, perigee is north of the equator; and if ω is between 180° and 360°, perigee is south of the equator. Luckily, we have the trusty $\vec{e}$ vector to tell us exactly where perigee is. If we look at just the $\hat{K}$ component of $\vec{e}$ (e_K), we can tell if perigee is in the Northern or Southern Hemisphere (positive e_K for Northern, negative e_K for Southern). We can write this as a logic statement

If $e_K \geq 0$ then $0° \leq \omega \leq 180°$

If $e_K < 0$ then $180° < \omega < 360°$

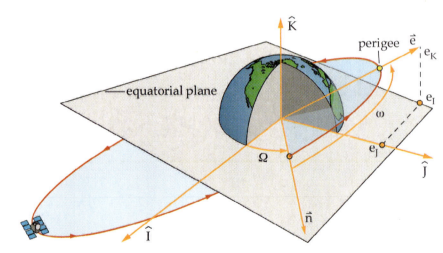

Figure 5-23.   Quadrant Check for the Argument of Perigee, ω. We check the quadrant for the argument of perigee, ω, by looking at the $\hat{K}$ component of the eccentricity vector, $\vec{e}$. If e_K is greater than zero, perigee lies above the equator; thus, ω is between 0° and 180°. If e_K is less than zero, perigee lies in the Southern Hemisphere; and, ω is between 180° and 360°.

Finding True Anomaly, ν

Finally, it's time to find out "what's nu." That is, we must find true anomaly, ν, to locate the spacecraft's position in its orbit. We define ν as the angle between perigee, $\vec{e}$, and the position vector, $\vec{R}$, as shown in Figure 5-24.

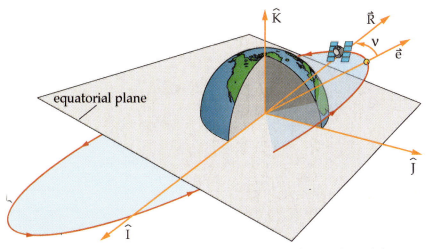

Figure 5-24.   Finding True Anomaly, ν. We find the true anomaly, ν, as the angle between the eccentricity vector, ê , and the spacecraft's position vector, $\vec{R}$.

We can start from our last point of reference, the perigee direction (using the eccentricity vector again), and measure to the position vector, $\vec{R}$. Applying our dot product relationship one last time, we arrive at

$$\nu = \cos^{-1}\left(\frac{\vec{e}\cdot\vec{R}}{eR}\right) \qquad (5\text{-}12)$$

where

ν = true anomaly (deg or rad)

$\vec{e}$ = eccentricity vector (unitless, points at perigee)

$\vec{R}$ = position vector (km)

e = magnitude of $\vec{e}$ (unitless)

R = magnitude of $\vec{R}$ (km)

To sort out the quadrant for this angle, we want to tell whether the spacecraft is heading away from or toward perigee. Recall from Chapter 4 our discussion of the flight-path angle, φ. If φ is positive, it's gaining altitude and heading away from perigee ("the houses are getting smaller"). If φ is negative, it's losing altitude and heading toward perigee ("the houses are getting bigger"), as seen in Figure 5-25.

So all we have to do is find the sign on φ. No problem. Remember that φ is the angle between the local horizontal and the spacecraft's velocity vector. By applying a little bit of trigonometry, we can show that the sign of the quantity $(\vec{R}\cdot\vec{V})$ is the same as the sign of φ! Thus, if we know $(\vec{R}\cdot\vec{V})$, we know what's nu, ν. Written as a logic statement, this idea boils down to

If $(\vec{R}\cdot\vec{V}) \geq 0$ (φ ≥ 0) then 0° ≤ ν ≤ 180°

If $(\vec{R}\cdot\vec{V}) < 0$ (φ < 0) then 180° < ν < 360°

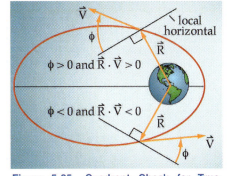

Figure 5-25.   Quadrant Check for True Anomaly, ν. To resolve the quadrant for true anomaly, ν, check the sign on the flight-path angle, φ. If φ is positive, the spacecraft is moving away from perigee, so true anomaly is between 0 and 180°. If φ is negative, the spacecraft is moving toward perigee, so true anomaly is between 180° and 360°.

173

Note that, if $(\vec{R} \cdot \vec{V}) = 0$ ($\phi = 0$), then we don't know if $\nu = 0°$ or $180°$. In this case, we have to compare the magnitude of $\vec{R}$ to the perigee and apogee radii, R_p and R_a. $R_p = a (1 - e)$ and $R_a = a (1 + e)$. If $|\vec{R}| = R_p$, then $\nu = 0°$. If $|\vec{R}| = R_a$, then $\nu = 180°$. Now we've been through all the steps needed to convert an obscure set of $\vec{R}$ and $\vec{V}$ vectors into the extremely useful COEs. Example 5-1 (parts 1, 2, and 3) goes through the entire process with some real numbers.

▮▮ Section Review

Key Terms

ascending node vector, $\vec{n}$
eccentricity vector, $\vec{e}$

Key Equations

$$\varepsilon = \frac{V^2}{2} - \frac{\mu}{R}$$

$$\vec{e} = \frac{1}{\mu}\left[\left(V^2 - \frac{\mu}{R}\right)\vec{R} - (\vec{R} \cdot \vec{V})\vec{V}\right]$$

$$i = \cos^{-1}\left(\frac{\hat{K} \cdot \vec{h}}{Kh}\right)$$

$$\vec{n} = \hat{K} \times \vec{h}$$

$$\Omega = \cos^{-1}\left(\frac{\hat{I} \cdot \vec{n}}{In}\right)$$

$$\omega = \cos^{-1}\left(\frac{\vec{n} \cdot \vec{e}}{ne}\right)$$

$$\nu = \cos^{-1}\left(\frac{\vec{e} \cdot \vec{R}}{eR}\right)$$

Key Concepts

➤ You can compute all six classic orbital elements (COEs) for an orbit using just one position vector, $\vec{R}$, and velocity vector, $\vec{V}$—a "snap shot" of the spacecraft's location at any point in time

➤ You can find semimajor axis, a, using the magnitudes of $\vec{R}$ and $\vec{V}$ to determine the orbit's specific mechanical energy, ε, which is a function of its semimajor axis

➤ You can find eccentricity computing the magnitude of the eccentricity vector, $\vec{e}$, which points at perigee

➤ Because all the remaining COEs are angles, you can find them using simple vector dot products. In general, for any two vectors $\vec{A}$ and $\vec{B}$, the angle between them, θ, is

$$\theta = \cos^{-1}\left(\frac{\vec{A} \cdot \vec{B}}{AB}\right)$$

➤ Because inclination, i, is the angle between the unit vector, $\hat{K}$, and the specific angular momentum vector, $\vec{h}$, you find it using the dot product relationship.
  • Remember: $0 \leq i \leq 180°$

➤ Because right ascension of the ascending node, Ω, is the angle between the principal direction, $\hat{I}$, and the ascending node ($\vec{n}$), you can find Ω using the dot product relationship.
  • Remember: If $n_J \geq 0$, then $0° \leq \Omega \leq 180°$
       If $n_J < 0$, then $180° < \Omega < 360°$

➤ Because the argument of perigee, ω, is the angle between the ascending node, $\vec{n}$, and perigee (represented by the eccentricity vector, $\vec{e}$), you can find it using the dot product relationship.
  • Remember: If $e_K \geq 0$, then $0° \leq \omega \leq 180°$
       If $e_K < 0$, then $180° < \omega < 360°$

➤ Because true anomaly, ν, is the angle between perigee (represented by the eccentricity vector, $\vec{e}$) and the spacecraft's position vector, $\vec{R}$, you can find it using the dot product relationship.
  • Remember: If $(\vec{R} \cdot \vec{V}) \geq 0 (\phi \geq 0)$, then $0° \leq \nu \leq 180°$
       If $(\vec{R} \cdot \vec{V}) < 0 (\phi < 0)$, then $180° < \nu < 360°$

Example 5-1 (Part 1)

Problem Statement

Space Operations Officers at Air Force Space Command have given you this set of position ($\vec{R}$) and velocity ($\vec{V}$) vectors for a new European Space Agency (ESA) satellite.

$$\vec{R} = 8228\ \hat{I} + 389.0\ \hat{J} + 6888\ \hat{K}\ \text{km}$$

$$\vec{V} = -0.7000\ \hat{I} + 6.600\ \hat{J} - 0.6000\ \hat{K}\ \text{km/s}$$

Determine the size (semimajor axis) and shape (eccentricity) for this satellite's orbit.

Problem Summary

Given: $\vec{R} = 8228\ \hat{I} + 389\ \hat{J} + 6888\ \hat{K}\ \text{km}$

$\vec{V} = -0.7000\ \hat{I} + 6.600\ \hat{J} - 0.6000\ \hat{K}\ \text{km/s}$

Find: a and e

Conceptual Solution

1) Determine magnitudes of the vectors, $\vec{R}$ and $\vec{V}$

2) Solve for the semimajor axis, a

Determine the orbit's size using the relationships shown in Equation (5-3) and Equation (5-1).

$$\varepsilon = \frac{V^2}{2} - \frac{\mu}{R}$$

$$\varepsilon = -\frac{\mu}{2a}$$

3) Solve for the eccentricity vector, $\vec{e}$, and its magnitude, e.

$$\vec{e} = \frac{1}{\mu}\left[\left(V^2 - \frac{\mu}{R}\right)\vec{R} - (\vec{R}\cdot\vec{V})\vec{V}\right]$$

Analytical Solution

1) Determine magnitudes of the vectors, $\vec{R}$ and $\vec{V}$

$$R = \sqrt{(8228)^2 + (389)^2 + (6888)^2} = 10{,}738\ \text{km}$$

$$V = \sqrt{(-0.7000)^2 + (6.600)^2 + (-0.6000)^2}$$
$$= 6.664\ \text{km/s}$$

2) Solve for the semimajor axis

$$\varepsilon = \frac{V^2}{2} - \frac{\mu}{R}$$

$$\varepsilon = \frac{(6.664\ \text{km/s})^2}{2} - \frac{3.986\times10^5\ \text{km}^3/\text{s}^2}{10{,}738\ \text{km}}$$

$$= -14.916\ \text{km}^2/\text{s}^2$$

$$\varepsilon = -\frac{\mu}{2a}$$

$$a = -\frac{\mu}{2\varepsilon} = \frac{-3.986\times10^5\ \text{km}^3/\text{s}^2}{2(-14.916\ \text{km}^2/\text{s}^2)} = 1.336\times10^4\ \text{km}$$

3) Solve for the eccentricity vector, $\vec{e}$, and its magnitude, e.

$$\vec{e} = \frac{1}{\mu}\left[\left(V^2 - \frac{\mu}{R}\right)\vec{R} - (\vec{R}\cdot\vec{V})\vec{V}\right]$$

We can start by finding the dot product between $\vec{R}$ and $\vec{V}$.

$$\vec{R}\cdot\vec{V} = (8228)(-0.7000) + (389.0)(6.600) +$$
$$(6888)(-0.6000) = -7325\ \text{km}^2/\text{s}$$

$$\vec{e} = \frac{1}{3.986\times10^5\,\text{km}^3/\text{s}^2} \times$$

$$\left[\left(\left(6.664\frac{\text{km}}{\text{s}}\right)^2 - \frac{3.986\times10^5\dfrac{\text{km}^3}{\text{s}^2}}{10{,}738\ \text{km}}\right)\vec{R} - \left(-7325\frac{\text{km}^2}{\text{s}}\right)\vec{V}\right]$$

$$\vec{e} = (2.5088\times10^{-6})[(7.288)\vec{R} - (-7325)\vec{V}]$$

$$\vec{e} = (1.8284\times10^{-5})\left[8228\ \hat{I} + 389.0\ \hat{J} + 6888\ \hat{K}\right] -$$
$$(-0.018377)\,[-0.7000\ \hat{I} + 6.600\ \hat{J} - 0.6000\ \hat{K}]$$

$$\vec{e} = 0.15044\ \hat{I} + 0.0071125\ \hat{J} + 0.12594\ \hat{K}$$
$$- \left[0.012864\ \hat{I} - 0.12129\ \hat{J} + 0.011026\ \hat{K}\right]$$

$$\vec{e} = 0.1376\ \hat{I} + 0.1284\ \hat{J} + 0.1149\ \hat{K}$$

Now that we have the eccentricity vector, we can solve for the magnitude, which tells us the shape of the orbit.

$$e = \sqrt{(0.1376)^2 + (0.1284)^2 + (0.1149)^2} = 0.2205$$

Interpreting the Results

We can see that the semimajor axis of this orbit is 13,360 km (8302 mi.). Because e = 0.2205 ($0 < e < 1$), we also know the orbit is an ellipse (not a circle, parabola, or hyperbola).

Example 5-1 (Part 2)

Problem Statement

Using the same position and velocity vectors as in Example 5-1 (Part 1), determine the inclination of the orbit.

Problem Summary

Given: $\vec{R} = 8228\,\hat{I} + 389.0\,\hat{J} + 6888\,\hat{K}$ km

$\vec{V} = -0.7000\,\hat{I} + 6.600\,\hat{J} - 0.6000\,\hat{K}\ \dfrac{km}{s}$

Find: i

Conceptual Solution

1) Solve for the specific angular momentum vector, $\vec{h}$, and its magnitude, h.

$$\vec{h} = \vec{R} \times \vec{V}$$

2) Solve for the inclination angle, i.

$$i = \cos^{-1}\left(\frac{\hat{K}\cdot\vec{h}}{Kh}\right) = \cos^{-1}\left(\frac{\hat{K}\cdot\vec{h}}{h}\right)$$

Analytical Solution

1) Solve for the specific angular momentum vector, $\vec{h}$, and its magnitude, h.

$$\vec{h} = \vec{R} \times \vec{V}$$

$$\vec{R} \times \vec{V} = \begin{vmatrix} \hat{I} & \hat{J} & \hat{K} \\ 8228 & 389.0 & 6888 \\ -0.7000 & 6.600 & -0.6000 \end{vmatrix} \frac{km^2}{s}$$

$$\vec{h} = [(389.0)(-0.6000) - (6.600)(6888)]\hat{I} -$$

$$[(8228)(-0.6000) - (-0.7000)(6888)]\hat{J}+$$

$$[(8228)(6.600) - (-0.7000)(389.0)]\hat{K}\ \frac{km^2}{s}$$

$$\vec{h} = [(-233.4) - (45,460.8)]\hat{I} -$$

$$[(-4936.8) + (4821.6)]\hat{J} +$$

$$[(54,304.8) + (272.3)]\hat{K}\ \frac{km^2}{s}$$

$$\vec{h} = -45,694.2\,\hat{I} + 115.2\,\hat{J} + 54,577.1\,\hat{K}\ \frac{km^2}{s}$$

$$h = \sqrt{(45,694.2)^2 + (115.2)^2 + (54,577.1)^2}$$

$$= 71,180.3\ \frac{km^2}{s}$$

2) Solve for the inclination angle, i.

$$i = \cos^{-1}\left(\frac{\hat{K}\cdot\vec{h}}{Kh}\right) = \cos^{-1}\left(\frac{\hat{K}\cdot\vec{h}}{h}\right)$$

$$\hat{K}\cdot\vec{h} = h_K = 54,577.1$$

$$i = \cos^{-1}\left(\frac{h_K}{h}\right) = \cos^{-1}\left(\frac{54,577.1}{71,180.3}\right) = \cos^{-1}(0.7667$$

At this point you need to pull out your calculator an take the inverse cosine of 0.76674 (unless you ca figure things like that in your head!). **But be carefu** When you take inverse trigonometric functions you calculator gives you only *one* of the possible corre angles. For an inverse cosine you must subtract th result from 360° to get the second possible answer. Fc our result from above we get two possible answers fc inclination

$$i = 39.94° \text{ or } (360°\text{-}39.94°) = 320.1°$$

To resolve this ambiguity, we must return to th definition of inclination. Because i must be between C and 180°, our answer must be i = 39.94°.

Interpreting the Results

The inclination of this orbit is 39.94°.

Example 5-1 (Part 3)

Problem Statement

Using the same position and velocity information from Example 5-1 (Part 1), determine the right ascension of the ascending node, argument of perigee, and true anomaly.

Problem Summary

Given: $\vec{R} = 8228 \, \hat{I} + 389.0 \, \hat{J} + 6888 \, \hat{K}$ km

$\vec{V} = -0.7000 \, \hat{I} + 6.600 \, \hat{J} - 0.6000 \, \hat{K} \, \dfrac{km}{s}$

Find: Ω, ω, and v

Conceptual Solution

Solve for the ascending node vector, $\hat{n}$, and its magnitude, n.

$$\vec{n} = \hat{K} \times \vec{h}$$

Solve for the right ascension of the ascending node, Ω. Do a quadrant check.

$$\Omega = \cos^{-1}\left(\frac{\hat{I} \cdot \vec{n}}{In}\right) = \cos^{-1}\left(\frac{\hat{I} \cdot \vec{n}}{n}\right)$$

Solve for the argument of perigee, ω. Do a quadrant check.

$$\omega = \cos^{-1}\left(\frac{\vec{n} \cdot \vec{e}}{ne}\right)$$

Solve for the true anomaly angle, v. Do a quadrant check.

$$v = \cos^{-1}\left(\frac{\vec{e} \cdot \vec{R}}{eR}\right)$$

Analytical Solution

Solve for the ascending node vector, $\hat{n}$, and its magnitude, n.

$$\vec{n} = \hat{K} \times \vec{h} = \begin{vmatrix} \hat{I} & \hat{J} & \hat{K} \\ 0.0 & 0.0 & 1.0 \\ -45,694.2 & 115.2 & 54,577.1 \end{vmatrix}$$

$$= -115.2 \, \hat{I} - 45,694.2 \, \hat{J} + 0 \, \hat{K}$$

Solving for the magnitude of $\vec{n}$, we get n = 45,694.3 km^2/s

2) Solve for the right ascension of the ascending node angle, Ω. Do a quadrant check.

$$\Omega = \cos^{-1}\left(\frac{\hat{I} \cdot \vec{n}}{In}\right) = \cos^{-1}\left(\frac{\hat{I} \cdot \vec{n}}{n}\right)$$

$$\hat{I} \cdot \vec{n} = n_I = -115.2$$

$$\Omega = \cos^{-1}\left(\frac{-115.2}{45,694.3}\right) = \cos^{-1}(-0.0025211)$$

$$\Omega = 90.14° \text{ or } (360° - 90.14°) = 269.9°$$

Once again we must decide which of the two possible answers is correct. Using the logic we developed earlier, we check n_J.

If $n_J \geq 0$ then $0° \leq \Omega \leq 180°$

If $n_J < 0$ then $180° < \Omega < 360°$

Because $n_J = -45,694.2$, $n_J < 0$, which means $180° < \Omega < 360°$. Thus, we can conclude that $\Omega = 269.9°$.

3) Solve for the argument of perigee angle, ω. Do a quadrant check.

$$\omega = \cos^{-1}\left(\frac{\vec{n} \cdot \vec{e}}{ne}\right)$$

$$\vec{n} \cdot \vec{e} = (-115.2)(0.1376) + (-45,694.2)(0.1284) + (0.0)(0.1149) = -5882.99$$

$$\omega = \cos^{-1}\left[\frac{-5882.99}{(45,694.3)(0.2205)}\right] = \cos^{-1}(-0.58389)$$

$$125.7° \text{ or } (360° - 125.7°) = 234.3°$$

To resolve which ω is right, we use the logic developed earlier and check the value of e_K.

If $e_K \geq 0$ then $0° \leq \omega \leq 180°$

If $e_K < 0$ then $180° < \omega < 360°$

Because $e_K = +0.1149$, $e_K > 0$, which means $0° \leq \omega \leq 180°$. Thus, we can conclude that $\omega = 125.7°$

Example 5-1 (Part 3) Continued

4) Solve for the true anomaly angle, ν. Do a quadrant check.

$$\nu = \cos^{-1}\left(\frac{\vec{e} \cdot \vec{R}}{eR}\right)$$

$$\vec{e} \cdot \vec{R} = (0.1376)(8228) + (0.1284)(389.0) +$$
$$(0.1149)(6888) = 1974.05$$

$$\nu = \cos^{-1}\left[\frac{1974.05}{(0.2205)(10,738)}\right] = \cos^{-1}(0.83373)$$

$$\nu = 33.52° \text{ or } (360° - 33.52°) = 326.48°$$

Finally, we resolve one last ambiguity between possible answers. Using our logic developed earlier, we check the sign on $(\vec{R} \cdot \vec{V})$.

If $(\vec{R} \cdot \vec{V}) \geq 0$ $(\phi > 0)$ then $0° \leq \nu \leq 180°$

If $(\vec{R} \cdot \vec{V}) < 0$ $(\phi < 0)$ then $180° < \nu < 360°$

The value for $(\vec{R} \cdot \vec{V})$ we found earlier was –7325. Because $(\vec{R} \cdot \vec{V}) < 0$, $180° < \nu < 360°$. Therefore, we can conclude that $\nu = 326.5°$

Interpreting the Results

We started with

$$\vec{R} = 8228 \; \hat{I} + 389.0 \; \hat{J} + 6888 \; \hat{K} \text{ km}$$

$$\vec{V} = -0.7000 \; \hat{I} + 6.600 \; \hat{J} - 0.6000 \; \hat{K}\frac{km}{s}$$

We found the following COEs

a = 13,360 km	Ω = 269.9°
e = 0.2205	ω = 125.7°
i = 39.94°	ν = 326.5°

5.3 Spacecraft Ground Tracks

▰ In This Section You'll Learn to...

☞ Explain why spacecraft ground tracks look the way they do

☞ Use ground tracks to describe why certain types of missions use specific types of orbits

☞ Use ground tracks to determine the inclination and period for direct orbits

The six classic orbital elements (COEs) allow us to visualize an orbit from space. Now let's beam back to Earth to see orbits from our perspective on the ground.

Many spacecraft users need to know what part of Earth their spacecraft is passing over at any given time. For instance, remote-sensing satellites must be over precise locations to get the coverage they need. As we'll see, we can learn a lot about a spacecraft's orbit and mission by examining the track it makes along Earth.

To understand ground tracks, imagine you're driving from San Francisco to Omaha. To get there, you go east out of San Francisco on Interstate 80 for a couple thousand miles. If you have a road map of the western United States, you can trace your route on the map by drawing a meandering line along I-80, as shown in Figure 5-26. This is your ground track from San Francisco to Omaha.

Now imagine you're taking the same trip in an airplane. You can trace your air route on the same map, but because you don't need roads, this ground track is nearly a straight line.

A spacecraft's ground track is similar to these examples. It's a trace of the spacecraft's path over Earth's surface. But it's more complicated because the spacecraft goes all the way around (more than 40,000 km or 25,000 mi.) during each orbit and Earth spins on its axis at more than 1600 km/hr (1000 m.p.h.) at the equator at the same time, as we show in Figure 5-27.

So what does a ground track look like? To make things easy, let's start by pretending Earth doesn't rotate. (Try not to get dizzy—we'll turn the rotation back on soon.) Picture an orbit above this non-rotating Earth. The ground track follows a great circle route around Earth. A *great circle* is any circle that "slices through" the center of a sphere. For example, lines of longitude, as shown in Figure 5-28, are great circles, because they slice through Earth's center, but lines of latitude are not great circles (except for 0° latitude at the equator), because they don't. An orbital trace must be a great circle because the spacecraft is in orbit around Earth's center; thus, the orbital plane also passes through Earth's center.

When we stretch Earth onto a flat-map projection (called a Mercator projection), the ground track looks a little different. To visualize how this flattening affects the ground-track shape, imagine Earth as a soda can. A

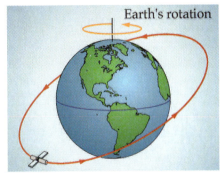

Figure 5-26.   Car and Airplane Ground Tracks. Ground tracks for a trip by car and air from San Francisco to Omaha.

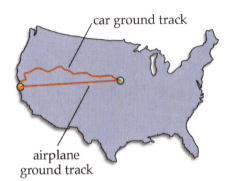

Figure 5-27.   Earth and Spacecraft Motion. The Earth spins on its axis at nearly 1600 km/hr (1000 m.p.h.) at the equator, while a spacecraft orbits above it.

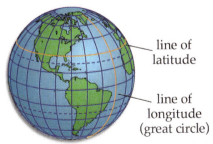

Figure 5-28.   Great Circles. A great circle is any circle around a sphere which bisects it (cuts it exactly in half). Lines of longitude are great circles whereas lines of latitude (except for the equator) are not.

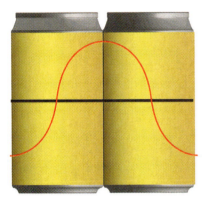

Figure 5-29.   Orbiting around a Soda Can. Imagine an orbit around a soda can. It draws a circle around the can. When we flatten the can, the line looks like a sine wave.

trace of the orbit on the soda can is shown in Figure 5-29. It looks like a circle slicing through the center of the can. But what if we were to flatten the can and look at the orbital trace, as shown in Figure 5-29? It looks like a sine wave!

Now imagine yourself on the ground watching the spacecraft pass overhead. Because we stopped Earth from rotating, the ground track will always stay the same, and the spacecraft will continue to pass overhead orbit after orbit, as shown in Figure 5-30. Even if we change the size and shape of the orbit, the ground track will look the same.

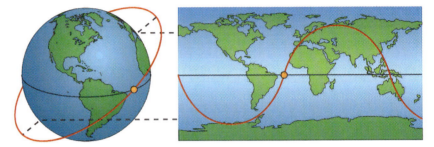

Figure 5-30.   An Orbit's Ground Track for a Non-Rotating Earth. For a non-rotating Earth, the ground track of an orbit will continuously repeat.

But suppose we start Earth rotating again. What happens? The space-craft passes overhead on one orbit but appears to pass to the *west* of you on the next orbit. How can this be? Because the orbital plane is immovable in inertial space, the spacecraft stays in the same orbit. But you're fixed to Earth and as it rotates to the east, you move away from the orbit, making it look as if the spacecraft moved, as seen in Figure 5-31. Each ground track traces a path on Earth farther to the west than the previous one.

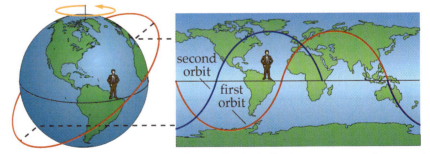

Figure 5-31.   A Normal Spacecraft Ground Track. As Earth rotates, successive ground tracks appear to shift to the west from an Earth-based observer's viewpoint.

Can we learn something about the orbit from all of this? Sure! Because Earth rotates at a fixed rate of about 15° per hr (360° in 24 hrs = 15°/hr) or 0.25° per minute, we can use this rotation as a "clock" to tell us the orbit's period. By measuring how much the orbit's ground track moves to the west from one orbit to the next, and we can establish a new parameter, *node displacement,* ΔN. We measure ΔN along the equator from one ascending node to the next and define it to be positive in the direction of the spacecraft's motion. Thus, the nodal displacement to the west during one orbit is the difference between 360° and ΔN.

We can put this ground track shift to work in finding the orbital period because the nodal displacement is simply Earth's rotation rate times the period of the orbit. For example, suppose the period of an orbit were two hours. Earth would rotate 30° (2 hr × 15°/hr) during one orbital revolution, producing a nodal displacement of 330° (360° − 30°). In terms of ΔN, we find the period from

$$\text{Period (hours)} = \frac{360° - \Delta N}{15°/\text{hr}} \text{ (for direct orbits)} \qquad (5\text{-}13)$$

[*Note:* As is, this equation applies only to direct orbits with a period less than 24 hours. For other orbits, the same concept applies but the equation changes. We'll only consider direct orbit ground tracks with periods less than 24 hours, so this equation will suffice.] If we can determine the period, we can also determine the orbit's semimajor axis using the equation for orbital period from Chapter 4.

$$P = 2\pi\sqrt{\frac{a^3}{\mu}} \qquad (5\text{-}14)$$

where

P = period (s)

π = 3.14159...(unitless)

a = semimajor axis (km)

μ = gravitational parameter $(km^3/s^2) = 3.986 \times 10^5 \ km^3/s^2$ for Earth

So, by finding ΔN from the ground track, we can find the period and then the semimajor axis. For example, in the ground track in Figure 5-32, ΔN is 330°. We find the orbital period using Equation (5-13) and the semimajor axis using Equation (5-14). But we must be careful to watch the units when using these equations.

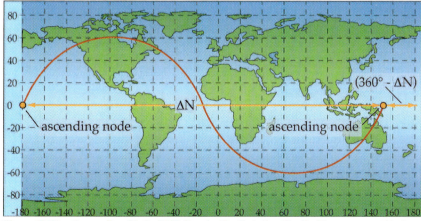

Figure 5-32.   Ascending Node Shift Due to the Rotating Earth. We measure ΔN along the equator from one ascending node to the next. It is positive in the direction of spacecraft motion. Thus, 360° − ΔN represents the amount Earth rotates during one orbit.

As the orbit's size increases, the semimajor axis gets bigger, so ΔN gets smaller. This happens because the spacecraft takes longer to make one revolution as Earth rotates beneath it (the bigger the semimajor axis, a, the longer the period). As the orbit gets bigger, the ΔN gets smaller, so the ground track appears to compress or "scrunch" together. Recall, we define a geosynchronous orbit as one with a period of approximately 24 hours. For such an orbit, the ΔN is 0°. This means the spacecraft's period matches Earth's rotational period. Thus, the orbit appears to retrace itself and form a figure 8, as shown in Figure 5-33, orbit D. If the orbit lies in the equatorial plane (has an inclination of 0°), the ground track will be just a dot on the equator, similar to orbit E, in Figure 5-33. A spacecraft with a period of 24 hours and an inclination of 0° is in a geostationary orbit. This name means the spacecraft appears stationary to Earth-based observers, making these orbits very useful for communication satellites. Once we point the receiving antenna at the satellite, we don't have to move the antenna as Earth rotates.

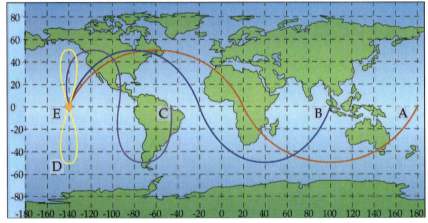

Figure 5-33.   Orbital Ground Tracks. Orbit A has a period of 2.67 hours. Orbit B has a period of 8 hours. Orbit C has a period of 18 hours. Orbit D has a period of 24 hours. Orbit E has a period of 24 hours.

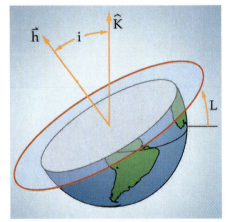

Figure 5-34.   Inclination Equals Highest Latitude, L. Because inclination relates the angle between the orbital plane and the equatorial plane, the highest latitude reached by a spacecraft equals its inclination (for direct orbits).

Besides using the ground track to determine an orbit's semimajor axis, we can also find its inclination. Imagine a spacecraft in a 50° inclined orbit. From our definition of inclination, we know in this case the angle between the equatorial plane and the orbital plane is 50°. What's the highest latitude the spacecraft will pass over directly? 50°! The highest latitude any spacecraft passes over equals its inclination. Let's see why.

Remember that latitude is the Earth-centered angle measured from the equator north or south to the point in question. But the orbital plane also passes through Earth's center, and the angle it forms with the equatorial plane is its inclination, as we show in Figure 5-34. Thus, for direct (prograde) orbits, when a spacecraft reaches its northernmost point, the point on Earth directly below it lies on the latitude line equal to the orbit's inclination.

In this way, we can use the ground track to tell us the orbit's inclination.

- For a direct orbit (0 < i < 90°), we find the northernmost or southernmost point on the ground track and read its latitude. This "maximum latitude" equals the orbit's inclination.

- For a retrograde orbit (90 < i < 180°), we subtract the maximum latitude from 180° to get the inclination

The Earth coverage a spacecraft's mission requires affects how we select the orbit's inclination. For example, if a remote-sensing spacecraft needs to view the entire surface during the mission, it needs a near polar inclination of about 90°. In Figure 5-35 we see several spacecraft ground tracks with the same period but with varying inclinations.

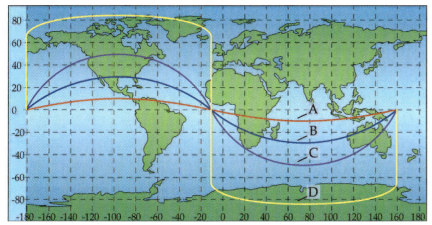

Figure 5-35. Changing Inclination. All four ground tracks represent orbits with a period of 4 hours. We can find the inclination of these orbits by looking at the highest latitude reached. Orbit A has an inclination of 10°. Orbit B has an inclination of 30°. Orbit C has an inclination of 50°. Orbit D has an inclination of 85°. (Note that Orbit D appears distorted, because ground distances elongate near the poles on a Mercator projection map.)

So far we've looked only at circular orbits. Now let's look at how eccentricity and the location of perigee affect the shape of the ground track. If an orbit is circular, its ground track is symmetrical. If an orbit is elliptical, its ground track is lopsided. That is, it will not look the same above and below the equator. Remember, a spacecraft moves fastest at perigee, so it travels farthest along its path near perigee, making the ground track look spread out. But, near apogee it's going slower, so the ground track is more scrunched. We show this effect in the two ground tracks in Figure 5-36. Orbit A has perigee in the Northern Hemisphere; Orbit B has perigee in the Southern Hemisphere.

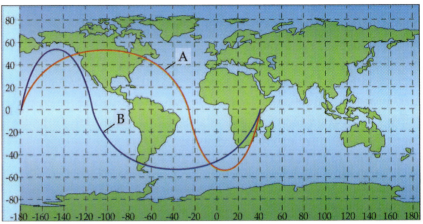

Figure 5-36.   Changing Perigee Location. Both ground tracks represent orbits with periods of 9.3 hours and inclinations of 50°. Both orbits are highly eccentric. Orbit A has perigee over the Northern Hemisphere. Orbit B has perigee over the Southern Hemisphere. If the mission objective is to get high-resolution photographs of locations in the United States, then orbit A has perigee properly positioned.

▬ Section Review

Key Terms

great circle
node displacement, ΔN

Key Equations

$$\text{Period (hours)} = \frac{360° - \Delta N}{15°/\text{hr}}$$

(for direct orbits)

$$P = 2\pi\sqrt{\frac{a^3}{\mu}}$$

Key Concepts

➤ A ground track is the path a spacecraft traces on Earth's surface as it orbits. Because a spacecraft orbits around Earth's center, the orbital plane slices through the center, so the ground track is a great circle.

➤ When the spherically-shaped Earth is spread out on a two-dimensional, Mercator-projection map, the orbital ground track resembles a sine wave for orbits with periods less than 24 hours

➤ Because orbital planes are fixed in inertial space and Earth rotates beneath them, ground tracks appear to shift westward during successive orbits

➤ From a ground track, you can find several orbital parameters

• Orbital period—by measuring the westward shift of the ground track

• Inclination of a spacecraft's orbit—by looking at the highest latitude reached on the ground track (for direct orbits)

• Approximate eccentricity of the orbit—nearly circular orbits appear symmetrical, whereas eccentric orbits appear lopsided

• Location of perigee—by looking at the point where the ground track is spread out the most

References

Bate, Roger R., Donald D. Mueller, and Jerry E. White. *Fundamentals of Astrodynamics.* New York, NY: Dover Publications, 1971.

Wertz, James R. and Wiley J. Larson. *Space Mission Analysis and Design.* Third edition. Dordrecht, Netherlands: Kluwer Academic Publishers, 1999.

Mission Problems

5.1   Orbital Elements

1   Why do we prefer classic orbital elements over a set of $\vec{R}$ and $\vec{V}$ vectors for describing an orbit?

2   How many initial conditions (ICs) do we need for solving the two-body equation of motion? Give an example of one set of ICs.

3   If a spacecraft has a high specific mechanical energy, what does this tell us about the size of the orbit? Why?

4   What is the specific mechanical energy, ε, of an orbit with a semimajor axis of 42,160 km?

5   What four things do classic orbital elements (COEs) tell us about a spacecraft's orbit and the spacecraft's position in the orbit?

6   What are the six COEs?

7   What are the two major categories of inclination relative to the Earth's motion?

8   How can we look only at the $\vec{R}$ and $\vec{V}$ vectors and tell if i = 0. What about i = 90° or 180°?

9   A Titan IV launches a spacecraft due south from Vandenburg AFB (34.6° N latitude, 120.6° W longitude). What's the most northerly point (latitude) the spacecraft can view directly below it on any orbit?

5.2   Computing Orbital Elements

10   After maneuvering, the spacecraft from the previous question has these orbital elements

semimajor axis = 8930 km
eccentricity = 0.3
inclination = 53°
right ascension of the ascending node = 165°
argument of perigee = 90°
true anomaly = 90°

a) Sketch a picture of the orbit

b) Is this a direct or retrograde orbit?

c) Where is the spacecraft located in its orbit?

d) Is perigee in the Northern or Southern Hemisphere?

e) Is this a circular, elliptical, parabolic, or hyperbolic orbit?

11   Discuss the use of inclination for different spacecraft missions.

12 Why don't we use vectors in the orbital and equatorial planes to measure inclination?

13 In general, how do we measure the alternate orbital elements when some of the reference vectors are undefined?

14 As a program manager for a major corporation, determine which alternate classic orbital elements (COEs) describe these orbits

a) Circular, $i > 0°$

b) Equatorial, $0 < e < 1$

c) Circular-equatorial

15 A spacecraft has these orbital elements

semimajor axis = 5740 km
eccentricity = 0.1
inclination = 53°
right ascension of the ascending node = 345°
argument of perigee = 270°
true anomaly = 183°

What is peculiar about this orbit?

16 A ground-based tracking station observes that a new Russian spacecraft has the following $\vec{R}$ and $\vec{V}$ vectors

$$\vec{R} = 7016 \,\hat{I} + 5740 \,\hat{J} + 638 \,\hat{K} \text{ km}$$

$$\vec{V} = 0.24 \,\hat{I} - 0.79 \,\hat{J} - 7.11 \,\hat{K} \text{ km/s}$$

a) What is the specific angular momentum of the spacecraft?

b) What is the spacecraft's inclination?

c) Calculate the ascending node vector.

d) What is the spacecraft's right ascension of the ascending node?

17 The above spacecraft was supposedly put into a Molniya type orbit.

a) Compute the eccentricity vector.

b) What is the spacecraft's argument of perigee? Is this the value needed for a Molniya orbit? See Table 5-4.

c) Where is the spacecraft? In other words, what is its true anomaly?

5.3 Spacecraft Ground Tracks

18 Given a non-rotating Earth, if the inclination stays the same but the orbital size increases or decreases, does the ground track change? Why or why not? Describe what, if anything, happens to the ground track.

19 Given a rotating Earth, if the inclination stays the same but the orbital size increases or decreases, does the ground track change? Why or why not? Describe what, if anything, happens to the ground track.

20 What type of ground track does a geostationary spacecraft have? How about a geosynchronous spacecraft with a 30° inclination?

21 Can we "hang" a reconnaissance satellite over Baghdad? Why or why not?

22 Sketch the ground track of a spacecraft with the following elements

period = 480 min
eccentricity = 0.0
inclination = 25°

a) What is the longitude shift of the ground track?

b) What is the highest and lowest latitude that the spacecraft's ground track reaches?

c) What shape does the ground track have? Is it symmetrical?

23 Given the ground track below of a direct-orbit, low-attitude spacecraft,

a) Identify the inclination

b) Determine the longitude shift and then compute the period of the orbit

c) Suggest a possible mission(s) for the spacecraft

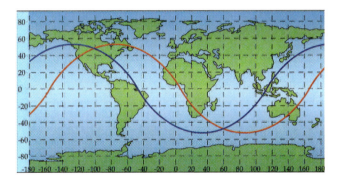

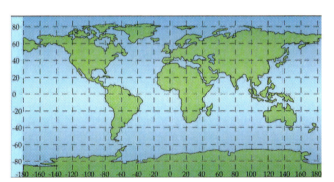

Notes

Mission Profile—IKONOS

With the launch of IKONOS, the world's first one-meter resolution, commercial, imaging satellite, on September 4, 1999, Space Imaging is set to deliver unprecedented images of Earth's surface from low-Earth orbit. With images from this system, farmers can precisely monitor the health of crops and estimate yields, scientists can look at environmentally sensitive areas and predict trends, and city planners can develop new housing communities.

Mission Overview

The mission of the IKONOS platform is to provide accurate and timely information to serve a variety of global needs, including environmental monitoring, mapping, infrastructure management, oceanographic and atmospheric research, agricultural monitoring, and others.

IKONOS Satellite. *(Courtesy of Space Imaging)*

Mission Data

✓ IKONOS was placed into orbit by a Lockheed Martin Athena II, a four-stage launch vehicle, with a lift-off weight of 121,000 kg (266,000 lb.)

✓ The IKONOS telescope has the equivalent resolving power of a 10,000 mm telephoto lens. Designed and built by Kodak, the telescope features three curved mirrors, each precisely configured to capture and focus high-resolution Earth imagery onto the imaging sensors at the focal plane. Two additional flat mirrors 'fold' the imagery across the inside of the telescope, thereby significantly reducing the telescope's length (from 10 m to about 2 m) and weight.

✓ To ensure the sharpest imagery possible, the surfaces of the three curved mirrors were polished to atomic-level accuracy. The primary mirror surface is so smooth, if it were enlarged to 160 km (100 mi.) in diameter, a car driven across its surface would not hit bumps any higher than 2 mm (0.08 in.)

✓ Each mirror was aligned in the telescope so precisely that the error is equivalent to placing a human hair under one end of a 6-m-long (20-ft.) wooden plank.

✓ The camera's Focal Plane Unit contains separate sensor arrays for simultaneous panchromatic (black and white) and multispectral (color) imaging. The panchromatic sensor array consists of 13,500, 12-micron-sized pixels—each about one-quarter the width of a human hair.

✓ The Digital Processing Unit compresses the digital image files from 11 bits per pixel (bpp) data to an average value of 2.6 bpp, at a speed of 115 million pixels per second. That's equivalent to capturing imagery simultaneously with 115 megapixel cameras, or enough to fill a photo CD every 17 seconds.

Mission Impact

The first IKONOS images show phenomenal detail. The Kodak camera is so powerful it can see objects less than 1 m square on the ground—enough to distinguish between a car and a truck. This capability from an orbital altitude of 680 km (400 mi.) represents a significant increase in image resolution over any other commercial, remote-sensing satellite system. With this new standard in place, our view of the world changes, and many industries can improve their productivity as a result.

IKONOS Image. This photo is of McNichols Arena and Mile High Stadium in Denver, Colorado. *(Courtesy of Space Imaging)*

For Discussion

- What industries may grow to rely on IKONOS high resolution images?
- How should governments encourage such innovation, yet control national security impact

Contributor

Douglas Kirkpatrick, the U.S. Air Force Academy

References

Space Imaging website prepared by their Public Relations office.

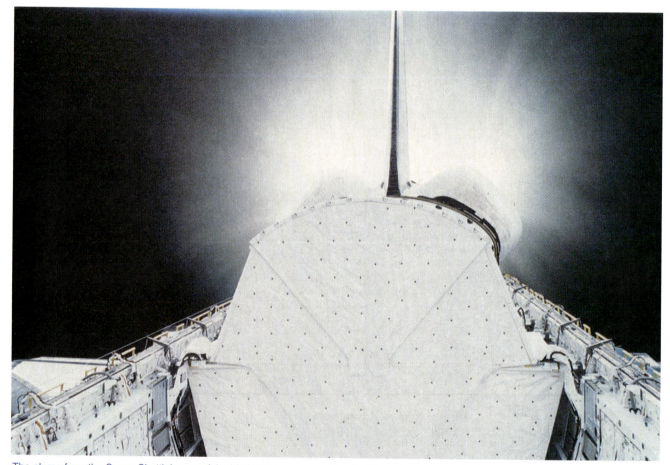

The plume from the Space Shuttle's powerful orbital maneuvering engines brightens the space at the back of the vehicle. *(Courtesy of NASA/Johnson Space Center)*

Maneuvering In Space

In This Chapter You'll Learn to...

- Explain the most energy-efficient means of transferring between two orbits—the Hohmann Transfer
- Determine the velocity change (ΔV) needed to perform a Hohmann Transfer between two orbits
- Explain plane changes and how to determine the required ΔV to accomplish them
- Explain orbital rendezvous and how to determine the required ΔV and wait time needed to start one

You Should Already Know...

- ☐ Basic orbital concepts (Chapter 4)
- ☐ Classic orbital elements (Chapter 5)

Outline

6.1 **Hohmann Transfers**

6.2 **Plane Changes**
   Simple Plane Changes
   Combined Plane Changes

6.3 **Rendezvous**
   Coplanar Rendezvous
   Co-orbital Rendezvous

Space...is big. Really big. You just won't believe how vastly, hugely, mind-boggling big it is. I mean, you may think it's a long way down the road to the chemist's [druggist's], but that's just peanuts to space.

Douglas Adams
The Hitch-hiker's Guide to the Galaxy
1979

Space Mission Architecture. This chapter deals with the Trajectories and Orbits segment of the Space Mission Architecture, introduced in Figure 1-20.

A spacecraft seldom stays very long in its assigned orbit. On nearly every space mission, there's a need to change one or more of the classic orbital elements at least once. Communication satellites, for instance, never directly assume their geostationary positions. They first go into a low-perigee (300 km or so) "parking orbit" before transferring to geosynchronous altitude (about 35,780 km). While this large change in semimajor axis occurs, another maneuver reduces the satellite's inclination from that of the parking orbit to 0°. Even after they arrive at their mission orbit, they regularly have to adjust it to stay in place. On other missions, spacecraft perform maneuvers to rendezvous with another spacecraft, as when the Space Shuttle rendezvoused with the Hubble Space Telescope to repair it (Figure 6-1).

Figure 6-1.   Shuttle Rendezvous with Hubble Space Telescope. In 1995 and again in 1999, the Space Shuttle launched into the same orbital plane as the Hubble Space Telescope. After some maneuvering, the Shuttle rendezvoused with and captured the telescope to make repairs. *(Courtesy of NASA/Johnson Space Center)*

As we'll see in this chapter, these orbital maneuvers aren't as simple as "motor boating" from one point to another. Because a spacecraft is always in the gravitational field of some central body (such as Earth or the Sun), it has to follow orbital-motion laws in getting from one place to another. In this chapter we'll use our understanding of the two-body problem to learn about maneuvering in space. We'll explain the most economical way to move from one orbit to another, find how and when to change a spacecraft's orbital plane, and finally, describe the intricate ballet needed to bring two spacecraft together safely in an orbit.

6.1 Hohmann Transfers

▬ In This Section You'll Learn to...

- ☞ Describe the steps in the Hohmann Transfer, the most fuel-efficient way to get from one orbit to another in the same plane
- ☞ Determine the velocity change (ΔV) needed to complete a Hohmann Transfer

One of the first problems faced by space-mission designers was figuring how to go from one orbit to another. Refining this process for eventual missions to the Moon was one of the objectives of the Gemini program in the 1960s, shown in Figure 6-2. Let's say we're in one orbit and we want to go to another. We'll assume for the moment that the initial and final orbits are in the same plane to keep things simple. We often use such coplanar maneuvers to move spacecraft from their initial parking orbits to their final mission orbits. Because fuel is critical for all orbital maneuvers, let's look at the most fuel-efficient way to do this, known as the Hohmann Transfer.

Figure 6-2.   The Gemini Program. During the Gemini program in the 1960s, NASA engineers and astronauts developed the procedures for all of the orbital maneuvers needed for the complex Lunar missions. Here the Gemini 6A command module rendezvous with the Gemini 7 command module. *(Courtesy of NASA/Johnson Space Center)*

In 1925 a German engineer, Walter Hohmann, theorized a fuel-efficient way to transfer between orbits. (It's amazing someone was thinking about this, considering artificial satellites didn't exist at the time.) This method, called the *Hohmann Transfer*, uses an elliptical transfer orbit tangent to the initial and final orbits.

To better understand this idea, let's imagine you're driving a fast car around a racetrack, as shown in Figure 6-3. The effort needed to exit the track depends on the off-ramp's location and orientation. For instance, if the off-ramp is tangent to the track, your exit is easy—you just straighten the wheel. But if the off-ramp is perpendicular to the track, you have to slow down a lot, and maybe even stop, to negotiate the turn. Why the difference? With the tangential exit you have to change only the magnitude of your velocity, so you just hit the brakes. With the perpendicular exit, you quickly must change the *magnitude* and *direction* of your velocity. This is hard to do at high speed without rolling your car!

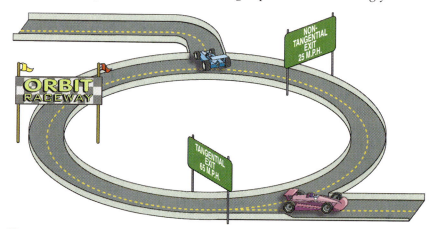

Figure 6-3.   Maneuvering. One way to think about maneuvering in space is to imagine driving around a racetrack. It takes more effort to exit at a sharp turn than to exit tangentially.

The Hohmann Transfer applies this simple racetrack example to orbits. By using "on/off-ramps" tangent to our initial and final orbits, we change orbits using as little energy as possible. For rocket scientists, saving energy means saving fuel, which is precious for space missions. By definition, we limit Hohmann Transfers to

- Orbits in the same plane (*coplanar orbits*)

- Orbits with their major axes (line of apsides) aligned (*co-apsidal orbits*) or circular orbits

- Instantaneous velocity changes (*delta-Vs* or ΔVs) tangent to the initial and final orbits

We'll deal with the problem of changing between orbital planes later. Co-apsidal orbits take this name because two elliptical orbits have their major axes (line of apsis) aligned with one another. Also, velocity changes are "instantaneous" because we assume that the time the engine fires is very short compared to the Hohmann Transfer time of flight. Tangential velocity changes mean the spacecraft changes its velocity vector magnitude but not direction to begin the Hohmann Transfer. Again at the end of the transfer the spacecraft changes its velocity vector magnitude but not its direction. To satisfy this tangential condition, the spacecraft must fire its thrusters or "burn" in a direction parallel to its velocity vector. These tangential ΔVs are the real secret to the Hohmann Transfer's energy savings.

Now let's look at what these velocity changes are doing to the orbit. By assuming all ΔVs occur nearly instantaneously (sometimes called an *"impulsive burn"*), we can continue to use the results from the two-body problem we developed in Chapter 4 to help us here. (Otherwise, we'd have to integrate the thrust over time and that would be too complex for this discussion.) Whenever we add or subtract velocity, we change the orbit's specific mechanical energy, ε, and hence its size, or semimajor axis, a. Remember these quantities are related by

$$\varepsilon = -\frac{\mu}{2a}$$
(6-1)

where

ε = specific mechanical energy (km^2/s^2)

μ = gravitational parameter = 3.986×10^5 (km^3/s^2) for Earth

a = semimajor axis (km)

If we want to move a spacecraft to a higher orbit, we have to increase the semimajor axis (adding energy to the orbit) by increasing velocity. On the other hand, to move the spacecraft to a lower orbit, we decrease the semimajor axis (and the energy) by decreasing the velocity.

During a space mission, we sometimes must transfer a spacecraft from one orbit (orbit 1) to another (orbit 2). It has to go into a *transfer orbit*, shown in Figure 6-4, on its way to orbit 2. To get from orbit 1 to the transfer orbit, we change the orbit's energy (by changing the spacecraft's velocity by an amount ΔV_1). Then, when the spacecraft gets to orbit 2, we must change its energy again (by changing its velocity by an amount ΔV_2). If we don't, the spacecraft will remain in the transfer orbit, indefinitely, returning to where it started in orbit 1, then back to orbit 2, etc. Thus, the complete maneuver requires two separate energy changes, accomplished by changing the orbital velocities (using ΔV_1 and ΔV_2).

Any ΔV represents a change from the present velocity to a selected velocity. For a tangential burn, we can write this as

$$\Delta V = |V_{selected} - V_{present}|$$

Notice we normally take the absolute value of this difference because we want to know the amount of velocity change, so we can calculate the energy and thus, the fuel needed. We're not concerned with the sign of the ΔV because we must burn fuel whether the spacecraft accelerates to reach a higher orbit or decelerates to drop into a lower orbit. If ΔV_1 is the change in velocity that takes the spacecraft from orbit 1 into the transfer orbit, then,

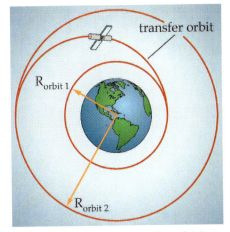

Figure 6-4. Getting From One Orbit to Another. The problem in orbital maneuvering is getting from orbit 1 to orbit 2. Here we see a spacecraft moving from a lower orbit to a higher one in a transfer orbit. If it doesn't perform the second ΔV when it reaches orbit 2, it will remain in the transfer orbit.

$$\Delta V_1 \,=\, \left|V_{\text{transfer at orbit 1}} - V_{\text{orbit 1}}\right|$$

where

ΔV_1 = velocity change to go from orbit 1 into the transfer orbit (km/s)

$V_{\text{transfer at orbit 1}}$ = velocity in the transfer orbit at orbit 1 radius (km/s)

$V_{\text{orbit 1}}$ = velocity in orbit 1 (km/s)

ΔV_2 is the change to get the spacecraft from the transfer orbit into orbit 2. Both of these ΔVs are shown in Figure 6-5.

$$\Delta V_2 \,=\, \left|V_{\text{orbit 2}} - V_{\text{transfer at orbit 2}}\right|$$

where

ΔV_2 = velocity change to move from the transfer orbit into orbit 2 (km/s)

We add the ΔV from each burn to find the total ΔV needed for the trip from orbit 1 to orbit 2.

$$\Delta V_{\text{total}} \,=\, \Delta V_1 + \Delta V_2 \qquad\qquad (6\text{-}2)$$

where

ΔV_{total} = total velocity change needed for the transfer (km/s)

When we cover the rocket equation in Chapter 14, we'll see how to convert this number into the amount of fuel required.

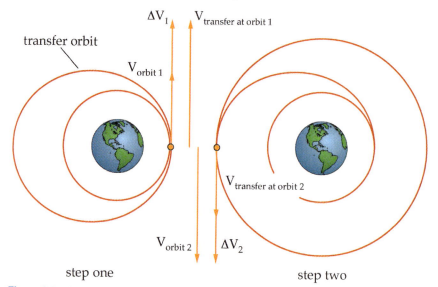

Figure 6-5.   Hohmann Transfer. Step 1: The first burn or ΔV of a Hohmann Transfer takes the spacecraft out of its initial, circular orbit and puts it in an elliptical, transfer orbit. Step 2: The second burn takes it from the transfer orbit and puts it in the final, circular orbit.

To compute ΔV_{total}, we use the energy equations from orbital mechanics. Everything we need to know to solve an orbital-maneuvering problem comes from these two valuable relationships, as you can see in Example 6-1. First, we need the specific mechanical energy, ε

$$\varepsilon = \frac{V^2}{2} - \frac{\mu}{R} \qquad (6\text{-}3)$$

where

ε = spacecraft's specific mechanical energy (km^2/s^2)

V = magnitude of the spacecraft's velocity vector (km/s)

μ = gravitational parameter (km^3/s^2) = $3.986 \times 10^5 \ km^3/s^2$ for Earth

R = magnitude of the spacecraft's position vector (km)

Then, we need the alternate form of the specific mechanical energy equation

$$\varepsilon = -\frac{\mu}{2a} \qquad (6\text{-}4)$$

where

ε = spacecraft's specific mechanical energy (km^2/s^2)

μ = gravitational parameter (km^3/s^2) = $3.986 \times 10^5 \ km^3/s^2$ for Earth

a = semimajor axis (km)

Let's review the steps in the transfer process to see how all this fits together. Referring to Figure 6-5,

- Step 1: ΔV_1 takes a spacecraft from orbit 1 and puts it into the transfer orbit

- Step 2: ΔV_2 puts the spacecraft into orbit 2 from the transfer orbit

To solve for these ΔVs, we need to find the energy in each orbit. If we know the sizes of orbits 1 and 2, then we know their semimajor axes ($a_{orbit\ 1}$ and $a_{orbit\ 2}$). The transfer orbit's major axis equals the sum of the two orbital radii, as shown in Figure 6-6.

$$2a_{transfer} = R_{orbit\ 1} + R_{orbit\ 2} \qquad (6\text{-}5)$$

Using the alternate equation for specific mechanical energy, we determine the energy for each orbit

$$\varepsilon_{orbit\ 1} = -\frac{\mu}{2a_{orbit\ 1}} \qquad (6\text{-}6)$$

$$\varepsilon_{orbit\ 2} = -\frac{\mu}{2a_{orbit\ 2}} \qquad (6\text{-}7)$$

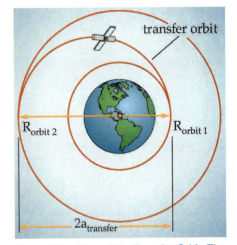

Figure 6-6.   Size of the Transfer Orbit. The major axis of the transfer orbit equals the sum of the radii of the initial and final orbits.

197

$$\varepsilon_{transfer} = -\frac{\mu}{2a_{transfer}} \qquad (6\text{-}8)$$

With the energies in hand, we use the main equation for specific mechanical energy, rearranged to calculate the orbits' velocities

$$V_{orbit\ 1} = \sqrt{2\left(\frac{\mu}{R_{orbit\ 1}} + \varepsilon_{orbit\ 1}\right)}$$

$$V_{orbit\ 2} = \sqrt{2\left(\frac{\mu}{R_{orbit\ 2}} + \varepsilon_{orbit\ 2}\right)}$$

$$V_{transfer\ at\ orbit\ 1} = \sqrt{2\left(\frac{\mu}{R_{orbit\ 1}} + \varepsilon_{transfer}\right)}$$

$$V_{transfer\ at\ orbit\ 2} = \sqrt{2\left(\frac{\mu}{R_{orbit\ 2}} + \varepsilon_{transfer}\right)}$$

Finally, we take the velocity differences to find ΔV_1 and ΔV_2, then add these values to get ΔV_{total}

$$\Delta V_1 = \left| V_{transfer\ at\ orbit\ 1} - V_{orbit\ 1} \right|$$

$$\Delta V_2 = \left| V_{orbit\ 2} - V_{transfer\ at\ orbit\ 2} \right|$$

$$\Delta V_{total} = \Delta V_1 + \Delta V_2$$

The Hohmann Transfer is energy efficient, but it can take a long time. To find the time of flight, look at the diagram of the maneuver. The transfer covers exactly one half of an ellipse. Recall that we find the total period for any closed orbit by

$$P = 2\pi\sqrt{\frac{a^3}{\mu}} \qquad (6\text{-}9)$$

So, the transfer orbit's time of flight (TOF) is half of the period

$$TOF = \frac{P}{2} = \pi\sqrt{\frac{a_{transfer}^3}{\mu}} \qquad (6\text{-}10)$$

where

TOF = spacecraft's time of flight (s)

P   = orbital period (s)

a   = semimajor axis of the transfer orbit (km)

μ = gravitational parameter $(km^3/s^2) = 3.986 \times 10^5\ km^3/s^2$ for Earth

Example 6-1 shows how to find time of flight for a Hohmann Transfer.

Now that we've gone through the Hohmann Transfer, let's step back to see what went on here. In the example, the spacecraft went from a low orbit to a higher orbit. To do this, it had to accelerate twice: ΔV_1 and ΔV_2. But notice the velocity in the higher circular orbit is less than in the lower circular orbit. Thus, the spacecraft accelerated twice, yet ended up in a slower orbit! Does this make sense?

ΔV_1 increases the spacecraft's velocity, taking the spacecraft out of orbit 1 and putting it into the transfer orbit. In the transfer orbit, its velocity gradually decreases as its radius increases, trading kinetic energy for potential energy, just as a baseball thrown into the air loses vertical velocity as it gets higher. When the spacecraft reaches the radius of orbit 2, it accelerates again, with ΔV_2 putting it into the final orbit. Even though the velocity in orbit 2 is lower than in orbit 1, the *total energy* is higher because it's at a larger radius. Remember, energy is the sum of kinetic plus potential energy. Thus, we use the spacecraft's rockets to add kinetic energy making it gain potential energy. Once it reaches orbit 2, it has higher total energy.

Section Review

Key Terms

co-apsidal orbits
coplanar orbits
delta-V, ΔV
Hohmann Transfer
impulsive burn
total energy
transfer orbit

Key Equations

$$\varepsilon = -\frac{\mu}{2a}$$

$$\varepsilon = \frac{V^2}{2} - \frac{\mu}{R}$$

$$2a_{transfer} = R_{orbit\ 1} + R_{orbit\ 2}$$

$$\varepsilon_{transfer} = -\frac{\mu}{2a_{transfer}}$$

$$TOF = \frac{P}{2} = \pi\sqrt{\frac{a_{transfer}^3}{\mu}}$$

Key Concepts

➤ The Hohmann Transfer moves a spacecraft from one orbit to another in the same plane. It's the simplest kind of orbital maneuver because it focuses only on changing the spacecraft's specific mechanical energy, ε.

➤ The Hohmann Transfer is the cheapest way (least amount of fuel) to get from one orbit to another. It's based on these assumptions

- Initial and final orbits are in the same plane (coplanar)

- Major axes of the initial and final orbits are aligned (co-apsidal)

- Velocity changes (ΔVs) are tangent to the initial and final orbits. Thus, the spacecraft's velocity changes magnitude but not direction

- ΔVs occur instantaneously—impulsive burns

➤ The Hohmann Transfer consists of two separate ΔVs

- The first, ΔV_1, accelerates the spacecraft from its initial orbit into an elliptical transfer orbit

- The second, ΔV_2, accelerates the spacecraft from the elliptical transfer orbit into the final orbit

Example 6-1

Problem Statement

Imagine NASA wants to place a communications satellite into a geosynchronous orbit from a low-Earth, parking orbit.

$R_{orbit\,1} = 6570$ km

$R_{orbit\,2} = 42,160$ km

What is the ΔV_{total} for this transfer and how long will it take?

Problem Summary

Given: $R_{orbit\,1} = 6570$ km
$R_{orbit\,2} = 42,160$ km

Find: ΔV_{total} and TOF

Problem Diagram

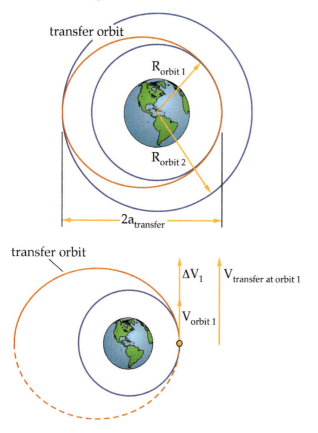

Conceptual Solution

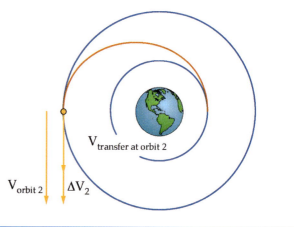

1) Compute the semimajor axis of the transfer orbit

$$a_{transfer} = \frac{R_{orbit\,1} + R_{orbit\,2}}{2}$$

2) Solve for the specific mechanical energy of the transfer orbit

$$\varepsilon_{transfer} = -\frac{\mu}{2a_{transfer}}$$

3) Solve for the energy and velocity in orbit 1

$$\varepsilon_{orbit\,1} = -\frac{\mu}{2a_{orbit\,1}}$$

$a_{orbit\,1} = R_{orbit\,1}$ (circular orbit)

$$\varepsilon = \frac{V^2}{2} - \frac{\mu}{R}$$

$$\therefore V_{orbit\,1} = \sqrt{2\left(\frac{\mu}{R_{orbit\,1}} + \varepsilon_{orbit\,1}\right)}$$

4) Solve for $V_{transfer\ at\ orbit\ 1}$

$$V_{transfer\ at\ orbit\ 1} = \sqrt{2\left(\frac{\mu}{R_{orbit\,1}} + \varepsilon_{transfer}\right)}$$

5) Find ΔV_1

$$\Delta V_1 = \left| V_{transfer\ at\ orbit\ 1} - V_{orbit\,1} \right|$$

Example 6-1 (Continued)

Solve for $V_{\text{transfer at orbit 2}}$

$$V_{\text{transfer at orbit 2}} = \sqrt{2\left(\frac{\mu}{R_{\text{orbit 2}}} + \varepsilon_{\text{transfer}}\right)}$$

Solve for the energy and velocity in orbit 2

$$\varepsilon_{\text{orbit 2}} = -\frac{\mu}{2a_{\text{orbit 2}}}$$

$a_{\text{orbit 2}} = R_{\text{orbit 2}}$ (circular orbit)

$$\varepsilon = \frac{V^2}{2} - \frac{\mu}{R}$$

$$\therefore V_{\text{orbit 2}} = \sqrt{2\left(\frac{\mu}{R_{\text{orbit 2}}} + \varepsilon_{\text{orbit 2}}\right)}$$

) Find ΔV_2

$$\Delta V_2 = \left|V_{\text{orbit 2}} - V_{\text{transfer at orbit 2}}\right|$$

) Solve for ΔV_{total}

$$\Delta V_{\text{total}} = \Delta V_1 + \Delta V_2$$

0) Compute TOF

$$TOF = \pi\sqrt{\frac{a_{\text{transfer}}^3}{\mu}}$$

Analytical Solution

) Compute the semimajor axis of the transfer orbit

$$a_{\text{transfer}} = \frac{R_{\text{orbit 1}} + R_{\text{orbit 2}}}{2} = \frac{6570 \text{ km} + 42{,}160 \text{ km}}{2}$$

$a_{\text{transfer}} = 24{,}365 \text{ km}$

?) Solve for the specific mechanical energy of the transfer orbit

$$\varepsilon_{\text{transfer}} = -\frac{\mu}{2a_{\text{transfer}}} = -\frac{3.986 \times 10^5 \dfrac{\text{km}^3}{\text{s}^2}}{2(24{,}365 \text{ km})}$$

$$\varepsilon_{\text{transfer}} = -8.1798\frac{\text{km}^2}{\text{s}^2}$$

(Note the energy is negative, which implies the transfer orbit is an ellipse; as we'd expect.)

3) Solve for energy and velocity of orbit 1

$$\varepsilon_{\text{orbit 1}} = -\frac{\mu}{2a_{\text{orbit 1}}}$$

$$= -\frac{3.986 \times 10^5 \dfrac{\text{km}^3}{\text{s}^2}}{2(6570 \text{ km})} = -30.33\frac{\text{km}^2}{\text{s}^2}$$

$$V_{\text{orbit 1}} = \sqrt{2\left(\frac{\mu}{R_{\text{orbit 1}}} + \varepsilon_{\text{orbit 1}}\right)}$$

$$\sqrt{2\left(\frac{3.986 \times 10^5 \dfrac{\text{km}^3}{\text{s}^2}}{6570 \text{ km}} - 30.33\frac{\text{km}^2}{\text{s}^2}\right)} = 7.789\frac{\text{km}}{\text{s}}$$

4) Solve for $V_{\text{transfer at orbit 1}}$

$$V_{\text{transfer at orbit 1}} = \sqrt{2\left(\frac{\mu}{R_{\text{orbit 1}}} + \varepsilon_{\text{transfer}}\right)}$$

$$\sqrt{2\left(\frac{3.986 \times 10^5 \dfrac{\text{km}^3}{\text{s}^2}}{6570\text{km}} - 8.1798\frac{\text{km}^2}{\text{s}^2}\right)}$$

$$V_{\text{transfer at orbit 1}} = 10.246\ \frac{\text{km}}{\text{s}}$$

5) Find ΔV_1

$$\Delta V_1 = \left|V_{\text{transfer at orbit 1}} - V_{\text{orbit 1}}\right|$$

$$\left|10.246\frac{\text{km}}{\text{s}} - 7.789\frac{\text{km}}{\text{s}}\right|$$

$$\Delta V_1 = 2.457\frac{\text{km}}{\text{s}}$$

6) Solve for $V_{\text{transfer at orbit 2}}$

$$V_{\text{transfer at orbit 2}} = \sqrt{2\left(\frac{\mu}{R_{\text{orbit 2}}} + \varepsilon_{\text{transfer}}\right)}$$

$$\sqrt{2\left(\frac{3.986 \times 10^5 \dfrac{\text{km}^3}{\text{s}^2}}{42{,}160 \text{ km}} - 8.1798\frac{\text{km}^2}{\text{s}^2}\right)}$$

$$V_{\text{transfer at orbit 2}} = 1.597\frac{\text{km}}{\text{s}}$$

Example 6-1 (Continued)

7) Solve for energy and velocity in orbit 2

$$\varepsilon_{orbit\ 2} = -\frac{\mu}{2a_{orbit\ 2}}$$

$$= -\frac{3.986 \times 10^5 \frac{km^3}{s^2}}{2(42,160\ km)} = -4.727 \frac{km^2}{s^2}$$

$$V_{orbit\ 2} = \sqrt{2\left(\frac{\mu}{R_{orbit\ 2}} + \varepsilon_{orbit\ 2}\right)}$$

$$\sqrt{2\left(\frac{3.986 \times 10^5 \frac{km^3}{s^2}}{42,160\ km} - 4.727 \frac{km^2}{s^2}\right)} = 3.075 \frac{km}{s}$$

8) Find ΔV_2

$$\Delta V_2 = \left|V_{orbit\ 2} - V_{transfer\ at\ orbit\ 2}\right|$$

$$\left|3.075 \frac{km}{s} - 1.597 \frac{km}{s}\right|$$

$$\Delta V_2 = 1.478 \frac{km}{s}$$

9) Solve for ΔV_{total}

$$\Delta V_{total} = \Delta V_1 + \Delta V_2 = 2.457 \frac{km}{s} + 1.478 \frac{km}{s}$$

$$= 3.935 \frac{km}{s}$$

10) Compute TOF

$$TOF = \pi \sqrt{\frac{a_{transfer}^3}{\mu}}$$

$$= \pi \sqrt{\frac{(24,365\ km)^3}{3.986 \times 10^5 \frac{km^3}{s^2}}}$$

$$TOF = 18,925\ s \cong 315\ min = 5\ hrs\ 15\ min$$

Interpreting the Results

To move the communication satellite from its low altitude (192 km) parking orbit to geosynchronous altitude, the engines must provide a total velocity change of about 3.9 km/s (about 8720 m.p.h.). The transfer will take five and a quarter hours to complete.

6.2 Plane Changes

In This Section You'll Learn to...

☛ Explain when to use a simple plane change and how a simple plane change can modify an orbital plane

☛ Explain how to use a plane change combined with a Hohmann Transfer to efficiently change an orbit's size and orientation

☛ Determine the ΔV needed for simple and combined plane changes

So far we've seen how to change the size of an orbit using a Hohmann Transfer. However, we restricted this transfer to coplanar orbits. As you'd expect, to change its orbital plane, a spacecraft must point its velocity change (ΔV) out of its current plane. By changing the orbital plane, it also alters the orbit's tilt (inclination, i) or its swivel (right ascension of the ascending node, Ω), depending on where in the orbit it does the ΔV burn. For plane changes, we must consider the direction and magnitude of the spacecraft's initial and final velocities.

To understand plane changes, imagine you're on a racetrack with off-ramps such as those on a freeway. If you want to exit from the track, you not only must change your velocity *within* its plane but also must go *above* or *below* the level of the track? This "out of plane" maneuver causes you to use even more energy than a level exit because you now have to accelerate to make it up the ramp or brake as you go down. Thus, out-of-plane maneuvers typically require much more energy than in-plane maneuvers.

Let's look at two types of plane changes—simple and combined. The difference between the two depends on how the orbital velocity vector changes. With a *simple plane change* only its direction changes, but to do a *combined plane change* we alter its direction and magnitude. We'll take on the simple plane change first.

Simple Plane Changes

Let's imagine we have a spacecraft in an orbit with an inclination, i, of 28.5°. (the inclination we'd get if we launched it due east from the Kennedy Space Center, as the Shuttle often does.) Assume we want to change it into an equatorial orbit (i = 0°). We must change the spacecraft's velocity to do this, but we want to change only the orbit's orientation, not its size. This means the velocity vector's magnitude stays the same, that is $\left|\vec{V}_{initial}\right| = \left|\vec{V}_{final}\right|$, but its direction changes.

How do we change just the direction of the velocity vector? Look at the situation in Figure 6-7. You can see we initially have an inclined orbit with a velocity $\vec{V}_{initial}$, and we want to rotate the orbit by an angle θ to reach a final velocity, $\vec{V}_{final}$. The vector triangle shown in Figure 6-7 summarizes

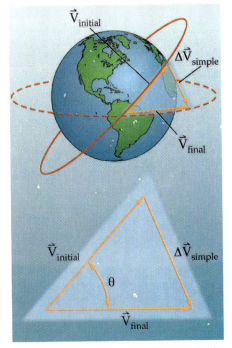

Figure 6-7.   Simple Plane Change. A simple plane change affects only the direction and not the magnitude of the original velocity vector.

this problem. It's an isosceles triangle (meaning it has two sides of equal length). Using plane geometry, we get a relationship for ΔV_{simple}—the change in velocity needed to rotate the plane

$$\Delta V_{simple} = 2\, V_{initial} \sin\left(\frac{\theta}{2}\right)$$

(6-11)

where

ΔV_{simple}	= velocity change for a simple plane change (km/s)
$V_{initial} = V_{final}$	= velocities in the initial and final orbits (km/s)
θ	= plane-change angle (deg or rad)

If we want to change only the orbit's inclination, we must change the velocity at either the ascending node or the descending node. When the ΔV occurs at one of these nodes, the orbit will pivot about a line connecting the two nodes, thus changing only the inclination.

We can also use a plane change to change the right ascension of the ascending node, Ω. This might be useful if we want a remote-sensing satellite to pass over a certain point on Earth at a certain time of day. When we consider a polar orbit (i = 90°), we see that a ΔV_{simple} at the North or South Pole changes just the right ascension of the ascending node, as illustrated in Figure 6-8. We can also change Ω alone for inclinations other than 90°. The trick is to perform the ΔV_{simple} where the initial and final orbits intersect. (Think of this maneuver as pivoting around a line connecting the burn point to Earth's center.) We won't go into the details of these cases because the spherical trigonometry gets a bit complicated for our discussion here.

The amount of velocity change a spacecraft needs to re-orient its orbital plane depends on two things—the angle it's turning through and its initial velocity. As the angle it's turning through increases, so does ΔV_{simple}. For example, when this angle is 60°, the vector triangle becomes equilateral (all sides equal). In this case, ΔV_{simple} equals the initial velocity, which is the amount of velocity it needed to get into the orbit in the first place! That's why we'd like the initial parking orbit to have an inclination as close as possible to the final mission orbit.

Also notice that ΔV_{simple} increases as the initial velocity increases. Therefore, we can lower ΔV_{simple} by reducing the initial velocity. For a circular orbit the velocity is constant throughout the orbit, but we know a spacecraft in an elliptical orbit slows down as it approaches apogee. Thus, if we can choose where to do a simple plane change in an elliptical orbit, we should do it at apogee, where the spacecraft's velocity is slowest. Remember our earlier analogy about changing speeds and directions on a racetrack. It's easier to change direction when we're going slower (even for a stunt driver). Example 6-2 demonstrates a simple plane change.

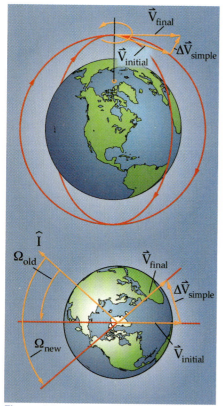

Figure 6-8. Changing Ω. A simple plane change as a spacecraft crosses the pole in a polar orbit (i = 90°) will change only the right ascension of the ascending node, Ω. Imagine the orbital plane pivoting about Earth's poles.

Combined Plane Changes

Suppose our spacecraft is in a low-altitude parking orbit with i = 28.5° and it needs to transfer to a geostationary orbit (R = 42,160 km, i = 0°). This transfer presents us with two separate problems: changing the size of the orbit and changing the orientation of the orbital plane. We might be tempted to tackle this problem in two parts—a Hohmann Transfer followed by a simple plane change. This two-part problem gets the job done in three separate ΔV burns. But we can do the job in two burns rather than three and save fuel. How? By combining the plane-change burn with one Hohmann Transfer burn to get a maneuver we call a *combined plane change*.

If we draw a diagram of this problem, as in Figure 6-9, we can see that $\Delta \vec{V}_{combined}$ is the vector sum of a simple plane change ($\Delta \vec{V}_{simple}$) and changing the orbit's size where $\Delta \vec{V}_{increase}$ is one of the two Hohmann Transfer burns. These three ΔVs form a triangle with $\Delta \vec{V}_{combined}$ as the third side. You may recall from plane geometry that the sum of any two sides of a triangle is greater than the third side. That is:

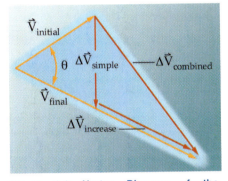

$$\left|\Delta \vec{V}_{simple}\right| + \left|\Delta \vec{V}_{increase}\right| > \left|\Delta \vec{V}_{combined}\right|$$

This means it's always cheaper (in terms of ΔV) to do a combined plane change than to do a simple plane change followed by one of the Hohmann Transfer burns.

To solve for the needed velocity change let's apply the ever-popular law of cosines to get

Figure 6-9. Vector Diagram of the Combined Plane Change. For a combined plane change, $\Delta \vec{V}_{combined}$ is always less than a simple plane change, $\Delta \vec{V}_{simple}$, followed by a tangential velocity increase, $\Delta \vec{V}_{increase}$.

$$\Delta V_{combined} = \sqrt{\left(\left|\vec{V}_{initial}\right|\right)^2 + \left(\left|\vec{V}_{final}\right|\right)^2 - 2\left|\vec{V}_{initial}\right|\left|\vec{V}_{final}\right|\cos\theta} \qquad (6\text{-}12)$$

where

$\Delta V_{combined}$ = velocity change for a combined plane change (km/s)

$\left|\vec{V}_{initial}\right|$ = magnitude of the velocity in the initial orbit (km/s)

$\left|\vec{V}_{final}\right|$ = magnitude of the velocity in the final orbit (km/s)

θ = plane-change angle (deg or rad)

By working through this equation, we find that it's cheaper to do a combined plane change at slower velocities (when farther from Earth) just as we found for the simple plane change. So what's the cheapest way to do a Hohmann Transfer with a plane change? For the case of going from a smaller to a larger orbit, we should begin the Hohmann Transfer (ΔV_1) at the lower altitude while keeping the same inclination. Then we should perform the combined plane change at apogee of the transfer orbit, completing the Hohmann Transfer and the plane change in one, combined ΔV burn.

Table 6-1 summarizes the four options for a transfer from a low-Earth orbit with an inclination of 28° to a geostationary orbit. It shows that starting the Hohmann Transfer closest to Earth and finishing with the combined plane change at apogee (Case 4) is the most economical in terms of ΔV. We base the results in Table 6-1 on the following example

Given: $R_{orbit\ 1} = 6570$ km
$R_{orbit\ 2} = 42{,}160$ km
$i_{orbit\ 1} = 28°$
$i_{orbit\ 2} = 0°$

Find: ΔV_{total}

Table 6-1.   Plane Change and Hohmann Transfer Options. Case 4 requires the least amount of ΔV.

Case 1	Case 2	Case 3	Case 4
Do a 28° inclination change using a simple plane change. Then do the Hohmann Transfer, ΔV_1 and ΔV_2.	Do the Hohmann Transfer, ΔV_1 and ΔV_2. Then do the 28° inclination change using a simple plane change.	Do a combined plane change at perigee of the transfer orbit. Do ΔV_2 of Hohmann.	Do ΔV_1 of Hohmann Transfer. Do combined plane change at apogee of transfer orbit.
$\Delta V_{simple} = 3.77$ km/s (in orbit 1)	$\Delta V_{Hohmann} = 3.94$ km/s	$\Delta V_{combined} = 4.98$ km/s (at perigee)	$\Delta V_1 = 2.46$ km/s
$\Delta V_{Hohmann} = 3.94$ km/s	$\Delta V_{simple} = 1.49$ km/s (in orbit 2)	$\Delta V_2 = 1.47$ km/s	$\Delta V_{combined} = 1.82$ km/s (at apogee)
$\Delta V_{total} = 7.70$ km/s	$\Delta V_{total} = 5.43$ km/s	$\Delta V_{total} = 6.46$ km/s	$\Delta V_{total} = 4.29$ km/s

▥ Section Review

Key Terms

combined plane change
simple plane change

Key Equations

$$\Delta V_{simple} = 2\ V_{initial} \sin\left(\frac{\theta}{2}\right)$$

$$\Delta V_{combined} =$$
$$\sqrt{(|\vec{V}_{initial}|)^2 + (|\vec{V}_{final}|)^2 - 2|\vec{V}_{initial}||\vec{V}_{final}|\cos\theta}$$

Key Concepts

➤ We need plane change maneuvers to move a spacecraft from one orbital plane to another

- Simple plane changes alter only the direction, not the magnitude, of the velocity vector for the original orbit

$$|\vec{V}_{initial}| = |\vec{V}_{final}|$$

  - A simple plane change at either the ascending or descending node changes only the orbit's inclination. On a polar orbit a simple plane change made over the North or South Pole changes only the right ascension of the ascending node. A simple plane change made anywhere else changes inclination and right ascension of the ascending node.

- A combined plane change alters the magnitude and direction of the original velocity vector

  - It's always cheaper (in terms of ΔV) to do a combined plane change than to do a simple plane change followed by a Hohmann Transfer burn

➤ It's always cheaper (in terms of ΔV) to change planes when the orbital velocity is *slowest*, which is at apogee for elliptical transfer orbits

Example 6-2

Problem Statement

Suppose a satellite is in a circular orbit at an altitude of 250 km. It needs to move from its current inclination of 28° to an inclination of 57°. What ΔV does this transfer require?

Problem Summary

Given: Altitude = 250 km

$i_{initial} = 28.0°$

$i_{final} = 57.0°$

Find: ΔV_{simple}

Conceptual Solution

1) Solve for the orbit's energy and velocity

$$\varepsilon = -\frac{\mu}{2a}$$

$$= -\frac{\mu}{2R} \quad \text{(circular orbit)}$$

$$\varepsilon = \frac{V^2}{2} - \frac{\mu}{R}$$

$$V = \sqrt{2\left(\frac{\mu}{R} + \varepsilon\right)}$$

2) Solve for the inclination change

$$\theta = \left|i_{final} - i_{initial}\right|$$

3) Find the change in velocity for a simple plane change

$$\Delta V_{simple} = 2\, V_{initial} \sin\frac{\theta}{2}$$

Analytical Solution

1) Solve for the energy and velocity of the orbit

$$\varepsilon = -\frac{\mu}{2R} = -\frac{3.986 \times 10^5 \frac{km^3}{s^2}}{2(6378 + 250 \ km)}$$

$$= -30.069 \ \frac{km^2}{s^2}$$

$$V_{initial} = \sqrt{2\left(\frac{\mu}{R} + \varepsilon\right)}$$

$$V_{initial} = \sqrt{2\left(\frac{3.986 \times 10^5 \frac{km^3}{s^2}}{6628 \ km} - 30.069 \frac{km^2}{s^2}\right)}$$

$$= 7.755\frac{km}{s}$$

2) Solve for the inclination change

$$\theta = \left|i_{final} - i_{initial}\right| = \left|57° - 28°\right|$$

$$\theta = 29°$$

Find ΔV for the simple plane change

$$\Delta V_{simple} = 2\, V_{initial} \sin\frac{\theta}{2} = 2\left(7.755\frac{km}{s}\right) \sin\frac{29°}{2}$$

$$\Delta V_{simple} = 3.88 \ km/s$$

Interpreting the Results

To change the inclination of the satellite by 29°, we must apply a ΔV of 3.88 km/s. This is 50% of the velocity we needed to get the satellite into space in the first place. Plane changes are *very expensive* (in terms of ΔV.)

6.3 Rendezvous

▰ In This Section You'll Learn to...

☛ Describe orbital rendezvous

☛ Determine the ΔV and wait time to execute a rendezvous

For the Hohmann Transfer and plane change maneuvers we described earlier in this chapter, we focused on how to move a spacecraft without considering where it is in relation to other spacecraft. However, several types of missions require a spacecraft to meet or *rendezvous* with another one, meaning one spacecraft must arrive at the same place at the same time as a second one. The Gemini program perfected this maneuver in the 1960s, as a prelude to the Apollo missions to the Moon, which depended on a Lunar-orbit rendezvous. Two astronauts returning from the Lunar surface had to rendezvous with their companion in the command module in Lunar orbit for the trip back to Earth. As another example, the Space Shuttle needs to rendezvous with the International Space Station routinely to transfer people and equipment. In this section, we'll examine two simple rendezvous scenarios between co-planar and co-orbital spacecraft.

─── *Astro Fun Fact* ───
Docking Drama on Gemini 8

The first man to walk on the Moon almost didn't survive his first flight into space! On March 16, 1966, NASA launched Gemini 8 from Cape Canaveral, Florida. The mission was to rendezvous and perform the first-ever docking with another spacecraft. A modified Agena upperstage (shown in the photo) had been launched 100 minutes before. Following a smooth rendezvous, astronauts Neil Armstrong and Dave Scott started the docking sequence and, for the first time, two spacecraft were joined together in orbit. Thirty minutes later, the two spacecraft began to spin uncontrollably. Thinking the problem was with the Agena, Armstrong undocked. Unfortunately, this made the spinning worse as the Gemini reached rates of 360° per second (60 r.p.m.). Nearing blackout, the two astronauts finally regained control by shutting down power to the primary thrusters and firing the re-entry thrusters. The mission was aborted early to a safe splashdown, but Neil Armstrong would fly again—next time to the Moon.

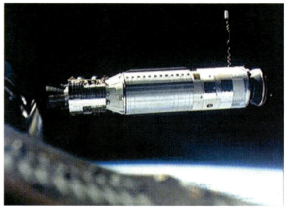

David Baker, PhD, The History of Manned Space Flight. New York, NY: Crown Publishers Inc., 1981.

(Courtesy of NASA/Johnson Space Center)

Coplanar Rendezvous

The simplest type of rendezvous uses a Hohmann Transfer between coplanar orbits. The key to this maneuver is timing. Deciding when to fire the engines, we must calculate how much to lead the target spacecraft, just as a quarterback leads a receiver in a football game. At the snap of the ball, the receiver starts running straight down the field toward the goal line, as Figure 6-10 shows. The quarterback mentally calculates how fast the receiver is running and how long it will take the ball to get to a certain spot on the field. When the quarterback releases the ball, it will take some time to reach that spot. Over this same period, the receiver goes from where he was when the ball was released to the "rendezvous" point with the ball.

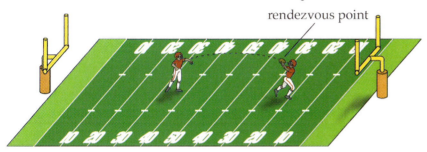

Figure 6-10. Orbital Rendezvous and Football. The spacecraft-rendezvous problem is similar to the problem a quarterback faces when passing to a running receiver. The quarterback must time the pass just right so the ball and the receiver arrive at the same place at the same time.

Let's look closer at this football analogy to see how the quarterback decides when to throw the ball so it will "rendezvous" with the receiver. Assume we have a quarterback who throws a 20-yard pass traveling at 10 yd/s and a wide receiver who runs at 4 yd/s. (Ironically, we use English units to describe American football.) How long must the quarterback wait from the snap (assuming the receiver starts running immediately) before throwing the ball? To analyze this problem, let's define the following symbols

$$V_{receiver} = \text{velocity of the receiver running down the field}$$
$$= 4 \text{ yd/s}$$
$$V_{ball} = \text{velocity of the ball}$$
$$= 10 \text{ yd/s}$$

We know the quarterback must "lead" the receiver; that is, the receiver will travel some distance while the ball is in the air. But how long will the ball take to travel the 20 yards from the quarterback to the receiver? Let's define

$$TOF_{ball} = \text{time of flight of the ball}$$
$$= \text{distance the ball travels}/V_{ball}$$
$$= 20 \text{ yd}/(10 \text{ yd/s})$$
$$= 2 \text{ s}$$

The lead distance is then the receiver's velocity times the ball's time of flight.

$$\alpha = \text{lead distance}$$
$$= V_{receiver} \times TOF_{ball}$$
$$= (4 \text{ yd/s}) \times 2 \text{ s}$$
$$= 8 \text{ yd}$$

This means the receiver runs an additional 8 yards down the field, while the ball is in the air. From this we can figure out how much of a head start the receiver needs before the quarterback throws the ball. If the receiver runs 8 yards while the ball is in the air, and the ball is being thrown 20 yards, the receiver then needs a head start of

$$\phi_{head\ start} = \text{head start distance needed by the receiver}$$
$$= 20 \text{ yd} - \alpha$$
$$= 20 \text{ yd} - 8 \text{ yd}$$
$$= 12 \text{ yd}$$

So before the quarterback throws the ball, the receiver should be 12 yards down the field. We can now determine how long it will take the receiver to go 12 yards down field.

$$\text{W.T.} = \text{wait time}$$
$$= \phi_{head\ start} / V_{receiver}$$
$$= 12 \text{ yd} / (4 \text{ yd/s})$$
$$= 3 \text{ s}$$

This is the time the quarterback must wait before throwing the ball to ensure the receiver will be at the rendezvous point when the ball arrives.

That's all well and good for footballs, but what about spacecraft trying to rendezvous in space? It turns out that the approach is the same as in the football problem. Let's look at the geometry of the rendezvous problem shown in Figure 6-11. We have a target spacecraft (say a disabled communication satellite that the crew of the Shuttle plans to fix) and an interceptor (the Space Shuttle). In this example, the target spacecraft is in a higher orbit than the Shuttle, but we'd take a similar approach if it were in a lower orbit. To rendezvous, the Shuttle crew must initiate a ΔV to transfer to the rendezvous point. But they must do this ΔV at just the right moment to ensure the target spacecraft arrives at the same point at the same time.

To see how to solve this problem, remember that the quarterback first had to know the velocities of the interceptor (the ball) and the target (the receiver). Because footballs move in nearly straight lines, their velocities are easy to see. However, for spacecraft in orbits, velocities aren't so straightforward. Instead of using a straight-line velocity (in meters per second or miles per hour), we use rotational velocity measured in radians per second or degrees per hour. We call this rotational velocity "angular velocity" and use the Greek letter small omega, ω, to represent it (not to be confused with the COE argument of perigee, ω). Because spacecraft

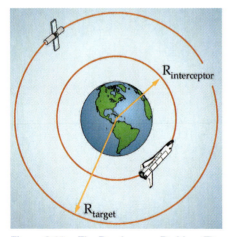

Figure 6-11.   The Rendezvous Problem. The Space Shuttle commander must do a Hohmann Transfer at precisely the right moment to rendezvous with another spacecraft.

move through 360° (or 2π radians) in one orbital period, we find their angular velocity from

$$\omega = \frac{2\pi(\text{radians})}{2\pi\sqrt{\dfrac{a^3}{\mu}}}$$

$$\boxed{\omega = \sqrt{\frac{\mu}{a^3}}} \qquad (6\text{-}13)$$

where

ω = spacecraft's angular velocity (rad/s)

μ = gravitational parameter $(\text{km}^3/\text{s}^2) = 3.986 \times 10^5\ \text{km}^3/\text{s}^2$ for Earth

a = semimajor axis (km)

For circular orbits, $a = R$ (radius), so this angular velocity is constant.

To solve the football problem, we had to find the ball's time of flight. For rendezvous in orbit, the time of flight is the same as the Hohmann Transfer's time of flight, which we found earlier

$$\text{TOF} = \pi\sqrt{\frac{a_{\text{transfer}}^3}{\mu}} \qquad (6\text{-}14)$$

where

TOF = interceptor spacecraft's time of flight (s)

π = 3.14159 . . . (unitless)

a_{transfer} = semimajor axis of the transfer orbit (km)

μ = gravitational parameter $(\text{km}^3/\text{s}^2) = 3.986 \times 10^5\ \text{km}^3/\text{s}^2$ for Earth

Finally, we need to get the timing right. In football, the quarterback must lead a receiver by a certain amount to get a pass to the right point for a completion. In rendezvous, the interceptor must lead the target by an amount called the *lead angle, α_{lead},* when the interceptor starts its Hohmann Transfer. This lead angle, shown in Figure 6-12, represents the angular distance covered by the target during the interceptor's time of flight. We find it by multiplying the target's angular velocity by the interceptor's time of flight.

$$\boxed{\alpha_{\text{lead}} = \omega_{\text{target}}\text{TOF}} \qquad (6\text{-}15)$$

where

α_{lead} = amount by which the interceptor must lead the target (rad)

ω_{target} = target's angular velocity (rad/s)

TOF = time of flight (s)

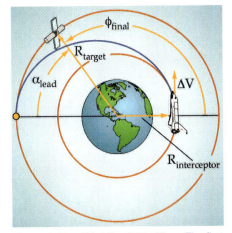

Figure 6-12. ΔV at the Right Time. The first ΔV of the rendezvous Hohmann Transfer starts when the interceptor is at an angle, ϕ_{final}, from the target.

We can now determine how big of a head start to give the target, just as a quarterback must give a receiver a head start before releasing the ball to complete a pass. For spacecraft, we call this the *phase angle, φ,* (Greek letter, small phi) measured from the interceptor's radius vector to the target's radius vector in the direction of the interceptor's motion. The interceptor travels 180° (π radians) during a Hohmann Transfer, so we can easily compute the needed phase angle, ϕ_{final}, if we know the lead angle.

$$\phi_{final} = \pi - \alpha_{lead} \tag{6-16}$$

where

ϕ_{final} = phase angle between the interceptor and target as the transfer begins (rad)

α_{lead} = angle by which the interceptor must lead the target (rad)

Chances are, when the interceptor is ready to start the rendezvous, the target won't be in the correct position, as seen in Figure 6-13. So what do we do? Just as a quarterback must wait a few seconds before releasing a pass to a receiver, the interceptor must wait until its position relative to the target is correct, as in Figure 6-12. But how long does it wait? To answer this we have to relate where the target is initially (relative to the interceptor), $\phi_{initial}$, to where the interceptor needs to be, ϕ_{final}, in time to begin the ΔV burn. Because the interceptor and target are moving in circular orbits at constant velocities, $\phi_{initial}$ and ϕ_{final} are related by

$$\phi_{final} = \phi_{initial} + (\omega_{target} - \omega_{interceptor}) \times \text{wait time} \tag{6-17}$$

Solving for wait time gives us

$$\text{wait time} = \frac{\phi_{final} - \phi_{initial}}{\omega_{target} - \omega_{interceptor}} \tag{6-18}$$

where

wait time   = time until the interceptor initiates the rendezvous (s)

$\phi_{final}, \phi_{initial}$ = initial and final phase angles (rad)

$\omega_{target}, \omega_{interceptor}$ = target and interceptor angular velocities (rad/s)

So far, so good. But if we look at the wait time equation, we see that wait time can be less than zero. Does this mean we have to go back in time? Luckily, no. Because the interceptor and the target are going around in circles, the correct angular relationship repeats itself periodically. When the difference between ϕ_{final} and $\phi_{initial}$ changes by 2π radians (360°), the correct initial conditions are repeated. To calculate the next available opportunity to start a rendezvous, we either add 2π to, or subtract it from, the numerator in Equation (6-18), whichever it takes to make the resulting wait time positive. In fact, we can determine future rendezvous opportunities by adding or subtracting multiples of 2π.

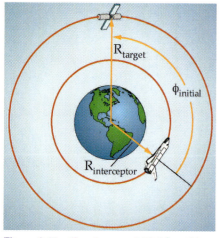

Figure 6-13. Rendezvous Initial Condition. At the start of the rendezvous problem, the target is some angle, $\phi_{initial}$, away from the interceptor.

Co-orbital Rendezvous

Another twist to the rendezvous problem occurs when the spacecraft are co-orbital, meaning the target and interceptor are in the same orbit, with one ahead of the other. Whenever the target is ahead, as shown in Figure 6-14, the interceptor must somehow catch the target. To do so, the interceptor needs to move into a waiting or *phasing orbit* that will return it to the same spot one orbit later, in the time it takes the target to move around to that same spot. Notice the target travels less than 360°, while the interceptor travels exactly 360°.

How can one spacecraft catch another one that's ahead of it in the same orbit? By *slowing down*! What?! Does this make sense? Yes, from specific mechanical energy, we know that if an interceptor slows down (decreases energy), it enters a smaller orbit. A smaller orbit has a shorter period, so it completes one full orbit (360°) in less time. If it slows down the correct amount, it will get back to where it started just as the target gets there.

To determine the right amount for an interceptor to slow down, first we find how far the target must travel to get to the interceptor's current position. If the target is ahead of the interceptor by an amount $\phi_{initial}$, it must travel through an angle, ϕ_{travel}, to reach the rendezvous spot, found from

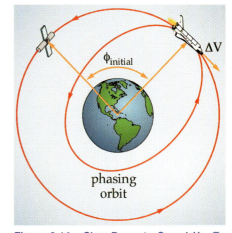

Figure 6-14.   Slow Down to Speed Up. To catch another spacecraft ahead of it in the same orbit, an interceptor slows down, entering a smaller phasing orbit with a shorter period. This allows it to catch the target.

$$\phi_{travel} = 2\pi - \phi_{initial} \qquad (6\text{-}19)$$

where

ϕ_{travel} = angle through which the target travels to reach the rendezvous location (rad)

$\phi_{initial}$ = initial angle between the interceptor and target (rad)

Now, if we know the angular velocity of the target, we can find the time it will take to cover this angle, ϕ_{travel}, by using

$$\text{TOF} = \frac{\phi_{travel}}{\omega_{target}} \qquad (6\text{-}20)$$

Remember we found the target's angular velocity from Equation (6-13)

$$\omega_{target} = \sqrt{\frac{\mu}{a_{target}^3}}$$

Because the time of flight equals the period of the phasing orbit, we equate this to our trusty equation for the period of an orbit, producing

$$\text{TOF} = \frac{\phi_{travel}}{\omega_{target}} = 2\pi\sqrt{\frac{a_{phasing}^3}{\mu}}$$

We can now solve for the required size of the phasing orbit

$$a_{phasing} = \sqrt[3]{\mu\left(\frac{\phi_{travel}}{2\pi\omega_{target}}\right)^2}$$

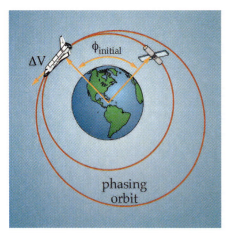

Figure 6-15.   Speed Up to Slow Down. If the target is behind the interceptor in the same orbit, the interceptor must speed up to enter a higher, slower orbit, thereby allowing the target to catch up.

where

$a_{phasing}$	= semimajor axis of the phasing orbit (km)
μ	= gravitational parameter $(km^3/s^2) = 3.986 \times 10^5 \; km^3/s^2$ for Earth
ϕ_{travel}	= angular distance the target must travel to get to the rendezvous location (rad)
ω_{target}	= target's angular velocity (rad/s)

Knowing the size of the phasing orbit, we can compute the necessary ΔVs for the rendezvous. The first ΔV slows the interceptor and puts it into the phasing orbit. The second ΔV returns it to the original orbit, right next to the target. These ΔVs have the same magnitude, so we don't need to calculate the second one.

We must also know how to rendezvous whenever the target is behind the interceptor in the same orbit. In this case, the angular distance the target must cover to get to the rendezvous spot is greater than 360°. Thus, the interceptor's phasing orbit for the interceptor will have a period greater than that of its current circular orbit. To get into this phasing orbit, the interceptor *speeds up*. It then enters a higher, slower orbit, allowing the target to catch up, as Figure 6-15 illustrates.

▦ Section Review

Key Terms

lead angle, α_{lead}
phase angle, ϕ
phasing orbit
rendezvous

Key Equations

$$\omega = \sqrt{\dfrac{\mu}{a^3}}$$

$$\alpha_{lead} = \omega_{target} TOF$$

$$\phi_{final} = \pi - \alpha_{lead}$$

$$wait \; time = \dfrac{\phi_{final} - \phi_{initial}}{\omega_{target} - \omega_{interceptor}}$$

$$a_{phasing} = \sqrt[3]{\mu \left(\dfrac{\phi_{travel}}{2\pi\omega_{target}} \right)^2}$$

Key Concepts

➤ Rendezvous is the problem of arranging for two or more spacecraft to arrive at the same point in an orbit at the same time

➤ The rendezvous problem is very similar to the problem quarterbacks face when they must "lead" a receiver with a pass. But because the interceptor and target spacecraft travel in circular orbits, the proper relative positions for rendezvous repeat periodically.

➤ We assume spacecraft rendezvous uses a Hohmann Transfer

➤ The lead angle, α_{lead}, is the angular distance the target spacecraft travels during the interceptor's time of flight, TOF

➤ The final phase angle, ϕ_{final}, is the "headstart" the target spacecraft needs

➤ The wait time is the time between some initial starting time and the time when the geometry is right to begin the Hohmann Transfer for a rendezvous

  • Remember, for negative wait times, we must modify the numerator in the wait time equation by adding or subtracting multiples of 2π radians

Example 6-3

Problem Statement

Imagine that an automated repair spacecraft in low-Earth orbit needs to rendezvous with a disabled target spacecraft in a geosynchronous orbit. If the initial angle between the two spacecraft is 180°, how long must the interceptor wait before starting the rendezvous?

$$R_{interceptor} = 6570 \text{ km}$$

$$R_{target} = 42{,}160 \text{ km}$$

Problem Summary

Given: $R_{interceptor} = 6570 \text{ km}$

$R_{target} = 42{,}160 \text{ km}$

$\phi_{initial} = 180° = \pi \text{ radians}$

Find: wait time

Problem Diagram

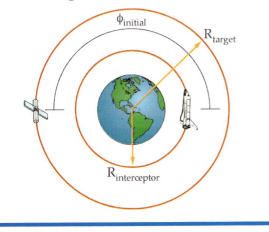

Conceptual Solution

Compute the semimajor axis of the transfer orbit

$$a_{transfer} = \frac{R_{interceptor} + R_{target}}{2}$$

2) Find the time of flight (TOF) of the transfer orbit

$$TOF = \pi\sqrt{\frac{a_{transfer}^3}{\mu}}$$

3) Find the angular velocities of the interceptor and target

$$\omega_{interceptor} = \sqrt{\frac{\mu}{R_{interceptor}^3}}$$

$$\omega_{target} = \sqrt{\frac{\mu}{R_{target}^3}}$$

4) Compute the lead angle

$$\alpha_{lead} = (\omega_{target})(TOF)$$

5) Solve for the final phase angle

$$\phi_{final} = \pi - \alpha_{lead}$$

6) Find the wait time

$$\text{Wait Time} = \frac{\phi_{final} - \phi_{initial}}{\omega_{target} - \omega_{interceptor}}$$

Analytical Solution

1) Compute the semimajor axis of the transfer orbit

$$a_{transfer} = \frac{R_{interceptor} + R_{target}}{2}$$

$$= \frac{6570 \text{ km} + 42{,}160 \text{ km}}{2}$$

$$a_{transfer} = 24{,}365 \text{ km}$$

2) Find the TOF of the transfer orbit

$$TOF = \pi\sqrt{\frac{a_{transfer}^3}{\mu}} = \pi\sqrt{\frac{(24{,}365 \text{ km})^3}{3.986 \times 10^5 \frac{\text{km}^3}{\text{s}^2}}}$$

$$TOF = 18{,}925 \text{ s} = 315 \text{ min } 25 \text{ s}$$

Example 6-3 Continued

3) Find the angular velocities of the interceptor and target

$$\omega_{interceptor} = \sqrt{\frac{\mu}{R_{interceptor}^3}} = \sqrt{\frac{3.986 \times 10^5 \frac{km^3}{s^2}}{(6570 \ km)^3}}$$

$$\omega_{interceptor} = 0.0012 \ rad/s$$

$$\omega_{target} = \sqrt{\frac{\mu}{R_{target}^3}} = \sqrt{\frac{3.986 \times 10^5 \frac{km^3}{s^2}}{(42,160 \ km)^3}}$$

$$\omega_{target} = 0.000073 \ rad/s$$

4) Compute the lead angle

$$\alpha_{lead} = (\omega_{target})(TOF)$$

$$= \left(0.000073 \frac{rad}{s}\right)(18,925 \ s)$$

$$\alpha_{lead} = 1.38 \ rad$$

5) Solve for the final phase angle

$$\phi_{final} = \pi - \alpha_{lead} = \pi - 1.38 \ rad$$

$$\phi_{final} = 1.76 \ rad$$

6) Find the wait time

$$wait \ time = \frac{\phi_{final} - \phi_{initial}}{\omega_{target} - \omega_{interceptor}}$$

$$wait \ time = \frac{1.76 \ rad - \pi}{0.000073 \frac{rad}{s} - 0.0012 \frac{rad}{s}}$$

$$wait \ time = 1225.9 \ s = 20.4 \ min$$

Interpreting the Results

From the initial separation of 180°, the intercepto must wait 20.4 minutes before starting the Hohman Transfer to rendezvous with the target.

References

Bate, Roger R., Donald D. Mueller, Jerry E. White. *Fundamentals of Astrodynamics.* New York, NY: Dover Publications, Inc., 1971.

Escobal, Pedro R. *Methods of Orbit Determination.* Malabar, FL: Krieger Publishing Company, Inc., 1976.

Kaplan, Marshall H. *Modern Spacecraft Dynamics and Control.* New York, NY: Wiley & Sons. 1976.

Vallado, David A. *Fundamentals of Astrodynamics and Applications.* New York, NY: McGraw-Hill Companies, Inc. 1997.

Mission Problems

6.1   Hohmann Transfers

1   What assumptions allow us to use a Hohmann Transfer?

2   What makes a Hohmann Transfer the most energy-efficient maneuver between coplanar orbits?

3   When going from a smaller circular orbit to a larger one, why do we speed up twice but end up with a slower velocity in the final orbit?

4   Why do we take the absolute value of the difference between the two orbital velocities when we compute total ΔV?

5   Suppose NASA wants to move a malfunctioning spacecraft from a circular orbit at 500 km altitude to one at 150 km altitude, so a Shuttle crew can repair it.

a) What is the energy of the transfer orbit?

b) What is the velocity change (ΔV_1) needed to go from the initial circular orbit into the transfer orbit?

c) What is the velocity change (ΔV_2) needed to go from the transfer orbit to the final circular orbit?

d) What is the time (TOF) required for the transfer?

6.2   Plane Changes

6   What orbital elements can a simple plane change alter?

7   For changing inclination only, where do we do the ΔV? Why?

8   Why do we prefer to use a combined plane change when going from a low-Earth parking orbit to a geostationary orbit rather than a Hohmann Transfer followed by a simple plane change?

9   Why does Case 3 in Table 6-1 (doing a combined plane change at perigee followed by ΔV_2 of the Hohmann Transfer) have a higher total ΔV than Case 2 (doing a Hohmann Transfer and then a simple plane change)?

10 A spacecraft deployed into a circular orbit, inclined 57° at 130 km altitude, needs to change to a polar orbit at the same altitude. What ΔV does this maneuver require?

11 Now that the spacecraft from Problem 10 is in a polar orbit, what ΔV will change the right ascension of the ascending node by 35°?

12 Suppose NASA wants to send a newly repaired spacecraft from its circular orbit at 150 km altitude (28° inclination) to a circular orbit at 20,000 km altitude (inclination of 45°).

a) What is the energy of the transfer orbit?

b) What is the velocity change (ΔV_1) needed to go from the initial circular orbit to the transfer orbit?

c) What is the combined plane change ΔV to go from the transfer orbit to the final circular orbit and change the inclination?

6.3   Rendezvous

13 Describe a rendezvous for an interceptor in a high orbit to a target spacecraft in a lower orbit. (Hint: draw a diagram and label the radii and angles)

14 Imagine you are in charge of a rescue mission. The spacecraft in distress is in a circular orbit at 240 km altitude. The Shuttle (rescue vehicle) is in a coplanar circular orbit at 120 km altitude. The Shuttle is 135° behind the target spacecraft.

a) What is the TOF of the Shuttle's transfer orbit to rendezvous with the target spacecraft?

b) What is the Shuttle's angular velocity? The target spacecraft's?

c) What is the lead angle?

d) What is the final phase angle?

e) How long must the Shuttle wait before starting the rendezvous maneuver?

15 In the above rescue mission, the Shuttle engines misfired, placing it in the same 240 km circular orbit as the target spacecraft, but 35° ahead of the target.

a) What is the TOF (and, therefore, the period) of the rendezvous phasing orbit?

b) What is the semimajor axis of the phasing orbit?

c) Compute the ΔV necessary for the Shuttle to move into the phasing orbit.

For Discussion

16 What extra steps must you add for a rendezvous between non-coplanar spacecrafts?

17 What types of space missions use rendezvous?

Mission Profile—Gemini

In December 1961, NASA let a contract to the McDonnell Corporation to build a "two-man spacecraft." This contract was the beginning of Project Gemini, the second U.S. human space program. NASA planners conceived it as an extension of the Mercury program to find solutions for many of the technical problems in a lunar mission. President Kennedy's goal of putting a human on the Moon by the end of the decade was a step closer to reality.

Mission Overview

The Gemini spacecraft carried two astronauts launched by a Titan 2 booster. Between April 1964 and November 1966, the program completed 10 launches with crews and two launches without crews. Major goals for the program included proving rendezvous and docking capabilities, extending the endurance of U.S. astronauts in space, and proving the ability to do extravehicular activity (EVA) or "spacewalking."

Mission Data

Gemini 1 and 2 were missions without crews to test the performance of the launch vehicle and spacecraft

Gemini 3 (Grissom, Young): First manual control of space maneuver and first manual re-entry

Gemini 4 (McDivitt, White): First U.S. citizen (White) to spacewalk. Eleven scientific experiments completed.

Gemini 5 (Conrad, Cooper): Seventeen scientific experiments completed

Gemini 6 (No crew): Failed at launch

Gemini 6A (Schirra, Stafford): Performed the first successful orbital rendezvous with Gemini 7

Gemini 7 (Borman, Lovell): Established an endurance record of 206 orbits in 330 hrs 36 mins, which was longer than any of the Apollo missions

Gemini 8 (Armstrong, Scott): Completed the first successful docking in space with an Atlas Agena upperstage. Failure of the spacecraft's attitude maneuvering system caused wild gyrations of the spacecraft–one of the worst emergencies of the program.

Gemini 9/9A (Cernan, Stafford): Failure of the target vehicle resulted in a delay and its redesignation as Gemini 9A two weeks later. Completed rendezvous with a new target but aborted docking because the docking apparatus had mechanical problems.

✓ Gemini 10 (Collins, Young): Rendezvoused and docked but used twice as much fuel as planned

✓ Gemini 11 (Conrad, Gordon): Docking achieved on first orbit

✓ Gemini 12 (Aldrin, Lovell): Conducted the first visual docking (due to a radar failure). More than five hours of EVA by Aldrin.

Astronaut Ed White makes the first U.S. space walk during the Gemini 4 mission in June, 1965. *(Courtesy of NASA/Johnson Space Center)*

Mission Impact

Gemini accomplished many "firsts" and showed human space-flight missions could overcome major problems. Despite some set-backs, the program succeeded beyond anyone's expectations and moved NASA toward more flexible operations.

For Discussion

- Do the lessons learned from Project Gemini affect how we work in space today?

- How might our current operations in space be different if we had not learned to walk in space or rendezvous and dock?

Contributor

Todd Lovell, the U.S. Air Force Academy

References

Baker, David. *The History of Manned Spaceflight*. New York: Crown, 1981.

Yenne, Bill. *The Encyclopedia of US Spacecraft*. New York, NY: Exeter, 1985.

Earth rise over the lunar horizon. *(Courtesy of NASA/Johnson Space Center)*

Interplanetary Travel

7

In This Chapter You'll Learn to...

- Describe the basic steps involved in getting from one planet in the solar system to another
- Determine the required velocity change, ΔV, needed for interplanetary transfer
- Explain how we can use the gravitational pull of planets to get "free" velocity changes, making interplanetary transfer faster and cheaper

You Should Already Know...

- ❏ Definition and use of coordinate systems (Chapter 4)
- ❏ Limits on the restricted two-body problem and its solution (Chapter 4)
- ❏ Definitions of specific mechanical energy for various conic sections (Chapter 4)
- ❏ How to use the Hohmann Transfer to get from one orbit to another (Chapter 6)
- ❏ Phasing for the rendezvous problem (Chapter 6)

Outline

7.1 Planning for Interplanetary Travel
Coordinate Systems
Equation of Motion
Simplifying Assumptions

7.2 The Patched-conic Approximation
Elliptical Hohmann Transfer between Planets—Problem 1
Hyperbolic Earth Departure—Problem 2
Hyperbolic Planetary Arrival—Problem 3
Transfer Time of Flight
Phasing of Planets for Rendezvous

7.3 Gravity-assist Trajectories

Greetings from the children of the planet Earth.

Anonymous greeting
placed on the Voyager spacecraft
in case it encounters aliens

Space Mission Architecture. This chapter deals with the Trajectories and Orbits segment of the Space Mission Architecture, introduced in Figure 1-20.

The wealth of information from interplanetary missions such as Pioneer, Voyager, and Magellan has given us insight into the history of the solar system and a better understanding of the basic mechanisms at work in Earth's atmosphere and geology. Our quest for knowledge throughout our solar system continues (Figure 7-1). Perhaps in the not-too-distant future, we'll undertake human missions back to the Moon, to Mars, and beyond.

How do we get from Earth to these exciting new worlds? That's the problem of interplanetary transfer. In Chapter 4 we laid the foundation for understanding orbits. In Chapter 6 we developed the Hohmann Transfer. Using this as a tool, we saw how to transfer between two orbits around the same body, such as Earth. The interplanetary transfer problem is really just an extension of the Hohmann Transfer. Only now, the central body is the Sun. In addition, as we'll see, we must be concerned with orbits around our departure and destination planets.

We'll begin by looking at the basic equation of motion for interplanetary transfer and then learn how we can greatly simplify the problem using a technique called the "patched-conic approximation." We'll see an example of how to use this simple method to plot a course from Earth to Mars. Finally, we'll look at gravity-assist or "slingshot" trajectories to see how we can use them for "free" ΔV, making interplanetary missions faster and cheaper.

Figure 7-1. Voyager Trajectory. Here we show an artist's concept of the Voyager spacecraft and their trajectories during their grand tours of the outer planets. *(Courtesy of NASA/Ames Research Center)*

7.1 Planning for Interplanetary Travel

▬ In This Section You'll Learn to...

- ☞ Describe the coordinate systems and equation of motion for interplanetary transfer
- ☞ Describe the basic concept of the patched-conic approximation and why we need it

To develop an understanding of interplanetary transfer, we start by dusting off our trusty Motion Analysis Process Checklist introduced in Chapter 4 and shown again in Figure 7-2. For our analysis, we'll deal with only the first three items on the checklist because looking at initial conditions, error analysis, and model testing would get far too involved for our simplified approach.

Coordinate Systems

Our first step in the Motion Analysis Process is to establish a coordinate system. When we developed the two-body equation of motion to analyze spacecraft motion around Earth in Chapter 4, two of our assumptions were

- There are only two bodies—the spacecraft and Earth
- Earth's gravitational pull is the only force acting on the spacecraft

So, for Earth-based problems, the Geocentric-equatorial frame is suitable. Once our spacecraft crosses a boundary into interplanetary space, however, Earth's gravitational pull becomes less significant and the Sun's pull becomes the dominant force. Therefore, because the Sun is central to interplanetary transfer, we must develop a sun-centered, or heliocentric coordinate system. By definition, *heliocentric* means the origin is the center of the Sun. In choosing a fundamental plane, we use the plane of Earth's orbit around the Sun, also known as the *ecliptic plane*. Next, because we need a principal direction, $\hat{I}$, fixed with respect to the universe, we bring the vernal equinox direction (♈) back for an encore performance. With the fundamental plane and principal direction chosen, we set the $\hat{J}$ axis in the ecliptic plane, 90° from the $\hat{I}$ axis in the direction of Earth's motion. Finally, the $\hat{K}$ axis is perpendicular to the ecliptic plane and it completes our right-handed system. Now we can relate any trajectory from Earth to another planet, or even to the edge of the solar system, to this *heliocentric-ecliptic coordinate system* defined in Figure 7-3.

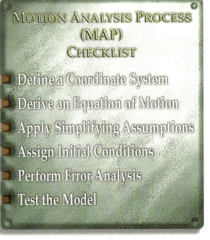

Figure 7-2. Motion Analysis Process Checklist (MAP). Apply the first three steps to learn about interplanetary travel.

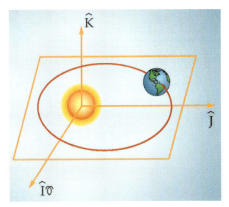

Figure 7-3. Heliocentric-ecliptic Coordinate System for Interplanetary Transfer. Origin— center of the Sun; fundamental plane—ecliptic plane (Earth's orbital plane around the Sun); principal direction—vernal equinox direction.

Equation of Motion

Now that we have a useful coordinate frame, the next step in the MAP checklist is to derive an equation to describe the motion of spacecraft around the Sun. We do this by returning to Newton's Second Law. First we must identify the forces a spacecraft will encounter while flying from Earth to another planet. As always, a spacecraft begins its mission under the influence of Earth's gravity, so that's the first force in our equation. When it gets far enough away from Earth, however, the Sun's gravitational pull begins to dominate. The Sun's gravity holds the spacecraft until it reaches the target planet, so we include that force in the equation. Finally, at journey's end, we must consider the gravitational attraction of the target planet. This attraction could range from Mercury's slight tug to Jupiter's immense pull, and we add it to the equation. As before, we can throw in "other" forces to cover anything we might have forgotten, such as solar pressure or pull from asteroids. When we consider all these forces, our equation of motion becomes pretty cumbersome

$$\sum \vec{F}_{external} = m\ddot{\vec{R}} = \vec{F}_{gravity\ Sun} + \vec{F}_{gravity\ Earth} + \vec{F}_{gravity\ target} + \vec{F}_{other} \quad (7\text{-}1)$$

Because this equation is so unwieldy, we have to make some simplifying assumptions to make our calculations more manageable.

Simplifying Assumptions

Thankfully, we can assume that the forces of gravity are much greater than all "other" forces acting on the spacecraft. This assumption leaves us with only the force of gravity, but gravity from three different sources!

$$\sum \vec{F}_{external} = m\ddot{\vec{R}} = \vec{F}_{gravity\ Sun} + \vec{F}_{gravity\ Earth} + \vec{F}_{gravity\ target} \quad (7\text{-}2)$$

Thus, as Figure 7-4, shows, we have a four-body problem—spacecraft, Earth, Sun, and target planet. Trying to solve for the spacecraft's motion under the influence of all these bodies could give us nightmares! Remember that gravity depends inversely on the distance (squared) from the central body to the spacecraft. To calculate all of these gravitational forces at once, we'd have to know the spacecraft's position and the positions of the planets as they orbit the Sun. This may not sound too tough, but the equation of motion becomes a highly non-linear, vector, differential equation that is very hard to solve. In fact, there's no closed-form solution to even a three-body problem, let alone one for four bodies. So how do we solve it? We use the old "divide and conquer" approach, taking one big problem and splitting it into three little ones. What kind of little problems can we solve? Two-body problems. For interplanetary transfers we call this approach the patched-conic approximation. The *patched-conic approximation* breaks the interplanetary trajectory into three separate regions and considers only the gravitational attraction on the spacecraft from one body in each region.

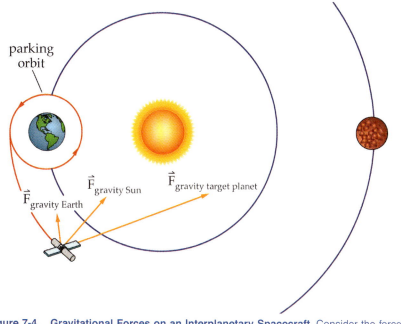

parking
orbit

$\vec{F}_{gravity\ Sun}$

$\vec{F}_{gravity\ target\ planet}$

$\vec{F}_{gravity\ Earth}$

Figure 7-4.   Gravitational Forces on an Interplanetary Spacecraft. Consider the forces on an interplanetary spacecraft as it makes its way from Earth to the target planet. We have the gravitational forces due to Earth, the Sun, and the target planet making it a four-body problem—Earth, Sun, target planet, and spacecraft.

By looking at the problem with respect to one attracting body at a time, we're back to our good-ol' two-body problem. Its equation of motion is

$$\ddot{\vec{R}} + \frac{\mu}{R^2}\hat{R} = 0 \qquad (7\text{-}3)$$

where
$\ddot{\vec{R}}$ = spacecraft's acceleration vector (km/s^2)
μ = gravitational parameter of the central body (km^3/s^2)
R = magnitude of the spacecraft's position vector (km)
$\hat{R}$ = unit vector in the $\vec{R}$ direction

As you may remember from Chapter 4, the solution to this equation describes a conic section (circle, ellipse, parabola, or hyperbola). Thus, the individual pieces of the spacecraft's trajectory are conic sections. By solving one two-body problem at a time, we "patch" one conic trajectory onto another, arriving at the patched-conic approximation. In the next section we'll see how all these pieces fit together.

■ Section Review

Key Terms

ecliptic plane
heliocentric
heliocentric-ecliptic
   coordinate system
patched-conic approximation

Key Concepts

➤ The coordinate system for Sun-centered or interplanetary transfers is the heliocentric-ecliptic system
  • The origin is the Sun's center
  • The fundamental plane is the ecliptic plane (Earth's orbital plane)
  • The principal direction ($\hat{I}$) is the vernal equinox direction

➤ Taken together, the interplanetary transfer problem involves four separate bodies
  • The spacecraft
  • Earth (or departure planet)
  • The Sun
  • The target or destination planet

➤ Because the four-body problem is difficult to solve, we split it into three, two-body problems using a method called the patched-conic approximation

7.2   The Patched-conic Approximation

▬ In This Section You'll Learn to...

- ☞ Describe how to solve interplanetary transfers with the patched-conic approximation
- ☞ Determine the velocity change (ΔV) needed to go from one planet to another
- ☞ Determine the time of flight for interplanetary transfer and discuss the problem of planetary alignment

As we introduced in the last section, the patched-conic approximation is a way of breaking the interplanetary trajectory into pieces (regions) we can handle, using methods we already know. By working within only one region at a time, we have to deal with the gravity from only one body at a time. In Figure 7-5 we see the three regions of the interplanetary transfer

- Region 1 (solved first)—*Sun-centered transfer from Earth to the target planet.* In this region, the Sun's gravitational pull dominates.

- Region 2 (solved second)—*Earth departure.* In this region, Earth's gravitational pull dominates.

- Region 3 (solved third)—*Arrival at the target planet.* In this region, the target planet's gravitational pull dominates.

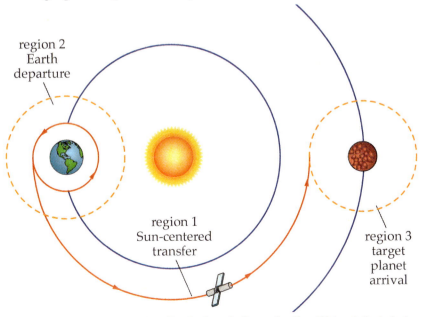

region 2
Earth
departure

region 1
Sun-centered
transfer

region 3
target
planet
arrival

Figure 7-5.   Three Regions of the Patched-conic Approximation. We break the trajectory for interplanetary transfer into three distinct regions in which the gravitational pull of only one body dominates the spacecraft's motion.

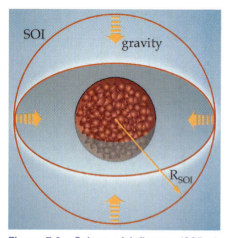

Figure 7-6. Sphere of Influence (SOI). A planet's SOI is the volume of space within which the planet's gravitational force dominates.

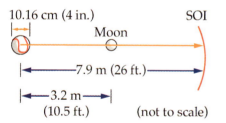

Figure 7-7. Earth's Sphere of Influence (SOI) Extends Well beyond the Orbit of the Moon. To put this in perspective, imagine if Earth were the size of a baseball; then the Moon would be 3.2 m (10.5 ft.) away and the SOI 7.9 m (26 ft.).

To deal with gravity from only one body at a time, we need to know how gravity operates in space. Any mass in space exerts a gravitational pull on other bodies. Newton's Law of Universal Gravitation describes this force as varying inversely with the square of the distance from the central body. Theoretically, a body's gravitational attraction reaches out to infinity, but practically, it's effective only within a certain volume of space called the body's *sphere of influence (SOI)*, as shown in Figure 7-6. For instance, within Earth's SOI, Earth's gravity dominates a spacecraft's motion. But at some point Earth's gravitational pull becomes insignificant and the pull of other bodies, such as the Moon and Sun, begins to dominate. The size of the SOI depends on the planet's mass (a more massive planet has a longer "gravitational reach") and how close the planet is to the Sun (the Sun's gravity overpowers the gravity of closer planets). To find the size of a planet's SOI, we use

$$R_{SOI} = a_{planet} \left(\frac{m_{planet}}{m_{Sun}} \right)^{\frac{2}{5}} \qquad (7\text{-}4)$$

where

R_{SOI} = radius of a planet's SOI (km)

a_{planet} = semimajor axis of the planet's orbit around the Sun (km)

m_{planet} = planet's mass (kg)

m_{Sun} = Sun's mass = 1.989×10^{30} kg

Earth's SOI is approximately 1,000,000 km in radius, well beyond the Moon's orbit but only a small fraction of the distance from Earth to the Sun (149.6 million km). To put this into perspective, imagine Earth being the size of a baseball, as in Figure 7-7. Its SOI would extend out 78 times its radius or 7.9 m (26 ft.). Appendix D.5 lists the sizes of the spheres of influence for other planets in the solar system.

To simplify the complex interactions between a spacecraft and the spheres of influence for the Earth, Sun, and target planet, we use the patched-conic approximation. By separately considering each of the regions, we set up three distinct two-body problems, solve them individually, and then "patch" them together to get a final solution. Our ultimate goal is to determine the total velocity change, ΔV_{total}, a spacecraft needs to leave Earth orbit and get into orbit around another planet. (In Chapter 14, we'll learn how to use this total ΔV requirement to determine the amount of rocket propellant needed for the trip.)

Let's use the patched-conic approach to analyze a down-to-Earth problem. Imagine you're driving along a straight section of highway at 45 m.p.h. Your friend is chasing you in another car going 55 m.p.h. A stationary observer on the side of the road sees the two cars moving at 45 m.p.h. and 55 m.p.h., respectively. But your friend's velocity with respect to you is only 10 m.p.h. (she's gaining on you at 10 m.p.h.) as illustrated in Figure 7-8.

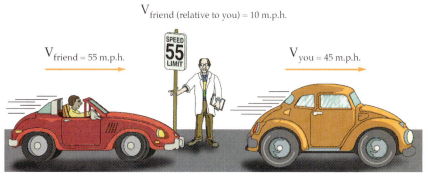

Figure 7-8.   Relative Velocity. From your perspective at 45 m.p.h., you see your friend at a speed of 55 m.p.h. gaining on you at a relative speed of 10 m.p.h.

Now suppose your friend throws a water balloon toward your car at 20 m.p.h. How fast is the balloon going? Well, that depends on the perspective. From your friend's perspective, it appears to move ahead of her car at 20 m.p.h. (ignoring air drag). From the viewpoint of the stationary observer on the side of the highway, your friend's car is going 55 m.p.h., and the balloon leaves her car going 75 m.p.h. What do you see? The balloon is moving toward you with a closing speed of 30 m.p.h. (10 m.p.h. closing speed for your friend's car plus 20 m.p.h. closing speed for the balloon, as shown in Figure 7-9.)

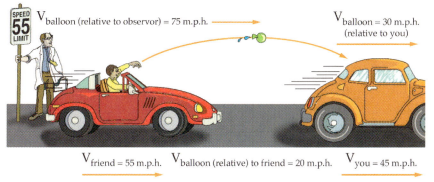

Figure 7-9.   Transfer from Car to Car. If your friend throws a water balloon at you at 20 m.p.h. (ignoring air drag) relative to your friend, it will be going 75 m.p.h. relative to a fixed observer and will appear to you to be gaining on you at 30 m.p.h.

By analyzing the balloon's motion, we see the three problems we use in a patched-conic approximation

- **Problem 1:** A stationary observer watches your friend throw a water balloon. The observer sees your friend's car going 55 m.p.h., your car going 45 m.p.h., and a balloon traveling from one car to the other at 75 m.p.h. The reference frame is a stationary frame at the side of the road. This problem is similar to Problem 1 of the patched-conic approximation (in region 1), where the Sun is similar to the observer and the balloon is similar to the spacecraft.

- **Problem 2:** The water balloon departs your friend's car with a relative speed of 20 m.p.h., as shown in Figure 7-9. The reference frame in this case is your friend's car. This problem relates to the patched-conic's Problem 2 (in region 2), where Earth is like your friend's car and the balloon is like the spacecraft.

- **Problem 3:** The water balloon lands in your car! It catches up to your car at a relative speed of 30 m.p.h. The reference frame is your car. This problem resembles the patched-conic's Problem 3 (in region 3), where the target planet is similar to your car and the balloon is still like the spacecraft.

Dividing interplanetary transfers into three problems requires us to keep track of velocities relative to a reference frame, which is different for each problem. In other words, the reference frame changes from one problem to the next. Thus, the spacecraft's velocity with respect to Earth isn't the same as its velocity with respect to the Sun. **This is a very important distinction to understand.** We find only one velocity that is common to Problems 1 and 2, and only one velocity that is common to Problems 1 and 3. This commonality allows us to "patch" the trajectories from the three regions together.

Elliptical Hohmann Transfer between Planets— Problem 1

The patched-conic approximation requires us to solve the Sun-centered problem first because the information from this solution allows us to solve the other two problems. We start by ignoring the Earth and target planet; then, we examine the interplanetary trajectory as though the spacecraft travels from Earth's orbit around the Sun to the target planet's orbit around the Sun. This part of an interplanetary trajectory is a Hohmann-transfer ellipse around the Sun (heliocentric). Because the Sun's gravity is the only force on the spacecraft, we use the heliocentric-ecliptic coordinate system for this part of the problem. Also, we assume the spacecraft starts and ends in a circular orbit. For most planets, this assumption is fine, because their orbital eccentricities are small.

For an interplanetary transfer, we follow nearly the same steps as for an Earth-centered Hohmann Transfer. For this problem, we assume the spacecraft has left Earth's sphere of influence (SOI) and is going off on its own independent orbit around the Sun, as shown in Figure 7-10. First, we find the spacecraft's initial velocity around the Sun. Because the spacecraft starts at Earth's radius from the Sun, its initial velocity with respect to the Sun is essentially the same as Earth's. (This isn't strictly true but it's close enough for a good approximation.) To find the spacecraft's initial velocity with respect to the Sun, we use the specific mechanical energy equation

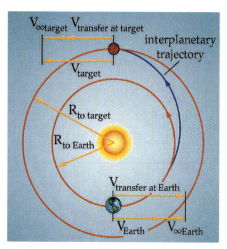

Figure 7-10.   Problem 1. To enter the heliocentric-elliptical transfer orbit, a spacecraft must have a velocity of $V_{transfer\ at\ Earth}$ relative to the Sun. To achieve this, it must change its current heliocentric velocity, V_{Earth}, by an amount, $V_{\infty\ Earth}$.

$$\varepsilon = \frac{V^2}{2} - \frac{\mu}{R}$$

(7-5)

where

ε = spacecraft's specific mechanical energy (km^2/s^2)

V = spacecraft's velocity (km/s)

μ = gravitational parameter of the central body (km^3/s^2)

$\quad$ = 1.327×10^{11} km^3/s^2 for our Sun

R = magnitude of the spacecraft's position vector (km)

and, in its alternate form (which we'll use soon), we have

$$\varepsilon = -\frac{\mu}{2a} \tag{7-6}$$

where

a = orbit's semimajor axis (km)

To find the specific mechanical energy the spacecraft would have if it stayed in the same orbit as Earth, just outside Earth's SOI, we use Equation (7-6) and Earth's major axis distance. Then we use this specific mechanical energy to determine its velocity at Earth's radius from the Sun, again, before it enters the Hohmann Transfer. We rearrange Equation (7-5) to get the relationship for orbital velocity

$$V = \sqrt{2\left(\frac{\mu}{R} + \varepsilon\right)} \tag{7-7}$$

We find the spacecraft's velocity around the Sun using Equation (7-7), being careful to use the correct quantity for each variable

$$V_{Earth} = \sqrt{2\left(\frac{\mu_{Sun}}{R_{to\ Earth}} + \varepsilon_{Earth}\right)} \tag{7-8}$$

where

V_{Earth} $\quad$ = Earth's orbital velocity with respect to the Sun (km/s)

μ_{Sun} $\quad$ = Sun's gravitational parameter = 1.327×10^{11} km^3/s^2

$R_{to\ Earth}$ = distance from the Sun to Earth (km)

$\quad$ = 1 astronomical unit (AU) (see Appendix B)

$\quad$ = 1.496×10^8 km (about 93 million statute miles)

ε_{Earth} $\quad$ = specific mechanical energy of Earth's orbit (km^2/s^2)

Notice we use μ of the Sun because we're referencing the spacecraft's motion to the Sun. V_{Earth} is not only Earth's velocity around the Sun, it's also a spacecraft's velocity with respect to the Sun while it's in orbit around Earth.

Next we find the velocity the spacecraft needs to enter the transfer ellipse. As before, we start with its specific mechanical energy

$$\varepsilon_{transfer} = -\frac{\mu_{Sun}}{2a_{transfer}} \tag{7-9}$$

231

where

$\varepsilon_{transfer}$ = spacecraft's specific mechanical energy in its heliocentric transfer orbit (km^2/s^2)

$a_{transfer}$ = semimajor axis of the transfer orbit (km)

We determine the semimajor axis ($a_{transfer}$) of the transfer orbit from

$$a_{transfer} = \frac{R_{to\ Earth} + R_{to\ target}}{2} \qquad (7\text{-}10)$$

where

$R_{to\ Earth}$ = radius from the Sun to Earth (km)

$R_{to\ target}$ = radius from the Sun to the target planet (km)

We use $R_{to\ Earth}$ and $R_{to\ target}$ because those radii mark the ends of the Hohmann Transfer ellipse, as shown in Figure 7-10. Then, we find the spacecraft's velocity on the transfer orbit at Earth's radius from the Sun by using

$$V_{transfer\ at\ Earth} = \sqrt{2\left(\frac{\mu_{Sun}}{R_{to\ Earth}} + \varepsilon_{transfer}\right)} \qquad (7\text{-}11)$$

where

$V_{transfer\ at\ Earth}$ = velocity the spacecraft needs at Earth's radius from the Sun to transfer to the target planet (km/s)

The difference between these two velocities, V_{Earth} and $V_{transfer\ at\ Earth}$, is the velocity relative to Earth which the spacecraft must have as it leaves Earth's SOI. For the patched-conic approximation, this velocity difference is the Earth-departure velocity, $V_{\infty\ Earth}$ or "V infinity at Earth." (Why "V infinity"? As we'll see in a bit, this is the spacecraft's velocity at an "infinite" distance from Earth.)

$$V_{\infty\ Earth} = |V_{transfer\ at\ Earth} - V_{Earth}| \qquad (7\text{-}12)$$

where

$V_{\infty\ Earth}$ = spacecraft's velocity "at infinity" with respect to Earth (km/s)

Let's review what all this means. The spacecraft, as it orbits Earth, goes around the Sun at V_{Earth} (the same as Earth's velocity with respect to the Sun). To enter a heliocentric transfer orbit to the target planet, the spacecraft needs to get from its orbit around Earth to a point beyond the SOI with enough velocity ($V_{transfer\ at\ Earth}$) with respect to the Sun. If the spacecraft leaves the SOI with $V_{\infty\ Earth}$, as calculated in Equation (7-12), it will have the correct velocity, as shown in Figure 7-11. We can relate this to the Hohmann Transfer by thinking of $V_{\infty\ Earth}$ as the ΔV_1 for the heliocentric transfer discussed in Chapter 6, even though no actual ΔV burn occurs here.

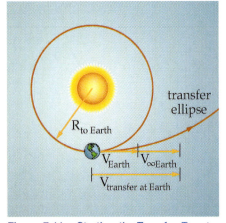

Figure 7-11.   Starting the Transfer. To enter the heliocentric transfer orbit, the spacecraft must change its velocity by an amount $V_{\infty\ Earth}$.

Now that we have $V_{\infty \; Earth}$, and we know it must add to V_{Earth} to equal $V_{transfer \; at \; Earth}$, we need to decide what direction our spacecraft must leave Earth's SOI. In Figure 7-11, $V_{\infty \; Earth}$ aligns nicely with V_{Earth}, so the vector addition creates $V_{transfer \; at \; Earth}$, which takes the spacecraft to an outer planet (further from the Sun than Earth is). To get the velocities to align so well, planners must ensure the spacecraft departs Earth's SOI ahead of Earth (aligned with Earth's velocity vector). If the $V_{\infty \; Earth}$ is not aligned with Earth's velocity vector, the $V_{transfer \; at \; Earth}$ won't be large enough, nor in the correct direction, to complete the Hohmann Transfer to the target planet.

Continuing with our heliocentric transfer, let's see what happens at the other end of the Sun-centered transfer, when the spacecraft approaches the target planet. Remember from the "big picture" of the Sun-centered Hohmann Transfer, the spacecraft coasts 180° around the Sun from Earth's SOI to the target planet's SOI. We can compute the spacecraft's velocity when it arrives in the target planet's region (region 3) from

$$V_{transfer \; at \; target} = \sqrt{2\left(\frac{\mu_{Sun}}{R_{to \; target}} + \varepsilon_{transfer}\right)} \qquad (7\text{-}13)$$

where

$V_{transfer \; at \; target}$ = spacecraft's velocity on the transfer orbit just outside the target planet's SOI (km/s)

μ_{Sun} = Sun's gravitational parameter (km^3/s^2)
= $1.327 \times 10^{11} \; km^3/s^2$

$R_{to \; target}$ = distance from the Sun to the target planet (km)

$\varepsilon_{transfer}$ = specific mechanical energy of the transfer orbit (km^2/s^2)

Notice here that the specific mechanical energy of the transfer ellipse remains constant from the first time we calculated it.

At the end of the Hohmann Transfer, the interplanetary spacecraft arrives at the target-planet's orbital radius with a velocity that is different from the target planet's circular velocity around the Sun. (The spacecraft is in an elliptical orbit that has a smaller semimajor axis than the target planet's.) Assuming the target planet arrives at the same time, a rendezvous, of sorts, occurs, where the spacecraft enters the target planet's SOI and is captured by the target planet's gravity. If the target planet isn't there at the same time, then the spacecraft misses the rendezvous, stays in its elliptical transfer orbit, and continues to orbit around the Sun. We'll assume we timed the transfer correctly, so the rendezvous occurs. To understand how the spacecraft arrives at the target planet, we must consider the velocities of the target planet and the spacecraft.

Let's look at the velocities first, then consider where on the SOI the spacecraft must arrive. We start with the target planet's specific mechanical energy

$$\varepsilon_{target} = -\frac{\mu_{Sun}}{2a_{target}}$$

(7-14)

where

ε_{target} = target planet's specific mechanical energy with respect to the Sun (km^2/s^2)

μ_{Sun} = Sun's gravitational parameter (km^3/s^2)
= $1.327 \times 10^{11} \, km^3/s^2$

a_{target} = target planet's semimajor axis (km)

Armed with the target planet's specific mechanical energy, we calculate its velocity with respect to the Sun, using

$$V_{target} = \sqrt{2\left(\frac{\mu_{Sun}}{R_{to \, target}} + \varepsilon_{target}\right)}$$

(7-15)

where

V_{target} = target planet's velocity around the Sun (km/s)

μ_{Sun} = Sun's gravitational parameter (km^3/s^2)

$R_{to \, target}$ = distance from the Sun to the target planet (km)

ε_{target} = target planet's specific mechanical energy (km^2/s^2)

We now know the heliocentric velocity the spacecraft has and the target planet's velocity. All that remains is to determine the difference between the two, which we call "V infinity target", $V_{\infty \, target}$.

$$V_{\infty \, target} = |V_{transfer \, at \, target} - V_{target}|$$

(7-16)

where

$V_{\infty \, target}$ = spacecraft's velocity "at infinity" with respect to the target planet (km/s)

$V_{transfer \, at \, target}$ = spacecraft's velocity on the transfer orbit just outside the target planet's SOI (km/s)

V_{target} = target planet's velocity around the Sun (km/s)

Using our Hohmann Transfer experience again, we can think of $V_{\infty \, target}$ as ΔV_2. Keep in mind, however, no rocket engine burn actually takes place here. If we take the perspective of an observer standing on the Sun, we see the spacecraft arriving at the target planet's radius with $V_{transfer \, at \, target}$ and the target planet moving with V_{target} with respect to the Sun. The difference, $V_{\infty \, target}$, is the spacecraft's velocity as it enters the target planet's SOI, as shown in Figure 7-12.

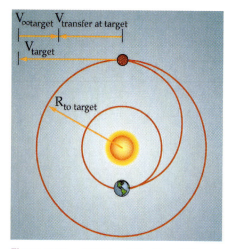

Figure 7-12.   Arriving at the Target Planet. From the Sun-centered perspective, the planet is traveling at V_{target} and the spacecraft at $V_{transfer \, at \, target}$. The difference, $V_{\infty \, target}$, is the speed with which the spacecraft enters the SOI.

Similar to the Earth-departure situation, the $V_{\text{transfer at target}}$ must align with the V_{target}, so that the $V_{\infty \text{ target}}$ moves the spacecraft correctly into the target planet's SOI and ultimately to a safe parking orbit. To do this for an outer-planet arrival, planners must ensure the spacecraft arrives ahead of the target planet's SOI, so the target planet can catch up to it and capture it. If the $V_{\infty \text{ target}}$ isn't aligned correctly, the spacecraft may travel directly toward the planet's surface (which some impacting probes do on purpose), or enter a hyperbolic arrival trajectory that misses its planned parking orbit.

For the other case (target planet is closer to the Sun than Earth is), the spacecraft must arrive behind the target planet's SOI, because it's velocity is higher than the planet's. The spacecraft then overtakes the planet and enters the SOI with $V_{\infty \text{ target}}$, and the planet's gravity pulls it in.

How and when do we actually fire our rockets to achieve $V_{\infty \text{ Earth}}$ and $V_{\infty \text{ target}}$? To find out, we need to examine the other two problems in the patched-conic approximation.

Astro Fun Fact
Pluto: A Planet or Not?

When Clyde Tombaugh discovered Pluto in 1930, it became the ninth known planet in our solar system. Admittedly, from February 7, 1979, to February 11, 1999, it was actually closer to the Sun than Neptune, but it is again further from the Sun than Neptune and will be until around 2219.

However, members of the International Astronomical Union (IAU) in January of 1999 considered assigning a minor planet number to Pluto. Due to the eccentric orbit, high inclination, and small size, the proposal considered classifying it as a Trans-Neptunian Object. Many people felt that this meant Pluto would be demoted to a lesser status. Following much debate and a barrage of e-mail, the IAU decided not to assign Pluto a minor planet number.

(Courtesy of the Association of Universities for Research in Astronomy, Inc./Space Telescope Science Institute)

CNN Interactive. "It's Official: No 'Demotion' for Planet Pluto." 3 February 1999.

Contributed by Scott R. Dahlke, the U.S. Air Force Academy

Example 7-1 (Part 1)

Problem Statement

The Jet Propulsion Lab (JPL) wants to send a probe from Earth ($R_{\text{to Earth}} = 1.496 \times 10^8$ km) to Mars ($R_{\text{to Mars}} = 2.278 \times 10^8$ km) to map landing sites for future manned missions. The probe will leave Earth from a parking orbit of 6697 km and arrive at Mars in another parking orbit of 3580 km.

- Part 1: What is the "extra" velocity the spacecraft needs to leave Earth ($V_{\infty\,\text{Earth}}$) and that it has at Mars ($V_{\infty\,\text{Mars}}$)?

- Part 2: What ΔV does it need in a parking orbit around Earth to begin the transfer?

- Part 3: What ΔV does it need to inject into a Mars parking orbit and what is the total mission ΔV?

Problem Summary—Part 1

Given: $R_{\text{to Earth}} = 1.496 \times 10^8$ km
  $R_{\text{to Mars}} = 2.278 \times 10^8$ km
  $R_{\text{park at Earth}} = 6697$ km
  $R_{\text{park at Mars}} = 3580$ km

Find: $V_{\infty\,\text{Earth}}$, $V_{\infty\,\text{Mars}}$ (Part 1)

Problem Diagram

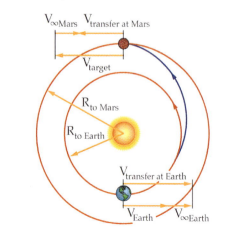

Conceptual Solution

Elliptical Hohmann Transfer—Problem 1

1) Find the semimajor axis of the transfer orbit, a_{transfer}

2) Find the energy of the transfer orbit, $\varepsilon_{\text{transfer}}$

3) Find the velocity of Earth around the Sun, V_{Earth}

4) Find the velocity in transfer orbit at Earth, $V_{\text{transfer at Earth}}$

5) Find the velocity at infinity near Earth, $V_{\infty\,\text{Earth}}$

6) Find the velocity of Mars around the Sun, V_{Mars}

7) Find the velocity in transfer orbit at Mars, $V_{\text{transfer at Mars}}$

8) Find the velocity at infinity near Mars, $V_{\infty\,\text{Mars}}$

Analytical Solution

1) Find a_{transfer}

$$a_{\text{transfer}} = \frac{R_{\text{to Earth}} + R_{\text{to Mars}}}{2}$$

$$= \frac{1.496 \times 10^8 \text{ km} + 2.278 \times 10^8 \text{ km}}{2}$$

$$a_{\text{transfer}} = 1.887 \times 10^8 \text{ km}$$

2) Find $\varepsilon_{\text{transfer}}$

$$\varepsilon_{\text{transfer}} = -\frac{\mu_{\text{Sun}}}{2a_{\text{transfer}}} = -\frac{1.327 \times 10^{11} \dfrac{\text{km}^3}{\text{s}^2}}{2(1.887 \times 10^8 \text{ km})}$$

$$\varepsilon_{\text{transfer}} = -351.6 \frac{\text{km}^2}{\text{s}^2}$$

(*Note:* negative energy because the transfer orbit is elliptical)

Example 7-1 (Part 1) Continued

) Find the velocity of Earth around the Sun, V_{Earth}

$$\varepsilon_{Earth} = -\frac{\mu_{Sun}}{2a_{Earth}} = -\frac{\mu_{Sun}}{2R_{to\,Earth}}$$

$$= -\frac{1.327 \times 10^{11} \dfrac{km^3}{s^2}}{2(1.496 \times 10^8 \text{ km})}$$

$$= -443.5 \text{ km}^2/s^2$$

(*Note:* Negative energy because Earth is in a circular orbit around the Sun)

$$V_{Earth} = \sqrt{2\left(\frac{\mu_{Sun}}{R_{to\,Earth}} + \varepsilon_{Earth}\right)}$$

$$= \sqrt{2\left(\frac{1.327 \times 10^{11} \dfrac{km^3}{s^2}}{1.496 \times 10^8 \text{ km}} - 443.5\frac{km^2}{s^2}\right)}$$

$$= 29.78 \text{ km}/s$$

) Find $V_{transfer\ at\ Earth}$

$$V_{transfer\,at\,Earth} = \sqrt{2\left(\frac{\mu_{Sun}}{R_{to\,Earth}} + \varepsilon_{transfer}\right)}$$

$$= \sqrt{2\left(\frac{1.327 \times 10^{11} \dfrac{km^3}{s^2}}{1.496 \times 10^8 \text{ km}} - 351.6\frac{km^3}{s^2}\right)}$$

$$= 32.72 \text{ km}/s$$

) Find $V_{\infty\,Earth}$

$$V_{\infty\,Earth} = |V_{transfer\,at\,Earth} - V_{Earth}|$$

$$= |32.72 \text{ km}/s - 29.78 \text{ km}/s|$$

$$= 2.94 \text{ km}/s$$

) Find the velocity of Mars around the Sun, V_{Mars}

$$\varepsilon_{Mars} = -\frac{\mu_{Sun}}{2a_{Mars}} = -\frac{\mu_{Sun}}{2R_{to\,Mars}}$$

$$= -\frac{1.327 \times 10^{11} \dfrac{km^3}{s^2}}{2(2.278 \times 10^8 \text{ km})}$$

$$= -291.3 \text{ km}^2/s^2$$

(*Note:* Negative energy)

$$V_{Mars} = \sqrt{2\left(\frac{\mu_{Sun}}{R_{to\,Mars}} + \varepsilon_{Mars}\right)}$$

$$= \sqrt{2\left(\frac{1.327 \times 10^{11} \dfrac{km^3}{s^2}}{2.278 \times 10^8 \text{ km}} - 291.3\frac{km^2}{s^2}\right)}$$

$$= 24.14 \text{ km}/s$$

7) Find $V_{transfer\ at\ Mars}$

$$V_{transfer\,at\,Mars} = \sqrt{2\left(\frac{\mu_{Sun}}{R_{to\,Mars}} + \varepsilon_{transfer}\right)}$$

$$= \sqrt{2\left(\frac{1.327 \times 10^{11} \dfrac{km^3}{s^2}}{2.278 \times 10^8 km} - 351.6\frac{km^2}{s^2}\right)}$$

$$= 21.49 \text{ km}/s$$

8) Find $V_{\infty\,Mars}$

$$V_{\infty\,Mars} = |V_{Mars} - V_{transfer\,at\,Mars}|$$

$$= \left|24.14\frac{km}{s} - 21.49\frac{km}{s}\right|$$

$$= 2.65 \text{ km}/s$$

Interpreting the Results

To leave Earth and enter an interplanetary trajectory to Mars, our probe needs to gain 2.94 km/s with respect to the Sun. The probe arrives at Mars' orbit going 2.65 km/s slower than Mars, so it enters the Mars SOI with a speed of 2.65 km/s. We'll see how to achieve these velocities in Parts 2 and 3.

Hyperbolic Earth Departure—Problem 2

Remember, we broke the interplanetary problem into three problems based on which attracting body (Sun, Earth, or target planet) was the major player in the spacecraft's trajectory. Problem 1 shows how to get a spacecraft from Earth's orbit to the target planet's orbit on a Sun-centered transfer ellipse. In Problem 2, we now back up to see how it gets from the Earth-centered trajectory to the Sun-centered one.

For this part of an interplanetary transfer, we assume Earth's gravity is the only force on the spacecraft and use the geocentric-equatorial coordinate system to describe its motion. From Problem 1, we know we want our spacecraft to leave Earth's sphere of influence (SOI) with some velocity we called $V_{\infty \text{ Earth}}$. To escape Earth's gravity, our spacecraft must be on a parabolic or hyperbolic trajectory (circular and elliptical orbits don't escape). But the parabolic trajectory wouldn't take it out of Earth's gravity; just to the SOI boundary (where R is almost 1,000,000 km), where it would have zero velocity relative to Earth when it got there.

$$V_{\text{relative to Earth}} = \sqrt{2\left(\frac{\mu}{R_{\text{"at infinity"}}} + \varepsilon\right)} = 0 \text{ (for a parabola)}$$

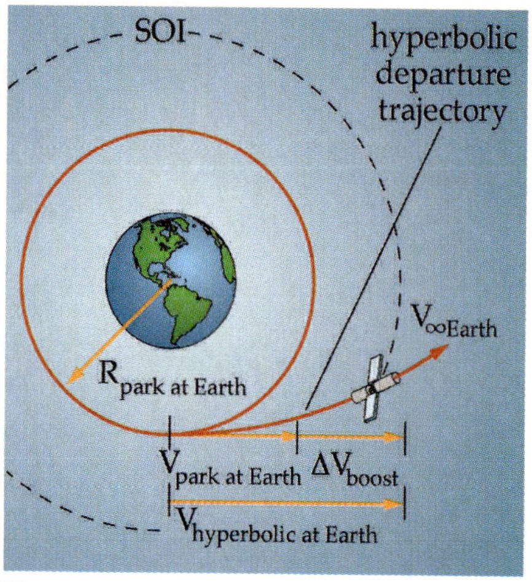

This parabolic trajectory would place our spacecraft in an orbit around the Sun exactly like Earth's (traveling 29.78 km/s), right on the SOI boundary. Relative to Earth, it would be stationary, so it wouldn't actually go anywhere. To illustrate this, imagine you're driving along the interstate at 55 m.p.h. If your friend pulls in front of you and sets his speed at 55 m.p.h., the relative velocity between the two cars is zero. This would be the case of the spacecraft's velocity relative to Earth at the SOI on a parabolic trajectory. Realize, of course, that a parabolic trajectory is a "special" case that we can't really achieve because of all the other forces (which we assumed away for the two-body problem) acting on our spacecraft. However, by understanding this special case, we should better understand the trajectory we really need–a hyperbolic trajectory.

If we put our spacecraft on the proper hyperbolic-departure trajectory when it leaves its low-Earth parking orbit, it will coast to the SOI boundary, arriving with the correct velocity, $V_{\infty \text{ Earth}}$. This is the velocity we computed in Problem 1 that takes the spacecraft on the heliocentric transfer to the target planet's orbital radius. We call $V_{\infty \text{ Earth}}$ the *hyperbolic excess velocity*, because our spacecraft leaves Earth's SOI with some "extra" velocity, as shown in Figure 7-13, the extra amount it needs to start its trip to the target planet.

Now that we know the spacecraft's velocity at the end of the Earth-centered hyperbolic-departure trajectory, we can work our way back to its velocity as it leaves the initial low-Earth parking orbit. For technical, as well as operational reasons, an interplanetary probe seldom launches directly into its transfer orbit from the launch pad. Instead, the launch vehicle first puts it into a circular, parking orbit close to Earth. This allows ground controllers time to check all the systems and wait for the right moment to ignite the upperstage rocket.

Figure 7-13. Escaping Earth. To escape Earth on a hyperbolic trajectory and arrive at the SOI with the required velocity to enter into the heliocentric transfer orbit, $V_{\infty \text{ Earth}}$, a spacecraft needs to increase its velocity in the parking orbit, $V_{\text{park at Earth}}$, by an amount ΔV_{boost}.

To determine our spacecraft's velocity on the hyperbolic-departure trajectory at the parking-orbit's radius, we must realize it has the same specific mechanical energy, ε, that it does at the SOI boundary. Remember, specific mechanical energy is a constant in the two-body problem. Knowing the excess velocity, $V_{\infty\,Earth}$, we calculate ε, using

$$\varepsilon_{\infty\,Earth} = \frac{V^2_{\infty\,Earth}}{2} - \frac{\mu_{Earth}}{R_{\infty\,Earth}} \qquad (7\text{-}17)$$

where

$\varepsilon_{\infty\,Earth}$ = specific mechanical energy on the hyperbolic-departure
    trajectory (km^2/s^2)

$V_{\infty\,Earth}$ = spacecraft's velocity at the SOI relative to Earth (km/s)

μ_{Earth} = Earth's gravitational parameter = $3.986 \times 10^5\ km^3/s^2$

$R_{\infty\,Earth}$ = "infinite" distance to the SOI from Earth's center (km)

Because the distance to the SOI is so big $(R_{\infty\,Earth} \approx \infty)$, the term for potential energy effectively is zero. Thus, the spacecraft's energy at the SOI becomes

$$\varepsilon_{\infty\,Earth} = \frac{V^2_{\infty\,Earth}}{2} \qquad (7\text{-}18)$$

(*Note:* This energy is positive, which, as we learned in Chapter 4, means the trajectory is hyperbolic).

We can now use this relationship to find the velocity the spacecraft must achieve at the parking-orbit's radius, R_{park}, to enter the hyperbolic-departure trajectory. Rearranging the specific mechanical energy equation for velocity we get

$$V_{hyperbolic\ at\ Earth} = \sqrt{2\left(\frac{\mu_{Earth}}{R_{park\ at\ Earth}} + \varepsilon_{\infty\,Earth}\right)} \qquad (7\text{-}19)$$

where

$V_{hyperbolic\ at\ Earth}$ = spacecraft's velocity on the hyperbolic-
    departure trajectory at the parking-orbit's
    radius (km/s)

μ_{Earth} = Earth's gravitational parameter
            = $3.986 \times 10^5\ km^3/s^2$

$R_{park\ at\ Earth}$ = parking orbit's radius (km)

Now we have the spacecraft's velocities at both ends of the hyperbolic-departure trajectory (at the SOI and at the parking orbit's radius). Next, we must find its velocity in the parking orbit before it accelerates onto the hyperbolic-departure trajectory, so that we know how much velocity change it needs. In its circular parking orbit, the spacecraft's velocity with respect to Earth is

$$V_{\text{park at Earth}} = \sqrt{\frac{\mu_{\text{Earth}}}{R_{\text{park at Earth}}}} \qquad (7\text{-}20)$$

where

$V_{\text{park at Earth}}$ = spacecraft's velocity in its parking orbit near Earth (km/s)

Finally, we have the spacecraft's velocity in its parking orbit and the velocity it needs at the parking orbit's radius to enter the hyperbolic-departure trajectory. The difference gives us the velocity change, ΔV_{boost}, that the upperstage rocket must provide to the spacecraft

$$\Delta V_{\text{boost}} = \left| V_{\text{hyperbolic at Earth}} - V_{\text{park at Earth}} \right| \qquad (7\text{-}21)$$

where

ΔV_{boost} = spacecraft's velocity change to go from its parking orbit around Earth onto its hyperbolic-departure trajectory (km/s)

$V_{\text{hyperbolic at Earth}}$ = spacecraft's velocity on its hyperbolic-departure trajectory at the parking-orbit's radius (km/s)

$V_{\text{park at Earth}}$ = spacecraft's velocity in its parking orbit around Earth (km/s)

ΔV_{boost} is the velocity change the spacecraft must generate to start its interplanetary journey. An attached upperstage—with a rocket engine, fuel tanks, and guidance system—normally provides this ΔV_{boost}. Once it applies the ΔV_{boost}, the spacecraft is on its way to the target planet!

Figure 7-13 shows the hyperbolic departure trajectory that the spacecraft must follow to depart Earth's SOI properly aligned with Earth's velocity vector. To get onto that trajectory, the spacecraft must do its ΔV_{boost} where the hyperbolic departure trajectory is tangent to the parking orbit. Doing the ΔV_{boost} at any other point in the parking orbit puts it on a hyperbolic trajectory that won't align with Earth's velocity vector, when it gets to the edge of the SOI. Any misalignment at the edge of Earth's SOI means a large error in the elliptical Hohmann Transfer, and probably a large miss distance at the target planet.

Let's recap how we "patch" Problems 1 and 2 together. We start the spacecraft in a circular parking orbit around Earth at a radius $R_{\text{park at Earth}}$ and a velocity $V_{\text{park at Earth}}$. We then fire the upperstage's rocket engines to increase the spacecraft's velocity by an amount, ΔV_{boost}, to give it a velocity, $V_{\text{hyperbolic at Earth}}$. This velocity puts it on a hyperbolic-departure trajectory away from Earth. Upon arrival at the SOI, the spacecraft has the necessary velocity, $V_{\infty \text{ Earth}}$, to escape Earth's gravity and enter a heliocentric-elliptical, transfer orbit. By design, our geocentric, hyperbolic trajectory blends smoothly into the heliocentric, elliptical orbit, so the two are now "patched" together. Notice that only one rocket-engine firing

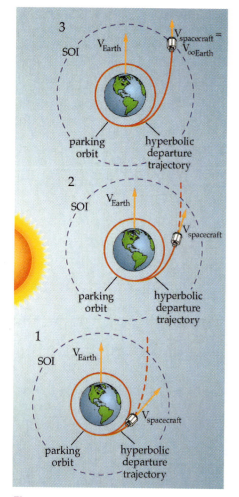

Figure 7-14. Hyperbolic Departure Trajectory. This sequence shows how a spacecraft departs the "front" edge of Earth's SOI to travel to a planet further from the Sun than Earth.

puts the spacecraft out of Earth's SOI and onto the Sun-centered transfer ellipse. Example 7-1 (Part 2) shows how to determine the ΔV_{boost}.

The last problem in the interplanetary transfer is the target-planet arrival, which looks a lot like the Earth-departure problem in reverse order.

Astro Fun Fact
Lagrange Points

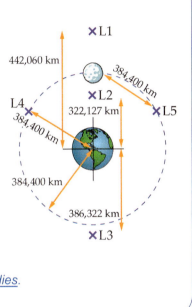

Five points (L1–L5) near the Earth and Moon are within both bodies' influence and revolve around Earth at the same rate as the Moon. French scientist Joseph Lagrange discovered these curious points in 1764 while attempting to solve the complex three-body problem. In his honor these points of equilibrium are now known as Lagrange Libration Points (LLP). LLPs exist for any multi-body system. The LLPs for the Earth-Moon system are shown at the right. These points would be an ideal location for future space stations or Lunar docking platforms, because they always keep the same relative position with respect to the Earth and Moon. In fact, in the Earth-Sun system, the Solar and Heliospheric Observatory launched in December, 1995, remains at the L2 libration point between the Earth and Sun.

Cousins, Frank W. The Solar System. New York, NY: Pica Press, 1972.

Szebehely, Victor. Theory of Orbits: The Restricted Problem of Three Bodies. Yale University, New Haven, CT: Academic Press, Inc., 1967.

Example 7-1 (Part 2)

Problem Summary—Part 2

Given: $R_{\text{to Earth}} = 1.496 \times 10^8$ km

$R_{\text{to Mars}} = 2.278 \times 10^8$ km

$R_{\text{park at Earth}} = 6697$ km

$R_{\text{park at Mars}} = 3580$ km

$V_{\infty\,\text{Earth}} = 2.94$ km/s (from Part 1)

Find: ΔV_{boost} (Part 2)

Problem Diagram

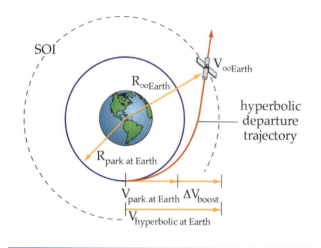

Conceptual Solution

Hyperbolic Earth Departure—Problem 2

1) Find the spacecraft's energy on its hyperbolic-escape trajectory, $\varepsilon_{\infty\,\text{Earth}}$

2) Find the spacecraft's velocity in the circular parking orbit around Earth, $V_{\text{park at Earth}}$

3) Find the spacecraft's velocity, $V_{\text{hyperbolic at Earth}}$, on the hyperbolic-escape trajectory at the parking orbit radius

4) Find the velocity change the spacecraft needs to enter the hyperbolic-escape trajectory, ΔV_{boost}

Analytical Solution

1) Find the spacecraft's energy on the hyperbolic-escape trajectory. Energy is the same everywhere on the trajectory, so we find it at the SOI, using information from Part 1.

$$\varepsilon_{\infty\,\text{Earth}} = \frac{V_{\infty\,\text{Earth}}^2}{2} = \frac{\left(2.94\frac{\text{km}}{\text{s}}\right)^2}{2} = 4.323\frac{\text{km}^2}{\text{s}^2}$$

(*Note:* positive energy on a hyperbolic trajectory)

2) Find $V_{\text{park at Earth}}$

$$V_{\text{park at Earth}} = \sqrt{\frac{\mu_{\text{Earth}}}{R_{\text{park at Earth}}}}$$

$$= \sqrt{\frac{3.986 \times 10^5 \frac{\text{km}^3}{\text{s}^2}}{6697 \text{ km}}}$$

$$= 7.71 \text{ km/s}$$

3) Find $V_{\text{hyperbolic at Earth}}$

$$V_{\text{hyperbolic at Earth}} = \sqrt{2\left(\frac{\mu_{\text{Earth}}}{R_{\text{park at Earth}}} + \varepsilon_{\infty\,\text{Earth}}\right)}$$

$$= \sqrt{2\left(\frac{3.986 \times 10^5 \frac{\text{km}^3}{\text{s}^2}}{6697\,\text{km}} + 4.323\frac{\text{km}^2}{\text{s}^2}\right)}$$

$$= 11.30 \text{ km/s}$$

4) Find ΔV_{boost}

$$\Delta V_{\text{boost}} = \left| V_{\text{hyperbolic at Earth}} - V_{\text{park at Earth}} \right|$$

$$= \left| 11.30\frac{\text{km}}{\text{s}} - 7.71\frac{\text{km}}{\text{s}} \right|$$

$$= 3.59 \text{ km/s}$$

Interpreting the Results

From the spacecraft's circular parking orbit around Earth, we must fire its upperstage engines to increase its velocity by 3.59 km/s, so it can enter a hyperbolic-departure trajectory, which starts it on its way to Mars.

Hyperbolic Planetary Arrival—Problem 3

At the other end of the spacecraft's heliocentric transfer ellipse, it arrives at the target planet's radius from the Sun. If we time the transfer ellipse correctly, the target planet is there to capture our spacecraft. If the target planet is farther from the Sun than Earth is, then the spacecraft arrives at apogee of its transfer ellipse ahead of, and moving slower than, the target planet. So, the target planet overtakes the spacecraft and the spacecraft enters the target planet's sphere of influence (SOI) in front of the planet. If the target planet is closer to the Sun than Earth is, then the spacecraft arrives at perigee of its transfer ellipse, behind, and moving faster than, the target planet. The spacecraft overtakes the planet in this case and enters the SOI from behind the planet.

To solve this third and final problem of the patched-conic approximation, we assume the target-planet's gravity is the only force on the spacecraft and use a coordinate frame similar to the geocentric-equatorial system (but centered at the target planet) to describe the spacecraft's motion. For this arrival problem, we need to find the spacecraft's velocity with respect to the planet, as it enters the SOI. The spacecraft's velocity on the heliocentric transfer ellipse is $V_{\text{transfer at target}}$, and the target planet's velocity with respect to the Sun is V_{target}. The difference is $V_{\infty \text{ target}}$, which is the spacecraft's velocity as it enters the SOI

$$V_{\infty \text{target}} = |V_{\text{transfer at target}} - V_{\text{target}}|$$
(7-22)

where

$V_{\infty \text{ target}}$ = spacecraft's velocity at the SOI with respect to the target planet (km/s)

$V_{\text{transfer at target}}$ = spacecraft's velocity in its heliocentric transfer orbit with respect to the Sun at the target planet (km/s)

V_{target} = target-planet's velocity with respect to the Sun (km/s)

The spacecraft's velocity, $V_{\infty \text{ target}}$, occurs at another boundary in the patched conic and represents its excess hyperbolic velocity at the target planet's SOI. Just as it left Earth on a hyperbolic trajectory, it arrives at the target planet on a hyperbolic trajectory, as shown in Figure 7-15. Applying the same energy technique, we can find the spacecraft's energy at an "infinite" radius from the target planet

$$\varepsilon_{\infty \text{ target}} = \frac{V_{\infty \text{ target}}^2}{2}$$
(7-23)

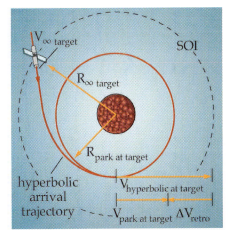

Figure 7-15.   Arriving at the Target Planet. If the spacecraft does nothing as it approaches the target planet, it will swing by on a hyperbolic trajectory and depart the SOI on the other side. To slow down enough to be captured into orbit at a radius R_{park}, it must change its velocity by an amount ΔV_{retro}.

where

$\varepsilon_{\infty \, target}$ = spacecraft's specific mechanical energy on its hyperbolic-arrival trajectory (km^2/s^2)

$V_{\infty \, target}$ = spacecraft's velocity at the SOI with respect to the target planet (km/s)

In a reflection of the way it left Earth, the spacecraft coasts on the hyperbolic-arrival trajectory and then performs a ΔV or "burn" at an assigned radius from the target planet ($R_{park \, at \, target}$). This maneuver moves it into a circular parking orbit around the target planet. Solving for this velocity

$$V_{hyperbolic \, at \, target} = \sqrt{2 \left(\frac{\mu_{target}}{R_{park \, at \, target}} + \varepsilon_{\infty \, target} \right)} \qquad (7\text{-}24)$$

where

$V_{hyperbolic \, at \, target}$ = spacecraft's velocity when it reaches the parking orbit altitude (km/s)

μ_{target} = target-planet's gravitational parameter (km^3/s^2)

$R_{park \, at \, target}$ = radius of the parking orbit from the target planet (km)

$\varepsilon_{\infty \, target}$ = spacecraft's specific mechanical energy on its hyperbolic-arrival trajectory (km^2/s^2)

If we didn't change the spacecraft's velocity, it'd speed around the planet and out into space on the other leg of the hyperbolic trajectory. To avoid this, it does a ΔV_{retro} to enter a circular parking orbit at the assigned radius. To compute how large the velocity change must be, we find the parking orbit velocity and subtract it from the spacecraft's hyperbolic arrival velocity at the parking orbit's radius.

$$V_{park \, at \, target} = \sqrt{\frac{\mu_{target}}{R_{park \, at \, target}}} \qquad (7\text{-}25)$$

where

$V_{park \, at \, target}$ = spacecraft's velocity in its parking orbit around the target planet (km/s)

The velocity change, ΔV_{retro}, to enter the parking orbit is

$$\Delta V_{retro} = |V_{park \, at \, target} - V_{hyperbolic \, at \, target}| \qquad (7\text{-}26)$$

where

ΔV_{retro} = spacecraft's velocity change required to go from its hyperbolic-arrival trajectory to its parking orbit around the target planet (km/s)

$V_{hyperbolic \, at \, target}$ = spacecraft's velocity on its hyperbolic-arrival trajectory at the parking-orbit's radius (km/s)

Figu ... hat the spacecraft
must f ... aligned to finally
descei ...). As it follows the
traject ... gain speed, which
it mu ... ere the hyperbolic
trajec ... ΔV$_{retro}$ at any other
poin ... it in an unplanned
orbit ... l for the mission.
N ... V$_{boost}$, to accelerate
our ... Earth, and ΔV$_{retro}$, to
decel ... arget planet. The total
velocity ... vide for the mission is
then

$$\Delta V_{mission} = \Delta V_{boost} + \Delta V_{retro} \qquad (7\text{-}27)$$

where

$\Delta V_{mission}$ = total velocity change required for the mission (km/s)

ΔV_{boost} = spacecraft's velocity change required to go from its parking orbit around Earth onto a hyperbolic-departure trajectory (km/s)

ΔV_{retro} = spacecraft's velocity change required to go from its hyperbolic-arrival trajectory to its parking orbit around the target planet (km/s)

The propulsion system on the spacecraft must provide $\Delta V_{mission}$ to leave the circular parking orbit around Earth and arrive into a circular parking orbit around the target planet. Example 7-1 (Part 3) shows how to calculate ΔV_{retro}.

To review, interplanetary flight involves connecting or "patching" three conic sections to approximate a spacecraft's path. We fire its rocket engines once to produce ΔV_{boost}, which makes it depart from its circular parking orbit around Earth on a hyperbolic trajectory, arriving at Earth's SOI with some excess velocity, $V_{\infty\ Earth}$. This excess velocity is enough to put it on a heliocentric, elliptical transfer orbit from Earth to the target planet. After traveling half of the ellipse, it enters the target planet's SOI, arriving at the target planet on a hyperbolic trajectory. We then fire its engines a second time to produce ΔV_{retro}, which captures it into its circular parking orbit. Table 7-1 summarizes the three regions of the problem and the necessary equations.

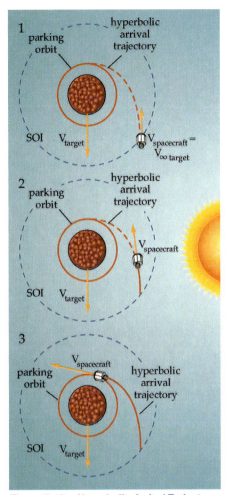

Figure 7-16.   Hyperbolic Arrival Trajectory. This sequence shows how a spacecraft enters the target planet's SOI and descends to its final parking orbit.

Table 7-1.   Summary of Interplanetary Transfer Problem.

Region	Reference Frame	Energy	Velocities
1: From Earth to the target planet (elliptical trajectory)	Heliocentric-ecliptic	$\varepsilon_{Earth} = -\dfrac{\mu_{Sun}}{2a_{Earth}}$ $\varepsilon_{transfer} = -\dfrac{\mu_{Sun}}{2a_{transfer}}$ $a_{transfer} = \dfrac{R_{to\ Earth} + R_{to\ target}}{2}$ $\varepsilon_{target} = -\dfrac{\mu_{Sun}}{2a_{target}}$	$V_{Earth} = \sqrt{2\left(\dfrac{\mu_{Sun}}{R_{to\ Earth}} + \varepsilon_{Earth}\right)}$ $V_{target} = \sqrt{2\left(\dfrac{\mu_{Sun}}{R_{to\ target}} + \varepsilon_{target}\right)}$ $V_{transfer\ at\ Earth} = \sqrt{2\left(\dfrac{\mu_{Sun}}{R_{to\ Earth}} + \varepsilon_{transfer}\right)}$ $V_{transfer\ at\ target} = \sqrt{2\left(\dfrac{\mu_{Sun}}{R_{to\ target}} + \varepsilon_{transfer}\right)}$ $V_{\infty\ Earth} = \left|V_{transfer\ at\ Earth} - V_{Earth}\right|$ $V_{\infty\ target} = \left|V_{target} - V_{transfer\ at\ target}\right|$
2: Departure from Earth (hyperbolic trajectory)	Geocentric-equatorial	$\varepsilon_{\infty\ Earth} = \dfrac{V_{\infty\ Earth}^2}{2}$	$V_{\infty\ Earth} = $ from above $V_{hyperbolic\ at\ Earth} = \sqrt{2\left(\dfrac{\mu_{Earth}}{R_{park\ at\ Earth}} + \varepsilon_{\infty\ Earth}\right)}$ $V_{park\ at\ Earth} = \sqrt{\dfrac{\mu_{Earth}}{R_{park\ at\ Earth}}}$ $\Delta V_{boost} = \left|V_{hyperbolic\ at\ Earth} - V_{park\ at\ Earth}\right|$
3: Arrival at the target planet (hyperbolic trajectory)	Planet-centered equatorial	$\varepsilon_{\infty\ target} = \dfrac{V_{\infty\ target}^2}{2}$	$V_{\infty\ target} = $ from above $V_{hyperbolic\ at\ target} = \sqrt{2\left(\dfrac{\mu_{target}}{R_{park\ at\ target}} + \varepsilon_{\infty\ target}\right)}$ $V_{park\ at\ target} = \sqrt{\dfrac{\mu_{target}}{R_{park\ at\ target}}}$ $\Delta V_{retro} = \left|V_{park\ at\ target} - V_{hyperbolic\ at\ target}\right|$

Example 7-1 (Part 3)

Problem Summary—Part 3

Given: $R_{to\ Earth} = 1.496 \times 10^8$ km

$R_{to\ Mars} = 2.278 \times 10^8$ km

$R_{park\ at\ Earth} = 6697$ km

$R_{park\ at\ Mars} = 3580$ km

$\mu_{Mars} = 43{,}050$ km^3/s^2

$V_{\infty\ Mars} = 2.65$ km/s (from Part 1)

$\Delta V_{boost} = 3.59$ km/s (from Part 2)

Find: $\Delta V_{mission}$ (Part 3)

Problem Diagram

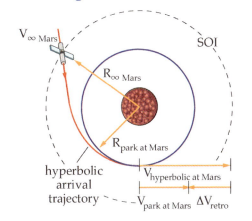

Conceptual Solution

Hyperbolic Planetary Arrival—Problem 3

1) Find the spacecraft's specific mechanical energy on its hyperbolic-arrival trajectory, $\varepsilon_{\infty\ Mars}$

2) Find its velocity on the hyperbolic-arrival trajectory at its parking-orbit's altitude, $V_{hyperbolic\ at\ Mars}$

3) Find its velocity in the circular, parking orbit, $V_{park\ at\ Mars}$

4) Find its velocity change, ΔV_{retro}, needed to enter its circular parking orbit,

5) Find its total velocity change for the mission, $\Delta V_{mission}$

Analytical Solution

1) Find the spacecraft's specific mechanical energy at Mar's SOI, using information from Part 1

$$\varepsilon_{\infty\ Mars} = \frac{V_{\infty\ Mars}^2}{2} = \frac{\left(2.65\frac{km}{s}\right)^2}{2} = 3.51\frac{km^2}{s^2}$$

2) Find $V_{hyperbolic\ at\ Mars}$

$$V_{hyperbolic\ at\ Mars} = \sqrt{2\left(\frac{\mu_{Mars}}{R_{park\ at\ Mars}} + \varepsilon_{\infty\ Mars}\right)}$$

$$= \sqrt{2\left(\frac{43{,}050\frac{km^3}{s^2}}{3580\ km} + 3.51\frac{km^2}{s^2}\right)}$$

$$= 5.57\ km/s$$

3) Find $V_{park\ at\ Mars}$

$$V_{park\ at\ Mars} = \sqrt{\frac{\mu_{Mars}}{R_{park\ at\ Mars}}}$$

$$= \sqrt{\frac{43{,}050\ km^3/s^2}{3580\ km}}$$

$$= 3.47\ km/s$$

4) Find ΔV_{retro}

$$\Delta V_{retro} = \left|V_{park\ at\ Mars} - V_{hyperbolic\ at\ Mars}\right|$$

$$= \left|3.47\frac{km}{s} - 5.57\frac{km}{s}\right|$$

$$= 2.10\ km/s$$

5) Find $\Delta V_{mission}$

$$\Delta V_{mission} = \Delta V_{boost} + \Delta V_{retro}$$

$$= 3.59\ km/s + 2.10\ km/s$$

$$= 5.69\ km/s$$

Interpreting the Results

To enter a circular parking orbit around Mars, our spacecraft needs to slow down by 2.10 km/s. Thus, to go from a parking orbit of 6697 km radius at Earth to a parking orbit of 3580 km radius at Mars requires a total velocity change from its rockets of 5.69 km/s.

Transfer Time of Flight

So far we've spent all our time figuring how much ΔV our spacecraft needs to get between planets. But before we launch it, we'd like to know how long the trip will take. The heliocentric, Hohmann Transfer ellipse approximates the time for our interplanetary journey, so, we use one-half the period of the transfer orbit to determine the time of flight

$$\boxed{TOF = \pi\sqrt{\frac{a_{transfer}^3}{\mu_{Sun}}}} \qquad (7\text{-}28)$$

where

TOF	= spacecraft's time of flight (s)
π	= 3.14159 . . . (unitless)
$a_{transfer}$	= semimajor axis of the transfer ellipse (km)
μ_{Sun}	= Sun's gravitational parameter (km^3/s^2)
	= 1.327×10^{11} km^3/s^2

(This does neglect the hyperbolic departure and arrival trajectories, but those times are insignificant compared to the long journey around the Sun.) Using information on a trip to Mars presented in Example 7-1 (Part 1, 2, and 3), we can determine the time of flight.

$$a_{transfer} = 1.887 \times 10^8 \text{ km}$$

$$TOF = \pi\sqrt{\frac{a_{transfer}^3}{\mu_{Sun}}} = \pi\sqrt{\frac{(1.887 \times 10^8 \text{ km})^3}{1.327 \times 10^{11}\frac{km^3}{s^2}}}$$

$$= 2.235 \times 10^7 \text{ s}$$
$$= 6208 \text{ hours}$$

$$TOF = 258.7 \text{ days or about 8.5 months}$$

That's a long time to be stuck in a tiny spaceship. Very long missions, such as a trip to Mars, would put significant demands on mission planners to sustain the crew by protecting them from the space environment and providing life support. Unfortunately, with current propulsion technology, these are the challenges we face when planning manned, planetary missions.

Phasing of Planets for Rendezvous

Another problem for an interplanetary transfer is finding the proper phasing of the planets for the transfer. Recall from Chapter 6 how we related the rendezvous problem to the way a quarterback synchronizes the flight of a football to a receiver's future position down field. Similarly, a spacecraft going to Mars must find the planet there when it arrives! The way we solve this problem is the same as the rendezvous of two spacecraft in Earth orbit. We need to find the lead angle, α_{lead}, and then the final phase angle, ϕ_{final}. For our Mars example, the angular velocity, ω, is

$$\omega = \sqrt{\frac{\mu}{R^3}} \qquad (7\text{-}29)$$

$$\omega_{Mars} = \sqrt{\frac{\mu_{Sun}}{R_{to\ Mars}^3}} = \sqrt{\frac{1.327 \times 10^{11} \dfrac{km^3}{s^2}}{(2.278 \times 10^8\ km)^3}} = 1.06 \times 10^{-7} \frac{rad}{s}$$

We then use the TOF from Equation (7-28) to calculate the lead angle:

$$\alpha_{lead} = \omega\ TOF \qquad (7\text{-}30)$$

$$\alpha_{lead} = (\omega_{Mars})(TOF) = \left(1.06 \times 10^{-7} \frac{rad}{s}\right)(2.235 \times 10^7\ s)$$

$$\alpha_{lead} = 2.37\ rad = 135.8°$$

So the final phase angle is

$$\phi_{final} = 180° - \alpha_{lead} \qquad (7\text{-}31)$$

$$\phi_{final} = 180° - 135.8° = 44.2°$$

A final phase angle of 44.2° means that, when we start the spacecraft on its interplanetary Hohmann Transfer, Mars needs to be 44.2° (0.7714 rads) ahead of Earth, as shown in Figure 7-17. If, for example, Earth were 50° (0.8727 rads) behind Mars, the phase angle, $\phi_{initial}$, is 50° (0.8727 rad), not 44.2° (0.7714 rad). Using the wait time equation we developed in Chapter 6, we solve for the amount of time to wait before we need to have our spacecraft ready to go.

$$\boxed{\text{Wait time} = \left(\frac{\phi_{final} - \phi_{initial}}{\omega_{Mars} - \omega_{Earth}}\right)} \qquad (7\text{-}32)$$

$$\text{Wait time} = \frac{0.7714\ rad - 0.8727\ rad}{\left(1.060 \times 10^{-7} \dfrac{rad}{s}\right) - \left(1.994 \times 10^{-7} \dfrac{rad}{s}\right)}$$

$$= 1.088 \times 10^6\ s$$

$$= 302.1\ hrs$$

$$= 12.59\ days$$

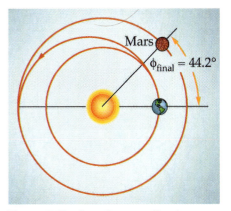

Figure 7-17. Interplanetary Rendezvous. A spacecraft launched from Earth to rendezvous with Mars should be 44.2° behind Mars at launch.

So, for this example, we would have to wait more than 12 days before the planets were phased properly for the spacecraft to launch.

If this were a manned mission, we'd have to worry not only about getting to Mars but also getting back. We know from Chapter 6 that the proper configuration for rendezvous recurs periodically, but we'd like to know how long we need to wait between opportunities. This wait time between successive opportunities is called the *synodic period*. Using the 2π relationship we discussed in Chapter 6, we can develop a relationship for synodic period as

$$\text{Synodic Period} = \frac{2\pi}{|\omega_{\text{Earth}} - \omega_{\text{target planet}}|} \quad (7\text{-}33)$$

For a trip to Mars using a Hohmann Transfer, the proper alignment between the two planets repeats itself about every two years. This means if we have a spacecraft sitting on the launch pad ready to go and we somehow miss our chance to launch, we must wait two more years before we get another chance! Appendix D.5 gives the synodic periods between Earth and the other planets.

▤ Section Review

Key Terms

hyperbolic excess velocity
sphere of influence (SOI)
synodic period

Key Equations

$$\epsilon = \frac{V^2}{2} - \frac{\mu}{R}$$

$$\epsilon = -\frac{\mu}{2a}$$

$$\Delta V_{\text{mission}} = \Delta V_{\text{boost}} + \Delta V_{\text{retro}}$$

$$\text{TOF} = \pi \sqrt{\frac{a_{\text{transfer}}^3}{\mu_{\text{Sun}}}}$$

$$\text{Wait time} = \left(\frac{\phi_{\text{final}} - \phi_{\text{initial}}}{\omega_{\text{Mars}} - \omega_{\text{Earth}}} \right)$$

See Table 7-1 for other Key
   Equations

Key Concepts

➤ The patched-conic approximation breaks interplanetary transfer into three regions and their associated problems

- Problem 1: From Earth to the target planet. This is a heliocentric transfer on an elliptical trajectory from Earth to the target planet

  - The velocity needed to change from Earth's orbit around the Sun to the elliptical transfer orbit is $V_{\infty \text{ Earth}}$

  - The velocity needed to change from the elliptical transfer orbit to the target planet's orbit around the Sun is $V_{\infty \text{ target}}$

- Problem 2: Earth departure. The spacecraft leaves Earth's vicinity on a hyperbolic trajectory

  - Earth's sphere of influence (SOI) defines an imaginary boundary in space within which Earth's gravitational pull dominates. When a spacecraft goes beyond the SOI, it has effectively left Earth. Earth's SOI extends to about 1,000,000 km.

  - To begin interplanetary transfer, a spacecraft needs a velocity relative to Earth of $V_{\infty \text{ target}}$ at the SOI. It achieves this velocity with ΔV_{boost}, which accelerates it from its circular, parking-orbit's velocity to its hyperbolic, departure-trajectory velocity

- Problem 3: Arrival at the target planet. The spacecraft arrives at the target planet on a hyperbolic trajectory

  - The spacecraft's velocity at the SOI, relative to the planet, is $V_{\infty \text{ target}}$

  - The spacecraft coasts on its hyperbolic-arrival trajectory from the SOI to its circular, parking orbit radius

  - To enter its circular, parking orbit around the target planet, the spacecraft performs a ΔV_{retro} burn

➤ Table 7-1 summarizes all equations for the interplanetary transfer

Continued on next page

▓ Section Review (Continued)

Key Concepts (Continued)

➤ Practically speaking, a spacecraft begins the interplanetary transfer in a parking orbit around Earth and ends in a final mission orbit around the target planet. To transfer between these two orbits, we must fire the spacecraft's engines twice to get two separate velocity changes

- First burn: ΔV_{boost} transfers the spacecraft from a circular parking orbit around Earth to a hyperbolic-departure trajectory with respect to Earth. This trajectory "patches" to an elliptical orbit around the Sun, taking the spacecraft to the target planet.

- Second burn: ΔV_{retro} slows the spacecraft from its hyperbolic-arrival trajectory with respect to the target planet to a final mission orbit around the target planet

➤ The total change in velocity the rocket must provide for the mission is the sum of ΔV_{boost} and ΔV_{retro}

➤ The time of flight (TOF) for an interplanetary transfer is approximately one-half the period of the transfer ellipse

➤ To ensure the target planet is there when the spacecraft arrives, we must consider the planets' phasing

- The phasing problem for interplanetary transfer is identical to the rendezvous problem from Chapter 6

➤ The synodic period of two planets is the time between successive launch opportunities

7.3 Gravity-assist Trajectories

▦ In This Section You'll Learn to...

☞ Explain the concept of gravity-assist trajectories and how they can help spacecraft travel between the planets

In the previous sections we saw how to get to other planets using an interplanetary Hohmann Transfer. Even this fuel-efficient maneuver requires a tremendous amount of rocket propellant, significantly driving up a mission's cost. Often, we can't justify a mission that relies solely on rockets to get the required ΔV. For example, if the Voyager missions, which took a "grand tour" of the solar system, had relied totally on rockets to steer between the planets, they would never have gotten off the ground.

Fortunately, spacecraft can sometimes get "free" velocity changes using gravity assist trajectories as they travel through the solar system. This *gravity assist* technique uses a planet's gravitational field and orbital velocity to "sling shot" a spacecraft, changing its velocity (in magnitude and direction) with respect to the Sun.

Of course, gravity-assisted velocity changes aren't totally free. Actually, the spacecraft "steals" velocity from the planet, causing the planet to speed up or slow down ever so slightly in its orbit around the Sun. Gravity assist can also bend the spacecraft's trajectory to allow it to travel closer to some other point of interest. The Ulysses spacecraft used a gravity assist from Jupiter to change planes, sending it out of the ecliptic into a polar orbit around the Sun.

How does gravity assist work? As a spacecraft enters a planet's sphere of influence (SOI), it coasts on a hyperbolic trajectory around the planet and the planet pulls it in the direction of the planet's motion, thus increasing (or decreasing) its velocity relative to the Sun. As it leaves the SOI at the far end of the hyperbolic trajectory, the spacecraft has a new velocity and direction to take it to another planet.

Of course, as Isaac Newton said, for every action there is an equal but opposite reaction. So, the spacecraft also pulls the planet a small amount. But, because the spacecraft is insignificantly small compared to the planet, the same force that can radically change the spacecraft's trajectory has no significant affect on the planet. (Imagine a mosquito landing on a dinosaur; the dinosaur would never notice the landing, but the mosquito would!)

As we saw in the previous sections, a spacecraft's velocity depends on the perspective of the beholder. During a gravity assist, we want to change the spacecraft's velocity with respect to the Sun, putting it on a different heliocentric orbit, so it can go where we want it to.

Let's consider what's going on from the planet's perspective. As a spacecraft flies by on a hyperbolic trajectory, the planet pulls on it. If the spacecraft passes behind the planet, as shown in Figure 7-18, it's pulled in

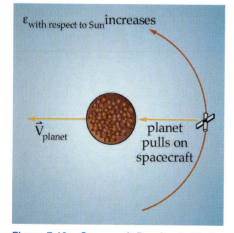

Figure 7-18.   Spacecraft Passing behind a Planet. During a gravity-assist maneuver, a spacecraft's energy will increase with respect to the Sun if it passes behind the planet.

the direction of the planet's motion and thus gains velocity (and hence energy) with respect to the Sun. This alters the spacecraft's original orbit around the Sun, as shown in Figure 7-19, sending it off to a different part of the solar system to rendezvous with another planet.

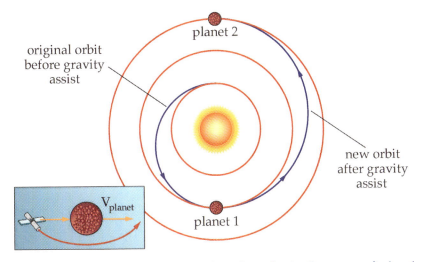

Figure 7-19.   Gravity Assist. During a gravity assist, a planet pulls a spacecraft, changing its velocity with respect to the Sun and thus altering its orbit around the Sun. The planet's orbit also changes, but very little.

When a spacecraft passes in front of a planet as in Figure 7-20, it's pulled in the opposite direction, slowing the spacecraft and lowering its orbit with respect to the Sun.

Gravity-assist trajectories often make the difference between possible and impossible missions. After the Challenger accident, the Galileo spacecraft's mission was in trouble. NASA officials banned liquid-fueled upperstages from the shuttle's payload bay, and solid-fuel upper stages simply weren't powerful enough to send it directly to Jupiter. Wisely, mission designers hit on the idea of going to Jupiter by way of Venus and Earth. They used one gravitational assist from Venus and two from Earth to speed the spacecraft on its way; hence, the name VEEGA (Venus, Earth, Earth Gravity Assist) for its new trajectory.

A gravity assist that changes the magnitude of a spacecraft's velocity is called *orbit pumping*. Using a planet's gravity to change the direction of travel is called *orbit cranking*. The gravity of Jupiter "cranked" the Ulysses solar-polar satellite out of the ecliptic plane into an orbit around the Sun's poles.

Realistically, mission requirements will constrain which planets we can use for gravity assist or whether it's even possible. For example, NASA used gravity assist for Voyager II's flights past Jupiter, Saturn, and Uranus (saving 20 years of trip time to Neptune), but couldn't use its flyby past Neptune to send it to Pluto. Doing so would have required a trip beneath Neptune's surface, which would have been a bit hard on the spacecraft! Instead, they used this last flyby to send Voyager II out of the

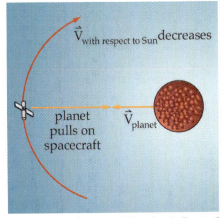

Figure 7-20.   Spacecraft Passing in Front of a Planet. During a gravity-assist maneuver, a spacecraft's energy will decrease with respect to the Sun if it passes in front of the planet.

solar system, where it travels today. In 1999, scientists received healthy transmissions from both Voyager spacecraft at distances of more than 11.2 billion km (6.9 billion mi.) and 8.7 billion km (5.4 billion mi.) for Voyager I and II, respectively. The messages take over 20 hrs, 40 min for Voyager I and 15 hrs, 40 min for Voyager II to travel to the spacecraft and return, at light speed.

Astro Fun Fact
Slowing Down Earth

Feeling a little slow today? Maybe you should. Following the Challenger accident, mission designers for the Galileo spacecraft had a problem. The high-energy Centaur upperstage they'd planned to use to boost the spacecraft to Jupiter wasn't available due to safety concerns. Instead, they had to use the safer, but less powerful, Inertial Upper Stage (IUS). Unfortunately, the IUS couldn't provide the necessary ΔV_{boost} to begin the transfer to Jupiter. Faced with this dilemma, they decided on a unique solution—not one, not two, but three gravity assists! To achieve the necessary ΔV, they planned to "steal" energy from Venus once and Earth twice, hence the name VEEGA—Venus, Earth, Earth Gravity Assist. The entire trajectory is in the figure on the right. Launched on October 18, 1989, Galileo

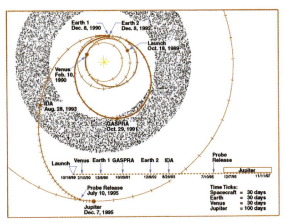

began its journey to Jupiter by going to Venus, flying by on February 10, 1990, and gaining 2.2 km/s. On December 8, 1990, Galileo returned to Earth on a hyperbolic trajectory that increased its velocity by another 5.2 km/s. Exactly two years later, it made a second pass by Earth, gaining the additional 3.7 km/s needed to take it to Jupiter, where it arrived in December 1995. But this extra ΔV from the gravity assists was not totally "free." Energy must be conserved, so for Galileo to speed up, Earth had to slow down. But don't worry, the result of both assists slowed Earth by a grand total of 4.3×10^{-21} km/s. That's about 13 cm (5 in.) in one billion years!

Information and diagram courtesy of NASA/Jet Propulsion Laboratory

---

▰ Section Review

Key Terms

gravity assist
orbit cranking
orbit pumping

Key Concepts

➤ Gravity-assist trajectories allow a spacecraft to get "free" velocity changes by using a planet's gravity to change a spacecraft's trajectory. This changes the spacecraft's velocity with respect to the Sun and slows the planet (but by a very small amount).

References

Bate, Roger R., Donald D. Miller, and Jerry E. White. *Fundamentals of Astrodynamics*. New York, NY: Dover Publications, Inc., 1971.

G.A. Flandro. *Fast Reconnaissance Missions to the Outer Solar System Utilizing Energy Derived from the Gravitational Field of Jupiter*. Astronautica Acta, Vol. 12 No. 4, 1966.

Jet Propulsion Laboratory Web Page, Voyager Home Page, 1998.

Wilson, Andrew. *Space Directory*. Alexandria, VA: Jane's Information Group Inc., 1990.

Mission Problems

7.1 Planning for Interplanetary Travel

1 How do we define the heliocentric-ecliptic coordinate frame?

2 Why can't we include all the appropriate gravitational forces for an interplanetary trajectory in a single equation of motion and solve it directly?

7.2 The Patched-conic Approximation

3 What does a planet's sphere of influence (SOI) represent?

4 What does the size of the SOI depend on?

5 What are the three regions used in the patched-conic approximation of an interplanetary trajectory and what coordinate frame does each use?

6 What "head start" does a spacecraft have in achieving the large velocities needed to travel around the Sun?

7 Why do we escape from Earth (or any planet) on a hyperbolic trajectory versus a parabolic trajectory?

8 A new research spacecraft is designed to measure the environment in the tail of Earth's magnetic field. It's in a circular parking orbit with a radius of 12,756 km. What ΔV will take the spacecraft to the boundary of Earth's SOI such that it arrives there with zero velocity with respect to Earth?

9 To continue the studies of Venus begun by the Magellan probe, NASA is sending a remote-sensing spacecraft to that planet.

Given:

$$R_{to\ Earth} = 1.496 \times 10^8 \text{ km}$$

$$R_{to\ Venus} = 1.081 \times 10^8 \text{ km}$$

$$R_{park\ at\ Earth} = 6600 \text{ km}$$

$$R_{park\ at\ Venus} = 6400 \text{ km}$$

$$\mu_{Earth} = 3.986 \times 10^5 \text{ km}^3/\text{s}^2$$

$$\mu_{Venus} = 3.257 \times 10^5 \text{ km}^3/\text{s}^2$$

$$\mu_{Sun} = 1.327 \times 10^{11} \text{ km}^3/\text{s}^2$$

a) Find the semimajor axis and specific mechanical energy of the transfer orbit.

b) Find $V_{\infty\ Earth}$

c) Find $V_{\infty\ Venus}$

d) Find ΔV_{boost}

e) Find ΔV_{retro}

f) Find $\Delta V_{mission}$

10 Find the time of flight (TOF) for the above mission to Venus.

11 Adjust Equation (7-4) and compute the radius of the moon's SOI relative to Earth.

Assume:

$$a_{Moon} = 3.844 \times 10^5 \text{ km and } \left(\frac{mass_{Moon}}{mass_{Earth}}\right) = \frac{1}{81.3}$$

12 What is the needed phase angle for the rendezvous of a flight from Earth to Saturn?

Given:

$$R_{to\ Earth} = 1.496 \times 10^8 \text{ km}$$

$$R_{to\ Saturn} = 1.426 \times 10^9 \text{ km}$$

7.3   Gravity-assist Trajectories

13 Explain how we can use gravity assist to get "free ΔV" for interplanetary transfers.

14 Is ΔV·from a gravity assist really "free?"

15 Can a flyby of the Sun help us change a spacecraft's interplanetary trajectory?

Projects

16 Write a computer program or spreadsheet to compute $\Delta V_{mission}$ for trips to all of the other eight planets in the solar system.

Mission Profile—Magellan

On May 4, 1989, the Magellan Space Probe launched from the Space Shuttle Atlantis—the first interplanetary spacecraft launched from a Space Shuttle and the first U.S. interplanetary mission since 1978. Magellan was designed to produce the first high-resolution images of the surface of Venus using a synthetic aperture radar (SAR). The project's manager was NASA, with coordinators at the Jet Propulsion Laboratory and the primary builders at Martin Marietta and Hughes Aircraft Company. By studying our neighboring planet with a spacecraft equipped with radar, researchers hope to learn more about how the solar system and Earth were formed.

Mission Overview

The primary objectives for Magellan were

✓ Map at least 70% of the surface of Venus using the SAR

✓ Take altitude readings of its surface, take its "temperature" (radiometry), and chart changes in its gravity field

Mission Data

✓ Launched by the Space Shuttle and placed into an interplanetary transfer orbit by the inertial upper-stage

✓ The interplanetary transfer orbit took Magellan around the Sun and placed it into a highly elliptical mapping orbit around Venus. The mapping orbit was elliptical to allow Magellan to map when at perigee and to transmit data to Earth near apogee.

✓ Magellan was built from left-over and back-up parts of previous satellites such as Voyager, Galileo, and Ulysses. Its configuration included

- Antennas with high, medium, and low gain plus a radar altimeter
- A forward equipment module, housing the communications and radar electronics and reaction wheels for attitude control
- A ten-sided spacecraft bus
- Two solar panels
- A propulsion module

✓ The main payload included the synthetic aperture radar and the radar altimeter

✓ To withstand harsh conditions close to the Sun, Magellan was covered in thermal blankets and heat-reflecting inorganic paint. It also used louvers to dissipate heat created by the SAR and other electronic components.

✓ The high-gain antenna provided communications and data gathering. Deep Space Network antennas on Earth received scientific data from the payload. This data then went to the Jet Propulsion Laboratory for interpretation. The Deep Space Network was also used to send commands to Magellan from Earth.

✓ Because Magellan used the high-gain antenna for radar mapping and data transmission to Earth, the spacecraft had to rotate four times in each orbit. A combination of reaction wheels, sun-sensors, and thrusters provided enough pointing accuracy for these maneuvers.

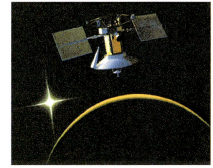

Magellan Mapper. Here's an artists' concept of the Magellan spacecraft mapping the Venus surface. *(Courtesy of NASA/Jet Propulsion Laboratory)*

Mission Impact

Magellan performed better than mission planners expected, mapping more than 95% of Venus's surface with images better than any of Earth. This single mission gathered more data than in all other NASA exploratory missions combined. Magellan's funding was cut for fiscal year 1993, the mission ended in May 1993.

For Discussion

- What information can we learn from studying Venus?
- How might the lessons learned from Magellan affect future unmanned space probes?

Contributors

Luciano Amutan and Scott Bell, the U.S. Air Force Academy

References

Wilson, Andrew. *Interavia Space Directory*. Alexandria, VA: Jane's Information Group, 1990.

Young, Carolynn (ed.). *The Magellan Venus Explorer's Guide*. Pasadena, CA: Jet Propulsion Laboratory, 1990.

What a dish! Antennas scan the heavens to monitor the paths of spacecraft through space. *(Courtesy of NASA/Jet Propulsion Laboratory)*

Predicting Orbits

8

Robert B. Giffen
Professor Emeritus, the U.S. Air Force Academy

▤ In This Chapter You'll Learn to...

- ☛ Determine the time of flight between two spacecraft positions within a given orbit
- ☛ Determine a spacecraft's future position using Kepler's Equation
- ☛ Describe the effects of perturbations on orbits and explain their practical applications
- ☛ Describe the overall problem of tracking spacecraft and predicting orbits

▤ You Should Already Know...

- ❏ The assumptions of the restricted two-body problem (Chapter 4)
- ❏ The two-body equation of motion and its solution (Chapter 4)
- ❏ The definition and use of the classic orbital elements, including argument of latitude, u (Chapter 5)

▤ Outline

8.1 **Predicting an Orbit (Kepler's Problem)**
Kepler's Equation and Time of Flight

8.2 **Orbital Perturbations**
Atmospheric Drag
Earth's Oblateness
Other Perturbations

8.3 **Predicting Orbits in the Real World**

What goes around comes around.

Anonymous

Space Mission Architecture. This chapter deals with the Trajectories and Orbits segment of the Space Mission Architecture, introduced in Figure 1-20.

By now you should have a pretty good feel for orbits: how they look, how they're defined, and how they're used. So far, everything we've done with orbits has been the result of a few basic equations developed in Chapter 4. Now it's time to take the next step. In this chapter we'll turn our attention to predicting orbits. To track spacecraft through space, we need to know where they are now and where they'll be later, so we can point our antennas at them to perform key functions. Although we can easily predict this motion when the orbit is a circle, the problem becomes more complicated when the orbit is an ellipse, and most orbits are at least slightly elliptical.

In this chapter we'll begin by looking at the orbital-prediction problem solved by Johannes Kepler over 300 years ago. We'll see how Kepler developed a method to analyze the motion of a spacecraft on an elliptical orbit. Next, we'll reexamine what we assumed about orbital motion in Chapter 4. We'll see that, in the "real world," our assumptions break down for low-Earth-orbiting spacecraft (Figure 8-1). We'll find that Earth's atmosphere and its non-spherical shape disturb or "perturb" orbits from the path we explained in Chapter 4. Finally, we'll combine Kepler's method with an understanding of how orbits are perturbed to predict orbits in the real world.

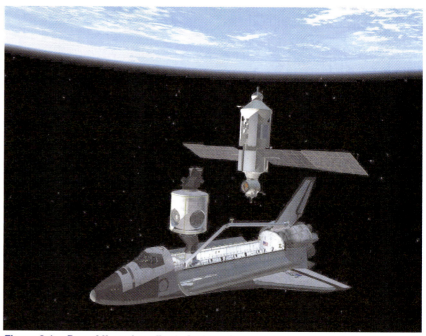

Figure 8-1.   Drag Affects Low-Earth Orbiting Spacecraft. Any objects in low-Earth orbit run into air molecules that slowly affect their orbits. *(Courtesy of NASA/Johnson Space Center)*

8.1 Predicting an Orbit (Kepler's Problem)

≡ In This Section You'll Learn to...

☛ Use Kepler's Equation to calculate a spacecraft's time of flight

☛ Use Kepler's Equation to predict a spacecraft's position at some future time

Let's look at the "big picture" of tracking and predicting orbits, shown in Figure 8-2. Imagine the Space Shuttle is in orbit and we've just received a position update on it from our tracking site located on an island in the Indian Ocean. The site provides us with the Shuttle's *range* (the distance from the tracking site), *azimuth* (the angle from true north), and *elevation* (the angle between the local horizon and the Shuttle). We then convert these observations into a position vector, $\vec{R}_{initial}$, and, with at least one more set of observations, we also find a velocity vector, $\vec{V}_{initial}$. Using the techniques developed in Chapter 5, we then convert $\vec{R}_{initial}$ and $\vec{V}_{initial}$ into a set of classic orbital elements (COEs).

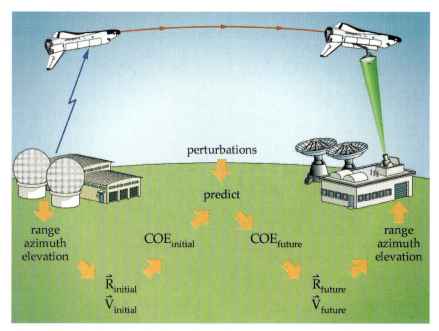

Figure 8-2. The Tracking Problem. To track and predict a spacecraft's orbit, we take tracking data, convert it first to $\vec{R}_{initial}$ and $\vec{V}_{initial}$ and then to COEs, and move these COEs to a future time (including perturbations). We then reconvert the COEs to $\vec{R}_{future}$ and $\vec{V}_{future}$ vectors, and, finally, to range, azimuth, and elevation angles.

For this example, we can assume the astronauts aboard the Shuttle are going to deploy a large mirror and we're going to bounce a low-power laser off the mirror's surface and back to the ground. To aim our laser, we have to know *precisely* when the Shuttle will pass over the test site and where it will be at that time. If all we know is the Shuttle's position and velocity some time in the past, we'll have to *predict* when it will be overhead. To predict or *propagate* any orbit into the future (COE_{future}), we have to develop a prediction method and understand how environmental factors such as drag and Earth's oblate shape affect this prediction. Once we know COE_{future}, we can then re-convert these updated orbital elements into $\vec{R}_{future}$ and $\vec{V}_{future}$, and then into range, elevation, and azimuth. These data will tell us how to point our laser.

In this section we'll determine the time of flight between two orbital positions. Then we'll predict a spacecraft's position within its orbit at some future time. We'll tackle the effects of drag and the oblate Earth in the next section.

Kepler's Equation and Time of Flight

In circular orbits, determining how long a spacecraft takes to get from an initial position to a future position is simple, because the spacecraft moves at a constant speed. Using Figure 8-3, let's define this angular speed as the *mean motion, n*, which tells us the mean, or average, speed on the orbit. We find the mean motion in terms of the period, P, and the radius, which, in the case of circular orbits, is the same as the semimajor axis, a. The mean motion is

$$n = \frac{\text{angle}}{\text{time}} = \frac{2\pi}{P} = \sqrt{\frac{\mu}{a^3}} \qquad (8\text{-}1)$$

where

n = spacecraft's mean motion (rad/s)

P = orbital period (s)

μ = central body's gravitational parameter (km^3/s^2)
  = $3.986 \times 10^5 \ km^3/s^2$ for Earth

a = semimajor axis (km)

Remember from Chapter 5, for a circular orbit we define a spacecraft's position within the orbit using the *argument of latitude, u*, the angular distance along the orbital path from the ascending node to the spacecraft. So, as Figure 8-3 shows, the time of flight in a circular orbit is

$$\text{Time of flight} = \text{TOF} = \frac{u_{future} - u_{initial}}{n} \qquad (8\text{-}2)$$

where

u_{future} = spacecraft's future argument of latitude (rad)

$u_{initial}$ = spacecraft's initial argument of latitude (rad)

n = spacecraft's mean motion (rad/s)

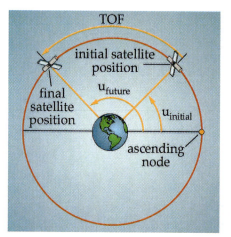

Figure 8-3.   Time of Flight on a Circular Orbit. To find the time of flight between two spacecraft positions in a circular orbit, divide the angle between these two positions by the orbital mean motion, n. Conversely, to find the spacecraft's location some time in the future, just multiply the time of flight by the mean motion and add the result to the initial position.

And the spacecraft's position at some future time equals its initial position plus the angle it travels through during the time of flight

$$u_{future} = u_{initial} + n(TOF) \qquad (8\text{-}3)$$

where

$$n(TOF) = \frac{angle}{time} \times time = angle$$

   In an elliptical orbit, however, the spacecraft motion is not uniform. We don't know how the true anomaly, v, changes with time because it doesn't change uniformly. Here's where Johannes Kepler came to the rescue. Remember from Chapter 2, he was trying to make his theories of planetary motion match Tycho Brahe's meticulous observations of the planet Mars. To solve this problem, he figured out how to move v to a time in the future and, conversely, given a future v, how to find how long Mars would take to get there.

   Kepler's approach was purely geometrical—he related motion on a circle to motion on an ellipse. He also defined a planet's mean motion, n, to be the average angular rate it travels in one orbital period

$$n = \sqrt{\frac{\mu}{a^3}} \qquad (8\text{-}4)$$

He then defined a new angle called the *mean anomaly, M*

$$M = nT \qquad (8\text{-}5)$$

where

M = mean anomaly (rad)

n = mean motion (rad/s)

T = the time since last perigee passage (s)

Mean anomaly is an angle that has no physical meaning and can't be drawn in a picture. We have to express it mathematically. The change in M between two points in the same orbit is

$$\boxed{M_{future} - M_{initial} = n(t_{future} - t_{initial}) - 2k\pi} \qquad (8\text{-}6)$$

where

M_{future}	= mean anomaly when the spacecraft is in the future position (rad)
$M_{initial}$	= mean anomaly when the spacecraft is in the initial position (rad)
$t_{future} - t_{initial}$	= time of flight (TOF) between two points in the orbit
t_{future}	= time when the spacecraft is in the final position (e.g., 3:47 A.M.)
$t_{initial}$	= time when the spacecraft is in the initial position (e.g., 3:30 A.M.)
k	= the number of times the spacecraft passes perigee during the TOF

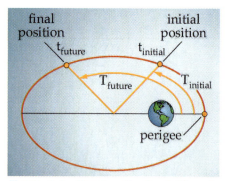

Figure 8-4. Keeping Track of Time in Kepler's Equation. We use T as the time since a spacecraft passed perigee (e.g., 30 min). Lower case "t" is the actual time (shown on a watch) that the spacecraft is at a particular location (e.g. 3:30 A.M.). Thus, $T_{initial}$ is the time elapsed since a spacecraft, located at an initial position, was at perigee (e.g., 20 min). And, $t_{initial}$ is the actual time when the spacecraft was at the initial position (e.g. 2:10 P.M.).

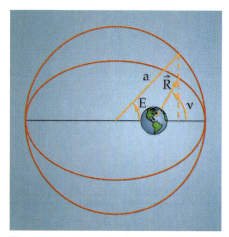

Figure 8-5. Eccentric Anomaly. We define the eccentric anomaly, E, geometrically by circumscribing an elliptical orbit with a circle and relating E to M, using the true anomaly, ν and the eccentricity, e.

Note that $M_{future} - M_{initial}$ must be greater than zero and is usually less than 2π, for convenience in working with angles. It's important to keep track of time when working these problems. Figure 8-4 shows the relationship between the time since last perigee passage, T, and the time of the clock, t.

To relate elliptical motion to circular motion, Kepler defined another new angle called the *eccentric anomaly, E*, so he could relate M to E and then E to ν geometrically, as seen in Figure 8-5. With all these things defined, Kepler was able to develop his now-famous equation, commonly called Kepler's Equation. (For this equation to work, all angles must be in radians.)

$$M = E - e \sin E \qquad (8\text{-}7)$$

where
M = mean anomaly (rad)
E = eccentric anomaly (rad)
e = eccentricity (unitless)

He then related E to ν using

$$\cos E = \frac{e + \cos v}{1 + e \cos v} \qquad (8\text{-}8)$$

where
ν = true anomaly (rad)

And ν to E through

$$\cos v = \frac{\cos E - e}{1 - e \cos E} \qquad (8\text{-}9)$$

Now we have the equations needed to solve two problems. The first problem, and the easier, is finding the time of flight between two points in an orbit. Given $v_{initial}$ and v_{future}, we simply go through the following steps

- Use Equation (8-8) to solve for $E_{initial}$ and E_{future}
- Use Equation (8-7) to solve for $M_{initial}$ and M_{future}
- Use Equation (8-6) to solve for the time of flight ($t_{future} - t_{initial}$)

Remember in Chapter 5, when we solved for the orbital elements, we had to check the quadrant when taking the inverse cosine? We have to do that here, too, but it's easier. Look at Figure 8-5 and you can see ν and E are always in the same half plane. It turns out that mean anomaly follows the same rule. This means if ν is between 0° and 180°, so are E and M.

Kepler's Equation and Future Position

The second problem we can solve using Kepler's method is far more practical. This involves determining a spacecraft's position at some future time, t_{future}, as shown in Figure 8-6. This second problem is trickier. We assume we know where the spacecraft is at time $t_{initial}$, so we also know $v_{initial}$. We start by finding $E_{initial}$, using Equation (8-8). Then we find $M_{initial}$ using Kepler's Equation (8-7).

$$M_{initial} = E_{initial} - e \; \sin E_{initial}$$

Now, because we know t_{future} (or we pick a future time), using Equation (8-6), we can find M_{future}.

$$M_{future} - M_{initial} = n(t_{future} - t_{initial}) - 2k\pi$$

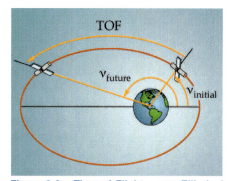

Figure 8-6. Time of Flight on an Elliptical Orbit. Predicting a spacecraft's future position at some time, t_{future}, is the second application of Kepler's Equation.

Great. We're now on our way to finding v_{future}, which tells us where the spacecraft will be. So we go to Kepler's Equation again to find E_{future}. Let's rearrange this equation and put E on the left side

$$E_{future} = M_{future} + e \; \sin E_{future} \qquad (8\text{-}10)$$

OOPS! We can't isolate E_{future} in this equation. Why? Because it's a variable in the sine function, which is a transcendental function (so are cosine and log). So, we call this a *transcendental equation*. In fact, almost every notable mathematician over the past 300 years has tried to find a direct solution to this form of Kepler's Equation without success. So we must resort to indirect methods to solve for E_{future}. The indirect approach we'll use is called *iteration*. To see how iteration works, think about the kids' game Twenty Questions. In this game, your partner thinks of a person, place, or thing and you must guess what he's thinking of. You get 20 questions (guesses) to which your partner can answer only "yes" or "no." In seeking the right answer, a good player will systematically eliminate all other possibilities until only the correct answer remains.

To demonstrate the power of iteration, let's look at an application of this process using another equation with a transcendental function

$$y = \cos y$$

Because we can't solve for y using algebra (we can't get the y out of the cosine function to put all the y's on the left side), we must iterate. We begin by taking a guess at the value for y in radians, and then we take the cosine to see how close we were. Then take this as the new value of y and use it for the next guess, and keep doing this iteration until the new y equals the old y (or is close enough, say, within 0.000001 radians of the old value).

Let's try it to see what the answer for y really is. Get out your calculator and use $\pi/4$ radians as your first guess for y. (Remember to set your calculator to use radians, not degrees.) Keep punching the cosine function button and you'll see the value slowly converges to 0.739085 radians (about 42.3°). Presto—you've now solved the transcendental equation y = cos y using iteration!

We can use this same iterative technique to solve Equation (8-10) for E_{future}. It turns out the values for M and E are always pretty close together, even for the most eccentric orbits, so let's use M_{future} for our first guess at E_{future}. Here's the algorithm

- Use M_{future} in radians for the first guess of E_{future}

- Solve Equation (8-10) for a new E_{future}

- Use this new E_{future} for the next guess for Equation (8-10)

- Keep doing the previous step until E_{future} doesn't change by much (less than about 0.0001 rad). At this point we say the solution has *converged*.

This brute force iteration method will solve Equation (8-10), but there are better methods. The most notable is Newton's Iteration Method.

Let's quickly summarize what we've learned. If we know where our spacecraft is in orbit at some point in time and we're interested in when it will reach some other point in the orbit, we can use Kepler's Equation to solve for the time of flight it takes to get there. The solution is very straight-forward. If, however, we know where it is and want to know where it'll be at some future time, we can use Kepler's Equation to find that location, only by iterating a transcendental equation for eccentric anomaly.

Astro Fun Fact
Astronomy vs. Astrology

Johannes Kepler was also a dabbler in astrology. From an early age, Kepler was interested in the study of how movements of the heavenly bodies affected people's lives. The beginnings of astrology go back to around 2000 B.C. in Babylonia. It purports to tell how people are affected by the positions of Earth, the planets (including the Sun and Moon), the zodiac, and the "houses" (similar to the zodiac except located on Earth). At first, Kepler told fortunes for family members, but later he made calendars of predictions to make more money. Into his third and final year of seminary, Kepler was appointed to become a math instructor at a Lutheran school in Graz, Austria, after the death of the professor. After arriving in Graz, his successful prediction of cold weather, peasant uprisings, and invasion by the Turks made Kepler's calendars a hot item. Kepler was split on his feelings about astrology. He referred to it as the "foolish little daughter of respectable astronomy" and stated "if astrologers sometimes do tell the truth, it ought to be attributed to luck." However, Kepler's firm belief in the harmony and sense of order in the universe kept him involved in astrology. He was also able to provide food for his family and pay the bills doing prediction calendars. Kepler was a bridge between the mysticism of astrology and the realism of astronomy.

<u>Dictionary of Scientific Biography</u>. New York, NY: Scribner's Sons, 1992.

Contributed by Steve Crumpton, the U.S. Air Force Academy

▦ Section Review

Key Terms

argument of latitude, u
azimuth
eccentric anomaly, E
elevation
iteration
mean motion, n
mean anomaly, M
range
transcendental equation

Key Equations

$$M_{future} - M_{initial} = n(t_{future} - t_{initial}) - 2k\pi$$

$$M = E - e \sin E$$

$$\cos E = \frac{e + \cos v}{1 + e \ \cos v}$$

$$\cos v = \frac{\cos E - e}{1 - e \ \cos E}$$

Key Concepts

➤ Kepler's Equation gives us the solution to two problems
  • Finding the time of flight between two known orbital positions
  • Finding a future orbital position, given the time of flight

➤ Mean motion, n, is the average angular speed of a spacecraft in orbit

➤ Mean anomaly, M, relates to mean motion through the time, T, since passing perigee

➤ Eccentric anomaly relates motion on an ellipse to motion on a circumscribed circle

➤ Spacecraft position defined by true anomaly, v, relates to eccentric anomaly

➤ Given a new spacecraft position, v, we can find E, M, and finally T. Or given T (or some future time, t_{future}), we can find M, solve for E using substitution and iteration, and then find v.

Example 8-1

Problem Statement

The Space Shuttle is conducting Spacelab experiments in a slightly elliptical orbit (e = 0.05) with a semimajor axis of 7000 km. Data from the Tracking and Data Relay Satellite tell us the Shuttle's current true anomaly is 270°. How long until it reaches a true anomaly of 50°?

Problem Summary

Given: $a = 7000$ km
$e = 0.05$
$\nu_{initial} = 270°$

Find: Time of flight from $\nu_{initial} = 270°$ to $\nu_{future} = 50°$

Problem Diagram

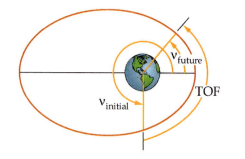

Conceptual Solution

1) Find mean motion, n

$$n = \frac{2\pi}{p} = \sqrt{\frac{\mu}{a^3}}$$

2) Find $E_{initial}$ and E_{future}

$$\cos E = \frac{e + \cos\nu}{1 + e\cos\nu}$$

3) Find $M_{initial}$ and M_{future}

$$M = E - e\sin E$$

4) Find time of flight

$$M_{future} - M_{initial} = n(t_{future} - t_{initial}) - 2k\pi$$

$$t_{future} - t_{initial} = \frac{M_{future} - M_{initial} + 2k\pi}{n}$$

Note from the diagram we must pass perigee once to get from $\nu_{initial}$ to ν_{final}

Analytical Solution

1) Find mean motion, n

$$n = \sqrt{\frac{\mu}{a^3}} = \sqrt{\frac{3.986 \times 10^5 \frac{km^3}{s^2}}{(7000 \text{ km})^3}}$$

$$= 0.001078 \text{ rad/s} = 14.82 \text{ rev/day}$$

2) Find $E_{initial}$, E_{future}

$$\cos E_{initial} = \frac{e + \cos\nu_{initial}}{1 + e\cos\nu_{initial}}$$

$$= \frac{0.05 + \cos 270°}{1.0 + 0.05\cos 270°} = 0.05$$

$E_{initial} = 87.13°$ or $272.87°$

Remember, ν, M, and E must all lie in the same half plane. Therefore, because $\nu_{initial} = 270°$, $E_{initial} = 272.87° = 4.762$ rad

$$\cos E_{future} = \frac{e + \cos\nu_{future}}{1 + e\cos\nu_{future}}$$

$$= \frac{0.05 + \cos 50°}{1.0 + 0.05\cos 50°} = 0.6712$$

$E_{future} = 47.84°$ or $312.16°$

$E_{future} = 47.84°$ (same half-plane as ν_{future}) = 0.835 rad

3) Find $M_{initial}$, M_{future}

$$M_{initial} = E_{initial} - e\sin E_{initial}$$

Note: Here $E_{initial}$ must be in **radians**

$$M_{initial} = 4.762 - 0.05\sin 4.762 = 275.727° = 4.812 \text{ rad}$$

Example 8-1 (Continued)

$M_{future} = E_{future} - e \sin E_{future}$

$M_{future} = 0.835 - 0.05\sin 0.835 = 45.716° = 0.798\,rad$

Find Time of Flight

$$t_{future} - t_{initial} = \frac{M_{future} - M_{initial} + 2k\pi}{n}$$

Because we must pass perigee once, $k = 1$

$$= \frac{(0.798 - 4.812 + 2\pi)\,rad}{0.001078\frac{rad}{s}}$$

$$= 2104.58\,s = 35.08\,min$$

Interpreting the Results

Using Kepler's equation, we found it takes about 35 minutes for the Space Shuttle to travel from $v_{initial} = 270°$ to $v_{future} = 50°$. It passes perigee one time so we must add in the factor 2π to the equation for time of flight. This answer makes sense because the period of a low-Earth orbit is about 90 minutes, and we're travelling 140°, or a little more than a third of the way around the orbit.

Example 8-2

Problem Statement

An Earth-observation satellite is in a slightly eccentric orbit (e = 0.05) with a semimajor axis of 7000 km and an inclination of 50°. If its current true anomaly is 270°, what will be the true anomaly six hours from now?

Problem Summary

Given: $a = 7000$ km
$e = 0.05$
$i = 50°$
$v_{initial} = 270°$

Find: The satellite's true anomaly six hours from now

Problem Diagram

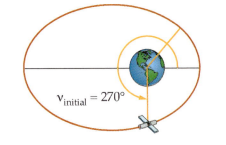

$v_{initial} = 270°$

Conceptual Solution

1) Find mean motion, n

$$n = \frac{2\pi}{P} = \sqrt{\frac{\mu}{a^3}}$$

2) Find $E_{initial}$

$$\cos E = \frac{e + \cos v}{1 + e \ \cos v}$$

3) Find $M_{initial}$

$$M = E - e \sin E$$

4) Move mean anomaly to the desired time

$$M_{future} = M_{initial} + n(t_{future} - t_{initial}) - 2k\pi$$

5) Solve for E_{future} using Kepler's equation (iterative solution required)

$$E = M + e \sin E$$

6) Find v_{future}

$$\cos v = \frac{\cos E - e}{1 - e \ \cos E}$$

Analytical Solution

1) Find mean motion, n

$$n = \sqrt{\frac{\mu}{a^3}} = \sqrt{\frac{3.986 \times 10^5 \ \frac{km^3}{s^2}}{(7000 \ km)^3}}$$

$$= 0.001078 \ rad/s = 14.82 \ rev/day$$

2) Find $E_{initial}$

$$\cos E_{initial} = \frac{e + \cos v_{initial}}{1 + e \ \cos v_{initial}}$$

$$= \frac{0.05 + \cos 270°}{1.0 + 0.05 \ \cos 270°} = 0.05$$

$E_{initial} = 87.13°, \ 272.87° = 272.87° = 4.762$ rad
Remember, v, M, and E must all lie in the same half plane.

3) Find $M_{initial}$

$M_{initial} = E_{initial} - e \sin E_{initial}$

Note: Here $E_{initial}$ must be in **Radians**

$M_{initial} = 4.762 - 0.05 \sin 4.762 = 4.812$ rad
$= 275.727°$

4) Move mean anomaly to the desired time

$M_{future} = M_{initial} + n(t_{future} - t_{initial}) - 2k\pi$

$M_{future} = 4.812 + (0.001078 \ rad/s)(6 \ hr \cdot 3600 \ s/hr) -$

$= 28.097$ rad

Because this value is greater than 2π, we need to subtract 2π until M_{future} is less than 2π.

Example 8-2 (Continued)

$M_{future} = 28.097 - 4(2\pi) = 2.964$ rad

Physically, this means our satellite passes perigee four times in the next six hours

Solve for E_{future} using Kepler's equation (iterative solution required)

First, guess $E_{future} = M_{future} = 2.964$ rad

$E_{future} = M_{future} + e \sin E_{future}$

$\qquad = 2.964 + 0.05 \sin 2.964 = 2.973$ rad

Now, use $E_{future} = 2.973$ rad for the next guess

$E_{future} = M_{future} + e \sin E_{future}$

$\qquad = 2.964 + 0.05 \sin 2.973 = 2.972$ rad

Again, use $E_{future} = 2.972$ rad next

$E_{future} = M_{future} + e \sin E_{future}$

$\qquad = 2.964 + 0.05 \sin 2.972 = 2.972$ rad
$\qquad = 170.3°$

Okay, within the accuracy we've chosen here, our solution has converged.

6) Find v_{future}

$$\cos v_{future} = \frac{\cos E_{future} - e}{1 - e \, \cos E_{future}} = \frac{\cos(2.972) - 0.05}{1 - 0.05 \, \cos(2.972)}$$

$v_{future} = 170.75°$ or $189.25° = 170.75°$
$\qquad$ (since $E_{future} = 170.3°$)

Interpreting the Results

In six hours our satellite will pass perigee four times and end up at $v = 170.75°$, which is about 100° behind where it is now. The period of our orbit is about 97 minutes, so in six hours we should go around the orbit about four times.

8.2 Orbital Perturbations

▬ In This Section You'll Learn to...

- ☞ Explain and determine how Earth's atmosphere changes a spacecraft's orbit
- ☞ Explain and determine how Earth's non-spherical shape changes a spacecraft's orbit
- ☞ Explain how sun-synchronous and Molniya orbits take advantage of Earth's non-spherical shape
- ☞ Describe other sources of orbital perturbations

In Chapter 4 we developed a general equation of motion for a spacecraft in orbit that looked like this

$$\sum \vec{F}_{external} = \vec{F}_{gravity} + \vec{F}_{drag} + \vec{F}_{thrust} + \vec{F}_{3rd\ body} + \vec{F}_{other} \quad (8\text{-}11)$$

We then assumed

- Gravity is the only force
- Earth's mass is much greater than the spacecraft's mass
- Earth is spherically symmetric with uniform density, so it could be treated as a point mass
- The spacecraft's mass is constant, so $\Delta m = 0$

These assumptions led us to the restricted two-body equation of motion

$$\ddot{\vec{R}} + \frac{\mu}{R^2}\hat{R} = 0 \quad (8\text{-}12)$$

The solution to this equation gave us the six classic orbital elements (COEs)

a—semimajor axis

e—eccentricity

i—inclination

Ω—right ascension of the ascending node

ω—argument of perigee

ν—true anomaly

Under our assumptions, the first five of these elements remain constant for a given orbit. Only the true anomaly, ν, varies with time as a spacecraft travels around its fixed orbit. What happens if we now change some of our original assumptions? Other COEs besides ν will begin to change as well. Any changes to these COEs due to other forces we call *perturbations*. To see which COEs change and by how much, let's look at our first assumption—gravity is the only force.

Atmospheric Drag

Gravity really isn't the only force acting on a spacecraft. In Chapter 3 we talked about the space environment and the effect Earth's atmosphere has on a spacecraft's lifetime. Recall that Earth's atmosphere gets thinner with altitude but still has some effect as high as 600 km (375 mi.). Because many important space missions are conducted in orbit with altitudes lower than 600 km, this very thin air causes drag on these spacecraft. Let's look at how drag affects their orbital elements.

Drag is a non-conservative force—it takes energy away from the orbit in the form of friction on the spacecraft. Because orbital energy is a function of semimajor axis, we expect the semimajor axis, a, to get smaller over time, due to drag. The eccentricity also decreases, since the orbit becomes more circular. Let's see why this is so. Imagine a spacecraft is in a highly elliptical orbit, as shown in Figure 8-7. The effect of drag will be most noticeable when the spacecraft goes through perigee. Each time, drag will slow the spacecraft slightly, like applying a small, negative ΔV. Recall from Chapter 6 that a negative ΔV applied at perigee will lower apogee. Thus, apogee for the orbit will gradually lower, making the orbit more circular, thus less eccentric.

Drag is very difficult to model because of the many factors affecting Earth's upper atmosphere and the spacecraft's attitude. Earth's day-night cycle, seasonal tilt, variable solar distance, and fluctuating magnetic field, as well as the Sun's 27-day rotation and 11-year cycle for Sun spots, make the modeling task nearly impossible. The force of drag, which we'll see in greater detail in Chapter 10, also depends on the spacecraft's coefficient of drag and frontal area, which can also vary widely, further complicating the modeling problem.

The uncertainty in these variables is the main reason Skylab decayed and burned in the atmosphere several years earlier than first predicted. For a given orbit, however, we can approximate how the semimajor axis and the eccentricity change with time, at least in the short term. Spacecraft-tracking organizations use complex techniques to determine these values and make them available on request for each spacecraft.

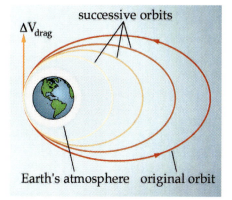

Figure 8-7. The Effects of Drag on an Eccentric Low-Earth Orbit. As the spacecraft passes through the atmosphere at perigee, the effect is like a series of small ΔVs, which lower apogee altitude, circularizing the orbit, until it decays and the spacecraft re-enters.

Earth's Oblateness

Our second assumption about a spacecraft's mass being much less than Earth's mass is still true, but what about the third assumption? Earth isn't really spherical. From space, it looks like a big, blue spherical marble, but when we look closer at the actual mass distribution, we find that Earth is actually kind of squashed. Thus, we can't really treat it as a pure point mass when we do very precise orbital modeling. We call this squashed shape *oblateness*. What does an oblate Earth look like? Imagine spinning a ball of gelatin and you can visualize how the middle (or equator) of the spinning gelatin would bulge out—Earth is fatter at the equator than at the poles. This bulge is often modeled by complex mathematics (which we won't do here) and is frequently referred to as the

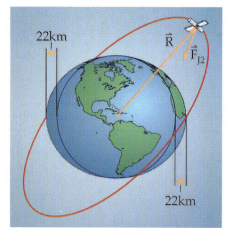

Figure 8-8. Earth's Oblateness. Earth's equatorial bulge, shown here greatly exaggerated, causes a slight shift in the direction that gravity pulls a spacecraft. The effect is a twisting force on the spacecraft's orbit.

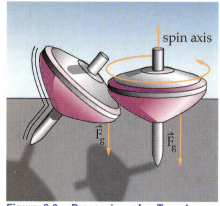

Figure 8-9. Precession of a Top. A non-spinning top will fall over if placed upright. But a spinning top wobbles about its spin axis due to the pull of gravity, $\vec{F}_g$. This motion is called precession.

J2 effect. J2 is a constant describing the size of the bulge in the mathematical formulas used to model the oblate Earth.

What effect does this have on an orbit? Let's look at Figure 8-8. Here the oblateness is shown very exaggerated; actually the bulge is only about 22 km thick. That is, Earth's radius is about 22 km larger along the equator than through the poles.

Let's reason what this bulge will do to the orbital elements. The force caused by the equatorial bulge is still gravity. Recall from Chapter 4 that gravity is a conservative force—the total mechanical energy of an orbit must be conserved. So, one of the constants of orbital motion we defined—the specific mechanical energy, ε—will not change. That means the semimajor axis, a, remains constant over the long term due to oblateness. It turns out the eccentricity, e, also doesn't change, although the explanation for this is beyond the scope of our discussion here. The bulge does pull on a spacecraft, so we expect the inclination to change due to oblateness, but it doesn't! Because the spacecraft is in orbit, the effect changes the right ascension of the ascending node, Ω, and moves the argument of perigee, ω, within the orbital plane. That's not very intuitive, but it's like a force acting on a spinning gyroscope, which we'll discuss in detail in Chapter 12. A similar analogy is a toy top. If you stand a non-spinning top on its point, gravity will cause it to fall over. But if you spin the top first, gravity still tries to make it fall over, but because of its angular momentum, it begins to swivel—this motion is called *precession*, as shown in Figure 8-9. Let's examine this effect on Ω and ω more closely.

How J2 Affects the Right Ascension of the Ascending Node, Ω

The effect of this equatorial bulge slightly perturbs the spacecraft because the gravitational force is no longer coming from Earth's exact center. This causes the plane of the orbit to precess (like the spinning top), resulting in a movement of the ascending node, $\Delta\Omega$. This motion is westward for direct orbits (inclination < 90°), eastward for retrograde orbits (inclination > 90°), and zero for polar orbits (inclination = 90°).

Figure 8-10 shows this *nodal regression rate,* $\dot{\Omega}$, as a function of inclination and orbital altitude. Let's look more closely at this figure. What it says is that the higher the spacecraft is, the less effect the bulge has on the orbit. This makes sense because the gravitational pull of the bulge decreases according to the inverse square law (μ/R^2). It also says that if the spacecraft is in a polar orbit (center of the graph), the bulge will have no effect. The greatest effect occurs at low altitudes with low inclinations. This makes sense, too, because the spacecraft is traveling much closer to the bulge during its orbit, and thus is pulled more by it. For low-altitude and low-inclination orbits, the ascending node can move as much as 9° per day (lower left corner and upper right corner of the graph).

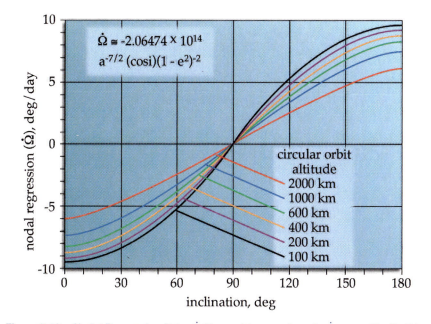

$$\dot{\Omega} \approx -2.06474 \times 10^{14}$$
$$a^{-7/2} (\cos i)(1 - e^2)^{-2}$$

circular orbit
altitude
2000 km
1000 km
600 km
400 km
200 km
100 km

Figure 8-10.   Nodal Regression Rate, $\dot{\Omega}$. The nodal regression rate, $\dot{\Omega}$, caused by Earth's equatorial bulge, reaches zero at an inclination of 90°. Positive numbers represent eastward movement; negative numbers represent westward movement. The less inclined an orbit is to the equator, the greater the effect of the bulge. The higher the orbit, the smaller the effect. An equation is given for finding $\dot{\Omega}$, where a = semimajor axis, i = inclination, and e = eccentricity.

How J2 Affects the Argument of Perigee, ω

Figure 8-11 shows how perigee location rotates for an orbit with a perigee altitude of 100 km, depending on the inclination, for various apogee altitudes. This *perigee rotation rate, $\dot{\omega}$*, is difficult to explain physically, but we can derive it mathematically from the equation for J2 effects on perigee location. With this perturbation, the major axis, or line of apsides, rotates in the direction of the spacecraft's motion, if the inclination is less than 63.4° or greater than 116.6°. It rotates opposite to the spacecraft's motion for inclinations between 63.4° and 116.6°.

Sun-synchronous and Molniya Orbits

The effects of Earth's oblateness on the node and perigee positions give rise to two unique orbits that have very practical applications. The first of these, the *sun-synchronous orbit*, takes advantage of eastward nodal progression at inclinations greater than 90°. Looking at Figure 8-10, we see at an inclination of about 98° (depending on the spacecraft's altitude), the ascending node moves eastward about 1° per day. Coincidentally, Earth also moves around the Sun about 1° per day (360° in 365 days), so at this sun-synchronous inclination, the spacecraft's *orbital plane* will always maintain the same orientation to the Sun, as shown in Figure 8-12. This means the spacecraft will always see the same Sun angle when it

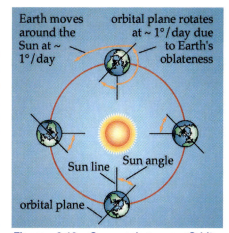

Figure 8-12. Sun-synchronous Orbits. Sun-synchronous orbits take advantage of the rate of change in right ascension of the ascending node, $\dot{\Omega}$, caused by Earth's oblateness. By carefully selecting the proper inclination and altitude, we can match the rotation of Ω with the movement of Earth around the Sun. In this case, the same angle between the orbital plane and the Sun will be maintained. Such orbits are useful for remote-sensing applications because shadows cast by targets on Earth stay the same.

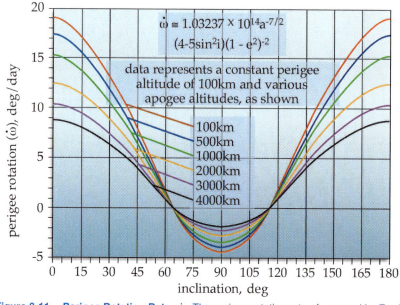

Figure 8-11.   Perigee Rotation Rate, $\dot{\omega}$. The perigee rotation rate, $\dot{\omega}$, caused by Earth's equatorial bulge depends on inclination and altitude at apogee. An approximate equation for finding $\dot{\omega}$ is given, where a = semimajor axis, i = inclination, and e = eccentricity.

passes over a particular point on Earth's surface. As a result, the Sun shadows cast by features on Earth's surface will not change when pictures are taken days or even weeks apart. This is important for remote-sensing missions such as reconnaissance, weather, and monitoring of Earth's resources, because they use shadows to measure an object's height and track other ones. By maintaining the same Sun angle day after day, observers can better track long-term changes in weather, terrain, and man-made features.

The second unique orbit is the *Molniya orbit*, named after the Russian word for lightning (as in "quick-as-lightning"). This is usually a 12-hour orbit with high eccentricity (about e = 0.7) and a perigee location in the Southern Hemisphere. The inclination is 63.4°—why? Because at this inclination the perigee doesn't rotate, as Figure 8-11 shows, so the spacecraft "hangs" over the Northern Hemisphere for nearly 11 hours of its 12-hour period before it whips "quick as lightning" through perigee in the Southern Hemisphere. Figure 8-13 shows the ground track for a Molniya orbit. The Russians used this orbit for their communication satellites because they didn't have launch vehicles large enough to put them into geosynchronous orbits from their far northern launch sites. Remember from Chapter 6, plane changes require large ΔVs and, as we'll see in Chapter 9, far northern launch sites don't get much extra kick from Earth's rotation. These Molniya orbits also better covered the polar and high latitudes above 80° north.

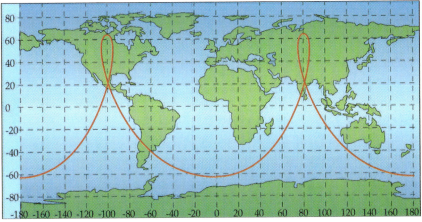

Figure 8-13.   Molniya Orbit. Molniya orbits take advantage of the fact that $\dot{\omega}$ due to Earth's oblateness is zero at an inclination of 63.4°. Thus, apogee stays over the Northern Hemisphere, covering high latitudes 11 hours out of the orbit's 12-hour period.

Other Perturbations

Other perturbative forces can affect a spacecraft's orbit and its orientation within that orbit. These forces usually are much smaller than the J2 (oblate Earth) and drag forces, but depending on the required accuracy, spacecraft planners may have to anticipate their effects. These forces include

- Solar radiation pressure—can cause long-term orbital perturbations and unwanted spacecraft rotation (as we'll discuss in Chapter 12)

- Third-body gravitational effects (Moon, Sun, planets, etc.)—can perturb orbits at high altitudes and on interplanetary trajectories

- Unexpected thrusting—caused by either outgassing or malfunctioning thrusters; can perturb orbits or cause spacecraft rotation

Astro Fun Fact
Russian Mirror

Russia has attempted a couple of new satellite experiments that could potentially improve the way of life on Earth. These experiments used the space-based mirror named Znamya, the Russian word for banner. The 25-meter mirror, made from Mylar™, reflects sunlight to locations on the night side of Earth. The light that reflects is 5–10 times brighter than the full Moon and illuminates an area on the ground roughly 4 km in diameter. Potential uses of this could be to provide illumination for night construction, night time disaster relief, and extended daylight in areas that have very little sunlight during the winter.

Beatty, J. Kelly. "Up in the Sky! It's a Bird, It's a Plane, It's Znamya!" Sky & Telescope Online. www.skypub.com, 1999.

Contributed by Scott R. Dahlke, the U.S. Air Force Academy

▰ Section Review

Key Terms

J2 effect
Molniya orbit
nodal regression rate, $\dot{\Omega}$
oblateness
perigee rotation rate, $\dot{\omega}$
perturbations
precession
sun-synchronous orbit

Key Concepts

➤ Perturbations resulting from small disturbing forces cause our two-body orbit to vary

➤ Atmospheric drag causes orbital decay by decreasing the semimajor axis, a, and the eccentricity, e

➤ Equatorial bulge of the oblate Earth (J2) causes the right ascension of the ascending node, Ω, and the argument of perigee, ω, to change in a predictable way

➤ We use oblateness perturbations to practical advantage in sun-synchronous and Molniya orbits

➤ Other perturbations may also have long-term effects on a spacecraft's orbit and orientation
  • Solar wind
  • Third body
  • Unexpected thrust

Example 8-3

Problem Statement

A remote-sensing satellite has the following orbital elements

$a_{initial} = 7303$ km

$e_{initial} = 0.001$

$i_{initial} = 50°$

$\Omega_{initial} = 90°$

$\omega_{initial} = 45°$

$\nu_{initial} = 0°$

U.S. Space Command has told you the semimajor axis is decreasing by 2 km/day. You would like nature to move the ascending node to a $\Omega = 30°$ before you command a spacecraft maneuver. You would also like to estimate your spacecraft's lifetime in case your orbital-correction thrusters stop operating. Assume your spacecraft will re-enter almost immediately if your semimajor axis drops below 6500 km. How long will it take for the ascending node to move to a position of 30°? How long can you expect your spacecraft to remain in orbit without further thrusting?

Problem Summary

Given: $a_{initial} = 7303$ km $\qquad \dot{a} = 2$ km/day

$e_{initial} = 0.001$ $\qquad$ Re-enter when

$i_{initial} = 50°$ $\qquad a = 6500$ km

$\Omega_{initial} = 90°$

$\omega_{initial} = 45°$

$\nu_{initial} = 0°$

Find: Time until $\Omega = 30°$

Time until $a = 6500$ km

Conceptual Solution

1) The orbit is essentially circular so you can use Figure 8-10 to find $\dot{\Omega}$, then

$$\text{wait time} = \frac{\Delta\Omega}{\dot{\Omega}}$$

2) Find decay time

$$\text{decay time} = \frac{\text{current } a - \text{minimum } a \text{ for decay}}{\text{decay rate}}$$

Analytical Solution

1) Using Figure 8-10, find $\dot{\Omega}$, then

$$\text{wait time} = \frac{\Delta\Omega}{\dot{\Omega}} = \frac{30° - 90°}{-5°/\text{day}} = 12 \text{ days}$$

2) Find decay time

$$\text{decay time} = \frac{\text{current } a - \text{minimum } a \text{ for decay}}{\text{decay rate}}$$

$$= \frac{7303 \text{ km} - 6500 \text{ km}}{2 \text{ km/day}} = 401.5 \text{ days}$$

Interpreting the Results

Earth's equatorial bulge will move the ascending node naturally to 30° in 12 days, thus saving precious fuel. The spacecraft will re-enter Earth's atmosphere in about 400 days if it can't thrust into a higher orbit.

8.3 Predicting Orbits in the Real World

▌▌▌ In This Section You'll Learn to...

☞ Combine what you've learned about Kepler's Problem and orbital perturbations to predict a spacecraft's future position

Let's put what we've learned about orbital perturbations together with the solution to Kepler's Problem and discuss in more detail how to predict a spacecraft's position in the real world. Assume we're tracking a spacecraft and have determined its orbital elements, $COE_{initial}$, at time $t_{initial}$.

Now let's step through the process of predicting the orbital elements, COE_{future}, at some time in the future, t_{future}. First, we need to find how these elements change with time due to the perturbations caused by atmospheric drag and the oblate Earth. We learned earlier that the oblate Earth (J2) affects Ω and ω, so we can use Figures 8-10 and 8-11 in Section 8.2 to find $\dot{\Omega}$ and $\dot{\omega}$. Inclination, i, isn't affected by either the oblate Earth or drag, so $i_{initial} = i_{future}$. We'll need to find out from our tracking organization (Figure 8-14) how drag affects our orbit's semimajor axis and eccentricity. They'll give us the time rate of change of the semimajor axis, $\dot{a}$, and the time rate of change in the eccentricity, $\dot{e}$.

Now we know how the first five elements change with time, so let's update them by multiplying the rate of change by the time interval and adding this to the initial value of the orbital element.

Figure 8-14. Tracking Site. We use data collected at radar tracking sites, such as this one at Millstone Hill, Massachusetts, to determine orbital perturbations due to drag. *(Reprinted with permission of MIT Lincoln Laboratory, Lexington, Massachusetts)*

$$a_{future} = a_{initial} + \dot{a}(t_{future} - t_{initial})$$

$$e_{future} = e_{initial} + \dot{e}(t_{future} - t_{initial})$$

$$i_{future} = i_{initial}$$

$$\Omega_{future} = \Omega_{initial} + \dot{\Omega}(t_{future} - t_{initial})$$

$$\omega_{future} = \omega_{initial} + \dot{\omega}\ (t_{future} - t_{initial})$$

(8-13)

where

$a_{initial}, a_{future}$	= initial and future values of semimajor axis (km)
$\dot{a}$	= time rate of change of semimajor axis (km/day)
$t_{initial}, t_{future}$	= initial and future time (days)
$e_{initial}, e_{future}$	= initial and future values of eccentricity
$\dot{e}$	= time rate of change of eccentricity (1/day)
$i_{initial}, i_{future}$	= initial and future values of inclination (deg)
$\Omega_{initial}, \Omega_{future}$	= initial and future values of the right ascension of the ascending node (deg)

$$\dot{\Omega} \qquad\qquad = \text{time rate of change of the right ascension of the ascending node (deg/day)}$$

$$\omega_{initial}, \omega_{future} = \text{initial and future values of the argument of perigee (deg)}$$

$$\dot{\omega} \qquad\qquad = \text{time rate of change of the argument of perigee (deg/day)}$$

We have only the true anomaly, v, left to update. We know true anomaly, v, changes with time, even without perturbations, but atmospheric drag also affects it, because drag decreases the semimajor axis, a, and therefore shortens the period. This means the spacecraft speeds up and so does the rate of change of v. As we learned in the last section, we need to use Kepler's equation to update v by using the mean anomaly, M, and the eccentric anomaly, E. Let's update the mean anomaly first, (using Equation (8-6)).

$$M_{future} - M_{initial} = n\,(t_{future} - t_{initial}) - 2k\pi$$

or

$$M_{future} = M_{initial} + n\,(t_{future} - t_{initial}) - 2k\pi \qquad (8\text{-}14)$$

where

$$M_{future}, M_{initial} = \text{future and initial values of the mean anomaly (rad)}$$

$$n \qquad\qquad\quad = \text{mean motion (rad/s)}$$

$$k \qquad\qquad\quad = \text{number of orbits from the initial position}$$

We know this equation works when there are no perturbations, but what happens when we add drag? What changes? Remember from Equation (8-1), we can find the mean motion

$$n = \frac{2\pi}{P} = \sqrt{\frac{\mu}{a^3}}$$

where

P = period (s)

μ = gravitational parameter (km^3/s^2)

a = semimajor axis (km)

Recall the semimajor axis, a, changes due to drag ($\dot{a}$). This means the mean motion, n, also changes, so we need to find $\dot{n}$. What value for n should we use in Equation (8-14) to solve for M_{future}? Let's look at how the mean motion changes. At $t_{initial}$, the mean motion is $n_{initial}$. At t_{future}, the mean motion is $n_{initial} + \dot{n}\,(t_{future} - t_{initial})$, so the *average mean motion*, $\bar{n}$, is

$$\bar{n} = \frac{n_{initial} + [n_{initial} + \dot{n}(t_{future} - t_{initial})]}{2}$$

$$= n_{initial} + \frac{\dot{n}}{2}(t_{future} - t_{initial}) \qquad (8\text{-}15)$$

where

$\bar{n}$ = average mean motion (rad/s)

$\dot{n}$ = time rate of change of mean motion (rad/s^2)

We just added the initial and future values for mean motion and divided by 2. If we now substitute this value of $\bar{n}$ for n into Equation (8-14), we get M_{future} based on the average mean motion for the time interval

$$M_{future} \;=\; M_{initial} + \bar{n}(t_{future} - t_{initial}) - 2k\pi \qquad (8\text{-}16)$$

We use the $2k\pi$ to subtract 2π from M_{future} until M_{future} is less than 2π, for convenience. Now we use Kepler's Equation (8-10) to find E_{future}

$$E_{future} = M_{future} + e_{future} \sin E_{future}$$

Remember, this is a transcendental equation, so we have to iterate to solve for E_{future}. Finally, we can use Equation (8-9) to solve for the true anomaly at t_{future}

$$\cos \nu_{future} \;=\; \frac{\cos E_{future} - e_{future}}{1 - e_{future} \cos E_{future}} \qquad (8\text{-}17)$$

We now have all the orbital elements (a_{future}, e_{future}, i_{future}, Ω_{future}, ω_{future}, and ν_{future}) for the future time, t_{future}. In real life we would then convert these elements back to position and velocity vectors, $\vec{R}$ and $\vec{V}$, and then back to range, azimuth, and elevation for our tracking site (Figure 8-15) to know where to point. Look at Figure 8-2 and you'll see we've just completed the entire tracking and prediction process.

Figure 8-15.   Eglin Phased Array Radar Site. Modern tracking sites don't have radar dishes, but instead, they electronically steer the radar beam from the face of a large concrete structure. *(Courtesy of the U.S. Air Force)*

In summary, we've looked at how our original restricted two-body problem changes by adding in the real-world effects of a non-spherical Earth and atmospheric drag. Then, using Kepler's Equation, we've learned to update our orbit to some future time, so we can use its position for our mission planning.

▦ Section Review

Key Terms

average mean motion, $\bar{n}$

Key Equations

$a_{future} = a_{initial} + \dot{a}(t_{future} - t_{initial})$

$e_{future} = e_{initial} + \dot{e}(t_{future} - t_{initial})$

$i_{future} = i_{initial}$

$\Omega_{future} = \Omega_{initial} + \dot{\Omega}(t_{future} - t_{initial})$

$\omega_{future} = \omega_{initial} + \dot{\omega}(t_{future} - t_{initial})$

$\bar{n} = \dfrac{n_{initial} + [n_{initial} + \dot{n}(t_{future} - t_{initial})]}{2}$

$\quad = n_{initial} + \dfrac{\dot{n}}{2}(t_{future} - t_{initial})$

$M_{future} = M_{initial} + \bar{n}(t_{future} - t_{initial}) - 2k\pi$

Key Concepts

➤ Using our knowledge of perturbations, we can update the orbital elements from time $t_{initial}$, to time t_{future}

➤ Drag causes the semimajor axis, and hence mean motion, to change with time

➤ To find v_{future}, we

- Determine average mean motion, $\bar{n}$
- Determine M_{future}

$$M_{future} = M_{initial} + \bar{n}(t_{future} - t_{initial})$$

- Compute E_{future} using iteration

$$E_{future} = M_{future} + e_{future}\sin E_{future}$$

- Check the E_{future} quadrant is the same as the M_{future} quadrant
- Solve for v_{future}

$$\cos v_{future} = \dfrac{\cos E_{future} - e_{future}}{1 - e_{future}\cos E_{future}}$$

- Check the v_{future} quadrant is the same as the E_{future} and M_{future} quadrants

Example 8-4

Problem Statement

A remote-sensing satellite has the following COEs and perturbations. Determine the COEs ten days from now.

$a_{initial} = 7000$ km $\quad \dot{a} = -0.7$ km/day

$\quad\quad\quad\quad\quad\quad\quad \dot{n} = 0.00003$ rad/day^2

$e_{initial} = 0.05 \quad\quad \dot{e} = -0.00003$/day

$i_{initial} = 50°$

$\Omega_{initial} = 90° \quad\quad \dot{\Omega} = -5.0°$/day

$\omega_{initial} = 45° \quad\quad \dot{\omega} = 4.0°$/day

$\nu_{initial} = 270°$

Problem Summary

Given: $a_{initial} = 7000$ km $\quad \dot{a} = -0.7$ km/day

$\quad\quad e_{initial} = 0.05 \quad \dot{n} = 0.00003$ rad/day^2

$\quad\quad i_{initial} = 50° \quad \dot{e} = -0.00003$/day

$\quad\quad \Omega_{initial} = 90° \quad \dot{\Omega} = -5.0°$/day

$\quad\quad \omega_{initial} = 45° \quad \dot{\omega} = 4.0°$/day

$\quad\quad \nu_{initial} = 270°$

Find: The satellite's COEs and position ten days from now

Problem Diagram

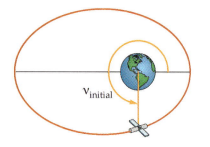

$\nu_{initial}$

Conceptual Solution

1) Update the orbital elements to the new time, t_{future}

$$a_{future} = a_{initial} + \dot{a}\,(t_{future} - t_{initial})$$

$$e_{future} = e_{initial} + \dot{e}\,(t_{future} - t_{initial})$$

$$i_{future} = i_{initial}$$

$$\Omega_{future} = \Omega_{initial} + \dot{\Omega}\,(t_{future} - t_{initial})$$

$$\omega_{future} = \omega_{initial} + \dot{\omega}\,(t_{future} - t_{initial})$$

2) Find $n_{initial}$

$$n_{initial} = \frac{2\pi}{p} = \sqrt{\frac{\mu}{a_{initial}^3}}$$

3) Find $E_{initial}$

$$\cos E_{initial} = \frac{e_{initial} + \cos\nu_{initial}}{1 + e_{initial}\,\cos\nu_{initial}}$$

4) Find $M_{initial}$

$$M_{initial} = E_{initial} - e_{initial}\sin E_{initial}$$

5) Find average mean motion, $\bar{n}$

$$\bar{n} = n_{initial} + \frac{\dot{n}}{2}(t_{future} - t_{initial})$$

6) Move mean anomaly to the desired time

$$M_{future} = M_{initial} + \bar{n}(t_{future} - t_{initial})$$

7) Solve for E_{future} using Kepler's Equation and iteration

$E_{future} = M_{future} + e_{future}\sin E_{future}$,
Check quadrant.

8) Find ν_{future}

$$\cos\nu_{future} = \frac{\cos E_{future} - e_{future}}{1 - e_{future}\,\cos E_{future}},$$
Check quadrant.

Example 8-4 (Continued)

Analytical Solution

1) Update the orbital elements

$a_{future} = a_{initial} + \dot{a}\,(t_{future} - t_{initial}) = 7000 \text{ km} -$
$\qquad (0.7 \text{ km}/\text{day})\,(10 \text{ days}) = 6993 \text{ km}$

$e_{future} = e_{initial} + \dot{e}\,(t_{future} - t_{initial}) = 0.05 -$
$\qquad (0.00003/\text{day})\,(10 \text{ days}) = 0.0497$

$i_{future} = i_{initial} = 50°$

$\Omega_{future} = \Omega_{initial} + \dot{\Omega}\,(t_{future} - t_{initial})$
$\qquad = 90° - (5°/\text{day})\,(10 \text{ days}) = 40°$

$\omega_{future} = \omega_{initial} + \dot{\omega}\,(t_{future} - t_{initial})$
$\qquad = 45° + (4°/\text{day})\,(10 \text{ days}) = 85°$

2) Find $n_{initial}$

$$n_{initial} = \sqrt{\frac{\mu}{a_{initial}^3}} = \sqrt{\frac{3.986 \times 10^5\, \dfrac{\text{km}^3}{\text{s}^2}}{(7000 \text{ km})^3}}$$

$\qquad = 0.001078 \text{ rad}/\text{s} = 14.82 \text{ rev}/\text{day}$

3) Find $E_{initial}$

$$\cos E_{initial} = \frac{e_{initial} + \cos\nu_{initial}}{1 + e_{initial}\cos\nu_{initial}}$$

$$= \frac{0.05 + \cos 270°}{1.0 + 0.05\cos 270°} = 0.05$$

$E_{initial} = 87.13°$, or $272.87°$

Remember, ν, E, and M must all lie in the same half plane. So, we choose $E_{initial} = 272.87°$, because $\nu_{initial} = 270°$. In radians, $E_{initial} = 4.762$ rad

4) Find $M_{initial}$

$M_{initial} = E_{initial} - e_{initial}\sin E_{initial}$

Note: Here $E_{initial}$ must be in radians

$M_{initial} = 4.762 - 0.05\sin 4.762 = 4.812 \text{ rad} = 275.73°$

Here, $M_{initial}$ is in the same half plane with $\nu_{initial}$ and $E_{initial}$.

5) Find the average mean motion, $\bar{n}$

$$\bar{n} = n_{initial} + \frac{\dot{n}}{2}(t_{future} - t_{initial})$$

$\bar{n} = 0.001078 \text{ rad}/\text{s}((86,400 \text{ s}/\text{day}) + \;)$

$$\frac{0.00003 \text{ rad}/\text{day}^2}{2} \times (10 \text{ day}) = 93.1394 \text{ rad}/\text{day}$$

6) Move the mean anomaly to the desired time

$M_{future} = M_{initial} + \bar{n}\,(t_{future} - t_{initial}) - 2k\pi$

$M_{future} = 4.812 \text{ rad} + 93.1394 \text{ rad}/\text{day} \times (10 \text{ day})$
$\qquad = 936.206 \text{ rad}$

Because this value is greater than 2π, we need to subtract 2π, repeatedly, until M_{future} is less than 2π.

$M_{future} = 0.01141 \text{ rad} = 0.6537°$

Physically, this means our satellite passes perigee 149 times in the next ten days

7) Solve for E_{future} using Kepler's Equation (iterative solution required)

First, guess $E_{future} = M_{future}$

$E_{future} = M_{future} + e_{future}\sin E_{future} = 0.01141 +$
$\qquad 0.0497\sin 0.01141 = 0.01198 \text{ rad}$

Now, use this value of E_{future} for your next guess

$E_{future} = M_{future} + e_{future}\sin E_{future} = 0.01141 +$
$\qquad 0.0497\sin 0.01255 = 0.01201 \text{ rad}$

Again, use this new value of E_{future} next

$E_{future} = M_{future} + e_{future}\sin E_{future} = 0.01141 +$
$\qquad 0.0497\sin 0.01201 = 0.01201 \text{ rad}$

Okay, within the accuracy we've chosen, our solution has converged.

8) Find ν_{future}

$$\cos\nu_{future} = \frac{\cos E_{future} - e_{future}}{1 - e_{future}\cos E_{future}}$$

$$= \frac{\cos 0.01201 - 0.0497}{1 - 0.0497\cos 0.01201}$$

$\cos\nu_{future} = 0.99992$, $\;\nu_{future} = 0.01262 \text{ rad} =$
$\qquad 0.7230°$

Interpreting the Results

In ten days, our satellite will make 149 trips around its orbit, which has been perturbed by drag and Earth's oblateness. The eccentricity and semimajor axis will decrease during this period. Inclination will remain unchanged while right ascension of the ascending node will regress 50° and argument of perigee will advance to 85°. True anomaly will be 0.7230°.

References

Bate, Roger R., Mueller, Donald D., and White, Jerry E., *Fundamentals of Astrodynamics.* New York: Dover Publications, Inc., 1971.

Mission Problems

8.1   Predicting an Orbit (Kepler's Problem)

1   Describe how to track and predict an orbit.

2   For a spacecraft in a circular, polar orbit with an altitude of 400 km, find its time of flight between the equator and the North Pole.

3   A spacecraft in an elliptical, polar orbit, with a semimajor axis of 6778 km and an eccentricity of 0.04, is at perigee as it passes the ascending node. How long will the spacecraft take to reach the North Pole?

4   For a spacecraft in a circular, polar orbit with an altitude of 400 km, find the spacecraft's location, argument of latitude, u, in 40 minutes if the spacecraft is at the ascending node.

5   A spacecraft in an elliptical, polar orbit, with a semimajor axis of 6778 km and an eccentricity of 0.04, is at perigee as it passes the ascending node. Find the true anomaly 40 minutes later.

8.2   Orbital Perturbations

6   Describe the effects of atmospheric drag on the orbits of spacecraft in low-Earth orbit.

7   Describe the effects of an oblate Earth on the orbits of spacecraft in low-Earth orbit.

8   List three other sources of orbital perturbations, besides atmospheric drag and the oblate Earth.

9   How long will a spacecraft in an elliptical orbit with a perigee altitude of 650 km take to decay to an altitude of 160 km, if the average decay rate is 0.6 km/day?

10   A spacecraft is in a circular orbit at an altitude of 200 km and an inclination of 28°. How far will the ascending node move in one day? In which direction will it move?

8.3   Predicting Orbits in the Real World

11   An Earth-resource satellite has the following orbital parameters

$a_{initial}$ = 6900 km $\dot{a}$ = –0.8 km/day

$e_{initial}$ = 0.03 $\dot{e}$ = –0.00004/day

$i_{initial}$ = 75° $\dot{i}$ = 0.0

$\Omega_{initial}$ = 90° $\dot{\Omega}$ = –3.0°/day

$\omega_{initial}$ = 45° $\dot{\omega}$ = –1.5°/day

$\nu_{initial}$ = 90° $\dot{n}$ = 0.00003 rad/day^2

Find the COEs of the satellite in 30 days.

Projects

12   Pick a clear night and try to observe spacecraft in low-Earth orbit right after sunset or right before sunrise (this is the only time they are illuminated by the Sun and not yet in Earth's shadow). Time their passage and approximate position in the sky and see if you can estimate their inclinations and right ascensions of ascending node.

Mission Profile—Skylab

on after the first successful lunar landing, NASA realized that without clear goals for its human space program, Congress and the public would quickly lose interest in funding future spaceflights. NASA proposed the Apollo Applications Program (AAP), which would use surplus Apollo hardware to bridge the gap between Apollo, and the proposed shuttle program, and a permanent space station. After cuts to Apollo (from ten lunar landings to six), the AAP became Skylab.

Mission Overview

Skylab was a prefabricated space station launched on a Saturn V. Three crews conducted long-duration human missions to the station that included solar and Earth science experiments. Future shuttle missions were to use it until a permanent station was built. Unfortunately, Skylab's orbit was so low that air drag reduced its velocity below minimum orbital speed, and the spacecraft was destroyed when it re-entered the atmosphere on July 11, 1979—two years before the Space Shuttle's first flight.

Mission Data

Skylab 1, May 14, 1973: Launch of Skylab station atop a Saturn V. Damaged during launch.

Skylab 2, May 25, 1973 (Crew: Conrad, Kerwin, Weitz): Mission lasted for 28 days and included extensive extra-vehicular activity to repair damage to the station, most notably to install a new heat shield.

Skylab 3, July 29, 1973 (Crew: Bean, Garriott, Lousma): Mission lasted for 59 days. Emphasis was on solar, Earth, and life sciences.

Skylab 4, November 16, 1973 (Crew: Carr, Gibson, Pogue): Mission lasted for 84 days (U.S. record). Observations of comet Kahoutek were a notable addition to the mission.

Mission Impact

Skylab was valuable because it was the United States' first space station. It helped scientists discover the effects of long spaceflights on astronauts, as well as valuable scientific data about the Earth and Sun. (Ironically, sunspot activity contributed to Skylab's orbital decay before it could be rescued).

Skylab in Orbit. *(Courtesy of NASA/Johnson Space Center)*

For Discussion

- What did we learn from Skylab?

- How will current proposals for a space station differ from Skylab?

Contributor

Todd Lovell, the U.S. Air Force Academy

References

Baker, David. *The History of Manned Spaceflight*. New York: Crown, 1981.

Yenne, Bill. *The Encyclopedia of U.S. Spacecraft*. New York: Exeter, 1985.

It's launch time for the Shuttle, Discovery, for STS-95, in October, 1998. *(Courtesy of NASA/Johnson Space Center)*

Getting To Orbit

9

with contributions from
Dr. Scott Dahlke, the U.S. Air Force Academy

In This Chapter You'll Learn to...

- Describe launch windows and how they constrain when we can launch into a particular orbit
- Determine when and where to launch, as well as the required velocity, to reach a specific orbit
- Demonstrate how mission planners determine when, where, and with what velocity to launch spacecraft into their desired orbits

You Should Already Know...

- ❏ Constants of orbital motion (Chapter 4)
- ❏ Definition of the Geocentric-equatorial Coordinate System (Chapter 4)
- ❏ Definitions of the classic orbital elements (Chapter 5)
- ❏ The difference between direct and retrograde orbits (Chapter 5)
- ❏ How to determine orbital inclination from the orbital ground track (Chapter 5)

Outline

9.1 **Launch Windows and Times**
   Launch Windows
   Launch Time

9.2 **When and Where to Launch**

9.3 **Launch Velocity**

If you don't know where you're going, you'll probably end up somewhere else.

Yogi Berra
former New York Yankees catcher

Space Mission Architecture. This chapter deals with the Trajectories and Orbits segment of the Space Mission Architecture, introduced in Figure 1-20.

F ew scenes are as spectacular as the launch of a Space Shuttle. The three main engines ignite, followed a few seconds later by the powerful solid-rocket motors. Then, the 20-story vehicle arcs into the clear Florida skies on a plume of gray exhaust. But how do we know when to light the engines, when to shut them off, and perhaps most important (at least for the people who live down range), where to point the thing? In this chapter we'll answer these questions. We'll first examine when the time is right to start the rockets for a journey into space. Then we'll dig into what velocity a spacecraft needs and in what direction we must launch it to get it into the desired mission orbit.

Figure 9-1.   NASA Views of Shuttle Launches. Upper left: The Shuttle, Endeavour, awaits launch time for STS-86, in September, 1997. Upper right: Endeavour clears the tower on STS-47, in September, 1992. Lower left: The Shuttle Columbia rises majestically off the pad for STS-90, in April, 1998. Lower right: Columbia climbs above the billowy smoke as it accelerates into orbit on STS-87, in November, 1997.

9.1 Launch Windows and Times

▤ In This Section You'll Learn to...

- ☛ Define a launch window
- ☛ Calculate time using Earth's rotation
- ☛ Explain the difference between the sidereal time we use to compute launch windows and the solar time we keep on our watches

For most space missions, the spacecraft must be placed into a particular orbit—polar orbits for remote-sensing, geostationary orbits for communication, and so on. To meet these requirements, launch team members need to launch it from a specific place at a particular time and in a particular direction. Let's see how we go about meeting these requirements.

Launch Windows

As part of mission planning we must select the exact orbit we want. The most common way to specify that orbit is to define a set of classic orbital elements that satisfy mission objectives. This may sound like no big deal, but trying to achieve a specific set of orbital elements along with other mission constraints can severely limit when and in what direction to launch. In some cases, certain orbital conditions may be impossible to achieve because of the launch-site location or launch vehicle capability.

We define the *launch window* to be the period of time when we can launch a spacecraft directly into a specified orbit from a given launch site. Notice we said "directly." We can always launch it into a parking orbit and then do a Hohmann Transfer and a plane change to put it in the desired orbit, but this is much more complicated and requires more fuel.

One way to understand launch windows is to think about bus schedules. Suppose you've made a date to meet a friend in a particular time and place and you need to catch a bus to get there. Only certain buses will get you to your meeting on time. If the schedule shows the bus you need leaves the bus stop at 11:13 A.M., you'd better be there at 11:13 A.M.— or close to it. If you miss this bus, you may have to wait quite a while until the next bus going your way comes along. The time and place specified for your meeting are similar to the desired orbital elements. The time the bus is scheduled to leave is similar to the launch window.

In our discussion, we define a launch window at an exact time—e.g., 11:13 A.M. In practice however, a launch window normally covers a period of time during which we can launch—usually several minutes or even hours around this exact time—just as a bus scheduled to leave at 11:13 A.M. could leave anytime between 11:10 A.M. and 11:15 A.M. and still arrive at its destination on time by adjusting its speed slightly. Mission planners have some flexibility in the orbital elements they can accept, and launch vehicles usually can steer enough to expand the length of the window.

291

Because a rocket must follow a trajectory governed by Newton's laws of motion, a launch window restricts a space launch much more than a bus schedule restricts passengers. Let's begin our investigation of launch windows by looking at the problem we face in relating what goes on in an orbit to what goes on at a launch site. As we learned in Chapter 4, orbital planes are fixed in inertial space, while Earth (and the launch site along with it) rotates *underneath* this orbital plane. As a result, a launch site at a particular point on Earth will intersect the orbital plane only periodically as Earth rotates. (In some cases, it may not intersect the orbital plane at all.) *When a launch site and an orbital plane intersect, we have a launch window and can launch directly into that orbit.*

Launch Time

In addition, we'll deal with only one major constraint on launch time— the physical alignment of the launch site and desired orbit. In practice, constraints on launch time include availability of the launch-site, tracking stations, weather, lighting conditions at runways in case of an abort (for human launches), and political considerations.

How do we know what time we can launch into an orbit? Because Earth's rotation under the spacecraft's intended orbit is periodic, by knowing this period, we can determine the launch window. To see why, it's helpful to think about a car driving around a two-mile, circular racetrack. Let's assume the car is one mile past the starting line, and the turnoff for the pit is 1/2 mile before the starting line, as seen in Figure 9-2. Using simple subtraction, we can see the car is 1/2 mile from the pit turnoff. If we also know the car's speed, we can easily determine how long before the car reaches the pit. If the car stays at a constant speed, we can also predict several laps in advance when it will reach the pit turn-off.

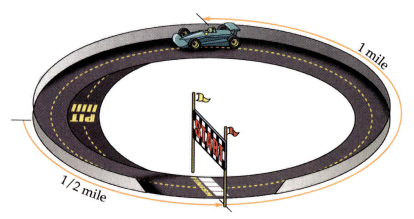

Figure 9-2.   Orbital Racetrack. To visualize how we relate periodic events for launch windows, we imagine a car going around a two-mile, circular racetrack. If the car is 1 mile past the starting line, and the pit turnoff is 1/2 mile before the starting line, then the car must be 1/2 mile from the pit. Given the car's speed, we can determine how long before the car reaches the pit.

What does all this have to do with launch windows? Well, launch windows also repeat periodically, so we can use almost the same approach to find when they occur. First, we must establish a reference direction, from which to measure the locations of the orbital plane and the launch site. Just as we used the starting line on the racetrack to reference the car's location, we can use the vernal equinox direction for our launch problem. Recall, the vernal equinox direction is the principal direction in the geocentric-equatorial coordinate system, which we use to describe the motion of Earth-orbiting spacecraft. Thus, it's a convenient reference from which to measure the angular distance between the orbital plane and our launch site, as Earth and the launch site rotate. (This distance is analogous to the distance between the race car and the pit in Figure 9-2.) By knowing this angular distance and Earth's rotation rate, we can determine opportunities to launch. In effect, we're using Earth as a big clock.

Establishing a launch window means we choose a particular clock (solar) time for launch. Mission planners reference the international clock time in Greenwich, England when establishing a launch time, so we need to understand how to measure solar time. People initially marked time in days because they didn't have a more refined measure. By observing shadows on a sundial, they could mark the time between the Sun's successive passages above a certain point, which we call an *apparent solar day*.

Because Earth's orbit around the Sun is slightly elliptical (e = 0.017), the apparent solar day's length varies a bit throughout the year. To compensate for this variation, we take the average of the lengths for one year to get a *mean solar day*. This is the time we see on our watches. There are 24 mean solar hours in a mean solar day.

Astro Fun Fact
Mean Solar Time

We calculate mean solar time by taking the annual average of the times between the Sun's successive passes over a point on Earth. This period of time must be averaged because Earth's orbit is slightly elliptical (which makes Earth's speed vary with its distance from the Sun.) As this method correlates with Earth's rotation it has some inherent inaccuracy. As you can imagine, many scientists and engineers want extreme accuracy regardless of the movement of the Earth and Sun. In response to this need, they developed a nearly perfect time-keeping device. This clock uses hydrogen atoms to produce a frequency which vibrates with amazing consistency. If this clock operated for a human's average life span, its error would be less than one microsecond. If it began operating when the solar system began, it would now be less than two minutes off the actual time!

Goudsmit, Samuel A. and Robert Claiborne, ed. Time. New York: Time Incorporated, 1966.

Cowan, Harrison J. Time and It's Measurement. Cleveland, OH: The World Publishing Company, 1958.

Contributed by Troy Kitch, the U.S. Air Force Academy

Because people around the world like to keep their time synchronized with the Sun, Earth has 24 time zones. To avoid confusion across time zones, we choose the Greenwich or 0° longitude line (Prime Meridian) as an international reference point. The local mean solar time at the Prime Meridian is the *Greenwich Mean Time (GMT)*, used by mission planners.

Unfortunately, when it comes to actually calculating when a particular launch site rotates under a specific orbit, solar time has a problem. We define our desired orbital elements with respect to the geocentric-equatorial coordinate system. Because this is an Earth-centered system, the Sun (and our solar time references) appear to move with respect to the system as Earth orbits around it. Instead of using solar time, then, we must define a new kind of time, *sidereal time*, which we measure as the angle between a longitude line and the vernal equinox. Sidereal means "related to the stars" and, as we're referencing a point way out in space, this is a good description. Just as we say that successive passages of the Sun over a given longitude is a solar day, so successive passages of the vernal equinox over a specific longitude is defined as a *sidereal day*. If the longitude we're using is the local longitude (say of our home town or a launch site), the time since the vernal equinox last passed over the local longitude is the *local sidereal time (LST)*.

Astro Fun Fact
Prime Meridian

(Courtesy of National Maritime Museum Greenwich, U.K.)

The Royal Observatory in Greenwich, England, was constructed in the late seventeenth century to track the movement of the Moon and location of the stars. This information (published in the Nautical Almanac of 1767) was essential for mariners and seamen who had no other way to locate their east-west position (longitude). They discovered their latitude by using this data coupled with an onboard device called a sextant. With the introduction of the railroads in the nineteenth century, a universal time standard became necessary. In England, local time had to be adjusted almost every time the train entered another town! The Greenwich site, with it's accurate time measurement and famous observatory, became the natural choice for a universal standard time. In time, the world accepted the Greenwich locale as 0° longitude, employing this Meridian as a timekeeping and navigation standard for land, air, space, and sea.

(Courtesy of Alan Palmer)

The Old Royal Observatory: The Story of Astronomy & Time. Courtesy British Royal Observatory, Greenwich, England.

Contributed by Troy Kitch, the U.S. Air Force Academy

To understand how this new way of measuring time works, imagine a polar orbit around Earth, as shown in Figure 9-3. As Earth rotates, the launch site comes under the orbital plane twice a day, twelve sidereal hours apart, at which times we can launch our spacecraft (our launch windows occur). So, at any given time, we can measure the angle from the launch site longitude to the orbital plane, we'll know how many sidereal hours until we can push the launch button.

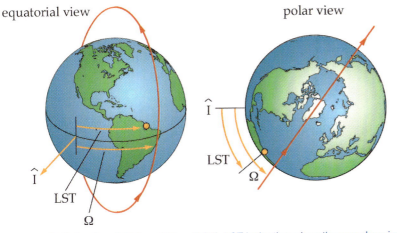

Figure 9-3.   **Defining Local Sidereal Time (LST).** LST is the time since the vernal equinox passed over a particular longitude line. For the launch site shown in the diagram, it's the time since the vernal equinox passed over the launch site longitude line.

We're used to telling clock time (related to the Sun) in hours, minutes, and seconds. For sidereal time, does it make sense to define "time" as an angle, since we're dealing with Earth's rotation? Why not? We can tell time just as easily in degrees as in hours, if we use Earth as a giant clock. Because Earth rotates 360° in 24 hours, it rotates 15°/hr (360° ÷ 24 hr). (Later we'll see it's actually a little more than 15°/hr) This explains why time zones around the world span about 15° of longitude. For example, we could say 1:00 A.M. is the same thing as 15° (Earth rotates 1 hour or 15° since midnight, which is 0 o'clock or 0°). 1:00 P.M., which is 1300 using a 24-hour clock, is 195° (15°/hr × 13 hr). A standard 12-hour clock face and the corresponding angles are in Figure 9-4. The relationship between time measured in hours and time measured in degrees is

$$\text{Time (degrees)} = \text{Time (hours)} \times \omega_{\text{Earth}}$$

where

ω_{Earth} = Earth's rotation rate = 15°/hr

Time in degrees may seem strange at first, but it allows us to find the launch time using the angle between the launch-site longitude and the orbital plane. As we'll see, to calculate launch time we also need another angle, the right ascension of the ascending node.

So, if our launch window is based on sidereal time, and we have only solar watches, how do we know when to punch the button to launch? We

Figure 9-4.   **Telling Time.** We can tell time in degrees as easily as in hours. 3:00 A.M. (0300) or 3 hours past midnight is 45° of Earth rotation. Similarly, noon (1200) is 180°, and 6:00 P.M. (1800), is 270°.

must understand the difference between solar and sidereal time. To visualize this difference, assume you want to track the position of the constellation Aquarius in the night sky (it will be the "age of Aquarius" in about 400 years). If, each night, we note on our watch when Aquarius rises in the sky, we'll see something strange: it appears about four minutes earlier than the previous night, and, eventually, it no longer appears in the night sky at all.

What's going on? Is our watch broken? No! The difference between solar time (on our watch) and sidereal time (referenced to the stars) causes this change in position relative to clock time. Eventually, Aquarius begins rising in the daytime, when we can't see it. To view it again, we have to wait through enough four-minute cycles for it to rise in the night sky once more (about six months).

Why doesn't this shift affect where we see the Sun at the same time each day? As Earth rotates, it also moves around the Sun in its orbit, causing an apparent shift in the Sun's position. Because Earth revolves 360° around the Sun in about 365 days, it moves slightly less than 1° per day along its orbit. Thus, it must rotate on its axis slightly more than 360° to bring the Sun successively over a given location. Marking one solar day in this way compensates for the apparent shift in the Sun's position. This also means one solar day is longer than one sidereal day

$$\text{1 mean solar day} = 1.0027 \text{ sidereal days}$$

23 hr 56 min 04 s mean solar time = 24 sidereal hours = 1 sidereal day

We illustrate the reason for this difference between a solar day and a sidereal day in Figure 9-5. Note that here we're using the vernal equinox direction as our fixed point of reference.

> Note: Earth's rotation rate, ω_{Earth}, is 360° in 24 sidereal hours or 15° per sidereal hour. There are 23 hr 56 min and 4 s of solar time in 24 sidereal hours. So, in terms of solar time ω_{Earth} is 15.04107° per mean solar hour.

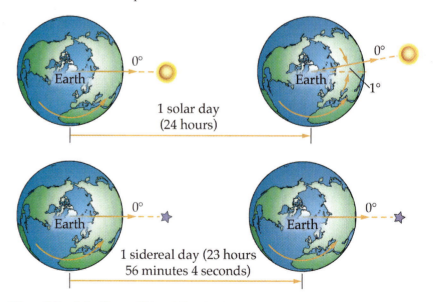

Figure 9-5.   Solar Versus Sidereal Day. A solar day is longer than a sidereal day because Earth rotates slightly more than 360° to bring the Sun back over a certain point, while Earth revolves around the Sun. The vernal equinox direction stays fixed in space, so Earth rotates exactly 360° for a sidereal day.

Unfortunately, converting from local sidereal time to local solar time is not simply a matter of subtracting four minutes. Although (fortunately) we don't have to worry about doing the conversion here, it requires us to know the exact position of the vernal equinox at some past solar time. We then have to "propagate," or account for Earth's rotation and revolution about the Sun, from that time to the sidereal time we're interested in. In practice, mission planners work in local sidereal time (converting it to Greenwich Mean Time [GMT] for launch controllers) and use precise launch-window geometry to calculate launch opportunities, as we'll see in the next section.

Section Review

Key Terms

apparent solar day
Greenwich Mean Time (GMT)
launch window
local sidereal time (LST)
mean solar day
sidereal day
sidereal time

Key Concepts

➤ A launch window is the period during which we can launch directly into a desired orbit from a particular launch site

➤ We can measure time in degrees as easily as in hours

➤ A mean solar day is the average time between the Sun's successive passages over a given longitude on Earth

  • Mean solar time is the time we keep on our clocks and watches

  • Greenwich Mean Time (GMT) is the mean solar time at Greenwich, England, which is on the Prime Meridian (0° longitude)

➤ We measure solar time with respect to the Sun. Because Earth revolves about the Sun, solar time isn't a good inertial time reference for launching spacecraft. Instead, we use sidereal time (having to do with the stars), with the vernal equinox direction as a reference

  • We define a sidereal day as the time between successive passages of the vernal equinox over a given longitude on Earth

  • Local sidereal time (LST) is the time since the vernal equinox was last over a given local longitude

➤ Earth must rotate slightly more than 360° to bring a given longitude back directly under the Sun, because Earth revolves about the Sun. Thus, to bring the Sun back over a given longitude, a solar day is slightly longer than a sidereal day.

9.2 When and Where to Launch

▆ In This Section You'll Learn to...

- ☛ Explain how many opportunities (launch windows) there are to launch from a given launch site into a specific orbit
- ☛ Draw a diagram representing launch-window geometry and use it to determine launch-window parameters

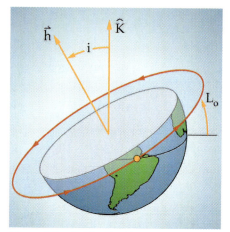

Figure 9-6.   Inclination Versus Latitude. An orbital plane slices through Earth's center and extends north and south to latitude lines equal to the orbital inclination. To launch directly from a given launch site on Earth, we must wait for the launch site to pass through the orbital plane.

Let's take a closer look at how we determine launch-window times and directions. But before we begin, we must remember one very important concept—*launch-window calculations always depend on geometry*! This means we *must* draw a picture to fully understand what is going on! In Chapter 5 we saw that, as an orbital plane slices through Earth's center, it extends north and south to latitude lines equal to its inclination, as shown in Figure 9-6. Thus, to launch a spacecraft directly from a given launch site into a given orbital plane, we must wait until the launch site rotates under the fixed orbital plane. Only at that point, do we have the correct geometry and can find the angles needed for a successful launch.

This means the orbit's inclination, i, must be equal to or greater than the launch site's latitude, L_o. In other words, a launch window can exist only for the following conditions

$$L_o \leq i \text{ (direct orbits)}$$

$$\text{or } L_o \leq 180° - i \text{ (in-direct or retrograde orbits)}$$

For example, we can't launch a spacecraft from the Kennedy Space Center (KSC) ($L_o = 28.5°$) *directly* into an equatorial orbit (i = 0°). Instead, we have to first launch it into an orbit with i = 28.5°, and then perform a plane change maneuver to reach i = 0°. But, as we saw in Chapter 6, plane changes require very expensive velocity changes (ΔV), so naturally we'd prefer to launch our spacecraft directly into its final orbital plane. Also, unless indicated, we'll focus here on prograde (direct) orbits (i < 90°). The geometry for retrograde orbits is the same, but the relationships are slightly different.

Assuming a launch window does exist ($L_o \leq i$ for a direct orbit), let's look at the number of launch windows we can have. Consider the two cases shown in Figure 9-7. In Case 1, the latitude equals the orbital inclination, that is $L_o = i$. This means the launch-site rotates to intersect the orbital plane only once daily, so we have only one launch opportunity per day. As Case 1 in Figure 9-7 shows, the launch site's latitude is tangent to the orbital plane at only the northern-most point under the plane. For example, if we want to launch into an orbit with i = 28.5° from KSC ($L_o = 28.5°$), we would have exactly one launch window per day. A polar view of this case is in Figure 9-8.

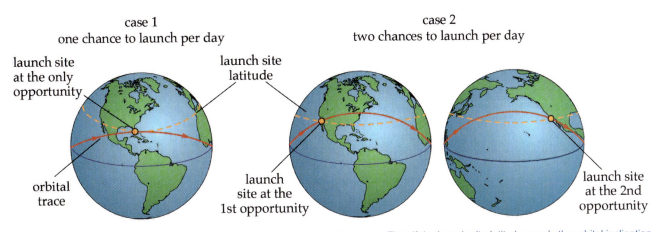

case 1
one chance to launch per day

case 2
two chances to launch per day

launch site
at the only
opportunity

launch site
latitude

launch
site at the
1st opportunity

orbital
trace

launch site
at the 2nd
opportunity

Figure 9-7.   Launch Opportunity. Chances to launch fall into two possible cases. First, if the launch-site latitude equals the orbital inclination, we have one chance per day. Second, if the launch-site latitude is less than the orbital inclination, we have two opportunities per day: once near the ascending node and once near the descending node. If the launch-site latitude is greater than the inclination, we have no opportunity to launch.

But what if the launch site's latitude is less than the orbit's inclination: $L_o < i$? This is illustrated as Case 2 in Figure 9-7. Remember the orbital plane is fixed in inertial space, and the launch site sits on the rotating Earth. As Earth rotates, it carries the launch site under the orbit twice each day. At those two times, windows are available to launch directly into the orbital plane at two locations—one near the ascending node and another near the descending node.

Of course, if the launch site's latitude is greater than the inclination, $L_o > i$, then the launch site never passes under the orbit, so no launch window exists.

Now, let's see how we determine when to launch. To do this, we use the local sidereal time (LST) for when the launch site is under the orbital plane (launch time). We call this the *launch-window sidereal time (LWST)*, which we measure from the vernal equinox direction ($\hat{I}$) to the point where the launch site passes through the orbital plane. So, whenever the local sidereal time (at the launch site) equals the launch-window sidereal time (LST = LWST), the correct geometry exists to launch the spacecraft into the desired orbit. In other words, if we're waiting for the 11:13 A.M. bus, and your watch says its 11:13 A.M., time to go!

Figure 9-9 illustrates the relationship between LWST and LST that we use to find the LWST for a particular opportunity. We start by drawing the orbital ground track on the sphere of the Earth. We then sketch in a dotted line showing the latitude for the launch site (L_o) and a symbol representing the launch site. We use Case 1 for the example. Notice the launch site latitude intersects the orbital trace at only one point, 90° from the ascending node. As Earth rotates, the launch site moves closer to this intersection.

To determine the LWST, again referring to Figure 9-9, we also need the orbit's right ascension of the ascending node, Ω, which we should already have, because it's one of the classic orbital elements that define this orbit.

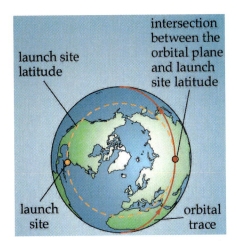

launch site
latitude

intersection
between the
orbital plane
and launch
site latitude

launch
site

orbital
trace

Figure 9-8.   Case 1. If we look down on Earth from the North Pole, we can sketch the latitude line of the launch site and the arc the orbital plane makes. If the orbital inclination equals the launch-site latitude, the two intersect each other at only one point. (Keep in mind the orbital plane is fixed, and Earth [and the launch site along with it] rotate underneath.) As Earth rotates, the launch site moves closer to the point where it will intersect the orbital plane.

So, the total angle from the vernal equinox to the launch point (LWST) for Case 1 is the sum of the right ascension of the ascending node and 90°. Thus, LWST = Ω + 90°.

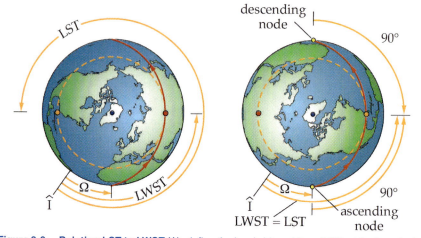

Figure 9-9.   Relating LST to LWST. We define the local sidereal time (LST) at the launch site as the angle from the vernal equinox direction, $\hat{I}$, to the launch site, as shown in the figure on the left. (Keep in mind the orbital plane is fixed and Earth [along with the launch site] rotates underneath.) The launch window opens when Earth has rotated enough to cause the launch site to intersect the orbital plane (LST = LWST), as shown in the figure on the right.

Thus, for Case 1, finding LWST is pretty straightforward, but for Case 2, things get a little more challenging. Because in this case the angle from the ascending node to either of the two points of intersection is *not* 90°, we must take a closer, 3-dimensional look at the geometry of the problem to find what these angles are.

We know that in Case 2 the orbital plane intersects the launch-site latitude at two points. One of these points is closer to the ascending node and we call it the *ascending-node opportunity*, while the other is closer to the descending node and we call it the *descending-node opportunity*.

We start by drawing the local longitude line on Earth's surface from the North Pole, through the launch site at the ascending-node opportunity, to the equator, as shown in Figure 9-10. This line crosses the orbital trace at the launch site's latitude and forms a triangle on Earth's surface with one side the launch site latitude (L_o), one side along the equator, and the third side along the orbital trace. Because we draw this triangle on the surface of a sphere, it's naturally called a *spherical triangle*. Spherical triangles are different from planar triangles. The sum of the angles in a spherical triangle can be more than 180° and we measure the sides as angles. By using the law of cosines for spherical trigonometry, we get a relationship between the two sides and two angles within the triangle. Calculating this triangle's sides provides the key to finding the launch-window sidereal time and the launch direction. (See Appendix A.1 for an explanation of spherical trigonometry.)

Figure 9-10.   Case 2. Taking a three-dimensional view of the problem, we can draw launch-site latitude and longitude lines on the globe to show where they intersect the orbital plane for the ascending-node opportunity. Notice this method forms a triangle on Earth's surface. Because Earth is a sphere, this is a spherical triangle, giving it properties different from planar triangles.

We define two auxiliary angles, α and γ, in this triangle. We call the first angle the *inclination auxiliary angle, α,* and define it at the ascending node between the equator and the ground trace of the orbit. Notice α equals inclination, i, for direct orbits. The second angle, called the *launch-direction auxiliary angle, γ,* we measure at the intersection of the ground trace and the longitude line. The side opposite γ, we call the *launch-window location angle, δ,* and measure it along the equator, between the node closest to the launch opportunity being considered and the longitude where the orbit crosses the launch-site latitude. The side opposite α is the *launch-site latitude, L_o.* Figure 9-11 shows another view of this auxiliary triangle. We can apply the law of cosines for spherical triangles to these angles to get an expression that leads to the value for γ

$$\cos\alpha = -\cos 90° \cos\gamma + \sin 90° \sin\gamma \cos L_o$$

$$\cos\alpha = \sin\gamma \cos L_o \qquad (9\text{-}1)$$

To find γ, we rearrange Equation (9-1)

$$\sin\gamma = \frac{\cos\alpha}{\cos L_o} \qquad (9\text{-}2)$$

where

γ = launch-direction auxiliary angle (deg or rad)

α = inclination auxiliary angle (deg or rad)

L_o = launch site's latitude (deg or rad)

The next important step is to define the triangle's sides. The side opposite γ is the launch-window-location angle, δ. We measure it along the equator, between the node closest to the launch opportunity and launch site's longitude. The other side we need is the one opposite α—the launch site's latitude, L_o, which we know. We find δ by using spherical trigonometry again to get

$$\sin\alpha \cos\delta = \cos\gamma \sin 90° + \sin\gamma \cos 90° \cos L_o \qquad (9\text{-}3)$$

Rearranging Equation (9-3), we get

$$\cos\delta = \frac{\cos\gamma}{\sin\alpha} \qquad (9\text{-}4)$$

where

δ = launch-window location angle (deg or rad)

γ = launch-direction auxiliary angle (deg or rad)

α = inclination auxiliary angle (deg or rad)

With all these auxiliary angle definitions in hand, we can find the launch-window sidereal time (LWST) for Case 2 (a direct orbit in the Northern Hemisphere), as shown previously in Figure 9-7. It always depends on the intended orbit's right ascension of the ascending node, Ω, and the launch-window location angle, δ. To find the window near the ascending node, use

$$LWST_{AN} = \Omega + \delta \qquad (9\text{-}5)$$

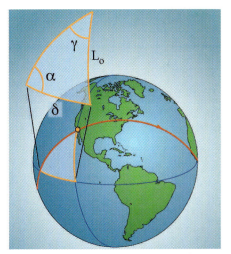

Figure 9-11. Auxiliary Triangle for a Launch Window. From the spherical triangle formed on Earth's surface for the ascending-node opportunity, we can see the right side is the launch-site latitude, L_o. The angle between the orbital trace and the equator is the inclination auxiliary angle, α. The angle between the orbital trace and the longitude line is the auxiliary angle, γ. The side opposite γ is the angle δ. (Note that in spherical triangles, sides are actually angles measured from Earth's center.)

To find it for the window near the descending node, use

$$LWST_{DN} = \Omega + (180° - \delta) \qquad (9\text{-}6)$$

If we know the current local sidereal time for the launch site, we can calculate how long to wait until it's time to launch by subtracting it from the launch-window sidereal time. For example, if the launch-window sidereal time is 1400 and the current local sidereal time is 1200, then we must wait two hours before the launch site comes under the orbital plane to open the launch window. (As we've shown in the previous section, these are sidereal hours, which are slightly shorter than the solar hours we keep on our watches.)

Finally, we need to define one more important angle—the *launch azimuth, β*. This angle tells us what direction to launch. We measure β from true north at the launch site, clockwise to the launch direction, as shown in Figure 9-12. Note that we measure β in the same way as magnetic heading on a compass, with north = 0°, east = 90°, south = 180°, and west = 270°. Also notice that for a launch at the ascending-node opportunity from a Northern Hemisphere site, β = γ. To find β for this case, use Equation (9-2) and set β = γ. For a launch at the descending-node opportunity, the spherical triangle is a mirror image of the one at the ascending node. For this case, calculate the launch-direction auxiliary angle, γ, using Equation (9-2), and then find the launch azimuth, β, using β = 180° − γ. Table 9-1 summarizes all these angles.

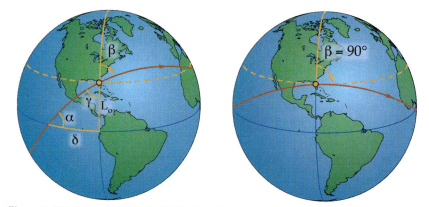

Figure 9-12.   Launch Azimuth, β. The launch azimuth, β, is the angle from true north (at the launch-site longitude line) clockwise to the launch direction. When only one chance to launch exists, the launch azimuth is 90° (due east), for direct orbits.

For sites in the Southern Hemisphere, the spherical triangles are turned 180°, because the launch site comes under the orbital plane at the descending-node opportunity before it reaches the ascending-node opportunity. But the equations for the launch azimuth corresponding to the two node opportunities remain the same

descending-node opportunity: β = 180° − γ

ascending-node opportunity: β = γ

By now you may be confused about the difference between α and LWST and all these other angles. To clear things up, you must remember that computing LWST and β depends on three choices

- Direct or retrograde orbit?
- Opportunity near the ascending node or descending node?
- Launch-site in Northern or Southern Hemisphere?

Table 9-1.   Key Angles for Launch Geometry.

Angle	Name	Definition
α	Inclination auxiliary angle	Measured from the equator to the ground track at the node—same as inclination for direct orbits
γ	Launch-direction auxiliary angle	Measured from the ground track to the launch-site local longitude line
δ	Launch-window location angle	Measured from the node closest to the launch opportunity to the launch-site local longitude line
β	Launch azimuth	Measured from due north clockwise to the launch direction
L_o	Launch site latitude	Measured from the equator to the launch site

As an example, let's look at a direct orbit near the descending node in the Northern Hemisphere with $i > L_o$. To see how to compute LWST for this situation, we begin by drawing two diagrams, as shown in Figure 9-13. One is in three dimensions showing the spherical launch geometry and the other is a two-dimensional polar view showing the relationship between LWST, Ω, and δ.

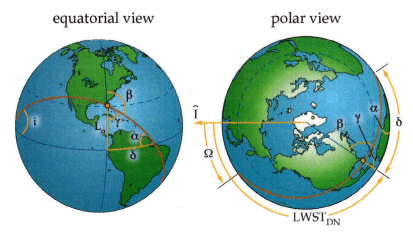

Figure 9-13.   Another Look at Launch Geometry. We can analyze the launch problem for a launch at the descending node opportunity from a launch site in the Northern Hemisphere. We begin by drawing the 3-D view showing the spherical Earth and the auxiliary triangle. We then sketch a 2-D polar view of the problem showing Ω and LWST in a polar view of Earth.

By inspection, we see the inclination auxiliary angle, α, equals the inclination, i. The launch azimuth, β, equals $180°$ minus the auxiliary angle, γ ($\beta = 180° - \gamma$). We find the launch-window location angle, δ, knowing α and γ and using Equation (9-4). To find LWST_{DN} (DN is descending node) for this situation, we look again at our diagram to see

$$\text{LWST}_{DN} = \Omega + (180° - \delta) \text{ for a direct orbit}$$
$$\text{and a Northern-hemisphere site}$$

Let's review where this came from. Ω is the part of LWST_{DN} between $\hat{\text{I}}$ and $\hat{\text{n}}$ (the ascending-node vector); adding $180°$ to this value takes us all the way to the descending node; then, we subtract δ to get the value for LWST_{DN} which we measure to the launch-site longitude.

In all, there are eight launch-window situations. Table 9-2 summarizes the four direct-orbit cases.

Table 9-2.   Summary of Direct-orbit Cases for Launch Opportunities.

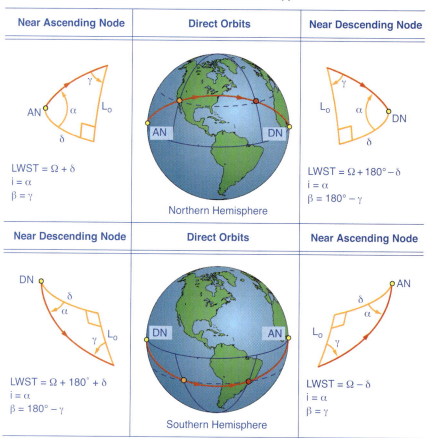

▦ Section Review

Key Terms

ascending-node opportunity
descending-node opportunity
inclination auxiliary angle, α
launch azimuth, β
launch-direction auxiliary angle, γ
launch-site latitude, L_o
launch-window sidereal time
(LWST)
launch-window location angle, δ
spherical triangle

Key Equations

$$\sin\gamma = \frac{\cos\alpha}{\cos L_o}$$

$$\cos\delta = \frac{\cos\gamma}{\sin\alpha}$$

$$\text{LWST}_{AN} = \Omega + \delta$$

$$\text{LWST}_{DN} = \Omega + 180° - \delta$$

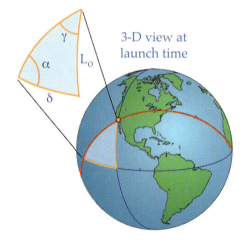

3-D view at
launch time

Key Concepts

➤ For a launch window to exist at a given launch site, the latitude of the launch site, L_o, must be less than or equal to the inclination of the desired orbit ($L_o \le i$).

➤ Computing launch-window sidereal time (LWST) and launch azimuth, β, depends on geometry. You must draw a diagram to clearly visualize all angles.

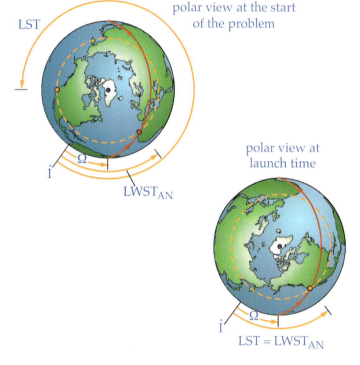

polar view at the start
of the problem

polar view at
launch time

- Launch-window geometry depends on spherical trigonometry
- After sketching the launch-window geometry, we can see an auxiliary triangle. Table 9-1 defines the auxiliary angles in this triangle.
- Launch-window sidereal time (LWST) is a function of the desired right ascension of the ascending node, Ω, and the launch-window location angle, δ
- Launch azimuth, β, is defined as the direction to launch from a given site to achieve a desired orbit. We measure β clockwise from due north at the launch site.

➤ Table 9-2 summarizes launch-window geometry for the four most common of the eight possible cases where $L_o < i$

Example 9-1

Problem Statement

Suppose the Space Shuttle will deploy an interplanetary probe bound for Saturn. The probe requires a parking orbit with a right ascension of the ascending node, Ω, of 195° and an inclination of 41°. If the current LST at the launch site is 0100, how long before the next launch window opens for a launch from Kennedy Space Center (28.5°N, 80°W)? What is the launch azimuth, β, for this opportunity?

Problem Summary

Given: LST = 0100, L_o=28.5°N, Ω = 195°, i= 41°

Find: Time in hours until the next launch window opens

Launch azimuth, β, for the next window

Problem Diagram

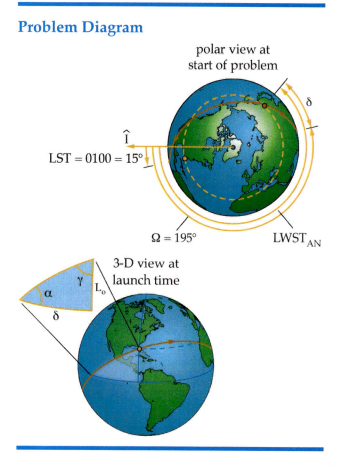

polar view at start of problem

$\hat{I}$

LST = 0100 = 15°

δ

$\Omega = 195°$

$LWST_{AN}$

3-D view at launch time

γ L_o

α

δ

Conceptual Solution

Time until the next launch window opens

1) By inspection, find the inclination auxiliary angle, α

$$\alpha = i$$

2) Knowing α and L_O, find the launch-direction auxiliary angle, γ

$$\sin\gamma = \frac{\cos\alpha}{\cos L_o}$$

3) Knowing γ, find the launch-window location angle, δ

$$\cos\delta = \frac{\cos\gamma}{\sin\alpha}$$

4) By drawing and inspecting launch-site geometry determine the relationship between LWST, Ω, and δ for the ascending-node opportunity

$$LWST_{AN} = \Omega + \delta$$

5) Solve for LWST for the ascending-node opportunity

6) By inspecting the launch-site geometry, determine the relationship between LWST, Ω, and δ for the descending-node opportunity

$$LWST_{DN} = \Omega + 180° - \delta$$

7) Solve for LWST for the descending-node opportunity

8) Determine which LWST is next

9) Determine the time until the next opportunity

10) Knowing γ and inspecting launch-site geometry for the next opportunity, find β

Example 9-1 (Continued)

Analytical Solution

1) By inspection, find the inclination auxiliary angle, α

$\alpha = i = 41°$

2) Knowing α and L_o, find the launch-direction auxiliary angle, γ

$$\sin\gamma = \frac{\cos\alpha}{\cos L_o} = \frac{\cos 41°}{\cos 28.5°} = 0.8588$$

$\gamma = 59.18°$

3) Knowing γ, find the launch-window location angle, δ

$$\cos\delta = \frac{\cos\gamma}{\sin\alpha}$$

$$\cos\delta = \frac{\cos 59.18°}{\sin 41°} = 0.7809$$

$\delta = 38.65°$

4) By inspecting the launch-site geometry, determine the relationship between LWST, Ω, and δ for the ascending-node opportunity

$LWST_{AN} = \Omega + \delta$

5) Solve for LWST for the ascending-node opportunity

$LWST_{AN} = 195° + 38.65° = 233.65°$

$$LWST_{AN} \text{ (hrs)} = \frac{LWST_{AN}(°)}{\omega_{Earth}} = \frac{233.65°}{15°/hr}$$

$= 15.577$ hr

0.577 hr = 34.6 min

so the time is 1534.6

6) By inspecting the launch-site geometry, determine the relationship between LWST, Ω, and δ for the descending-node opportunity

$LWST_{DN} = \Omega + (180° - \delta)$

7) Solve for LWST for the descending-node opportunity

$LWST_{DN} = 195° + (180° - 38.65°) = 336.35°$

$$LWST_{DN} \text{ (hrs)} = \frac{LWST_{DN}(°)}{\omega_{Earth}} = \frac{336.35°}{15°/hr}$$

$= 22.423$ hr

0.423 hr = 25.4 min

so the time is 2225.4

8) Determine which LWST is next in hours

Because LST = 0100 and $LWST_{DN}$ = 2225.4, we missed the descending-node opportunity (0100 is after 2225.4). Thus, we must wait for the ascending-node opportunity, and the next LWST is at 1534.6.

9) Determine the time until the next opportunity

The time until the next opportunity is the time from the current LST (0100 = 1.00) until $LWST_{AN}$ (1534.6 = 15.577) or 15.577 - 1.0 = 14.577 hours = 14 hours and 34.6 minutes.

10) Knowing γ and inspecting the launch-site geometry for the next opportunity, find β.

For the ascending-node opportunity, $\beta = \gamma$, thus $\beta_{AN} = 59.18°$

Interpreting the Results

We must wait 14 hours and 34.6 minutes until the ascending-node opportunity to launch the Space Shuttle into a parking orbit with $\Omega = 195°$ and $i = 41°$. We will launch when LST is 1534.6 hours, at an azimuth of 59.18° (northeast).

9.3 Launch Velocity

▬ In This Section You'll Learn to...

☞ Determine the total change in velocity a launch vehicle must deliver to put a spacecraft into a given orbit

The launch window tells us when (Launch Window Sidereal Time) and in what direction (azimuth, β) to launch our launch vehicle to achieve some desired orbit. Now let's examine how much velocity the launch vehicle must deliver to place a payload into this orbit.

During liftoff, a launch vehicle goes through four distinct phases on its way from the launch pad into orbit, as shown in Figure 9-14.

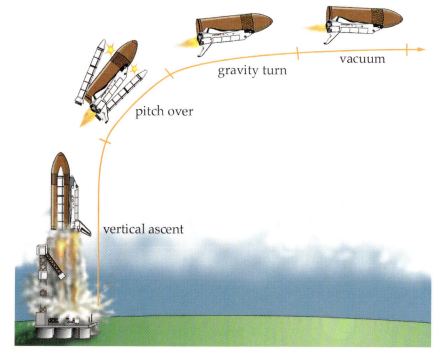

vacuum

gravity turn

pitch over

vertical ascent

Figure 9-14.   Phases of Launch Vehicle Ascent. During ascent a launch vehicle goes through four phases—vertical ascent, pitch over, gravity turn, and vacuum.

❏ Phase one—vertical ascent

• During this phase, it needs to gain altitude quickly to get out of the dense atmosphere that slows it down due to drag. In some cases, it also does a distinctive roll maneuver as it leaves the pad. This roll properly aligns its launch azimuth with the correct orbital plane.

❏ Phase two—pitch over

- After the launch vehicle has gained enough altitude, it must pitch over slightly so it can begin to gain velocity downrange (horizontally). As we showed in Chapter 4, this horizontal velocity keeps a spacecraft in orbit.

❏ Phase three—gravity turn

- During this phase, gravity pulls the launch vehicle toward horizontal, thereby also pulling its velocity vector toward horizontal

❏ Phase four—vacuum phase

- During this phase, the launch vehicle is effectively out of Earth's atmosphere and continues to accelerate to gain the necessary velocity to achieve orbit. In this final phase of powered flight, the control system works to deliver the vehicle to the desired burnout conditions: velocity ($\vec{V}_{burnout}$), altitude ($Alt_{burnout}$), flight-path angle ($\phi_{burnout}$), and downrange angle ($\theta_{burnout}$), as shown in Figure 9-15.

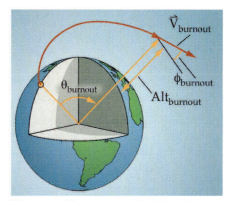

Figure 9-15. Launch Vehicle Burnout Conditions. During the vacuum phase the control system is trying to deliver the launch vehicle to a specified set of burnout conditions including velocity, flight-path angle, altitude, and downrange angle. *Note:* We measure $\phi_{burnout}$ between $V_{burnout}$ and the local horizontal.

In this section we'll look at these burnout conditions, specifically, burnout velocity ($\vec{V}_{burnout}$). As we'll see, this is a very important parameter that goes into the design and selection of a launch vehicle for a given mission, when a launch vehicle reaches the burn out conditions, it is in the intended mission orbit.

The velocity needed to get to orbit consists of the launch vehicle's burnout velocity and the tangential velocity that exists at its launch site due to Earth's rotation. First, let's consider what we mean by tangential velocity. Suppose we have an old-fashioned record player and we set weights on a spinning record—one near the center and one near the outside edge. We see that the weight farther from the center moves faster (has a larger tangential velocity) than the one near the center. The entire record is spinning at the same angular velocity (revolutions per minute), but the tangential velocity is higher at the larger radius, as seen in Figure 9-16.

We express the relationship among tangential velocity, angular rate, and radius as

$$V = R\,\omega \qquad (9\text{-}7)$$

where

V = instantaneous tangential velocity (m/s)

R = radius from the center of rotation (m)

ω = angular velocity (rad/s)

Figure 9-16. Tangential Velocity Increases with Radius. If you look at a spinning record you can see that the tangential velocity increases as you look farther out from the center.

The same thing happens to us because we're rotating with Earth. Even when we think we're motionless, we're actually moving fast. "Doesn't feel like it," you say? Well, Earth spins on its axis, so any point on it (except at the poles) has some tangential velocity due to this rotation. We don't feel it, but Earth's surface (and us along with it) moves eastward at nearly 1,600 kilometers per hour (1,000 m.p.h.)! A launch vehicle sitting on the launch pad is moving eastward too. Thus,

$$V_{\text{launch site}} = R\,\omega_{\text{Earth}}$$

where

$V_{\text{launch site}}$ = launch site's instantaneous tangential velocity (m/s)

R = radius from the spin axis to the launch site (m)

ω_{Earth} = Earth's angular velocity (rad/s)

= 15.04107 (°/hr) = 7.29212×10^{-5} (rad/s)

Note: Here we use a more exact rotation rate for Earth (15.04107°/hr) than we did in Section 9.1 (15°/hr).

In this case, the velocity is

$$V_{\text{at equator}} = (15.04107°/\text{hr})\,(6378.137 \text{ km})$$

Converting units, gives us

$$V_{\text{at equator}} = (0.0000729212 \text{ rad/s})\,(6378.137 \text{ km})$$

$$V_{\text{at equator}} = 0.4651 \text{ km/s}$$

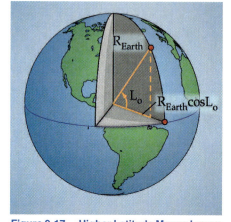

Figure 9-17. Higher Latitude Means Lower Velocity. The tangential velocity of a point on Earth's surface is a function of its latitude. At higher latitudes (either north or south) the perpendicular distance to the spin axis decreases, so the tangential velocity decreases.

The Boeing Sea Launch platform can be positioned at the equator to take full advantage of this velocity and the European Space Agency's site at Kourou, French Guyana, with a latitude of 4° N gets nearly all the effect. At other locations, we can't use Earth's radius directly to find tangential velocity. To find the radius from the spin axis to a site at other locations, we must multiply by the cosine of the latitude. Thus, as we increase latitude, L_o, up to 90°, the distance from the spin axis, R, decreases, and so the tangential velocity decreases, as shown in Figure 9-17.

$$V_{\text{launch site}} = (R_{\text{Earth}} \cos L_o)\omega_{\text{Earth}}$$

and because

$$V_{\text{at equator}} = R_{\text{Earth}}\,\omega_{\text{Earth}} = 0.4651 \text{ km/s}$$

we can say

$$V_{\text{launch site}} = (0.4651 \text{ km/s}) \cos L_o$$

Notice that, for higher latitudes, the radius from the spin axis is smaller, and so is the launch site's tangential velocity.

To fully describe a launch site's tangential velocity, we need to know its direction of motion, so we can put it into vector form. To express the velocity as a vector, we need to choose a new coordinate frame. Previously, we picked inertial frames because we were writing equations of motion and needed to apply Newton's laws. Now, because we want to know the velocity a launch vehicle must deliver from a given launch site, we pick an Earth-fixed reference, called a *topocentric-horizon frame*. As the name implies, the origin for this frame is at the launch site, with the horizontal (a plane tangent to Earth's surface at the launch site) as the fundamental plane. If we choose the vector pointing due south from the site as the principal direction ($\hat{S}$) and the straight-up or zenith direction ($\hat{Z}$) as the out-of-plane vector, the east direction ($\hat{E}$) completes the right-hand rule. We call this the *south-east-zenith (SEZ) coordinate system*, as shown in Figure 9-18. We can now express the velocity due to Earth's rotation as a vector in the eastward direction.

$$\vec{V}_{\text{launch site}} = (0.4651\,\text{km/s})\cos L_o\,\hat{E} \qquad (9\text{-}8)$$

So what does this mean in terms of putting a payload into orbit? Because a launch site has a tangential velocity in the eastward direction, it gives a launch vehicle a "head start" (assist) for launches in the easterly direction—into direct (prograde) orbits. The closer a launch site is to the equator, the greater assist the launch vehicle gets when launching eastward. For example, the European Space Agency's launch site at Kourou (4° N latitude) gives launch vehicles an assist of 0.464 km/s versus 0.4087 km/s for the Kennedy Space Center at 28.5° latitude. This means that for a given launch vehicle we can launch a larger payload from a launch site at a lower latitude. This seemingly small "free" velocity can actually be a big advantage, especially for large commercial spacecraft where every kilogram they save means longer lifetime on orbit generating revenue.

What about launching into a retrograde orbit? A launch vehicle won't get any help because it'd be launching in a westerly direction while the *launch-site velocity,* $\vec{V}_{\text{launch site}}$, is eastward. In fact, it's coming from behind because it has to make up this difference to get into orbit. Thus, it's more costly (in terms of velocity) to launch into a retrograde orbit.

To achieve some specified orbital elements, the spacecraft must reach a certain inertial velocity and altitude. To get to this condition, a launch vehicle must meet two primary objectives (Figure 9-19.)

- Increase altitude to orbital altitude (increase potential energy)
- Increase velocity to orbital velocity (increase kinetic energy)

In order to have a successful launch, it must provide enough velocity so that it can meet both of these objectives. Determining this velocity during launch planning is a very complex problem and to get accurate values, engineers usually use numerical integration in sophisticated trajectory modeling programs. This process often incorporates properties of the launch vehicle, atmospheric density models, and other factors. There may be significant differences between accurate numerical integration solutions and answers produced by the method below, but many of the basic concepts are similar.

To begin with, we initially define four velocities we need to find

- $\vec{V}_{\text{loss gravity}}$ = extra velocity needed to overcome gravity and reach the correct altitude

- $\vec{V}_{\text{burnout}}$ = inertial velocity needed at burnout to be in the desired orbit

- $\vec{V}_{\text{launch site}}$ = velocity of the launch pad due to Earth's rotation (which works for us or against us depending on whether we launch east or west)

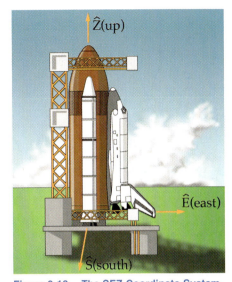

Figure 9-18.   The SEZ Coordinate System. We use the south, east, up (zenith) or SEZ coordinate system centered at the launch site as a tropocentric-horizon frame, from which to analyze launch-site velocity.

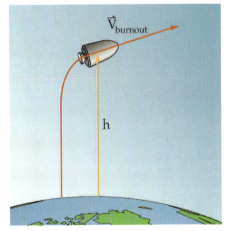

Figure 9-19.   Launch Vehicle's Two Primary Objectives. A launch vehicle must get to the desired altitude and have the inertial velocity to stay in orbit.

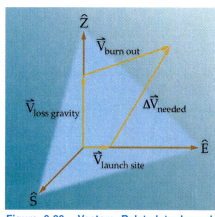

Figure 9-20. Vectors Related to Launch Velocity. $\vec{V}_{\text{loss gravity}}$ and $\vec{V}_{\text{burnout}}$ form the desired velocity. $\vec{V}_{\text{launch site}}$ and $\Delta\vec{V}_{\text{needed}}$ are the components that get the launch vehicle to the right velocity.

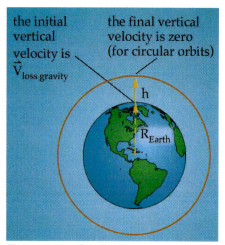

Figure 9-21. Gravity Losses. We use the gravity loss term to account for the energy needed to go from the launch altitude (usually Earth's surface) to orbital altitude.

- $\Delta\vec{V}_{\text{needed}}$ = total velocity change that the launch vehicle must generate to meet the mission requirements

We add the first two terms, $\vec{V}_{\text{loss gravity}}$ and $\vec{V}_{\text{burnout}}$, to get the total velocity desired. Together they show how the launch vehicle gets to the desired altitude and leaves the spacecraft with the desired inertial velocity. We add the second two terms, $\vec{V}_{\text{launch site}}$ and $\Delta\vec{V}_{\text{needed}}$, to show what the launch vehicle must deliver to reach these desired conditions. We show a graphical representation of these velocities in Figure 9-20. In practice, we calculate all the velocity terms except $\Delta\vec{V}_{\text{needed}}$. After we obtain these three values, we can determine $\Delta\vec{V}_{\text{needed}}$ from them.

We already saw how to find $\vec{V}_{\text{launch site}}$, so next let's tackle the gravity losses. Please note that this value is not really a "loss," from an energy standpoint. What occurs here is that the launch vehicle must effectively provide an additional vertical velocity component to overcome gravity. However, as the launch vehicle goes up, it trades this vertical velocity (kinetic energy) for potential energy. (Figure 9-21). To model this, we can assume the launch vehicle adds just enough vertical velocity (kinetic energy), so that by the time it reaches the desired altitude, it has the desired potential energy (i.e., no vertical velocity remains). In this approximation, the left hand side of Equation (9-9) represents the specific mechanical energy the launch vehicle needs at the launch altitude (usually at the launch pad) to reach the desired altitude with zero vertical velocity, and the right hand side represents the needed specific mechanical energy of the launch vehicle at the desired altitude.

$$\frac{V_{\text{loss gravity}}^2}{2} - \frac{\mu}{R_{\text{launch}}} = \frac{0^2}{2} - \frac{\mu}{R_{\text{burnout}}} \quad (9\text{-}9)$$

where

$V_{\text{loss gravity}}$ = velocity needed to reach the correct altitude (km/s)

R_{launch} = radius to the launcher (usually Earth's radius) (km)

R_{burnout} = radius to the burnout point (km)

We can solve this equation for $V_{\text{loss gravity}}$, but the answer is only a magnitude. What we need is a vector, so we must determine the direction. Because this component is solely for getting the correct altitude, the direction is "up" or in the zenith direction. Using this information, Equation (9-9), and some algebraic manipulation, we get

$$\vec{V}_{\text{loss gravity}} = \sqrt{\frac{2\mu(R_{\text{burnout}} - R_{\text{launch}})}{(R_{\text{launch}} \cdot R_{\text{burnout}})}} \hat{Z} \quad (9\text{-}10)$$

where

$\hat{Z}$ = unit vector in the zenith direction (unitless)

This equation gives us an approximation for the additional velocity the launch vehicle must provide to reach the desired orbital altitude. The greater the gravitational parameter, μ, the greater the amount of additional velocity it needs to reach orbit. Therefore, as we'd expect, it is cheaper, in terms of velocity, to reach orbit from the surface of a smaller planet like Mars than from Earth. This is sometimes called the *gravity well* that launch vehicles must fly out of to reach orbit.

Now we can switch our focus to the *burnout velocity, $\vec{V}_{burnout}$*. We use the velocity magnitude, $V_{burnout}$, the flight-path angle, ϕ, and the launch azimuth angle, β, to compute the SEZ components of the velocity at burnout (Figure 9-22 and Appendix C.8).

$$V_{burnout_{south}} = -V_{burnout}\cos\phi\cos\beta \tag{9-11}$$

$$V_{burnout_{east}} = V_{burnout}\cos\phi\sin\beta \tag{9-12}$$

$$V_{burnout_{zenith}} = V_{burnout}\sin\phi \tag{9-13}$$

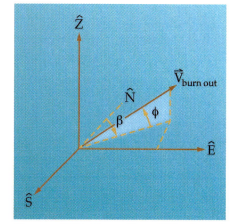

Figure 9-22. Converting Velocity at Burnout to SEZ Coordinates. We use the velocity magnitude, flight-path angle, and launch azimuth to perform this conversion.

where

$V_{burnout_{south,\,east,\,zenith}}$ = components of the burnout velocity in the south, east, and zenith directions (km/s)

$V_{burnout}$ = magnitude of the velocity from the launch (km/s)

ϕ = flight-path angle of spacecraft at burnout (deg or rad) = 0° for circular orbits

β = launch azimuth (deg or rad)

We now have enough information to compute the *velocity needed, $\Delta\vec{V}_{needed}$*. From Figure 9-20 we can show that

$$\Delta\vec{V}_{needed} = \vec{V}_{loss\,gravity} + \vec{V}_{burnout} - \vec{V}_{launch\,site} \tag{9-14}$$

In terms of the individual components, we have

$$\Delta V_{needed_{south}} = 0 + V_{burnout_{south}} - 0 \tag{9-15}$$

$$\Delta V_{needed_{east}} = 0 + V_{burnout_{east}} - V_{launch\,site} \tag{9-16}$$

$$\Delta V_{needed_{zenith}} = V_{loss\,gravity} + V_{burnout_{zenith}} - 0 \tag{9-17}$$

where

$\Delta V_{needed_{south,\,east,\,zenith}}$ = components of the velocity needed by the launch vehicle to get the spacecraft from the launch site to orbit in the south, east, and zenith components (km/s)

$V_{burnout_{south,\,east,\,zenith}}$ = components of the burnout velocity in the south, east, and zenith directions (km/s)

$V_{launch\,site}$ = velocity of the launch pad due to Earth's rotation (km/s)

$V_{\text{loss gravity}}$ = velocity needed to reach the correct altitude (km/s)

The magnitude of the needed velocity is

$$\Delta V_{\text{needed}} = \left| \Delta \vec{V}_{\text{needed}} \right| = \sqrt{(\Delta V_{\text{needed}_{\text{south}}})^2 + (\Delta V_{\text{needed}_{\text{east}}})^2 + (\Delta V_{\text{needed}_{\text{zenith}}})^2}$$

(9-18)

This is the velocity that the launch vehicle needs to provide to get the spacecraft into its orbit. This estimate accounts for the potential energy that it must gain to reach the correct altitude, provides for the kinetic energy that it must gain for the desired velocity at burnout, and accounts for the motion of the launch pad.

In practice, launch vehicles must also overcome significant air drag, back pressure, and steering losses. They incur drag losses as they pass through the atmosphere. Back pressure losses result from operating a rocket engine in an atmosphere. And steering losses happen when they have to correct for winds and other disturbances that take them off their planned trajectory. Adding an extra term ΔV_{losses}, compensates for these last few losses and gives a new value that we'll call *design velocity,* ΔV_{design}. The value, ΔV_{losses}, varies depending on the launch vehicle and mission, but a rough estimate is 1.5 km/s.

$$\Delta V_{\text{design}} = \Delta V_{\text{needed}} + \Delta V_{\text{losses}}$$

(9-19)

where

ΔV_{design} = design velocity the launch vehicle must deliver to reach the desired orbit (km/s)

ΔV_{needed} = velocity needed by the launch vehicle to get the spacecraft from the launch site to orbit (km/s)

ΔV_{losses} = velocity losses during ascent due to drag, back pressure, and steering $\cong 1.5$ km/s

This is the velocity we must design our launch vehicle to provide. As we'll see in Chapter 12, the launch vehicle continually compares its desired velocity with its actual velocity and makes corrections to ensure that it satisfies the desired burnout conditions.

As we'll explore in Chapter 14, ΔV_{design} is a critical requirement when it comes to sizing a launch vehicle's propulsion subsystem and staging options.

Section Review

Key Terms

burnout velocity, $\vec{V}_{burnout}$

design velocity, ΔV_{design}

gravity well

launch-site velocity, $\vec{V}_{launch\,site}$

south-east-zenith (SEZ) coordinate system

topocentric-horizon frame

velocity needed, $\Delta \vec{V}_{needed}$

Key Equations

$$\vec{V}_{loss\,gravity} = \sqrt{\frac{2\mu(R_{burnout} - R_{launch})}{(R_{launch} \cdot R_{burnout})}} \hat{Z}$$

$$V_{burnout_{south}} = -V_{burnout}\cos\phi\cos\beta$$

$$V_{burnout_{east}} = V_{burnout}\cos\phi\sin\beta$$

$$V_{burnout_{zenith}} = V_{burnout}\sin\phi$$

$$\Delta\vec{V}_{needed} = \vec{V}_{loss\,gravity} + \vec{V}_{burnout} - \vec{V}_{launch\,site}$$

$$\Delta V_{needed_{south}} = 0 + V_{burnout_{south}} - 0$$

$$\Delta V_{needed_{east}} = 0 + V_{burnout_{east}} - V_{launch\,site}$$

$$\Delta V_{needed_{zenith}} = V_{loss\,gravity} + V_{burnout_{zenith}} - 0$$

$$\Delta V_{needed} = \left| \Delta\vec{V}_{needed} \right| =$$

$$\sqrt{(\Delta V_{needed_{south}})^2 + (\Delta V_{needed_{east}})^2 + (\Delta V_{needed_{zenith}})^2}$$

$$\Delta V_{design} = \Delta V_{needed} + \Delta V_{losses}$$

Key Concepts

➤ We design a launch vehicle to go from a given launch site and deliver a spacecraft of a certain size into a specified orbit. It does this in four phases
  • Vertical ascent
  • Pitch over
  • Gravity turn
  • Vacuum

➤ Because Earth is rotating eastward, a launch vehicle sitting on the launch pad already has some velocity in the eastward direction. Thus,
  • A launch vehicle has a "head start" for launching into direct orbits
  • A launch vehicle must overcome Earth's rotation to get into a retrograde orbit
  • The velocity of a launch site depends on the launch-site's latitude and is in the eastward direction

➤ To determine the velocities needed to get into orbit, we define the topocentric-horizon coordinate system (or SEZ), as shown in Figure 9-18

➤ Launch vehicles must meet two primary objectives
  • Increase altitude to orbital altitude
  • Increase velocity to orbital velocity

➤ Four velocities help us analyze what a launch vehicle must deliver

  • $\vec{V}_{loss\,gravity}$ = extra velocity needed to overcome gravity and reach the correct altitude

  • $\vec{V}_{burnout}$ = inertial velocity needed at burnout to be in the desired orbit

  • $\vec{V}_{launch\,site}$ = velocity of the launch pad due to Earth's rotation (which works for us or against us depending on whether we launch east or west)

  • $\Delta\vec{V}_{needed}$ = total velocity change that the launch vehicle must generate to meet the mission requirements

➤ In practice, launch vehicles also encounter significant air drag, back pressure, and steering losses
  • So, ΔV_{design} is the velocity we must design the launch vehicle to deliver. $\Delta V_{design} = \Delta V_{needed} + \Delta V_{losses}$

Example 9-2

Problem Statement

Suppose you have to design a launch vehicle to place a spacecraft into a circular orbit, at 400 km altitude, and an inclination of 28.5°. Plan to launch from the Kennedy Space Center, Florida, and use 1.5 km/s for ΔV_{losses}. Compute the ΔV_{design} for this launch.

Problem Summary

Given: $h_{burnout} = 400$ km
Circular orbit
$\Delta V_{losses} = 1.5$ km/s
Launch site: Kennedy Space Center, $L_o = 28.5^o$

Find: ΔV_{design}

Conceptual Solution

1) Determine $V_{burnout}$ from the velocity of a circular orbit at 400 km altitude

$$V_{burnout} = V_{circular\ orbit} = \sqrt{\frac{\mu_{Earth}}{R_{circular}}}$$

2) Determine the launch azimuth, β, and burnout flight-path angle, ϕ

Use the launch-site latitude, L_o, and orbital inclination to decide on the launch azimuth, β. Use the flight-path angle of circular orbits to determine the burnout flight-path angle, ϕ.

3) Compute the launch-site velocity

$$V_{launch\ site} = 0.4651 \cos L_o$$

4) Compute the gravity-loss velocity that will result in the correct burnout altitude

$$\vec{V}_{loss\ gravity} = \sqrt{\frac{2\mu(R_{burnout} - R_{launch})}{(R_{launch} \cdot R_{burnout})}}\ \hat{Z}$$

5) Compute the SEZ components of the burnout velocity, $\vec{V}_{burnout}$

$V_{burnout\ south} = -V_{burnout} \cos \phi \cos \beta$
$V_{burnout\ east} = V_{burnout} \cos \phi \sin \beta$
$V_{burnout\ zenith} = V_{burnout} \sin \phi$

6) Compute the SEZ components of the needed velocity change, $\vec{V}_{needed}$

$$\Delta\vec{V}_{needed} = \vec{V}_{loss\ gravity} + \vec{V}_{burnout} - \vec{V}_{launch\ site}$$

$\Delta V_{needed\ south} = V_{loss\ gravity\ south} + V_{burnout\ south}$
$\quad - V_{launch\ site\ south}$

$\Delta V_{needed\ east} = V_{loss\ gravity\ east} + V_{burnout\ east}$
$\quad - V_{launch\ site\ east}$

$\Delta V_{needed\ zenith} = V_{loss\ gravity\ zenith} + V_{burnout\ zenith}$
$\quad - V_{launch\ site\ zenith}$

$\Delta V_{needed} =$

$$\sqrt{(\Delta V_{needed\ south})^2 + (\Delta V_{needed\ east})^2 + (\Delta V_{needed\ zenith})^2}$$

7) Compute the launch vehicle's design velocity accounting for drag and steering losses

$$\Delta V_{design} = \Delta V_{needed} + \Delta V_{losses}$$

Analytical Solution

1) Determine $V_{burnout}$ from the velocity of a circular orbit at 400 km altitude

$$\vec{V}_{loss\ gravity} = \sqrt{\frac{2\mu(R_{burnout} - R_{launch})}{(R_{launch} \cdot R_{burnout})}}\ \hat{Z} =$$

$$\sqrt{\frac{3.986 \times 10^5 \frac{km^3}{s^2}}{(6378\ km + 400\ km)}} = \sqrt{\frac{3.986 \times 10^5 \frac{km^3}{s^2}}{6778\ km}}$$

$$= \sqrt{58.808 \frac{km^3}{s^2}} = 7.669 \frac{km}{s}$$

2) Determine the launch azimuth, β, and burnout flight-path angle, ϕ

The launch site latitude, L_o, is 28.5° and the orbital inclination is 28.5°, so the launch vehicle must launch due east, which means, $\beta = 90^o$. The flight-path angle for circular orbits is 0°, so the burnout flight-path angle must also be 0°.

3) Compute the launch-site velocity

$$V_{launch\ site} = 0.4651 \cos L_o = 0.4651 \cos 28.5^o$$
$$= 0.4651\ (0.87882) = 0.4087\ km/s$$

Example 9-2 (Continued)

Compute the gravity-loss velocity that will result in the correct burnout altitude

$$\vec{V}_{loss\ gravity} = \sqrt{\frac{2\mu h}{(R_{Earth})(R_{Earth} + h)}}\ \hat{Z} =$$

$$\sqrt{\frac{2\left(3.986 \times 10^5 \frac{km^3}{s^2}\right)(400\ km)}{6378\ km\ (6378\ km + 400\ km)}}\ \hat{Z} = 2.716\ \hat{Z}\ km/s$$

Compute the SEZ components of the burnout velocity, $\vec{V}_{burnout}$

$$V_{burnout\ south} = -V_{burnout} \cos\phi \cos\beta = -7.669 \cos 0° \cos 90° = 0\ km/s$$

$$V_{burnout\ east} = V_{burnout} \cos\phi \sin\beta = 7.669 \cos 0° \sin 90° = 7.669\ km/s$$

$$V_{burnout\ zenith} = V_{burnout} \sin\phi = 7.669 \sin 0° = 0\ km/s$$

Compute the SEZ components of the needed velocity change, $\Delta\vec{V}_{needed}$

$$\Delta V_{needed\ south} = V_{loss\ gravity\ south} + V_{burnout\ south} - V_{launch\ site\ south} = 0 + 0 - 0 = 0$$

$$\Delta V_{needed\ east} = V_{loss\ gravity\ east} + V_{burnout\ east} - V_{launch\ site\ east} = 0 + 7.669\ km/s - 0.4087\ km/s = 7.260\ km/s$$

$$\Delta V_{needed\ zenith} = V_{loss\ gravity\ zenith} + V_{burnout\ zenith} - V_{launch\ site\ zenith} = 2.716\ km/s + 0 - 0 = 2.716\ km/s$$

$$\Delta V_{needed} =$$

$$\sqrt{(\Delta V_{needed\ south})^2 + (\Delta V_{needed\ east})^2 + (\Delta V_{needed\ zenith})^2} =$$

$$\Delta V_{needed} = \sqrt{(0)^2 + \left(7.260 \frac{km}{s}\right)^2 + \left(2.716 \frac{km}{s}\right)^2}$$

$$= \sqrt{60.08 \frac{km}{s}} = 7.751\ \frac{km}{s}$$

7) Compute the launch vehicle's design velocity, accounting for drag and steering losses

$$\Delta V_{design} = \Delta V_{needed} + \Delta V_{losses} = 7.751\ km/s + 1.5\ km/s = 9.251\ km/s$$

Interpreting the Results

The launch vehicle for the mission needs a ΔV_{design} of 9.251 km/s to reach the required burnout conditions and overcome the losses due to gravity, drag, and steering.

▰ References

Bate, Roger R., Donald D. Mueller, and Jerry E. White. *Fundamentals of Astrodynamics.* New York: Dover Publications, Inc., 1971.

▰ Mission Problems

9.1   Launch Windows and Time

1   What is a launch window?

2   How do mission planners specify a desired orbit so a spacecraft can do its mission?

3   What is local sidereal time (LST)? Draw a diagram to illustrate your answer. What is meant by "sidereal"?

4   Why do we use sidereal rather than solar time for computing launch windows?

5   What is the difference between solar and sidereal time? Draw a diagram to illustrate which is longer and why.

6   How does local sidereal time (LST) change as Earth rotates? How does right ascension of the ascending node, Ω, change as Earth rotates?

7   If LST is 45°, what is it in hours, minutes, and seconds? Draw a diagram to illustrate this time.

8   If your current location has rotated 50° past the vernal equinox direction, what is your LST in hours, minutes, and seconds?

9.2   When and Where to Launch

9   What do we mean when we say orbital planes are fixed in inertial space?

10   If an orbit has an inclination of 45°, what is the highest northern latitude it will pass over? The highest southern latitude?

11   Define right ascension of the ascending node, Ω.

12   Mission planners want to launch the Space Shuttle from Kennedy Space Center ($L_o = 28.5°$) into an orbit with an inclination of 28.5°. How many launch windows will there be each day? Draw a diagram to illustrate this case. How would this change if the desired inclination were 57°? Draw a diagram to illustrate this case.

13   Define launch-window sidereal time (LWST)? What is the difference between $LWST_{AN}$ and $LWST_{DN}$? Draw a diagram to illustrate your answers. How does LWST change as Earth rotates?

14 Mission planners want to launch the Space Shuttle from Kennedy Space Center ($L_o = 28.5°$) into an orbit with an inclination of 28.5° and a right ascension of ascending node of 45°.

a) What is the LWST for this launch, in degrees?

b) What is the LWST for this launch in hours, minutes, seconds?

c) If the current LST at Kennedy Space Center is 1200 hrs, how long until the launch window opens?

15 Why do we need to determine angles on a spherical auxiliary triangle to determine LWST for cases with two opportunities per day?

16 Sketch the spherical auxiliary triangle we use to compute launch windows and define all angles.

17 Mission planners at the European Space Agency want to launch their Ariane 4 launch vehicle from French Guyana ($L_o = 4°$ N) into a low-Earth orbit with an inclination of 30° and a right ascension of ascending node of 135°. LST at the launch site is 1430.

a) How many launch opportunities will there be per day?

b) Draw a 2-D polar view of this launch geometry.

c) Draw a 3-D side view of this launch geometry.

d) Draw the auxiliary triangle for the ascending-node and descending-node opportunities.

e) What is the inclination auxiliary angle, α, for the ascending- and descending-node opportunities?

f) What is the launch-direction auxiliary angle, γ, for the ascending- and descending-node opportunities?

g) What is the launch-window location angle, δ, for both opportunities?

h) What is the $LWST_{AN}$?

i) How long until the $LWST_{AN}$?

j) What is the launch azimuth, β, for the ascending node?

k) What is the $LWST_{DN}$?

l) How long until the $LWST_{DN}$?

m) What is the launch azimuth, β, for the descending node?

18 Mission planners want to launch the Proton launch vehicle from Baikonur cosmodrome ($L_0 = 51°$ N) into an orbit with an inclination of 63° and a right ascension of ascending node of 270°. If the LST is 0945, when will the next launch window open? What direction will they launch?

19 Australian launch planners are designing a new launch site for the northern coast ($L_0 = 13°$ S).

a) What is the lowest inclination for an orbit from this site?

b) If planners want to use this site to resupply the Space Station ($i = 51.6°$ and $\Omega = 35°$), how much time elapses between launch opportunities each day?

9.3   Launch Velocity

20 What are the four phases of launch vehicle ascent?

21 How fast is Kennedy Space Center ($L_0 = 28.5°$ N) moving?

22 Explain why eastward launches get a "head start" but westward launches don't.

23 Define the following

a) $\vec{V}_{loss\ gravity}$

b) $\vec{V}_{burnout}$

c) $\vec{V}_{launch\ site}$

d) $\Delta \vec{V}_{needed}$

e) ΔV_{losses}

f) ΔV_{design}

g) The SEZ coordinate system

24 Mission planners want to launch a new spacecraft to monitor hurricanes in the Pacific Ocean. The launch vehicle will launch from Kennedy Space Center into a circular orbit at an altitude of 800 km, with an inclination of 28.5° and right ascension of ascending node of 25°.

a) Find the SEZ components of $\vec{V}_{burnout}$.

b) Compute $\vec{V}_{launch\ site}$.

c) Compute $\Delta \vec{V}_{needed}$ in SEZ coordinates.

d) What is the magnitude of $\Delta \vec{V}_{needed}$?

e) If we assume ΔV_{losses} are 1.0 km/s for this launch, what is ΔV_{design}?

25 Launch-process teams are preparing the new Ariane V launch vehicle for launch from French Guyana ($L_0 = 4°$ N). For propellant loading, what ΔV_{design} does it need to achieve a circular orbit at 500 km, with an inclination of 4°, if the total losses from drag are 0.8 km/s?

Mission Profile—Salyut

he of the most ambitious goals of space exploration is tablishing a permanent human presence beyond the dle of Earth. From 1971–1986, the Soviet Salyut ogram logged a staggering number of human hours space, established procedures for resupplying smonauts, and laid the foundation of technology and perience for eventually colonizing of space.

ission Overview

om 1971 to 1986, the former Soviet Union designed, ilt, launched, and operated seven Salyut space tions of increasing capability. Cosmonauts conducted periments in life sciences, astronomy, Earth observa-n, and materials processing. They developed proce-ires for automated resupply, extravehicular activities VA or space walks), and in-flight maintenance and pair.

ission Data

Salyut 1: Launched April 19, 1971. Occupied for more than 23 days beginning June 16, 1971. Cosmo-nauts Dobrovolsky, Volkov, and Patsaiev died returning to Earth due to a sudden loss of cabin pressure.

Salyut 2: Launched April 3, 1973. The station broke up and decayed shortly after launch. No cosmo-nauts ever visited it.

Salyut 3: Launched June 25, 1974. Cosmonauts Popvich and Artyukhin spent more than 15 days onboard. It re-entered in January 1975 after seven months of occupied and unoccupied operation.

Salyut 4: Launched December 26, 1974. Cosmonauts Goubarev and Grechko logged more than 29 days onboard followed by Klimauk and Sevastianov with a record-setting 63-day stay. The unmanned Soyuz 20 mission demonstrated automated rendezvous and docking. The station deorbited on February 3, 1977.

Salyut 5: Launched June 22, 1976. Cosmonauts Volynov and Jolobov spent more than 49 days in the station. Soyuz 23 failed to dock. Soyuz 24 with Gorbatko and Glazkov, successfully docked in Feb-ruary 1977, and they spent 17 days onboard. The station deorbited August 8, 1977.

Salyut 6: Launched September 29, 1977. Signifi-cantly advanced over its predecessors, it consisted of five modules and two docking ports. The crews used a shower and self-contained space suits.

Salyut 6 was home to 33 cosmonauts who kept it occupied for 676 days—far longer than Skylab's 171 days. Deorbited July 1982.

✓ Salyut 7: Launched April 19, 1982. Ten crews spent a total of 812 days onboard. It received more than 37,000 kg (81,400 lb.) of cargo from twelve Progress and three Cosmos vehicles. Cosmonauts Dzhanibekov and Savinykh rescued the station after a nearly catastrophic power failure. Salyut 7 was visited one last time by Kizim and Soloviev to salvage equipment for use onboard the new Mir space station in 1986. Salyut 7 finally succumbed to atmospheric drag, re-entering in 1991.

Salyut 7 was the capstone of the Soviet Salyut program. *(Courtesy of Planeta Publishers, Moscow, Russia)*

Mission Impact

The Salyut program showed the importance of incre-mental improvement in a total program, each success setting the stage for the next.   Building on their Salyut experience, in 1986 the former Soviets launched the Mir station, which they can expand by adding on modules.

For Discussion

• The U.S. built the Space Shuttle while the former Soviets built Salyut. Which was the better choice? Why?

• How could the experience gained during Salyut apply toward planning a manned Mars mission?

References

Wilson, Andrew (ed.) *Space Directory 1990-91*. Coulsdon, U.K.: Jane's Information Group, 1990.

Rycroft, Michael (ed.) *The Cambridge Encyclopedia of Space*. New York, N.Y.: Cambridge University Press, 1990.

The Space Shuttle orbiter streaks into the atmosphere blazing a fiery trail through the sky. *(Courtesy of NASA/Johnson Space Center)*

Returning from Space: Re-entry

10

▨ Outline

10.1 Analyzing Re-entry Motion
Trade-offs for Re-entry Design
The Motion Analysis Process
Re-entry Motion Analysis in
Action

10.2 Options for Trajectory Design
Trajectory and Deceleration
Trajectory and Heating
Trajectory and Accuracy
Trajectory and the Re-entry
Corridor

10.3 Options for Vehicle Design
Vehicle Shape
Thermal-protection Systems

10.4 Lifting Re-entry

▨ In This Chapter You'll Learn to...

☛ Describe the competing design requirements for re-entry vehicles

☛ Describe the process for analyzing re-entry motion

☛ Describe the basic trajectory options and trade-offs in re-entry design

☛ Describe the basic vehicle options and trade-offs in re-entry design

☛ Describe how a lifting vehicle changes the re-entry problem

▨ You Should Already Know...

❑ The motion analysis process checklist (Chapter 4)

❑ Conservation of energy (Chapter 4)

❑ Newton's Second Law of Motion (Chapter 4)

❑ Basic concepts of calculus (Appendix A.3)

❑ Basic approach to interplanetary transfer (Chapter 7)

All around him glows the brilliant orange color. Behind, visible through the center of the window is a bright yellow circle. He sees that it is the long trail of glowing ablation material from the heat shield, stretching out behind him and flowing together. "This is Friendship 7. A real fireball outside!"

Astronaut, John Glenn
during re-entry of Mercury-Atlas 6
February 20, 1962
[Voas, 1962]

Space Mission Architecture. This chapter deals with the Trajectories and Orbits segment of the Space Mission Architecture, introduced in Figure 1-20.

Walking along the shore of a tranquil lake on a sunny, spring day, most of us have indulged in one of life's simplest pleasures: skipping stones. When the wind is calm, the mirror-like surface of the water practically begs us to try our skill. Searching through pebbles on the sandy bank, we find the perfect skipping rock: round and flat and just big enough for a good grip. We take careful aim, because we want the stone to strike the water's surface at the precise angle and speed that will allow its wide, flat bottom to take the full force of impact, causing it to skip. If we have great skill (and a good bit of luck), it may skip three or four times before finally losing its momentum and plunging beneath the water. We know from experience that, if the rock is not flat enough or its angle of impact is too steep, it'll make only a noisy splash rather than a quiet and graceful skip.

Returning from space, astronauts face a similar challenge. Earth's atmosphere presents to them a dense, fluid medium, which, at orbital velocities, is not all that different from a lake's surface. They must plan to hit the atmosphere at the precise angle and speed for a safe landing. If they hit too steeply or too fast, they risk making a big "splash," which would mean a fiery end. If their impact is too shallow, they may literally skip off the atmosphere and back into the cold of space. This subtle dance between fire and ice is the science of atmospheric re-entry.

In this chapter we explore the mission requirements of vehicles entering an atmosphere—whether returning to Earth or trying to land on another planet. We consider what engineers must trade in designing missions that must plunge into dense atmospheres (Figure 10-1). When we're through, you may never skip rocks the same way again!

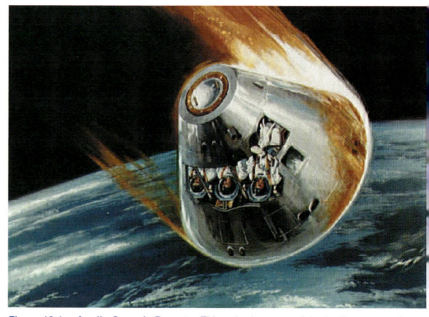

Figure 10-1.   Apollo Capsule Re-entry. This artists' concept of the Apollo re-entry shows that air friction causes the capsule to glow red hot. The astronauts inside stay cool, thanks to the protective heat shield. *(Courtesy of NASA/Johnson Space Center)*

10.1 Analyzing Re-entry Motion

▰ In This Section You'll Learn to...

☞ List and discuss the competing requirements of re-entry design

☞ Define a re-entry corridor and discuss its importance

☞ Apply the motion analysis process (MAP) checklist to re-entry motion and discuss the results

☞ Describe the process for re-entry design and discuss its importance

Trade-offs for Re-entry Design

All space-mission planning begins with a set of requirements we must meet to achieve mission objectives. The re-entry phase of a mission is no different. We must delicately balance three, often competing, requirements

- Deceleration
- Heating
- Accuracy of landing or impact

The vehicle's structure and payload limit the maximum deceleration or "g's" it can withstand. (One "g" is the gravitational acceleration at Earth's surface—9.798 m/s^2.) When subjected to enough g's, even steel and aluminum can crumple, like paper. Fortunately, the structural g limits for a well-designed vehicle can be quite high, perhaps hundreds of g's. But a fragile human payload would be crushed to death long before reaching that level. Humans can withstand a maximum deceleration of about 12 g's (about 12 times their weight) for only a few minutes at a time. Imagine eleven other people with your same weight all stacked on top of you. You'd be lucky to breathe! Just as a chain is only as strong as its weakest link, the maximum deceleration a vehicle experiences during re-entry must be low enough to prevent damage or injury to the weakest part of the vehicle.

But maximum g's aren't the only concern of re-entry designers. Too little deceleration can also cause serious problems. Similar to a rock skipping off a pond, a vehicle that doesn't slow down enough may literally bounce off the atmosphere and back into the cold reaches of space.

Another limitation during re-entry is heating. The fiery trail of a meteor streaking across the night sky shows that re-entry can get hot! This intense heat is a result of friction between the speeding meteor and the air. How hot can something get during re-entry? To find out, think about the energies involved. The Space Shuttle in orbit has a mass of 100,000 kg (220,000 lb.), an orbital velocity of 7700 m/s (17,225 m.p.h.), and an altitude of 300 km (186 mi.). We can find its total mechanical energy, E, using the relationship we developed in Chapter 4.

$$E = \frac{1}{2}mV^2 + mgh \qquad (10\text{-}1)$$

where

E = total mechanical energy (kg m^2/s^2 = joule)

m = mass (kg)

V = velocity (m/s)

g = acceleration due to gravity (m/s^2) = 8.94 m/s^2

h = altitude (m)

Substituting the above values and converting to standard units of energy, we get

$$E = 3.23 \times 10^{12}\ \text{joules} = 3.06 \times 10^9\ \text{Btu}$$

Let's put this number in perspective by recognizing that heating the average house in Colorado takes only about 73.4 × 10^5 Btu/year. So, the Shuttle has enough energy during re-entry to heat the average home in Colorado for 41 years!

The Shuttle has kinetic energy due to its speed of 7700 m/s and potential energy due to its altitude. It must lose all this energy in only about one-half hour to come to a full stop on the runway (at Earth's surface). But, remember, energy is conserved, so where does all the "lost" energy go? It converts to heat (from friction) caused by the atmosphere's molecules striking its leading edges. This heat makes the Shuttle's surfaces reach temperatures of up to 1477° C (2691° F). We must design the re-entry trajectory, and the vehicle, to withstand these high temperatures. As we'll see, not only do we have to contend with the total heating during re-entry, but the peak heating rate as well.

The third mission requirement is accuracy. Beginning its descent from over 6440 km (4000 mi.) away, the Space Shuttle must land on a runway only 91 m (300 ft.) wide. However, the re-entry vehicle (RV) of an Intercontinental Ballistic Missile (ICBM) has even tighter accuracy requirements. To meet these constraints, we must again adjust the trajectory and vehicle design.

On the other hand, if a vehicle can land in a larger area, the accuracy constraint becomes less important. For example, the Apollo missions required the capsules to land in large areas in the Pacific Ocean—much larger landing zones than for an ICBM's RV payload. Thus, the Apollo capsule was less streamlined and used a trajectory with a shallower re-entry angle. In all cases, designers adjust the trajectory and vehicle shape to match the accuracy requirement.

As you can see from all these constraints, a re-entry vehicle must walk a tightrope between being squashed and skipping out, between fire and ice, and between hitting and missing the target. This tightrope is actually a three-dimensional *re-entry corridor*, shown in Figure 10-2, through which a re-entry vehicle must pass to avoid skipping out or burning up.

The size of the corridor depends on the three competing constraints—deceleration, heating, and accuracy. For example, if the vehicle strays

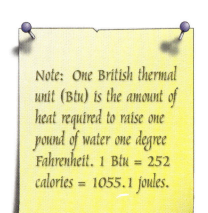

Note: One British thermal unit (Btu) is the amount of heat required to raise one pound of water one degree Fahrenheit. 1 Btu = 252 calories = 1055.1 joules.

below the lower boundary (undershoots), it will experience too much drag, slowing down rapidly and heating up too quickly. On the other hand, if the vehicle enters above the upper boundary (overshoots), it won't experience enough drag and may literally skip off the atmosphere, back into space. If designers aren't careful, these competing requirements may lead to a re-entry corridor that's too narrow for the vehicle to steer through!

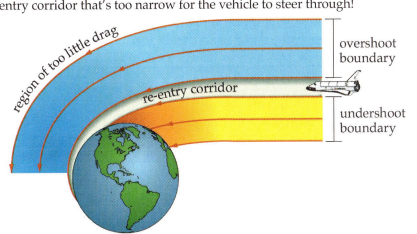

Figure 10-2.   Re-entry Corridor. The re-entry corridor is a narrow region in space that a re-entering vehicle must fly through. If the vehicle strays above the corridor, it may skip out. If it strays below the corridor, it may burn up.

Whereas the above three constraints determine the re-entry corridor's size, the vehicle's control system determines its ability to steer through the re-entry corridor. In this chapter we concentrate on describing what affects the corridor's size. We'll discuss limits on the control system in Chapter 12.

The Motion Analysis Process

Imagine one of Earth's many, small, celestial companions (say, an asteroid) wandering through space until it encounters Earth's atmosphere at more than 8 km/s, screaming in at a steep angle. Initially, in the upper reaches of the atmosphere, there is very little drag to slow down the massive chunk of rock. But as the meteor penetrates deeper, the drag force builds rapidly, causing it to slow down dramatically. This slowing is like the quick initial deceleration experienced by a rock hitting the surface of a pond. At this point in the meteor's trajectory, its heating rate is also highest, so it begins to glow with temperatures hot enough to melt the iron and nickel within. If anything is left of the meteor at this point, it will continue to slow down but at a more leisurely pace. Of course, most meteors burn up completely before reaching our planet's surface.

The meteor's velocity stays nearly constant through the first ten seconds, when the meteor is still above most of the atmosphere. But things change rapidly over the next ten seconds. The meteor loses almost 90% of its velocity—almost like hitting a wall. With most of its velocity lost, the deceleration is much lower—it takes 20 seconds more to slow down by another 1000 m/s.

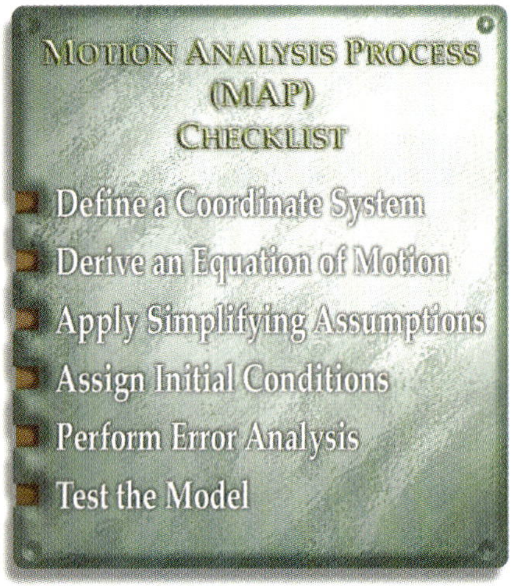

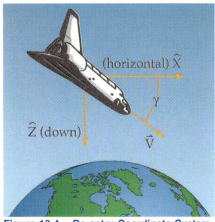

MOTION ANALYSIS PROCESS (MAP) CHECKLIST

■ Define a Coordinate System
■ Derive an Equation of Motion
■ Apply Simplifying Assumptions
■ Assign Initial Conditions
■ Perform Error Analysis
■ Test the Model

Figure 10-3. Motion Analysis Process (MAP) Checklist. This checklist is the same one we introduced in Chapter 4.

Figure 10-4. Re-entry Coordinate System. Our re-entry-coordinate system uses the center of the vehicle at the start of re-entry as the origin. The orbital plane is the fundamental plane, and the principal direction is down. The re-entry flight-path angle, γ, is the angle between local horizontal and the velocity vector.

Of course, unlike the meteor, in establishing a trajectory for a re-entry vehicle, we must keep the vehicle intact. Thus, we must trade deceleration, heating, and accuracy to calculate the correct trajectory for each vehicle. But first, to understand these trade-offs, we need to understand the motion of re-entering objects.

Before we can see how to juggle all these re-entry constraints, we need to develop a way of analyzing re-entry motion, to see how various trajectories and vehicle shapes affect its re-entry. Whether it's a rock hitting the water or a spacecraft hitting the atmosphere, we still have a dynamics problem, one we can solve by applying our trusty motion analysis process (MAP) checklist, as shown in Figure 10-3.

First on the list is defining a coordinate system. We still need an inertial reference frame (so Newton's laws apply), which we call the re-entry coordinate system. To make things easy, we place the origin of the *re-entry coordinate system* at the vehicle's center of mass *at the start of re-entry*. We then analyze the motion with respect to this fixed center.

The fundamental plane is the vehicle's orbital plane. Within this plane, we can pick a convenient principal direction, which points "down" to Earth's center. (By convention, the axis which points down is the $\hat{Z}$ direction.) We define the $\hat{X}$ direction along the local horizontal in the direction of motion. The $\hat{Y}$ direction completes the right hand rule. However, because we assume all motion takes place in plane, we won't worry about the $\hat{Y}$ direction. Figure 10-4 shows the re-entry coordinate system.

We also define the *re-entry flight-path angle, γ,* which is the angle between the local horizontal and the velocity vector. (Note this angle is the same as the orbital flight-path angle, ϕ, used earlier, but re-entry analysts like to use gamma, γ, instead, so we play along.) Similar to ϕ, a re-entry flight-path angle below the horizon (diving toward the ground) is negative, and a flight-path angle above the horizon (climbing) is positive.

Next, we derive an equation of motion. To do this, we brainstorm what forces could possibly affect a re-entering spacecraft. Of course there's gravity (it always seems to get involved) and, because it's travelling through the dense atmosphere, just like an airplane, it must also contend with lift and drag. Finally, we throw in "other" forces to cover all our bases. We show a vehicle with all these forces in Figure 10-5. Summing all of these forces, we get

$$\sum \vec{F}_{\text{external}} = \vec{F}_{\text{gravity}} + \vec{F}_{\text{drag}} + \vec{F}_{\text{lift}} + \vec{F}_{\text{other}} \qquad (10\text{-}2)$$

We now apply Newton's Second Law, which, in equation form, is

$$\sum \vec{F}_{\text{external}} = m\vec{a} \qquad (10\text{-}3)$$

Once again, we have a rather complicated equation to solve. So, it's time for some assumptions to bail us out. To make our lives easier, let's assume

- The re-entry vehicle is a point mass

- Drag is the dominant force—all other forces, including lift and gravity, are insignificant. (We'll see why this is a good assumption later.)

For meteors entering the atmosphere, the lift force is almost zero. Even for the Space Shuttle, lift is relatively small when compared to drag. For these reasons, we can assume for now that our vehicle produces no lift, thus, $\vec{F} = 0$. (Actually, the lift generated by the Space Shuttle is enough to significantly alter its trajectory, as we'll see in Section 10.2. However, this assumption will greatly simplify our analysis and allow us to demonstrate the trends in re-entry design.) Thus, we can assume gravity doesn't affect the vehicle and the vehicle produces no lift.

Looking at Figure 10-5, we can see that drag acts in the direction opposite the vehicle's motion. Because it has magnitude and direction, we must apply a little trigonometry, using the re-entry flight-path angle, γ, to resolve the components of the drag vector in the $\hat{X}$ and $\hat{Z}$ directions

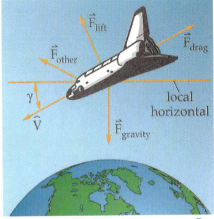

Figure 10-5.   Significant Forces on a Re-entry Vehicle. A re-entry vehicle could potentially encounter lift, drag, gravity, and other forces.

$$\sum \vec{F}_{external} = (-F_{drag}\cos\gamma)\hat{X} + (F_{drag}\sin\gamma)\hat{Z} \qquad (10\text{-}4)$$

Next we make some assumptions about drag. Drag on a vehicle depends on the *dynamic pressure, $\bar{q}$* ("q-bar"). $\bar{q}$ describes the effect of traveling through a fluid (air) with a density, ρ, at a velocity, V.

$$\bar{q} = \frac{\rho V^2}{2} \qquad (10\text{-}5)$$

where
$\bar{q}$ = dynamic pressure on the vehicle (N/m^2)
ρ = atmospheric density (kg/m^3)
V = vehicle's velocity (m/s)

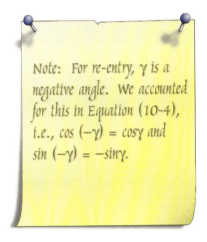

Note: For re-entry, γ is a negative angle. We accounted for this in Equation (10-4), i.e., $\cos(-\gamma) = \cos\gamma$ and $\sin(-\gamma) = -\sin\gamma$.

We can then describe drag using a unique property of vehicle shape—*coefficient of drag, C_D*. Engineers compute and validate this quantity using wind tunnels. Combining it with the dynamic pressure and the cross-sectional area of the vehicle, A, we can describe the drag force as

$$F_{drag} = \bar{q}\,C_D\,A = \frac{1}{2}\rho\,V^2\,C_D\,A \qquad (10\text{-}6)$$

where
F_{drag} = drag force on a vehicle (N)
C_D = drag coefficient (unitless)
A   = vehicle's cross-sectional area (m^2)
ρ = atmospheric density (kg/m^3)
V   = vehicle's velocity (m/s)

We can now simplify Equation (10-4) even more, to get

$$\sum \vec{F}_{external} = (-\bar{q}C_D A\cos\gamma)\hat{X} + (\bar{q}C_D A\sin\gamma)\hat{Z} \qquad (10\text{-}7)$$

Figure 10-6. Comparing Ballistic Coefficients. A sack of potatoes and a skydiver have about the same ballistic coefficient (BC).

If we divide both sides by the vehicle's mass to get the vehicle's acceleration, $\vec{a}$, we notice the result has $C_D A/m$ in both terms

$$\vec{a} = \left(-\bar{q}\frac{C_D A}{m}\cos\gamma\right)\hat{X} + \left(\bar{q}\frac{C_D A}{m}\sin\gamma\right)\hat{Z} \qquad (10\text{-}8)$$

where
$\vec{a}$ = vehicle's acceleration (m/s^2)
m = vehicle's mass (kg)
γ = vehicle's flight-path angle (deg)

Ever since engineers began to analyze the trajectories of cannon balls, this quantity $(C_D A/m)$ has had a special significance in describing how an object moves through the atmosphere. By convention, engineers invert this term and call it the *ballistic coefficient, BC.*

$$BC = \frac{m}{C_D A} \qquad (10\text{-}9)$$

where
BC = vehicle's ballistic coefficient (kg/m^2)
m = vehicle's mass (kg)
C_D = vehicle's drag coefficient (unitless)
A = vehicle's cross-sectional area (m^2)

From Equation (10-8), we can see the magnitude of the deceleration from drag, $|\vec{a}|$, is inversely related to BC

$$|\vec{a}| = \frac{\bar{q}}{BC} \qquad (10\text{-}10)$$

This relationship means that as BC goes up, deceleration goes down and vice versa.

Let's take a moment to see what BC really represents. Suppose a 60 kg (150 lb.) skydiver and a 60 kg (150 lb.) sack of potatoes fall out of an airplane at the same time (same mass, same initial velocity). If the skydiver and the potatoes have about the same mass, m; cross-sectional area, A; and drag coefficient, C_D, they have the same BC. Thus, the drag force on each is the same, and they fall at the same rate, as shown in Figure 10-6. What happens when the skydiver opens his parachute? He now slows down significantly faster than the sack of potatoes. But what happens to his BC? His mass stays the same, but when his chute opens, his cross-sectional area and C_D increase dramatically. When C_D and area increase, his BC goes down compared to the sack of potatoes, slowing his descent rate, as shown in Figure 10-7. From this example, we see that *an object with a low BC slows down much quicker than an object with a high BC.* In everyday terms, we would say a light, blunt vehicle (low BC) slows down much more rapidly than a heavy, streamlined (high BC) one, as shown in Figure 10-8.

Figure 10-7. Changing BC. With his parachute open, the skydiver greatly increases his area, A, and drag coefficient, C_D, thus decreasing his ballistic coefficient, BC, and slowing down much faster than the potatoes.

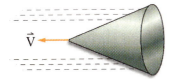 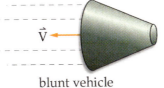

streamlined vehicle
(high ballistic coefficient)

blunt vehicle
(low ballistic coefficient)

Figure 10-8.   Blunt Versus Streamlined Vehicles. A light, blunt vehicle (low BC) slows down much more rapidly due to drag than a heavy, streamlined (high BC) one.

Now that we have the re-entry equation of motion, we can turn our attention to the next item on the MAP checklist—Initial Conditions (ICs). These ICs are especially important for re-entry. The initial re-entry velocity, $V_{re-entry}$, and the initial re-entry flight-path angle, γ, determine most of the conditions experienced during re-entry. Determining what these ICs should be involves many trade-offs for trajectory and vehicle designers. For re-entry analysis, we'll concentrate on the effects of these ICs and not spend any time on Error Analysis or Testing the Model. So, this concludes the MAP checklist. Let's look at how we can use what we've learned about re-entry motion.

Astro Fun Fact
Dinosaurs and Meteors

Every day, 400 tons of micrometeorite dust hit Earth in the form of minute cosmic particles. Yet, this did not explain what geologist Walter Alvarez discovered in Italy in the late 1970s. He unearthed a half-inch layer of clay deposited 65 million years ago. He named this layer the K-T layer, as the clay lay between the Cretaceous and Tertiary Time periods. Later, a technique called neutron activation found this deposit contained thirty times the normal amount of iridium, an element rare on Earth but abundant in meteors. This evidence led to the theory that a massive meteor collision with Earth caused the extinction of dinosaurs. The theory, officially called the K-T theory of extinction, appears viable. Possible sites for the meteor's impact include a 190-mile-wide crater off the coast of South America, as well as, an unknown-sized crater 3500 feet below ground on the Yucatan peninsula (indicated by geographical surface features). Why then, you may ask, did other species survive such an enormous catastrophe? While no scientific explanation can yet answer this question, many scientists believe it may simply have been another event in the natural selection process—survival of the fittest!

Evans, Barry. <u>The Wrong Way Comet and Other Mysteries of Our Solar System</u>. Blue Ridge Summit: Tab Books, 1992.

Contributed by Troy Kitch, the U.S. Air Force Academy

Re-entry Motion Analysis in Action

Because the equation of motion we developed for re-entry in Equation (10-8) is still quite complicated, let's take some time to see how we can use it. We need to understand how the acceleration equation affects a vehicle's velocity and, in turn, its position during re-entry.

If we give an object a constant acceleration, we can determine its velocity after some time, t, from

$$\vec{V}_{final} = \vec{V}_{initial} + \vec{a}t \tag{10-11}$$

where

$\vec{V}_{final}$ = final velocity (m/s)
$\vec{V}_{initial}$ = initial velocity (m/s)
$\vec{a}$ = acceleration (m/s^2)
t   = time (s)

The final position of the object is

$$\vec{R}_{final} = \vec{R}_{initial} + \vec{V}_{initial}\, t + \frac{1}{2}\vec{a}t^2 \tag{10-12}$$

Unfortunately, a re-entry vehicle's acceleration isn't constant. Notice in Equation (10-8) that drag deceleration is a function of velocity, and the velocity changes due to drag! This equation is another example of a transcendental function similar to the one we discussed in Chapter 8. We can't solve these equations for a vehicle's position directly. How do we deal with it, then? We use a method first developed by Isaac Newton—numerical integration. Sound complicated? Actually it's not that bad. We assume that over some small time interval, Δt, the acceleration *is* constant (a good assumption if Δt is small enough). This allows us to use the velocity and position equations for constant acceleration during that time interval. By adding the acceleration effects during each time interval, we can determine the cumulative effect on velocity and position. (Of course this means lots of calculations, so it's best to use a computer. We could either write a new computer program or use the built-in flexibility of a spreadsheet. We did all the analysis in this chapter using a spreadsheet.)

Let's start by applying this numerical analysis technique to the motion of the meteor entering the atmosphere, as we discussed earlier. Recall, its velocity is pretty much constant initially, while it is high in the thin atmosphere. But then, it hits a wall as the atmosphere thickens and it slows rapidly. The results of the numerical integration for this example are shown in Figure 10-9. We can see from the graph what we expected to find from our discussion. Notice in the figure that the velocity stays nearly constant through the first ten seconds, when the meteor is still above most of the atmosphere. But conditions change rapidly over the next ten seconds. The meteor loses about 90% of its velocity—almost like hitting a wall. With most of its velocity lost, the vehicle decelerates much more slowly—it takes 20 seconds more to slow down by another 1000 m/s.

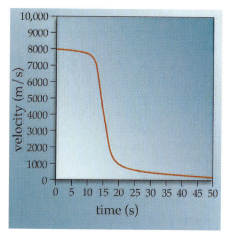

Figure 10-9. Meteor Re-entering the Atmosphere. Notice how abruptly a meteor slows down—similar to a rock hitting the surface of a pond.

We now have a precise mathematical tool to analyze re-entry character-istics. We can use this tool to balance all the competing mission require-ments by approaching them on two broad fronts

- Trajectory design, which includes changes to
  - Re-entry velocity, $V_{re-entry}$
  - Re-entry flight-path angle, γ
- Vehicle design, which includes changes to
  - Vehicle size and shape (BC)
  - Thermal-protection systems (TPS)

Trajectory design involves changing the re-entry initial conditions, defined by the vehicle's velocity as it enters the effective atmosphere. These initial conditions are the *re-entry velocity, $V_{re-entry}$*, and re-entry flight-path angle, γ. Vehicle design includes changing the vehicle's shape to alter the BC or designing a thermal-protection systems (TPS) to deal with re-entry heating.

As seen in Figure 10-10, re-entry design requires iteration. Mission requirements affect the vehicle design. The design drives deceleration, heating, and accuracy. These parameters, in turn, affect trajectory options, which may change the vehicle design, and so on. In practice, we must continually trade between trajectory and vehicle design, until we reach some compromise vehicle that meets mission requirements. In the next few sections, we'll explore trajectory options and vehicle design in greater detail.

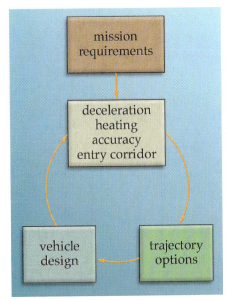

Figure 10-10.   Re-entry Design. Re-entry design begins with mission requirements. Then engineers must work the trade-offs between vehicle design, deceleration, heating, accuracy, re-entry corridor, and trajectory options.

▦ Section Review

Key Terms

ballistic coefficient, BC
coefficient of drag, C_D
dynamic pressure, $\bar{q}$
re-entry coordinate system
re-entry corridor
re-entry flight-path angle, γ
re-entry velocity, $V_{re\text{-}entry}$

Key Equations

$$\bar{q} = \frac{\rho V^2}{2}$$

$$F_{drag} = \bar{q}\, C_D\, A = \frac{1}{2}\rho\, V^2\, C_D\, A$$

$$\vec{a} = \left(-\bar{q}\frac{C_D A}{m}\cos\gamma\right)\hat{X} + \left(\bar{q}\frac{C_D A}{m}\sin\gamma\right)\hat{Z}$$

$$BC = \frac{m}{C_D A}$$

Key Concepts

➤ We must balance three competing requirements for re-entry design

- Deceleration
- Heating
- Accuracy

➤ We base the re-entry coordinate system on the

- Origin—vehicle's center of gravity at the beginning of re-entry
- Fundamental plane—vehicle's orbital plane
- Principal direction—down

➤ To analyze re-entry trajectories, we must use numerical integration with the following assumptions

- Re-entry vehicle is a point mass
- Drag is the dominant force—all other forces, including gravity and lift, are insignificant

➤ Ballistic coefficient, BC, quantifies an object's mass, drag coefficient, and cross-sectional area and predicts how drag will affect it

- Light, blunt vehicle—low BC—slows down quickly
- Heavy, streamlined vehicle—high BC—doesn't slow down quickly

➤ To balance competing requirements, we tackle the re-entry-design problem on two fronts

- Trajectory design—changes to re-entry velocity, $V_{re\text{-}entry}$, and re-entry flight-path angle, γ
- Vehicle design—changes to a vehicle's size and shape (BC) and thermal-protection systems (TPS)

10.2 Options for Trajectory Design

▰▰▰ In This Section You'll Learn to...

- ☞ Explain how changing the re-entry velocity and flight-path angle affects deceleration and heating rates
- ☞ Determine the maximum deceleration and the altitude at which this deceleration occurs for a given set of re-entry conditions
- ☞ Determine the maximum heating rate and the altitude at which this rate occurs for a given set of re-entry conditions
- ☞ Explain how changing the re-entry velocity and flight-path angle affects accuracy and size of the re-entry corridor

Depending on the mission and vehicle characteristics, planners can do only so much with the re-entry trajectory. For example, the amount of propellant the Space Shuttle can carry for the engines in its orbital maneuvering system (OMS) limits how much it can alter velocity and flight-path angle at re-entry. Re-entry conditions for ICBM re-entry vehicles, depend on the velocity and flight-path angle of the booster at burnout. In either case, we must know how the re-entry trajectory affects a vehicle's maximum deceleration, heating, and accuracy, as well as the re-entry corridor's size.

Trajectory and Deceleration

As we showed with our meteor example in the last section, a vehicle re-entering from space takes time to make its way into the denser layers of the atmosphere. Deceleration builds gradually to some maximum value, a_{max}, and then begins to taper off. To see how varying the re-entry velocity and angle affects this maximum deceleration, let's apply our numerical tool to the re-entry equation of motion we developed in the last section. We begin by keeping all other variables constant and change only the initial re-entry velocity, $V_{re-entry}$, to see its effect on a_{max}. We can plot the deceleration versus altitude for various re-entry velocities, if we set the following initial conditions

   Vehicle mass = 1000 kg

   Nose radius = 2 m

   Cross-sectional area = 50.3 m^2

   $C_D = 1.0$

   BC = 19.9 kg/m^2

   Re-entry flight-path angle, $\gamma = 45°$

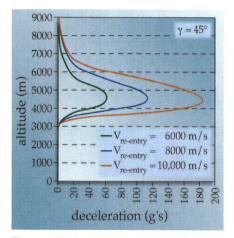

Figure 10-11. Deceleration Profiles for Various Re-entry Velocities. For a given re-entry flight-path angle, the higher the re-entry velocity, the greater the maximum deceleration.

Figure 10-11 shows that a higher re-entry velocity means greater maximum deceleration. This should make sense, if we think again about skipping rocks. The harder we throw a rock at the water (the higher the $V_{re\text{-}entry}$), the bigger the splash it will make (greater a_{max}). Without going into a lengthy derivation, we can find the vehicle's maximum deceleration, and the altitude at which it occurs, from

$$a_{max} = \frac{V_{re\text{-}entry}^2 \beta \sin\gamma}{2e} \tag{10-13}$$

$$Altitude_{a_{max}} = \frac{1}{\beta}\ln\left(\frac{\rho_o}{BC\,\beta\,\sin\gamma}\right) \tag{10-14}$$

where

a_{max} = vehicle's maximum deceleration (m/s^2)

$V_{re\text{-}entry}$ = vehicle's re-entry velocity (m/s)

β = atmospheric scale height, a parameter used to describe the density profile of the atmosphere $= 0.000139\ m^{-1}$ for Earth

γ = vehicle's flight-path angle (deg or rad)

e = base of the natural logarithm $= 2.7182...$

$\ln$ = natural logarithm of the quantity in parentheses

ρ_o = atmospheric density at sea level $= 1.225\ kg/m^3$

BC = vehicle's ballistic coefficient (kg/m^2)

Notice the maximum deceleration depends on the re-entry velocity and flight-path angle, but the altitude of a_{max} depends only on the flight-path angle (see Equation 10-14). So, as Figure 10-11 shows, no matter what the velocity, the altitude of a_{max} will be the same for a given flight-path angle.

Now that we know how $V_{re\text{-}entry}$ affects deceleration, let's look at the other trajectory parameter—flight-path angle, γ. Keeping the same initial conditions and fixing the re-entry velocity at 8 km/s, we can plot the deceleration versus altitude profiles for various re-entry flight-path angles.

In Figure 10-12, we show that the steeper the re-entry angle the more severe the peak deceleration. Once again, this should make sense from the rock-skipping example, in which a steeper angle causes a bigger splash. In addition, we show that a vehicle with a steeper re-entry angle plunges deeper into the atmosphere before reaching the maximum deceleration.

Now let's look at the amount of maximum deceleration (in g's) for varying re-entry velocities and flight-path angles. Notice the maximum deceleration is over 160 g's! Because the acceleration from gravity is defined as 1 g, we can conclude the dominant force on a vehicle during re-entry is drag. This justifies our earlier decision to ignore gravity.

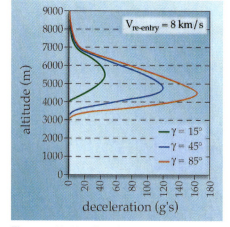

Figure 10-12. Deceleration Profile for Various Re-entry Flight-Path Angles. For a given velocity, the higher the re-entry flight-path angle (steeper the re-entry) the greater the maximum deceleration experienced.

Trajectory and Heating

Earlier, we described *why* a re-entry vehicle gets hot—all the orbital energy it starts with must go somewhere (conservation of energy). Before looking at *how* the vehicle gets hot, let's review how heat transfers from one place to another by radiation, conduction, and convection. *Radiation* or *radiative heat transfer*, discussed in Chapter 3, involves the transfer of energy from one point to another through electromagnetic waves. If you've ever held your hand in front of a glowing space heater, you've felt radiative heat transfer.

Conduction or *conductive heat transfer* moves heat energy from one point to another through some physical medium. For example, try holding one end of a metal rod and sticking the other end in a hot fire. Before too long the end you're holding will get HOT (ouch)! The heat "conducts" along the metal rod.

Finally, *convection* or *convective heat transfer* occurs when a fluid flows past an object and transfers energy to it or absorbs energy from it (depending on which object is hotter). This is where we get the concept of "wind chill." As a breeze flows past us, heat transfers from our body to the air, keeping us cool.

So what's all this have to do with a re-entering vehicle? If you've ever been on a ski boat, plowing at high speeds through the water, you may have noticed how the water bends around the hull. At the front of the boat, where the hull first meets the water, a bow wave forms so the moving boat never appears to run into the still water. This bow wave continues around both sides of the boat, forming the wake of turbulent water that's so much fun to ski through.

A spacecraft re-entering the atmosphere at high speeds must plow into the fluid air, much like the boat. Because of the extremely high re-entry speeds, even the wispy upper atmosphere creates a profound effect on a vehicle. In front of the re-entering spacecraft, a bow wave of sorts forms. This *shock wave* results when air molecules bounce off the front of the vehicle and then collide with the incoming air. The shock wave then bends the air flow around the vehicle. Depending on the shape of the vehicle, the shock wave can either be attached or detached. If the vehicle is stream-lined (high BC, like a cone), the shock wave may attach to the tip and transfer a significant amount of heat, causing localized heating at the attachment point. If the vehicle is blunt (low BC, like a rock), the shock wave will detach and curve in front of the vehicle, leaving a boundary of air between the shock wave and the vehicle's surface. Figure 10-13 shows both types of shock waves.

So how does the vehicle get so hot? As the shock wave slams into the air molecules in front of the re-entering vehicle, they go from a cool, dormant state to an excited state, acquiring heat energy. (To see why, strike a metal object, such as a nail, with a hammer many times and feel the object get hot.) Similar to the energetic re-entry vehicle, transferring energy to air molecules, the hammer converts its kinetic energy into heat, which it transfers to the metal object upon contact.

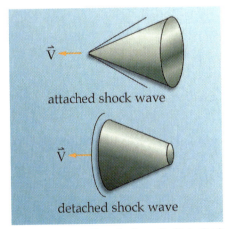

attached shock wave

detached shock wave

Figure 10-13. Attached and Detached Shock Waves. As a vehicle plows into the atmosphere from space a shock wave forms out in front. This shock wave attaches to streamlined vehicles (high BC) but detaches from blunt vehicles (low BC).

These hot air molecules then transfer some of their heat to the vehicle by convection. Convection is the primary means of heat transfer to a vehicle entering Earth's atmosphere at speeds under about 15,000 m/s. (For a re-entry to Mars or some other planet with a different type of atmosphere, this speed will vary.) Above this speed, the air molecules get so hot they begin to transfer more of their energy to the vehicle by radiation.

Without going into all the details of aerodynamics and thermo-dynamics, we can quantify the *heating rate, $\dot{q}$* ("q dot" or rate of change of heat energy) a re-entry vehicle experiences. We express this quantity in watts per square meter, which is heat energy per unit area per unit time. It's a function of the vehicle's velocity and nose radius, and the density of the atmosphere. Empirically, for Earth's atmosphere, this becomes approximately

$$\dot{q} \cong 1.83 \times 10^{-4} \; V^3 \sqrt{\frac{\rho}{r_{nose}}} \qquad (10\text{-}15)$$

where

$\dot{q}$ = vehicle's heating rate (W/m^2)

V = vehicle's velocity (m/s)

ρ = air density (kg/m^3)

r_{nose} = vehicle's nose radius (m)

Returning to our numerical analysis of a generic re-entry vehicle with the same initial conditions as before, we can plot heating rate, $\dot{q}$, versus altitude for various re-entry velocities. In Figure 10-14 we show that the maximum heating rate increases as the re-entry velocity goes up. We can find the altitude and velocity where the maximum heating rate occurs using

$$\text{Altitude}_{\dot{q}_{max}} = \frac{1}{\beta}\ln\left(\frac{\rho_o}{3BC\,\beta\,\sin\gamma}\right) \qquad (10\text{-}16)$$

where

β = atmospheric scale height = 0.000139 m^{-1} for Earth

ρ_o = atmospheric density at sea level = 1.225 kg/m^3

BC = vehicle's ballistic coefficient (kg/m^2)

γ = vehicle's flight-path angle (deg or rad)

and

$$V_{\dot{q}_{max}} \approx 0.846 \; V_{re\text{-}entry} \qquad (10\text{-}17)$$

where

$V_{\dot{q}_{max}}$ = vehicle's velocity when it reaches maximum heating rate (m/s)

$V_{re-entry}$ = vehicle's re-entry velocity (m/s)

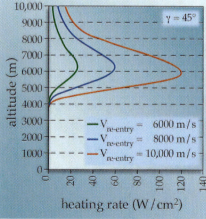

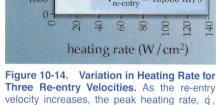

Figure 10-14.   Variation in Heating Rate for Three Re-entry Velocities. As the re-entry velocity increases, the peak heating rate, $\dot{q}$, also increases.

From Equation (10-17), we learn that the velocity for the maximum heating rate is about 85% of the re-entry velocity.

We can also vary the re-entry flight-path angle, γ, to see how it affects the maximum heating rate. Let's use a re-entry velocity of 8 km/s again. Keeping all other initial conditions the same and varying γ, we can plot $\dot{q}$ versus altitude for various re-entry flight-path angles, as shown in Figure 10-15.

Notice the correlation between steepness of re-entry and the severity of the peak heating rate. Recall from our earlier discussion that the steeper the re-entry the deeper into the atmosphere the vehicle travels before reaching maximum deceleration. This means the steeper the re-entry angle, the more quickly the vehicle reaches the ground, creating an interesting dilemma for the re-entry designer

- Steep re-entry angles cause high maximum heating rates but for a short time
- Shallow re-entry causes low maximum heating rates but for a long time

A steep re-entry causes a very high heating rate but for a brief time, so the overall effect on the vehicle may be small. On the other hand, shallow re-entries lead to much lower heating rates. However, because heating continues longer, the vehicle is more likely to "soak up" heat and be damaged.

To understand this difference, imagine boiling two pots of water. For the first pot we build a fire using large, thick logs. They'll build up a low, steady heating rate, lasting for a long time. Under the second pot we place an equal mass of wood but in the form of sawdust. The sawdust will burn much faster than the logs but will also burn out much more quickly. Which option will boil the water better? Because the logs burn at a lower heat rate but for much longer, the water is more likely to soak up this heat and begin to boil. The sawdust burns so fast that the pot can't absorb it quickly enough, so most of its heat simply escapes into the air.

This example underscores the importance of considering the heating rate, $\dot{q}$, along with the total heat load, Q. *Total heat load, Q,* is the total amount of thermal energy (J/m^2) the vehicle receives. We find Q by integrating or summing all the $\dot{q}$'s over the entire re-entry time. As we've already seen, $\dot{q}$ varies with re-entry velocity. Q also varies with velocity but *not* with flight-path angle. This makes sense when we consider the heat results from mechanical energy dissipating during re-entry, which is independent of re-entry angle. This means, the higher the re-entry velocity, the higher the total heat load, as shown in Figure 10-16. Thus, although the peak heating rate varies with flight-path angle, the total heat load for a given re-entry velocity is constant.

Again, we face an acute engineering dilemma for manned re-entry vehicles. We'd like a shallow re-entry to keep the maximum deceleration low (don't crush the crew), but this means a greater risk of soaking up the re-entry heat. Fortunately (for the crew), we have ways to deal with this heat energy, as we'll see in the next section.

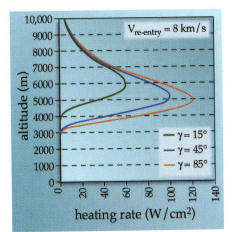

Figure 10-15.   Variation in Heating Rate at Different Re-entry Flight-Path Angles. The steeper the re-entry angle, γ, [Equation (10-16)] the higher the peak heating rate, $\dot{q}$.

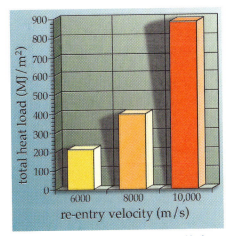

Figure 10-16.   Total Heat Load for Various Re-entry Velocities. The higher the re-entry velocity, the greater the total heat load, Q.

Trajectory and Accuracy

Next, we can look at how trajectory affects accuracy. Consider what the atmosphere does to a re-entering vehicle. Drag and lift forces perturb its trajectory from the path it would follow under gravity alone. When we modeled these effects, we used several parameters to quantify how the atmosphere affects the vehicle. Whether we're modeling the density, ρ, or the drag coefficient, C_D, the values we use are, at best, only close to the real values and, at worst, mere approximations. Thus, the actual trajectory path will be somewhat different, so when we try to aim at a particular target we might miss!

To reduce these atmospheric effects, and improve our accuracy, we want a trajectory that spends the least time in the atmosphere. So we choose a high re-entry velocity and a steep re-entry angle. But as we've just seen, this increases the severity of deceleration and heating. Thus, to achieve highly accurate re-entry for ICBMs, we build these vehicles to withstand extremely high g forces and peak heating. Manned vehicles, on the other hand, accept lower accuracy to get much lower peak deceleration and heating.

Trajectory and the Re-entry Corridor

From the definition of re-entry corridor, we can think of the upper or overshoot boundary as the "skip out" boundary. A vehicle entering the atmosphere above this boundary risks bouncing off the atmosphere and back into space. While hard to quantify exactly, this boundary is set by the minimum deceleration needed to "capture" the vehicle. Changes to re-entry velocity or flight-path angle don't move this boundary significantly. Therefore, we can change the size of the re-entry corridor most effectively by tackling the lower or undershoot boundary.

As we've just shown, maximum deceleration and maximum heating rate, the two parameters that set the undershoot boundary, increase directly with increased re-entry velocity, $V_{re\text{-}entry}$, or re-entry flight-path angle, γ, (steeper re-entry). Most programs limit maximum deceleration and maximum $\dot{q}$ to certain values. Thus, we could still expand the re-entry corridor by decreasing $V_{re\text{-}entry}$ or γ. This change would give us a larger margin for error in planning the re-entry trajectory and relieve requirements placed on the control system. Unfortunately, for most missions, $V_{re\text{-}entry}$ and γ are set by the mission orbit and are difficult to change significantly without using rockets to perform large, expensive ΔVs. Therefore, as we'll see in the next section, our best options for changing the re-entry corridor size lie in the vehicle design arena.

Table 10-1 summarizes how trajectory options affect deceleration, heating, accuracy, and re-entry-corridor size.

Table 10-1.   Trajectory Trade-offs for Re-entry Design. Notice that maximum deceleration and maximum heating rates vary directly with velocity and re-entry flight-path angle. For a constant velocity, altitudes for maximum deceleration and maximum heating rate vary inversely with flight-path angle. For a constant re-entry flight-path angle, altitudes for maximum deceleration and maximum heating rate are independent of velocity. total heat load varies directly with velocity and is independent of re-entry flight-path angle.

Parameter	Maximum Deceleration	Altitude of Maximum Deceleration	Maximum Heating Rate	Altitude of Maximum Heating Rate	Accuracy	Corridor Width
Re-entry velocity, $V_{re\text{-}entry}$ (constant γ)						
High	High	Same	High	Same	High	Narrow
Low	Low	Same	Low	Same	Low	Wide
Re-entry flight-path angle, γ (constant $V_{re\text{-}entry}$)						
Steep	High	Low	High	Low	High	Narrow
Shallow	Low	High	Low	High	Low	Wide

▰ Section Review

Key Terms

conduction
conductive heat transfer
convection
convective heat transfer
heating rate, $\dot{q}$
radiation
radiative heat transfer
shock wave
total heat load, Q

Key Equations

$$a_{max} = \frac{V_{re\text{-}entry}^2 \beta \sin\gamma}{2e}$$

$$\text{Altitude}_{a_{max}} = \frac{1}{\beta}\ln\left(\frac{\rho_o}{BC\ \beta\ \sin\gamma}\right)$$

$$\text{Altitude}_{\dot{q}_{max}} = \frac{1}{\beta}\ln\left(\frac{\rho_o}{3BC\ \beta\ \sin\gamma}\right)$$

$$V_{\dot{q}_{max}} \approx 0.846\ V_{re\text{-}entry}$$

Key Concepts

➤ We can meet re-entry mission requirements on the trajectory front by changing
  - Re-entry velocity, $V_{re\text{-}entry}$
  - Re-entry flight-path angle, γ

➤ Increasing re-entry velocity increases
  - Maximum deceleration, a_{max}
  - Maximum heating rate, $\dot{q}_{max}$

➤ Compared to the drag force, the gravity force on a re-entry vehicle is insignificant

➤ Increasing the re-entry flight-path angle, γ, (steeper re-entry) increases
  - Maximum deceleration, a_{max}
  - Maximum heating rate, $\dot{q}_{max}$

➤ The more time a vehicle spends in the atmosphere, the less accurate it will be. Thus, to increase accuracy, we use fast, steep re-entry trajectories.

➤ To increase the size of the re-entry corridor, we decrease the re-entry velocity and flight-path angle. However, this is often difficult to do.

➤ Table 10-1 summarizes the trajectory trade-offs for re-entry design

10.3 Options for Vehicle Design

▥ In This Section You'll Learn to...

☞ Discuss two ways to determine the hypersonic drag coefficient for a given vehicle shape

☞ Discuss the effect of changing the ballistic coefficient on deceleration, heating rate, and re-entry-corridor width

☞ Discuss three types of thermal-protection systems and how they work

Once we've exhausted all trajectory possibilities, we can turn to options for vehicle design. Here, we have two ways to meet mission requirements

• Vehicle size and shape
• Thermal-protection systems (TPS)

In this section, we'll look at both of these.

Vehicle Shape

The re-entry vehicle's size and shape help determine the ballistic coefficient (BC) and the amount of lift it will generate. Because adding lift to the re-entry problem greatly complicates the analysis, we'll continue to assume we're dealing only with non-lifting vehicles. In the next section, we'll discuss how lift affects the re-entry problem.

The hardest component of BC to determine for re-entry vehicles is the drag coefficient, C_D, which depends primarily on the vehicle's shape. At low speeds, we could just stick a model of the vehicle in a wind tunnel and take specific measurements to determine C_D. But at re-entry speeds approaching 25 times the speed of sound, wind tunnel testing isn't practical because no tunnels work at those speeds. Instead, we must create mathematical models of this hypersonic flow to find C_D. The most accurate of these models requires us to use high-speed computers to solve the problem. This approach is now a specialized area of aerospace engineering known as *computational fluid dynamics (CFD)*.

Fortunately, a simpler but less accurate way will get us close enough for our purpose. We can use an approach introduced more than 300 years ago called *Newtonian flow*. Yes, Isaac Newton strikes again. Because Newton looked at a fluid as simply a collection of individual particles, he assumed his laws of motion must still work. But they didn't at low speeds. Centuries later, however, Newton was vindicated when engineers found his model worked quite well for flow at extremely high speeds. So the grand master of physics was right again—but only for certain situations. Figure 10-17 summarizes these two approaches to analyzing fluid dynamics. Using Newton's approach, we can calculate C_D and thus find BC. We show three examples using this approach for three simple shapes in Table 10-2.

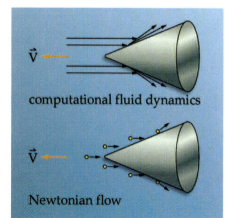

Figure 10-17.   Computational Fluid Dynamics (CFD) Versus Newtonian Flow. In CFD, high-speed computers numerically model the fluid flow. Newton's approach models the fluid flow as many individual particles impacting the vehicle.

Table 10-2. Examples of Estimating BC Using Newton's Approach.

Shape	Example Values	Estimated Ballistic Coefficient
Sphere	$D = 2$ m $C_D = 2.0$ $m = 2094$ kg (Assumes density = 500 kg/m^3)	BC ≅ 333 kg/m^2
Cone	$l = 3.73$ m $\delta_c = 15° =$ cone half angle $r_c = 1$ m = cone radius $C_D \cong 2\,\delta_c^2 = 0.137$ $m = 1954$ kg (Assumes density = 500 kg/m^3)	BC ≅ 4543 kg/m^2
Blunted cone	$l = 3.04$ m $\delta_c = 15° =$ cone half angle $r_c = 1$ m = cone radius $r_n = 0.304$ m $m = 1932$ kg (Assumes density = 500 kg/m^3) $C_D = (1 - \sin^4\delta_c)\left(\dfrac{r_n}{r_c}\right)^2$ $\quad + 2\sin^2\delta_c\left[1 - \left(\dfrac{r_n}{r_c}\right)^2\cos^2\delta_c\right]$ $C_D \cong 0.188$	BC ≅ 3266 kg/m^2

Effects of Vehicle Shape on Deceleration

Now that we have a way to find BC, we can use the numerical tool we used in the last section to see how varying BC changes a re-entry vehicle's deceleration profile. Let's start by looking at three very different vehicles entering Earth's atmosphere at an angle of 45° and a velocity of 8000 m/s. Notice something very interesting in Figure 10-18: the maximum deceleration, a_{max}, is the same in all cases! But the altitude of a_{max} varies with BC. This is what we'd expect from Equations (10-13) and (10-14). The higher the BC (the more streamlined the vehicle), the deeper it plunges into the atmosphere before reaching a_{max}. This means a streamlined vehicle spends less time in the atmosphere and reaches the ground long before a blunt vehicle.

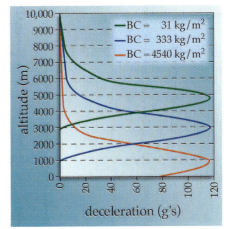

Figure 10-18. Deceleration Profiles for Various Ballistic Coefficients (BC). Note that, regardless of shape, all the vehicles experience the same maximum deceleration but at different altitudes.

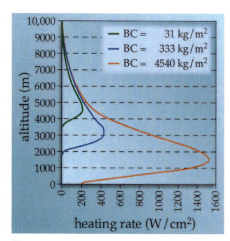

Figure 10-19. Heating Rate Profiles for Various Ballistic Coefficients (BC). Streamlined vehicles have a much higher maximum heating rate, lower in the atmosphere, than blunt vehicles.

Effects of Vehicle Shape on Heating Rate

Now let's see how varying BC affects the maximum heating rate. In Figure 10-19, notice the maximum heating rate is much more severe for the high-BC (streamlined) vehicle and occurs much lower in the atmosphere. The shape of the shock wave surrounding each vehicle causes this difference. Remember the nature of shock waves for blunt and streamlined vehicles, shown in Figure 10-13. Blunt vehicles have detached shock waves that spread the heat of re-entry over a relatively large volume. Furthermore, the air flow near the surface of blunt vehicles tends to inhibit convective heat transfer. Thus, the heating rate for blunt vehicles is relatively low.

Streamlined vehicles, on the other hand, have attached shock waves. This situation concentrates a large amount of heat near the sharp tip causing it to reach very high temperatures—hot enough to melt most materials. In addition, the heat around the vehicle stays in a smaller volume, and the air flow near the surface doesn't inhibit heat transfer as well. As a result, the overall heating rate is higher as illustrated in Figure 10-20. For these reasons, "needle-nosed" vehicles (like you see in some science fiction movies) aren't very practical. In practice, even relatively streamlined vehicles have slightly rounded noses to keep the tips from burning off.

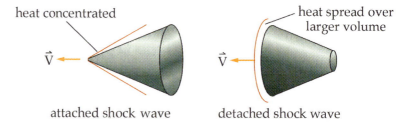

Figure 10-20. Shock Waves and Heating. For streamlined vehicles (high BC), the shock wave is attached, concentrating heat at the tip. For blunt vehicles (low BC), the shock wave is detached, spreading the heat over a larger volume.

Effects of Vehicle Shape on Accuracy

As we've seen, a more streamlined (high-BC) vehicle reaches maximum deceleration much lower in the atmosphere than a blunt (low-BC) vehicle; thus, it reaches the ground more quickly. We know from earlier discussion that the atmosphere can greatly decrease re-entry accuracy, so we want our vehicle to spend as little time in the atmosphere as possible. As a result, we want a streamlined vehicle for better accuracy, even though we must accept more severe heating rates. As we'll see, thermal-protection systems can deal with this heating.

Effects of Vehicle Shape on the Re-entry Corridor

We already said that the re-entry corridor's upper or overshoot boundary depends on the minimum deceleration for atmospheric capture. Variations in vehicle shape don't affect this end of the corridor significantly. However, we can change the lower or undershoot boundary by changing the limits on deceleration or heating rate. But maximum deceleration is independent of the BC, so a vehicle's shape doesn't affect this boundary either. On the other hand, as we've seen, decreasing the BC can dramatically decrease the maximum heating rate. Thus, when the corridor's lower boundary is set by the maximum heating rate, decreasing BC can be helpful. This decrease expands the re-entry corridor and gives us more margin for navigational error.

Table 10-3 summarizes how vehicle shape affects re-entry parameters.

Table 10-3. Summary of Ballistic Coefficient (BC) Trade-offs for Re-entry Design.

Ballistic Coefficient (BC)	Maximum Deceleration	Altitude of Maximum Deceleration	Maximum Heating Rate	Altitude of Maximum Heating Rate	Accuracy	Corridor Width
High (streamlined)	Same	Low	High	Low	High	Narrow
Low (blunt)	Same	High	Low	High	Low	Wider

Thermal-protection Systems

As you know by now, during re-entry, things get hot. How do we deal with this massive heat accumulation without literally burning up? We use specially formulated materials and design techniques called thermal-protection systems (TPS). We'll look at three approaches to TPS

- Heat sinks
- Ablation
- Radiative cooling

Heat Sinks

Engineers first dealt with the problem of massive re-entry heating for ICBMs, in the 1950s. Initially, they couldn't get rid of the heat, so they decided to spread it out and store it in the re-entry vehicle, instead. In other words, they created a *heat sink*—using extra material to absorb the heat, keeping the peak temperature lower.

To see how a heat sink works, let's consider what happens when we put a five-liter pan and a ten-liter pan of water over a fire. Which pan will boil first? The five-liter pan will because less water is storing the same amount of heat, so the water heats faster. Similarly, a vehicle with less material will heat faster during re-entry. Thus, whenever a vehicle faces a fixed amount of heat energy (such as for a given set of re-entry conditions), designers can lower the peak temperature by increasing the volume of its material to "soak up" more heat.

The heat sink, although heavy, was a simple, effective solution to re-entry heating of early ICBMs. These missions used high re-entry angles, giving better accuracy, because the vehicle traveled more quickly through the atmosphere. Thus, the heat sink had to absorb heat for a relatively short period. Unfortunately, for a given launch vehicle, as designers increased a heat sink's mass, they had to drastically constrain the available payload mass. Because payload is what they were trying to put on target, they had to consider alternatives to the simple, but heavy, heat sink.

Ablation

How do you keep your sodas cold on a hot day at the beach? You put them in a cooler full of ice. At the end of the day, the ice is gone, and only cold water remains. Why don't you just fill your cooler with cold water to start with? Because ice at 0° C (32° F) is "colder" than water at the same temperature! Huh? When ice goes from a solid at 0° C to a liquid at the same temperature, it absorbs a lot of energy. By definition, 1 kilocalorie of heat energy will raise the temperature of one liter of water by 1° C. (1 kilocalorie = 1 food calorie, those things we count every day as we eat candy bars.) But to melt 1 kg of ice at 0° C to produce one liter of water at the same temperature requires 79.4 kilocalories! This phenomenon, known as the *latent heat of fusion*, explains why your sodas stay colder on ice.

So what does keeping sodas cold have to do with a re-entry vehicle? Surely we're not going to wrap it in ice? Not exactly, but pretty close! A re-entry-vehicle designer can take advantage of this concept by coating the vehicle's surface with a material having a very high latent heat of fusion, such as carbon or ceramics. As this material melts or vaporizes, it soaks up large amounts of heat energy and protects the vehicle. This melting process is known as *ablation*.

Ablation has been used on the warheads of ICBMs and on all manned re-entry vehicles, such as the Apollo capsule shown in Figure 10-21, until the time of the Space Shuttle. Russia's manned vehicles still use this process to protect cosmonauts during re-entry. But ablation has one major drawback. By the time the vehicle lands, part of it has disappeared! This means we must either build a new vehicle for the next mission or completely refurbish it. To get around this problem, engineers, faced with designing the world's first reusable spaceship, devised a new idea—radiative cooling.

Figure 10-21.   Ablative Cooling. The bottom side of the Apollo re-entry capsule shown here was coated with a ceramic material that literally melted away during re-entry. As it melted, it took away the fierce heat and kept the astronauts safe and comfortable. *(Courtesy of NASA/Johnson Space Center)*

Radiative Cooling

Stick a piece of metal in a very hot fire and, before long, it will begin to glow red hot. Max Planck first explained this process. When you apply heat to an object, it will do three things—transmit the heat (like light through a pane of glass), reflect it (like light on a mirror), or absorb it (like a rock in the Sun). If an object absorbs enough heat, it warms up and, at the same time, radiates some of the heat through *emission*. This emission is what we see when a metal piece begins to glow. If heat energy continues to strike the object, it heats until the energy emitted balances the energy absorbed. At this point, it's in *thermal equilibrium*, where its temperature levels off and stays constant.

The amount of energy emitted per square meter, E, is a function of the object's temperature and a surface property called emissivity. *Emissivity, ε*, is a unitless quantity $(0 < ε < 1.0)$ that measures an object's relative ability to emit energy. A perfect black body would have an emissivity of 1.0. We determine the energy emitted using the Stefan-Boltzmann relationship

$$E = \sigma \varepsilon T^4 \qquad\qquad (10\text{-}18)$$

where

E = object's emitted energy (W/m^2)

σ = Stefan-Boltzmann constant = $5.67 \times 10^{-8}\ W/m^2\ K^4$

ε = object's emissivity $(0 < \varepsilon < 1.0)$ (unitless)

T = object's temperature (K)

If an object being heated has a high emissivity, it will emit almost as much energy as it absorbs. This means it reaches thermal equilibrium sooner, at a relatively low temperature. This process of reducing equilibrium temperatures by emitting most of the heat energy before a vehicle's structure can absorb it is known as *radiative cooling*. However, even for materials with extremely high emissivities, equilibrium temperatures during re-entry can still exceed the melting point of aluminum.

The high temperatures of re-entry pose two problems for us in finding materials for radiative cooling. First, we must select a surface-coating material that has a high emissivity and a high melting point, such as a ceramic. Second, if we place this surface coating directly against the vehicle's aluminum skin, the aluminum would quickly melt. Therefore, we must isolate the hot surface from the vehicle's skin with very efficient insulation having a high emissivity.

This artful combination of a surface coating on top of a revolutionary insulator describes the, now famous, Shuttle tiles. The insulation in these tiles is made of a highly refined silicate (sand). At the points on the Shuttle's surface where most of the heating takes place, a special coating gives the tiles an emissivity of about 0.8, as well as their characteristic black color, as shown in Figure 10-22.

Figure 10-22.   Shuttle Tiles. Space Shuttle tiles composite material has high emissivity and is an efficient high-temperature insulator. *(Courtesy of NASA/Johnson Space Center)*

Astro Fun Fact
Shuttle Tiles

(Courtesy of NASA/Johnson Space Center)

We know re-entry gets hot. For the Space Shuttle, temperatures can exceed 1247° C (2300° F). But the aluminum skin of the Shuttle doesn't reach its maximum temperature of 350° until almost 20 minutes <u>after</u> landing, thanks to perhaps the greatest technical advance of the Shuttle program—tiles. Designed to withstand the aerodynamic loads of ascent and re-entry, temperature extremes of over 1350° C (2400° F), and repeated use, they're one of the most unique materials ever invented. To cover the complex contours of the Shuttle surface, over 30,000 individually machined tiles fit together like a big jigsaw puzzle. Each tile has two pieces—a white silica fiber structure (basically highly refined sand) covered by the characteristic black coating made of reaction-cured glass (RCG). During re-entry, the RCG dissipates 90% of the heat energy in radiation back to the atmosphere while the white silica fiber structure insulates the inner aluminum skin and bears the brunt of the aerodynamic forces.

Refractory Composite Insulation, LI900, LI2200, FRCI, How It Works..., Lockheed Missiles & Space Company, Sunnyvale, CA.

▰ Section Review

Key Terms

ablation
computational fluid dynamics (CFD)
emission
emissivity, ε
heat sink
latent heat of fusion
Newtonian flow
radiative cooling
thermal equilibrium

Key Equations

$$E = \sigma \varepsilon T^4$$

Key Concepts

➤ We can meet mission requirements on the design front by changing
  - Vehicle size and shape, BC
  - Vehicle thermal-protection systems (TPS)

➤ Increasing the vehicle's ballistic coefficient, BC,
  - Doesn't change its maximum deceleration, a_{max}
  - Increases its maximum heating rate, $\dot{q}$

➤ There are three types of thermal-protection systems
  - Heat sinks—spread out and store the heat
  - Ablation—melts the vehicle's outer shell, taking heat away
  - Radiative cooling—radiates a large percentage of the heat away before the vehicle can absorb it

Example 10-1

Problem Statement

Long-range sensors determine a re-entry capsule is emitting 45,360 W/m^2 of energy during re-entry. If the emissivity of the capsule's surface is 0.8, what is its temperature?

Problem Summary

Given: $E = 45{,}360$ W/m^2

$\varepsilon = 0.8$

Find: T

Conceptual Solution

Solve Stefan-Boltzmann relationship for T

$$E = \sigma \varepsilon T^4$$

$$T = \sqrt[4]{\frac{E}{\sigma \varepsilon}}$$

Analytical Solution

1) Solve Stefan-Boltzmann equation for T

$$T = \sqrt[4]{\frac{E}{\sigma \varepsilon}}$$

$$T = \sqrt[4]{\frac{45{,}360\dfrac{W}{m^2}}{\left(5.67 \times 10^{-8}\dfrac{W}{m^2 K^4}\right)(0.8)}}$$

$$T = 1000 \text{ K}$$

Interpreting the Results

During re-entry, the capsule's surface reached 1000 K. With the surfaces' emissivity, this means 45,360 W/m^2 of energy is emitted. Imagine 450 100-watt light bulbs in a 1 m^2 area!

10.4 Lifting Re-entry

▉ In This Section You'll Learn to...

- ☛ Discuss the advantages offered by lifting re-entry
- ☛ Explain aerobraking and discuss how interplanetary missions can take advantage of it

Figure 10-23. An Astronaut's View of Landing. The Space Shuttle uses the lift from its wings to guide it to a pin-point landing on a tiny runway. This photograph shows the pilot's view of the landing strip at Edwards Air Force Base. *(Courtesy of NASA/Johnson Space Center)*

In Sections 10.1 through 10.3, we assumed the force of lift on our re-entering vehicle was zero, so we could use a straightforward equation of motion to investigate the trade-offs between re-entry characteristics. Adding lift to the problem takes it beyond the scope of our simple model but gives us more flexibility. For example, we can use the lifting force to "stretch" the size of the corridor and allow a greater margin of error in re-entry velocity or angle. Controlling lift also improves accuracy over a strictly ballistic re-entry. We can change the vehicle's *angle of attack* (angle between the vehicle's nose and its velocity vector) to improve lift, making the vehicle fly more like an airplane than a rock. This allows the pilot or onboard computer to guide the vehicle directly to the desired landing area, as shown in Figure 10-23.

The Space Shuttle is a great example of a lifting-re-entry vehicle. About one hour before landing, re-entry planners send the Shuttle crew the necessary information to do a deorbit burn. This burn changes the Shuttle's trajectory to re-enter the atmosphere by establishing a –1° to – 2° re-entry flight-path angle. After this maneuver, the Shuttle is on "final approach." Because it has no engines to provide thrust in the atmosphere, it gets only one chance to make a landing!

Preparing to hit the atmosphere (just like a skipping stone), the Shuttle rotates its nose to a 40° angle of attack. This high angle of attack exposes it's wide, flat bottom to the atmosphere. At an altitude of about 122,000 m (400,000 ft.), the re-entry interface takes place. Here the atmosphere begins to be dense enough for the re-entry phase to begin. From this point, more than 6400 km (4000 mi.) from the runway, the Shuttle will land in about 45 minutes! Figure 10-24 shows a graph of the Shuttle's re-entry profile.

Throughout re-entry, the Shuttle rolls to change lift direction in a prescribed way, keeping maximum deceleration well below 2 g's. These roll maneuvers allow the Shuttle to use its lift to steer toward the runway. In contrast, Apollo and Gemini capsules had minimal lifting ability, so they re-entered much more steeply and didn't roll much, so they endured up to 12 g's. Figure 10-25 compares these re-entry profiles.

Another exciting application of lifting re-entry is *aerobraking*, which uses aerodynamic forces (drag and lift) to change a vehicle's velocity and, therefore, its trajectory. In Chapter 7 we explored the problem of interplanetary transfer, and we saw that to get from Earth orbit to another planet required us to start the spacecraft's engines twice: ΔV_{boost} to get it on its way and ΔV_{retro} to capture it into orbit around the target planet.

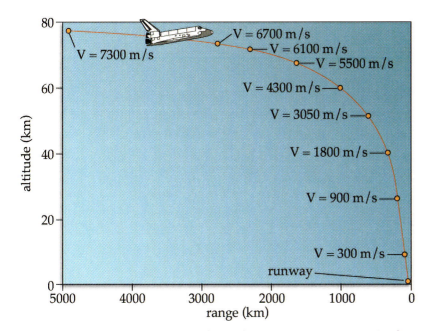

Figure 10-24.   Re-entry Profile for the Space Shuttle. This graph shows the Space Shuttle's altitude and velocity profile for a typical re-entry.

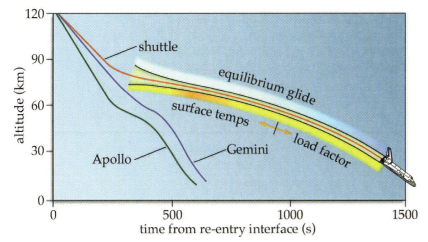

Figure 10-25.   Re-entry Profiles for the Shuttle Versus Gemini and Apollo. This graph shows the difference between re-entry profiles for Apollo, Gemini, and the Space Shuttle. Notice Gemini and Apollo re-entered much more steeply than the Space Shuttle. The Shuttle's re-entry profile must stay within a tight corridor between equilibrium glide, which insures it will slow enough to avoid skipping out and not over shoot the runway, and surface temperature/load factor requirements, which determine maximum heating and deceleration.

But if the target planet has an atmosphere, there's another option. Instead of using engines to slow the spacecraft enough to enter a parking orbit, we can plan the hyperbolic approach trajectory to take it right into the atmosphere and then use drag to do the equivalent of the ΔV_{retro} burn. We then use its lift to pull it back out of the atmosphere before it crashes

Figure 10-26. Aerobraking Concept. This artist's concept shows a heat shield that could be used for aerobraking at Mars or Earth. *(Courtesy of NASA/Goddard Space Flight Center)*

into the planet! By getting this "free" ΔV, we can save an enormous amount of fuel. Calculations show that using aerobraking, instead of conventional rocket engines, is almost ten times more efficient. This efficiency could mean a tremendous savings in the amount of material that must be put into Earth orbit to mount a mission to Mars. Figure 10-26 shows an artist's conception of an aerobraking vehicle. In his novel *2010: Odyssey Two*, Arthur C. Clarke uses aerobraking to capture a space ship into orbit around Jupiter. The movie made from this novel dramatically depicts the aerobraking maneuver.

Figure 10-27 shows an aerobraking scenario. On an interplanetary transfer, the spacecraft approaches the planet on a hyperbolic trajectory (positive specific mechanical energy with respect to the planet). During aerobraking, it enters the atmosphere at a shallow angle to keep maximum deceleration and heating rate within limits. Drag then reduces its speed enough to capture it into an orbit (now it has negative specific mechanical energy with respect to the planet). To "pull out" of the atmosphere, it changes its angle of attack, lift. Basically, the vehicle dives into the atmosphere, and then "bounces" out. In the process it loses so much energy that it is now captured into orbit. This atmospheric encounter now leaves the vehicle on an elliptical orbit around the planet. Because periapsis is now within the atmosphere, the vehicle would re-enter if it took no other actions. Finally, it completes a single burn, much smaller than ΔV$_{retro}$, is completed to put the vehicle into a circular parking orbit well above the atmosphere.

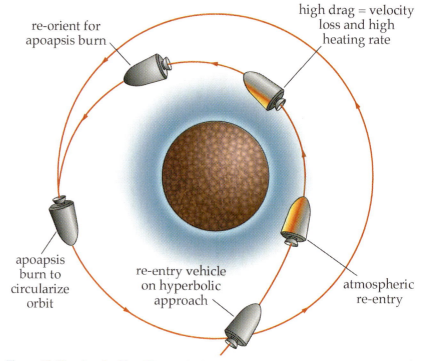

Figure 10-27.   Aerobraking. The aerobraking maneuver allows a vehicle to get "free" ΔV by diving into the atmosphere and using drag to slow down.

The Mars Global Surveyor spacecraft, shown in Figure 10-28, was the first interplanetary spacecraft designed to take advantage of aerobraking. it was initially captured into a relatively high orbit around Mars, and, over the course of several months, it used aerobraking to lower itself to the final mission orbit, saving many kilograms of precious propellant.

Figure 10-28.   Mars Global Surveyor. The Mars Global Surveyor spacecraft was the first interplanetary mission that was designed to use aerobraking to lower itself into its final mission orbit. *(Courtesy of NASA/Jet Propulsion Laboratory)*

▦ Section Review

Key Terms

aerobraking
angle of attack

Key Concepts

➤ Applying lift to the re-entry problem allows us to stretch the size of the re-entry corridor and improve accuracy by flying the vehicle to the landing site.

➤ The Space Shuttle is a good example of a lifting-re-entry vehicle. It uses its lift to keep re-entry deceleration low and fly to a pinpoint runway landing.

➤ Aerobraking can significantly decrease the amount of mass needed for interplanetary transfer. During an aerobraking maneuver, the vehicle dives into the target planet's atmosphere, using drag to slow enough to be captured into orbit.

References

Chapman, Dean. *An Analysis of the Corridor and Guidance Requirements for Supercircular Entry Into Planetary Atmospheres.* NASA TR R-55, 1960.

Concise Science Dictionary. Oxford: Oxford University Press, U.K. Market House Books, Ltd., 1984.

Eshbach, Suoder, (ed.). *Handbook of Engineering Fundamentals.* 3rd edition. New York, NY: John Wiley & Sons, Inc., 1975.

Entry Guidance Training Manual. ENT GUID 2102, NASA Mission Operations Directorate, Training Division, Flight Training Branch, NASA/Johnson Space Center, Houston, TX, December 1987.

Regan, Frank J. *Reentry Vehicle Dynamics.* AIAA Education Series, J.S. Przemieniecki series ed. in chief. New York, NY: American Institute of Aeronautics and Astronautics, Inc., 1984.

Tauber, Michael E. *A Review of High Speed Convective Heat Transfer Computation Methods.* NASA Technical Paper 2914, 1990.

Tauber, Michael E. *Atmospheric Trajectories.* Chapter for AA213 Atmospheric Entry. NASA/Ames Research Center, Stanford University, 1990.

Tauber, Michael E. *Hypervelocity Flow Fields and Aerodynamics.* Chapter for AA213 Atmospheric Entry. NASA/Ames Research Center, Stanford University, 1990.

Voas, Robert B. *John Glenn's Three Orbits in Friendship 7.* National Geographic, Vol. 121, No. 6. June 1962.

Mission Problems

10.1   Analyzing Re-entry Motion

1   What are the three competing re-entry requirements?

2   Where does all the heat generated during re-entry come from?

3   Why would increasing the ability of a re-entry vehicle to withstand higher g forces not necessarily increase the maximum deceleration requirement for the mission?

4   What is the re-entry corridor? Define its upper and lower boundaries.

5   Describe the re-entry coordinate system.

6   What are the potential forces on a re-entry vehicle? What is the dominant force during re-entry? Why?

7   Define ballistic coefficient and describe how a blunt versus a streamlined shape affects how a body will slow due to drag.

8   What two approaches can we use to balance competing re-entry requirements?

10.2   Options for Trajectory Design

9   Describe re-entry design.

10 To save fuel, Venture Star (X-33) engineers want to increase the velocity and the flight-path angle for re-entry. How will this affect the maximum deceleration and maximum heating rate? The altitudes for maximum deceleration and maximum heating rate?

11 Contact lenses being manufactured in space are returned in a re-entry capsule to Earth for distribution and sale. If the re-entry velocity is 7.4 km/s and the re-entry flight-path angle is $10°$, determine the maximum deceleration it will experience and at what altitude? The capsule's BC is 1000 kg/m^2.

12 For the same capsule in Problem 11, determine the altitude of maximum heating rate and the velocity at which this occurs.

13 What is a shock wave?

10.3   Options for Vehicle Design

14 In what two ways can we determine the hypersonic drag coefficient for a vehicle?

15 Mission planners for a manned Mars spacecraft face two different re-entry vehicles. Vehicle A has a high BC; vehicle B has a low BC. Assuming the re-entry velocity and flight-path angle are the same for both vehicles, explain any differences in deceleration profile.

16 Compare the advantages and disadvantages the three types of thermal-protection systems.

17 What is latent heat of fusion and how does it relate to ablation?

18 During re-entry, a meteor reaches a temperature of $1700°$ K. If its emissivity is 0.25, how much energy is emitted per square meter?

10.4 Lifting Re-entry

19 What are the advantages offered by a lifting-re-entry vehicle?

20 How does the Space Shuttle use lift to reach the runway?

21 A vehicle attempting to aerobrake into orbit around Mars needs to achieve an equivalent ΔV_{retro} of 2 km/s. If the entire aerobraking maneuver lasts for 10 minutes, what is the average drag force (in g's) we must attain?

Mission Profile—Space Shuttle

On April 12, 1981, the world's first reusable space ship rocketed into the Florida skies with astronauts John Young and Robert Crippen aboard. The successful flight of STS-1 (Space Transportation System mission 1) heralded a new era which promised to make access to space routine.

Mission Overview

The Space Shuttle, or Space Transportation System as it's sometimes called, is the most complex flying vehicle ever constructed. It has three main parts—the winged orbiter, the external tank (ET), and a pair of solid-rocket boosters (SRBs). The orbiter houses the crew compartment with avionics, payload bay, three Space Shuttle main engines (SSMEs), two orbital maneuvering system (OMS) engines, and 44 reaction control system (RCS) thrusters for attitude control. The ET is a big gas tank holding 790,000 kg (1.58×10^6 lb.) of liquid hydrogen and liquid oxygen fed to the three SSMEs on the orbiter through large interconnect valves. The SRBs provide the necessary thrust to get the entire system off the pad at lift-off. A typical Shuttle mission divides into three phases—ascent, on-orbit, and re-entry.

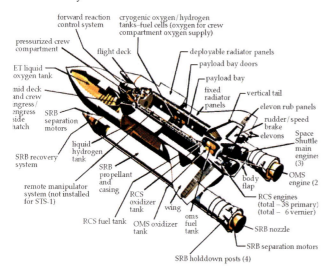

The Space Shuttle has three main parts—the orbiter, external tank, and solid rocket boosters—with numerous subsystems in each. *(Courtesy of NASA/Johnson Space Center)*

Mission Data

✓ Shuttle Ascent

- T –8 seconds: The three main engines ignite. As they throttle up to 104% capacity, generating five million newtons of thrust (1.125×10^6 lbf.), the entire vehicle pitches forward slightly. If the onboard computers detect any engine problems, they shut down all three and Mission Control scrubs the mission for the day.

- T –0 seconds: As the vehicle rocks back to upright at T – 0, the mighty SRBs ignite. Each applies a force of more than 11.8 million newtons (2.65×10^6 lbf.) to the vehicle, causing it to almost leap off the pad. When the SRBs ignite, there's no way to stop them. At lift-off, the Shuttle system has a gross mass of more than two million kg (4.4×10^6 lb.).

- T +60 seconds: The three main engines throttle down to 65% to minimize loads, as the vehicle flies through "Max-Q," the region of maximum dynamic pressure. After Max-Q, the engines throttle up to 104%. [*Note:* The reason the SSMEs can exceed 100% has to do with engine calibration data established early in the Shuttle's development. 100% is simply a benchmark value which it can safely exceed by 4%.]

- T +120 seconds: The two SRBs burn out and are jettisoned to parachute into the ocean, where barge crews recover them and return them to shore to be refurbished for future missions.

- T +480 seconds: SSMEs again throttle down to stay below three g's on the vehicle and crew.

- T +500 seconds: Main engine cut-off (MECO). The ET jettisons and falls into the atmosphere to burn up over the Indian Ocean.

- At MECO, the orbiter is not yet in orbit as perigee is well within the atmosphere. The OMS engines must fire at least once to establish a safe orbit.

✓ On-Orbit

- Once on-orbit, the Shuttle uses its OMS engines to change orbits and rendezvous with satellites or to achieve the correct parking orbit and deploy payloads. Typical Shuttle orbits are nearly circular at about 300 km (186 mi.) altitude with an inclination of 28.5° or 57°.

✓ Re-entry

- De-orbit burn: The re-entry phase starts with the de-orbit burn of 100 m/s (328 ft./s) over the Indian Ocean about one hour before landing. This burn lowers the Shuttle's orbit for a controlled re-entry into the atmosphere.

- De-orbit coast: The crew orients the vehicle to present the wide, flat bottom to the atmosphere at a 40° angle of attack.

- Re-entry interface (RI): RI takes place about 30 minutes before landing, at an altitude of 122,000 m (400,000 ft.) more than 8300 km (5158 mi.) up range from the landing site. At this altitude, the atmosphere becomes dense enough for aerodynamic forces to be significant. Throughout re-entry, the guidance, navigation, and control system must manage the Shuttle's energy to guide it through the narrow re-entry corridor, bleed off enough energy to land safely (but not too much, so it can overfly the landing site), and maintain acceptable heating levels.

- TACAN acquisition: At about 40,000 m (131,000 ft.), the Shuttle's navigation system begins processing data from ground-based Tactical Air Navigation (TACAN) stations—the same ones used by military and commercial aircraft—to update its position and velocity and ensure it's on course for the runway.

- HAC intercept: At about 20,000 m (65,600 ft.) the Shuttle intercepts the heading alignment cone (HAC) for a wide (up to 270°) turn to align with the runway centerline.

- On glide-slope: The Shuttle seems to dive at the runway, flying a 19° glide slope, much steeper than a convention airliner's 3°.

- Touchdown: The Shuttle touches down at 98 m/s (218 m.p.h.) compared to only 67 m/s (150 m.p.h.) for an airliner.

Mission Impact

As of 1999, the Shuttle has flown over 100 missions. From satellite deployment and retrieval to satellite repair and scientific experiments, the Shuttle has proven its flexibility. However, the emotional impact of the Challenger accident, caused by burn-through in a booster, showed how fragile the complex Shuttle system is and showed that access to space has a long way to go before we consider it routine.

On Space Shuttle mission STS-49, astronauts grabbed and boosted the Intelsat VI satellite. *(Courtesy of NASA/Johnson Space Center)*

For Discussion

- Has the Space Shuttle been a good investment for the U.S.?

- Given the inherent dangers of space flight, was the public's strong reaction to the Challenger accident justified?

- How would you design a next-generation Shuttle to replace the current one?

References

Space Shuttle System Summary, SSV80-1, Rockwell International Space Systems Group, May 1980.

Even the design of a small satellite constellation, such as the one shown in this artists' conception, represents a significant effort in space mission engineering. *(Courtesy of Maarten Meerman, Surrey Satellite Technology, Ltd., U.K.)*

Space Systems Engineering

11

In This Chapter You'll Learn to...

- Describe the systems engineering process and apply it to designing space missions
- Describe how payload requirements drive the rest of the spacecraft design
- Identify the major spacecraft subsystems and their associated performance budgets

You Should Already Know...

- ❏ Elements of a space mission architecture (Chapter 1)
- ❏ Basic requirements for maneuvering in orbit (Chapter 6)
- ❏ Concepts of vector mathematics (Appendix A.2)
- ❏ Calculus concepts (Appendix A.3)
- ❏ Kepler's Laws of Planetary Motion (Chapter 2)

Outline

11.1 Space Mission Design
The Systems Engineering
  Process
Designing Payloads and
  Subsystems
The Design Process

11.2 Remote-sensing Payloads
The Electromagnetic Spectrum
Seeing through the
  Atmosphere
What We See
Payload Sensors
Payload Design

Man must rise above the Earth—to the top of the atmosphere and beyond—for only then will he fully understand the world in which he lives

Socrates ca. 450 B.C.

Space Mission Architecture. This chapter deals with the Spacecraft segment of the Space Mission Architecture, introduced in Figure 1-20, and how it relates to the other mission elements.

We all use space systems. As you kick back in your easy chair, remote control in hand, the cable TV shows that you watch are beamed around the world using communication spacecraft. Weather forecasts on the evening news depend on up-to-the-minute images from weather spacecraft that patiently track the motion of clouds and storms around the globe. In the last several chapters, we've focused all of our attention on the trajectories spacecraft follow through space. You've seen how to select the proper orbit for their mission, launch them from a given launch site, and help them re-enter the atmosphere for a smooth landing on Earth.

But orbits are only part of the story. In this chapter, we'll step back and take a look at the "big picture" of space mission design. We'll discover how designers use the tried-and-true systems engineering process to translate user needs into fully-integrated space missions. We'll learn how these needs form the basis for mission requirements and constraints from which we decide how all of the elements of a mission will fit together. From these top-level requirements, we'll learn how to design the payload and individual subsystems that make up the spacecraft (Figure 11-1). As we'll discover, the payload is one of the most important drivers of any mission. With this in mind, we'll turn our attention to an important class of space payloads, remote sensing, to learn the basic principles governing their limitations and design.

This chapter lays the groundwork for future chapters that delve in greater detail into the design of spacecraft subsystems, launch vehicles and the other elements of a space mission architecture. While you may not be able to build your own spacecraft in your garage when you're done, you should develop a greater appreciation for the challenge of designing missions to work in the final frontier.

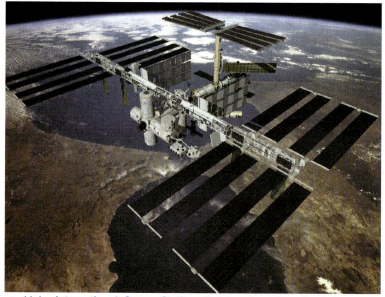

Figure 11-1. International Space Station. Building the International Space Station depends on a careful space systems engineering process. This starts with well-defined requirements needed to tie together the dozens of countries that participate in designing, building, and operating it. *(Courtesy of NASA/Johnson Space Center)*

11.1 Space Mission Design

In This Section You'll Learn to...

- ☛ Describe the systems engineering process
- ☛ Apply the systems engineering process to designing space missions
- ☛ Describe how the payload requirements drive the rest of the spacecraft design
- ☛ Identify the major spacecraft subsystems and their associated performance budgets

If you've ever built a tree house or even planned a big party, you've probably applied a systematic approach to translate your needs into a final product. In this section we'll look closely at this process to understand its component steps and see how we can apply it to design anything from a backyard BBQ platform to an on-orbit observation platform. We'll begin by looking at the basic steps in the process, then we'll see how we can apply it to the design of complex space missions.

The Systems Engineering Process

All design problems begin with a need. Ancient astronomers needed a means to track celestial events, so they designed Stonehenge. Henry Ford needed a car that appealed to the average driver, so he designed the Model-T. NASA needed a vehicle to launch astronauts to the Moon, so they designed the Saturn V, shown in Figure 11-2. Over the centuries, engineers have developed a well-tested process for translating simply-stated needs into complex systems. We call this process *systems engineering*.

To see how the systems engineering process works in action, let's look at a simple weekend project—building a backyard deck. Imagine you wake up one bright, sunny Saturday and take an appraising look at your backyard. The lawn looks good, the hedges are trimmed, its almost perfect. You only need one thing to finish it—a deck. A nice deck or patio would give you a great place to put the BBQ and a nice location for that new hot tub (which you'll decide you need next weekend).

The right approach to this project would be to resist the urge to drive to the hardware store, buy a load of lumber, grab your power saw, and start filling the yard with sawdust. The right approach would be to think before you act, to start by carefully defining what it is you really need. A general statement of your requirement may look something like this

Requirement: A flat, dry area in your backyard, roomy enough to house your BBQ and entertain friends.

Figure 11-2. Saturn V Launch Vehicle. NASA scientists and engineers designed and built the Saturn V launch vehicle in response to a need to take astronauts safely to the Moon. *(Courtesy of NASA/Johnson Space Center)*

361

Figure 11-3. Systems Engineering in Action. Any project, large or small, can benefit from the systems engineering process, even something as simple as a backyard deck. *(Courtesy of Decks U.S.A.)*

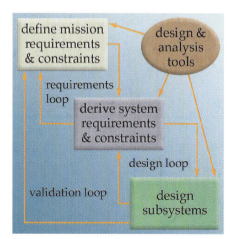

Figure 11-4. The Systems Engineering Process. The systems engineering process is the fundamental technique we can use to design anything from a backyard BBQ platform to space platforms. The process goes through several steps. Each step draws upon design and analysis tools. Note that the process is iterative. Between each step in the process, there are loops that take us back to review decisions in the previous step.

Along with this general requirement, you could probably define some constraints as well. For example, you could specify your budget for the project (say, less than $10,000), some basic ideas on when you'd like it to be done (in time for the summer) and its overall quality (you'd like it to last at least as long as you'll live in the house).

With this basic understanding of your needs on paper, you can then start to derive more specific requirements about what your deck will look like (Figure 11-3). You'll probably want to do a bit of research on pre-fabricated "deck kits" you can buy off-the-shelf or maybe some basic plans from a "How to Build a Deck in a Weekend" book. Once you shop around a bit, you may find that the deck you really want is way over your budget, or will take too long to build. Or, the deck that is in budget may not be sturdy enough to survive more than one season of BBQ parties. At that point, you may need to revisit your initial requirements to see if you should expand your budget or relax your other requirements.

Once you've made this trade-off between what you want and what you can really afford, you can start specifying your deck characteristics (size, shape, materials, etc.). With these decisions made, you can finally get down to the business of designing what it will look like and how you will build it. You'll need to make some detailed drawings, specify the amount and type of lumber to use, types of nails and bolts and all the other construction details. As you do this, you may find that some of the specifications aren't possible given the available materials or backyard conditions. This may mean a few trips "back to the drawing board" as you modify your original specifications.

Finally, you finish a design you are happy with. At this point, before you actually break ground and start getting your hands dirty, it's a good idea to review this massive project to make sure it's really what you need. With that decided and maybe a few minor design adjustments, the fun begins and your project gets underway (watch out for flying sawdust!).

Now that we've seen a little bit of the systems engineering process in action, let's take a closer look at it. Figure 11-4 shows the generic systems engineering process. In our deck example, we started by defining our basic "mission" requirements and constraints. Then we derived requirements for what the deck, the "system," would look like. Once we shopped for materials, we had to trade-off requirements versus what was realistically available, given our budget and time constraints. With the basic deck requirements decided, we then specified details about the design and construction of the foundation, sub-floor, railing, and other "subsystems." Along the way, we used a number of tools (spreadsheets, design books, rulers and other "Design and Analysis Tools") to help us make decisions.

The backyard deck example helps to introduce the systems engineering process. Now, we can apply it to space systems. In the following sections, we'll step through each phase of the process in more detail to see how the same process can help us to design complex spacecraft with down-to-Earth benefits.

Defining Mission Requirements and Constraints

Yogi Berra once said, "if you don't know where you're going, you'll probably end up someplace else!" We begin the systems engineering process by defining top-level mission requirements and constraints. Before we set out to solve any problem, we want to make sure we're solving the *right* problem. The best way to do this is to clearly state what we really want—our requirements. For this reason, the first and most important, step in the systems engineering process is to define the mission requirements. We need to ask ourselves

- "What end result do we want to achieve or accomplish?"
- "What is our ultimate objective?"
- In space terms, "What is the mission?"

The more clear and specific our requirements, the easier it is chart a course to achieve them. Requirements communicate what we need to others and to ourselves as well. Clearly-defined requirements also give us an opportunity to eliminate distracters, to separate what's important from what's not. Often we get sidetracked from our ultimate goal by competing objectives that, on closer analysis, really aren't that important anyway! Equally important, requirements give us a means of evaluating (and measuring) our performance, by comparing where we are with where we want to go. We have a compass heading we can continually refer to: "Are we headed in the right direction?"

There are two important qualities of requirements to remember: they should be *clear* and *simple*. Clearly stated requirements tell us exactly what we want. Ambiguity can lead to misunderstandings and disaster. Requirements should also be simple. As General George S. Patton, one of the greatest tank commanders of WW II, was fond of saying, "Never tell people how to do things. Tell them what to do and they'll surprise you with their ingenuity." By applying the "keep it simple" rule to requirements at the beginning of a program, we're far more likely to have a useful product at the end.

Applying this first step to space missions, we see that all space missions begin with a specific need, such as the need to economically transmit television shows or to provide world-wide navigation. Only with this need clearly in mind, can we begin to develop space missions to satisfy it.

To help us define space mission requirements, we begin by developing a mission statement. As we introduced in Chapter 1, the *mission statement* clearly and simply lays out

- The mission objective—why we do the mission
- The mission users—who will benefit from and use the information
- The operations concept—how all the mission elements fit together

The *mission objective* defines the purpose of the mission and what services or information it will deliver to users. *Users* are the customers who give us the reason (and usually money!) for the mission. Once we know why and who, the next question is how, which leads us to the operations concept. The *operations concept* describes how people, systems, and all the elements of the mission architecture will interact to satisfy the mission requirements.

In Chapter 1, we introduced the space mission architecture, shown in Figure 11-5. This architecture shows the basic elements of any space mission and how they relate to each other. In Chapters 4 through 10, we've focused on only one important element—trajectories and orbits. Now we turn our attention to defining the rest of the elements. During this first step of the space systems engineering process, we focus on the heart of the mission architecture by deciding what form our mission will take and how the various elements will interact. The architecture we define, and the various systems and subsystems that underlies it, must ultimately satisfy our mission need (Figure 11-6).

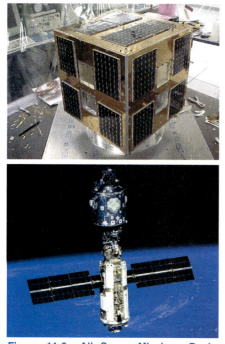

Figure 11-6.   All Space Missions Begin with a Need. Whether we're launching a tiny microsat (FalconSAT above) or a massive space station (ISS below), we must define a need to start the systems engineering process. *(Courtesy of USAF Academy and NASA/ Johnson Space Center)*

Figure 11-5.   Space Mission Architecture. The elements of a space mission perform the functions that satisfy the mission requirements.

Figure 11-7.   Detecting Forest Fires. The FireSat mission is motivated by the need to detect and contain the damage done by forest fires. *(Courtesy of the National Forest Service)*

To see the space systems engineering process in action, let's begin by describing a mission need that will serve as an example to guide our discussion in this and subsequent chapters. We'll pick a need with global impact, one that concerns all of us in some way: detecting forest fires. Every year forest fires devastate tens of thousands of hectares (acres) of valuable timber or wilderness resources and wildlife habitat, as shown in Figure 11-7. As Smokey the Bear reminds us, "Only *you* can prevent forest fires." So, while space assets can do little to prevent the human carelessness that often starts fires, space assets *can* help us track and put them out faster after they start.

For this example mission, we need to notify world-wide forest services in time for them to contain fires before they spread out of control. We'll call this mission *FireSat*. An example FireSat mission statement could be

- Mission objective—detect and locate forest fires worldwide and provide timely notification to users

- Users—U.S. Forest Service and other national and international agencies responsible for fighting forest fires

- Operations Concept—there are a number of possible operations concepts for this type of mission. For this example, let's pick a concept that relies on a number of spacecraft in low-Earth orbit to detect and locate the fires. The system will communicate this information to users through the internet. We'll control the entire mission via a single, dedicated ground station (Figure 11-8).

While our discussion so far has focused mainly on requirements (what we want), an equal, and sometimes more important, mission aspect to define is mission constraints—limits on what we can do. Even though the spacecraft will operate in a vacuum, the mission design cannot. We design all missions in the "real world" and they must conform to a variety of economic, technical and political constraints. We'll examine the impact of all of these in more detail in Chapter 16. Purely political constraints aside (such as constraints on which congressional district to build it in), systems engineering constraints typically fall into three general categories: cost, schedule, and performance.

The bottom-line program cost is an easily recognized constraint to anyone who operates on a budget. In the realities of the post-cold war aerospace communities, budget constraints have become one of the biggest drivers of modern space missions. More and more often, mission planners use a dedicated approach known as *design-to-cost* that is specifically oriented toward developing effective design solutions within a cost-constrained environment.

Along with cost, schedule can also be a significant constraint on any engineering solution. Often, missions must conform to a specific schedule to meet a particular launch window or simply to ensure the required spacecraft is on station in time to service paying customers. Naturally, cost and schedule go hand-in-hand. "Time is money" is a well-known economic principle. Typically, the longer a mission takes to get off the ground, the more expensive it will be.

Balanced against cost and schedule is performance. It's not difficult to imagine that the cheapest and fastest solution may not be the best technical approach. Systems engineers must design a spacecraft composed of various subsystems that work together reliably to accomplish the mission. They may use off-the-shelf parts or design new-technology components for each of the subsystems. The combined performance of these subsystems determine the success of the overall mission. On the other hand, the "best" technical solution offering the greatest performance may be expensive and time consuming to implement.

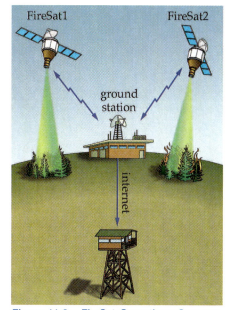

Figure 11-8.   FireSat Operations Concept. This operations concept uses two FireSat satellites in low-Earth orbit, one ground station, and an Internet link that passes forest fire information to users.

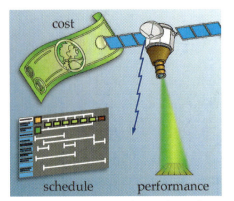

Figure 11-9. Mission Constraints: Cost, Schedule, and Performance. Cost, schedule, and performance comprise the 3-dimensional trade-space that all missions must operate within. Systems engineers must constantly trade these competing objectives to achieve a well-balanced solution.

Define Mission Requirements and Constraints

- Define the mission statement
  - State the mission objective
  - Identify users
  - Create the operations concept
- Identify the mission constraints
  - Cost
  - Schedule
  - Performance

Figure 11-10. The First Phase of the Systems Engineering Process. This phase focuses on defining mission requirements and constraints.

Figure 11-11. Real-world System Constraints. Small satellites can often take advantage of low-cost "piggyback" opportunities on large launch vehicles, such as, the Ariane. We show four of them here, attached to the Ariane Structure for Auxiliary Payloads. But we don't get something for nothing. By taking advantage of this cheap ride into space, the mission accepts a number of system constraints on launch time, orbit, etc. *(Courtesy of Surrey Satellite Technology, Ltd., U.K.)*

Cost, schedule, and performance represent the three-dimensional "trade-space" in which all space missions, and any other problems for that matter, are constrained as Figure 11-9 illustrates. Whether we're deciding what to eat for dinner (fast food, cheap food, or health food) or designing a global communications network, we must constantly trade these three constraints, one against the other. Later we'll see how this trade-off takes place between mission requirements and system requirements. Figure 11-10 summarizes the activities that take place during the first phase of the space systems engineering process.

With the mission requirements and constraints defined, the next phase in the systems engineering process translates these basic, top-level requirements into more specific system requirements. We'll see how system requirements develop from mission requirements next.

Deriving System Requirements and Constraints

While mission requirements focus on the big picture items (the reasons for and results of the project), systems requirements focus on the individual elements of the system architecture to describe in more detail what we expect of each for mission success. Naturally, the elements are closely interrelated. The launch vehicle depends on the size of the spacecraft and the mission orbit. The operations network depends on the number and types of spacecraft and the amount of data transmitted to users. The mission orbit depends on the altitude and inclination to satisfy the mission requirements.

In the "real world," few space missions begin with a totally blank sheet of paper. Typically, at least one of the mission elements is completely defined, or severely constrained, by economic, political or other factors at the outset. For example, we may design a mission to take advantage of a specific launch opportunity, in which case the launch vehicle, range of achievable orbits, and hard limits on spacecraft mass are defined before we begin system design, as shown in Figure 11-11.

Even if there are no specific limits on the launch vehicle, there are practical limits on it because there are only a relatively small number of options available "off the shelf," and few missions can afford to build their own launch vehicle from scratch. Typically, mission operations are also constrained at the beginning of mission design in some way. While very large programs, such as the Iridium global telephone network, can justify designing and building a dedicated operations network from scratch, most missions, such as NASA or U.S. Military missions, must use existing facilities.

Generally, the most unconstrained element of a space mission is the spacecraft. For this reason, we'll focus our discussion of the space systems engineering process on it. Recall from Chapter 1 that we conceptually divide the spacecraft into two functionally different parts—the payload and the spacecraft bus. The *payload* consists of the sensors or other instruments that perform the mission. The *bus* is a collection of subsystems designed to support the payload. Payload requirements are usually the biggest drivers of the spacecraft configuration. For this reason, the payload is the first system, for which we must derive requirements.

We start deriving payload requirements by returning to the mission statement defined earlier. Somewhere in the description of the objectives, users, and operations concept should be an indication of the mission subject. We define the *subject* of the mission to be a natural or manufactured object or phenomena that the payload will sense or interact with. We characterize the subject by such things as its color, size, shape, temperature, chemical composition, or frequency. Only after we know what type of subject we're dealing with, can we lay out clear and simple payload requirements.

Returning to our FireSat mission example, the obvious subject for this mission is forest fires. After all, they're what the mission objective states we should detect and locate. But what kind of forest fires? How big or how hot? What particular characteristics of forest fires should the payload detect? These questions may sound trivial, but to the payload designer, they can be very important. After all, we don't want to send out a forest-fire alert every time someone starts a campfire. On the other hand, we don't want to ignore a multi-acre blaze that may be out of control. We must design FireSat to respond to the *right kind* of fires. (Figure 11-12)

With further analysis, we could also specify other characteristics of such a fire, including the amount of heat and smoke it gives off. All of these will be important when it comes to designing the type of sensor or sensors the payload will employ to detect and locate fires. We'll learn more about how sensors do this in the Section 11.2.

Once we know these payload requirements, the rest of the mission elements fall into place. The type of payload greatly determines the mission orbit. Payload mass, volume, power and other requirements determine the basic size and mass of the spacecraft. If they are not already constrained, these system requirements determine, or at least limit the options, on the launch vehicle and operations systems. Figure 11-13 summarizes the steps during this second phase of the space systems engineering process. In the next section, we revisit the decisions made during the previous phase in light of these results to see if there is room for trade-offs.

Trading Requirements

While the shortest distance between two points on a map is a straight line, this is seldom the case for systems engineering. Few successful projects take a linear course to their final design. Instead, we write mission requirements and constraints knowing they may change. Often, the results from detailed analysis of the system requirements may make us rethink some of them. Quite often, program sponsors unknowingly tie mission designers' hands with an ambiguous or overcomplicated requirement. So, we use the "requirements loop," a necessary and continuous "process within a process," to re-evaluate requirements, based on new information.

For example, the FireSat mission objectives require us to "detect and locate forest fires worldwide and provide timely notification to users." How fast is "timely?" A preliminary analysis of the number of spacecraft needed to provide global coverage may show that to provide truly instant

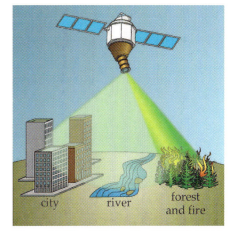

Figure 11-12.   Defining the Right Subject. Looking down from space, a spacecraft's payload must detect or interact with the right subject for the mission.

Derive the System Requirement

- Review the constraints on mission architecture (launch vehicle, orbit, operations, etc.)
- Identify and characterize the mission subject
- Derive payload requirements
- Derive orbital requirements
- Determine basic spacecraft size and mass
- Identify the potential launch vehicle(s)
- Derive operations systems requirements

Figure 11-13.   Derived System Requirements. In the second phase of the systems engineering process, we must derive system-level requirements.

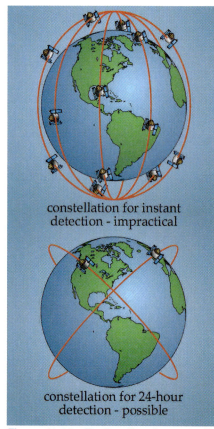

constixtuation for instant
detection - impractical

constellation for 24-hour
detection - possible

Figure 11-14. Trading Requirements. By trading mission requirements versus system requirements, a mission that is impractical or too expensive may become doable and affordable.

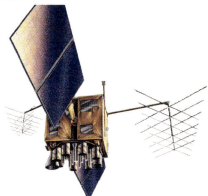

Figure 11-15. Defining Payload Requirements. In considering the payload requirements for the GPS satellite, engineers had to define support requirements for power, temperature, data handling, and communication that affected the overall system design. *(Courtesy of the U.S. Air Force)*

notification the manufacturing, launch, and operations costs are prohibitive. However, a single spacecraft may be able to detect and notify within three days—too much time to be useful. This analysis would mean mission designers and sponsors would need to work together to more clearly define the requirement for "timely" coverage. By refining this requirement somewhat, say by redefining "timely" to be "within 24 hours," a mission that is too expensive or impossible to accomplish may become affordable and doable, as illustrated in Figure 11-14.

Designing Payloads and Subsystems

After we fully derive the system requirements, and, where necessary, redefine the mission requirements that drove them, we can finally move to the last phase of the space systems engineering process—payload and subsystem design. This phase is the real nuts and bolts of the mission. Here we roll up our sleeves and design detailed specifications for the payload and its supporting subsystems, including drawings, mechanical and electrical interfaces, and even nuts and bolts. Let's begin by looking briefly at payload requirements, then consider the subsystems that make up the spacecraft bus.

The Payload

In the previous phase of the space systems engineering process, we derived the overall payload requirements in terms of the subject with which it must interact. During this final phase of the process, we take a more detailed look at the payload to design the components that make this interaction possible. The GPS mission, shown in Figure 11-15, is just one example of how the payload drives the other subsystem requirements.

Continuing with our FireSat mission scenario, we now know that the subject is a forest fire. But how do we detect such a fire? We know fires generate heat, light, and smoke. Fortunately, we can build electronic devices, or sensors, to detect each of these attributes of our subject. Imagine you're sitting around a campfire on a clear, cool night. You can feel the heat from the fire. You can see its light, with sparks and ashes dancing in the air, and smell the smoke. If we put these kinds of sensors on a spacecraft, they become the spacecraft's payload.

The payload could consist of a single, simple camera to detect light from the fire, or include a collection of several sensors, each tuned to detect a particular characteristic such as its light, heat, or smoke. We'll explore how sensors remotely detect these phenomena in greater detail in Section 11.2. For now, we simply need to understand that the number and type of sensors chosen, and how they work together to form the spacecraft's payload, determine how we design the rest of the spacecraft to support it. As we design the payload, we generate a number of "spin-off" requirements for the spacecraft bus that dictate

- Where and how precisely the spacecraft must point

- The amount of data the bus must process and transmit

- How much electrical power it needs

- The acceptable range of operating temperatures

- The payload's volume and mass

The subsystems that make up the spacecraft bus must satisfy all of these payload requirements. Typically, mission designers define these requirements in terms of performance budgets. Just as your household budget determines the amount of money you have to spend on a given activity (such as going to the movies), subsystem performance budgets specify the amount of velocity change, electrical power or other limited resource that it must "spend" to accomplish some activity (such as getting to the correct orbit or turning on the payload). In the remainder of this section, we'll look at spacecraft subsystems and their performance budgets in more detail.

The Spacecraft Bus

The spacecraft bus exists solely to support the payload, with all the necessary housekeeping to keep it healthy and safe. Perhaps the best way to visualize the relationship between the payload and bus is to picture a common, everyday, school bus, such as the one in Figure 11-16.

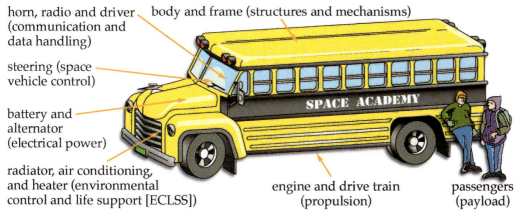

horn, radio and driver (communication and data handling)

body and frame (structures and mechanisms)

steering (space vehicle control)

battery and alternator (electrical power)

radiator, air conditioning, and heater (environmental control and life support [ECLSS])

engine and drive train (propulsion)

passengers (payload)

Figure 11-16.   A Spacecraft "Bus." The major functions performed on a spacecraft are also performed on a school bus.

A school bus is designed to take its payload—students—to and from school, and all of its subsystems support this goal. To design a school bus, we must understand the specific needs of the payload. For this example, we have to know things such as

- How far and how fast the students need to go, so we have a big enough engine and plenty of gas

- How many students there are, so we know how big to make the bus

- How warm to keep the bus, so the students don't freeze or overheat

In addition to knowing these things to directly support the payload, the school bus designer must keep in mind other requirements to support the overall mission. This leads to the design of other important subsystems such as the steering system and the horn and the radio.

Now that we've shown how a school bus works to support its payload, let's apply the analogy to understand the specific subsystems that comprise a spacecraft bus. Figure 11-17 shows an exploded view of the Magellan spacecraft with its payload and many of its subsystems. In this final phase of the space systems engineering process, we want to understand the critical requirements that drive the subsystem design.

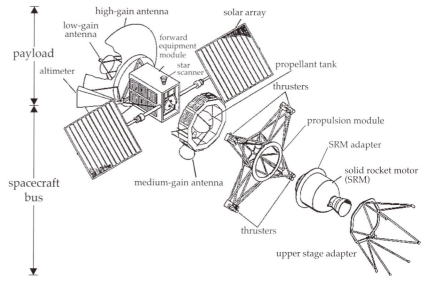

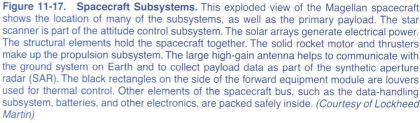

Figure 11-17.   Spacecraft Subsystems. This exploded view of the Magellan spacecraft shows the location of many of the subsystems, as well as the primary payload. The star scanner is part of the attitude control subsystem. The solar arrays generate electrical power. The structural elements hold the spacecraft together. The solid rocket motor and thrusters make up the propulsion subsystem. The large high-gain antenna helps to communicate with the ground system on Earth and to collect payload data as part of the synthetic aperture radar (SAR). The black rectangles on the side of the forward equipment module are louvers used for thermal control. Other elements of the spacecraft bus, such as the data-handling subsystem, batteries, and other electronics, are packed safely inside. *(Courtesy of Lockheed Martin)*

Figure 11-18.   Attitude and Orbit Control. To accurately point at ground stations, the Tracking and Data Relay Satellite relies on its attitude and orbit control subsystem (AOCS). The required mission pointing accuracy translates into attitude and orbit control budgets that drive the AOCS design. *(Courtesy of Hughes Space and Communications Company)*

Steering—Spacecraft Control. School bus drivers must know what route to take, where they are, and how to steer the bus to get where they need to go. This is all part of controlling their vehicle. On a spacecraft, there is a similar subsystem that "steers" the vehicle to control its attitude and orbit. A spacecraft has to have the right attitude, or orientation, to point cameras and antennas at targets on Earth. (Experience has shown there's nothing worse than a spacecraft with a bad attitude!) For attitude control, rockets or other devices rotate it around its center of mass. Also, a spacecraft must be able to reach and maintain its operational orbit. To do orbital control it uses rockets that perform the maneuvers described in Chapter 6. (Figure 11-18)

The subsystem that controls the spacecraft's attitude and orbit we naturally call the *attitude and orbit control subsystem (AOCS)*. We derive the critical performance budgets for the AOCS from the payload and other system requirements. We can split these into an attitude-control budget and an orbital-control budget. The *attitude-control budget* is the total angular momentum that the spacecraft may use during the mission lifetime. The *orbital-control budget* is the total ΔV needed to reach and maintain the operational orbit.

In addition to the raw numbers in the performance budgets, AOCS designers need to know how the spacecraft will determine its attitude and orbit and what mechanisms it will use to control it. In Chapter 12 we'll describe in greater detail how control systems use sensors, rockets and other tools to keep spacecraft pointed in the right direction and settled into the right orbit.

The Horn, Radio, and Driver—Communication and Data Handling.

On school buses, drivers have a demanding (and often thankless job). They have to

- Monitor the activities of dozens of overactive little "payloads" in the back of the bus (using their eyes), while

- Keeping the passengers comfortable (using the heater/air conditioner), while

- Keeping track of the bus's location (using their eyes and a map), while

- Steering the bus on the road and dodging road hazards (using the steering wheel), while

- Communicating location with the central office (using the radio), while

- Announcing their frustrations to other busses and cars (using the horn), while

- Getting their bus where it needs to go (using the engine, transmission, and wheels)

We can easily identify the driver on the school bus, in the front, behind the steering wheel. On a spacecraft, the driver is less easy to pick out, but the role is just as important. If we look at the list of tasks the driver needs to perform, we can see they interact with the payload and all other subsystems in the spacecraft, as well as users and operators on the ground. This interaction is in the form of communicating and handling data from other parts of the bus and the outside world. For this reason, the "driver" of the spacecraft bus is called the *communication and data-handling subsystem (CDHS)*.

The school bus drivers take in information about the route and the bus's performance, and processes this information through their brain to decide how to use it. On a spacecraft, the CDHS consists of computers to collect, process, and store data and radios to communicate to the outside world. (Figure 11-19)

Figure 11-19. Communication and Data-handling Subsystem (CDHS). The CDHS collects and processes data from onboard and from the ground, decides on courses of action, and executes commands. Here a NASA astronaut performs these functions as he talks to mission control while analyzing mission data. *(Courtesy of NASA/Johnson Space Center)*

A spacecraft may have several computers that work in the same way as a typical home computer. The size and complexity of onboard systems for data handling depend on the volume of commands and data received, stored, processed, and transmitted, as well as the degree of autonomy built into the vehicle. In design terms, we define these *data budgets* similar to specifications for our personal computer: speed (instructions per second e.g. "333 MIPS") and storage (e.g. "Megabytes" or "Gigabytes").

Radios on a spacecraft are not that different from the AM/FM radios we listen to every day. In fact, a spacecraft may have several radios to

- Allow controllers to keep track of where the spacecraft is, check how it's doing, and tell it what to do
- Send mission data back to users
- Relay data sent from ground stations

Analysis of the communication requirements produces a *link budget* that specifies parameters for the type of radio and the amount of data. Engineers use the data budgets and link budgets, along with other mission and subsystem requirements to design the CDHS. We'll learn more about how they do this when we explore the fundamentals of communication and data-handling subsystems in more detail in Chapter 13.

Battery and Alternator—Electrical Power. To start the school bus in the morning, the driver uses electrical power stored in the battery. Once the engine is running, the alternator keeps the battery charged and provides electrical power to run the lights, the radio, and other electrical components. Just like the school bus, spacecraft depend on electrical power to keep components running.

The electrical power on a spacecraft is no different from the electrical power used to run your television. Unfortunately, in space, there's no outlet to plug into, and an extension cord to Earth would have to be very long! Therefore, a spacecraft must produce its own electrical power from some energy source, just as the school bus uses chemical energy released from burning gasoline to run the engine that turns the alternator. In Chapter 13, we'll see how the electrical power subsystem (EPS) converts some energy, such as solar energy, into usable electrical power and stores it to run the entire spacecraft. We add the electrical power requirements for each of the other bus systems to determine the total electrical power budget. This is the biggest driver of the EPS design.

Radiator, Air Conditioner, and Heater—Environmental Control and Life Support. A school bus engine would overheat without a radiator to keep it cool. The radiator circulates a mixture of water and antifreeze through the engine block, where it heats up and then transfers the heat to the cooler air that rushes through the radiator. If you've ever ridden on a school bus on a cold winter morning or a hot summer day, you also know the bus's heating and air conditioning systems can be life savers! All these subsystems are designed to keep the passengers comfortable (Figure 11-20).

Figure 11-20. Environmental Control and Life-support Subsystem (ECLSS). To keep the payload healthy and happy, the ECLSS must maintain the temperature within an acceptable operating range and provide air, water, food, and other amenities, if necessary. Here, astronaut Cady Coleman tends to a mouse-ear plant on Columbia's flight deck *(Courtesy of NASA/Johnson Space Center)*

Similarly, a spacecraft must regulate the temperature of its components to keep them from getting too cold or too hot. In addition, on the Shuttle and the International Space Station, astronauts must be protected from the harsh space environment. They need a breathable atmosphere at a comfortable temperature, humidity, and pressure, along with water and food to sustain life. Unmanned spacecraft don't have these tight requirements on passenger comfort, however, careful temperature control is still critical to keep payloads functioning normally.

The job of the spacecraft *environmental control and life-support subsystem (ECLSS)* is to provide the required temperature, atmosphere and other conditions necessary to keep the payload (including astronauts!) healthy and happy. Rather than specifying budgets, we typically state ECLSS design requirements in terms of *acceptable operating ranges*. For example, a payload sensor may have a temperature range of 0° to 50° C, or an astronaut may need an atmospheric pressure of 1 ±0.1 bar (14.5 ±1.5 p.s.i.). In Chapter 13 we'll learn how a spacecraft moderates temperatures and other parameters to keep vital components and passengers working smoothly.

Figure 11-21. Spacecraft Structure. Here we see a technician working on the NICMOS remote-sensing instrument. It's structure holds everything together and supports it during launch. *(Courtesy of Ball Aerospace & Technologies Corporation)*

The Body and Frame—Structures and Mechanisms. Holding the school bus together are all sorts of beams, struts, and fasteners. These components comprise the basic structure of the bus and support the loads it feels as it bounces along the road. The school bus structure literally holds it together. When the driver steps on the gas pedal, the bus accelerates, forcing passengers back into their seats. When a spacecraft blasts into orbit, it experiences a similar force as the launch vehicle accelerates it into space. The spacecraft structure must be sturdy enough to handle all these high loads and hold all the other subsystems in place.

In addition to these static structures, spacecraft also have many mechanisms that crank, extend, bend, and turn to deploy solar panels, extend antennas, or perform other functions. In Chapter 13, we'll see the kinds of structures needed to hold spacecraft together, learn some of the basic principles of structural analysis, and discover the types of mechanisms needed to complete the mission, as we look at the structures and mechanisms subsystem. (Figure 11-21)

The Engine and Drive Train—Propulsion Subsystem. Finally, the bus has an engine and drive train. These supply torque to the wheels, moving the bus where the driver wants it to go. This torque, combined with the friction of the wheels on the road, causes the bus to move forward just as Newton's Third Law of Motion (*for every action there is an equal and opposite reaction*) describes. In space we take advantage of Newton's Third Law by using rockets to expel mass in one direction causing the spacecraft to move in the other. Large rockets on launch vehicles produce the necessary thrust to get the spacecraft into orbit. (Figure 11-22) Once there, smaller rockets in its propulsion subsystem produce thrust to maneuver it between orbits and control its attitude.

The design of the propulsion subsystem depends to a large extent on the attitude and orbital-control budgets determined by the AOCS. But we must consider other factors as well, such as the number and size of

Figure 11-22. Propulsion Subsystem. The propulsion subsystem consists of the rockets that take a spacecraft where it needs to go and maintain its orbit. Here, an engineer inspects the Shuttle main engines prior to STS-51. *(Courtesy of NASA/Johnson Space Center)*

maneuvers and how quickly they must be performed. From these requirements, we can derive an equally important design parameter for the propulsion system, the *propellant budget*. No one would start a long bus journey without a full tank of gas. Similarly, designers must carefully determine how much "gas" or propellant the spacecraft must have to satisfy mission requirements. In Chapter 14 we'll explore rocket science in greater detail to learn how rockets take spacecraft where we want them to go in space and how much propellant they need to get there.

The Design Process

Now that we understand the scope of the subsystem design task, we can step back to look at the process needed to complete the spacecraft design. The challenge is to painstakingly design every mechanical and electrical detail of the payload, along with every bus subsystem, and carefully define the interfaces between them. Once again, requirements drive the process. For example, the mission concept and payload determine the attitude and orbital-control budgets. These budgets affect design decisions about how the control system measures and changes the spacecraft's orientation and position in space. These results determine the mission's velocity change, ΔV, and propellant budgets that drive the propulsion system design. The communication and data-handling subsystem must receive and process commands to move the spacecraft and reorient it. While the propulsion subsystem provides the ΔV to change the spacecraft's orbit, the structure must bear the loads this maneuver generates. The environmental control and life-support subsystem must dissipate the additional heat that sunlight on the spacecraft creates. Finally, all subsystems need electrical power to meet these demands. During the "design loop," of the space systems engineering process, these detailed subsystem design issues must be constantly traded with overall system requirements.

Figure 11-23 shows the interdependence involved with the design of all the subsystems. Notice all subsystems link together in a continuous chain. Spacecraft designers must be very conscious of this interrelationship between the payload and the bus subsystems. During the design process, seemingly innocent changes to the design of the electrical power subsystem, for example, may have profound effects on the performance of other subsystems. Thus, the design process is, by its very nature, an iterative one. That is, changes to one subsystem lead to changes in another which, in turn, lead to changes in another and. . . well, you get the idea.

We can never achieve a perfect design. One of the most important, practical rules of engineering design is *"Better Is the Enemy of Good."* We can design a perfectly good system to death in an attempt to constantly reach for a "better" design. Time and money (two sides of the same coin) are not infinite. At some point we have to quit designing and go with the design we have. The best we can hope for is to iterate the process enough times to reach a design that satisfies the mission performance requirements on time and within budget.

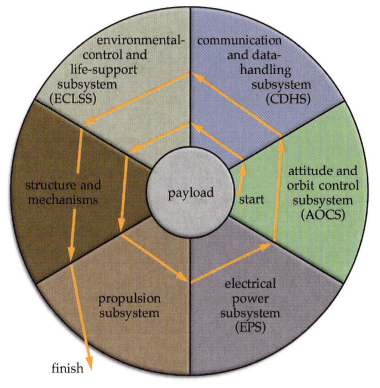

Figure 11-23.   The Spacecraft Design Process. Here we show how the interdependence of all the spacecraft subsystems. When we adjust the design of one subsystem, we are likely to have to adjust some, or all, of the other subsystems.

It is also important to keep in mind that the subsystem design is only one part of the problem. After we have a design, we have to build the spacecraft. All too often, initial designs satisfy mission requirements, but are difficult or impossible to build! Throughout the design process, there needs to be good communication between the engineer that designs the part and the technician who must make it. *Design-for-manufacturing* principles force us to focus on issues such as types of materials, number of parts, commonality of components and other details that can make the crucial difference between a useful subsystem that can be economically produced and one that simply can't be built.

Design and Analysis Tools

With perhaps billions of dollars or even human lives at stake, programmatic and system design decisions made throughout the space systems engineering process can't rely on gut feelings alone. We must support all decisions, big or small, with hard data. How do we characterize the mission subject? What payload will detect it? What is the spacecraft's mass? Will the launch vehicle deliver it to the right orbit? To help make these decisions, mission planners and systems engineers have a wide variety of design and analysis tools in their toolkit. These range from

Figure 11-24.   Mission Analysis. Computer-based mission analysis tools allow engineers to rapidly model mission scenarios and trade-off system requirements. *(Courtesy of Analytical Graphics, Inc.)*

simple "back of the envelope" calculations using a pencil, paper, and calculator (or spreadsheet) to complex computer simulations requiring hours of run time.

Computer-based tools have increasingly become indispensable to modern space systems engineering. From initial mission definition through requirements trade-off and subsystem design iterations, these sophisticated tools allow designers to quickly assess the far ranging impact of a single decision on the entire mission architecture. Off-the-shelf packages that support routine analysis and design tasks are now widely available to rapidly perform complex orbital analysis (Figure 11-24) or complete detailed engineering drawings.

Ironically, as these design and analysis tools become more sophisticated and widely used, it becomes even more important that mission designers understand the fundamental principles on which they're based. Because the ultimate responsibility for the mission success (or failure) rests with the mission planners, they must carefully act as a "reality check" on all analysis results. The "garbage in, garbage out" rule always applies. The designers must understand the assumptions behind the analysis in order to assess the reasonableness of the results. If the spacecraft breaks, we can't blame it all on the computer simulation!

Validating the Design

One of the biggest challenges during the design phase is keeping the entire mission in perspective. Too often, people responsible for specific subsystems get so involved in designing their own small piece of the mission that they lose sight of how their decisions affect other subsystems and the overall mission performance. Figure 11-25 offers a humorous look at how different subsystem designers often view the spacecraft.

Of course, we don't want subsystem managers to look at the spacecraft this way. To design a spacecraft to meet the objectives on time and within budget, everyone involved with a project must keep one eye clearly focused on the overall mission—the big picture. Only then can they make effective trade-offs between competing requirements to produce the best result. For this reason, we build the "Validation Loop" into the systems engineering process. This loop forces us to constantly compare subsystem design decisions to the original mission requirements. The most cleverly designed subsystem is useless if it doesn't support the payload. The most high-tech spacecraft is useless if it doesn't satisfy the user's needs.

Figure 11-26 summarizes the entire space systems engineering process we've explored in this chapter. As we learned, one of the biggest drivers of the design process is the payload. Because payloads are so important to mission planning and to the ultimate design of the entire spacecraft, let's continue by looking at the physical limitations on their design in the next section.

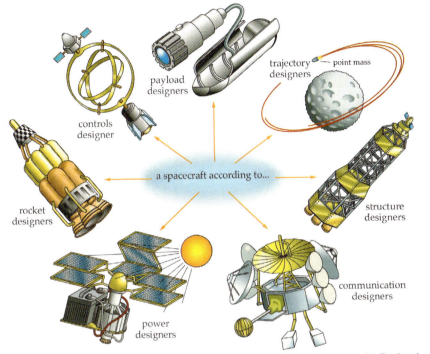

Figure 11-25.   The Spacecraft According to the Communication, Controls, Payload, Power, Rocket, Structure, and Trajectory Designers. Sometimes individual subsystem designers get so focused on their subsystem they lose sight of the overall mission.

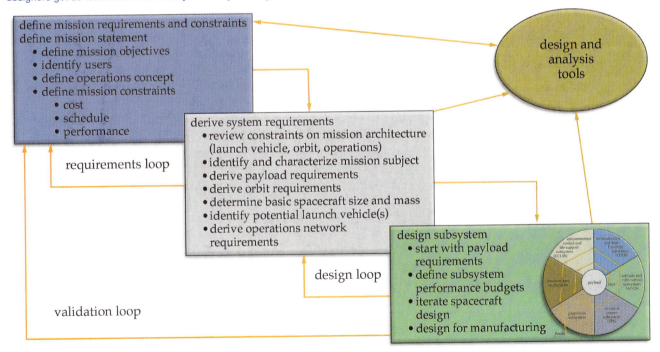

Figure 11-26.   Space Systems Engineering Process. By following this process, systems engineers design spacecraft that meet mission requirements while staying within budget and schedule constraints.

377

▬ Section Review

Key Terms

acceptable operating ranges
attitude and orbit control
    subsystem (AOCS)
attitude-control budget
bus
communication and data-handling
    subsystem (CDHS)
data budgets
design-for-manufacturing
design-to-cost
environmental control and life-
    support subsystem (ECLSS)
link budget
mission objective
mission statement
operations concept
orbital-control budget
payload
propellant budget
subject
systems engineering
users

Key Concepts

➤ We can apply the systems engineering process, illustrated in Figure 11-4, to any project from a backyard BBQ platform to a space platform. It consists of

- Three major steps
    - Define mission requirements and constraints
    - Derive system requirements and constraints
    - Design subsystems

- Three "loops" that form processes within the process
    - Requirements loop—verify derived requirements match overall mission requirements and constraints
    - Design loop—verify subsystem designs meet system requirements and constraints
    - Validation loop—verify the overall system design meets mission requirements and constraints

- Design and analysis tools include computer-based and other techniques to calculate subsystem specifications and simulate trade-offs

➤ During the first step of the space systems engineering process we define mission requirements and constraints. This involves

- Define mission statement
    - State the mission objective—why we do the mission
    - Identify mission users—who will benefit from or use the information produced by the mission
    - Create the operations concept—how will all the mission elements fit together
    - Identify mission constraints (cost, schedule, and performance)

➤ During the second step of the space systems engineering process, we derive the system requirements

- Review the constraints on mission architecture (launch vehicle, orbit, operations, etc.)
- Identify and characterize the mission subject—the subject of a mission is "what" the spacecraft payload will sense or interact with
- Derive payload requirements
- Derive orbital requirements
- Determine basic spacecraft size and mass
- Identify potential launch vehicle(s)
- Derive the operations-network requirements

Continued on next page

Section Review (Continued)

Key Concepts (Continued)

➤ The spacecraft bus provides all the housekeeping functions needed to run the payload and get data to users

➤ The attitude and orbit control subsystem (AOCS) controls the spacecraft's attitude, so it can point in the right direction, and controls the spacecraft's position and velocity, so it can get where it needs to go

  • The attitude budget is the total angular momentum the spacecraft may use during the mission lifetime

  • The orbital-control budget is the total ΔV needed to reach and maintain the mission's operational orbit

➤ The communication and data-handling subsystem (CDHS) consists of computers and radios needed to process mission data, as well as send and receive information from operators on the ground

  • The data budgets refer to the total amount of data the CDHS must process and store

  • The link budget refers to the amount of data that must be communicated to/from the ground station

➤ The electrical power subsystem (EPS) converts energy from some source into usable electrical power to run the other subsystems and the payload. The spacecraft power budget is the sum total of all mission electrical power requirements.

➤ The environmental control and life-support subsystem (ECLSS) controls

  • Temperature for all onboard hardware

  • The environment, including air and water, for fragile human payloads

  • Requirements are specified in terms of operating ranges, e.g., temperature between $0°$ and $50°$ C

➤ The structural subsystem holds together all the other subsystems and payload and withstands launch and mission loads. A variety of mechanisms may also deploy and retract throughout the mission.

➤ The propulsion subsystem provides the torque and ΔV needed by the AOCS. The propellant budget is the sum total of propellant needed throughout the mission.

➤ The spacecraft design process (as illustrated in Figure 11-23) is inherently iterative as changes made in one subsystem ultimately effect the design of all other subsystems

➤ Figure 11-26 illustrates the space systems engineering process

Example 11-1

Problem Statement

Environmental damage from forest fires can destroy valuable resources and wildlife habitats. To contain fires before they rage out of control, forest services worldwide need timely warning of new fires. You have been given the job of program manager for the FireSat mission to detect and locate fires that exceed 4 hectares (40,000 m^2 or ~10 acres) and notify U.S. and other national forest services worldwide within 24 hours. This is a low-cost demonstration mission to prove the usefulness of a space-based forest fire system. This will be a "bare bones" mission with a total budget of $10M for all spacecraft and 5 years of operation. Operations will be conducted from an existing ground station in Colorado Springs, Colorado, U.S.A., with mission data distributed to users through the Internet. Launches will be donated, but must be "piggy back" opportunities on the new Falcon launch vehicle. These donated opportunities allow spacecraft designers to use spare launch capacity to place small spacecraft into 500-km altitude circular, polar orbits (near the sunsynchronous orbit of the primary payload). Launches are limited to "nanosatellites" up to 15 kg in mass with dimensions of 0.30 × 0.30 × 0.30 m. Each launch can accommodate 3 nanosatellites and there will be 2 launch opportunities available beginning in 2 years (a total of up to 6 spacecraft). To satisfy the mission sponsors (the people with the money), the mission must be operational within 3 years. Complete, on a conceptual basis only, the first two phases of the space systems engineering process for this mission.

Problem Summary

Given: Need to detect and locate forest fires (>4 hectares) worldwide and provide information to forest services within 24 hours. Launch opportunities for nanosatellites (up to 15 kg with dimensions of 0.30 × 0.30 × 0.30 m) into 500-km altitude, circular, polar orbits. Total budget of $10M. The mission must be operational within 3 years.

Find: 1) Define the mission requirements and constraints

2) Derive the system requirements and constraints

Conceptual Solution

There are practically an infinite number of possible mission designs to satisfy this need. We'll go through one possible solution that will serve as the basis for subsequent example problems.

1) Define mission requirements and constraints

Stepping through the process outlined in Figure 11-26

Define the mission statement

- Define the mission objective—detect and locate forest fires (>4 hectares) worldwide and inform the users within 24 hours

- Identify the users—national forest services in the U.S. and worldwide

- Define the operations concept—Assume the mission will need a six-satellite constellation to provide 24-hour notification to users. All operations will take place from an existing ground station in Colorado Springs, Colorado. Spacecraft will collect and store mission data onboard and relay it to the ground station when they pass overhead. Notify the users through the Internet.

Define the mission constraints

- Cost—<$10M

- Schedule—first 3 spacecraft ready for launch in 2 years

- Performance—minimum to detect >4 hectares from 500-km circular, polar orbit and relay data to the ground station

Example 11-1 (Continued)

) Derive the system requirements and constraints

We can now look at each of the major systems that comprise the mission architecture to see what requirements and constraints are already known

- Launch vehicle—constrained by mission management to be the Falcon launch vehicle. No further requirements can be defined at this point. Specific requirements of the launch vehicle will come from the launch vehicle provider.

- Orbit—constrained by the launch vehicle to 500-km, circular, polar. No further requirements can be defined.

- Operations network—constrained by mission management to be the existing facility in Colorado Springs, Colorado. No further requirements can be defined at this point. Specific requirements will come from the ground station operators.

- Spacecraft—constrained by the launch vehicle to be <15 kg with dimensions of 0.30 × 0.30 × 0.30 m. 5 year onboard lifetime. Derived requirements include

  – Payload capable of detecting forest fire >4 hectares from the mission orbit

  – Spacecraft bus must support the payload

  – Spacecraft bus must communicate mission data to existing operations network

Interpreting the Results

The space systems engineering process is a powerful tool that helps us translate general mission requirements and constraints into more specific system and subsystem requirements. From these, we can start to design the hardware to do the mission. In this scenario, much of the mission architecture was "pre-defined" by the mission constraints. This is not unusual for typical missions, especially low-cost ones. For our FireSat mission, we have described a system of 6 satellites in 500-km circular, polar orbits. Each spacecraft must be less than or equal to 15 kg and not exceed the volume of a cube 0.30 m on a side. The spacecraft must have a payload capable of detecting forest fires >4 hectares ($40,000 \text{ m}^2$) and relaying this information each time it passes over the single ground station in Colorado Springs, Colorado. Users will receive this information from the ground station through the Internet. The entire mission must be operational in 3 years at a total cost (excluding launch) of $10M. For other example problems, in this and subsequent chapters, we'll continue this scenario to see how the systems engineering process is used as we define more details of the mission.

11.2 Remote-sensing Payloads

▰ In This Section You'll Learn to...

- ☞ Identify the elements of a remote-sensing system
- ☞ Describe and compute important parameters of electromagnetic radiation
- ☞ Use Wien's Law and the Stefan-Boltzmann equation to analyze an object's temperature versus the wavelength of its emitted radiation
- ☞ Identify the two types of remote-sensing payloads and describe their basic functions

As you read this book and listen to the sounds around you, your eyes and ears act as sensors to help you perceive your surroundings. Your eyes see all the colors of the rainbow—light. Your ears detect tiny disturbances in the pressure of the air around you—sound. Payloads are the "eyes" and "ears" of a spacecraft. Payloads detect objects on Earth and in space. They also "talk" to other spacecraft and Earth-based ground stations. Looking and listening from space involves sensing portions of the *electromagnetic (EM) spectrum*. All space missions—communications, navigation, weather, and remote-sensing—rely on the EM spectrum to collect data and interact with other elements of the mission.

To further illustrate the basic process of payload design that we began in the last section, we'll concentrate on *remote sensing*—a broad category of missions designed to detect and monitor subjects on Earth, on other planets, or far out in space. Remote sensing missions monitor the global environment, observe the weather, spy on enemies, search for natural resources, assess agricultural yields, observe the universe, and even detect forest fires, like the one shown in Figure 11-27, and communicate this information to users.

Electromagnetic energy provides the information necessary for remote sensing. This energy is either emitted by a subject, e.g., the Sun, or reflects off the subject, e.g., ships at sea. For example, if a spacecraft needs to detect an airplane in flight, the EM radiation originates at the Sun. This radiation first travels through space and through the atmosphere. It then reflects off the airplane, carrying information back through the atmosphere for sensors on the spacecraft to detect. Computers on the spacecraft process the detected radiation, forming an image of the airplane, which it relays to users on the ground. Figure 11-28 shows how these components of a remote-sensing system all work together. To learn more about these systems, let's start by looking closer at the EM spectrum.

Figure 11-27. Forest Fires from Space. NASA astronauts captured this photograph of a forest fire. The FireSat mission would need other types of remote-sensing payloads to automatically detect them. *(Courtesy of NASA/ Johnson Space Center)*

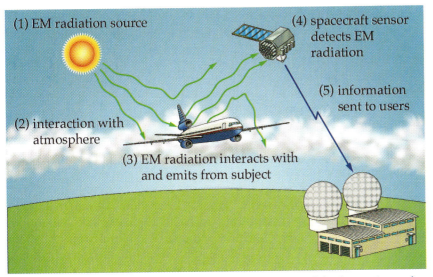

Figure 11-28.   Components of a Remote-sensing Mission. A basic remote-sensing system includes an EM radiation source, atmospheric interaction, a subject, a sensor, and an information link to users.

The Electromagnetic Spectrum

The colors of the rainbow (red, orange, yellow, green, blue, indigo, and violet or "Roy G. Biv" if you need an easy way to remember them) represent the range or spectrum of light that is visible to the human eye. However, this spectrum is only a tiny fraction of all possible colors. Colors of light beyond the visible range we refer to as *bands* of electromagnetic radiation. The light we see with our eyes is just one small part of the entire EM spectrum. Figure 11-29 shows the entire EM spectrum with key parts identified. We use the visible portion of the spectrum to take conventional pictures using reflected light. We use the infrared part of the spectrum to sense heat emitted by a subject. Finally, we use radio wavelengths to transmit voice, television, and radio signals. Most space missions use several different parts of the EM spectrum.

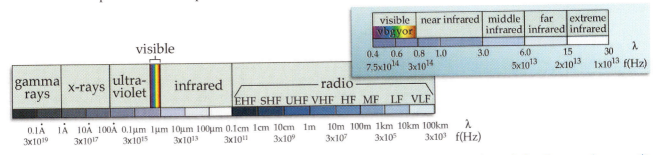

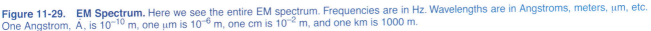

Figure 11-29.   EM Spectrum. Here we see the entire EM spectrum. Frequencies are in Hz. Wavelengths are in Angstroms, meters, μm, etc. One Angstrom, Å, is 10^{-10} m, one μm is 10^{-6} m, one cm is 10^{-2} m, and one km is 1000 m.

Electromagnetic radiation is a fickle phenomenon. We can describe it as either particle or wave motion. Sometimes it's useful to think about EM radiation as waves, such as the ripples on a pond after we've dropped in a rock. EM waves spread out from the source, traveling at the speed of light: 300,000 km/s (186,000 mi./s). These waves can be detected by sensors (like our eyes) and interpreted by computers (like our brains). Other times it's more useful to think of radiation as *photons*—tiny, massless bundles of energy emitted from the source at the speed of light. This dual nature of radiation (sometimes called the "wave-icle" theory) can, of course, lead to some confusion. For the most part, except where noted, we'll think of radiation as waves.

Because payloads can use different parts of the EM spectrum, we need to understand how we describe these different types of radiation. Let's look at the EM wave shown in Figure 11-30. The distance from crest to crest is called the *wavelength, λ*. The number of waves that go by in one second is the *frequency, f*. So, if 10 waves or cycles go by per second, the frequency is 10 cycles/s or 10 Hertz (1 cycle/s = 1 Hz). The time it takes for just one wave to go by is 1/f. That is, if the frequency is 10 Hz, one cycle takes 1/10 second. Now, because we know that distance is just speed multiplied by time, we can relate wavelength, speed, and frequency as

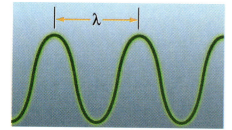

Figure 11-30.   Wavelength. We describe EM radiation in terms of its wavelength (λ)—the distance between crests of the waves.

$$\lambda = \frac{c}{f} \tag{11-1}$$

where

λ = wavelength (m)

c = speed of light in a vacuum = 3×10^8 m/s

f = frequency (Hz)

Energy in EM radiation relates to the number of waves that hit an object during a given time. Thus, ten waves hitting in one second (10 Hz) delivers twice the energy of five waves (5 Hz). This energy relationship can be expressed as

$$Q = hf \tag{11-2}$$

where

Q = energy (joules, J)

f = frequency (Hz)

h = Planck's constant = 6.626×10^{-34} J · s

Equation (11-2) implies more energy is available at higher frequencies, so higher-frequency waves have more energy than lower-frequency waves. For example, radio frequencies have much less energy than X-rays. That's why we can walk through radio waves all day long without worry, but must take special precautions around medical X-ray equipment. As we discussed in Chapter 3, this high energy radiation can have harmful effects on the human body.

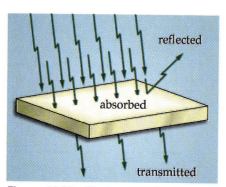

Figure 11-31.   Electromagnetic Radiation Interacting with an Object. When EM radiation hits an object, the energy is either transmitted, reflected, or absorbed.

Whenever EM radiation hits an object, one of three things can happen, as Figure 11-31 illustrates. The object transmits, reflects, or absorbs the energy. *Transmitted energy* passes directly through an object, (like light through a pane of glass). *Reflected energy* bounces off the object (like light off a mirror). *Absorbed energy* adds to the internal energy of the object (usually in the form of heat, like a hot pavement on a sunny day). From the principle of conservation of energy discussed in Chapter 4, we can say that the sum of the transmitted, reflected, and absorbed energy must equal the total energy that initially hits the object. The amount of energy transmitted, reflected, or absorbed depends on the wavelength (or frequency) of the incident energy and the properties of the material. In the following sections, we'll explore the importance of transmission, reflection, and absorption to remote sensing missions.

Seeing through the Atmosphere

On a crystal clear night, we can see the light from stars thousands of light years away. On nights like these, the sky appears completely transparent. In other words, it transmits all the radiation that hits it. But we know that on a cloudy day, we can't see the Sun, so not all the light is being transmitted. However, even on a cloudy day it can be warm and we can still get sunburned. How can this be? It turns out that the atmosphere is selectively transparent to different wavelengths of radiation. On a cloudless day, we see light from the Sun, feel its heat, and get sunburned from its ultraviolet (UV) radiation. On a cloudy day, some of the light and heat are blocked, but some UV radiation still gets through, so we can still get sunburned. Clouds consist of water droplets that block certain wavelengths of radiation. Even on a clear day, different molecules in Earth's atmosphere, such as ozone, block various wavelengths, such as, certain harmful frequencies of UV. This protects living things on Earth from the harmful effects of radiation, as discussed in Chapter 3.

For a remote-sensing system, we want to collect the EM radiation reflected by or emitted from objects on Earth's surface, while our spacecraft is above the atmosphere in space. We also need to use radio waves to communicate to and from the surface, again through the atmosphere. Thus, we must know which wavelengths the atmosphere readily blocks and those that easily pass through it. Figure 11-32 shows the percentage of each wavelength of EM radiation that passes through the atmosphere.

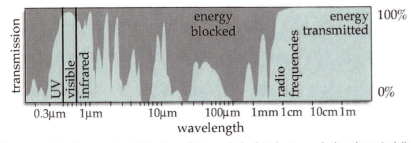

Figure 11-32.   Atmospheric Windows. This graph depicts the transmission characteristics of Earth's atmosphere at various wavelengths. Notice that certain parts of the EM spectrum—visible light, heat, and radio waves—get through the atmosphere while other wavelengths are blocked. [1979, Lillesand]

Notice in Figure 11-32 that some wavelengths (such as visible light) are completely transmitted while others are almost completely blocked. We see that our spacecraft instruments have access to Earth from space through various windows of transmission. *Atmospheric windows* are the wavelengths that are 80%–100% transmitted through the atmosphere. The most notable atmospheric windows are the visible, infrared, and radio wavelengths. Using the visible and infrared windows, spacecraft instruments can peer through the atmosphere to sense properties of objects on Earth from space. We use the radio-frequency window to pass television and radio signals from studios on the ground through satellites to our living rooms. Now that we've shown how transmission of EM radiation allows us to see through the atmosphere, let's take a closer look at what we're seeing.

What We See

After EM radiation makes it through the atmosphere, it hits a subject (a field of grass, a patch of dirt, or other object), which either reflects or absorbs it. Reflected light is what we see. If we look out a window on a sunny day, the grass is green because it absorbs all light frequencies except green, which it reflects. The dirt is brown because it absorbs all frequencies except those that mix to make the color we call brown. In conversation, we say "the grass is green" and the "the dirt is brown," but in reality the grass is every color *except* green and dirt is every color *except* brown! Because objects reflect different wavelengths of EM radiation, measuring the amount and type of radiation reflected tells us many things about it. Just by the color, we can conclude that green grass is alive and brown grass is dead. More detailed analysis of the reflected energy can tell us about soil properties, moisture content, types of grass, and many other important details.

As noted earlier, absorbed energy causes an object to heat up. However, some objects are already hot. A piece of metal heated in a fire starts to glow red-hot, as shown in Figure 11-33. This red color is not due to reflected energy, but due to emitted energy. Everything above the temperature of absolute 0 K (–273° C or –460° F) emits some EM radiation. It's easy to imagine that the Sun, at 6000 K, emits energy we call sunshine. However, Earth, at a mere 300 K, emits radiation in the form of "Earth shine," as well. You too, at a robust 310 K (98.6° F), emit EM radiation (and you thought it was just your glowing personality). Because everything has some temperature, everything emits some EM radiation.

Objects emit EM radiation at different wavelengths depending on their material properties and temperature. The classic explanation for this phenomenon is that thermal radiation begins with accelerated, charged particles near the surface of an object. These charges then emit radiation like tiny little antennas. The thermally excited charges can have different accelerations, which explains why an object emits energy at many different wavelengths.

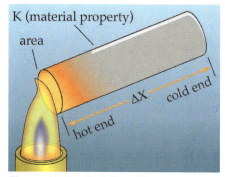

Figure 11-33. Red Hot. When a piece of metal reaches a certain temperature, it begins to glow red hot. This represents the emittance of EM radiation due to temperature.

Max Planck (1858–1947) refined this explanation and helped to usher in the field of quantum physics. He postulated that objects emit energy in tiny bundles or "quanta" called photons. Planck used a black-body as his perfect case. A *black body* is an object that absorbs and re-emits all of the radiation that strikes it. He developed a model that related the amount of power given off at specific wavelengths as a function of an object's temperature. Remember the glowing red-hot piece of metal? Using Planck's relationship, we can relate the distribution of wavelengths of this glow directly to its temperature. The curves in Figure 11-34 show the wavelengths of black-body radiation for the Sun and Earth.

Notice in Figure 11-34 that the peak output for the Sun is in the visible region. Thus, our eyes take advantage of the most abundant type of radiation from the Sun. How convenient! The relationship between temperature and wavelength of peak emission had been noted earlier by Wilhelm Wien (1864–1928). Wien's Displacement Law relates the wavelength (and hence the frequency) of maximum output for a given black-body radiator to its temperature.

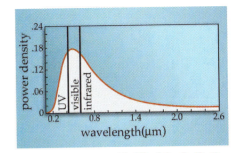

Figure 11-34. Planck's Black-body Radiation Curve for the Sun. The hotter the object, such as the Sun, the more EM radiation it emits at shorter wavelengths (higher frequencies and hence higher energy). As this graph illustrates, the Sun at a robust 6000 K emits energy over a range of wavelengths that peak in the visible bands. UV refers to ultraviolet [1979, Lillesand].

$$\lambda_m = \frac{2898}{T}$$

(11-3)

where

λ_m = wavelength of the maximum output (μm)
T   = object's temperature (K)

For the Sun at 6000 K, λ_m is 0.483 μm, which is in the middle of the range of visible light (0.39–0.74 μm). Just what we'd expect!

Astro Fun Fact
The Greenhouse Effect

The property of the atmosphere that makes it transparent to some wavelengths but opaque to others, causes the greenhouse effect. The greenhouse effect is a result of near-infrared radiation at short wavelengths being allowed into the atmosphere. When this radiation hits the ground and ocean, it's reflected or re-radiated as longer-wave radiation, such as far infrared, which can't penetrate the atmosphere. Thus, energy is "trapped" in the atmosphere as heat. Without the greenhouse effect, the effective temperature of Earth would be about 0° F (–18° C). Thus, the greenhouse effect is a very useful, natural phenomenon which keeps Earth warm enough to support life. The controversy over the greenhouse effect has to do with man-made things that block Earth's window that allows thermal radiation (heat) to escape. Certain man-made and naturally occurring compounds, such as chlorofluorocarbons (CFCs) and methane, block this window. As a result, Earth may heat up even more than we want it to. NASA's Earth Science Enterprise spacecraft will observe the greenhouse effect to determine whether human actions, such as pollution and deforestation, or natural phenomena, such as volcanos, may lead to more global warming.

Masters, Prof. Gil. Environmental Science and Technology. Autumn 1989. Course Reader, CE170. Stanford University.

Using Wien's law, we can determine the best frequency to use to see a particular subject. For example, if we need to detect the hot plume from a rocket's exhaust, we'd need to know its temperature and then design a sensor tuned to the frequency of maximum output for that temperature.

We can also determine the total power output of black-body radiation. To do this, we use the Stefan-Boltzmann equation developed by Joseph Stefan (1853–1893) and theoretically derived by Ludwig Boltzmann (1844–1906).

$$E = \varepsilon \sigma T^4 \qquad (11\text{-}4)$$

where

E = object's energy per square meter (W/m^2)

ε = object's emissivity ($0 < \varepsilon < 1$)

σ = Stefan-Boltzmann constant = $5.67 \times 10^{-8}\ W/m^2\ K^4$

T = object's temperature (K)

This equation estimates the total amount of energy available over all wavelengths for a specified temperature. Note that the energy output of a black body goes up as the fourth power of temperature. So, if we double an object's temperature, its energy output increases 16 times! Recall we used the Stefan-Boltzmann equation in Chapter 10 to estimate the amount of energy emitted by a vehicle re-entering the atmosphere. Later, in Chapter 13, we'll use this same relationship to analyze thermal control for spacecraft.

Payload Sensors

So far our discussion has focused mainly on the physics involved with EM radiation used for remote sensing. Now let's look at the technology involved with making workable sensor systems. To observe a subject on the ground, a spacecraft sensor must

1) *Look at it*—move the sensor to point at the subject

2) *See it*—collect EM radiation from it

3) *Convert it*—transform EM radiation into usable data

4) *Process it*—turn data into usable information

Let's start with step one. If someone asks you what time it is, you need to move your head or your eyes to focus on your watch or the clock on the wall. Before you can see the clock, you must first *look* at it. Spacecraft sensors have a similar limitation. Before they can see the subject, they must first point at it. Sensor scanning mechanisms can vary widely in complexity. Some sensors simply stare at the ground and use the spacecraft's own motion over Earth to scan the area beneath it. In this case, information collection is limited by the field-of-view of the sensor, as shown in Figure 11-35. *Field-of-view (FOV)* is the angular width that a sensor can see. Our eyes, for example, have an angular field-of-view of about 130°. This means we can detect objects out to 65° on either side of

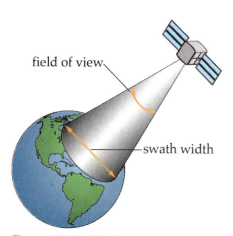

Figure 11-35. Field-of-View (FOV). The FOV of a satellite determines how much of Earth it can scan at any one time.

where we look. The edge of this range is called our *peripheral vision*. There, we can sense an object only if it's moving. To really see something, we use the muscles around our eyes to move them in the direction of the object of interest. Similarly, many spacecraft use mirrors to move the image in a sweeping pattern to look at objects of interest. The speed at which this motion takes place is called the *scan rate*.

The second important function of a sensor system is to collect incident radiation and focus it on a detector. This is exactly what our eyes and cameras do. In our eyes, the lens collects light and focuses it on our retina. In an ordinary camera, a lens gathers light from the subject and focuses it onto photographic film, as shown in Figure 11-36. We use two important parameters to describe how our eyes, a telescope, or any sensor works: focal length and aperture. *Focal length, fl*, is the distance from the lens to the detector. The *aperture, D*, is the diameter of the lens or antenna used. The aperture gathers light either by *refraction*, using an ordinary lens, or by *reflection*, using a mirror.

Cameras, small telescopes, and our eyes rely on refraction. Large telescopes, such as the Hubble and a radio frequency antenna use reflection. For simplicity, we'll consider only the refraction case. A diagram of a simple space-based telescope is shown in Figure 11-37. The radius of the detector is r_d and the distance from the lens to the subject is the height, h. For telescopes, we're most often interested in the magnification factor it gives us—how close are things made to look that are far away? We define *magnification* as the ratio of focal length to height or detector radius to the ground image radius, R_g (the distance across the ground that the instrument can "see.")

$$\frac{fl}{h} = \frac{r_d}{R_g} = \text{magnification} \qquad (11\text{-}5)$$

Realize that twice the ground image radius, R_g, gives us the sensor *swath width* (width of the field of view across the ground).

$$\text{Swath width} = 2\,R_g$$

Knowing the focal length and the detector radius, we can also determine the sensor field of view (FOV).

$$\text{FOV} = 2\tan^{-1}\left(\frac{r_d}{fl}\right) \qquad (11\text{-}6)$$

The size of the aperture, and the wavelength of radiation we're interested in, determines the smallest object we can see—the *resolution*. If we have perfect eyesight, our vision is "20/20." This measurement means that at a distance of about 6.1 m (20 ft.) we can read all the letters on a specific line of an eye chart. Opticians have determined the resolution limit of the human eye and they compare our vision to this standard. For optical systems, like our eyes, cameras, and telescopes, we express the *angular resolution, θ*, as an angle that is a function of wavelength, λ, and aperture diameter, D

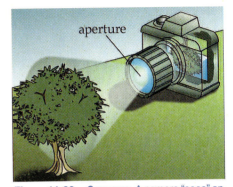

Figure 11-36.   Cameras. A camera "sees" an image by collecting light reflected from the subject through the lens and focusing it on the detector—photographic film.

Figure 11-37.   Simple Space-based Telescope. For an Earth-observing telescope, the detector radius and focal length determine the field-of-view, the ground image radius and detector radius set the magnification, and the aperture diameter and wavelength determine the resolution.

$$\theta = \frac{1.22\lambda}{D} \tag{11-7}$$

where

θ = instrument's angular resolution (rads)

λ = wavelength of radiation sensed (m)

D = instrument's aperture diameter (m)

This equation says the smallest angle we can detect between points is directly proportional to the wavelength we select and indirectly proportional to the size of the lens aperture we use. From geometry, shown in Figure 11-38, we determine the smallest linear dimension our instrument can distinguish at a given distance by multiplying the angular resolution by two times the distance from the sensor to the object. Note that the resolution tells us the smallest detectable object, so the smaller the resolution, the better

$$Res = 2\theta h = \frac{2.44\lambda h}{D} \tag{11-8}$$

where

Res = instrument's resolution (m)

θ = angular resolution or beamwidth (rads)

h = distance between the lens and the viewed object (m)

λ = wavelength of radiation sensed (m)

D = instrument's aperture diameter (m)

The human eye, for example, has an aperture diameter of about 1 cm. Taking a wavelength in the middle of the visible range of 0.5 μm, the resolution limit is about 0.2 mm at 3.2 m (20 ft.).

These relationships tell us the limits on resolution due to optics. However, a lens only collects energy and focuses it on a detector. The detector still must convert this energy into a useful form. This is the third function of a sensor system. The complexity of the detector depends on the nature of the radiation being sensed. Most sensors operate in the visible or infrared (IR) parts of the spectrum. The simplest type of detector is conventional photographic film. The film is coated with chemicals that react to the incident radiation, forming an image when developed. The first spy satellites took images using conventional film and then dropped the film canisters back to Earth, where the U.S. Air Force C-119 and C-130 airplanes caught them in mid-air! (Figure 11-39)

Since then, visual and IR detectors have been produced using semiconductor materials similar to solar cells. In an array of these solid-state detectors each one generates an electrical current proportional to the incident radiation. A computer samples the entire array and processes the voltage from each cell, producing the final image. The most commonly used solid-state detector is a *charge-coupled device (CCD)*. These are now used in virtually all video cameras. The big advantages of CCDs are they

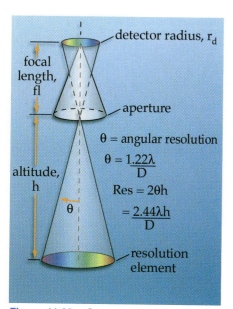

Figure 11-38.   Ground Resolution Versus Angular Resolution. Ground resolution (size of resolution element) is two times the angular resolution times the altitude, h.

Figure 11-39.   C-119 Box Car Catches a Discoverer Canister. Through careful planning and a bit of luck, aircrews could catch a film canister that was jettisoned by a remote-sensing satellite, after its parachute deployed. *(Courtesy of the U.S. Air Force)*

don't wear out and they produce a digital, electronic image that computers can easily store, process, or enhance.

Infrared detectors are similar to visible-wavelength detectors, but they sense a different part of the spectrum—heat. If you close your eyes and hold your hand close to a fire, you can feel the heat and use it to move your hand directly to the fire without peeking. The reason you can detect the fire is that your hand is much cooler than the heat from the fire. For spacecraft infrared sensors to be sensitive enough to detect rather small amounts of infrared energy from space, they must be very cold—about 77 K to 120 K (–383° F to –340° F). To keep them this cold, spacecraft rely on liquid nitrogen (80 K) coolers or active refrigeration techniques. The Landsat satellites use sensitive IR sensors, cooled by cryogenic coolers, to image vegetation. Figure 11-40 shows a Landsat infrared image.

Most people are familiar with optical sensors, such as cameras; however, many astronomers use sensors in another part of the spectrum— radio frequencies. Scientists are interested in the radio waves emitted by distant stars, as well as unique interstellar radio sources such as pulsars. Radio telescopes are simply radio receivers designed to "tune in" to these stellar radio programs. Other telescopes detect the high-energy gamma and X-rays emitted in violent supernovas or mysterious black holes. Regardless of the frequencies used, all types of astronomy benefit greatly by having instruments in space, where faint signals are not blocked or attenuated by the atmosphere space-based sensors, such as the Chandra X-ray Telescope shown in Figure 11-41 and with greater resolution than their Earth-based counterparts.

So far, we've only looked at sensors that detect either emitted radiation or reflected radiation that originated at some other source, e.g., sunlight. Sensors of this type are called *passive sensors* since they do not directly interact with the subject. Passive sensors (Figure 11-42) work well if the subject of interest emits enough energy to detect or we only want to take pictures in daylight. However, when the spacecraft moves into Earth's shadow or clouds move over the subject, the mission pauses temporarily until the subject is again clearly visible in sunlight. To work around this problem, and enable all-weather imaging, another type of sensor, called an *active sensor*, is available. These sensors transmit their own radiation that reflects off the subject and returns to the sensor.

When you're groping across a very dark room, your eyes, which are passive sensors, can't detect enough reflected or emitted EM radiation from the objects in the room. So you hit your toe on that darn coffee table again. To avoid another toe injury, you could use a flashlight to illuminate the objects in the room and allow your eyes to see them. In this way, you are effectively acting as an active-sensor system by shining EM radiation on an object you want to see and then detecting the reflected energy with your sensors (eyes).

Of course, we don't actually put big spotlights in space to shine down on the night side of the planet. But we do "shine" EM radiation from other parts of the spectrum, such as radar. If you've ever known anyone pulled over for speeding (surely you'd never do such a thing), you're

Figure 11-40. Infrared Sensors. This infrared image of Mount Saint Helens volcano taken by Landsat shows vegetation in red and rocks in green and blue. Remote-sensing images of this type are especially useful for monitoring natural disasters such as volcanos. *(Courtesy of NASA/ Goddard Space Flight Center)*

Figure 11-41. Space-based Sensor. Telescopes, such as the Chandra X-ray Observatory benefit from being in space, above the atmosphere, where they can see further with greater resolution. *(Courtesy of NASA/Chandra X-ray Center/Smithsonian Astrophysical Observatory)*

Figure 11-42. Passive Sensors. Your eye is a passive sensor. It detects EM radiation other objects reflect or emit.

familiar with the police radar gun. The radar gun is an active sensor. It transmits EM radiation in the radar frequencies at speeding cars. The radiation reflects off the cars back to a sensor on the gun. The gun then measures this reflected radar energy to determine the car's speed.

Radar can do many other things besides enforce speed limits. Air traffic controllers use radar to track aircraft. Airplanes use onboard radars to create an image of the terrain below them, even if they can't see the ground through the clouds. This allows the aircrew to navigate by comparing terrain features to maps. Large ground-based radars are an important part of a mission operations network to determine the position and velocity of spacecraft.

Because radar is an active sensor, it works day or night and doesn't depend on sunlight, as optical sensors do. The reflected radar signal reveals much about the topography of the ground being imaged, as well as its composition (such as soil type and presence of subsurface features like ancient river beds). Space-based radar allows us to measure terrain features accurately to construct a 3-D picture of a planet's surface.

Resolution is still an issue with active sensors. Because resolution relates directly to the wavelength of the signal, shorter wavelengths yield better resolution than longer wavelengths. Optical sensors use wavelengths on the order of 0.5 μm, while radar systems operate at about 240,000 μm. Thus, for optical and radar systems with the same size aperture, the optical system has almost one-half million times better resolution. For conventional radar to have the same resolution as an optical system, we must increase the size of the radar's aperture. A conventional radar operating at a wavelength of 240,000 μm would need an aperture of more than 3900 km (6200 mi.) to get the same resolution as an optical system with a mere 1 m (3.28 ft.) aperture! Obviously, an aperture this size is impractical. Instead, we've developed signal-processing techniques that can make the electronics think the aperture is much larger than it really is. This is the basis for a *synthetic aperture radar (SAR)*. SARs have been successful in remote sensing of Earth and Venus. The SAR on the Magellan spacecraft used a very high-resolution (around 150 m or 492 ft.) to give us a detailed map of more than 98% of Venus, as shown in Figure 11-43.

The final function of a sensor system is to process the data from the detector into useful information. One of the many tasks for computers on spacecraft is to interface with sensors to process images for storage and/or transmission to users. As we'll learn in Chapter 13, the communication and data-handling subsystem (CDHS) can be quickly overwhelmed if it tries to store every possible image coming from a sensor. For this reason, "smart" sensors have built-in computers to process the raw signal coming from the detector. This task may involve screening images to sort good ones from bad ones. The processor may also need to compress the data for easier storage in limited computer space. Fortunately, advances in computer technology to support video cameras and internet teleconferencing have made many of these computer, image-processing operations routine.

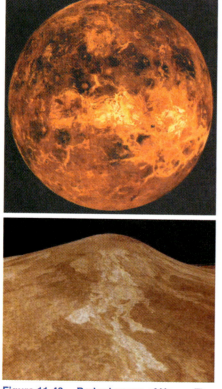

Figure 11-43. Radar Images of Venus. The Magellan spacecraft used a synthetic aperture radar to pierce the dense clouds and return highly accurate images of the surface of Venus. *(Courtesy of NASA/Jet Propulsion Laboratory)*

Payload Design

In this section we've reviewed important fundamental principles of EM radiation and learned how they relate to remote-sensing payloads. Space mission design generally, and spacecraft design specifically, grow from the payload requirements. When it comes to space missions, the payload is the star of the show. In Section 11.1 we described the entire space systems engineering process. There we learned that two important steps involved with deriving systems requirements are

1) Identify and characterize the mission subject
2) Define the payload requirements

Based on our discussion in this section, we can now break down these steps into more specific actions, as shown in Figure 11-44 for remote-sensing payloads.

With the payload requirements determined, we can now turn our attention to designing the rest of the spacecraft. The performance budgets needed by the payload will help us design the necessary subsystems. In the following chapters we'll dig into the spacecraft subsystems in greater detail and explore the basic principles that govern their function and design.

Identify and Characterize the Mission Subject

- Define the wavelength/frequency of EM radiation that will best characterize the subject
- Define the required minimum resolution
- Design the sensor system to observe the subject
  - Passive vs. active sensor
  - Aperture diameter
  - Focal length
- Derive the performance budgets for supporting subsystems
  - Attitude and orbital control budgets
  - Electrical power budget
  - Thermal control ranges
  - Data budget
  - Link budget
  - Launch and other structural loads

Figure 11-44. Payload Design Process. We follow these steps when designing a remote-sensing mission.

▬▬ Section Review

Key Terms

absorbed energy
active sensor
angular resolution, θ
aperture, D
atmospheric windows
bands
black body
charge-coupled device (CCD)
electromagnetic (EM) spectrum
field-of-view (FOV)
focal length, fl
frequency, f
magnification
passive sensors
peripheral vision
photons
reflected energy

Key Concepts

➤ Sensors are the spacecraft's "eyes" and "ears." Spacecraft "look" and "listen" from space using the EM spectrum.

➤ EM radiation can be thought of as waves of energy emanating from some source like ripples in a pond

- EM radiation is classified by wavelength—the distance between crests of the waves
- The shorter the wavelength, the higher the frequency—the number of waves or cycles per second
- The higher the frequency, the greater the energy of the radiation

➤ Certain frequencies of EM radiation are blocked by Earth's atmosphere, while other frequencies pass through. The range of frequencies that pass through are known as atmospheric windows. Engineers must be careful to select frequencies that can penetrate the atmosphere for remote-sensing.

Continued on next page

▦ Section Review (Continued)

Key Terms (Continued)

reflection
refraction
remote sensing
resolution
scan rate
swath width
synthetic aperture radar (SAR)
transmitted energy
wavelength, λ

Key Equations

$$\lambda_m = \frac{2898}{T}$$

$$E = \varepsilon \sigma T^4$$

$$\frac{fl}{h} = \frac{r_d}{R_g} = magnification$$

$$FOV = 2\tan^{-1}\left(\frac{r_d}{fl}\right)$$

$$\theta = \frac{1.22\lambda}{D}$$

$$Res = 2\theta h = \frac{2.44\lambda h}{D}$$

Key Concepts (Continued)

➤ Any object with a temperature greater than 0 K emits EM radiation: Planck described how an object's temperature determines its varying radiation frequencies

 • Wien's law tells us the frequency at which the maximum amount of black-body radiation will take place, depending on an object's temperature

 • The Stefan-Boltzmann relationship describes how much energy an object will emit, depending on its temperature

➤ To observe a subject on the ground, spacecraft sensors must

 • Look at it—move the sensor to point at the subject

 • See it—collect EM radiation from it

 • Convert it—transform EM radiation into usable data

 • Process it—turn data into usable information

➤ Sensor resolution refers to the smallest object a sensor can detect. Resolution depends on the

 • Sensor's wavelength

 • Distance to the subject

 • Diameter of the sensor's aperture

➤ Remote-sensing payloads use passive and active sensors

 • Passive sensors detect energy reflected or emitted from a subject

 • Active sensors shine EM radiation at the subject and then detect the energy reflected

➤ During the payload design process, we start by identifying and characterizing the subject, which involves

 • Defining the wavelength/frequency of EM radiation that will best characterize the subject

 • Defining the required minimum sensor resolution

 • Designing the sensor system to observe the subject

  – Passive versus active sensor

  – Aperture diameter

  – Focal length

 • Derive performance budgets for the supporting subsystems

Example 11-2

Problem Statement

You've been named payload manager for the FireSat mission described in Example 11-1. You know the mission subject is a forest fire >40,000 m². Assuming the average forest fire has a temperature of 1160 K, characterize the subject by determining the best wavelength/frequency to monitor and the required sensor resolution. The following system requirements, relevant to the payload design, have already been defined in Example 11-1

- Orbit constrained by the launch vehicle to 500-km, circular, polar. No further requirements can be defined.

- Spacecraft-constrained by launch vehicle to be <15 kg with dimensions of 0.30 × 0.30 × 0.30 m

Problem Summary

Given: Forest fires 1 km²
T = 1160 K

Find: Resolution, λ_{max}, f_{max}

Conceptual Solution

1) Determine the necessary payload resolution

Since the minimum size of a forest fire is given to be 40,000 m² (10 acres), the logical answer would be a linear resolution of 200 m (200 m × 200 m = 40,000 m²). Regardless of the actual dimensions of the fire, if the total area is greater than 40,000 m², it will exceed 200 m in at least one dimension.

2) Determine the wavelength of maximum emission for a 1160 K forest fire using Equation (11-3).

$$\lambda_{max} = \frac{2898(\mu m K)}{T(K)}$$

3) Convert the wavelength to frequency using Equation (11-1).

$$c = 3.0 \times 10^8 \frac{m}{s}$$

$$f = \frac{c}{\lambda_{max}}$$

Analytical Solution

1) Determine the minimum sensor resolution.

No further analysis is required, $Res_{min} = 200$ m

2) Determine the wavelength of maximum emission using Equation (11-3)

$$\lambda_{max} = \frac{2898\ (\mu m\ K)}{T(K)} = \frac{2898 \times 10^{-6} m\ K}{1160\ K}$$

$$\lambda_{max} = 2.50 \times 10^{-6} m$$

3) Convert the wavelength to frequency using Equation (11-1)

$$f_{max} = \frac{c}{\lambda_{max}} = \frac{3.0 \times 10^8 \frac{m}{s}}{2.50 \times 10^{-6} m}$$

$$f_{max} = 1.2 \times 10^{14} Hz$$

Interpreting the Results

The FireSat payload sensor must have a linear resolution of at least 200 m. At the expected temperature of 1160 K, forest fires will have a maximum energy emission at a wavelength of 2.50 μm, which corresponds to a frequency of 1.2 × 10¹⁴ Hz. According to Figure 11-29, this frequency is in the mid-infrared region. Fortunately, from Figure 11-32, we can see there is an atmospheric window at this wavelength that will allow us to see the fire.

Example 11-3

Problem Statement

Continuing the payload design problem for FireSat begun in Example 11-2, we want to determine critical parameters for the sensor and its operation. From Example 11-1, we know the operational altitude for the spacecraft will be 500 km. From Example 11-2, we know the required frequency and resolution. A review of available sensor materials reveals that Lead Selenide (PbSe) detectors, although somewhat expensive, are sensitive to this range of IR. A 1024 × 1024 pixel CCD PbSe array will serve as the sensor detector. Unfortunately, this detector must be cooled to at least 196 K to be effective, creating additional requirements for the other subsystems we'll see in Chapter 13. Assume the detector will have an effective radius of 1 cm. For the required sensor resolution of 200 m, determine the corresponding ground swath width and ground width along with sensor focal length. Then, find the minimum sensor aperture to achieve the required resolution. Finally, compute the sensor field of view.

Problem Summary

Given: $\lambda_{max} = 3.22 \times 10^{-6}$ m
Res = 200 m
pixels = 1024 × 1024
h = 500 km
$r_{detector} = 0.01$ m

Find: Swath width; sensor focal length, fl; minimum aperture, D; and field of view, FOV

Conceptual Solution

1) Find sensor swath width based on detector characteristics and resolution

swath width = (Res) (pixels)

2) Find ground radius

$$R_g = \frac{\text{swath width}}{2}$$

3) Solve Equation (11-5) for sensor focal length, fl

$$fl = \frac{r_{detector} h}{R_g}$$

4) Solve Equation (11-8) for the minimum sensor aperture

$$D = \frac{2.44 \lambda_{max} h}{\text{Res}}$$

5) Solve Equation (11-6) for the sensor field of view (FOV)

$$FOV = 2 \operatorname{atan}\left(\frac{r_{detector}}{fl}\right)$$

Analytical Solution

1) Find sensor swath width based on detector characteristics and resolution

swath width = (Res) × (pixels)
swath width = (200 m) × (1024)
swath width = 204.8 km

2) Find ground radius

$$R_g = \frac{\text{swath width}}{2} = \frac{204.8 \text{ km}}{2}$$
$$R_g = 102.4 \text{ km}$$

3) Solve Equation (11-5) for sensor focal length, fl

$$fl = \frac{r_{detector} h}{R_g} = \frac{(0.01 \text{ m})(500 \text{ km})}{102.4 \text{ km}}$$
$$fl = 0.049 \text{ m}$$

4) Solve Equation (11-8) for the minimum sensor aperture

$$D = \frac{2.44 \lambda_{max} h}{\text{Res}} = \frac{2.44(3.22 \times 10^{-6} \text{ m})(500 \text{ km})}{200 \text{ m}}$$
$$D = 19.6 \text{ mm}$$

5) Solve Equation (11-6) for the sensor field of view (FOV)

$$FOV = 2 \operatorname{atan}\left(\frac{r_{detector}}{fl}\right) = 2 \operatorname{atan}\left(\frac{0.01 \text{ m}}{0.049 \text{ m}}\right)$$
$$FOV = 0.404 \text{ rad}$$
$$FOV = 23.148 \text{ deg}$$

Interpreting the Results

We now have a preliminary design of the FireSat payload. To achieve 200 m linear resolution imaging 900 K forest fires emitting radiation in the $\lambda = 3.22$ μm range, the sensor will need a focal length of approximately 5 cm with a minimum lens aperture of about 2 cm. Of course, this lens must be transparent to the IR wavelength we're interested in. Fortunately, there are a variety of lens materials that are transparent to these wavelengths (such as CaF_2). With a focal length of 5 cm and an image plane with a radius of 1 cm, the sensor should easily fit in the available spacecraft volume. This sensor configuration will give the mission a field of view of just over 23° for a ground swath width of 204 km in a 500-km altitude orbit. In Chapters 12, 13, and 14 we'll analyze performance requirements for the spacecraft bus.

References

Elachi, Charles. *Scientific American.* December 1982. Vol. 247, No. 6. Radar Images of the Earth From Space.

Halliday, David and Robert Resnick. *Fundamentals of Physics.* 2nd edition. New York, NY: John Wiley and Sons, Inc., 1981.

Lillesand, Thomas A. and Ralph W. Diefer. *Remote Sensing and Image Interpretation.* New York, NY: John Wiley and Sons, Inc., 1979.

Masters, Prof. Gil. *Environmental Science and Technology.* Autumn 1989. Course Reader, CE170. Stanford University.

NASA-STD-3000. *Man-Systems Integration Standards.*

Serway, Raymond A. *Physics For Scientists and Engineers.* Vol. II. 1990. Saunders Golden Sunburst Series.

Stanford University. Spring 1990. AA141, Colloquium on Life in Space.

Wertz, James R. and Wiley J. Larson. *Space Mission Analysis and Design.* Third edition. Dordrecht, Netherlands: Kluwer Academic Publishers, 1999.

Mission Problems

11.1 Space Mission Design

1 Describe the systems engineering process and explain how it could be applied to an every day project.

2 What three things does the mission statement tell us?

3 What is the subject of a mission and how does it relate to the payload design?

4 Describe the mission objective and operations concept for our proposed FireSat. Why are these two things so important to overall mission design?

5 What is the subject of the proposed FireSat mission?

6 Describe the steps in the space systems engineering process and relate them to the FireSat mission. How do decisions made in this phase affect what will happen during the rest of the mission?

7 List the various parts of the spacecraft bus and describe functions for each one.

8 Describe what happens during preliminary spacecraft design. What are the key concerns for each subsystem? Why is this an iterative process?

11.2 Remote-sensing Payloads

9 Describe the primary elements for the remote-sensing system.

10 What is the difference between wavelength and frequency?

11 The wavelength of visible light is around 0.5 μm. What is its frequency?

12 Gamma rays have a wavelength of about 10–13 μm, AM radio waves have a wavelength of about 102 m. Which has more energy? Why?

13 What contribution did Max Planck make to our understanding of how objects emit radiation?

14 What is Wien's Law and why is it important?

15 How could you use the Stefan-Boltzmann relationship to determine an object's temperature?

16 Forest fires burn at about 538° C (1000° F). What is the wavelength of maximum power output for a forest fire? What is its energy output per square meter?

17 What are atmospheric windows and why are they important to selecting a spacecraft payload?

18 Using Figure 11-32, which frequency would be better to use for a remote-sensing spacecraft—7.5×10^{13} Hz or 1.03×10^{14} Hz? Why?

19 Describe the differences between passive and active sensors and give examples of each.

20 What are the main parts of a passive sensor?

21 What is sensor resolution?

22 Which sensor has better resolution, one with Res = 5 cm or one with Res = 10 cm? Why?

23 What is the angular resolution of an antenna with a 1 m aperture operating at a wavelength of 1 μm?

24 A remote-sensing payload operates in the visible part of the spectrum ($l = 0.5$ μm). If it has an aperture diameter of 1 m, what is the highest altitude orbit it should operate in to achieve a resolution of 10 m?

For Discussion

25 What may have been some of the mission requirements and constraints for the Hubble Space Telescope mission?

26 What systems did engineers have to plan when designing the satellites for global positioning (navigation)?

Projects

27 The Environmental Protection Agency is interested in detecting the amount of pollution flowing from the mouth of the Mississippi River into the Gulf of Mexico. Plan the space mission for this problem, identify the major trade-offs in selecting a payload, and discuss some of the considerations involved with spacecraft design.

28 Step through the space systems engineering process for a manned mission to Mars. Discuss how the mission objective and payload requirements affects the eventual spacecraft design.

Mission Profile—Landsat

The Earth Resources Technology Satellite (ERTS-1) was designed in the 1960s and launched in 1972. It was the first satellite designed specifically for broad-scale, repetitive observation of Earth's land areas. The program was renamed Landsat (land satellite) in 1975. Landsat developed as a cooperative, multi-agency, government project under NASA's direction. During the 1970's and 80's, Landsat transitioned to a commercial project under the private sector's control. In 1985, the Earth Observation Satellite Company (EOSAT) won a competitive bid to operate Landsat 4 and 5—to collect, archive, distribute, and sell Landsat images and to increase the user base.

Mission Overview

The objective of the Landsat satellite is to conduct remote sensing of Earth's resources, geology, and features made by humans and return this data to users on Earth. Landsat employs two primary sensors—the multi-spectral scanner and the thematic mapper. The scanner images in four distinct bands ranging from green to near infrared, with a spatial resolution of 80 m. The mapper images over a larger part of the EM spectrum than the scanner, with seven bands instead of four and a resolution of 30 m.

Mission Data

✓ Landsat launched in June, 1972. Designed for only a one-year mission, it wasn't retired from service until June, 1978, after returning more than 300,000 images of Earth.

✓ Landsat 2 launched in January, 1975 and remained in service until September, 1983.

✓ Landsat 3 launched in March, 1978 and retired in September, 1983

✓ Landsat 4 was an improved design of 1, 2, and 3, and launched in July, 1982. Because Landsat 4 suffered a power distribution problem in March 1983, it operates at reduced power.

✓ Landsat 5, identical to Landsat 4, launched in March 1984, with a planned lifetime of three years. It was still in use nine years later, in 1993.

✓ All Landsat satellites operate in a near-polar, sun-synchronous orbit crossing the equator at 9:45 A.M. local time.

Mission Impact

For more than 24 years, the Landsat program has produced images of most of Earth's land masses, useful for analyzing long-term and quick-response changes. In the 1980s, with the improvement of computer technology to process Landsat images, the applications for this data have exploded. Landsat images have been used for many long-term environmental studies, such as disappearing tropical forests, expanding desert areas, and climatic changes, as well as Earth's response to natural disasters such as the explosion of Mount St. Helens and fires arising from the Midwestern floods. Landsat imagery has also been used for monitoring oil spills, identifying wildlife habitats, and measuring the growth of urban areas.

Investigators use Landsat images to track the destruction of forests. These Landsat images show Rondonia, Brazil, in 1975 (left) and 1986 (right). Settlers colonized the area and converted forest land to agricultural land, which shows on the right as a fishbone pattern radiating from the highway. *(Courtesy of NASA/Goddard Space Flight Center)*

For Discussion

- How have we benefited from commercial remote-sensing technologies?
- What are the disadvantages of the Landsat sensors? How could the sensors be improved?
- How can we use Landsat data to better manage Earth resources?

Contributors

Steve McGregor and Mark Hatfield, U.S. Air Force Academy

References

American Society of Remote Sensing. *Manual of Remote Sensing.* Virginia: The Sheridan Press, 1983.

Campbell, James B. *Introduction to Remote Sensing.* New York, NY: Guilford Press, 1987.

EOSAT Corp., Landsat Data Users Notes, published quarterly.

The manned maneuvering unit (MMU) gives astronauts the freedom to soar through space as a one-person spaceship. *(Courtesy of NASA/Johnson Space Center)*

Space Vehicle Control Systems

12

In This Chapter You'll Learn to...

- Describe the elements of and uses for control systems
- Explain the elements of space vehicle attitude determination and control subsystems and describe various technologies currently in use
- Explain the elements of space vehicle navigation, guidance, and control subsystems and how they work together to deliver a vehicle to a desired point in space

You Should Already Know...

- ❑ Effects of the space environment on spacecraft (Chapter 3)
- ❑ Newton's Laws of Motion (Chapter 4)
- ❑ The principle of conservation of momentum (Chapter 4)
- ❑ Components and functions of the spacecraft bus (Chapter 11)

Outline

12.1 Control Systems

12.2 Attitude Control
Having the Right Attitude
Attitude Dynamics
Disturbance Torques
Spacecraft Attitude Sensors
Spacecraft Attitude Actuators
The Controller

12.3 Orbit Control
Space Vehicle Dynamics
Navigation—The Sensor
Rockets—The Actuators
Guidance—The Controller

The Earth is a cradle of the mind, but we cannot live forever in a cradle.

Konstantin E. Tsiolkovsky
father of Russian cosmonautics

magine you're a one-person spacecraft, flying the manned maneuvering unit out of the Space Shuttle's payload bay, as shown in Figure 12-1. Your mission is to fly to a crippled satellite and install a new black box. You must somehow manipulate the joy sticks in your hands to control your position, velocity, and orientation so you're lined up with the access panel on the spacecraft. How should you rotate? In which direction should you fire your thrusters? Do you speed up or slow down? While this may sound like a fun scenario for a video game, we must answer these questions for nearly all spacecraft. In this chapter we'll begin by examining the basics of any control system and then see how we can apply this process to rotate and move a spacecraft through space.

Figure 12-1. Space Vehicle Control. An astronaut flying the manned-maneuvering unit (MMU) must carefully control rotation, position, and velocity to accomplish the mission (and not get lost in space!). *(Courtesy of NASA/Johnson Space Center)*

12.1 Control Systems

In This Section You'll Learn to...

- ☞ Describe the elements of a system
- ☞ Explain the difference between open-loop and closed-loop control systems and give examples of each
- ☞ Describe the steps in the control process
- ☞ Apply block diagrams to describe the functions of control systems

A *system* is any collection of things that work together to produce something. Systems have inputs (what goes in), outputs (what comes out), and some process in between that turns the inputs into outputs. In electronic systems, the inputs and outputs are called *signals*. The part of the system that performs the process is typically called the *plant*. The plant is usually an "equal opportunity" processor that will respond to either precisely calculated inputs or random environmental inputs or both.

To simplify our discussion of systems, we like to illustrate them using *block diagrams* where lines represent input and output signals and boxes represent the plant or other components. Figure 12-2 shows the simplest type of system block diagram. For space vehicle applications, the success or failure of the mission depends on the output of various subsystems. Therefore, we're most interested in a specific class of systems called *control systems*. Control systems are everywhere. If you've ever flushed a toilet, driven a car, or turned up the thermostat on a frigid winter night, you've used a control system.

Figure 12-2. System Block Diagram. All systems take some input (or inputs), perform some process in the "plant," to produce an output (or outputs). We illustrate the functions of systems using block diagrams where input and output signals are shown as arrows and the plant, or other components, are shown as boxes.

To understand how we use control systems, let's look at a problem we're all familiar with—heating a house. In the old days, people heated their homes only with fireplaces. They started with some desired result "It's too cold in here, let's warm things up!" and decided what action to take—"Throw some logs on the fire!" The logs burned, providing heat, and the house warmed up. We can draw a simple diagram for this whole process, as shown in Figure 12-3. As you can see, the input (logs) go into the fireplace and burn, which produces the output (heat).

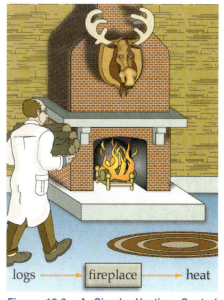

Figure 12-3. A Simple Heating Control System. With the simplest type of heating control system, we throw logs on the fire and wait for heat to come out, warming the room. Unfortunately, such a simple system is "open-looped," so there is no guarantee we'll get the right amount of the desired output.

Unfortunately, this simple control system has one major drawback. If you put on too many logs, the room can get too hot. And, when you went to sleep at night, there was no way to ensure that the house would still be warm in the morning. A system that can't dynamically adjust the inputs based on what's actually happening is an *open-loop control system*. Of course, people lived with this kind of heating system for thousands of years (and still do). But eventually we got tired of waking to a cold house and invented the modern home-heating system. Let's see what makes this modern control system an improvement.

On cold winter nights, we turn up the dial on the thermostat to a desired temperature and wait for the heat. After some time, the furnace reaches its operating temperature, turns on its fan, and the room temperature starts to rise. When the house reaches the desired temperature, the furnace shuts off. Simple, right? But what's really going on here?

As with any control system, we have some desired result—a house at 20° C (75° F). This desire is what we tell the thermostat when we set the dial. For this example, the heating-control system has different jobs. First, it measures the current temperature in the house using a thermometer. In control-system lingo, the thermometer is a *sensor*, because it measures the output of the system. Next, the control system decides what to do using the "brains" of the thermostat. The "brain" of the thermostat is called the *controller*, because it compares the sensor output to the desired output and decides what type of input the system needs. If it's cold outside, the "environmental inputs" will eventually reduce the temperature in the house to less than 20° C (75° F). When this happens, the controller knows to turn on the furnace. Similarly, if the temperature is greater than 20° C (75° F), it knows to turn off the furnace. Finally, the furnace carries out the thermostat's decision by providing heat. We call the furnace an *actuator*, because it takes commands from the controller and produces the required output for the system. Figure 12-4 illustrates the various pieces of this control system.

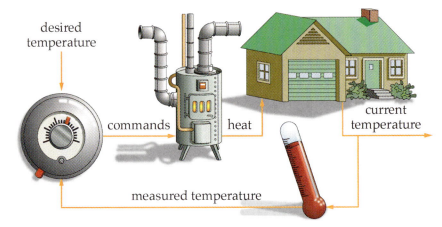

desired temperature

commands   heat

current temperature

measured temperature

Figure 12-4.   The Modern Home-heating System. A modern home-heating system constantly measures the temperature and decides when to turn the furnace on or off.

404

We call this type of system a *feedback control system* or a *closed-loop control system*, because we have a thermometer (sensor), which constantly measures the current temperature and feeds this information back to the thermostat (controller). With this information, the thermostat produces the correct furnace (actuator) commands to maintain the house at 20° C (75° F), in spite of environmental inputs (close the window please!). Again, we can draw a simple block diagram to illustrate what's going on in this system, as shown in Figure 12-5.

In this simple example we've identified the four basic tasks all control systems must do

- *Understand* the system's behavior—how the plant will react to inputs, including environmental inputs, to produce outputs. This is also known as the *plant model*.

- *Observe* the system's current behavior—using sensors

- *Decide* what to do—the job of the controller

- *Do* it—using actuators

Closed-loop control systems, such as the heating system described above, are in cars, planes, spacecraft, and even the human body. They are extremely useful because, unlike open-loop systems, they can make a system do what we want even in the face of random environmental inputs.

On space vehicles, control systems are an integral part of virtually all payloads and subsystems. Recall the payload and bus design issues discussed in Chapter 11. For example, a remote sensing payload may need to control

- Exposure

- Aperture settings

- Lens cover mechanisms

- Imaging time and duration

To support the payload, each subsystem in the spacecraft bus needs to control something as well

- Momentum (angular and linear)—attitude and orbit control subsystem (AOCS)

- Data (bits and bytes)—communication and data-handling subsystem (CDHS)

- Power (current, voltage, distribution)—electrical power subsystem (EPS)

- Internal environment (temperature, air, water, food, waste)— environmental control and life-support subsystem (ECLSS)

- Loads (bending, twisting, shaking)—structures and mechanisms

- Rocket thrust (valves, pressure, temperature)—propulsion subsystem

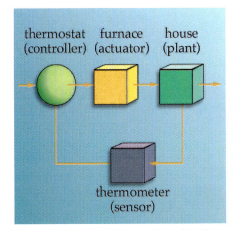

Figure 12-5.   Block Diagram for a Heating-control System. It's easier to represent the elements of a control system using blocks.

In the rest of this chapter, we'll focus our attention on momentum control—angular and linear—the job of the *attitude and orbit control subsystem (AOCS)*. This is the "steering" function we described for the school bus in Chapter 11. As you can imagine, in operation the two functions of controlling angular and linear momentum often overlap. (In practice, it's difficult to just change angular momentum without having some effect on linear momentum—try spinning a frisbee without moving it!) However, for purposes of discussion here, we'll keep the two functions (angular and linear momentum control) separate. In Chapters 13 and 14 we'll return to the other subsystems to see their control systems in action.

▬ Section Review

Key Terms

actuator
attitude and orbit control
   subsystem (AOCS)
block diagrams
closed-loop control system
control systems
controller
feedback control system
open-loop control system
plant
plant model
sensor
signals
system

Key Concepts

➤ All systems take some input and perform some process to produce an output. Inputs and outputs are called signals and the element performing the process is called the plant. We can best illustrate systems using block diagrams.

➤ The simplest type of control system is open-loop. Input produces an output. Unfortunately, open-loop systems can't dynamically adjust inputs to control outputs.

➤ Feedback control systems, also called closed-loop control systems can better assure we get our desired output because it can sense outputs (what we get), compare them to desired outputs (what we want), and adjust inputs as needed. Closed-loop control systems accomplish this in four steps

  • *Understand* the system's behavior—how the plant will react to inputs, including environmental inputs, to produce outputs. This is also known as the plant model.

  • *Observe* the system's current behavior—using sensors

  • *Decide* what to do—the job of the controller

  • *Do* it—using actuators

➤ Virtually all spacecraft payloads and subsystems rely on closed-loop systems to control

  • Imaging, communicating, and operating other missions—payloads

  • Momentum (angular and linear)—attitude and orbit control subsystems (AOCS)

  • Data (bits and bytes)—communication and data-handling subsystem (CDHS)

  • Power (current, voltage, distribution)—electrical power subsystem

  • Internal environment (temperature, air, water, food, waste)—environmental control and life-support subsystem (ECLSS)

  • Loads (bending, twisting, shaking)—structures and mechanisms

  • Rocket thrust (valves, pressure, temperature)—propulsion subsystem

12.2 Attitude Control

▤ In This Section You'll Learn to...

- ☛ Explain and apply important concepts in attitude dynamics to the problem of space vehicle control
- ☛ Describe key elements and technologies used in space-vehicle attitude determination and control subsystems and explain them using block diagrams

We'll begin our discussion of the space vehicle attitude and orbit control subsystem (AOCS) by focusing on the attitude part of the problem. Attitude defines a vehicle's orientation in space. For example, if we want a spacecraft to take pictures of a particular spot on Earth, we need to align the payload so it points at the spot. In this case, we'd need to control the spacecraft's attitude so it points "down," toward Earth. In space terms, we say, "down toward Earth" is the nadir direction. The opposite direction, away from Earth toward space, is the zenith direction. Similarly, launch vehicles need to control their attitude to steer into the correct orbit and keep forces aligned along the long axis where they are strongest. However, in this section, we'll focus primarily on the unique problems for spacecraft. Because this function is so important, it is sometimes given a separate name—*attitude determination and control subsystem (ADCS)*. In this section, we'll refer to it by that name. Regardless of the name given to the subsystem, its job is the same—keep its spacecraft pointed in the right direction.

In the last section, we learned that all closed-loop control systems have the same basic components and functions, as shown in Figure 12-5. In this section, our "desired state" is the specific attitude a vehicle needs to do its mission. We start by defining this desired attitude. Then, we move on to the first function of any control system—understanding system behavior. We explore attitude dynamics to understand the basic principles that govern a vehicle's angular momentum. We also see how various phenomena in the space environment affect a spacecraft's attitude. After this introduction, we'll turn our attention to attitude sensors to learn how we use instruments to "look out the window" to determine a spacecraft's orientation in space. Before looking at attitude controllers, we'll first examine the types of attitude actuators available to designers. With these in mind, we'll finally consider the controller and see how the entire subsystem fits together (Figure 12-6).

Having the Right Attitude

Before we go too far, let's review some basic terms used to describe attitude. Recall that when we described the motion of an object in Chapter 4, we referenced it to some coordinate system. To describe

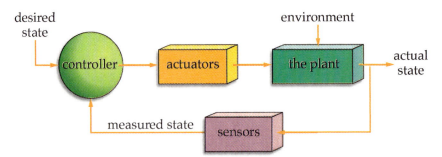

Figure 12-6.   Closed-loop Control System. All closed-loop control systems have the same basic elements. The desired state is one input to the controller. It compares this state to the actual state from the sensors. By comparing the difference between these two input signals, it decides on specific commands to send to the actuators. Actuator changes, along with environmental inputs, affect the final output of the plant. System sensors detect and measure this output.

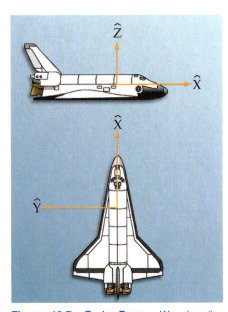

Figure 12-7.   Body Frame. We describe attitude in terms of rotations in degrees (or radians) around one or more of the body-centered axes from the body frame. For airplanes and vehicles like the Space Shuttle, the $\hat{X}$-direction points out the nose, the $\hat{Y}$-direction out the left wing, and the $\hat{Z}$-direction (out of the page) completes the right hand rule. For box-shaped spacecraft without a "nose" or wings, designers pick convenient, preferred directions through the center of mass to define the body frame.

attitude we must do the same thing, but we're now interested in rotation rather than translation. For this reason, we define attitude in terms of angles instead of distances. Attitude is described as an angular rotation with respect to a body-centered coordinate frame, called the *body frame*, where $\hat{X}$ points out the nose, $\hat{Y}$ out the left wing, and $\hat{Z}$ out the top, as shown in Figure 12-7. It is usually given as roll, pitch, and yaw angles, where *roll* is a rotation about the $\hat{X}$ axis, *pitch* is a rotation about the $\hat{Y}$ axis, and *yaw* is a rotation about the $\hat{Z}$ axis, as shown in Figure 12-8. Obviously, box-shaped spacecraft don't have noses or wings. Instead, designers define preferred directions through the center of mass in a body-centered system and then they define roll, pitch, and yaw angles with respect to it.

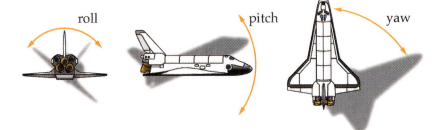

Figure 12-8.   Describing Attitude. We describe space-vehicle attitude in terms of roll, pitch, and yaw angles around the axes of the body frame.

Now that we showed how to describe a spacecraft's attitude, how do we determine whether it has the right attitude? We must first know what attitude it needs. Typically, we describe attitude-control requirements in terms of accuracy and rate of attitude change. To understand what we mean by attitude or pointing accuracy, let's pretend we're trying to point a laser beam at a target, as shown in Figure 12-9. Our ability to keep the beam on the target depends on the size of the target and the steadiness of our hand. It should make sense that the smaller the target, the more steady our hand must be to maintain the laser on target. Hopefully, even as our hand wavers, the beam will tend to stay within a cone, more or less, centered on the target. The angular size of this cone defines *pointing* or *attitude accuracy, ψ*. For a spacecraft trying to point an antenna at a ground station on Earth, for example, the control system must be accurate enough to keep the radio beam focused over the receiver antenna.

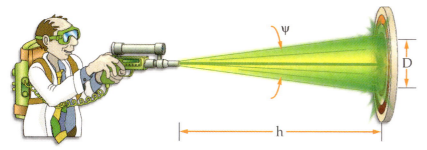

Figure 12-9.   Attitude Accuracy. In this example, the shooter is pointing a laser beam at a dinner-plate sized target. As his hand wavers, the beam describes a cone, more or less, centered on the target. The angular size of this cone, ψ, defines attitude or pointing accuracy.

To get a better feel for what we mean by attitude or pointing accuracy, let's pretend we're trying to keep the laser pointer focused on a target about the size of a 25 cm (10 in.) dinner plate, as shown in Figure 12-9. We know the pointing accuracy, ψ, and the distance to the target, h. To find the apparent diameter of the target we can hit, D, we use

$$D = h\,\psi \qquad\qquad (12\text{-}1)$$

where

D = approximate diameter of target (m)

h = distance to target (m)

ψ = pointing accuracy (rad)

Table 12-1 shows the required pointing accuracy to stay focused on the dinner plate at various distances. Now let's put this in space terms. A remote-sensing spacecraft passing directly overhead at an altitude of 500 km (310 mi.), for example, would need about 0.003° of accuracy to point a laser range finder directly at a house (D = 26 m or 85 ft.). Fortunately, pointing a laser beam is a worst-case scenario because the narrow beam has a very narrow field of view. Remote sensing missions using optical or infrared cameras typically have lenses with fields of view of several degrees or more, depending on the application. To give the widest possible coverage, communication missions will often design antennas with very wide fields of view. The actual requirement for spacecraft pointing, then, depends on the subject, the sensor's field of view, and other factors, such as timing and viewing angles.

The rate of attitude change is also important to consider when defining attitude control requirements. For example, a remote-sensing spacecraft may need to shift its attention between various targets on the ground. To shift attention means it must rapidly change its attitude to focus on a new point of interest. *Slew rate* is the angular speed (in degrees, or radians, per second) describing how fast the spacecraft can change its attitude.

Table 12-1.   Pointing Accuracy.

To Point at a Target the Size of a Dinner Plate (25 cm or 10 in. diameter) at this Distance...	The Pointing Accuracy Needs to be...
1.4 m (4.6 ft.)	10°
14 m (46 ft.)	1°
140 m (460 ft.)	0.1°
1400 m (0.87 mi.)	0.01°

Now that we understand more about describing attitude, let's start to see how we control it by first trying to understand attitude dynamics. We'll then turn our attention to environmental factors affecting attitude that spacecraft must deal with.

Attitude Dynamics

As we know, all spinning objects—tops, yo-yo's, ice skaters, and even spacecraft—follow Newton's Laws of Motion. Recall from Chapter 4 that a spinning mass has angular momentum, which is a function of its shape and mass distribution, along with its rate of spin. Notice, for example, that a compact object with all the mass concentrated near the center of mass spins much easier than an object that has a lot of mass located far from the center of mass. As Figure 12-10 shows, this is why figure skaters bring their arms in to spin faster and extend their arms to slow down. The distribution of mass describes an object's *mass moment of inertia, I*. By knowing the mass moment of inertia, I, and the object's angular velocity, $\vec{\Omega}$, we can find its angular momentum, $\vec{H}$, in pretty much the same way we found its linear momentum.

$$\boxed{\vec{H} = I\vec{\Omega}}$$

(12-2)

where

$\vec{H}$ = angular momentum $(kg \cdot m^2/s)$

I   = mass moment of inertia $(kg \cdot m^2)$

$\vec{\Omega}$ = angular velocity (rad/s)

We also know from Chapter 4 that to change an object's momentum we must apply a force. We explained this using Newton's Second Law, which we express as

$$\vec{F} = \frac{d\vec{p}}{dt} = \dot{\vec{p}}$$

(12-3)

where

$\vec{F}$ = force (N)

$\vec{p}$ = momentum $(kg \cdot m/s)$

$\dot{\vec{p}}$ = time rate of change of momentum $\left(\dfrac{kg \cdot m/s}{s}\right)$

Figure 12-10. Changing Mass Moment of Inertia. Figure skaters change their moments of inertia to vary their rate of spin. For the same total angular momentum, they will spin faster by bringing their arms in (lower moment of inertia) and slower by extending their arms (higher moment of inertia).

410

When we kick a football or serve a volleyball, it's not hard to see that applying a force to a mass changes its velocity. But how do we apply force to a rotating mass? If we push on spinning ice skaters they'll start moving in a straight line across the ice while continuing to spin. What if we want to change only their rate or direction of spin but not move them anywhere? Then we must apply a torque. A *torque* is a twisting force that results when we try to rotate an object, such as when we use a wrench to turn a bolt. We apply a force some distance away from the bolt, producing a torque, as shown in Figure 12-11. A torque in one direction tightens the bolt. A torque in the other direction loosens it.

Mathematically, we define the direction of torque as the vector cross product of the applied force's position vector with the force vector. In other words, we use the good-old right-hand rule: we point the fingers of our right hand in the direction of the twist and our thumb points in the torque-vector direction. In Figure 12-11, we have a force applied to the end of a wrench. The torque vector, $\vec{T}$, points into the bolt. We can compute the torque using

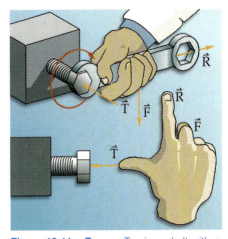

Figure 12-11.   Torque. Turning a bolt with a wrench is a good example of applying a torque. We find the direction of torque using the right hand rule. By wrapping the fingers of our right hand in the direction of spin, our thumb points in the direction of the torque vector. In this case, the torque direction is into the bolt, as we'd expect.

$$\boxed{\vec{T} = \vec{R} \times \vec{F}} \qquad (12\text{-}4)$$

where

$\vec{T}$ = torque (N · m)

$\vec{R}$ = distance from the center of mass to the point where the force is applied (m)

$\vec{F}$ = applied force (N)

According to this relationship, we can get more torque with the same force by simply applying the force farther from the center of rotation. Aristotle knew of this effect when he bragged he could move the Earth if given "a fulcrum, a long enough staff, and a place to stand." We don't have to move the Earth to see this effect. All we have to do is push open a door. If we push at the edge of the door, far from the hinges, the door swings right open. If we push on the door right next to the hinges, it's much harder to move.

Returning to Newton's Second Law, we can now see how to relate torque and angular momentum. Just as force equals the time rate of change of linear momentum, torque is the time rate of change of angular momentum. In other words, if we apply a torque to an object, its angular momentum will change. We can express this as

$$\vec{T} = \frac{d\vec{H}}{dt} = \dot{\vec{H}} \qquad (12\text{-}5)$$

where

$\vec{T}$ = torque (N · m)

$\vec{H}$ = angular momentum (kg · m^2/s)

$\dot{\vec{H}}$ = time rate of change of angular momentum $\left(\dfrac{\text{kg} \cdot \text{m}^2/\text{s}}{\text{s}} \right)$

This relationship tells us something we already learned back in Chapter 4. When torque is zero, angular momentum stays constant. Later, we'll see we can use this basic principle to give us accurate attitude sensors, as well as efficient actuators to control attitude. For now let's see how we can use it to analyze how attitude works. Remember that if we apply a force to an object, it will accelerate. Similarly, if we apply a torque to a free-floating object, it will start to spin faster and faster. That is, it will experience angular acceleration, $\vec{\alpha}$. Thus,

$$\boxed{\vec{T} = \dot{\vec{H}} = I\vec{\alpha}}$$ (12-6)

where

I = mass moment of inertia ($kg \cdot m^2$)

$\vec{\alpha}$ = angular acceleration ($rads/s^2$)

As we know from our discussion of linear motion, as something accelerates over time, it acquires velocity. If we drop a ball, it accelerates, gains velocity, and falls faster with time. Similarly, when an object has angular acceleration over time, it gains angular velocity, $\vec{\Omega}$. Thus, to determine a spacecraft's attitude, described by an angle θ (an amount the spacecraft rotated from its previous attitude), we must look at how long it accelerates and how long it moves at some angular velocity. In other words, by applying a torque to a non-spinning, free-floating object (such as a spacecraft), we create angular acceleration, leading to angular velocity and hence a change in angular position. The model for this aspect of spacecraft behavior is the block diagram in Figure 12-12.

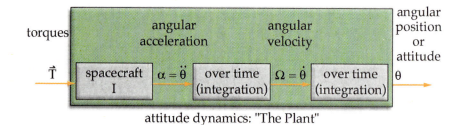

attitude dynamics: "The Plant"

Figure 12-12.   Block Diagram of Attitude Dynamics. A torque applied to a spacecraft (or any object for that matter) causes an angular acceleration. Over time, this acceleration increases angular velocity, changing its attitude.

When we apply a torque to a non-spinning spacecraft, predictable things happen. For example, when we turn a screw with a screwdriver, it rotates in the way we'd expect. But if the spacecraft is spinning when we apply the torque, the dynamics get more complicated. As we know, a spinning object has angular momentum. If we apply a torque *parallel* to the angular momentum direction, it causes angular acceleration and angular velocity changes. However, if we apply the torque in a direction other than parallel to the angular momentum vector, something quite different happens.

In Figure 12-13, we have a spinning disk with a force couple (torque) applied to it. You might expect the mass to begin to rotate in the same direction you're torquing it, or clockwise as you look down on it. But that's not what happens! The mass begins to rotate counter-clockwise about an axis that comes out of the page! This phenomenon is known as *precession*.

For the disk shown in Figure 12-14, the $\vec{H}$ vector will begin to move (or precess) toward the $\vec{T}$ vector (things would behave differently if it were a different shape). This precession occurs around a third vector called the precession vector, $\vec{\omega}$, and is at right angles to both $\vec{H}$ and $\vec{T}$. For a constant torque, the precession rate is also constant; it doesn't accelerate, as we'd expect! As we'll see, knowing how a spacecraft gains angular velocity and precesses helps us determine how to apply forces to adjust it's attitude. Note that the direction of precession depends on *how* mass is distributed in the object—its mass moment of inertia. Analyzing *why* it precesses the way it does is beyond the scope of this book.

There is one more important result of the interaction between a spinning spacecraft and an applied torque. We all know that spinning footballs and spinning rifle bullets travel further and faster than non-spinning ones. This is because spin makes them more stable and resistant to outside torques. The faster they spin, the more stable they become. This stability is referred to as an object's *gyroscopic stiffness*. The mathematical explanation for what makes a spinning object "stiffer" than a not-spinning object is beyond the scope of our discussion here. However, as we'll see, we can make use of this fact to keep spacecraft pointed where we want them.

Disturbance Torques

So why can't we just stick a satellite out in space with the desired attitude and forget about it? As we know from Chapter 3, space can be a nasty place. Over time, if we do nothing, environmental effects called *disturbance torques* will drive a spacecraft away from its original attitude. These torques are extremely small (in most cases, they literally couldn't kill a fly). But just as tiny drops of water can wear away mountains over time, these torques can eventually rotate even very large spacecraft. We're concerned with four main sources of disturbance torques

- Gravity gradient
- Solar-radiation pressure
- Earth's magnetic field
- Atmospheric drag

We'll go through each of these before turning our attention to the problem of attitude determination.

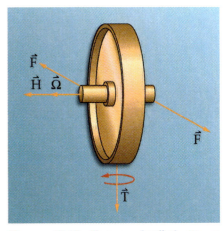

Figure 12-13. Torque Applied to a Spinning Disk. Here, the angular momentum vector, $\vec{H}$, points to the left. As we apply a force couple to the spinning disk, into and out of the page, it creates a torque, $\vec{T}$, that points down.

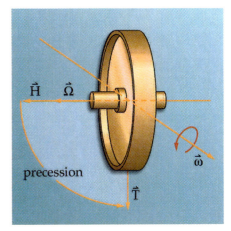

Figure 12-14. Precession of a Spinning Disk. When we apply a torque to the spinning disk, it begins to precess by rotating around an axis 90° from both the torque and angular momentum axes. In general, $\vec{H}$ tends to move towards $\vec{T}$. Using the right-hand rule, curling your fingers from $\vec{H}$ to $\vec{T}$ allows you to predict the direction of the precession axis with your thumb.

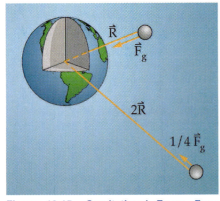

Figure 12-15. Gravitational Force. From Newton's Law of Universal Gravitation, we know that gravitational attraction decreases with the square of the distance between two objects. Thus, if we double the distance, the gravitational force is only 1/4 as strong.

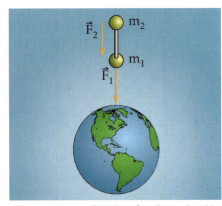

Figure 12-16. Gravity Gradient. In this simplified, dumb bell shaped spacecraft, we show that the gravitational force on the lower part is slightly greater than the force on the upper part, $\vec{F}_1 > \vec{F}_2$. The same effect happens in more conventional shaped spacecraft due to differences in internal mass distribution.

Gravity-gradient Torque

Gravity-gradient torque results from the difference in gravitational force exerted on different parts of a spacecraft. Recall from Chapter 4 that Newton said the force of gravity on an object varies inversely with the square of the distance from the central body.

$$\vec{F}_g = \frac{-\mu \, m}{R^2}\hat{R} \qquad (12\text{-}7)$$

where

$\vec{F}_g$ = force of gravity (N)

μ = gravitational parameter of the central body (km^3/s^2)

m = mass of the object (kg)

R = distance from the object to the central body (km)

$\hat{R}$ = unit vector in the $\vec{R}$ direction (dimensionless)

Thus, we show in Figure 12-15, if one object is twice as far from Earth as a second object, the gravitational force will be one-fourth as large. This is easy to visualize if the difference in distances from the central body is very great, but it works the same way for very small differences. Imagine we have a dumbbell-shaped spacecraft in Earth orbit. If the dumbbell is hanging vertically, as in Figure 12-16, the lower mass (m_1) will have a slightly greater gravitational force on it than the higher mass (m_2).

If m_2 is directly above m_1, nothing interesting happens. However, if the dumbbell gets displaced slightly off vertical, as in Figure 12-17, this slight difference between the gravitational forces on the two masses will create a torque on the spacecraft that will tend to restore it to vertical. This is fine if you want it to be vertical, but if you don't, your control system must constantly fight against this torque. We can estimate the magnitude of this torque using

$$T_g = \left(\frac{3\mu}{2R^3}\right)|I_Z - I_Y|\sin(2\theta) \qquad (12\text{-}8)$$

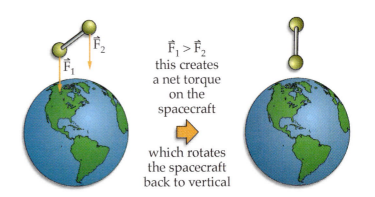

Figure 12-17. Gravity-gradient Torque. The slight difference in gravitational force between the upper and lower part of a spacecraft will tend to rotate the spacecraft to vertical, with its long axis pointed to Earth.

where

T_g = gravity-gradient torque (Nm)

μ = gravitational constant $(km^3/s^2) = 3.986 \times 10^5\ km^3/s^2$ for Earth

I_Z = spacecraft moment of inertia about the $\hat{Z}$ axis (where we assume $I_X = I_Y$ and $I_Z >> I_X$) $(kg\ m^2)$

I_Y = spacecraft moment of inertia about the $\hat{Y}$ axis $(kg\ m^2)$

θ = angle between the body $\hat{Z}$ axis and the local vertical

This equation tells us three important things about the gravity-gradient effect

- It decreases with the cube of the distance, e.g., by going from a 500 km altitude orbit (R = 6878 km) to a 1000 km (R = 7378 km) altitude orbit the torque reduces by almost 20%

- It depends on the difference between moments of inertia in the $\hat{Z}$ axis and $\hat{X} - \hat{Y}$ plane, thus for a homogenous spacecraft with $I_x = I_y = I_z$ the effect is zero

- It depends on the angle between the $\hat{Z}$ axis and local vertical. The greater the angle from local vertical, the greater the torque

Later in this section we'll see how we can turn this sometimes annoying effect to our advantage.

Solar Radiation Pressure

Another source of disturbance torque for spacecraft comes from the Sun. The Sun exerts an ever-so-slight force called *solar-radiation pressure* on exposed surfaces. We're used to being warmed by the Sun, tanned by the Sun, and even burned by the Sun, but pushed by the Sun? Yes. One way to think about sunlight (or any light for that matter) is as tiny bundles of energy called *photons*. In one of those seeming paradoxes of modern physics, we say these photons are massless (thus, they can travel at the speed of light), but they do have momentum. As photons strike any exposed surface, they transfer this momentum to the surface. Why can't we feel this force when we hold our hand up to the Sun? Because this force is very, very small. We can estimate the force using

$$F = \left(\frac{Fs}{c}\right) A_s (1 + r) \cos I \qquad (12\text{-}9)$$

where

F   = force on a surface (N)

Fs = solar constant = $1358\ W/m^2$ at Earth's orbit around the Sun

c   = speed of light = $3 \times 10^8\ m/s$

A_s = illuminated surface area (m^2)

r   = surface reflectance (where r = 1 for a perfect reflector and 0 for a perfect absorber) (unitless)

I   = incidence angle to the Sun (deg)

The force exerted on even a very large spacecraft with ten square meters of surface (assuming perfect reflectance) is only 9×10^{-5} N (2×10^{-5} lb$_f$)! We assume this force acts at the center of pressure for the surface. The moment arm is the distance from the center of pressure to the spacecraft's center of mass. We find the resulting torque by multiplying this force times the moment arm ($T = F \times d$). Even with a 1 m moment arm, the resulting solar pressure torque is only 9×10^{-5} Nm. So why worry about it? Over time, even this tiny force, acting unevenly over different parts of the spacecraft, especially large areas like solar panels, can cause problems for spacecraft needing precise pointing. In Chapter 14 we'll see how we harness this small force to propel large solar sails.

Magnetic Torque

A third source of disturbance torque comes from Earth's magnetic field. As we learned in Chapter 3, because of the impact of charged particles in space, the surface of a spacecraft can develop a charge of its own giving it a distinct *dipole*—north/south ends, like a compass. Just as a compass needle rotates to align with Earth's magnetic field, the dipole-charged spacecraft will similarly try to rotate as it passes through the magnetic field. The size of this *magnetic torque* depends on the spacecraft's effective magnetic dipole and the local strength of Earth's magnetic field. We estimate this from

$$T = D\,B \qquad (12\text{-}10)$$

where

T = torque on a spacecraft (Nm)

D = spacecraft dipole (amp-m^2)

B = local magnetic field strength (tesla). This varies with altitude (R = distance to Earth's center) and latitude. Earth's magnetic moment, M, is approximately 7.99×10^{15} tesla-m^3. At the poles, $B = 2M/R^3$. At the equator, $B = M/R^3$. (tesla = kg/amp-s^2)

Magnetic torque is a big concern for operators of small satellites in low, polar orbit but hardly noticeable for large spacecraft in geostationary orbit. Later, we'll see how we can create a large dipole on purpose to use this torque as an attitude actuator.

Aerodynamic Drag

The last disturbance torque we have to worry about is drag. As we saw in Chapter 3, in low-Earth orbit the atmosphere applies a drag force to a vehicle, eventually causing it to re-enter the atmosphere and burn up. In Chapter 10 we introduced the drag force as

$$F_{drag} = \frac{1}{2}\,\rho\,V^2\,C_D\,A \qquad (12\text{-}11)$$

where

F_{drag} = force of drag (N)

ρ = atmospheric density (kg/m^3)

V = velocity (m/s)

C_D = drag coefficient (unitless)

A = impacted area (m^2)

Because parts of a spacecraft may have different drag coefficients (solar panels, for example, act like big sails), drag forces on different parts of the spacecraft can also differ. This difference, along with the distance between the center of pressure (where the drag acts) and the center of mass, causes a *drag torque*. A spacecraft designer can do little to prevent this torque (short of moving the spacecraft to a higher orbit), so again the control system must be designed to deal with it.

Spacecraft Attitude Sensors

As we've seen, an essential element of closed-loop control systems is a device that can watch what's happening to the system and report this information to the controller. In other words, we need a sensor. Sensors are the control system's "eyes and ears." Sensors observe the system to determine attitude and transform these observations into signals that the controller processes.

All of us have a built-in attitude-sensor system. As we know from our discussion of the human vestibular system in Chapter 3, we use fluid flowing over tiny hairs in our inner ear, along with information from our eyes to detect changes in our attitude. For example, they sense if we're standing up or falling over. If we suddenly tilt our head to the side, these sensors detect this motion. If our body violently moves or shakes (like when we ride a roller coaster), our eyes and inner ear can get "out of synch," leading to motion sickness. Fortunately, spacecraft don't get sick from all their rotating, but they do need good attitude sensors. So let's take a look at a spacecraft's eyes and ears.

To understand how sensors help spacecraft determine their attitude, pretend you're flying the Space Shuttle in low-Earth orbit and need to point the nose at some spot on the surface. You're in the commander's seat facing toward the nose. To point the nose at the surface, you must first determine where you're currently pointed. How can you do this? The obvious answer is to look out the window at some reference. Let's say you look out and see the Sun out the left-hand window. Would this tell you all you need to know? Unfortunately, no. A single reference point tells you your current attitude in only two dimensions. In other words, you'd know that the left wing is pointed at the Sun and the nose is pointed perpendicular to the Sun. But the nose could point in various directions and still be perpendicular to the Sun, so what do you do?

To determine your attitude in three dimensions, you need another reference. If you could see some known star out the front window you'd know your orientation with respect to two reference points—the Sun and a star. Knowing the angle between the Sun and the Earth and between a known star and Earth, you could determine how to change your attitude to point the nose at Earth. Let's look at how we can apply this technique for attitude determination.

"Looking out the Window"

When pilots fly along in their airplanes, the easiest way for them to determine attitude is to look out the window (if the weather is good). If the ground is down and the sky is up, they're flying upright. Similarly, the simplest way for a spacecraft to determine its attitude is to just "look out the window." One important class of attitude sensors works the same way as the remote sensing payloads discussed in Chapter 11 (on some spacecraft, the payload can actually serve both functions). Recall, to look at a subject, a remote sensing system must perform the following functions

- Look at it—scan the sensor to point at the subject
- See it—collect EM radiation from it
- Convert it—transform EM radiation into a usable data
- Process it—turn data into usable information

When it comes to remote sensing for attitude determination, there are three primary subjects available for reference—Earth, the Sun, and the Stars. This gives us three classes of "out the window" sensors

- Earth sensors
- Sun sensors
- Star sensors

All of these sensors work in pretty much the same way as other remote sensing devices. Typically, they are attached to the spacecraft so the spacecraft must rotate to bring the subject into the sensor field of view or rely on "targets of opportunity" that will routinely go in and out of the field of view. Similar to a telescope or camera, EM radiation from the primary subject enters through a lens and focuses on solid state detectors, such as the charged-coupled devices described in Chapter 11. The accuracy of the sensor depends on how precisely it can discriminate the target, or portion of the target, and how much onboard processing it can accomplish.

In low-Earth orbit, Earth fills a big portion of the sky. *Earth sensors* can roughly indicate the "down" direction by simply discriminating where Earth is with respect to the rest of the sensor's field of view. At geostationary altitude, the angular radius of Earth is about 10°, so a sensor that can find Earth is at least accurate to within that amount. To use Earth as a more accurate method of attitude determination, a sensor must focus only on one small portion. Conveniently, sensors can detect Earth's horizon by focusing on a narrow band of EM radiation emitted by carbon dioxide, CO_2, in the atmosphere, as shown in Figure 12-18. These Earth-horizon sensors can be as much as ten times more accurate than a simple Earth detector.

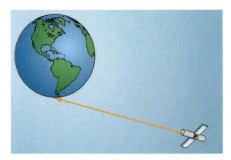

Figure 12-18.   Earth Sensors. As their name implies, Earth sensors use Earth as a target for determining spacecraft attitude. Sensors focus either on the gross direction of Earth or on narrower (and more accurate) parts of Earth, such as the horizon.

Sun sensors, the most widely used spacecraft attitude sensors, are similar in function to Earth sensors. As the name implies, a sun sensor finds the Sun and determines its direction with respect to the spacecraft body frame, as shown in Figure 12-19. By their nature, Earth and Sun sensors can give accurate information about attitude in only two-dimensions. For example, this means an Earth or Sun sensor can measure pitch and roll relative to the horizon, but not yaw; or pitch and yaw but not roll, etc.

A more accurate 2-axis reference is a *star sensor*. As Figure 12-20 shows, star sensors measure a spacecraft's attitude with respect to known star locations. Then they compare these measurements to accurate maps of the brightest stars stored in the spacecraft's memory. The angle between the known star's position and a reference axis on the spacecraft, θ, then helps determine the spacecraft's inertial attitude. By using two or more star sensors located around a spacecraft (or by taking multiple measurements with the same sensor), the system can determine its attitude in 3-dimensions.

As we mentioned, each of these sensors provides only a 2-D reference. To determine attitude in 3-dimensions, we often use two or more sensors together. For example, onboard computers can combine data from an Earth sensor with Sun-sensor data to get a 3-D, accurate fix. As we'll see, all of these sensors can also work in conjunction with a spacecraft's "ears"—gyroscopes and magnetometers.

Gyroscopes

Gyroscopes, like our inner ear, can determine attitude and changes in attitude, directly, without needing to look out the window. The simplest type of gyroscope is a spinning mass. As we know, any spinning mass has angular momentum that is conserved. By using this fundamental principle, we can use the gyroscope to detect spacecraft's angular motion. Two basic principles of gyros make them useful as attitude sensors

- With no torques, their angular momentum is conserved—they always point in the same direction in inertial space

- With torque applied, they precess in a predictable direction and amount

When a mass starts to spin, its angular momentum vector remains stationary in inertial space, unless acted on by an outside torque. For example, let's spin a gyroscope at 6:00 A.M. (see Figure 12-21) with its angular momentum vector pointed at some convenient inertial reference—say, a star just above the eastern horizon (somewhere to the right side of the page). We can then observe how conservation of angular momentum works to keep the gyro always pointed in the same inertial direction, as long as no torque affects it.

In this case, the angular momentum vector, $\vec{H}$, appears to "track" the star because the star is essentially fixed in inertial space. As the gyro sits in its stand, it looks like it's rotating throughout the day. Actually, the stand is moving as Earth rotates. The gyro remains stationary in inertial space.

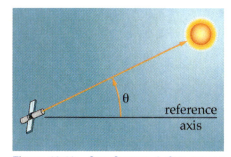

Figure 12-19. Sun Sensor. A Sun sensor determines spacecraft attitude by finding the direction of the Sun with respect to the body frame. Like Earth sensors, this sensor can only give a 2-dimensional fix on attitude without another point of reference.

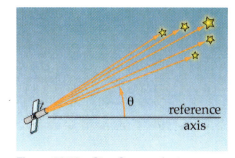

Figure 12-20. Star Sensor. A star sensor determines a spacecraft's attitude with respect to the known orientation of certain, bright stars.

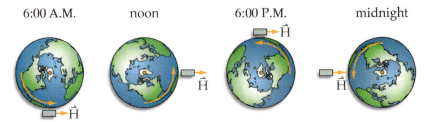

6:00 A.M.   noon   **6:00 P.M.**   midnight

Figure 12-21. Conservation of Angular Momentum. A spinning mass, such as a gyroscope, has angular momentum that is naturally conserved. If we spin a freely rotating gyro pointing east at 6 A.M., in this polar view of Earth, it appears to rotate as the day goes on. Actually, the Earth-bound observer rotates with Earth, but the gyro stays pointing in the same direction in inertial space.

Museums often demonstrate this principle with huge pendulums suspended on long cables. The swinging pendulum's plane remains fixed in inertial space, but as Earth turns, the pendulum path appears to move, knocking over dominos spaced around it to the delight of the crowds.

The second basic principle of gyroscopes relates to their strange motion in response to an applied torque. Earlier, we called this motion precession—rotation with constant angular velocity in a direction 90° from the direction of the applied torque.

Knowing these two basic principles, let's see how we can use a gyro to sense attitude. Because its angular momentum vector stays constant in inertial space, it provides a constant reference for inertial direction. One way to measure rotation with respect to the reference is to isolate the gyro from torques by mounting it on a *gimbal* (hinged brackets that allow it to rotate freely or that allow the mounting box to rotate freely around the stationary gyro). We then mount the gimbal on a platform in a spacecraft and measure the spacecraft's rotation by measuring how much the spacecraft rotates with respect to the stationary gyro.

Astro Fun Fact
Foucault Pendulum

For centuries, scientists have known that Earth rotates on its axis, but until 1851, no one had proved it. That year, a young French physicist names Leon Foucault (1819–1868) demonstrated that a simple pendulum appeared to change its swing axis over the course of a day. He surmised that since there were no rotational forces to cause the axis change, then Earth must rotate under the arcing pendulum. Inspiration for the idea came from an experiment with a long steel rod that he twanged while it was spinning in a lathe. He noted that while it continued to spin in the lathe, its vibration axis stayed in the same plane. He tried his first pendulum experiment in his home, using a 5-kg brass ball (bob) on a 2-m wire. It worked, so he lengthened the wire to 11 m and moved it to the Paris Observatory. Because of growing public interest, he was invited to show all of Paris in the Pantheon, where he hung a 28-kg bob on a 67-m wire. Important witnesses saw the pendulum move slowly clockwise with each swing, thus proving that Earth rotated counterclockwise.

Robin, William. Leon Foucault. Scientific American. July, 1998.

Contributed by Douglas Kirkpatrick, the U.S. Air Force Academy

Another way to measure a spacecraft's rotation is to strap a gyro directly to the spacecraft. Then, when the gyro (or the spacecraft) rotates around an axis perpendicular to the spin vector, the resulting torque will cause the gyro to precess. By measuring this precession angle and rate, the system can compute the amount and direction of the spacecraft's rotation and thus determine its new attitude.

Newer types of gyroscopes, called *ring-laser gyroscopes*, don't use these principles of a spinning mass. They use principles associated with laser light! A ring-laser gyro consists of a circular cavity containing a closed path, through which two laser beams shine in opposite directions (it's all done with mirrors). As the vehicle rotates, the path lengths traveled by the two beams change, causing a slight change in the frequency of both beams. By measuring this frequency shift, the system can compute the vehicle's rate of rotation. By integrating this rate over time, it can determine the amount of rotation and hence the vehicle's new orientation. Ring-laser technology offers similar or better accuracy, with greater reliability than the old style spinning-mass gyros.

Magnetometers

Another means of measuring attitude directly uses Earth's magnetic field. A *magnetometer* is basically a fancy compass that measures the direction of the magnetic field and its strength. Earlier, when we discussed the disturbance torque caused by the magnetic field, we indicated its strength varies with the cube of radius (R^3) and by a factor of two between the pole and equator. By comparing the measured direction and strength of the local field with a high fidelity model of Earth's field, the sensor can determine the orientation of the spacecraft with respect to Earth. An engineering drawing of a magnetometer is shown in Figure 12-22.

To see how this works, think about a compass needle. It's usually just a lightweight magnet that can rotate freely. If you've ever played with magnets, you know that one side of a magnet will readily attract and stick to another magnet, while the opposite side will repel it. With magnets, opposites attract and likes repel, so the north pole of a magnet attracts the south pole of another magnet. The lightweight magnet rotating freely in a compass tries to do the same thing. The north end of the compass tends to point at Earth's North Pole, and suddenly, you're no longer lost!

Use of magnetometers is limited by the strength of the field, making them more useful in low-Earth orbit than at geostationary altitude. The sensor accuracy depends on the accuracy of the field model. Even so, they offer a relatively inexpensive sensor that can deliver an independent reference from the other sensors we've discussed.

Global Positioning System (GPS)

The newest attitude sensor to emerge on the scene is the "differential" *Global Positioning System (GPS)*. GPS is a constellation of 24 satellites in high Earth orbit (12-hour period) designed and deployed by the U.S. Air Force to provide world-wide position, velocity, and time information.

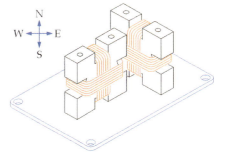

Figure 12-22.   Magnetometers. A magneto-meter functions as a highly accurate compass that measures the direction and strength of the local magnetic field. By comparing this mea-surement to a model of Earth's field, it can determine an accurate estimate of the current attitude. *(Courtesy of Maarten Meerman, Surrey Satellite Technology, Ltd., U.K.)*

Clever engineers figured that by placing two GPS receivers some distance apart on a vehicle, and carefully measuring the difference between the two signals, they could determine a vehicle's attitude. This attitude determination technique has the potential to offer a relatively inexpensive, independent system for low-Earth orbiting spacecraft.

Spacecraft Attitude Actuators

After we've determined what our spacecraft's attitude is, we need to know how to change it. For example, we may need to compensate for disturbance torques or rotate the spacecraft to point at a new subject. As we've seen, applying a torque changes a vehicle's attitude. That's why we need actuators. Actuators provide "torque on demand" to rotate a spacecraft as needed to take pictures, downlink data, or meet other mission requirements.

Many types of attitude actuators are available to spacecraft designers. Just as several different types of sensors often work together to accurately measure attitude, typically two or more types of actuators combine to apply torque to achieve a desired attitude. For simplicity, we'll discuss each type of actuator separately.

We can conceptually divide actuator types into two general classes, passive and active. *Passive actuators* operate more or less open loop. In other words, after the spacecraft is in the desired attitude, passive actuators will keep it there with little or no additional torques needed. *Active actuators*, on the other hand, require continuous feedback and adjustment. As you would expect, passive actuators typically can't reach the same level of accuracy as active ones; however, in many cases, they're good enough. We'll look at three types of passive actuators

- Gravity-gradient stabilization
- Spin stabilization
- Dampers

and three types of active actuators

- Thrusters
- Magnetic torquers
- Momentum-control devices

Gravity-gradient Stabilization

The first type of passive actuator takes advantage of the gravity-gradient disturbance torque discussed earlier. We can exploit this "free" torque to keep a spacecraft oriented in a local vertical, or "downward," orientation. Fortunately, a spacecraft doesn't have to be shaped like a dumbbell to take advantage of this effect. For example, why do we see only one face of the Moon and never the mysterious "dark side?" Because of uneven distribution of mass within the Moon's crust, it's in a gravity-gradient-stabilized attitude with respect to Earth. However, to maximize

Figure 12-23. Gravity-gradient Stabilization. Some spacecraft take advantage of the gravity-gradient torque to keep them oriented in a local vertical, or "downward," attitude. Usually, they maximize this effect by deploying a small mass at the end of a very long boom. This artist's conception of the PicoSAT spacecraft shows it to scale with a 6 m long deployable boom with a small mass on the end. *(Courtesy of Surrey Satellite Technology, Ltd., U.K.)*

the effect of this cheap and reliable attitude actuator, spacecraft will usually deploy weighted booms to create a more dumbbell-like shape. Figure 12-23 shows an artist's conception of the PicoSAT spacecraft using a 6 m deployable boom.

Gravity-gradient attitude control offers a simple, reliable, inexhaustible (as long as there's gravity) system with no moving parts. However, it has a few drawbacks

- Two axes control only—pitch and roll but not yaw
- Limited accuracy—depending on the spacecraft moments of inertia, downward pointing accuracy is only about ±10°
- Only effective in low-Earth orbit—because gravity varies with the square of the distance, it's not very effective beyond LEO

Despite these disadvantages, gravity-gradient-controlled spacecraft have been used effectively on a variety of missions.

Spin Stabilization

Earlier we saw that a spinning mass has unique gyroscopic properties. A *spin-stabilized* spacecraft takes advantage of the conservation of angular momentum to maintain a constant inertial orientation of one of its axes. Because the angular momentum vector, $\vec{H}$, of a spinning mass is fixed in inertial space, the spacecraft tends to stay in the same inertial attitude, as shown in Figure 12-24.

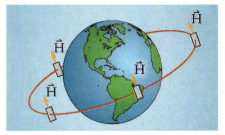

Figure 12-24.   Spin Stabilization. A spinning spacecraft keeps its angular momentum vector fixed in inertial space.

Perhaps the best example of a spin-stabilized satellite is Earth. The spinning Earth is essentially a giant gyroscope. Earth's $\vec{H}$ vector points out of the North Pole. This $\vec{H}$ stays fixed in inertial space (except for a minor wobble), always pointed at the same place in the sky. When we observe the motion of the stars at night, we see they all appear to rotate around one star—the North Star. This occurs because Earth's $\vec{H}$ vector points at the North Star!

Spin stabilization is useful, as long as we want our spacecraft to stay pointed in the same inertial direction. However, usually we're more interested in non-inertial pointing. For example, spin stabilization isn't very useful for pointing $\vec{H}$ at Earth, as illustrated in Figure 12-25. For this reason, we mostly use it only during spacecraft deployment, when the natural gyroscopic stiffness we discussed earlier is useful to maintain a known orientation until the spacecraft is free from the launch vehicle. This spin is usually maintained through the first major maneuver, providing a stiff, stable platform during a rocket firing. During high-thrust, orbit-insertion rocket firings, spin stabilization is often the only technique that can efficiently keep the spacecraft stable.

Figure 12-25.   Spin Stabilization Isn't Much Good for Earth Pointing. Because spin stabilized spacecraft have fixed pointing with respect to inertial space, they aren't a good choice for Earth-pointing missions. During part of the orbit, they may point toward Earth but during other parts of the orbit, they'll point away.

One way to avoid Earth-pointing limitations of spin stabilization is to use a dual-spin system. *Dual-spin systems* also take advantage of the constant angular momentum vector of a spinning mass. These systems consist of an inner cylinder called the "de-spun" section, surrounded by an outer cylinder that is spinning at a high rate. The outer cylinder provides overall spacecraft stability. The word "de-spun" is actually a misnomer.

In fact, the "de-spun" section does spin, but at a much slower rate than the outer section. To allow for antenna and sensor pointing, the "de-spun" section spins at a rate to keep them pointed at Earth. For example, if a spacecraft is in geostationary orbit, the de-spun section rotates at "orbit rate" or once every 24 hours, keeping antennas or other sensors focused on Earth, as shown in Figure 12-26.

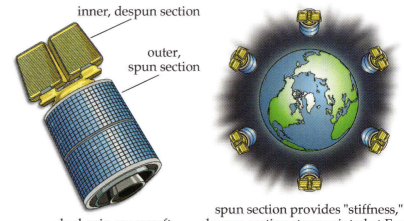

inner, despun section

outer, spun section

dual-spin spacecraft

spun section provides "stiffness," despun section stays pointed at Earth

Figure 12-26.   Dual-spin Spacecraft. A dual-spin spacecraft uses the inherent stiffness of a spinning outer section with a "de-spun" inner section that can independently point at Earth. The de-spun section turns at the orbital rate to keep sensors and antenna pointed at Earth.

Of course, the need for independently spinning sections makes dual-spin spacecraft much more complex. Electrical and other connections must run from the spun to the "de-spun" sections. Highly reliable bearings must allow the two sections to spin at different rates with little friction. Even with these inherent technical challenges, dual-spin has been a popular control option for large, geosynchronous, communication spacecraft, such as the one shown in Figure 12-27.

Dampers

As mentioned earlier, we seldom use a single type of attitude actuator alone. A damper is another actuator usually used in combination with others for a complete system. Generally speaking, a *damper* is a device that changes angular momentum by absorbing energy. We know momentum is constant only as long as energy stays constant. If we add or take away energy, momentum changes. As a spacecraft attitude actuator, dampers absorb unwanted momentum. Where does it go? When we hit the brakes in a car, the linear momentum "goes" into heat produced by friction between the brake pads and the disks or drums. Similarly, attitude dampers use friction or other means to convert angular momentum energy into other forms.

One simple type of momentum damper consists of a small ball in a circular tube filled with highly viscous fluid, as illustrated in Figure 12-28. As a spacecraft rotates, some of its momentum is contained in the ball that moves inside the tube. Friction between the ball and the fluid in the

Figure 12-27.   Dual-spin Communication Spacecraft. Large geosynchronous communication spacecraft, such as the Satellite Business Systems spacecraft, shown here, make good use of dual-spin attitude control. *(Courtesy of Hughes Space and Communications Company)*

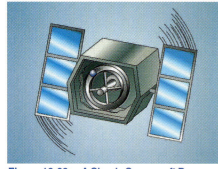

Figure 12-28.   A Simple Spacecraft Damper. Dampers "absorb" unwanted angular momentum by converting the energy into friction, in much the same way as the brakes in a car turn linear momentum into heat through friction. A ball inside a circular tube filled with a viscous fluid is one type of damper. As the spacecraft rotates, the ball moves through the fluid. The resistance produces heat, dissipating the angular motion.

tube converts some of the momentum into heat that slowly dissipates throughout the spacecraft. Over time, the spacecraft can use this simple technique to absorb mechanical energy, slowing it's rotation. Dampers are usually designed and oriented to reduce rotation about a specific axis. In this way, designers often use them in spinning spacecraft to remove unwanted "wobbles" in the spin axis.

Thrusters

All of the actuators we've discussed so far are passive, in that, once put in motion, they can more or less function in an open-loop mode, with little or no additional inputs. Now we'll turn our attention to active actuators. Thrusters are perhaps the simplest type of active actuator to visualize. *Thrusters* are simply rockets that rely on "brute force" to rotate a spacecraft. By applying a balanced force with a pair of rockets on opposite sides of a spacecraft, we can produce a torque, as shown in Figure 12-29. By varying which thruster pair we use and how much force we apply, we can rotate a spacecraft in any direction.

Placing thrusters as far from the satellite's center of mass as possible gives them a larger moment arm and allows them to exert a greater torque for a given force. This is evident by looking at Equation (12-4), where we see that the bigger R is, the more torque is delivered from the same force. However, as we learned earlier, because of precession, when a spacecraft is already spinning, any applied torque in a direction other than the spin axis causes the spacecraft to rotate at constant velocity about an axis perpendicular to the torque direction.

The biggest advantage of using thrusters is that they can produce a well-defined "torque on demand," allowing the spacecraft to slew quickly from one attitude to another. Unfortunately, the amount of propellant a spacecraft can carry limits their use. For short missions, such as those flown by the Space Shuttle, this limit is no problem. For longer missions (months or years), designers use thrusters only as a backup and for other purposes we'll discuss later. We'll explore basic principles of rocket science and propulsion system technologies in greater detail in Chapter 14.

Magnetic Torquers

A *magnetic torquer* is another type of actuator that takes advantage of a naturally occurring torque in the space environment. Earlier we looked at the magnetic disturbance torque caused by the interaction of the spacecraft's magnetic field due to surface charging with Earth's magnetic field. We can use this effect in an active mode by creating powerful onboard magnets and switching them on and off as needed to rotate "against" Earth's magnetic field. Magnetic torquers are simply electromagnets produced by running an electrical current through a loop of wire onboard. Like a compass needle, this electromagnet tries to align with Earth's magnetic field, dragging the rest of the spacecraft with it, as seen in Figure 12-30.

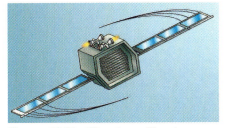

Figure 12-29. Thrusters. Thrusters are rockets that apply a force some distance away from the center of mass, causing a torque that rotates the spacecraft.

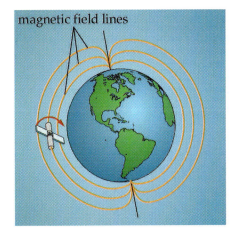

Figure 12-30. Magnetic Torquers. A magnetic torquer is an active spacecraft attitude actuator that takes advantage of the natural torque caused by Earth's magnetic field interacting with a magnet; it's the same effect that rotates a compass needle. Onboard, the system switches electromagnets on and off as needed, pushing "against" the magnetic field, producing the necessary torque.

Magnetic torquers offer a relatively cheap and simple way to control a spacecraft's attitude. Furthermore, because they need only electrical power to run, they're inexhaustible—unlike thrusters. Unfortunately, their effectiveness depends directly on the strength of Earth's magnetic field, so they become less useful in higher orbits. Also, because the field strength varies by a factor of two between the equator and the poles, they are most useful in highly inclined orbits. Even so, they are an important secondary means of attitude control used on many LEO spacecraft.

Momentum-control Devices

The most common actuator for spacecraft attitude control is a family of systems that all rely on angular momentum. These *momentum-control devices* actively vary the angular momentum of small, rotating masses within a spacecraft to change its attitude. How can this work? If you stand on a turntable, holding a spinning bicycle wheel at arm's length, you can cause yourself to rotate by tilting the bicycle wheel to the left or right. This works because total angular momentum of a system is always conserved. As the bicycle wheel rotates one way, you rotate another to compensate, keeping the total angular momentum constant, as you can see in Figure 12-31.

Figure 12-31.   Bicycle Wheels in Space? You can do a simple experiment to see one way spacecraft control their attitude. By standing on a turntable holding a spinning bicycle wheel, as shown in the left-hand photograph, you can change direction (your attitude) by applying a small torque to the wheel by slightly tilting the wheel to one side, as shown in the right-hand photograph.

Let's look at where this attitude change comes from. From Equation (12-2), we know angular momentum is the product of an object's mass moment of inertia, I, and its angular velocity, $\vec{\Omega}$.

$$\vec{H} = I\vec{\Omega} \qquad (12\text{-}2)$$

Note that a large mass (high I) spinning at a relatively slow speed (low $\vec{\Omega}$) can have exactly the same angular momentum as a small mass (low I) spinning at a much higher rate (high $\vec{\Omega}$). If we consider a spacecraft and all mass inside it to be one system, we can control where the spacecraft points by changing the angular momentum (rate and direction of spin) of a small spinning mass inside. There are three approaches to momentum-control devices currently in wide use

- Biased momentum systems—"Momentum wheels" that typically rely on a single wheel with a large, fixed momentum to provide overall stiffness. The wheel's speed gradually increases to absorb disturbance torques.

- Zero-bias systems—"Reaction wheels" that rely on three or more wheels, normally with little or no initial momentum. Each wheel spins independently to rotate the spacecraft and absorb disturbance torques.

- Control-moment gyroscopes— rely on three or more wheels, each with a large, fixed momentum. The wheels are mounted on gimbals, rotating the wheels about their gimbals changes the spacecraft orientation.

Biased momentum systems are the simplest type of momentum control device. In operation, these systems use one or two constantly spinning *momentum wheels*, each with a large, fixed momentum. (They are "biased" toward having a particular, set momentum, hence the name). Because they are always spinning, they give the spacecraft a large angular momentum vector, fixed in inertial space. This is exactly the same concept used by spin-stabilized spacecraft, discussed earlier. Only, in this case, instead of spinning the whole spacecraft, we only spin a small wheel inside the spacecraft to achieve the same effect, as illustrated in Figure 12-32.

In contrast, *reaction wheels* are a type of *zero-bias system*, because their normal momentum is at or near zero (no bias). Typically, an attitude control system uses at least three separate reaction wheels, oriented at right angles to each other, as seen in Figure 12-33. Often, a fourth wheel is skewed to the other three for redundancy. When the spacecraft needs to rotate to a new attitude, or to absorb a disturbance torque, the system spins one or more of these wheels. To see how this works, let's step through what happens to the relationship between a reaction wheel and the overall spacecraft momentum.

First of all, realize that without any external torque, the total angular momentum of the spacecraft (including the reaction wheels) is conserved (and usually maintained at or near zero). Thus, the angular momentum of the spacecraft plus the angular momentum of the reaction wheels must add up to a constant vector quantity. Now, imagine one of the reaction wheels begins to spin using a motor. As the wheel's spin rate increases, its angular momentum also increases. But, the total angular momentum of the wheel and spacecraft must always sum to a constant value. So what happens to the spacecraft?

Figure 12-32. Biased Momentum Systems. We use momentum wheels in biased momentum systems. They typically rely on a single wheel with a large, fixed ("biased") momentum to provide overall stiffness. The wheel speed gradually increases to absorb disturbance torques. *(Courtesy of Ball Aerospace & Technologies Corporation)*

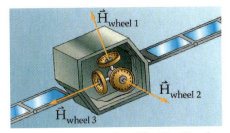

Figure 12-33. Reaction Wheels. Reaction wheels are part of a zero-bias system that uses three independent wheels, one along each axis, normally with zero or near zero momentum. To rotate the spacecraft or absorb disturbance torques, one or more wheels begin to spin. Often, designers add a fourth wheel, skewed with respect to the other three, for redundancy.

Let's look at a more specific example to get a better idea. We can express the total angular momentum of the spacecraft (including reaction wheels) as

$$\vec{H}_{TOT} = \vec{H}_{S/C} + \vec{H}_{RW} \qquad (12\text{-}12)$$

where

$\vec{H}_{TOT}$ = total angular momentum of the spacecraft (kg m^2/s)

$\vec{H}_{S/C}$ = angular momentum of just the spacecraft (kg m^2/s)

$\vec{H}_{RW}$ = angular momentum of the reaction wheels (kg m^2/s)

[*Note:* This relationship is vector addition, so $\left|\vec{H}_{TOT}\right| \neq \left|\vec{H}_{S/C}\right| + \left|\vec{H}_{RW}\right|$!]

If a reaction wheel spins faster, its angular momentum increases by an amount $\Delta\vec{H}_{RW}$. Because the total angular momentum must stay constant, the spacecraft's angular momentum *must* automatically decrease to compensate by an amount $\Delta\vec{H}_{S/C}$. The vector increase in the reaction wheel's momentum must exactly equal the decrease in the spacecraft's momentum, or $\Delta\vec{H}_{RW} = -\Delta\vec{H}_{S/C}$, to keep a constant total. Figure 12-34 shows these relationships. To conserve momentum, the spacecraft must either slow its rotation or start rotating in the opposite direction. In either case, the spacecraft's attitude has changed simply by spinning a small mass faster inside.

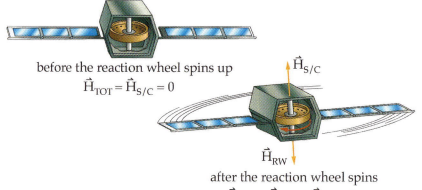

before the reaction wheel spins up

$$\vec{H}_{TOT} = \vec{H}_{S/C} = 0$$

$\vec{H}_{S/C}$

$\vec{H}_{RW}$

after the reaction wheel spins

$$\vec{H}_{TOT} = \vec{H}_{S/C} + \vec{H}_{RW} = 0$$

Figure 12-34.   Reaction Wheels in Operation. The total angular momentum of a spacecraft system is the sum of the spacecraft's momentum plus the momentum of each reaction wheel. In this example, we start with a non-rotating spacecraft that has zero total angular momentum. To rotate the spacecraft in one direction, a reaction wheel is spun up in the opposite direction, such that the total angular momentum of the system stays constant.

Figure   12-35.   Reaction   Wheels   for Accurate   Pointing. The Hubble Space Telescope observes many interstellar objects at such long distances that it must point very accurately. To be this accurate, it relies on very accurate reaction wheels. *(Courtesy of NASA/ Johnson Space Center)*

Three reaction wheels can deliver precise control of a spacecraft's attitude in all three axes. Unfortunately, as with any mechanism with moving parts, they can be complex, expensive, and have a limited operational lifetime. Despite these limitations, they remain the primary choice for attitude control on large, modern spacecraft requiring very accurate pointing, such as the Hubble Space Telescope shown in Figure 12-35.

The final type of momentum-control device is the *control-moment gyroscope (CMG)*. A CMG consists of one or more spinning wheels, each mounted on gimbals that allow them to rotate freely in all directions. Recall that reaction wheels change momentum by changing magnitude only (spinning faster or slower). CMGs change momentum by changing their magnitude *and* direction (physically rotating the spinning wheel). Again, because the total angular momentum of the system must be conserved, as the momentum of a CMG changes in one way, the spacecraft will rotate in the other to compensate. CMGs provide pointing accuracy equivalent to reaction wheels but offer much higher slew rates and are especially effective on very large platforms, such as the Skylab Space Station shown in Figure 12-36.

Figure 12-36.   A Control Moment Gyroscope (CMG) in Space. CMGs can vary the magnitude and direction of their angular momentum, allowing for much higher slew rates and making possible efficient attitude control on very large platforms, such as the Skylab Space Station shown here. *(Courtesy of NASA/Johnson Space Center)*

One important limitation of all momentum control devices is that there is a practical limit to how fast a given wheel can spin. In operation, all of these systems must gradually spin faster and faster to rotate the spacecraft and absorb disturbance torques. Eventually, a wheel will be spinning as fast as it can, without damaging bearings or other mechanisms. At this point, the wheel is "saturated," meaning it has reached its design limit for rotational speed. When this happens, the wheels must "de-saturate" through a process known as "momentum dumping." *Momentum dumping* is a technique for decreasing the angular momentum of a wheel by applying a controlled torque to the spacecraft. The wheel can absorb this torque in a way that allows it to reduce its rate of spin. Of course, this means the spacecraft needs some independent means of applying an external torque. For this reason, on all spacecraft using momentum control devices, designers use either magnetic torquers or thrusters (or both) to allow for momentum dumping.

The Controller

So far, we've looked at the dynamics of rotating systems to understand how torque affects a spacecraft's angular momentum, including the environmental sources for disturbance torques. We then looked at the various types of sensors used to measure attitude. Finally, we discussed the different types of actuators, passive and active, used to generate torques that allow us to freely change a spacecraft's attitude. Now we can put the entire attitude determination and control subsystem together by looking at the "brains" of the operation—the controller.

The controller's job is to generate commands for the actuators to make the spacecraft point in the right direction based on mission requirements for accuracy and slew rate. To use the information from sensors and continuously adjust actuator commands, the controller must be smart. It has to know what's happening and decide what to do next. To do this right, the controller has to keep track of

- What's happening now
- What may happen in the future
- What happened in the past

Knowing what's happening now is pretty easy—the controller simply asks the sensors to find the current attitude. It then compares this to the desired attitude. The difference between the measured and desired attitude is the *error signal*. Based on this error signal, the controller steers in the direction of the proper orientation. That is, if the attitude is 10° off, the controller commands a 10° change. This is known as *proportional control* and is used in some form in virtually all closed-loop control systems.

However, predicting what's going to happen and remembering what's happened in the past can be just as important. For example, if you need to stop at a stop sign, you need to know not only where you are, but also how fast you're going, so you can hit the brakes in time. Similarly, to hit the desired attitude, the spacecraft controller must monitor the attitude rate, as well as the current attitude. For you calculus buffs, you may recognize this rate of change calculation as a derivative. In this case, by knowing the rate of change or "speed" of attitude, the controller can more accurately determine how to command the actuators to achieve better accuracy. This process is called *derivative control*.

Sometimes we can be more precise by keeping track of how close we're getting to the desired result. One way to do this is for the controller to monitor the angular difference between the measured and desired attitude, $\Delta\theta$. When the spacecraft reaches the desired attitude, this difference, $\Delta\theta$, will be zero. If the system stops commanding the actuators at this point, the attitude will immediately begin to drift due to disturbance torques. A really smart controller, however, won't just look at the instantaneous $\Delta\theta$. Instead, it would keep a running tally, summing the $\Delta\theta$ over time. The result would always be some value other than zero and would tell the controller how much torque to add in a "steady-state" mode to compensate for the disturbance torques. In calculus, this process is called integration, so we call this type of control *integral control*. Designers use it for highly accurate pointing.

Regardless of the exact scheme used, the controller combines its memory with its current measurements and an ability to predict future behavior to decide how to command the actuators. We can now complete a block diagram of a spacecraft's attitude-control system in Figure 12-37.

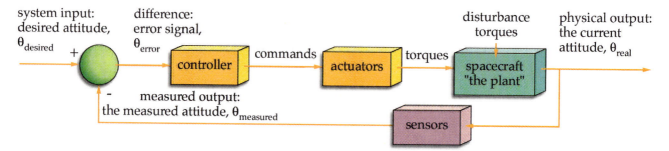

Figure 12-37. **Attitude Determination and Control Subsystem (ADCS).** A complete ADCS (the attitude part of an AOCS) includes a controller, actuators, the spacecraft ("the plant"), and sensors that work together to maintain or change spacecraft attitude in response to changing mission requirements.

Section Review

Key Terms

active actuators
attitude accuracy, ψ
attitude determination and
    control subsystem (ADCS)
biased momentum systems
body frame
control-moment gyroscope
    (CMG)
damper
derivative control
dipole
disturbance torques
drag torque
dual-spin systems
Earth sensors
error signal
gimbal
Global Positioning System
    (GPS)
gravity gradient torque
gyroscopic stiffness
gyroscopes
integral control
magnetic torque
magnetic torquer
magnetometer
mass moment of inertia, I
momentum-control devices
momentum dumping
momentum wheels
passive actuators
photons
pitch
pointing accuracy
precession
proportional control
reaction wheels
ring-laser gyroscopes
roll
slew rate
solar-radiation pressure
spin stabilized

Key Concepts

➤ To understand a spacecraft's behavior or how it reacts to inputs, we must understand the model of system dynamics based on linear and rotational laws of motion. To rotate a spacecraft we must recognize that

- Angular momentum is always conserved
- A torque describes the direction of a force couple applied to a system
- A torque applied to a non-spinning object (or applied parallel to the direction of spin for a spinning object) causes angular acceleration, which leads to angular velocity and hence, change in angular orientation
- A torque applied in a direction other than the direction of spin for a spinning object will cause precession. This means it will begin to rotate at constant angular velocity about an axis perpendicular to the torque direction.

➤ A Spacecraft experiences many environmental disturbance torques that, over time, work to change its attitude. These include

- Gravity gradient
- Magnetic
- Solar-radiation pressure
- Atmospheric drag

➤ Sensors determine a spacecraft's attitude

- Sun sensors, horizon sensors, and star sensors are the "eyes" of the spacecraft, determining attitude by "looking out the window." They function in much the same way as remote sensing payloads.
- Gyroscopes are the "inner ears" of spacecraft. They can directly sense changes in attitude because a spinning mass has two important properties
  - The angular momentum of a spinning mass is constant
  - Torque applied to a spinning mass causes precession
- Ring-laser gyros measure the changing frequency of light to detect attitude changes
- Magnetometers measure the direction and magnitude of Earth's magnetic field to determine attitude
- Differential Global Positioning System (GPS) measures the difference between signals received at two or more locations on a spacecraft to determine attitude

Continued on next page

▰ Section Review (Continued)

Key Terms (Continued)

star sensors
Sun sensors
thrusters
torque
yaw
zero-bias system

Key Equations

$$\vec{H} = I\vec{\Omega}$$

$$\vec{T} = \vec{R} \times \vec{F}$$

$$\vec{T} = \dot{\vec{H}} = I\vec{\alpha}$$

$$\vec{H}_{TOT} = \vec{H}_{S/C} + \vec{H}_{RW}$$

Key Concepts (Continued)

➤ Applying torques to the spacecraft requires spacecraft actuators. Actuators are either passive or active.

➤ Passive attitude actuators include

- Gravity-gradient stabilization

- Spin stabilization

- Dampers

➤ Active attitude actuators include

- Thrusters

- Magnetic torquers

- Momentum control devices

  – Zero-bias systems—momentum wheels

  – Bias momentum systems—reaction wheels

  – Control moment gyros

➤ The controller decides what commands to send to active actuators based on current and historical data from sensors and an understanding of spacecraft rotational properties

Example 12-1

Problem Statement

Based on your experience in designing the FireSat payload (as described in Example Problems 11-1, 11-2, and 11-3), you've now been put in charge of the attitude control system. For the FireSat constellation to provide continuous coverage, the center of the sensor's FOV must not deviate from nadir by more than ±100 km. Determine the corresponding attitude accuracy required. Given this accuracy, complete a conceptual design of the FireSat attitude control subsystem using the simplest available techniques. Draw a simple block diagram for your resulting subsystem.

Problem Summary

Given: D = 100 km
        h = 500 km (from Example 11-1)

Find: ψ, conceptual design of FireSat
        control system

Conceptual Solution

1) Determine ψ, for the given D and h, using Equation (12-1)

$$D = h\psi$$

$$\psi = \frac{D}{h}$$

2) Complete a conceptual design of the FireSat attitude control system using the simplest techniques.

3) Draw a simple block diagram for the subsystem.

Analytical Solution

1) Determine ψ, given D and h

$$\psi = \frac{D}{h} = \frac{100 \text{ km}}{500 \text{ km}}$$

$$\psi = 0.2 \text{ rad}$$
$$\psi = 11.459°$$

2) Complete a conceptual design of the FireSat attitude control system, using the simplest techniques.

There are almost an infinite number of possible control schemes we could use to achieve modest pointing accuracy. Given that the spacecraft will operate in LEO and must be primarily nadir pointing, the simplest technique would be gravity-gradient stabilization. The addition of a short boom (2 m–3 m) with a small mass at the tip (~1 kg) should provide sufficient gravity-gradient torque to keep the spacecraft nadir pointing. To ensure the payload is pointing "right side down," a small magnetorquer could also be added. Three-axis attitude determination can be done using a simple sun sensor (2 axis) plus a magnetometer (3-axis).

3) Draw a simple block diagram for the subsystem.

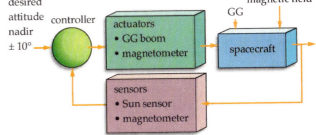

Interpreting the Results

Given the computed attitude accuracy and the conceptual design of the FireSat attitude control system, we can present the system block diagram, as shown above. As team players in the overall spacecraft design, we need to consider the corresponding requirements this particular subsystem will have on other spacecraft subsystems. For example, the use of a gravity-gradient boom will mean the structures and mechanisms must be able to accommodate its storage during launch and successful deployment on orbit. The use of magnetorquer will place additional demands on the electrical power subsystem.

12.3 Orbit Control

▀▀▀ In This Section You'll Learn to...

- ☞ Describe key elements and technologies used in space-vehicle orbit control and explain them using block diagrams
- ☞ Apply orbit control concepts to the problem of delivering a launch vehicle to orbit

In the last section, we looked at the problem of how to point a space vehicle in the right direction—the "A" of the vehicle's attitude and orbit control subsystem (AOCS). In this section, we look at the separate, but related, problem of getting a vehicle in the right position with the right velocity—the "O" part of the AOCS. This function is so important that it is often given the separate name *navigation, guidance, and control (NGC) subsystem*. In this section we'll refer to it by that name. Like any closed-loop control system, it uses the same four steps we defined in Section 12.1.

- *Understand* the system's behavior—how the plant reacts to inputs, including environmental inputs, to produce outputs. This is also known as the *plant model*.
- *Observe* the system's current behavior—using sensors
- *Decide* what to do—the job of the controller
- *Do* it—using actuators

Let's look at how the separate functions of a NGC subsystem relate to these steps. Observing and measuring a vehicle's current position and velocity using various sensors and computational techniques is the problem of navigation. Understanding how the vehicle behaves and deciding what steering commands it needs is called *guidance*. The guidance system is the controller. Finally, implementing the guidance commands using various actuators is the problem of *control*.

As we did with our discussion of attitude control in the last section, let's start by taking a closer look at the system dynamics that governs vehicle position and velocity. Then we'll look at how we keep track of position and velocity. After a brief look at actuators, we'll see how the whole subsystem fits together.

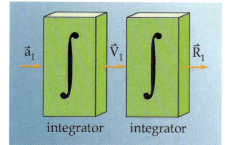

Figure 12-38. Space Vehicle Dynamics. We describe all dynamics for a vehicle's velocity and position using Newton's Laws. From them we know that the longer an object accelerates, the faster it goes. If it goes at a velocity for a certain time, it will reach a new position. From a calculus standpoint, acceleration over time (integration) produces a velocity change, and velocity over time (integration) produces a position change.

Space Vehicle Dynamics

The basic principles that explain how to move a spacecraft, once again, stem from Newton's Laws of Motion. If we apply force to a mass, it accelerates. The longer it accelerates, the faster it goes. If it goes at some velocity for a certain time, it'll reach a new position. Figure 12-38 illustrates this basic concept from a calculus standpoint. Accelerating a vehicle over time (integration) produces a velocity change. Holding a velocity over time (integration) produces a position change. This is the basis for modeling space vehicle dynamics. Designers know that to reach a certain velocity in a certain direction, a vehicle must produce a force in the opposite direction.

Navigation—The Sensor

Navigation is the problem of knowing where a vehicle is and where it's going. Within the NGC subsystem is a group of sensors working in concert to figure this out. The two primary navigation concerns are knowing a vehicle's velocity vector, $\vec{V}$, and its position vector, $\vec{R}$, with respect to an inertial reference. Using a Global Positioning System (GPS) receiver, we can learn these two things directly. However, without a GPS receiver, we have to resort to other, internal means. To determine velocity and position, we only have to measure acceleration along each of our three axes. Starting from a known position and velocity, we apply the basic principles from Newton's Laws and work forward in time (integrate) to determine our vehicle's new velocity and position. All we need is some way to measure inertial acceleration. Let's see how we can build something to do that.

When we stomp on the accelerator in our car, it responds by applying a contact force to everything attached to it (including us), moving the car forward and throwing us back in our seat. We feel this contact force on our body and in our inner ear. Onboard a spacecraft, we use this same effect to detect and measure acceleration. A device that can sense acceleration due to contact forces is called an *accelerometer*. An accelerometer can be as simple as a free-floating mass suspended between two springs. If we suddenly accelerate a box containing the mass and springs by applying a contact force, the springs on one side will compress as the inertia of the mass resists the acceleration. We illustrate this principle in Figure 12-39.

Notice we said that an accelerometer measures only contact forces. But gravity also accelerates the vehicle. How do we measure it? We can't. Because gravity acts on all bodies, and we can't "shield" a mass from its effect, there is no known way to build a device to sense gravity directly! Fortunately, armed with Newton's good ol' Law of Universal Gravitation, we can compute it. Recall from Chapter 4 that we find acceleration due to gravity from

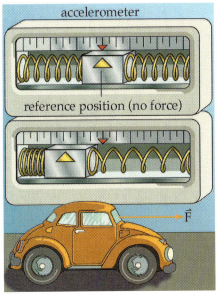

Figure 12-39. Accelerometers. Accelerometers measure acceleration. You experience this basic principle, which makes an accelerometer work, every time you push on the accelerator in your car. You feel yourself pushed back in your seat. Accelerometers use masses that are similarly displaced when subjected to contact forces. By measuring this displacement, you can determine the magnitude of the applied force and, hence, the acceleration on the vehicle.

$$\boxed{\vec{a}_g = \ddot{\vec{R}}_g = -\frac{\mu}{R^2}\hat{R}} \qquad (12\text{-}13)$$

where

$\vec{a}_g = \ddot{\vec{R}}_g$ = acceleration due to gravity (m/s²)

μ = gravitational parameter of the central body (m³/s²)

R = magnitude of the position vector (m)

$\hat{R}$ = unit vector in the direction of $\vec{R}$

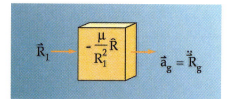

Figure 12-40.   A Simple Gravity Computer. We can determine gravitational acceleration, $\vec{a}_g = \ddot{\vec{R}}_g$, by knowing our vehicle's position, $\vec{R}_I$, and using Newton's Law of Universal Gravitation as a "gravity computer."

So if we have a "gravity computer," as shown in Figure 12-40, we can compute $\ddot{\vec{a}}_g$, as long as we know the inertial position ($\vec{R}_I$). But wait a minute—that's what we were looking for in the first place! Are we going in circles? Actually, we are, but we'll see how it all works in just a bit.

We now have the gravitational forces computed from Newton's Law (the "gravs") and the contact, or non-gravitational forces ("grav-nots") measured directly by the accelerometer. Because these constitute all possible accelerations on the vehicle, the total inertial acceleration is the vector sum of the two.

$$\vec{a}_I = \vec{a}_g + \vec{a}_N \qquad (12\text{-}14)$$

where

$\vec{a}_I$ = inertial acceleration (m/s^2)

$\vec{a}_g$ = acceleration due to gravity (m/s^2)

$\vec{a}_N$ = acceleration due to non-gravitational forces (m/s^2) (e.g., lift, drag, thrust, etc.)

Astro Fun Fact
Gravity Probe B "As Accurate as It Gets"

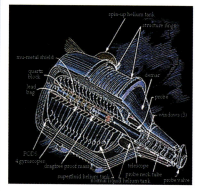

Due to launch in 2001, Gravity Probe B is a relativity experiment using a small satellite in a 644-km (400-mi.), polar orbit. NASA and Stanford University are developing the experiment to test two extraordinary, unverified predictions of Einstein's general theory of relativity. The experiment will very precisely measure tiny changes in the spin direction of four, extraordinarily precise gyroscopes. They will measure how Earth's rotation drags space-time around with it (frame-dragging), and how Earth's presence warps space and time (geodetic effect). These effects, though small for Earth, have far-reaching implications for the nature of matter and the structure of the universe.

The Gravity Probe B experiment also requires a reference telescope sighted on HR8703 (IM Pegasus), a binary star in the constellation Pegasus. With all the gyro spin directions also pointing toward HR8703, the frame-dragging and geodetic effects appear as spin-axis movement at right angles to the orbital plane. To measure a spin-axis movement of about 42 milliarc-seconds for frame-dragging, the gyroscopes must be stable to 10^{-11} deg/hr—a million times better than the best inertial-navigation gyroscopes. To build such gyroscopes, engineers and physicists electrically suspend 3.8 cm (1.5 in.) spheres made of fused quartz, thinly coated with niobium metal. They spin them at 10,000 r.p.m. and freeze them to 1.8 K. In the orbital free fall, and magnetically shielded, the gyroscopes are completely isolated from outside disturbances. They even compensate for atmospheric drag.

The thin niobium metal coating creates a weak magnetic field around each sphere as it spins, so that an extremely sensitive instrument can sense any spin-axis movement. With this extraordinary satellite, scientists hope to measure the general relativity effects called frame-dragging and the geodetic effect.

Stanford University website, History of Gravity Probe B topic. No author indicated.

Contributed by Douglas Kirkpatrick, the U.S. Air Force Academy

Note that this is a vector equation. Thus, the accelerometer needs to know the direction and magnitude of the acceleration caused by contact forces. This means the system must also know the attitude of the accelerometer that measures the acceleration. To find attitude, it needs an attitude sensor. Earlier, we discussed using gyroscopes as attitude sensors. Armed with an accelerometer and a gyroscope, we can then determine a_N, as shown in Figure 12-41.

Now that we have both $\vec{a}_g$ and $\vec{a}_N$, we can add them to get $\vec{a}_I$. We do this by building a little "position and velocity computer," or *inertial navigation system*, that applies principles of feedback-control systems to converge on an accurate solution. All we need are the initial conditions of the vehicle (say, at lift-off), and the navigation system does the rest. We show the elements of such an "inertial navigation system" in Figure 12-42.

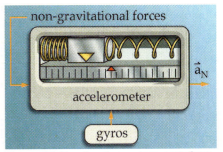

Figure 12-41.   Accelerometers and Gyroscopes. To determine the direction as well as, the magnitude of the non-gravitational accelerations, $\vec{a}_N$, we use accelerometers in conjunction with gyroscopes.

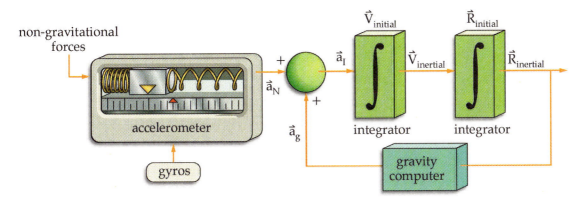

Figure 12-42.   Inertial Navigation System. An inertial navigation system is the primary "sensor" of an NGC subsystem. It uses accelerometers, gyroscopes, and knowledge of Newton's Laws to determine the current inertial position and velocity.

Rockets—The Actuators

As Mr. Newton said, "to create acceleration we need to apply a force." The force to move launch vehicles and spacecraft comes from rockets. Currently, rockets are the only actuator available to NGC subsystems (we'll explore some other futuristic options in Chapter 14). *Thrust* is the force produced by a rocket. Of course, the resulting acceleration depends on the amount of thrust applied and the mass of the vehicle (remember, $\vec{F} = m\vec{a}$). But a vehicle can reach the same total velocity and position change by using a high thrust rocket for a short time or a low thrust rocket for a long time. In Chapter 14, we'll look at rockets and propulsion subsystems in more detail to see how they work. For the purpose of the NGC subsystem, it simply has to know how much thrust to expect, then decide which direction to apply it, when, and for how long.

Guidance—The Controller

We can now look at the NGC subsystem in action by looking at the guidance portion—the controller. Since we control position by controlling velocity, let's look at just that. We start by checking our vehicle's *current* velocity, $\vec{V}_{current}$ (current velocity with respect to an inertial reference frame), by asking the navigation system. The guidance system then compares this to its desired velocity $\vec{V}_{desired}$, as provided by pre-defined mission requirements or sent to it by ground controllers. The difference between these two is the additional velocity needed ($\Delta\vec{V}_{needed}$) to achieve the desired velocity. Figure 12-43 shows how the guidance system finds $\Delta\vec{V}_{needed}$ by subtracting the other two velocity vectors.

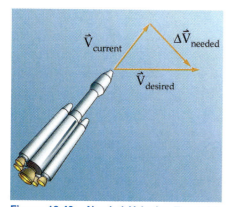

Figure 12-43. Needed Velocity. The NGC subsystem subtracts the velocity it has, $V_{current}$, from the velocity it wants, $V_{desired}$, to determine the additional velocity needed, ΔV_{needed}. Using this value, it computes the necessary steering and rocket commands to produce the extra velocity.

$$\Delta\vec{V}_{needed} = \vec{V}_{desired} - \vec{V}_{current} \qquad (12\text{-}15)$$

where

$\Delta\vec{V}_{needed}$ = velocity change needed to reach the desired velocity vector (m/s)

$\vec{V}_{desired}$ = desired velocity vector (m/s)

$\vec{V}_{current}$ = spacecraft's current velocity vector (m/s)

This equation gives the NGC subsystem a simple algorithm to use. (More complicated schemes give somewhat more accurate results.) Figure 12-44 shows the entire process for controlling the vehicle's position and velocity. In Example 12-2, you can see how this whole system works together to determine the velocity a launch vehicle needs during the final phase of powered flight to reach the desired burnout velocity.

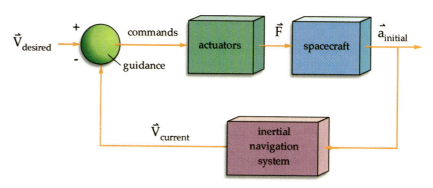

Figure 12-44. Complete Navigation, Guidance, and Control (NGC) Subsystem. The block diagram shows all the elements of an NGC subsystem working together. Given the desired velocity, $\vec{V}_{desired}$, from mission requirements, and the current velocity, $V_{current}$, from the navigation subsystem, it computes guidance commands for the actuators (rockets). These commands produce a force (thrust) on the vehicle causing acceleration.

Astro Fun Fact
Accelerometers and Seat Belts

Most of us use accelerometers every day and we hardly notice it! The most common place you come in contact with accelerometers is with the seat belt in your automobile. If you've been in an accident, you know that the seat belt latches instantly and holds you in the car seat upon impact. The simple accelerometer in the seat belt mechanism detects the rapid deceleration and causes the seat belt to hold. Two types of accelerometers are in most cars. The first is an electromechanical sensor that is a metal ball held in place by a magnet. An impact interrupts the magnetic pull, releases the metal ball, and activates the seat belt. The second is electric. It uses a glass rod that moves horizontally and, when enough deceleration is placed on the system, emits a radio frequency signal which activates the seat belt. Accelerometers can protect you every time you're in a car.

Adler, U. Automotive Handbook. Stuggart: Robert Bosch GmbH, p. 628–629, 1986.

Contributed by Michael A. Banks, the U.S. Air Force Academy

Section Review

Key Terms

accelerometer
control
guidance
navigation, guidance, and
    control (NGC) subsystem
inertial navigation system
plant model
thrust

Key Equations

$$\vec{a}_g = \ddot{\vec{R}}_g = -\frac{\mu}{R^2}\hat{R}$$

$$\vec{a}_I = \vec{a}_g + \vec{a}_N$$

$$\Delta\vec{V}_{needed} = \vec{V}_{desired} - \vec{V}_{current}$$

Key Concepts

➤ The navigation, guidance, and control (NGC) subsystem is another name for the orbital part of an attitude and orbit control subsystem (AOCS). It maintains and changes a vehicle's position and velocity. As with all closed-loop control systems, it must

- *Understand* the system's behavior—how the plant will react to inputs, including environmental inputs, to produce outputs. This is also known as the plant model.
- *Observe* the system's current behavior—using sensors
- *Decide* what to do—the job of the controller
- *Do* it—using actuators

➤ To understand the dynamics of a vehicle, we must understand Newton's Laws of Motion and know that a force applied over time causes an acceleration. Acceleration over time causes a velocity change, and velocity over time changes position.

➤ The navigation system is the "sensor." It uses various other sensors to determine current position and velocity. It combines sensing with Newton's Law of Universal Gravitation to form an inertial-navigation system.

➤ Rockets are the primary actuators for NGC subsystems. They produce a force on a vehicle called thrust.

➤ The guidance system decides what commands to send to the actuators by comparing the current position and velocity to the desired position and velocity

Example 12-2

Problem Statement

The new Falcon launch vehicle on its maiden flight has launched nearly due south ($\beta = 188°$) from the Vandenberg launch site in California to put a small remote sensing spacecraft into Sun-synchronous orbit. Mission planners want the Falcon to achieve a velocity at engine cut off of 7613 m/s, with a flight-path angle of 2°. The azimuth at burnout should be the same as the launch azimuth. During the last navigation cycle, four seconds ago, the navigation system placed the Falcon at an altitude of 455 km with the following velocity vector in the topocentric-horizon (SEZ) frame: $V_Z = 946.1$ m/s; $V_E = -56.4$ m/s; $V_S = 7434.5$ m/s. During the four seconds since the last navigation update, the inertial measurement units detected the following acceleration: –0.4 g's in the east direction; 1.2 g's in the zenith direction; 4.4 g's in the south direction. What are the three components of the velocity-desired vector ($\vec{V}_{desired}$) that the guidance system should compute so the launch vehicle can reach the desired burnout conditions?

Problem Summary

Given: $g = 9.798$ m/s^2
$\beta_{BO} = 188°$
$\Delta t = 4$ s
$V_{BO} = 7613$ m/s
$\phi_{BO} = 2°$
Altitude = 455 km
$V_{S_{initial}} = 7434.5$ m/s
$V_{E_{initial}} = -56.4$ m/s
$V_{Z_{initial}} = 946.1$ m/s
$\mu = 3.986 \times 10^5$ km^3/s^2
Acceleration$_S$ = 4.4 g's
Acceleration$_E$ = –0.4 g's
Acceleration$_Z$ = 1.2 g's

Find: $V_{desired}$

Conceptual Solution

1) Determine the desired velocity at burnout, $\vec{V}_{BO}$, using the method from Chapter 9

$$V_{BO_S} = -V_{BO}\cos(\phi_{BO})\cos(\beta_{BO})$$

$$V_{BO_E} = V_{BO}\cos(\phi_{BO})\sin(\beta_{BO})$$

$$V_{BO_Z} = V_{BO}\sin(\phi_{BO})$$

2) Find the local acceleration due to gravity, $\vec{a}_g$

$$\vec{a}_g = -\frac{\mu}{R^2}\hat{R} = -\frac{\mu}{R^2}\hat{Z}$$

3) Compute the non-gravitational acceleration, $\vec{a}_N$, over the navigation cycle in m/s^2

$$a_{N_S} = Accel_S(g)$$

$$a_{N_E} = Accel_E(g)$$

$$a_{N_Z} = Accel_Z(g)$$

4) Compute the inertial acceleration, $\vec{a}_I$, during the navigation cycle

$$\vec{a}_I = \vec{a}_g + \vec{a}_N$$

5) Find the change in velocity, $\Delta\vec{V}$, during the navigation cycle

$$\Delta\vec{V} = \vec{a}_I\Delta t$$

6) Determine the current inertial-velocity vector

$$\vec{V}_{current} = \vec{V}_{initial} + \Delta\vec{V}$$

7) Determine $\Delta\vec{V}_{needed}$

$$\Delta\vec{V}_{needed} = \vec{V}_{BO} - \vec{V}_{current}$$

Analytical Solution

1) Determine the desired velocity at burnout, $\vec{V}_{BO}$, using the method from Chapter 9

$$V_{BO_S} = -V_{BO}\cos(\phi_{BO})\cos(\beta_{BO})$$
$$= -7613 \text{ m/s} \cos(2°)\cos(188°)$$
$$V_{BO_S} = 7534.32 \text{ m/s}$$

$$V_{BO_E} = V_{BO}\cos(\phi_{BO})\sin(\beta_{BO})$$
$$= 7613 \text{ m/s} \cos(2°)\sin(188°)$$
$$V_{BO_E} = -1058.88 \text{ m/s}$$

$$V_{BO_Z} = V_{BO}\sin(\phi_{BO})$$
$$= 7613 \text{ m/s} \sin(2°)$$
$$V_{BO_Z} = 265.69 \text{ m/s}$$

Example 12-2 (Continued)

■ Find the local acceleration due to gravity, $\vec{a}_g$

$\hat{Z} = 1$ in the Z direction
$R = 6378 \text{ km} + \text{Alt} = 6378 \text{ km} + 455 \text{ km}$
$R = 6833 \text{ km}$

$$\vec{a}_g = -\frac{\mu}{R^2}\hat{Z} = -\left(\frac{3.986 \times 10^5 \frac{\text{km}^3}{\text{s}^2}}{(6833 \text{ km})^2}\right)\hat{Z}$$

$\vec{a}_g = -8.54 \text{ m/s}^2$ in the $\hat{Z}$ direction

) Compute the non-gravitational acceleration, $\vec{a}_N$, over the navigation cycle in m/s²

$a_{N_S} = \text{Accel}_S(g) = 4.4 \text{ g's} \left(9.798 \frac{\text{m/s}^2}{\text{g}}\right)$

$a_{N_S} = 43.11 \text{m/s}^2$

$a_{N_E} = \text{Accel}_E(g) = -0.4 \text{ g's} \left(9.798 \frac{\text{m/s}^2}{\text{g}}\right)$

$a_{N_E} = -3.92 \text{m/s}^2$

$a_{N_Z} = \text{Accel}_Z(g) = 1.2 \text{ g's} \left(9.798 \frac{\text{m/s}^2}{\text{g}}\right)$

$a_{N_Z} = 11.76 \text{m/s}^2$

.) Compute the inertial acceleration, $\vec{a}_I$, during the navigation cycle

$a_I = a_g + a_N$

$a_{I_S} = 0 + a_{N_S} = 0 + 43.11 \text{ m/s}^2$

$a_{I_S} = 43.11 \text{m/s}^2$

$a_{I_E} = 0 + a_{N_E} = 0 + (-3.92 \text{ m/s}^2)$

$a_{I_E} = -3.92 \text{m/s}^2$

$a_{I_Z} = a_g + a_{N_Z} = -8.54 \text{ m/s}^2 + 11.76 \text{ m/s}^2$

$a_{I_Z} = 3.22 \text{m/s}^2$

5) Find the change in velocity, $\Delta\vec{V}$, during the navigation cycle

$\Delta V = a_I(\Delta t)$

$\Delta V_S = a_{I_S}(\Delta t) = 43.11 \text{ m/s}^2 \text{ (4 s)}$

$\Delta V_S = 172.44 \text{ m/s}$

$\Delta V_E = a_{I_E}(\Delta t) = -3.92 \text{ m/s}^2 \text{ (4 s)}$

$\Delta V_E = -15.68 \text{ m/s}$

$\Delta V_Z = a_{I_Z}(\Delta t) = 3.22 \text{ m/s}^2 \text{ (4 s)}$

$\Delta V_Z = 12.88 \text{ m/s}$

6) Determine the current inertial velocity vector, $\vec{V}_{current}$

$\vec{V}_{current} = \vec{V}_{initial} + \Delta\vec{V}$

$V_{current_S} = V_{S_{initial}} + \Delta V_S = 7434.5 \text{ m/s} + 172.44 \text{ m/s}$

$V_{current_S} = 7606.94 \text{ m/s}$

$V_{current_E} = V_{E_{initial}} + \Delta V_E = -56.4 \text{ m/s} + (-15.68 \text{ m/s})$

$V_{current_E} = -72.08 \text{ m/s}$

$V_{current_Z} = V_{Z_{initial}} + \Delta V_Z = 946.1 \text{ m/s} + 12.88 \text{ m/s}$

$V_{current_Z} = 958.98 \text{ m/s}$

7) Determine $\Delta\vec{V}_{needed}$

$\Delta\vec{V}_{needed} = \vec{V}_{BO} - \vec{V}_{current}$

$\Delta V_{needed_S} = V_{BO_S} - V_{current_S}$
$= 7534.32 \text{ m/s} - 7606.94 \text{ m/s}$

$\Delta V_{needed_S} = -72.62 \text{ m/s}$

$\Delta V_{needed_E} = V_{BO_E} - V_{current_E}$
$= 1058.88 \text{ m/s} - (-72.08 \text{ m/s})$

$\Delta V_{needed_E} = 1130.96 \text{ m/s}$

$\Delta V_{needed_Z} = V_{BO_Z} - V_{current_Z}$
$= 265.69 \text{ m/s} - 958.98 \text{ m/s}$

$\Delta V_{needed_Z} = -693.29 \text{ m/s}$

$$\Delta V_{needed} = \sqrt{\left(-72.62\frac{\text{m}}{\text{s}}\right)^2 + \left(1130.96\frac{\text{m}}{\text{s}}\right)^2 + \left(-693.29\frac{\text{m}}{\text{s}}\right)^2}$$

$\Delta V_{needed} = 1328.5 \text{ m/s}$

Interpreting the Results

The launch vehicle is slightly off course and needs to make some final adjustments to its velocity to reach the desired burnout conditions. It needs to gain an additional 1130.96 m/s in the downrange or east direction and −693.29 m/s in the up or vertical direction to get the correct flight-path angle (2°) at MECO. It also needs to correct its extra southerly velocity by 72.62 m/s. The total ΔV_{needed} is 1328.5 m/s.

References

Asimov, Isaac. *Asimov's Biographical Encyclopedia of Science and Technology.* Garden City, NJ: Doubleday and Company, Inc., 1972.

Chetty, P.R.K. *Satellite Power Systems: Energy Conversion, Energy Storage, and Electronic Power Processing.* George Washington University Short Course 1507. October, 1991.

Chetty, P.R.K. *Satellite Technology and Its Applications.* THB Professional and Reference Books, New York, NY: McGraw-Hill, Inc., 1991.

Gere, James M., Stephen P. Timoshenko. *Mechanics of Materials.* Boston, MA: PWS Publishers, 1984.

Gonick, Larry and Art Huffman. *The Cartoon Guide to Physics.* New York, NY: Harper Perennial, 1991.

Gordon, J.E. *Structures: Why Things Don't Fall Down.* New York, NY: Da Capp Press, Inc., 1978.

Holman, J.P. *Thermodynamics.* New York, NY: McGraw-Hill Book Co., 1980.

Pitts, Donald R. and Leighton E. Sissom. *Heat Transfer.* Schaum's Outline Series. New York, NY: McGraw-Hill, Inc., 1977.

Wertz, James R. and Wiley J. Larson. *Space Mission Analysis and Design.* Third edition. Dordrecht, Netherlands: Kluwer Academic Publishers, 1999.

Wertz, James R. [ed.] *Spacecraft Attitude Determination and Control.* Netherlands: D. Reidel Publishing Co., Kluwer Group, 1986.

Mission Problems

12.1   Control Systems

1   What are block diagrams and why are they useful?

2   Define the four steps in control and apply them to some everyday process such as hitting a baseball with a bat.

3   What is a plant model? What is the plant model for the baseball and bat example?

4   How do spacecraft use control?

5   What do sensors do in a control system? What sensors do you use to hit a baseball with a bat?

6   Draw a block diagram for the baseball-and-bat system. Assume you are the controller and the bat is the actuator.

12.2   Attitude Control

7   Give an example of how you can use roll, pitch, and yaw to describe the attitude a remote-sensing spacecraft needs to point a camera at a target on Earth.

8   What is the difference between attitude accuracy, slew, and slew rate?

9   A spacecraft is not spinning. If a 2.0 N thruster fires 1.0 m from the spacecraft's center of mass, perpendicular to the center of mass, how much torque does it generate and how will the angular momentum change? If the spacecraft's moment of inertia is $1000 \text{ kg} \cdot \text{m}^2$, what angular acceleration will the spacecraft experience?

10 Describe how gyroscopic precession works. What is gyroscopic stiffness?

11 What are disturbance torques and how do they affect a spacecraft? Which ones do spacecraft designers worry about most?

12 Suppose a burst of solar wind disturbed the attitude of your gravity-gradient-stabilized spacecraft, and knocked it 30° off of vertical. The spacecraft specifications show the moments of inertia are $I_z = 100 \text{ kg m}^2$, $I_x = I_y = 20 \text{ kg m}^2$. It's in a 500-km altitude polar orbit. Find how much torque is available from the gravity gradient to restore your spacecraft to vertical.

13 After designing the solar panels for the electrical power subsystem on a LEO spacecraft, you want to know how much solar radiation pressure they will have. So, for solar panels that are 20 m² in area, assume a perfect reflectance ($\rho = 1.0$) and compute the worst case ($I = 0°$) force on the panels. Use Fs = 1358 W/m² for the solar constant.

14 What is the force of air drag on a spacecraft with a cross-sectional area of 10 m² in a low-Earth, circular orbit at 200 km? Assume the air density is $2.53 \times 10^{-10} \text{ kg/m}^3$ and $C_D = 1.0$.

15 Which of the spacecraft's sensors allow it to "look out the window" to determine attitude and position?

16 What unique properties of spinning masses make them useful as gyroscopic attitude sensors?

17 What is a ring-laser gyro and how is it different from a spinning mass?

18 How do magnetometers determine a spacecraft's attitude and what limits their use to low-Earth orbits?

19 How does differential GPS determine a spacecraft's attitude?

20 What are the three main passive attitude actuators used on spacecraft?

21 Describe the potential spacecraft design impacts for using gravity-gradient control.

22 List the advantages and disadvantages of control strategies using gravity gradient, spin, and dual spin.

23 How can thrusters be used to torque a spacecraft?

24 How do magnetic torquers work?

25 What are the three momentum-control devices and how do they work? What are their main differences?

26 What does a controller do and what types of things does it have to know?

27 Draw the block diagram of a complete attitude determination and control subsystem and discuss all of the components.

12.3   Orbit Control

28 Define the functions of navigation, guidance, and control.

29 How do accelerometers work? What do they measure?

30 The Space Shuttle is on its way into orbit and is currently at an altitude of 40 km. What gravitational acceleration is it experiencing?

31 Describe the purpose and various parts of an inertial-navigation system.

32 What type of actuators are used on spacecraft?

33 What is the basic relationship between needed velocity, desired velocity, and current velocity used by the booster controller?

34 A Titan IV rocket has been launched due east from the Kennedy Space Center (28.5° N, 75° W). Mission designers want the booster to reach a burnout velocity of 8220 m/s, with a flight-path angle of 5°. The azimuth at engine cut off should be the same as the launch azimuth. During the last navigation cycle five seconds ago, the navigation system placed the booster at an altitude of 60 km with the following velocity vector in the topocentric-horizon (SEZ) frame: $V_S = -5.7$ m/s, $V_E = 436.2$ m/s, $V_Z = 544.1$ m/s. During the five seconds since the last navigation update, the inertial measurement units detected the following acceleration

- –0.6 g's in the south direction

- 6.4 g's in the east direction

- 4.2 g's in the zenith direction

What are the three components of the $\vec{\Delta V}_{needed}$ vector that the guidance system should compute so the Titan IV can achieve the desired burnout conditions?

Mission Profile—GPS

For centuries, mariners relied on the stars to tell them their location at sea. Today, ships, airplanes, and even spacecraft can still look to the skies to determine where they are. But instead of a crude reckoning based on known star positions, they can now achieve unprecedented accuracy by looking at a man-made constellation of stars, better known as NAVSTAR or the Global Positioning System (GPS).

Mission Overview

The NAVSTAR Global Positioning System is a space-based radio navigation system which, for any number of users, provides extremely accurate position and velocity data anywhere on Earth to those equipped with GPS receivers.

Mission Data

✓ The elements of the mission architecture are

- Objective—provide world-wide navigation reference
- User—United States armed forces and anyone with a receiver
- Operations concept—a user receives signals from four (or more) satellites (either simultaneously or sequentially) to determine their three-dimensional position and velocity

✓ The receiver calculates range information by computing the difference between the current receiver time and the time transmitted in the pulse train, and then multiplying this time by the speed of light.

✓ By using four (or more) signals, the receiver can mathematically eliminate its own clock errors (which are significant), and rely solely on the time kept onboard the satellites by atomic clocks. These clocks would lose or gain only one second every 160,000 years if they were not updated every day.

- Spacecraft—NAVSTAR GPS. The payload consists of two cesium and two rubidium atomic clocks and a communication package to broadcast GPS system time and the spacecraft's position information to users.
- Space operations—five unmanned stations monitor the GPS block II satellites at Hawaii, Ascension Island, Diego Garcia, Kwajalein Atoll, and Schriever AFB in Colorado. Schriever AFB also serves as the Master Control Station.
- Booster—Delta II

- Communication network—antennas at the five monitor stations

GPS Block II-F. The next generation GPS satellites will improve accuracy and availability of the signal. *(Courtesy of the U.S. Air Force)*

Mission Impact

During Operation Desert Storm, almost every element of the coalition used the GPS system. Supporting everything from precision air strikes to lights-out delivery of meals to foot soldiers in the field, the GPS proved itself a true "force multiplier." Receivers were in such demand that soldiers bought civilian sets out of their own pockets so they could navigate in the featureless deserts of the Middle East. Because there's no limit to the use of GPS signals, the volume and scope of its applications are also limitless. Cars are now being equipped with "moving maps," which use GPS and digital maps stored on compact disks (CDs). GPS may soon be a household word on par with even radio and television.

For Discussion

- What other applications can you think of for GPS?
- What are its potential limitations?
- Could we use it for interplanetary or interstellar navigation? Why or why not?
- The United States government provides this service to the world basically free of charge. Is this the right thing to do?

Contributor

Kirk Emig, the U.S. Air Force Academy

References

Logsdon, Tom. *The Navstar Global Positioning System.* New York, NY: VanNostrand Reinhold, 1992.

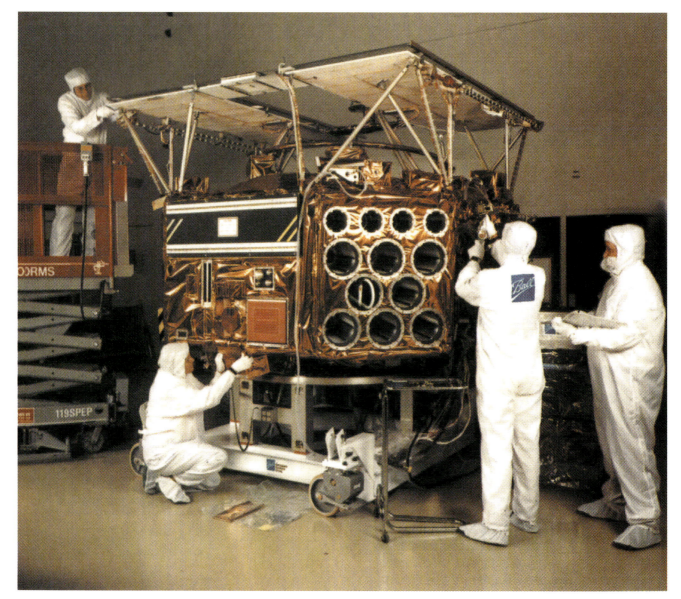

Technicians in a clean room prepare the combined release and radiation effects satellite (CRRES) for its journey into space. *(Courtesy of Ball Aerospace & Technologies Corporation)*

Spacecraft Subsystems

13

In This Chapter You'll Learn to...

☛ Describe the main functions and requirements of the communication and data-handling subsystem (CDHS)

☛ Describe the main functions and requirements of the electrical power subsystem (EPS)

☛ Describe the main functions and requirements of the environmental control and life-support subsystem (ECLSS)

☛ Describe the main functions and requirements of the spacecraft structures and mechanisms

You Should Already Know...

❏ The space systems engineering process (Chapter 11)

❏ Basic functions of the spacecraft bus's subsystems (Chapter 11)

❏ Effects of the space environment on spacecraft (Chapter 3)

❏ Black-body radiation, Stefan-Boltzmann relationship (Chapter 11)

❏ Basic elements of a control system (Chapter 12)

Outline

13.1 Communication and Data-handling Subsystem (CDHS)
System Overview
Basic Principles
Systems Engineering

13.2 Electrical Power Subsystem (EPS)
Basic Principles
Systems Engineering

13.3 Environmental Control and Life-support Subsystem (ECLSS)
System Overview
Basic Principles of Thermal Control
Basic Principles of Life Support
Systems Engineering

13.4 Structures and Mechanisms
System Overview
Basic Principles
Systems Engineering

You would make a ship sail against the winds and currents by lighting a bonfire under her deck...I have no time for such nonsense.

Napoleon commenting
on Fulton's steamship

Space Mission Architecture. This chapter deals with the Spacecraft segment of the Space Mission Architecture, introduced in Figure 1-20.

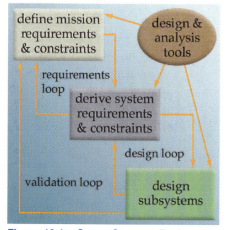

Figure 13-1.   Space Systems Engineering Process. The space systems engineering process, introduced in Chapter 11, allows us to turn basic requirements into real system hardware.

B eing a spacecraft is a lot of work. It gets packed into the nose cone of a rocket sitting on tons of explosives. It gets blasted into orbit on a bumpy ride that subjects it to many g's of acceleration. Then it gets dumped into the cold vacuum of space to fend for itself. It spends the rest of its life keeping the payload happy, supplying it with electrical power, keeping it not-too-hot and not-too-cold, pointing its sensors in the right direction, and processing its data.

In this chapter, we'll see how the spacecraft does all this. In Chapter 11, we presented the *space systems engineering process*, shown again in Figure 13-1, and reviewed the basic functions of the spacecraft bus. This process gave us a systematic way to use general mission requirements, such as "build a small satellite to look for forest fires," to develop specific payload and subsystem designs. In Chapter 12, we looked at control systems and how they manage a spacecraft's angular momentum, position, and velocity in the attitude and orbit control subsystem (AOCS). In this chapter, we'll continue to explore the other subsystems that are on the spacecraft bus. Here we'll see how to apply the systems engineering process to control and manage the information flow, electrical power, and heat on a spacecraft, as well as it's structural design. We'll focus on the basic principles for design and analysis tools, as well as the "design loop" and "verification loop" issues.

We'll start by looking at the flow of information to see how the communication and data-handling subsystem (CDHS) uses radios and computers to collect, store, send, and receive vital data from payloads and ground controllers. Next, we'll look at the flow of electrical power to see how the electrical power subsystem (EPS) creates, stores and distributes electricity to power-hungry components. All this power, together with the Sun and other sources, produces lots of heat. We'll see how the environmental control and life-support subsystem (ECLSS) creates a cozy working space for all the subsystems, astronauts, and other payloads. Finally, we'll look at how everything fits together. While not a "subsystem" in the conventional sense, we'll see how structures and mechanisms literally form the backbone of the spacecraft, providing a framework for integrating all the other subsystems and payloads. So let's kick the tires and take a peek under the hood to see how a spacecraft really works.

13.1 Communication and Data-handling Subsystem (CDHS)

▰ In This Section You'll Learn to...

- ☛ Describe the inputs, outputs, and basic processes within the spacecraft communication and data-handling subsystem (CDHS)

- ☛ Explain the basic principles used to communicate and handle data

- ☛ Apply the space systems engineering process to the design and testing of the CDHS

The most important commodity that "flows" within a spacecraft and down to the ground is information. After all, the need for the unique payload information motivated the mission originally. Monitoring information on the health and status of the spacecraft is vital to the continued success of the mission. The *communication and data-handling subsystem (CDHS)* processes all this information. It serves as the spacecraft's ears, brain, and mouth. Its "ears" gather raw data from the payload, other subsystems, and ground controllers. Its "brain" processes this data and translates it into a useful form that can be passed between onboard systems and to the ground. Its "mouth" communicates this information in a meaningful way to users in the mission control center.

We introduced the spacecraft's communication and data-handling subsystem (CDHS) in Chapter 11. This subsystem acquires, stores, and transfers all information for the spacecraft. The communication part of this subsystem transfers data and operating commands among the spacecraft, ground stations, and other spacecraft. The data handling part collects payload or housekeeping data on all the subsystems and stores it until it can be transmitted to the ground.

In this section, we'll start by looking at the system-level inputs and outputs of the CDHS. Then we'll briefly review some of the basic principles of communication and data handling needed to do its job. We'll explain the process for translating data into a form that can be communicated from one place to another. Then we'll look at the major components of any data-handling subsystem and the software requirements needed to make it all work together. Finally, we'll bring the pieces together to understand some of the fundamental issues of systems engineering that drive the requirements and testing of the CDHS.

System Overview

Throughout this chapter, and in Chapters 14 and 15, we'll continue to apply the systems approach to understanding how things work on a spacecraft we began in Chapter 12. Recall the major characteristic of any system is that it has input, performs some process, and produces output.

Figure 13-2 illustrates the inputs and outputs of the CDHS. As the figure shows, the CDHS is somewhat unique in that inputs and outputs flow in both directions. Starting from the left, the CDHS collects raw data from the payload and the other spacecraft systems. Within the data handling portion, the CDHS processes this raw data to translate it into valuable mission information (images, measurements, etc.) or important information on subsystem performance (temperature, voltage, etc.).

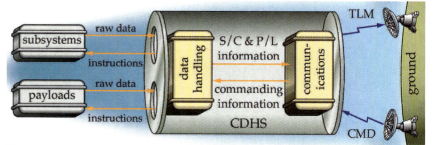

Figure 13-2. Communication and Data-handling Subsystem (CDHS). The CDHS is a spacecraft's "ears," "brain," and "mouth." It gathers data from the payload and other subsystems and the ground controllers and translates it into useful information that is passed to other "customers."

Note we make a distinction here between "data" and "information." *Data* can be any collection of facts. For example, 23° C, or 14 volts. These facts become useful *information* only when we put them in a specific context, e.g., the temperature of the solar panels at 12:34 GMT was 23° C or the battery voltage after eclipse was 14 volts.

After we assemble this information, the next step is to communicate it to eager users on the ground. The communication portion of the CDHS is simply a radio that takes the "script" written by the data handling piece and broadcasts it to the ground. The payload and spacecraft health and status information that the controllers receive on the ground for analysis is called *telemetry* (from the Latin for "far-measurement").

Now working from right-to-left in Figure 13-2, notice that the communications portion collects commands radioed from the ground (e.g., "take an image over Washington, D.C.") and translates them into specific commands that the data handler can interpret. The CDHS then breaks this information into specific sets of instructions to the payload and other subsystems, which execute the command (e.g., rotate the spacecraft 14° and take a picture at 09:47 GMT). Next, we'll look at how the CDHS performs all these processes.

Basic Principles

Whether we're speaking, writing, reading, or thinking, we're processing information. Biologists and psychologists are only beginning to understand the complex processes our brains use to collect, store, retrieve, and assemble new information. This powerful tool that sits between our ears performs many of these same functions that happen within the components we call computers and radios in a spacecraft. All CDHS design and analysis tools depend on some basic principles of operation for these two components. To understand these principles, we'll start by looking at the communication function.

Communication involves translating information into a form that we can move from point A to point B. On a fundamental level, this is the same process that happens in a computer. The primary difference is the amount of "translating" that occurs. In fact, on modern spacecraft, it can be very difficult to draw a line between where computing ends and communication begins. Both of these functions have blended together in the same basic hardware and software. However, for purposes of discussion here, we'll break the CDHS into two parts—communicating and data handling.

Communicating

On May 24, 1844, Samuel Morse (1791–1872) made history by sending a simple message, "What Hath God Wrought," from Washington, D.C., to Baltimore, Maryland, through a wire strung between the two cities. The Information Age was dawning. Within a few years, telegraph lines spanned the country, linking far-flung cities. With the invention of radio communication by Guglielmo Marconi (1874–1937) in 1895, the telegraph slowly became obsolete. However, the techniques pioneered by the telegraph are very useful in illustrating the basic process of communication still used today.

In a simple telegraph system, a single wire connects two stations. When an electromechanical switch closes at either end, an electric pulse travels along the wire and another electromechanical switch at the other end detects it. Given the obvious limitations of this fairly crude system, how can we devise a useful communication technique?

Realize that in the absence of any communication, there is no activity in the wire. It is silent. In engineering terms, we call the default condition of our communication system the *carrier signal*. The carrier signal literally "carries" the information with it, so it doesn't contain any information itself. By prior agreement, we decide this electrical signal will always have the same amplitude, phase and frequency. We communicate by laying the message onto the carrier to get a signal to transmit. For a telegraph, the carrier signal of the system is a silent wire (zero amplitude, phase and frequency). When we add a message, it's like shouting across a quiet room, it gets noticed. The specific technique we use to blend the message signal with the carrier signal is called *modulation*. For telegraph, the modulation options are pretty limited. We can only turn the pulse on or off, or vary its width (shorter or longer in duration). Even so, Morse devised a simple but effective modulation scheme that still bears his name: Morse code.

Morse assigned a specific series of "dots" (short pulses) and "dashes" (long pulses) to each letter of the alphabet and numbers, 0–9. Thus, operators could translate any message into Morse code and transmit it using the telegraph. Figure 13-3 illustrates how the signal strength, or *amplitude*, of the message signal—dots or dashes—is always the same, but the pulse width varies over time to deliver the message (now, commonly called "pulse width modulation").

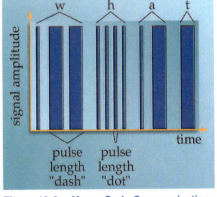

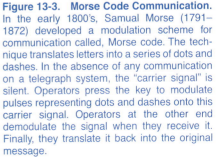

Figure 13-3.   Morse Code Communication. In the early 1800's, Samual Morse (1791–1872) developed a modulation scheme for communication called, Morse code. The technique translates letters into a series of dots and dashes. In the absence of any communication on a telegraph system, the "carrier signal" is silent. Operators press the key to modulate pulses representing dots and dashes onto this carrier signal. Operators at the other end demodulate the signal when they receive it. Finally, they translate it back into the original message.

With the invention of radio, the carrier signal was freed from the confines of wires and broadcast far and wide through the atmosphere. These carrier signals are simply specific frequencies (naturally called *radio frequencies*, or simply *rf*) of the electromagnetic (EM) spectrum we discussed in Chapter 11. A wide variety of modulation schemes have been developed to overlay messages on these carrier frequencies, including the well known amplitude modulation (AM) and frequency modulation (FM) we can tune our car radios to.

Physical, technical, and legal limitations dictate allowable spacecraft communication frequencies. Atmospheric windows, discussed in Chapter 11, place physical limitations on which frequencies can pass undisturbed through the atmosphere. Technical requirements such as antenna size, available power, data transmission rate, and other spacecraft and ground system specifications also constrain the range of frequencies that we can use for a certain mission. In Chapter 15, we'll look at this aspect of the communication problem and the trade-offs between the spacecraft and ground system. Finally, international licensing of spacecraft frequencies by the International Telecommunication Union (ITU) and the World Administrative Radio Conference (WARC) dictate what frequencies we *can* use that may be different than the ones we *want* to use. In the competitive spacecraft-communication market, a license to use specific frequencies can be an extremely valuable company asset. We'll learn more about these regulatory agencies in Chapter 16.

Figure 13-4 shows the basic components of the communication portion of the CDHS. Payload and spacecraft information is coded onto the carrier signal by a *modulator* before being amplified and sent to the spacecraft antenna for broadcast to the ground. The receiver antenna collects commands from the ground, then an amplifier boosts the weak signal, and a *demodulator* translates (decodes) it, before sending it to the data handler.

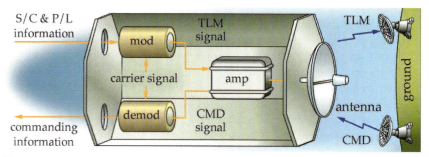

Figure 13-4.   Communication Components. The communication portion of the CDHS consists of modulators (MOD), demodulators (DEMOD), amplifiers (amp), and antennas. Payload and spacecraft information modulates onto the carrier signal, then is amplified and broadcast to the ground system through the antenna. The spacecraft collects commands from ground stations (using its antenna), amplifies the signal, then demodulates it to produce instructions for the data handling portion of the CDHS.

A home computer handles access to the Internet in much the same way, using a combination modulator/demodulator called a "*MODEM*." Later in this section, we'll look at a simple example of how this process works to transmit information.

Handling Data

To put the message signal in a form to transmit, the modulator must combine the message signal with the carrier signal. In mathematical terms, the two are multiplied together to form the transmitted signal. To reach that point, we must first have useful information to send. To get the information, the data handler must first manipulate the payload and subsystem data, a process that involves adding, subtracting, multiplying, and dividing data signals. These functions form the heart of data handling, which is the sole function of computers. In almost every way, the data-handling hardware (and, even software) in spacecraft is identical to a home personal computer (PC), similar to the one shown in Figure 13-5.

Modern computers labor under the same basic limitations faced by Samuel Morse. Similar to data from a telegraph, data within a computer can only travel as electronic pulses that are either on or off. However, instead of Morse code, we treat these on/off pulses as 1's and 0's and apply binary arithmetic. (Normal mathematics uses a decimal or base-10 system. Binary mathematics employs a base-2 system.) Each 1 or 0 is called a *bit*. Just as letters combine to form words, bits of data combine into groups called *bytes*. Depending on the computer, 8, 16, 32, or more bits form a *word*.

Millions of transistors packed onto computer chips complete the mathematical operations within the computer hardware. The more densely we pack these transistors, the more operations they can complete in a given time. We measure this throughput or processing speed in cycles per second (or Hertz, abbreviated Hz) and sometimes call it "clock speed" on home computers.

Figure 13-6 illustrates the three main components of the data handling portion of the CDHS. *Input/output (I/O)* devices interact with the "outside world" to collect and distribute data to other users. "Thinking" takes place within the *central processing unit (CPU)*. This is the main processor in the computer and is similar (and in some cases, even identical) to the Pentium III™ or other processors that power a home PC.

When it needs additional information, the CPU can access and use the information stored in memory. Memory comes in many different forms. *Read only memory (ROM)* stores fixed programs and data that don't change throughout the mission (unless overwritten by operators or engineers). This is non-volatile memory that remains constant even if the computer shuts off. Within a home computer, ROM stores basic instructions that allow the computer to turn on and talk to input devices.

For other long-term memory storage, spacecraft use non-volatile memory that stores large amounts of data such as specific programs or payload data. Originally, bulky tape drives did this work, however, solid state memory, similar to memory cards used with computers and digital cameras, has replaced these mechanical storage options, for the most part on modern spacecraft.

Computers rely on *random access memory (RAM)* for short-term storage. Information flows from long-term storage into RAM where the CPU can access it at electronic speeds. Unfortunately, RAM usually holds much

Figure 13-5. Personal Computers and Spacecraft. The communication and data-handling subsystem (CDHS) has many of the same components as a personal computer. It communicates using modulators/demodulators (MODEMs). Data handling happens in the CPU. Both of those systems run operating-system and application software. *(Courtesy of the U.S. Air Force Academy)*

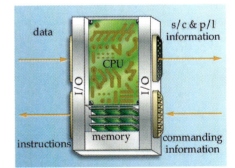

Figure 13-6. Data Handling. The data handling portion of the CDHS is the "brains" of the operation. It collects and distributes data through input/output devices. "Thinking" takes place in the central processing unit (CPU). Mass memory modules store data for later retrieval.

less information than tape drives or other long-term memory storage devices. In addition, RAM is volatile memory. That is, all information stored in RAM vanishes when power shuts off.

Software controls the actions of all these tiny data manipulators. *Operating-system software* works at the most basic level to run the entire computer and to schedule events. On home computers, this operating system is the basic set of instructions that allows the computer to start up and run (e.g. Windows™ or Apple™ OS). *Application software* on your PC handles specific tasks such as word processing or spreadsheets. On a spacecraft, a separate application program may run soon after deployment, to spin the spacecraft or extend antennas. Other application software programs control sensors, fire thrusters, and perform other specific tasks.

We can't over-emphasize the importance of software to a space mission. It controls virtually every action on a spacecraft (as well as in the operational control center). The success or failure of a mission can depend on a single line of computer code. To insure success, mission managers typically must perform rigorous software tests. However, as any computer user knows, no software is ever completely bug free. For this reason, smart mission designers build in the flexibility to reprogram their spacecraft "on the fly," from the ground system, after launch. This allows them to fix bugs as needed and upgrade the computer functions throughout the life of the mission. For example, when the Voyager spacecraft had trouble pointing it's camera after flying by Saturn, ground controllers had to extensively modify the programs from more than three billion km (two billion miles) away to get it working again.

Astro Fun Fact
Space Shuttle Computers

Computers sometimes lockup and need to be reset. However, in systems where computers control almost everything and life is at risk, protection against lockups has to be incorporated into the system. The Space Shuttle is a prime example of such a situation. To protect against computer errors during the trip to orbit, four computers, all running the same software, compare answers with each other. If one computer disagrees with the other three, it is voted out of the system and is ignored from then on. Two of the remaining computers can also vote out the third computer, if it doesn't agree with them.

If, along the way, the voting gets to the point where the system can't decide which answer is right, the astronauts have another option available. A fifth computer, running different software, can be brought on-line as a backup in order to finish the ascent to orbit.

NASA Johnson Space Center. "Space Shuttle News Reference." "Data Processing System Training Manual." TD292.

Contributed by Scott R. Dahlke, the U.S. Air Force Academy

454

Software and hardware are susceptible to the space environment. As we discussed in Chapter 3, cosmic rays, in the form of charged particles, can cause a variety of problems, including *single event upsets (SEU)* that cause bits stored in memory to flip (0's change to 1's). Figure 13-7 shows the pattern of SEU's recorded by the UoSAT spacecraft. Notice the majority of these events take place in a region known as the *South Atlantic Anomaly*, where Earth's magnetic field dips closer to the surface than normal. Single event latch-ups effect the system at the hardware level, changing the action within a single transistor. As engineers pack more transistors together to increase performance, the hardware becomes increasingly susceptible to these stray particles. One major focus of space electronics research is to develop radiation tolerant (rad-hard) components for future mission applications.

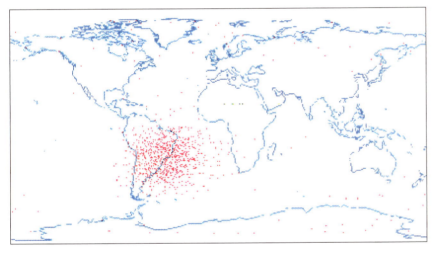

Figure 13-7. Single Event Upsets. This map illustrates the pattern of Single Event Upsets (SEU) recorded by the UoSAT spacecraft. Notice most of events take place over the South Atlantic Anomaly. *(Courtesy of the Surrey Space Centre, U.K.)*

CDHS Example

To see how data handling and communication work together in action, let's look at the simple example of measuring the temperature on a solar panel and preparing the value to send to interested ground controllers. Earlier we defined data as any collection of facts. But data can be one of two forms: analog or digital. *Analog data* is continuous within any range. The real world is analog, meaning, there are an infinite number of numbers between 0 and 1; and events take place in a continuous flow. Humans are analog creatures. We see and hear over a continuous range of sights and sounds. Temperature is another example of analog data. A solar panel doesn't feel compelled to be exactly 23.4° C or 23.5° C. It can be any temperature in between.

Unfortunately, as we discussed earlier, computers think in 1's and 0's. They perceive the world in terms of digital data. *Digital data* has discrete

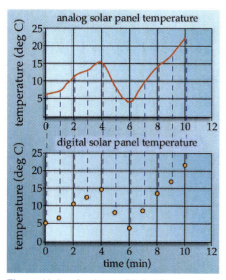

Figure 13-8.   Analog to Digital Conversion. Data can be either analog or digital. Here we show the analog output from a thermometer measuring solar panel temperature. To "digitize" this data, we sample it at a specific rate to return a set of discrete digital data.

values within any range. So, one of the first tasks of the CDHS is to convert the real-world analog data into computer-world digital data that it can easily manipulate. The most common way to do this conversion is to sample the analog signal at discrete times to produce digital values at those times. Figure 13-8 illustrates *sampling* an analog temperature output to produce a digital representation.

Sampling is a science in its own right. The *Nyquist criteria* (developed by Harry Nyquist in 1928) dictates that to accurately represent an analog signal, we must sample it at least twice as often as its highest frequency output. In other words, if the temperature varies once per minute, we must sample it at least every 30 seconds to accurately represent it.

To continue our example, let's look at a single data point. From the graph, we can say that at exactly two minutes, the digital temperature is 11° C. Therefore, to make the temperature data into useful information, we must include the time tag. Thus, we must keep the two numbers—time and temperature—together to make sense. The analog-to-digital (A/D) conversion task can take place within the sensor, or as part of the I/O of the data handling portion of the CDHS. Any additional processing to format the data, or collect it into "packets" that we send to ground controllers, we assign to the CPU, based on existing software.

Converting our example data into 6-bit binary "words" we have

- 2 = 000010 (representing time tag of 2 min)

- 11 = 001011 (representing temperature of 11° C)

The CDHS subsystem then transfers these words to the communication portion of the subsystem. Here, each word modulates onto the chosen carrier frequency. Figure 13-9 illustrates a crude example of how the temperature could modulate onto a carrier frequency using amplitude modulation (in reality, the frequencies used are far too high to easily illustrate here). By changing, the pulse width (like Morse code), frequency, and other parameters of the carrier frequency, we have a wide variety of coding techniques available in addition to simple AM.

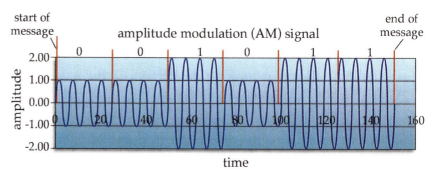

Figure 13-9.   Amplitude Modulation. By laying the message signal (001011) onto the carrier frequency, we can see a crude representation of pulse amplitude modulation. While the amplitude stays below the maximum carrier amplitude (1.0) the message is "0." When the amplitude exceeds this value (2.0) the message is "1." In reality, the frequencies used are much, much higher (10^9 Hz range) and thus impossible to illustrate on paper.

Systems Engineering

The basic principles we've reviewed form the basis for CDHS design and analysis tools. We apply these tools within the design loop of the systems engineering process. As we've described, information is composed of data. The shear volume of data that the system must collect, process, store, and communicate is one of the biggest drivers of the CDHS design. Typically, the payload produces the bulk of mission data. Remote-sensing missions especially can accumulate large amounts of data to process, store, and send to users.

To define the CDHS requirements, we must first determine

- The communication frequencies available
- Other limitations on system design (mass, power, volume)
- The total amount of data
- The amount of onboard processing
- The amount of time available for downlink to ground stations

Physical and legal restrictions limit the available communication frequencies. Technology (and cost!) limits how much performance we can pack into the available mass and volume. As with all system designs, we have a wide variety of trade-offs to consider when it comes to building the CDHS. For example, the more processing it does onboard, the more sophisticated the data-handling components must be. The advantage is that additional onboard processing could significantly decrease the total amount of data that it must send to the ground system, making the communication design simpler and the ground operations lower in cost.

Two important parameters that emerge from this design trade-off study are the total amount of data collected and processed per orbit (the data budget), and the required communication data rate, part of the overall link budget. We can represent the *data budget* in terms of the total amount of data generated in Mbytes/orbit. We express the *data rate* as the number of bits per second that must be transmitted during a given pass over the ground station. The *link budget* accounts for the data rate and other factors that contribute to the basic ability for the spacecraft and ground system to communicate effectively. In Examples 13-1, and 13-2 we'll show how we can use the data budget and data rate to help us design the onboard CDHS for the FireSat mission. In Chapter 15, we'll look at the ground-system side of the problem to determine antenna size, power, and other parameters go into determining the overall mission link budget.

After we design and build the subsystem, we must finally test it. Recall, the verification loop is in the systems engineering process to make sure what we build is really what we want. Prior to launch, we subject our CDHS hardware and software to a variety of tests. Typically, we test each component separately to ensure it meets subsystem specifications. The ultimate test of the CDHS takes place during *integrated testing*, when we combine the subsystem with other subsystems and the payloads on the spacecraft bus. One critical test measures the antenna's radiation pattern inside a vast *anechoic chamber*, as shown in Figure 13-10.

Figure 13-10.   CDHS Testing. During integrated testing, the spacecraft's antenna radiation pattern is measured inside a large anechoic chamber, such as the one shown here. *(Courtesy of Surrey Satellite Technology, Ltd., U.K.)*

During this test phase, we fully test onboard software to verify all subsystems are "go" and ground controllers can send commands and receive telemetry as expected. As you can imagine, all these computers and radios need electrical power to operate. We'll look at where all this power comes from in the next section.

▮ Section Review

Key Terms

amplitude
anechoic chamber
analog data
application software
bit
bytes
carrier signal
central processing unit (CPU)
communication and data-
    handling subsystem (CDHS)
data
data budget
data rate
demodulator
digital data
information
input/output (I/O)
integrated testing
link budget
MODEM
modulation
modulator
Nyquist criteria
operating system software
radio frequencies (rf)
random access memory
    (RAM)
read only memory (ROM)
sampling
single event upsets (SEU)
South Atlantic Anomaly
space systems engineering
    process
telemetry
word

Key Concepts

➤ The space systems engineering process (Figure 13-1) is the foundation for space mission design

- It helps us translate requirements and constraints into a workable subsystem design

- Along the way, we iterate requirements in the "requirements loop" and the design in the "design loop"

- Throughout the process, we rely on various design and analysis tools

➤ The communication and data-handling subsystem (CDHS) acts as the ears, brain, and mouth of the spacecraft

- It has two primary portions—communication and data handling

- It translates raw data from the payloads and subsystems into useful information and sends it to the ground system

- Commands from ground controllers generally request information from and give specific instructions to individual subsystems

➤ Communicating is the process of laying, or modulating, an information signal onto a carrier signal

- The communication portion of the CDHS sends payload and subsystem information to the ground system and receives commanding information

- It consists of modulators/demodulators ("MODEMs"), amplifiers, and antennas

- Physical, technical, and legal limitations dictate allowable communication frequencies

➤ The data handling portion of the CDHS is very similar to what is in a home computer

- Computers "think" in 1's and 0's, called bits

- Bits are collected into groups called bytes

- Data handling consists of input/output devices (I/O), the central processing unit (CPU), and various types of memory (RAM, ROM, and long-term storage)

Continued on next page

Section Review (Continued)

Key Concepts (Continued)

➤ Software drives the data handling function

- Operating system software runs the computer and schedules events
- Application software executes specific tasks

➤ Data handling is susceptible to space environmental effects such as single even upsets (SEU) and single event latch-ups

➤ To translate real-world analog data into computer-world digital data we must sample the analog signal. The Nyquist criteria dictate that to obtain an accurate representation of the original signal, we must sample at least twice as often as the lowest frequency of the signal.

➤ The CDHS systems engineering process is driven by

- The communication frequencies available
- Other limitations on system design (mass, power, volume, etc.)
- The total amount of onboard data
- The amount of onboard processing
- The amount of time available for downlink to ground stations

➤ Two important CDHS quantities are the data budget and the data rate

  - The data budget tells us the total amount of data to collect, process, and store
  - The data rate tells us how fast the spacecraft must send data to the ground

- During CDHS testing, we execute all the software to simulate mission conditions and we test the communication portion in large anechoic chambers

Example 13-1

Problem Statement

Using the FireSat payload requirements we developed in Chapter 11, we must now derive the data and link budgets for the spacecraft CDHS. Recall, we chose a 1024 × 1024-pixel detector for the FireSat payload. Assume that we will need 8 bits to accurately record each pixel. To minimize data collection, we know we only need to image over land. This means the sensor will only be active about 30% of each orbit. The 500-km orbit will have a period of 90 minutes. To insure we achieve the required coverage, we will need to collect an image about every 30 seconds. To minimize the amount of data sent to the ground station, we'll rely on onboard processing to review and reject images with low probability of having a forest fire. Analysis indicates this will eliminate about 95% of the images taken. All of the remaining images must be down-linked during a 15-min pass over the mission ground station in Colorado. To allow additional margin for problems, we must assume that as much as 3 orbits worth of data must be saved and downloaded during a pass. Given these assumptions, determine the maximum amount of data that must be stored onboard and the minimum data rate needed to download the data to the ground.

Problem Summary

Given: Number of Pixels = 1024 × 1024, 8 bits per pixel, 8 bits per word
P_{orbit} = 90 min
Image every 30 s
Imaging over 30% of each orbit, save 5% of images per orbit
Store up to 3 orbits worth of data
Pass time = 15 min

Find: Maximum data to store (maximum data) and minimum data rate (data rate min)

Conceptual Solution

1) Determine data per image
Data per image = (pixels wide) (pixels long) (bits per pixel)

2) Determine images collected per orbit
Images saved per orbit = (orbital period) (image collection rate) (percent sensor active) (percent images not rejected)

3) Determine maximum data stored
Max data bits = (number of orbits) (images saved per orbit) (data per image)
$$\text{max data bytes} = \frac{\text{max data bits}}{8}$$

4) Determine minimum data rate
$$\text{min data rate} = \frac{\text{max data bits}}{\text{pass time}}$$

Analytical Solution

1) Determine data per image
data per image = (1024) (1024) (8)
data per image = 8.389×10^6 bits

2) Determine images collected per orbit
$$\text{images per orbit} = (90 \text{ min})\left(\frac{2}{\text{min}}\right)(0.30)(0.05)$$
images per orbit = 2.7
Round up to 3 images saved per orbit

3) Determine maximum data stored
max data bits = (3 orbits) (3 images per orbit) $(8.389 \times 10^6$ bits per image)
max data bits = 7.55×10^7 bits
$$\text{max data bytes} = \frac{\text{max data bits}}{8} = \frac{7.55 \times 10^7 \text{bits}}{8}$$
max data bytes = 9.437×10^6 bytes

4) Determine minimum data rate
$$\text{min data rate} = \frac{7.55 \times 10^7 \text{ bits}}{(15 \text{ min})\left(\frac{60 \text{ s}}{\text{min}}\right)}$$
min data rate = 8.389×10^4 bits/s

Interpreting the Results

Given our assumed sensor duty cycle, FireSat will collect about 9 images during 3 orbits of operation. We must design our CDHS to store at least 9.5 MBytes of data before downloading the payload data to the operational control center. To accomplish this, the CDHS needs a minimum data rate of 8.39×10^4 bits/s.

13.2 Electrical Power Subsystem (EPS)

▬ In This Section You'll Learn to...

- ☞ Describe the basic functions of the spacecraft electrical power subsystem (EPS)
- ☞ Define basic concepts and determine parameters of electrical power
- ☞ Identify the main energy sources available to spacecraft, their applications and limitations
- ☞ Discuss the power supply and power conditioning and distribution portions of the EPS
- ☞ Apply the systems engineering process to designing and testing the EPS

Like all modern devices, payloads and subsystems need electrical power to operate. Unfortunately, in space there are no convenient wall outlets, and an extension cord would be much too long! The job of the electrical power subsystem (EPS) is to take some convenient energy source, either one the spacecraft packs along or finds along the way, and convert its energy into usable electrical power to run the entire spacecraft. Similar to any other system, the EPS has inputs and outputs. As Figure 13-11 illustrates, the EPS takes in raw energy from a convenient source, usually the Sun, and converts it into electrical power in a form that other equipment on the spacecraft can use.

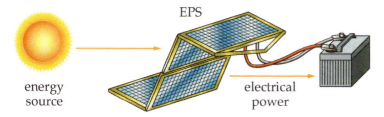

Figure 13-11. Electrical Power Subsystem (EPS). The spacecraft EPS takes in energy from some convenient source, usually the sun, and converts it's energy into electrical power in a form that the payload and other subsystems can readily use.

In this section we'll focus primarily on the basic principles of the EPS operation. We'll start by reviewing some basic concepts in electricity, then turn our attention to spacecraft energy sources. With this background, we can better appreciate the more detailed system functions. Finally, we'll return to the systems engineering process to review some of the design and testing issues important to building an effective EPS.

Basic Principles

To apply the design and analysis tools in the EPS systems engineering process, we need to know about the energy source (or sources) it uses, and the functions of the subsystem. However, to fully appreciate these topics involves knowing about volts, watts, circuits and other concepts in electronics. So let's start there.

Electricity and Circuits

In Chapter 4 we discussed basic concepts that describe the motion of bodies, something we can all relate to at a basic level. But understanding electricity is a bit less intuitive. After all, most of us have never seen an electron, even though we use them whenever we flip a switch to light a room. There are six terms and concepts we must first understand before starting to examine electrical power subsystems. These are

- Charge
- Coulomb's Law
- Voltage
- Current
- Resistance
- Power

To describe these concepts, we build upon some of the concepts of mass and motion we described in Chapter 4 and draw some analogies. Let's start with charge. *Charge* is the basic unit of electricity. It's similar in concept to mass. Recall from Chapter 4 that mass is a basic property of matter that describes its behavior in terms of how much stuff it has or how much gravitational force it generates. We know that applying a force, F, to a mass causes it to accelerate an amount, a, according to Newton's famous second law, F = ma.

Similarly, charge is a basic property of matter, for which we have laws of behavior. Charge comes in two flavors—positive (+) and negative (–). If a particle lacks charge or has an equal number of positive and negative charges, we say it is *neutral*. For mass, the basic unit is the gram. For charge, the basic unit is the *coulomb (C)*. One electron has 1.6×10^{-19} coulombs of negative charge. One proton has exactly the same amount of positive charge.

Due to their nature, opposite charges attract, and like charges repel. This force of attraction or repulsion between charges is called an *electrostatic force*. We show this force in Figure 13-12. We quantify it using Coulomb's Law.

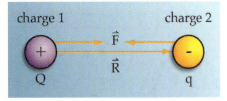

Figure 13-12.   Electrostatic Force. Coulomb's Law of electrostatic force is very similar to Newton's Law of Universal Gravitation.

Coulomb's Law. The force of attraction (or repulsion) between two charges is directly proportional to the amount of each charge and inversely proportional to the square of the distance between them.

Realize that this looks a lot like Newton's Law of Universal Gravitation. We can write a similar equation for electrostatic force

$$\vec{F} = \left(\frac{K\,Qq}{R^2}\right)\hat{R} \tag{13-1}$$

where

$\vec{F}$ = electrostatic force vector between charges 1 and 2 (N)

K = constant (9×10^9) $(kg\,m^3/C^2s^2)$

Q = value of charge 1 (coulombs, C)

q = value of charge 2 (coulombs, C)

R = distance between charges (m)

$\hat{R}$ = unit vector in the $\vec{R}$ direction

Just as mass accelerates in a gravitational field, a particle with charge (electron, proton, or ion) accelerates in an *electric field*. Imagine that a charged particle is like a water droplet. Under pressure, the water droplet flows through a hose. We define *current, i*, as the rate at which charges flow through a given area, such as the cross section of a wire. The unit for current is *amperes (amps)*, which equals the charges flowing per second.

$$amps = charges/time\ (coulombs/s)$$

$$i = \frac{dQ}{dt} \tag{13-2}$$

where

i   = current (amps)

dQ = flow of charges (coulomb)

dt = time (s)

In electricity, potential or voltage is similar to the concept of potential energy from mechanics. Remember that we must expend energy to move an object against a gravitational field. For example, when we climb a flight of stairs, we expend energy that is stored as potential energy. If we were to jump out of a window at the top of the stairs, this potential energy would transform into kinetic energy, because of the gravitational field, as we plummet to the ground. Similarly, we describe *electrical potential* as the energy an electric field can transmit to a unit charge. Returning to our garden hose analogy, it's similar to water pressure: the greater the pressure, the faster the water flows. We use the unit of *volts, V,* to describe this relationship.

$$energy\ per\ charge = volts = joules/coulomb$$

For example, a 12-volt battery in a car can deliver 12 joules of energy to each coulomb of charge. A battery with higher voltage has more potential and can deliver more energy per coulomb of charge. We use the terms *potential* and *voltage* interchangeably.

In electricity, *resistance, R,* tries to prevent charges from flowing. Just as a kink in a hose slows the water flow, a resistor slows the flow of current. The unit of resistance is the *ohm, Ω.* Different materials have different resistance. A material with very low electrical resistance is a *conductor.* Naturally, a material with very high resistance is a *resistor.*

Ohm's Law. *Relates the current to the voltage pushing the charge and the amount of resistance to that push.*

$$i = \frac{V}{R}$$ (13-3)

where

i = current (amps)
V = voltage (volts)
R = resistance (ohms)

This equation tells us that current is directly proportional to voltage and inversely proportional to resistance. Does this make sense? For a given resistance, the higher the voltage, the greater the energy delivered to a unit charge; thus, more charges flow, and current increases. It's as if we increase the pressure in a garden hose. The more pressure on the water (higher voltage), the more water flows (higher current). Furthermore, the higher the resistance, the lower the flow of charges and the lower the current (put a kink in the garden hose and less water flows).

Another important parameter is power, which is the amount of energy delivered per unit time. The unit of power is the *watt, W,* and is defined as one joule of energy per second.

$$1 \text{ W} = 1 \text{ J/s}$$

Power also equals the product of the voltage and the current. This makes sense if we look at what voltage and current represent.

$$V = \text{energy/charge, and } i = \text{charge/time so,}$$

$$P = (\text{energy/charge}) \times (\text{charge/time}) = \text{energy/time}$$

Thus,

$$P = Vi$$ (13-4)

where
P = power (W)
V = voltage (volts)
i = current (amps)

Unlike water in a hose, however, current can only flow in a closed loop called a *circuit.* For example, when we turn on a lamp, we close a switch and establish a circuit of current. With the circuit closed, electrons flow through the light bulb. Resistance within the filament of the bulb causes it to heat up and glow, emitting light. We generally define the flow of current in a circuit to be from positive to negative, as seen in Figure 13-13.

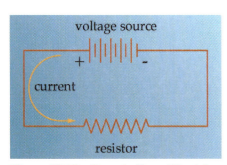

Figure 13-13. Current Flow. We define current as the flow of charge around a closed path. Current can either be direct, DC, or alternating, AC.

The negative terminal of the circuit is also called the ground or Earth. Traditionally, this terminal connects to the ground to dissipate energy into it.

Current is either *direct current, DC,* or *alternating current, AC.* With DC, the current always flows in one direction. For AC, the direction of current flow switches back and forth at some cyclic rate. On Earth, the standard current in the U.S. is 60 Hz (meaning 60 cycles per second) AC at 110V. Standard European AC is 50 Hz at 230V. AC is more efficient for very high power demands and allows for transporting power over long distances through high-voltage transmission lines with lower losses. However, spacecraft most often use DC because most energy sources produce DC current, transmission distances are short, and standard spacecraft components run on DC.

Now that we've reviewed some of the basic principles of electricity, let's explore the energy sources available to spacecraft and see how to use them.

Energy Sources

As we learned at the beginning of the section, the sole input to the EPS is energy from an energy source. Next, we look at the most common spacecraft energy sources, starting with the Sun.

Solar Energy. The Sun is the energy source that drives life on Earth, so it's an obvious energy source for Earth-orbiting spacecraft, as well. We can convert solar energy to electrical energy either indirectly or directly. Indirect methods concentrate sunlight to produce heat that drives a working fluid to run a generator. Direct means use incoming solar photons to create a flow of electrons.

By far, the most common way is direct conversion using solar cells, more technically called *photovoltaic, PV,* cells, similar to the ones in Figure 13-14. When sunlight shines on a solar cell, electrical current flows. But how? *Solar cells* consist of a thin wafer of silicon, gallium arsenide, or other semiconductor crystal. As photons strike the cell, they transmit their energy to the atoms in the cell, freeing some electrons. These electrons then move across tiny junctions of silicon and phosphorous, or similar materials, within the cell. This movement decreases the resistance in the cell, and the freed electrons start to flow. Recall, a flow of electrons (charges) is a current. We have electricity! Due to manufacturing limitations, the maximum size of a single solar cell is fairly small. Therefore, we must wire hundreds or even thousands of individual cells together to form a complete solar array.

Currently, the efficiency of widely available solar cells is quite modest. We define the *solar cell efficiency, η,* as the percentage of incident solar energy that converts to electrical energy. Although laboratory specimens have converted energy at efficiencies over 30%, typical production silicon cells provide only around 15%, while more expensive gallium-arsenide cells approach 20%. Even so, this means only 20% of the solar energy that strikes the surface converts to electrical energy. The rest of the solar energy either reflects or remains as waste heat.

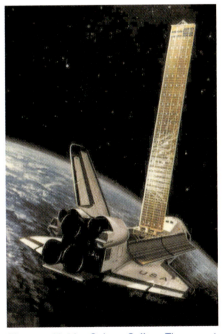

Figure 13-14. Solar Cells. The most common source of spacecraft energy is the sun. Solar cells (or photovoltaic, PV, cells) convert solar energy directly into electricity. Here's an artist's concept of the Solar Array Experiment. *(Courtesy of NASA/Johnson Space Center)*

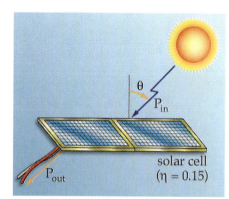

Figure 13-15.   Angle of Incidence. A photovoltaic cell takes incident solar energy and converts it to electrical power at some efficiency (usually around 18%).

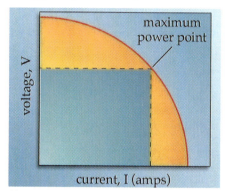

Figure 13-16.   Solar Cell Current/Voltage (I/V) Curve. All solar cells exhibit a unique I/V curve that describes how power output varies with voltage and current. The point of maximum power output is at the "knee" of the curve.

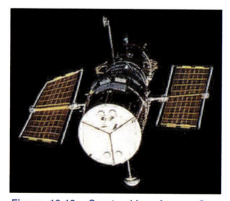

Figure 13-18.   Sun-tracking Arrays. Sun-tracking solar arrays, like the ones shown on the Hubble Telescope, get higher efficiency by maintaining a zero angle of incidence—that is, the sunlight always is perpendicular to the surface. *(Courtesy of NASA/Johnson Space Center)*

Practically speaking, only the component of solar energy hitting the solar cell perpendicular to the surface transforms into electrical energy. We use the *angle of incidence, θ,* to determine how much solar energy hits a solar panel. This angle lies between a line perpendicular to the cell and the Sun line. Figure 13-15 shows solar energy striking a solar cell and the angle of incidence. Thus, the total power output of a solar cell depends on the intensity of the solar energy (in W/m^2) input, its efficiency, and the angle of incidence.

With this information, we can express the output power density (P_{out}) as

$$P_{out} = P_{in} \eta \cos \theta \qquad (13\text{-}5)$$

where

P_{out} = solar cell's output power density (W/m^2)
P_{in} = solar input power density (W/m^2)
η = solar cell's energy-conversion efficiency (typically < 0.25)
θ = incidence angle (deg or rad)—perpendicular to surface = 0°

As we showed in Equation (13-4), power is a function of the current and voltage. All solar cells have a distinctive current/voltage curve (I/V curve), as shown in Figure 13-16. Notice from this curve there is a definite trade-off between current and voltage output. The shape of this curve for a given solar array can change due to temperature, age, and other factors. We can reach the maximum power output by adjusting current and voltage to stay on the "knee" of the curve.

Spacecraft solar arrays can be mounted in one of several ways. The simplest way is to attach all the cells to the outside of the spacecraft. This is called a *body-mounted solar array*. These arrays are relatively simple and typically used on spacecraft with spin or dual-spin stability, such as the IntelSat spacecraft, shown in Figure 13-17. Unfortunately, for body-mounted arrays, the Sun shines on less than half of the cells at one time. The spacecraft's body shades the rest or they have an angle of incidence so high that they produce very little effective power. For a cylindrical spacecraft, less than one-third (actually it's $1/\pi$) of the spacecraft's total surface can generate electricity at any time.

We can overcome this problem by having the solar arrays actively track the Sun, keeping the angle of incidence near zero and the resulting power output near maximum. To do this, we must make the arrays movable and build a control system to keep them constantly pointed at the Sun. Figure 13-18 shows a typical sun-tracking array as used on the Hubble Telescope.

Besides angle of incidence, several other environmental factors can degrade solar cell performance

- Temperature
- Radiation and charged particles
- Eclipses

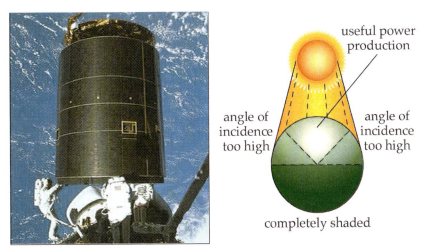

Figure 13-17.   Body-mounted Arrays. The IntelSat spacecraft, shown here being repaired by astronauts, uses body-mounted solar arrays covering the entire cylindrical surface. As the diagram shows, approximately 1/3 of the solar cells produce power at any one time as the spacecraft spins, due to shading and high incidence angle. *(Courtesy of NASA/Johnson Space Center)*

Solar cells are very sensitive to temperature, being most efficient at low temperatures and losing efficiency at higher temperatures. Typical solar cells lose from 0.025% to 0.075% of their efficiency per °C as the temperature increases above 28°C. For example, an array that's 15% efficient at 28° C would be only about 14.75% efficient at 38° C. This means that some type of thermal control is very important for solar arrays.

Solar cells and their cover glass are also extremely sensitive to the radiation and charged particles in space. As radiation and particles hit the arrays, the materials begin to degrade. Depending on the orbit, solar arrays can lose up to 30% of their effectiveness over ten years. For this reason, spacecraft designers deliberately over-design the size of the solar arrays for the start of the mission (*beginning-of-life, BOL*), knowing that over the lifetime of the mission the power output levels gradually decrease. Thus, beginning-of-life power must be high enough, so that, by the end of the mission (*end-of-life, EOL*), power is still available to run all spacecraft systems.

One other complication of relying on solar energy is that orbiting spacecraft periodically pass through Earth's shadow (enter eclipse), as shown in Figure 13-19. During these eclipse periods, the incident solar energy drops to zero, and the solar cells stop producing power. The duration of an orbital eclipse depends mostly on the spacecraft's altitude, which determines the *Earth's angular radius, ρ,* as shown in Figure 13-20 (we can ignore the effect of inclination to find a good worst-case approximation).

Figure 13-19. Earth's Shadow. Most spacecraft pass into Earth's shadow once each orbit blocking input to their solar cells.

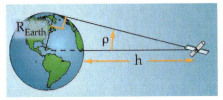

Figure 13-20.   Earth's Angular Radius. As viewed from a spacecraft in orbit, Earth has an angular radius, ρ, which depends on the spacecraft's altitude, h.

$$\rho = \sin^{-1}\left(\frac{R_{Earth}}{h + R_{Earth}}\right) \qquad (13\text{-}6)$$

where

ρ   = Earth's angular radius viewed from space (deg)

h   = orbital altitude (km)

R_{Earth} = Earth's radius = 6378 km

We can then find the total time in eclipse during one orbit by computing the fraction of the orbital period spent in Earth's shadow. (*Note:* This gives us an approximate maximum eclipse time. Actual eclipse times will be shorter due to orbital inclination and other factors.)

$$TE = \frac{2\rho}{360°} \times P \qquad (13\text{-}7)$$

where

TE = maximum time of eclipse (min)

ρ   = Earth 's angular radius (deg)

P   = orbital period (min)

For example, a spacecraft in low-Earth orbit (say at an altitude of 550 km or 342 mi.) experiences 15 eclipses per day (one per orbit) or about 5500 per year. Each lasts up to 36 minutes (depending on the inclination). In contrast, a geostationary orbit (at 35,000 km [21,749 mi.] altitude and 0° inclination) experiences only 90 eclipses per year, with a maximum of 72 minutes.

Of course, if less solar energy goes in, less electrical power comes out. The intensity of solar energy decreases with the square of the distance from the Sun. Near Earth, the density of this incident solar power is about 1358 W/m^2. In comparison, near Venus, it is 2596 W/m^2, and at Mars, it decreases to only 585 W/m^2. Farther from the Sun, near Jupiter, the solar energy is a mere 50 W/m^2. Because the energy is so low, solar power becomes impractical for spacecraft going to Jupiter or any of the other outer planets. The solar arrays or collectors would be far too large to be practical. For these missions, such as the Galileo mission to Jupiter shown in Figure 13-21, we must turn to other energy sources, primarily nuclear, as we'll see later.

Chemical Energy. Anther common energy source for space missions is chemical energy. We'll look at two types of chemical-energy systems— batteries and fuel cells.

Most of us have used batteries in our portable stereos and watches. Batteries are the most common means for converting chemical energy to electrical energy. How do they work? We can demonstrate their basic operating principle by sticking two different types of metal nails (one steel, one copper) into a lemon and connecting them with a conducting wire. If we measure the voltage and current in the wire, we'll find a small potential exists and current is flowing. In this case, the citric acid of the lemon acts as an *electrolyte*—a fluid containing charged particles in the form of ions. The two nails become the *electrodes*—conductors that emit or

Figure 13-21.   Too Far from the Sun. Missions to the outer planets, such as the Galileo mission to Jupiter shown here, are too far from the Sun to rely on solar energy. So, they must rely on radioisotope thermoelectric generators. *(Courtesy of NASA/Jet Propulsion Laboratory)*

collect electrons. One nail collects positive ions (called the *anode*), and the other collects negative ions (the *cathode*).

All types of batteries work using this same basic principle. As electrical energy goes into the battery (charging), a chemical reaction takes place in the electrolyte, storing the electrical energy as chemical energy. When we stop charging and close the circuit, the process reverses and the chemical reaction produces electrical energy (discharging).

Spacecraft can use either primary or secondary batteries. *Primary batteries*, like those in portable stereos, provide the sole source of electrical power. They're designed for a single, short use and can't be recharged. For very short space missions (less than one month), primary batteries can sometimes do the job. It's easier to load enough primary batteries to last throughout a brief mission than to provide some way to recharge them. The FalconGold spacecraft, shown in Figure 13-22 had a mission life of only a few weeks, so it relied on primary batteries as its sole source of power.

Figure 13-22.   FalconGold. The FalconGold spacecraft built at the USAF Academy had a mission duration of only a few weeks. Thus, it was able to rely on primary batteries as the sole source of electrical power. *(Courtesy of the U.S. Air Force Academy)*

However, most space missions last many months, or years, so secondary batteries are much more common. *Secondary batteries* discharge and recharge repeatedly, similar to a car battery. This cycle makes them an ideal backup source of electrical power for solar cells, because they can charge while the spacecraft is sunlit, then discharge during eclipse periods.

One of the most common types of secondary batteries used onboard spacecraft is virtually the same as the rechargeable batteries found in video camcorders—*nickel-cadmium, NiCd*. NiCd batteries can be discharged and recharged thousands of times, and they also have high energy density. This means they can store a relatively high amount of electrical energy for their weight—about 25–30 W · hr/kg.

Nickel-hydrogen, NiH$_2$, batteries, recently developed for space applications, have an even better energy density. Figure 13-23 shows a photograph of NiH$_2$ batteries.

A battery's lifetime depends on how deeply it discharges and the total number of charge/discharge cycles it experiences. *Depth-of-discharge (DOD)* is the percentage of total stored energy that is removed from a battery during a discharge period. The smaller the DOD, the more times a battery can be cycled before it eventually dies. For example, a NiCd battery can be cycled more than 20,000 times at a 25% DOD, but only about 800 times at 75% DOD, before it is unable to hold a charge. This same effect occurs with car batteries. If you repeatedly leave your lights on all day while the car is parked (causing a DOD of nearly 100%), your battery will wear out much faster than the manufacturer's advertised lifetime. Eventually, no matter how long you try to charge it, it just won't hold a charge.

The total number of charge/discharge cycles that batteries undergo depends on the number of eclipse cycles. Lower orbits have shorter periods and experience more eclipses during their missions than spacecraft in higher orbits. Because of this, low-Earth orbit spacecraft need more total battery capacity, meaning more batteries (with nearly 5500 cycles per year), versus those in geostationary orbit (with only 90 cycles per year).

Figure 13-23.   NiH$_2$ Batteries. The top view shows a single nickel-hydrogen (NiH$_2$) battery cell. The bottom view shows a set of NiH$_2$ cells wired to provide total energy storage requirements. *(Courtesy of Eagle Picher Technologies, LLC)*

Batteries also have stringent thermal requirements. Because of the nature of the chemical reactions in the electrolyte, extremely high or extremely low temperatures can greatly affect their ability to hold a charge. Again, car batteries illustrate this point. If you've ever experienced the frustration of a dead car battery on a bitter, cold winter morning, you know how battery charge can decline when exposed to extreme cold. Nominal operating temperature ranges for NiCd batteries, for example, are about 10° C to 30° C.

We size batteries based on the total amount of energy they can hold. We usually express this in terms of watt-hours (W·hrs). Remember that W = energy/time, so W · hr is really (energy/time) × time = energy. As part of Example 13-2, we will determine the total battery capacity for the FireSat mission.

Fuel cells are the second type of chemical-energy system used on spacecraft. When two highly reactive compounds mix in a carefully controlled environment, their chemical reaction can produce a current. Fuel cells have been used on all U.S. manned spacecraft since Gemini (except for Skylab), because these missions have lasted two weeks or less and required power of 1000 W or more. Typically, fuel cells use gaseous hydrogen, H_2, and oxygen, O_2, as the reactants. From elementary chemistry, we know that these two compounds combine to produce H_2O (water). Thus, using fuel cells for manned missions gives an added bonus—drinking water as a by-product of electrical power production. (Drink it and you could get a real jolt!) Figure 13-24 shows a typical fuel cell.

The main drawback of fuel cells is the limited amount of H_2 and O_2 that we can practically place on a mission. As missions last longer than a month, solar cells or other power options become much more practical.

Nuclear Energy. As mentioned earlier, for missions beyond the orbit of Mars, solar energy becomes impractical. Batteries are also impractical because of the mission length. For these applications, mission planners turn to nuclear energy.

On Earth, thermal-cycle fission systems produce electrical power by driving generators with steam created by the heat from nuclear fission. So far, space applications of nuclear power have been far more modest. The primary nuclear power source for space missions has been radioisotope thermoelectric generators (RTGs).

Radioisotope thermoelectric generators (RTGs) use the heat generated by the decay of radioactive isotopes to produce electricity. Isotopes of uranium or plutonium, for example, naturally decay into non-radioactive elements over time. The time it takes for one half of the material to decay is called its *half-life*. Half-lives of some radioactive isotopes can last tens, hundreds, or even thousands of years. As the radioactive material decays, it gives off large amounts of heat. In an enclosed container, the temperature can reach many hundreds of degrees.

Figure 13-25 shows a cut-away view of the RTG used for the Galileo spacecraft. To turn this by-product heat into electricity, RTGs use thermocouples attached to the outside of the heated chamber. *Thermocouples* are metallic strips formed from two unlike metals. As one

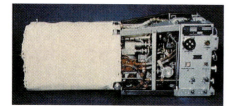

Figure 13-24. Fuel Cells. Fuel cells, like these shown for the Space Shuttle, combine gaseous hydrogen (H_2) and oxygen (O_2) to produce electrical power. Drinking water is a useful by-product. *(Courtesy of NASA/Johnson Space Center)*

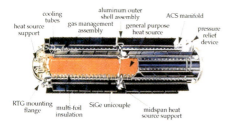

Figure 13-25. Radioisotope Thermoelectric Generator (RTG). RTGs like this one used on the Galileo spacecraft produce electrical power by converting the heat from radioactive decay into electricity using thermocouples. *(Courtesy of NASA/Jet Propulsion Laboratory)*

side of the thermocouple heats, the difference in resistance between the two metals causes free electrons to flow, producing a current. Various industries use thermocouples—often as thermometers—because we can easily compute surface temperatures by measuring the current produced. By placing thermocouples against the contained heat source, engineers can get an RTG to sustain a flow of electricity for many years. For example, RTGs on the Voyager spacecraft are still delivering some power after over twenty years in space.

RTGs have two major drawbacks over other power options—expense and political concerns. Radioactive material is, of course, expensive. But, to design, construct, and test a self-contained RTG power system, housed in a containment shell that can survive in the event of a launch disaster is even more expensive. (Typically, RTGs cost $15,000/W versus solar cells at $3000/W).

Any use of radioactive material brings public scrutiny and invites political concerns. Due to safety concerns, activists sought a court injunction to block the launch of the Galileo spacecraft that used an RTG. Fortunately, mission planners were able to convince the courts that they designed the RTG to safely contain the radioactive material during any

Astro Fun Fact
How Safe are RTGs?

Because RTGs contain radioactive material such as uranium or plutonium, their launch poses obvious public health concerns. To ensure maximum safety, engineers design RTGs to withstand every conceivable launch failure and then they test them to precisely quantify the level of risk for a launch. During the testing program, hundreds of thousands of computer simulations look at how an RTG could be damaged due to booster explosions, accidental re-entry, or impact with the ground. In addition to the computer simulations, engineers shoot bullets traveling 684 m/s (1530 m.p.h.) at the RTG containment cases, burn them with solid rocket propellant, hurl them at high-velocity at concrete and steel, and shoot SRB fragments at them using gas guns and rocket sleds. They also propel fuel elements, clad in iridium, with high velocity at concrete, steel, and sand to ensure their integrity in an accident. The results from all these tests indicate that the probability of exposing even one person

(Courtesy of NASA/Kennedy Space Center)

to 6.37×10^{-3} REM is 16/10,000 (more than one hundred times less than simply living one year at sea level). When they consider the exposure to significant dosages, the probabilities enter the realm of the incredibly small. In comparison, the probability of 1000 fatalities from a dam failure during the same time period of the launch is 1/100. So, are there risks to launching RTGs? Yes, but the risks are extremely small and well understood. It's much more likely someone will be injured driving a car to protest a launch than in an RTG accident.

Executive Summary of the Final Safety Analysis Report for the Ulysses Mission, Prepared for U.S. Department of Energy Office of Special Applications, General Purpose Heat Source Radioisotope Thermoelectric Generator Program, ULS-FSAR-006, NUS Corporation, March, 1990.

conceivable launch disaster and the mission proceeded. Because of these inherent drawbacks—expense and politics—we use RTGs only for missions that preclude using any other type of power system. Future, more ambitious missions to colonize the planets may look for more extensive use of nuclear energy, for electrical power, and for propulsion, as we'll describe in Chapter 14.

System Functions

Now that we've described some of the basic principles of electrical power and the energy sources available, we can focus our attention on the system functions in a bit more detail.

Figure 13-26 shows a more detailed view of a typical EPS that relies on solar energy. In this example, the system converts raw solar energy into DC electrical power within the solar panels, as described earlier. We divide the other EPS functions into two stages: power supply, and power conditioning and distribution. First, we'll describe the functions of each of these stages, then look at a simple example of an EPS in action.

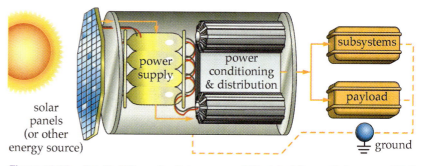

Figure 13-26.   Detailed View of a Solar-powered Electrical Power Subsystem (EPS). An EPS using solar energy first converts the energy to electricity using solar panels. The EPS then distributes the power to the subsystems and payloads. To safely complete the circuit, everything must be tied to a common ground.

Power Supply. In the first stage, the power supply regulates the power output from the solar panels. There are two ways to do this: direct energy transfer and peak power tracking. With *direct energy transfer*, the most common way, the output voltage from the solar cells is constant, and the total current (hence total power) varies depending on the operating location on the I/V curve, as shown in Figure 13-16. This method means that in some situations (such as when the solar panels first come into sunlight after eclipse, when they're coolest) more power will be available than the system can handle. The system simply wastes this extra power by running it through a resistor (heater). Wasting this small amount of energy is considered worth the convenience of having a standard output voltage.

When power is at a premium, especially for very small spacecraft, we can use a peak power tracking system. As the term implies, for *peak power tracking*, the system constantly adjusts the current and voltage to stay on the "knee" of the I/V curve (Figure 13-16), so that it maximizes the power output.

Whatever the total power output from the panels, the control system must condition it, in terms of voltage and current, for supply to the spacecraft bus. When the bus has any excess power, it charges the batteries. During eclipse, of course, the batteries supply all of the power. Figure 13-27 illustrates these functions within the power supply portion of the EPS and shows the flow of electrical power.

Power Conditioning and Distribution. In the second stage, the EPS further conditions bus power for supply to the individual subsystem users and distributes it according to specific mission needs, as shown in Figure 13-28. Let's consider the conditioning task first. The bus voltage may be a standard 28V, while the CDHS may need only a 5V-power supply. In this case, the EPS needs to drop the voltage before supplying power to this user.

Note that this conditioning function can take place either within the EPS (a "centralized" system) or within each subsystem ("decentralized"). The decentralized approach is somewhat more power efficient, but places extra demands on each subsystem design. However, for purposes of our discussion, we'll focus on a centralized approach to conditioning and distributing electrical power.

Once its produced, the power must go to users. However, users can't simply take what they want. First of all, this approach could endanger the entire spacecraft. Second, it is simply greedy! As Figure 13-28 shows, before being distributed, the power must pass through some protection circuits. In our homes, all power passes through fuses or circuit breakers before going throughout the house. If any individual user starts drawing too much current (for example, if you have too many appliances plugged into the same outlet), the fuse burns through or the circuit breaker trips, cutting power to that circuit. On a spacecraft, fuses are difficult to replace, so we use power switches or solid-state breakers that ground commands can reset.

Pre-determined mission requirements and commands from the ground determine the allocation of power to individual users. For example, a mission using synthetic aperture radar, such as the Magellan spacecraft used to map Venus, had very high payload power demands during short, specific phases of the mission. During those times, power diverted from less important functions to ensure the payload was adequately supplied.

One final system design issue to consider is grounding. Notice in Figure 13-26 that the EPS, the payload, and all of the subsystems tie to a common ground terminal. Electrical power will always take the easiest path to the ground. During the EPS design phase, proper attention to spacecraft grounding is important to prevent unwanted "ground loops" that can lead to power drains and subsystem damage.

Example. Let's look at a simple example to see how the various principles of EPS work together. Imagine an average-sized spacecraft with 1-m^2 sun-tracking arrays in Earth orbit. Analyzing the EPS functions mainly involves "bookkeeping" the total energy input versus the electrical energy output.

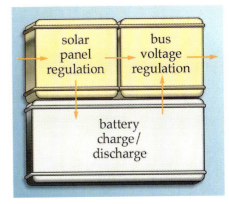

Figure 13-27. Power Supply. The power supply portion of the EPS regulates power output from the solar cells, charges the batteries, and conditions power for further distribution.

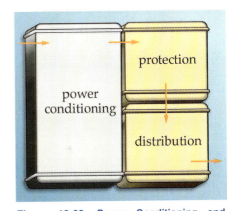

Figure 13-28. Power Conditioning and Distribution. The power conditioning and distribution portion of the EPS conditions power for individual subsystems as needed, provides circuit protection in the form of fuses and breakers, and distributes power to onboard users.

Figure 13-29 illustrates a possible power flow and distribution for our example spacecraft. The area of the arrays, their efficiency, and the incidence angle determines the maximum power produced. By tracking current and voltage, the EPS power supply can keep this power maximized.

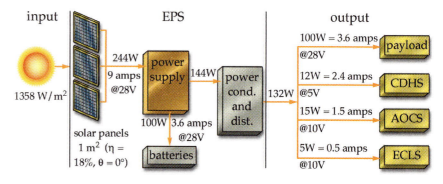

Figure 13-29.   Electrical Power Subsystem (EPS) Example. This example illustrates how the EPS converts energy into electrical power and rations it to onboard users.

Now, some of this power must go to run the spacecraft and some to charge the batteries for use during eclipses. If about 40% goes to charge the batteries, the remainder is available to run the bus. Some of this power the EPS uses for its internal requirements. This amount represents overhead, to keep the whole system running. If we start with 244 W, for our example, this leaves 132 W for the payload and subsystems. However, different users can have different delivery requirements. The payload, for example, may use the standard bus voltage of 28 V, while the CDHS needs its 12 W at 5 V for a total current demand of 2.4 amps. As the example shows, we must account for all inputs and outputs.

Systems Engineering

After this electrifying review of EPS basics and system functions, we finish our discussion by looking at some of the design and testing issues associated with EPS systems engineering. In Chapter 11, we said the overall spacecraft *power budget* drives the EPS design. Let's start by looking closer at what this budget represents.

As our earlier example illustrated, we can build the *total system power budget* by adding the individual demands of each power "customer" in the spacecraft. Typically, the payload is one of the neediest customers, followed by the other subsystems. However, this isn't necessarily the final power output, for which we should design the EPS. Recall from our discussion of photovoltaic systems, that environmental factors such as radiation and charged particles can degrade their power by as much as 4% per year. Thus, we should size the total system power budget to supply the end-of-life (EOL) power needs. Then, we can factor in the planned mission lifetime to determine the required beginning-of-life (BOL) power that we must design the EPS to produce.

All of these budgets represent static, maximum amounts. In reality, to rely on these amounts alone would be similar to computing your house power demand by assuming that all of your appliances (TV, stereo, washing machine, dryer, etc.) were on at the same time (not usually the case!). Instead, we must consider the dynamic demands of the mission. This effort includes understanding the duty cycle of the batteries, payload, and subsystems. *Duty cycle* defines the use timeline for individual components. For example, a SAR payload may require very large amounts of power, but only during short periods of the mission. Rather than designing the solar array size to meet this peak power demand, mission designers could oversize the batteries and use them to supplement direct solar power during payload operations.

Due to eclipse periods, power input also varies widely throughout an orbit. If our spacecraft has body-mounted arrays, the power output can also vary widely even while in sunlight. For these reasons, designers make a distinction between peak power and orbit average power. *Peak power* refers to the maximum power output from the EPS when the arrays are in full sunlight. *Orbit average power*, a more useful measure of available power, is the EPS power output averaged over a single orbit. In Example 13-2 we'll see how all these requirements effect the overall EPS design for the FireSat mission.

After we consider all these design factors, and build the final system, it will require extensive testing to validate that the design meets the overall mission objectives. As with the CDHS, we test individual components of the EPS separately and then together in successive stages of assembly. During integrated testing, we can put the spacecraft in a large solar simulation chamber, as shown in Figure 13-30, to measure output from the solar arrays and ensure the EPS works well with the other subsystems making electrical demands.

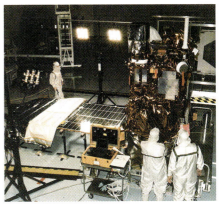

Figure 13-30.   Spacecraft EPS Test in Solar Chamber. Here we show solar panels from the Mars Global Surveyor being subjected to simulated sunlight inside a large solar chamber at Kennedy Space Center. *(Courtesy of NASA/ Kennedy Space Center)*

Section Review

Key Terms

alternating current, AC
amperes (amps)
angle of incidence, θ
anode
beginning-of-life (BOL)
body-mounted solar array
cathode
charge
circuit
conductor
coulomb, C
current, i
depth-of-discharge (DOD)
direct current, DC

Key Concepts

➤ The spacecraft's electrical power subsystem (EPS) converts energy, such as solar energy, into usable electrical power to run the spacecraft

➤ Charge, current, voltage, and power are some of the basic quantities of electrical power

- Charge is the basic unit of electricity; it can be positive or negative

- Coulomb's Law describes the electrostatic force between charges in an electric field

- Voltage, or potential, describes the amount of energy that is delivered to a unit of charge

- Current is the rate at which charges flow through a given area

- Power is energy delivered per unit time

Continued on next page

■ Section Review (Continued)

Key Terms (Continued)

direct energy transfer
duty cycle
Earth's angular radius, ρ
electric field
electrical potential
electrodes
electrolyte
electrostatic force
end-of-life (EOL)
half-life
neutral
nickel-cadmium, NiCd
nickel-hydrogen, NiH_2
ohm, Ω
orbit average power
peak power
peak power tracking
photovoltaic, PV
potential
power budget
primary batteries
radioisotope thermoelectric
    generators (RTGs)
resistor
resistance, R
secondary batteries
solar cells
solar cell efficiency, η
thermocouples
total system power budget
voltage
volts, V
watt, W

Key Equations

$$i = \frac{V}{R}$$

$$P = Vi$$

$$P_{out} = P_{in}\eta\cos\theta$$

$$\rho = \sin^{-1}\left(\frac{R_{Earth}}{h + R_{Earth}}\right)$$

$$TE = \frac{2\rho}{360°} \times P$$

Key Concepts (Continued)

➤ Spacecraft energy sources include the Sun, chemical energy, and nuclear energy

➤ Solar energy systems primarily rely on solar cells, also called photovoltaic, PV, cells that convert sunlight directly into electricity

➤ Chemical-energy systems use either batteries or fuel cells

   • Spacecraft batteries are either primary or secondary

      – Primary batteries are the sole source of power. They don't recharge and are for short, one-shot missions.

      – Secondary batteries provide backup power and can recharge many times

   • The lifetime of secondary batteries depends on the depth of discharge (DOD) and the number of cycles

   • Fuel cells produce power by combining two reactants such as oxygen and hydrogen. Most United States manned missions have used them, but they have limited lifetimes.

➤ Nuclear energy sources for spacecraft are primarily radioisotope thermoelectric generators (RTGs). These have been almost exclusively used for missions to the outer solar system where sunlight is too week to be a practical energy source. The primary disadvantages of nuclear energy sources are expense and political concerns.

➤ There are two stages within the EPS process: power supply, and power conditioning and distribution

➤ Mission planners must design to a beginning-of-life (BOL) system power output that will give the necessary end-of-life (EOL) power. In addition, they must consider the battery, payload, and subsystem duty cycles, as well as the orbit average power produced.

Example 13-2

Problem Statement

As the FireSat program manager, your task is to complete preliminary design calculations for the electrical power subsystem (EPS). Analysis of the spacecraft bus indicates the following power allocation requirements for each subsystem

- CDHS: 8W
- AOCS: 3W
- ECLSS: 0 W (passive thermal control)
- EPS: 0.5 W

At this time, we're still completing the final power requirements for the payload. We know the sensor, including CCDs and electronics, will consume 7.5 W. In addition, it will use a solid state thermoelectric cooler, but we haven't selected a final component yet. You can assume the operating voltage for the cooler will be 5V, and you can model the device as a 2.0Ω resistor. Fortunately, at most, the cooling system and payload will operate 30% of each orbit.

From Example 11-1, we know to plan for a 5 year mission in a 500-km low-Earth orbit (with a solar input of 1358 W/m²). Fortunately, the spacecraft will be able to use gallium arsenide solar cells at 28% efficiency. Assume that at this relatively low altitude solar panels will lose 3% of their efficiency per year. Given these system constraints, the payload duty cycle and assuming other subsystems will operate continuously, compute the average orbit power required, which will be the same as the mission end-of-life power. From here, determine if the planned solar array area that will cover four sides of the 0.3 m cubic spacecraft will provide sufficient electrical power for the demand (assume a maximum of one panel fully illuminated at any time). Finally, determine the required battery capacity (in W · hr) to meet only subsystem demands (no payload operations) during eclipse at a 25% depth of discharge.

Problem Summary

Given: Payload and subsystem power budgets with duty cycles
payload power = 7.5 W
cooler voltage = 5 V
cooler resistance = 2.0 Ω

Operating 30% of the orbit
payload duty = 0.2
CDHS power = 8 W
AOCS power = 3 W
ECLS power = 0 W
EPS power = 0.5 W
Array design data
$\eta_{cell} = 0.28$
$h_{mission} = 500$ km
$R_{Earth} = 6378$ km
mission life = 5 yrs

$$\mu_{Earth} = 3.986 \times 10^5 \frac{km^3}{s^2}$$

array degrade = 0.03/yr
panel width = 0.3 m
panel height = 0.3 m
DOD = 0.25

$$solar\ input = 1358 \frac{W}{m^2}$$

Find: Average orbit power required = required end-of-life power
Actual end-of-life power
Minimum solar array area compared to actual planned area
Battery capacity

Conceptual Solution

1) Determine the total peak payload operating power, including the solid-state cooler

$$cooler\ current = \frac{cooler\ voltage}{cooler\ resistance}$$

cooler power = (cooler current) (cooler voltage)
peak payload power = payload power + cooler power

2) Determine the average payload power required

average payload power = (peak payload power) (payload duty)

3) Determine the continuous subsystem power required

subsystem power = CDHS power + AOCS power + ECLS power + EPS power

Example 13-2 (Continued)

4) Find Earth's angular radius at the mission altitude

$$\rho = asin\left(\frac{R_{Earth}}{h_{mission} + R_{Earth}}\right)$$

5) Find the orbit's semimajor axis and period

$$a_{mission} = R_{Earth} + h_{mission}$$

$$P = 2\pi\sqrt{\frac{a_{mission}^3}{\mu_{Earth}}}$$

6) Find the maximum time of eclipse (TE) and the time in sunlight (TS)

$$TE = \frac{2\rho}{360 \text{ deg}}P$$

$$TS = P - TE$$

7) Find the energy needed during eclipse

energy eclipse = (subsystem power) (TE)

8) Find the required battery capacity

$$\text{batter capacity} = \frac{\text{energy eclipse}}{DOD}$$

9) Determine the actual solar array area exposed to sunlight

array area = (spacecraft width) (spacecraft height)

10) Determine the beginning-of-life peak array power

array power BOL = (sunlit array area) (η_{cell}) (solar input)

11) Determine the end-of-life array power

array power EOL = (sunlit array area) (η_{cell})
$[(1.0 - \text{array degrade})^{\text{mission life}}]$ solar input

12) Determine the orbit's average power produced

$$\text{array orbit average power} = \frac{(\text{array power EOL})(TS)}{P}$$

13) Determine the required orbit average power

required orbit average power = average payload power + subsystem power

14) Determine if the planned array size is sufficient for the mission

Compare if the array power EOL ≥ orbit average power

15) Check the energy "bookkeeping"

energy input EOL = (array power EOL) (TS)
energy output = (subsystem power) (P) + (peak payload power) (P) (0.2)

Analytical Solution

1) Determine the total peak payload operating power including the solid-state cooler

$$\text{cooler current} = \frac{\text{cooler voltage}}{\text{cooler resistance}} = \frac{5 \text{ V}}{2\Omega}$$

cooler current = 2.5 A
cooler power = (cooler current) (cooler voltage)
= (2.5 A) (5 V)
cooler power = 12.5 W
peak payload power = payload power + cooler power = 7.5 W + 12.5 W
peak payload power = 20.0 W

2) Determine the average payload power required

average payload power = (peak payload power) (payload duty) = (20.0 W) (0.3)
average payload power = 6.0 W

3) Determine the continuous subsystem power required

subsystem power = CDHS power + AOCS power + ECLSS power + EPS power = 8 W + 3 W+ 0 W + 0.5 W)
subsystem power = 11.5 W

4) Find Earth's angular radius at mission altitude

$$\rho = asin\left(\frac{R_{Earth}}{h_{mission} + R_{Earth}}\right)$$

$$= asin\left(\frac{6378 \text{ km}}{500 \text{ km} + 6378 \text{ km}}\right)$$

$$\rho = 68.0 \text{ deg}$$

5) Find the orbit's semimajor axis and period

$$a_{mission} = R_{Earth} + h_{mission} = 6378 \text{ km} + 500 \text{ km}$$

$$a_{mission} = 6.878 \times 10^3 \text{ km}$$

$$P = 2\pi\sqrt{\frac{a_{mission}^3}{\mu}} = 2\pi\sqrt{\frac{(6878 \text{ km})^3}{3.986 \times 10^5 \dfrac{km^3}{s^2}}}$$

$$P = 94.6 \text{ min}$$

Example 13-2 (Continued)

Find the maximum time of eclipse (TE) and the time in sunlight (TS)

$$TE = \frac{2\rho}{360 \text{ deg}} P = \frac{2(68.0°)}{360°}(94.6 \text{ min})$$

TE = 35.7 min
TS = P − TE = 94.6 min − 35.7 min
TS = 58.9 min

Find the energy needed during eclipse

energy eclipse = (subsystem power) (TE) =
$$\frac{(11.5\text{W})(35.7 \text{ min})}{60 \text{ min}/\text{hr}}$$
energy eclipse = 6.84 W · hr = 2.463×10^4 J

) Find the required battery capacity

$$\text{battery capacity} = \frac{\text{energy eclipse}}{\text{DOD}} = \frac{6.84\text{W} \cdot \text{hr}}{0.25}$$

battery capacity = 27.4 W · hr

) Determine the actual solar array area exposed to sunlight

sunlight array area = (spacecraft width)
(spacecraft height) = (0.3 m) (0.3 m)
sunlight array area = 0.09 m^2

0) Determine the beginning-of-life peak array power

array power BOL = (sunlight array area) (η_{cell})
(solar input) = (0.09 m^2) (0.28) (1358 W/m^2)
array power BOL = 34.2 W

1) Determine the end-of-life array power

array power EOL = (sunlight array area) (η_{cell})
[(1.0 − array degrade)$^{\text{mission life}}$] solar input =
(0.09 m^2) (0.28) (1.0 − 0.03)^5 (1358 W/m^2)
array power EOL = 29.4 W

2) Determine the array orbit average power produced

array orbit average power = $\frac{\text{(array power EOL)(TS)}}{P}$

$$= \frac{(29.4 \text{ W})(58.9 \text{ min})}{94.6 \text{ min}}$$
array orbit average power = 18.3 W

3) Determine the required orbit average power

required orbit average power = average payload
power + subsystem power = 6.0 W + 11.5 W
required orbit average power = 17.5 W

14) Determine if the planned array size is sufficient for the mission

array orbit average power – required orbit
average power = 18.3 W − 17.5 W = 0.8 W

Our array size gives us a 0.8 W margin, so the planned array size is sufficient.

15) Check energy "bookkeeping"

energy input EOL = (array power EOL) (TS) =

$$(29.4 \text{ W}) (58.9 \text{ min}) \left(\frac{1 \text{ hr}}{60 \text{ min}}\right)$$

energy input EOL = 28.9 W · hr = 1.04×10^5 J

energy output = (subsystem power) (P) + (peak payload power) (P) (0.2) =

$$(11.5 \text{ W} \cdot 94.6 \text{ min} \cdot \frac{1 \text{ hr}}{60 \text{ min}}) +$$

$$(20.0 \text{ W} \cdot 94.6 \text{ min} \cdot \frac{1 \text{ hr}}{60 \text{ min}} \cdot 0.3) =$$

18.1 + 9.46

energy output = 27.56 W · hr = 9.92×10^4 J

Energy input is greater than energy output. The bookkeeping checks and the array's sizes are confirmed to be sufficient.

Interpreting the Results

By using body-mounted arrays, the available surface area on our cubical spacecraft will be sufficient to provide adequate end-of-life power for our assumed payload duty cycle and subsystem power requirements. We are left with about 0.8 W additional power margin to allow for extra payload operations or other contingencies, as needed. The batteries will need at least 27.4 W · hr capacity to handle the load during eclipse at 25% DOD.

13.3 Environmental Control and Life-support Subsystem (ECLSS)

▰ In This Section You'll Learn to...

☞ Explain the primary function of the environmental control and life-support subsystem—thermal control and life support

☞ List the primary sources of heat for a spacecraft

☞ Describe the three basic means of heat transfer—conduction, convection, and radiation—and how to use them on a spacecraft

☞ Describe the various ways to control heat outside and inside a spacecraft

☞ Explain how, from the standpoint of the life support, we view humans as systems with inputs and outputs

☞ Apply the space systems engineering process to the design and testing of the ECLSS

Figure 13-31. Healthy and Happy "Pay-loads." The job of the environmental control and life-support subsystem (ECLSS) is to provide a comfortable environment for sub-systems and payloads, including astronauts, to live and work. *(Courtesy of NASA/Johnson Space Center)*

As we know from Chapter 3, space is a rough place to live and work—for humans and machines. For spacecraft to survive, and even thrive, we need some way to keep the payload and all the subsystems onboard (including the crew) healthy and happy, as shown in Figure 13-31. Providing a livable environment in the harshness of space is the function of the environmental control and life-support subsystem (ECLSS).

We can divide the tasks of the ECLSS conceptually into two problems—thermal control and life support. In this section, we'll focus primarily on the thermal control problem, the major concern for unmanned spacecraft. Then we'll introduce some of the additional complications caused by placing fragile humans onboard. Finally, we'll look at some of the major systems engineering issues of ECLSS design.

System Overview

A spacecraft, orbiting peacefully in space, is perhaps the ultimate example of an isolated system. We can quite easily analyze everything that goes in and out. One of those things is heat. If more heat goes into a spacecraft, or is produced internally, than leaves it, then it's temperature increases. If more heat leaves than goes in, it begins to cool. Because sensitive equipment and payloads (including humans) can't survive wide temperature swings, we need to balance the heat flow in, plus the heat generated internally, with the heat flow out. We'd prefer the spacecraft's average temperature to stay nearly constant, a condition we call *thermal equilibrium*.

The primary job of the thermal control portion of the ECLSS is to regulate and control the amount of heat that gets in, out, and moves around inside of a spacecraft. Just as the furnace and air conditioner do in our homes, the thermal-control subsystem regulates and moderates the spacecraft's temperature. To maintain thermal equilibrium, the ECLSS must balance inputs and outputs, as well as internal heat sources. This means the heat coming in plus the heat produced internally must equal the total heat ejected.

Heat Out = Heat In + Internal Heat (for thermal equilibrium)

Figure 13-32 shows a simple view of the thermal control inputs and outputs, and illustrates the flow of heat between payloads and subsystems.

Basic Principles of Thermal Control

Later we'll see what techniques the thermal control subsystem uses to control this heat flow. But first, we'll review some internal and external heat sources for spacecraft. Following this, further discussion of thermal control methods requires an understanding of basic thermodynamics, so we'll review some of the principles of heat transfer.

Heat Sources in Space

Typically, the biggest problem for spacecraft thermal control is removing heat. For the most part, we must maintain temperatures inside even unmanned spacecraft at normal room temperature (20° C or about 70° F). In some cases, specific payloads may have more demanding requirements. Infrared sensors, for example, require refrigeration units to supercool them to 70 K (–193° C or –316° F).

As Figure 13-33 shows, heat in space comes from three main sources

- The Sun
- Earth
- Internal sources

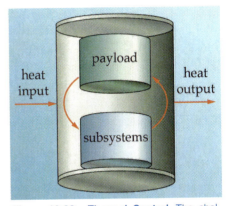

Figure 13-32.   Thermal Control. The challenge of spacecraft thermal control is to balance the heat input plus the heat produced internally with the heat output to maintain thermal equilibrium.

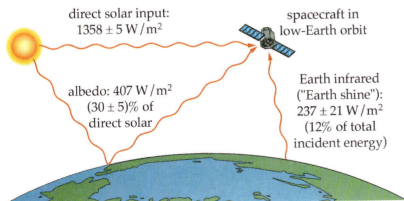

direct solar input:
$1358 \pm 5 \text{ W/m}^2$

spacecraft in low-Earth orbit

albedo: 407 W/m^2
$(30 \pm 5)\%$ of direct solar

Earth infrared ("Earth shine"):
$237 \pm 21 \text{ W/m}^2$
(12% of total incident energy)

Figure 13-33.   Heat Sources in Space. A spacecraft gets heat from the Sun, Earth (reflected and emitted energy), and from within itself.

Near Earth, the biggest source of heat for orbiting spacecraft is the Sun—about 1358 W/m². We all know how hot we get standing in the Sun on a summer day. For a satellite in space, the Sun's heat is much more intense because there is no atmosphere to absorb radiant energy and moderate the temperature. On the side facing the Sun, the surface of a spacecraft can reach many hundreds of degrees Kelvin. On the side away from the Sun, the temperature can plunge to a few degrees Kelvin.

So what is the temperature in space? Earth is about 300 K, while temperatures in space range from 900–1300 K. Sounds hot, but is it? On Earth we measure temperature using a thermometer. The fluid in the thermometer expands when heated by air molecules brushing by it. Temperature is proportional to the velocity and number of the molecules. In space, the molecules are traveling faster but there aren't very many of them. So while the temperature appears higher in space, the effect on people and materials is much less than the equivalent temperature on Earth.

For satellites in low-Earth orbit, Earth is also an important source of heat because of two effects. The first results from sunlight reflecting off Earth—called *albedo*. It accounts for as much as 407 W/m², or 30% of the direct solar energy on a spacecraft. Another important source is "Earth shine," or the infrared energy Earth emits directly, as a result of its temperature. This accounts for another 237 W/m² or about 12% of the incident energy on a spacecraft.

Internal sources also add heat. Electrical components running onboard and power sources such as radioisotope thermoelectric generators (RTGs) produce waste heat. If you've ever placed your hand on the top of your television after it's been on a while, you know how hot it can get. In your living room, the heat from the television quickly distributes throughout the room because of small air currents. Otherwise, your television would overheat and be damaged. Unfortunately, it's not so easy to keep spacecraft temperatures balanced onboard. We'll see what methods are available for moving heat around, next.

Heat Transfer

Recall from Chapter 3 that heat transfers from one point to another through

- Conduction
- Convection
- Radiation

Let's start with conduction. If you hold one end of a long metal rod and put the other end into a fire, as in Figure 13-34, what happens? You get burned! The heat from the fire somehow flows right along the metal rod. When heat flows from hot to cold through a physical medium (in this case the rod), we call it *conduction*, and we experience it every day. It's the reason we put insulation in the walls of our home to prevent heat from the inside being conducted outside (and vice versa in the summer time).

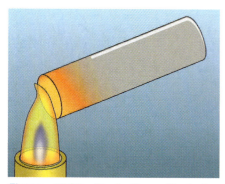

Figure 13-34. Conduction. Conduction occurs when heat flows through a physical medium from a hot point to a cooler point.

We can describe the amount of heat transfer due to conduction using the Fourier Law (developed by J. B. J. Fourier [1768–1830]).

$$q = -kA\frac{\Delta T}{\Delta X}$$ (13-8)

where

q   = heat energy conducted per unit time (W)

k   = thermal conductivity of the material (W / K m)

A   = cross-sectional area of the material (m^2)

ΔT = temperature difference between two sides of the material (K)

ΔX = distance between "hot" and "cold" points in the material (m)

Similar to the flow of electrons discussed in the last section, this relationship indicates heat will flow faster if the material is a better heat *conductor* (high k), such as metal rather than wood. It will also flow faster if a larger area is available, if the temperature difference is greater, or if the distance is smaller. We use this principle to insulate the walls in our homes by building thick walls (large ΔX) and filling them with poor heat conductors (insulators) that have low k.

The second method of heat transfer is convection. If you've ever boiled water, you've used convection. Let's look at how water boils in a pot on the stove, as illustrated in Figure 13-35. Water on the bottom of the pot, nearest the heat source, gets hot first, through conduction, directly from the heat source. As the water gets hot, it expands slightly, making it a bit less dense than the water above it. At the same time, gravity pulls heavier material to the bottom of the pot. Thus, the cooler, denser water at the top of the pot displaces the warmer, less dense water at the bottom. Once on the bottom, this cooler water also heats, expands, and rises. A convection current then continues as water flows past the heat source, driven by the force of gravity, until it reaches thermal equilibrium (heat boils out the top at the same rate the stove adds heat on the bottom). Unlike conduction, which relies on heat flow *through* a solid medium, *convection* transfers heat to a fluid medium flowing past a heat source. Obviously, if the fluid is flowing, something must make it flow. Convection relies on gravity, or some other force, to push the fluid past the heat source.

In the free-fall environment of space, there are no forces to cause cooler water to replace the warmer water (everything free falls together). For convection to work in space, we must supply the force to move the fluid. For example, Russian spacecraft have long relied on forced convection to cool their spacecraft electronics. The components are in a large pressure vessel filled with nitrogen at about 1 bar pressure (14.7 p.s.i.). They use fans to circulate the nitrogen around the vessel, cooling the electronics. Figure 13-36 shows one of the Russian *Meteor* spacecraft that relies on this method of thermal control.

Figure 13-35. Convection. Convection occurs when some driving force, such as gravity, moves the medium (usually a liquid or gas) past a heat source.

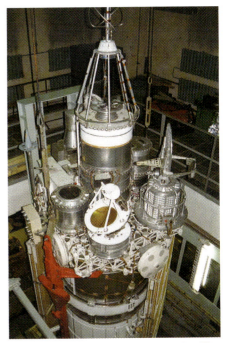

Figure 13-36. Meteor Spacecraft. The Russian Meteor spacecraft is able to use convective cooling of onboard electronics by sealing everything in a large pressure vessel and using fans to circulate nitrogen. *(Courtesy of A. Koorbanoff and C. Maag)*

The final method of heat transfer is radiation. If you've ever basked in the warm glow of an electric space heater, you've felt the power of heat transfer by radiation. *Radiation* is the means of transferring energy (such as heat) through space. More specifically, radiative heat transfer occurs through electromagnetic (EM) radiation. As we described in Chapter 11, EM radiation is the waves (or particles) emitted from an energy source. Recall that a red-hot piece of metal acts as a blackbody radiator, meaning the intense heat energy causes it to emit EM radiation. In this case, the frequency of the EM radiation is in the visible (red) portion of the EM spectrum. We use Stefan-Boltzmann's Law to describe the heat-power transfer by radiation.

$$q = \sigma \varepsilon A T^4 \qquad (13\text{-}9)$$

where

q = heat-power transfer per unit time (W)

σ = Stefan-Boltzmann's constant (5.67×10^{-8} W/m^2 K^4)

ε = emissivity ($0 < e < 1$)

A = surface area of the body (m^2)

T = black body temperature (K)

This relationship tells us that, as the temperature of a black body increases, the amount of heat power it emits increases by the fourth power of the temperature. Thus, if we double the temperature, the amount of energy emitted will increase sixteen times.

As Figure 13-37 shows, when radiation strikes a surface, the material reflects, absorbs, or transmits it. Reflected radiation is the same as reflected light from a mirror. This type of radiation basically bounces off the surface. We use the symbol, ρ, to identify the *reflectivity* of a surface (don't confuse this ρ with Earth's angular radius from Section 13.2). We work with reflectivity as a percentage; that is, $\rho = 0.3$ means that an object reflects 30% of the radiation that hits it.

Absorbed radiation is energy the surface captures, just as a sponge soaks up water. Absorbed radiation eventually causes the surface temperature to rise. We use the symbol, α, to identify absorptivity. We also work with *absorptivity* as a percentage; that is, $\alpha = 0.5$ means an object absorbs 50% of the radiation that hits it.

Transmitted radiation is energy that passes right through (the same as visible light passes through a pane of glass). We use the symbol τ to quantify transmissivity. *Transmissivity* too is a percentage; that is, $\tau = 0.2$ means an object transmits 20% of the radiation that hits it.

Because of the conservation of energy, all of the radiation must go somewhere. So the sum of the reflected, absorbed, and transmitted radiation energy equals the incoming energy. Another way of looking at this is

$$\tau + \alpha + \rho = 1 \qquad (13\text{-}10)$$

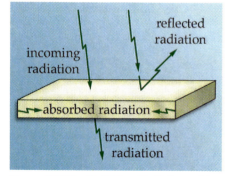

Figure 13-37.   Radiation. Radiation striking a surface is either reflected, absorbed, or transmitted.

where

τ = transmissivity $(0 < \tau < 1)$
α = absorptivity $(0 < \alpha < 1)$
ρ = reflectivity $(0 < \rho < 1)$

As an object absorbs energy, the kinetic energy of individual molecules increases and the object gets hotter. As Figure 13-38 shows, all objects above absolute zero (0 K) emit radiation. But not all materials emit heat with the same efficiency. We call a material's ability to emit heat its *emissivity, ε.* A pure black body has an emissivity of 1.0. The black tiles on the Space Shuttle have a very high emissivity ($\varepsilon = 0.8$).

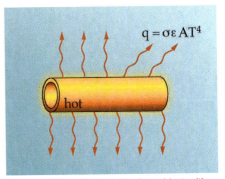

Figure 13-38.   Emissivity. Any object with a temperature above 0 K (meaning basically everything in the universe) emits (EM) radiation per the Stephan-Boltzmann relationship. The greater the emissivity of a material the more energy it emits at a given temperature.

Methods for Spacecraft Thermal Control

Now that we've explored the heat transfer options, let's see how we use them to maintain spacecraft thermal equilibrium. As we said earlier, we must manage heat coming into and out of the spacecraft (external thermal control), as well as heat generated inside (internal thermal control).

There are two basic approaches to thermal control. The easiest method is passive thermal control. *Passive thermal control* is an open-loop means of controlling the spacecraft's temperature, by carefully designing the entire system to regulate heat input and output, and creating convenient heat conduction paths. The nice thing about passive thermal control is, once it gets going, it requires no additional control inputs. Unfortunately, some systems have too much heat to control or the environment is too unpredictable. In these situations, we resort to closed-looped, active thermal control. *Active thermal control* employs working fluids, heaters, pumps, and other devices to move and eject heat. Next, we'll look at methods for external thermal control.

External Thermal Control. The challenge of external thermal control is to manage the flow of heat into and out of a spacecraft. Let's start with the problem of heat input.

We know the primary external heat sources are the Sun and Earth. Our first line of defense is to carefully control the amount of heat absorbed by spacecraft surfaces. Realize that in the vacuum of space, the side facing the Sun gets terribly hot and the side facing space gets terribly cold. One of the simplest ways to balance this temperature differential is to slowly rotate the spacecraft around an axis perpendicular to the Sun. In this "barbecue" mode, a spacecraft surface alternately heats when facing the Sun and then cools when facing the cold of space, maintaining a moderate surface temperature without hot spots. The Apollo spacecraft used this method all the way to the Moon and back. This is one method of active thermal control of heat input.

The barbecue mode can help to equalize the heat hitting the surface. The next challenge is to control the amount that gets absorbed. As we've seen, depending on the material, it absorbs, reflects, or transmits incident radiation. By changing the type of surface coatings on the spacecraft, we can control its total absorptivity and emissivity, thus its equilibrium temperature. We can change the ratio of heat absorbed to emitted (α/ε)

by carefully selecting materials to keep the surface temperature at a desired level.

Instead of actually coating the metallic surface of the spacecraft structure, we place various types of multi-layer insulation (MLI) on top of the structure. MLI consists of alternating sheets of polymer material, such as Mylar™ or Kapton™. Kapton™ often looks like gold foil and is similar to the "space blankets" sold in sporting goods stores to keep you warm during emergencies (another great spin-off from space technology!). We can apply MLI or simply Mylar™ or Kapton™ adhesive tape to surfaces to vary the amount of heat absorbed by different areas and insulate the subsystems underneath. Typically, we can meet nearly 85% of a spacecraft's thermal-control demands through passive means, by simply choosing the right surface coatings and insulation. Figure 13-39 shows MLI used for thermal control on the outside of the Upper Atmospheric Research Satellite (UARS).

Figure 13-39. Upper Atmospheric Research Satellite (UARS). We can meet nearly 85%–100% of a spacecraft's thermal-control demands by choosing the right coatings and insulation. Here we see foil wrapping used on the UARS. *(Courtesy of NASA/Goddard Space Flight Center)*

We've looked at controlling heat flow into the spacecraft. Now let's consider how to control the flow out of the spacecraft. In space, surrounded by a vacuum, using conduction or convection to eject heat is possible, but not too convenient. For example, we could transfer the heat to some fluid, such as water, and then dump it overboard. The Space Shuttle uses this method to remove excess heat with a device called a *flash evaporator*. Water pumps around hot subsystems cool them by convection, and then vent overboard. Unfortunately, this method works only as long as they have extra water onboard. This is one type of active thermal control that, for long missions, is impractical.

So the most effective long-term method for ejecting heat is by radiation. To radiate heat, we must design special surface areas on the spacecraft with low absorptivity and very high emissivity (low α/ε). These special areas then readily emit any heat concentrated near them. These surfaces are called *radiators*. Radiators are similar to "heat windows" that allow hot components on the inside of a spacecraft to radiate their heat into the cold of space. Often a radiator is simply a section of glass coating over a particularly hot section of the spacecraft. This greatly increases the emissivity of that part of the spacecraft so more heat radiates away. The radiators on the Space Shuttle are quite evident on the inside of the payload bay doors, as shown in Figure 13-40.

Internal Thermal Control. Inside the spacecraft we have different problems. Often the trouble is not having too much or too little heat, but, instead, it's having the heat (or lack of it) in the right place. Each subsystem has different thermal requirements, and we must keep them all happy. Some components, such as propellant lines and tanks, need to stay warm to prevent freezing. Others, such as high-power payloads, need active cooling.

The complexity of the internal thermal control techniques depends on two things—how fast we need to move the heat and how much heat we need to move. To remove modest amounts of heat from spacecraft components when time isn't critical, the easiest way is simply to establish a heat-conduction path from the hot component to a passive external

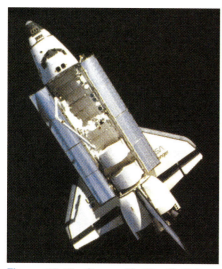

Figure 13-40. Space Shuttle Radiators. Radiators, like the ones on the inside of the Space Shuttle's payload bay doors, are areas of low absorptivity and high emissivity that radiate heat transferred to them. *(Courtesy of NASA/Johnson Space Center)*

radiator. This path can be as simple as connecting the two with a piece of heat-conducting metal. This is another form of passive thermal control.

As the amount of heat and the urgency to remove it increases, we need more complex, active thermal-control methods. One of the simplest of these is to use heat pipes. *Heat pipes* are tubes closed at both ends, filled with a working fluid, such as ammonia, as shown in Figure 13-41. When one end of the pipe is close to a heat source, the fluid absorbs this heat and vaporizes. Gas pressure forces the heated vapor to the cold end of the pipe where the heat passes out of the pipe via conduction. As the vapor loses its heat, it re-condenses as a liquid. It then flows back to the other end along a wick—just as liquid flows through a candlewick.

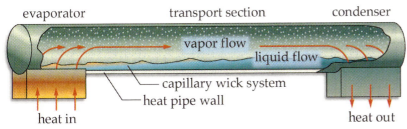

Figure 13-41.   Heat Pipe. Heat pipes employ some liquid with a low boiling point inside a hollow tube. As the liquid absorbs heat at the hot end, it vaporizes and carries the heat to the cool end. There it re-condenses and "wicks" back to the hot end.

Heat pipes offer a simple, open-loop active thermal control technique. The big cooling advantage comes from the latent heat absorbed when liquids vaporize. What do we mean by this? If you heat water on the stove, how hot can it get? Only about 100° C (212° F). No matter how long it's heated, it can reach only this temperature—the boiling point of H_2O. When you add more heat to the water, it changes phase (vaporizes) from a liquid to a gas (steam). Steam is not limited to 100° C; it can get much hotter and thus, can store more heat. *Latent heat of vaporization* is the principle of storing additional heat in a liquid as it changes phase. If you look at the graph of energy input versus temperature for water (or almost any substance for that matter) in Figure 13-42, you'll see where this latent heat comes in. As the fluid in a heat pipe vaporizes, it absorbs a large amount of heat due to this phenomenon.

Another simple method for removing heat is to use paraffin or some other phase-change material with a relatively low melting point to remove heat from a component during times of peak thermal demand. As the paraffin absorbs heat, it melts. When the component is no longer in use and stops producing heat, the melted paraffin conducts or radiates this heat to other parts of the spacecraft. Eventually, the thermal control system must eject the heat by radiation. As the paraffin cools, it solidifies and is ready for use during the next peak demand cycle. This thermal control method tends to be very reliable because it has no moving parts and the paraffin essentially never wears out. What makes this method so efficient is the same principle that makes ice a good thing to put in your cooler—latent heat of fusion. *Latent heat of fusion* is the same basic idea as

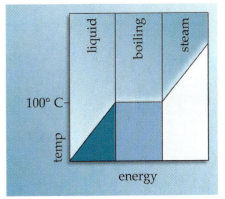

Figure 13-42.   Latent Heat of Vaporization. As heat is added to a liquid, such as water, its temperature increases linearly until it reaches the boiling point. Then, the temperature of the water stays constant as more heat is added. This additional heat needed to change the phase of the substance from liquid to steam is known as latent heat of vaporization.

Figure 13-43. Defense Support Program (DSP) Spacecraft. From geosynchronous altitude the DSP spacecraft needs supercool IR sensors to detect missile launches. *(Courtesy of NASA/Goddard Space Flight Center)*

Figure 13-44. Cyro-cooler. Sophisticated cryogenic coolers, such as the one shown here, keep IR sensors at extremely low temperatures (<20 K). *(Courtesy of the U.S. Air Force Research Laboratory)*

latent heat of vaporization, but uses melting instead of boiling. As the ice melts in your cooler, it takes heat out of your sodas. We applied this principle in Chapter 10 to remove heat through ablation from spacecraft re-entering the atmosphere.

In many cases, heat pipes or paraffin can't do the job, so we must resort to more complex, closed-looped, thermal-control methods. This is especially true for infra-red (IR) sensors, used on some remote-sensing missions, such as the Defense Support Program spacecraft, as shown in Figure 13-43. To be sensitive to minute changes in background temperature representing a missile launch seen from 36,000 km (22,370 mi.) away, these IR sensors must be super-cooled to 70 K (–200° C) or less. Older systems used liquid helium stored in the equivalent of large Thermos™ bottles called *Dewar flasks*. This method had a limited lifetime. After the liquid helium evaporated, the mission was over. Modern systems still use liquid helium, but actively cool it on the spot, using cryogenic pumps similar in concept to the pumps found in kitchen refrigerators. Advances in design and analysis of these components have drastically reduced their size, while increasing their operational lifetime to over ten years, making long-term IR remote sensing more practical and cost-effective. We show an example of an advanced cryo-cooler unit in Figure 13-44.

Basic Principles of Life Support

As we described in Chapter 3, space is a hostile place. Charged particles, solar radiation, vacuum, and free fall are all potentially harmful, even fatal, to unprepared humans. A spacecraft's life support system provides a "home away from home" for space travelers. To understand its requirements, we can consider humans as just another system (self-loading baggage), similar to the payload or the electrical power subsystem. As such, humans have various needs that the life-support subsystem must provide. In this section, we'll focus on the basic principles that drive those needs.

In Chapter 3 we introduced the long-term physiological challenges of living in the space environment from charged particles, radiation, and free fall. Here we'll focus on the much shorter-term problem of staying alive (and even living comfortably) in space for days or years at a time. We already looked at the problem of thermal control. Like any other "payload," humans have their own, specific temperature range, where they function best. Now we'll look at the basic necessities to keep humans alive for even a short time in space. These include

- Oxygen—at the right pressure
- Water—for drinking, hygiene, and humidity
- Food
- Waste management

From a systems perspective, we can look at human requirements in terms of inputs and outputs. Figure 13-45 shows the amount of oxygen, water, and food an average person needs for minimum life support and the amount of waste he or she produces. Let's look at each of these requirements to better see what the life-support subsystem must deliver.

Oxygen

At sea level, we breathe air at a pressure of 101 kPa (14.7 p.s.i.). Of this, 20.9% is oxygen (O_2), 78.0% is nitrogen (N_2), 0.04% is carbon dioxide (CO_2), and the rest consists of various trace gasses, such as argon. During respiration, our lungs take in all of these gasses but only the oxygen gets used. Our bodies use it to "burn" other chemicals as part of our metabolism. Within the lungs, O_2 transfers to the blood in exchange for a metabolic by-product, CO_2. By exhaling, we dump CO_2 back into the air around us. On Earth, plants eventually absorb this waste CO_2 and exchange it for O_2, and the process continues. In space, it's not that simple.

To provide a breathable atmosphere in space, the life-support subsystem must provide O_2 at a high enough partial pressure to allow for comfortable breathing. *Partial pressure* refers to the amount of the total pressure accounted for by a particular gas. At sea level, the partial pressure of O_2 (PPO_2) is 20.9% of 101 kPa (14.7 p.s.i.), or about 21 kPa (3.07 p.s.i.). After becoming acclimated, people living at high altitudes (above 2000 m or about 6560 ft.) show little discomfort with PPO_2 of 13.8 kPa (2.0 p.s.i.) or less. The Space Shuttle's life support system maintains a PPO_2 close to Earth's sea level standard at 22 ± 1.7 kPa (3.2 ± 0.25 p.s.i.).

Besides keeping the PPO_2 high enough, we must also not let it get too high. Breathing oxygen at too high of a partial pressure is literally toxic. This is a problem which scuba divers must also avoid during deep dives. We consider a PPO_2 of less than 48 kPa (7 p.s.i.) safe.

Besides providing adequate oxygen to breathe, we must consider other trade-offs. We want the PPO_2 to be low enough so it doesn't create a fire hazard in the crew cabin. This was the problem during the Apollo 1 accident, shown in Figure 13-46. At that time, the cabin atmosphere was pure oxygen. This led to the untimely deaths of three astronauts, when a wiring problem caused a fire during a routine ground test. The pure oxygen atmosphere let the fire spread much more rapidly than it would have in a normal O_2/N_2 atmosphere. Since then, cabin atmospheres in U.S. manned spacecraft have contained a mixture of oxygen and nitrogen to decrease this fire hazard.

The correct mixture and pressure of gasses is also important for thermal control. Convective heat transferred into the cabin atmosphere also cools electronic components, so atmospheric composition and circulation must support that function.

A final concern for cabin air is the astronauts preparing to leave the spacecraft for an extravehicular activity (EVA or space walk). Because of design limitations, Shuttle space suits operate at 29.6 kPa (4.3 p.s.i.). To avoid potential decompression problems, the astronauts reduce the Shuttle pressure to 70.3 kPa (10.2 p.s.i.) 12 hours before a planned EVA.

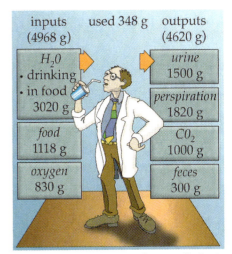

Figure 13-45.   The Human System. Similar to any other system, humans take some amount of input, process it, and produce output. Here we see the approximate daily food, water, and oxygen requirements for an astronaut and the corresponding urine, perspiration, CO_2, and feces produced. (Adapted from Nicogossian, et al and Chang, et al)

Figure 13-46.   Apollo 1 Disaster. The Apollo 1 fire that claimed the life of astronauts Grissom, White, and Chaffee, was caused by the use of a pure oxygen environment inside the capsule. *(Courtesy of NASA/Johnson Space Center)*

Even then, they must breathe pure oxygen for 3 – 4 hours before the EVA, to purge nitrogen from their bodies. Otherwise, the nitrogen could form bubbles in the blood, causing a potentially deadly problem known as "the bends," a condition that scuba divers must also carefully avoid. However, given the relative infrequency of EVAs, most astronauts consider these procedures a minor inconvenience.

Where does the air for the life-support subsystem come from? For Space Shuttle missions, tanks hold liquid oxygen and liquid nitrogen. As liquids, they need much less volume than as gasses. The life support system warms the liquids and then evaporates them into gases at the correct partial pressures. This process also replenishes air that vents during space walks or leaks out.

It is important to realize that astronauts need an efficient, closed-looped control system to monitor and maintain a safe atmosphere. Sensors constantly monitor the pressure and composition of cabin air and alert the crew and ground controllers to any problems before they can become a health hazard.

Astro Fun Fact
Cosmonauts Rescue
Crippled Space Station

Power out, life support systems failing, temperature dropping. . .sounds like an episode of Star Trek. But that's the situation faced by cosmonauts Vladimir Dzhanibekov and Viktor Savinykh when they blasted off from the Baikonur Cosmodrome on June 16, 1985, on a mission to rescue the crippled Salyut 7 space station. With the station's automated rendezvous equipment shut down, the crew had to make a tricky manual docking. Once attached to the station, they cautiously opened the valve into the station to sample its atmosphere. Luckily, the station still held pressure, but it was cold! "Ice was everywhere," Dzhanibekov later commented, "on the instruments, control panels, windows. Mold from past occupations was frozen to the walls." Because their suit thermometers only went down to 0° C, the crew ingeniously decided to spit on the walls and time how long it took to freeze. Using this crude estimate, they determined the station was at −10° C (14° F). The crew's first order of business was to recharge the station batteries so heat, light, and ventilation could be restored. Even bundled up "like babies in a Moscow winter," they found they could stand the cold for only about 40 minutes at a time before retreating into the refuge of their Soyuz capsule. Other problems plagued them as well. Without ventilators to circulate the air, carbon dioxide from their breath hung around their heads, causing headaches and sluggishness. After nearly 24 hours of constant work, they switched on the power and "suddenly the lights turned on and ventilators started whirring. . .the station was saved."

Canby, Thomas Y. Are the Soviets Ahead in Space? National Geographics Vol. 170, No. 4. Washington, D.C.: National Geographics Society, October 1986.

Water and Food

With an understanding of the air supply system, we can now turn our attention to one of the simpler pleasures of life—eating and drinking. We normally eat and drink without much concern for the total mass that we consume (we're much more concerned with calories!). For space missions, every gram taken to orbit represents a huge cost, so we want very little waste. On the other hand, because we can't call out for pizza, we must carry enough water and food for any contingency. Thus, we must fully understand the crew's needs when we design this piece of the ECLSS.

Astronauts need water onboard for many reasons. As a minimum, humans need about two liters of drinking water per day (about 2 kg or 4.4 lb.) to stay alive. We also need another liter or so of water for food preparation and re-hydration. Besides this minimal amount of water to maintain life, astronauts need water for personal hygiene (washing, shaving, etc.), as well as, doing the dishes and washing clothes. All told, this can add up to more than 20 liters per person per day.

We also need food. The average human needs about 29 calories per kilogram of body weight per day, to maintain their present weight. This means a typical 70 kg (154 lb.) astronaut needs at least 1972 calories per day—more for days when strenuous EVAs occur.

Space food has come a long way since astronauts ate peanut butter and drank Tang from tubes during the Gemini missions. Nowadays, astronaut food isn't much different from what we're used to on Earth. Figure 13-47 shows an astronaut "sitting down" to a healthy meal with a tray strapped to her thigh. Planners use the recommended daily allowances (RDA) of carbohydrates, protein, fat, vitamins, and minerals that we read about on food labels. To conserve mass and volume, they dehydrate or freeze-dry much of the food and then rehydrate it on orbit. For short-term missions, astronauts take fresh fruit and other perishable items—storage space and mass permitting. Depending on how they package it and the total calories needed, they must plan as much as 2 kg of food per person per day.

Where does all this food and water come from? For U.S. manned flights during Apollo and on the Space Shuttle, ample water comes as a by-product of the fuel cells used to produce electrical power, discussed in Section 13.2. Thus, astronauts have had the luxury of using as much water as they want. They simply dump wastewater overboard into the vacuum of space. Menu planners help them order all the necessary food for their mission. For extended missions lasting several months or more, such as those flown on the Russian Mir space station, unmanned re-supply spacecraft are launched every few months with more groceries.

Figure 13-47.   Food in Space. Food astronauts eat isn't all that different from Earth food. However, to save space, more dehydrated and freeze-dried foods are often used. Above you can see a typical meal tray used on the Shuttle. *(Courtesy of NASA/Johnson Space Center)*

Waste Management

Humans produce waste in the form of urine, feces, and CO_2, simply as a by-product of living. Collecting and disposing of this waste in an effective and healthy manner is one of the biggest demands on the life-support subsystem. Urine and feces pose health risks, as well as odor problems. CO_2 poses a subtler problem. Unless they remove it from the air, its concentration builds up, eventually causing increased heart and respiratory rates, a change in the body acidity, and other health complications.

One of the most commonly asked questions of the entire space program is, "How do you go to the bathroom in space?" Collecting urine and feces in a free-fall environment is a challenge. In the early days of the space program, designers subjected those dashing young astronauts with the "right stuff" to a most humbling experience. They collected urine and feces using inconvenient and messy methods, euphemistically called *intimate-contact devices*. Because all the U.S. astronauts were male at that time, they collected urine using a roll-on cuff placed over the penis and connected to a bag. They collected feces using a simple diaper, or, even more messy, a colostomy-type bag taped or placed over the buttocks. They either dumped urine and feces overboard (to eventually burn up in the atmosphere) or returned them to Earth for analysis and disposal.

The Skylab program ushered in a new era in free-fall toilets. For the first time, intimate contact devices were no longer necessary. An advanced version of this system is now on the Space Shuttle, as shown in Figure 13-48. However, the free-fall toilet created (and still creates) considerable challenge to engineers. We tend to take for granted all the work that gravity does for us every time we go to the bathroom. On Earth, urine and feces fall away from our bodies; in orbit, it's a different story. In free fall, us, our urine, and feces are all falling at the same rate. As a result, waste isn't compelled to move away from us, so it tends to float next to our body in a smelly blob (or, as one anonymous astronaut put it "those little guys just don't want to leave home!"). To get around this problem, engineers use forced air to create a suction, pulling urine and feces into a waste-collection system for disposal. Unfortunately, this method doesn't work nearly as well as good ol' gravity, but at least it's a vast improvement over the older methods.

In comparison, removing CO_2 from the air is much simpler and far less messy. On the Space Shuttle, canisters containing charcoal and lithium hydroxide (LOH) filter the air. The LOH chemically reacts with the CO_2, trapping it in the filter. The charcoal absorbs odors and other contaminants, as well. The crew must change these canisters periodically during the flight as we show one astronaut doing in Figure 13-49. On Skylab, for missions lasting up to 84 days, the crew filtered CO_2 using a molecular sieve, which they then "baked out" and re-used.

Figure 13-48.   Shuttle Toilet. The toilet used by astronauts on the Space Shuttle compensates for the free-fall environment. On Earth, gravity does all the work; in free fall, forced air creates a suction to draw waste away from the body. *(Courtesy of NASA/Johnson Space Center)*

Figure 13-49.   "Scrubbing CO_2." Lithium hydroxide canisters, like the ones being changed here, remove carbon dioxide (CO_2) from the air on the Space Shuttle. *(Courtesy of NASA/Johnson Space Center)*

Closed-loop Life Support

From a systems point of view, astronaut and cosmonaut life support systems have been largely open loop in nature. Certainly, there are closed-loop aspects that monitor temperature and cabin atmosphere, but all of the system inputs (air, water, food) are eventually thrown away as waste.

For future long-term missions to the Moon or Mars, lasting many months or years, it probably won't be practical to take along all the supplies or rely on re-supply missions. Instead, we need to establish a closed system that can reclaim and recycle water and other waste. Such closed-loop systems could recycle urine, feces, and CO_2 to provide water and food to the crew, as illustrated in Figure 13-50. While this may not sound appetizing, it promises to greatly reduce the mass they need to pack along for very long missions. Scientists are investigating life-support subsystems that can effectively reclaim and recycle water, the heaviest item. One limited approach to this idea is to reclaim so called "gray" water (used for washing and rinsing) and reuse it for purposes other than drinking.

Other scientists are looking beyond such limited systems to ones that will fully recycle nearly everything onboard and provide all the oxygen, water, and food crew members need for missions lasting for years. Such systems could eventually make it much easier for astronauts to eat, drink, breath, and even go to the bathroom. Unfortunately, such systems are still far in the future. Until then, these pioneers in the high frontier must accept some austere conditions.

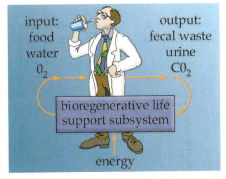

Figure 13-50. Bioregenerative Life-support Subsystems. For long space missions to Mars or beyond, bioregenerative life support subsystems may be needed. These systems allow us to "close the loop" and recycle all human waste into food, water, and oxygen.

Systems Engineering

Now that you've warmed up to the concepts of thermal control and life support, we can return our attention to the systems engineering challenges of ECLSS design. We'll look briefly at the inputs to the design process, then review some of the testing requirements spacecraft must endure before launching into the harshness of space.

Requirements and Constraints

As described in Chapter 11, we kick off the space systems engineering process by defining mission and system-level requirements and constraints. Of course, the biggest driver for the ECLSS design is whether the mission includes a fragile human payload. With astronauts onboard, we must include all of the necessary life support systems that we discussed earlier—water, air, food, and waste management. To total these requirements, we multiply the daily human requirements by the number of astronauts and the total mission duration. This gives us a starting point for designing the ECLSS *life-support budget*.

Without humans to worry about, the "LS" part of ECLSS goes away and environmental control remains. This system must mainly keep components at the right temperature. Unlike mission data or electrical power, the thermal budget is a collection of acceptable operating temperature ranges for each subsystem. System engineers must look at component placement, conduction paths, and thermal input and output to ensure each subsystem stays comfortable.

Analysis and Design

At the beginning of the section, we introduced the concept of thermal equilibrium. Recall that to maintain this state, the heat output must balance the heat input plus internal heat. The biggest challenge of the thermal control system design is meeting this requirement under all mission conditions. Keeping things warm isn't a problem in full sunlight.

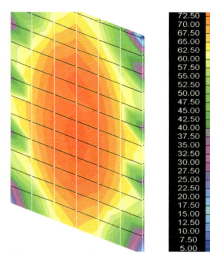

Figure 13-52.   Spacecraft Thermal Analysis Results. This solar panel shows the distribution of temperatures (°C) as a result of a computer thermal simulation. *(Courtesy of Surrey Satellite Technologies, Ltd., U.K.)*

Figure 13-53.   Thermal Cycling. These tests screen components to ensure they'll function in the space environment by subjecting them to a series of thermal cycles—hot-cold-hot-cold. Here we show the Picosat microsatellite in a small thermal cycling chamber. *(Courtesy of Surrey Satellite Technology, ltd., U.K.)*

Figure 13-54.   Thermal Vac Facility. Prior to flight, individual subsystems, and entire spacecraft, go through thermal cycling tests inside large vacuum chambers to simulate the effects of the space environment. *(Courtesy of Hughes Space and Communications, Co.)*

Keeping things cool isn't a big problem in the darkness of space. But keeping them cool in the sun and warm in the shade can be a big problem.

During the design loop, thermal control engineers must stay closely involved with the other subsystem designers to understand their requirements and carefully analyze where all the heat will go. One method for managing heat flow is to create a detailed thermal model of the entire system, by dividing it into a series of nodes. A *thermal node* is any payload, subsystem or even part of the structure that has unique thermal properties to consider. We carefully define the thermal properties of each node (heat input, absorptivity, emissivity, etc.). We then connect all of the nodes, in a virtual sense, as illustrated in Figure 13-51, and calculate the equilibrium temperature using a complex computer simulation. We show the output of one such simulation in Figure 13-52. The results give designers a good indication of potential thermal problems, and help them design passive and active thermal control techniques to take care of them.

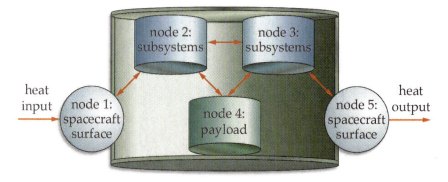

Figure 13-51.   Spacecraft Thermal Analysis Techniques. To model spacecraft thermal control, engineers define a series of interconnected nodes on the spacecraft. By defining the unique thermal properties of each node and understanding the heat conduction and radiation between nodes, they can use computer analysis techniques to determine overall system temperatures.

Testing

Finally, when we finish a design, we must validate it against the original requirements. For thermal control, this involves a testing process that subjects the spacecraft to the simulated vacuum and temperature conditions of space. To test temperature extremes, we perform thermal cycling tests (hot-cold-hot-cold, etc.) on individual components, subsystems, and the entire spacecraft, as shown in Figure 13-53. These tests subject components to the wide temperature extremes they'll see on orbit. These tests also determine if thermal cycling may cause them to fail.

In addition to relatively inexpensive thermal-cycling tests, we add a strong vacuum to the thermal cycling, using large, more expensive thermal/vacuum facilities, as shown in Figure 13-54. These "thermal-vac" tests create realistic heat transfer situations by eliminating convective cooling processes (no air). If the spacecraft survives these grueling series of tests, we certify the ECLSS ready to fly!

▰ Section Review

Key Terms

absorptivity, α
active thermal
    control
albedo
conduction
conductor
convection
Dewar flasks
emissivity, ε
flash evaporator
heat pipes
intimate-contact
    devices
latent heat of fusion
latent heat of
    vaporization
life-support budget
partial pressure
passive thermal
    control
radiation
radiators
reflectivity, ρ
thermal equilibrium
thermal node
transmissivity, τ

Key Equations

$$q = -KA\frac{\Delta T}{\Delta X}$$

$$q = \sigma\varepsilon A T^4$$

$$\tau + \alpha + \rho = 1$$

Key Concepts

➤ A spacecraft's environmental control and life-support subsystem (ECLSS) has two primary tasks—environmental control (primarily temperature) and life support

➤ Thermal control balances heat input, internal heat, and heat output to maintain thermal equilibrium

➤ Sources of heat for a spacecraft include

- The Sun

- Earth (for spacecraft in low-Earth orbits)—from albedo and "Earth shine"

- Internal sources—such as electrical components

➤ Heat can transfer between two points in three ways

- Conduction—heat transfer through a solid medium, the Fourier Law

- Convection—heat transfer to a flowing fluid

- Radiation—heat transfer by EM radiation, the Stephan-Boltzmann Law

➤ All EM radiation striking a surface must be either reflected, absorbed or transmitted. Reflectivity, ρ, absorptivity, α, and transmissivity, τ, describe the percentage of each for a given surface. Once absorbed, the energy can re-radiate based on the surface emissivity, ε.

➤ Spacecraft thermal control regulates external heat input/output as well as internal heat flow

- Passive thermal control uses open-loop methods, such as surface coatings, multi-layer insulation (MLI), and conduction paths, to control overall temperature.

- Active, thermal-control techniques, such as heaters, heat pipes, and cryogenic coolers use some power and/or some working fluid to control heat in specific locations

➤ Life support keeps humans alive in space. Humans need

- Oxygen at the right partial pressure

- Water and food

- Methods for waste disposal (CO_2, urine, and feces)

➤ The ECLSS systems engineering process is driven by life-support budgets and individual operating temperature ranges for each payload and subsystem

- Engineers conduct thermal analysis by simulating each component as a series of nodes with specific thermal properties

- Spacecraft testing involves subjecting components and entire systems to thermal cycling, as well as combined, thermal-vacuum facilities

Example 13-3

Problem Statement

Prior to thermal-vacuum-facility testing for FireSat, engineers want to check their passive thermal design for the spacecraft. The structure is cubic shaped, 0.3 m on an edge, with solar panels on the four sides. Manufacturer specifications for the panels tell us their transmissivity is zero ($\tau = 0$), their reflectivity ($\rho = 0.05$), and their absorptivity ($\alpha = 0.95$). Panel emissivity is 0.85. The top and bottom square sections of the spacecraft are covered in multi-layer insulation (MLI) providing an effective emissivity of 0.0 and absorptivity of 0.0 (perfect insulator). From Example 13-2, we know that the spacecraft bus needs 11.5 W of power to operate. During payload operations, an additional 20 W peak power is consumed. Determine the equilibrium temperature for the spacecraft with full sunlight on one solar panel during payload operations, and in eclipse (no payload operation during eclipse). You can ignore the input of "Earth shine" (237 W/m^2) on the nadir-facing panel since it is covered in MLI.

Problem Summary

Given: spacecraft width = 0.3 m
spacecraft height = 0.3 m
no. panels = 4
panel $\rho = 0.05$
panel $\alpha = 0.95$
panel $\varepsilon = 0.85$
MLI $\varepsilon = 0.0$
solar input = 1358 W/m^2
Stefan-Boltzmann's constant

$$\sigma = 5.67 \times 10^{-8} \frac{\text{W}}{\text{m}^2\text{K}^4}$$

MLI $\alpha = 0.0$
payload power = 20.0 W
subsystem power = 11.5 W

Find: spacecraft equilibrium temperature during full-sun payload operations and in eclipse (no payload) operations.

Conceptual Solution

1) Find the total area and solar input area of the panels

total array area = (spacecraft width) (spacecraft height) (no. panels)

$$\text{sunlit area} = \frac{\text{total array area}}{4}$$

2) Find the total sunlit energy input during payload operations

$q_{\text{input sunlight}}$ = (sunlight area) (solar input) (panel absorptivity)

$q_{\text{internal sunlight}}$ = subsystem power + payload power

$q_{\text{in sun}} = q_{\text{input sunlight}} + q_{\text{internal sunlight}}$

3) Set q_{in} equal to q_{out} and solve for the sunlit side's equilibrium temperature

$q_{\text{in sun}} = q_{\text{out}}$

$q_{\text{out}} = (\sigma)$ (panel emissivity) (total array area) $(T_{\text{sunlit}})^4$

$$T_{\text{sunlit}} = \left(\frac{q_{\text{in sun}}}{(\sigma)(\text{panel } \varepsilon)(\text{total array area})} \right)^{\frac{1}{4}}$$

4) Set q_{out} equal to $q_{\text{internal eclipse}}$ and solve for the eclipse equilibrium temperature

$q_{\text{internal eclipse}}$ = subsystem power

$q_{\text{out}} = q_{\text{internal eclipse}}$

$$T_{\text{eclipse}} = \left(\frac{q_{\text{out}}}{(\sigma)(\text{panel } \varepsilon)(\text{total array area})} \right)^{\frac{1}{4}}$$

Analytical Solution

1) Find the total area and solar input area of the panels

total array area = (spacecraft width) (spacecraft height) (no. panels) = (0.3 m) (0.3 m) (4)

total array area = 0.36 m^2

$$\text{sunlit area} = \frac{\text{total array area}}{4} = \frac{0.36 \text{m}^2}{4}$$

sunlit area = 0.09 m^2

Example 13-3 (Continued)

3) Find the total sunlight energy input during payload operations (assuming one side of cube is in direct sun and ignoring albedo and Earth IR)

$q_{\text{input sunlight}}$ = (sunlit area) (solar input) (panel absorptivity) = (0.09 m^2) (1358 W/m^2) (0.95)

$q_{\text{input sunlight}}$ = 116.11 W

$q_{\text{internal sunlight}}$ = subsystem power + payload power = 11.5 W + 20.0 W

$q_{\text{internal sunlight}}$ = 31.5 W

$q_{\text{in sun}} = q_{\text{input sunlight}} + q_{\text{internal sunlight}}$
$\qquad$ = 116.11 W + 31.5 W

$q_{\text{in sun}}$ = 147.61 W

4) Set q_{in} equal to q_{out} and solve for the sunlit side's equilibrium temperature

$q_{\text{in sun}} = q_{\text{out}}$

q_{out} = (σ) (panel emissivity) (total array area) $(T_{\text{sunlit}})^4$

$$T_{\text{sunlit}} = \left(\frac{q_{\text{in sun}}}{(\sigma)(\text{panel emissivity})(\text{total array area})}\right)^{\frac{1}{4}}$$

$$= \left(\frac{147.61\,\text{W}}{\left(5.67 \times 10^{-8}\,\dfrac{\text{W}}{\text{m}^2\text{K}^4}\right)(0.85)(0.36\,\text{m}^2)}\right)^{\frac{1}{4}}$$

T_{sunlit} = 303.71 K

$T = T_{\text{sunlit}} - 273.15$ K

$T = 30.56°$ C

4) Set q_{out} equal to $q_{\text{internal eclipse}}$ and solve for the eclipse equilibrium temperature

$q_{\text{internal eclipse}}$ = subsystem power = 11.5 W

$q_{\text{out}} = q_{\text{internal eclipse}}$

$$T_{\text{eclipse}} = \left(\frac{q_{\text{out}}}{(\sigma)(\text{panel emissivity})(\text{total array area})}\right)^{\frac{1}{4}}$$

$$= \left(\frac{11.5\,\text{W}}{\left(5.67 \times 10^{-8}\,\dfrac{\text{W}}{\text{m}^2\text{K}^4}\right)(0.85)(0.36\,\text{m}^2)}\right)^{\frac{1}{4}}$$

T_{eclipse} = 160.45 K

$T = T_{\text{eclipse}} - 273.15$ K

$T = -112.70°$ C

Interpreting the Results

Between full sunlight with payload operations and eclipse, the spacecraft will experience a widely varying equilibrium temperature between 303 K (31° C) and 160 K (–113° C). Fortunately, these represent the worst-case extremes in temperature. In reality, the actual spacecraft temperature will stay somewhere between these extremes. However, from the analysis, it would appear that there may be a tendency for the spacecraft to be on the cold side, depending on the actual length of each eclipse. This may require special attention to specific spacecraft components, particularly those that are susceptible to extreme cold.

13.4 Structures and Mechanisms

▬ In This Section You'll Learn to...

- ☞ Discuss the main functions of a spacecraft's structures and mechanisms
- ☞ Discuss and apply basic principles of structural mechanics
- ☞ Apply the space systems engineering process to the spacecraft structural design
- ☞ Explain the importance of engineering drawings for communicating the design process

To design the perfect house, an architect must understand the needs (and budget) of its future occupants and balance these against the physical demands of the location and its environment. Similarly, space mission architects must balance all of the requirements and constraints defined during the space systems engineering process to achieve a final structural design that supports the mission. Traditionally, we don't think of a structure as a system, with inputs, processes, and outputs. Instead, we look at it according to how it holds together the rest of the subsystems and payloads to create a functional space system.

The structure holds all the other subsystems together during the stresses, strains, and vibrations of launch and orbital operations. In addition, the structure supports all the *mechanisms* that push, pull, extend, or pivot to do various tasks. Without a strong enough structure, the spacecraft might collapse before it gets into space. Without well-designed mechanisms, the spacecraft couldn't function after it gets there.

In many cases, the success of the entire mission may depend on a single bolt holding (or giving way) at the right time. For example, solar arrays normally store folded for launch and then deploy after the spacecraft reaches orbit, as shown in Figure 13-55. If the solar panels don't extend, the spacecraft won't produce power, and the mission will fail!

Figure 13-55. Structures and Mechanisms. To deploy solar panels, like the ones shown in this photograph being extended from the Shuttle bay, engineers must carefully design the structure to take the necessary forces, and hinge mechanisms to ensure smooth opening at exactly the right point in the mission. *(Courtesy of NASA/Johnson Space Center)*

In this section we'll explore what structures do, review some basic mechanics to see how they work, and then examine some of the important systems engineering issues associated with their design. We'll show how the structural design serves as the backbone of a spacecraft and as a focal point for integrating all the other subsystems with the payload.

System Overview

Typically, the spacecraft structure accounts for about 10%-20% of the spacecraft's total dry weight. *Dry weight* is the total spacecraft weight minus propellant. We can think of the structure as having two parts. The *primary structure* maintains the physical integrity of the spacecraft. It carries most of the loads the spacecraft must withstand. A *load* is any force pushing, pulling, or vibrating the structure. In Figure 13-56 we show engineers assembling the primary structure of a sensitive instrument. The *secondary structure* accommodates the mechanical configuration of the payload and subsystems. It holds together all the mechanisms, cameras, wires, pipes, doors, and brackets inside and outside the spacecraft. Figure 13-57 shows engineers installing brackets to hold an antenna.

Figure 13-57.   Secondary Structure. Secondary structure includes all wires, pipes, and brackets needed to house and connect the payload and subsystems. Here, we show a skilled engineer installing the antenna to the UoSAT-12 spacecraft. *(Courtesy of Surrey Satellite Technology, Ltd., U.K.)*

Figure 13-56.   Primary Structure. In this photograph, engineers are assembling the primary structure of the near infrared camera and multi-object spectrometer (NICMOS). The primary structure carries the majority of loads during launch and operations. *(Courtesy of Ball Aerospace & Technologies Corporation)*

Although mechanisms account for only a small fraction of the mass of a typical spacecraft, their success (or failure) usually weighs heavily on the minds of mission designers and operators. We classify spacecraft mechanisms generally by the number of times they need to work. Low-cycle mechanisms must work a few times at most, to separate the spacecraft from the launch vehicle, deploy antennas or booms, or open isolation valves. Since most of these actions occur only once in a mission, designers often use pyrotechnic actuators for these mechanisms because of their high reliability, relatively low cost, and ease of integration. Pyrotechnic actuators use explosive charges that can cut bolts and open (or close) valves. High-cycle mechanisms must work repeatedly throughout the mission. They include motors that rotate and point antennas or solar arrays, rocket propellant control valves, as well as reaction wheels, gyroscopes, and any other moving parts.

To understand how structures and mechanisms meet all the various mission requirements, we can start by looking at a snapshot of the systems engineering process for a spacecraft structure. As Figure 13-58 illustrates, all of the requirements and constraints—from the mission level, systems level, and from other subsystems—feed into the structural design process. This design is a balance between the mechanical configuration demanded by the subsystems and payloads, and the sometimes-competing requirements for mechanical behavior imposed by the launch environment and mission profile.

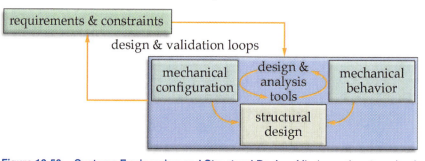

Figure 13-58.   Systems Engineering and Structural Design. Mission and systems-level requirements and constraints feed into the structural design process. During the systems engineering process for the spacecraft structure, we must balance the requirements of mechanical configuration (mass, volume, layout) imposed by the payload and subsystems with the equally important requirements for mechanical behavior dictated by the launch vehicle and mission profile to achieve a final structural design. Analysis and design tools provide the dialog between these two parallel efforts and help us converge on a final design.

As we learned in the last two chapters, and in this one, after we define the mission, we identify the requirements and constraints on each element of the mission architecture. These system requirements flow down to the spacecraft payload and its subsystems in terms of specific demands for mass, volume, field of view, and physical layout. In the structural design process, we must blend these requirements together, along with the simple need to tie all these components together with power and data lines, to produce a workable, even elegant, *mechanical configuration*.

But that's only half the story. Even a house of cards can have an elegant mechanical configuration. To perform the mission, a spacecraft's structure must first survive launch. Launch vehicle requirements, along with the mission profile, dictate how severely the launch environment pushes, pulls, shakes, and thermal cycles the structure. Taken together, these requirements determine the structure's required *mechanical behavior*—how it responds to the mission environment.

The final *structural design* emerges by trading these sometimes-conflicting demands of mechanical configuration and mechanical behavior. To do the trade-offs, designers extensively use analysis and design tools, as we described throughout the systems engineering process. As we said, the systems engineering process is iterative. After they reach a preliminary design, the engineers must go back through the design loop to ensure it meets the needs of the payload and subsystems and back through the "validation loop" to ensure it satisfies overall mission requirements.

In this section, we'll look at the analysis tools needed to understand mechanical behavior, by reviewing some basic principles of structural mechanics. After that, we'll return to the systems engineering process to see how mission and system requirements and constraints drive mechanical behavior and configuration.

Basic Principles

Whether we're building a bridge, a car, or a spacecraft, structures work pretty much the same way. Basically, a structure is anything that carries a load. For example, our skeleton, muscles, and connective tissue form the structure of our body. This structure allows us to stand up in the pull of Earth's powerful, gravitational field. And it has enough built-in safety factors to allow us to run and jump—sometimes causing loads many times those due to gravity.

What kinds of loads must we design a structure to deal with? Basically, we can

- Push, pull, twist, or bend it—causing stress, strain, and shear
- Shake it—causing vibrations
- Change its temperature—causing thermal stress

Any or all of these things put demands on the structure in some way and can deform or break it, if it's not built strong enough. To understand this in more detail, let's look at how all of these loads affect a structure.

Types of Loads

Like most things in life, loads can either stay constant or vary with time. Constant, or relatively constant, loads are *static loads*. Loads that vary widely with time are *dynamic loads*. (There's another definition of dynamic load that results from lift and drag due to motion through air or water. Unless stated otherwise, we'll deal with time varying dynamic loads, here).

We can classify static and dynamic loads based on how they're applied to a structure's axis of orientation. The load types are axial, lateral, or torsional. *Axial loads* are those applied parallel to the longitudinal (long) axis of a structure. *Lateral loads* are applied parallel to the lateral (short) axis of a structure. Twisting loads apply a torque to a structure, so we call them *torsional loads*.

To see all of these loads in action, imagine you have a soda can, as shown in Figure 13-59. If you push on the can at both ends, you're applying an axial load. In this case, the load is one of *compression* or a *compressive load*. If you pull out on both ends you're again applying an axial load, but in this case it is one of *tension* or a *tensile load*. If you now attach one end of the can to a tabletop (glue it down) and push or pull on the side of the can, you're applying a lateral load. Notice that, in this case, one side of the can is in compression, while the other is in tension. Finally, if you twist the can, as though you were wringing a wet towel, you're applying a torsional load.

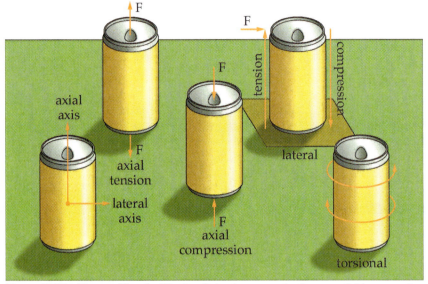

Figure 13-59.   Types of Loads. We classify loads on any structure as axial (compression or tension), lateral, or torsional.

When a structure receives axial or lateral loads, bending can occur. It's easiest to visualize bending for the case of lateral loads. Notice in Figure 13-60 that the lateral load causes a bending moment about the attachment point. A *bending moment, M,* results from a load applied at a distance away from the attachment point.

$$M = F \, d \tag{13-11}$$

where

M = bending moment on an object (N m)

F = force (N)

d = distance between the load and attachment points (m)

Bending moments can be especially dangerous because even a very small load, when applied over a great enough distance, can cause significant bending.

Stress, Strain, and Shear

Now that we've described a little bit about the types of loads on a structure, we can look at their effect. The fundamental question of structural analysis is, "will it break?" To answer this question we look at three parameters that quantify changes within a structure due to loads: stress, strain, and shear.

When we hear someone talk about stress, we may think about the panic before a big project's deadline. But for structures, stress has an entirely different meaning. When we apply a load to a structure, that load

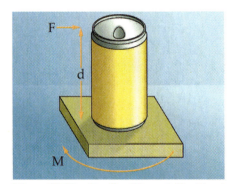

Figure 13-60.   Bending Moment. The lateral load applied to the fixed soda can causes a bending moment, M, about the attachment point.

essentially spreads over its entire cross-sectional area. We define *stress* as the applied load per unit area. We compute it using

$$\sigma = \frac{F}{A}$$ (13-12)

where

σ = stress (N/m^2)

F = applied load (N)

A = cross sectional area (m^2)

Notice the units on stress are the same as pressure—force per unit area. One way to visualize the effect of stress on a structure is to imagine inflating a car tire. The more air pumped in, the greater the pressure in the tire. Similarly, the greater the load applied to a given structure, the greater the stress. Depending on the direction of the applied load, stress is either compressive (pushing in) or tensile (pulling out).

As a structure undergoes stress, it begins to deform. Imagine pulling on a hunk of Play-Doh™. As you pull (applying an axial, tensile load), it stretches. This change in length, or deformation, due to an applied load is known as *strain*. We determine the strain in a structure by computing the ratio of a structure's change in length compared to its original length, as shown in Figure 13-61.

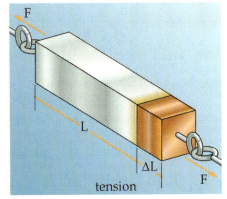

Figure 13-61.   Strain. Strain describes how an object changes in length due to stress.

$$\varepsilon = \frac{\Delta L}{L}$$ (13-13)

where

ε = strain (m/m)

ΔL = change in length (m)

L   = original length (m)

All types of loads—axial, lateral, and torsional—can cause a third effect within a structure, known as shear. When we apply transverse loads to our soda can, as shown in Figure 13-62, the magnitude of the internal force that results, we call *shear*. Shear also results from torsional loads. The resulting stress generated by a shearing force is *shear stress*.

You might think that a spacecraft, minding its own business in orbit, would be generally free from external loads. However, stress, strain, and shear can result from a variety of loads, including environmental factors, such as heating.

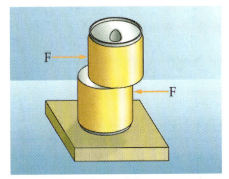

Figure 13-62.   Shear. When we apply a transverse load, as shown above, to a structure, the resulting shear causes shear stress. Depending on how they're applied, torsional or axial loads may also cause shear.

To introduce the effects of heating, we ask, "have you ever wondered where all those cracks in the sidewalk come from?" As most materials heat up, they expand. Similarly, as they cool, they contract. This deformation due to heat is no problem if they have plenty of space to expand into or aren't constrained from contracting. But in the sidewalk case, or for any constrained structure (i.e., locked in on both sides), this expansion and contraction produces stresses that eventually cause it to crack.

Of course, a similar problem can occur in a spacecraft structure. As it enters and exits Earth's shadow, it almost literally goes from fire to ice.

These frequent thermal cycles (remember LEO satellites orbit 16 times a day) can cause stresses in the structure. For example, when the Hubble Space Telescope, shown in Figure 13-63, was first put into orbit, engineers found that thermal expansion and contraction of the solar array structure, due to thermal cycling through eclipses each orbit, caused a small vibration in the entire structure. This vibration made focusing on distant objects more difficult.

We quantify thermal expansion and contraction by using an object's length and a material property called its *coefficient of thermal expansion, α*. We can then find the change in length, ΔL, from

$$\Delta L = \alpha(\Delta T)L \qquad (13\text{-}14)$$

where

ΔL = object's expansion or contraction (m)

α = object's coefficient of thermal expansion (1/°C)

ΔT = temperature change (°C)

L = object's original length (m)

Figure 13-63.   Thermal-induced Vibrations on the Hubble Space Telescope. When the Hubble Space Telescope was first deployed, operators noticed vibrations each time it went from eclipse into sunlight. These vibrations were caused by thermal contraction and expansion of the struts supporting the solar panels. The problem was eventually corrected during a Shuttle repair mission. *(Courtesy of NASA/Johnson Space Center)*

Using this change in length, we can find the corresponding strain by applying Equation (13-13).

Whether a structure breaks as a result of too much stress, strain, or shear stress depends on its material properties. In other words, the same amount of strain that breaks a steel girder wouldn't even faze a hunk of Play-Doh™. We'll look briefly at how we quantify these material properties later in this section.

Vibrations

So far, the load effects we've described are mainly from static loads. But getting into space is a very dynamic undertaking! Getting there can get a spacecraft all shook up. If you've ever ridden on an old wooden roller coaster or in the back of a pickup going down a bumpy road, you've experienced dynamic loads similar to launch. Dynamic loads that vary widely and randomly are called *vibrations*. We're concerned with two effects of these vibrations on a structure. The first effect is the cumulative load produced by applying random vibrations over time. These vibrations have the same effect we see if we bend a piece of metal back and forth many times even if we don't deform it much. Eventually, it breaks due to *fatigue*. Fortunately, rides into space are short enough that fatigue isn't much of a problem. The second concern with vibrations is far more serious. This effect describes how a given structure responds to vibrations of a particular frequency.

All objects (even you!) have a natural frequency. A *natural frequency* is the rate at which an object vibrates when given one sudden impulse and then left undisturbed. For example, when we pluck a guitar string, it vibrates at a particular natural frequency depending on the string's length, thickness, and other properties. When we vibrate an object at any frequency other than one of its natural frequencies, nothing interesting happens. It simply acts like any other load and we find the cumulative effect by summing all the individual effects of the vibrations (some positive and some negative). However, when we vibrate an object at its natural frequency, something very interesting happens—it begins to resonate. *Resonance* is the tendency for an object to vibrate with increased amplitude (higher peaks), due to a synchronized, applied, periodic force. An object actually has several natural frequencies. The most important part of resonance is its lowest natural frequency, called the *fundamental frequency*.

To understand resonance, think back to when you played on a swing, as shown in Figure 13-64. After giving yourself a little push to start, you pumped your legs at just the right time, gradually increasing your swing's amplitude. What you did was time your leg pumps to the frequency at which you were swinging. In other words, you provided a cyclic force at the same frequency as your effective natural frequency. This cyclic force allowed you to work with the swing instead of against it every time, eventually building up a large amplitude arc.

Resonance can be very powerful. Because the frequency of the input force matches the fundamental frequency of the structure, each vibration amplifies the structure's oscillation. Even a very small input, over a short time, can cause a structure to build up very large-amplitude oscillations. One of the classic examples of resonance is an opera singer who is able to shatter a glass with just the sound of her voice. Setting the pitch of her note to coincide with the fundamental frequency of the glass causes large-amplitude vibrations that eventually cause the glass to break.

A more dangerous resonance occurs during earthquakes. During the 1985 earthquake in Mexico City, scientists recorded the primary frequency of the forcing vibration to be about 0.5 Hz. After the quake, they found some of the most severe damage occurred in buildings between 10 and 14 stories tall, while taller and shorter buildings were relatively unaffected. Through some analysis, they discovered that the fundamental frequency of 10–14 story buildings is about 0.5 Hz!

To estimate the fundamental frequency of a structure we can assume it acts like a big spring (a pretty close approximation for most structures). The frequency then depends on its mass and a property called the spring constant. The *spring constant, k,* is a measure of the force it takes to compress a spring, or any structure, by a certain amount. Knowing this value, we can compute the fundamental frequency using

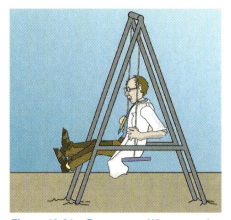

Figure 13-64. Resonance. When you play on a swing you pump your legs at just the right time to increase the amplitude at which you're swinging. In this way, the forcing vibration is in resonance with the natural frequency of the swing, so you go higher.

$$f_{fund} = \frac{1}{2\pi}\sqrt{\frac{k}{m}} \qquad (13\text{-}15)$$

where

f_{fund} = object's fundamental frequency (Hz = cycles/s)

k   = object's spring constant (N/m)

m   = object's mass (kg)

A structure with a high spring constant, k, (and correspondingly high fundamental frequency) is *stiff*. Thus, the spring constant is also a measure of a structure's *stiffness*. Note that the fundamental frequency is inversely proportional to mass. So, roughly speaking, the bigger the structure, the lower the fundamental frequency. For that reason, a spacecraft's primary structure generally has a much lower fundamental frequency than the bits and pieces that comprise the secondary structure.

There are two ways to avoid resonance. The first is to ensure that the fundamental frequency of the structure is different from the frequency of the forcing vibration. As we'll see, this is a mandatory requirement for spacecraft design. The second way is to include damping within the structure. *Damping* is a passive or active mechanism to dissipate vibrational energy, which limits the magnitude and duration of a structure's response to input forces. The shock absorbers in a car are a simple example of dampers that dissipate the vibrations caused by speed bumps and potholes, giving passengers a smoother ride.

Material Properties

The ultimate effect of stress, strain, vibration, bending, and thermal loading depends on what material we use in the structure—its material properties. When choosing what material to use to build a particular structure, we must consider a number of factors

- Mechanical properties— strength and stiffness
- Physical properties—stability in the space environment, magnetic properties, density, etc.
- Ease of fabrication—how easy is it to bend, cut, and weld?
- Cost

Let's start with mechanical properties. As we all know, different materials react differently to the same loads. For example, you could easily pull apart a hunk of Play-Doh™, but only Superman could do the same thing with a hunk of steel. The ultimate question of structural design is—will it break? To answer this we need to know a material's *strength*.

As we subject different materials to stress and strain, they behave in predictable ways. As we apply a load to most metals, they first begin to deform *elastically*. That is, they stretch but then return to their original size and shape when we remove the load. In this *elastic region*, the relationship between stress and strain is linear, that is, the plot of stress versus strain is a straight line. This elastic region remains linear until reaching a point that we call the *proportional limit*. If we remove the load before it reaches this point, the material will return to its original shape. If the load increases beyond the material's proportional limit, it will reach a *yield point*, where a *residual* (left over) *strain* of 0.2% will remain after we remove the load. That is, the material will be permanently 0.2% longer (or shorter) than when we started. Applying more load beyond the yield point eventually leads to the material failing or breaking. This amount of load is the *ultimate failure point*. The stress-strain curve in Figure 13-65 illustrates all of these relationships.

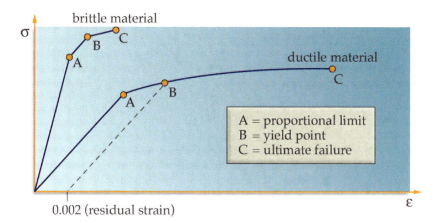

Figure 13-65. Stress-Strain Curves. Typical stress-strain curves show the difference between ductile and brittle materials.

We can describe the strength of a material in one of several ways, depending on what point on the stress-strain curve we're using. The *ultimate strength, F_u,* for example, tells us the stress, beyond which the material will break. The *yield strength* tells us the stress at the yield point. As you can image, this value varies widely between materials. Aluminum (6061-T6) has an F_u of about 290×10^6 N/m^2 (42×10^3 lb./in.^{2}) while Steel (17-4PH H1150z) has a value nearly three times larger, at 860×10^6 N/m^2 (125×10^3 lb./in.^{2}). Notice the complex designation after each type of material (e.g., 6061-T6), this refers to the particular grade. Properties can also vary widely between grades of the same material.

Any material that can withstand significant strain before failing, we say is *ductile*. For example, Play-Doh™ is ductile, because it stretches pretty far before it breaks. A material that breaks after only a small amount of strain, like glass, is *brittle*. (That's why it's called peanut brittle, not peanut ductile).

Earlier we learned how to determine the fundamental frequency for an entire structure. This gave us an idea of its stiffness. To determine the stiffness of a particular material we, once again, treat it like a spring. This allows us to apply Hooke's Law, named for English mathematician Robert Hooke (1635–1703), which describes the stress in a material in terms of its strain and a new parameter—modulus of elasticity, E.

$$\boxed{\sigma = E\varepsilon} \qquad (13\text{-}16)$$

where

σ = stress (N/m^2)

E = modulus of elasticity or Young's modulus (N/m^2)

ε = strain (m/m)

The *modulus of elasticity, E,* also called *Young's modulus* after English scientist Thomas Young (1773–1829), is a basic property of any material and describes the amount of deformation it undergoes for a given amount of stress. For example, steel has a modulus of elasticity of around 200×10^9 N/m^2, but for wood, the value is only about 2×10^6 N/m^2, telling us what we already knew—steel is stiffer than wood (but, now we know how much stiffer.). We can then use Young's modulus, along with other material properties, for selecting which material to use to build a particular structure, depending on the loads it will have.

In addition to mechanical properties, strength and stiffness, we must also consider a material's basic physical properties. In Section 13-3, we saw the importance of thermal conductivity for thermal control on a spacecraft. In this section, we've already used the coefficient of thermal expansion, α, to determine thermal strain. Understanding how a material behaves under different temperature conditions is extremely important, before selecting it for a given application. Failure to do so can have serious consequences. For example, the poor performance of rubber O-rings under extremely cold conditions led to the Space Shuttle Challenger disaster.

Other physical properties are also important to consider. In Chapter 3 we discussed many problems with living and working in the space environment. This environment subjects materials to vacuum and radiation. Recall, a vacuum can create a problem with materials called out-gassing. *Out-gassing* comes from minute quantities of gas trapped in a material leaking when vacuum conditions exist. Engineers must carefully screen materials for their out-gassing properties before selecting them. Otherwise, this leaking gas can damage lenses or cause shorts in electrical equipment. Radiation can also lead to long-term damage to materials, causing them to lose their mechanical properties, and making them more susceptible to failure.

We must also consider the magnetic properties of a material. Obviously, for some material applications, such as in electromagnets for magnetorquers, magnetic properties are a good thing. For other applications, such as in the tray that holds a magnetometer, magnetic properties are bad.

Finally, density is another basic physical property we must know before selecting a material. Because mass is usually a premium on space missions, we like to use the lightest materials possible. That's why aluminum (about 2800 kg/m^3; 174.8 lbm/ft.^{3}) is often best for primary spacecraft structures, instead of steel (about 7860 kg/m^3; 490.6 lbm/ft.^{3}).

In most cases, we assume a material's mechanical and physical properties are *homogeneous*; that is, it has the same properties throughout and in all directions. This is a pretty good assumption for common spacecraft materials, such as aluminum, but not all materials are homogeneous. Wood, the oldest known building material, for example, is generally not homogeneous. It has a grain and tends to be much stronger along the grain than across it. Ironically, modern material scientists re-discovered this ancient material property, or, more precisely, carefully applied it, in the form of composite materials.

Composite materials (or simply *composites*) are "designer" materials that can be carefully tailored to carry a specific load in a specific way. Like wood, composites can be very strong in a particular direction. What makes them so attractive for aerospace applications (as well as for cars, tennis rackets, and thousands of other everyday items) is that they can have the same strength as a metal, at a fraction of the weight of aluminum or other traditional materials. After all, if a structure needs strength only in one direction, why not build it strong only in that direction and save the additional weight? Figure 13-66 shows the composite structure used for the Space Technology Research Vehicle (STRV) spacecraft.

Another important material property to consider is ease of fabrication. Again, analysis may show one material handles the loads better than another, but if the better material is too difficult or dangerous to work with, it's not practical to use. Titanium, for example, is very strong and has good high temperature properties. However, it is much more difficult to machine than aluminum. Beryllium also has many advantageous material properties, but is toxic.

Ultimately, cost is one of the biggest drivers in deciding what type of material to use. Even if your analysis indicates it would be the best material, it's impractical to build a spacecraft out of solid gold! Thus, we must always try to balance the best material properties with reasonable costs. We'll see how this and all the competing mission requirements blend together next.

Figure   13-66.   Composite   Structures. Composites are "designer" materials that are made to be strong in a particular direction, so they are lighter for specific applications than aluminum or other homogeneous materials. The Space Technology Research Vehicle (STRV) spacecraft, built by the Defense Evaluation and Research Agency (DERA) in the United Kingdom, in cooperation with U.S. government agencies, used composite structures to reduce mass. *(Courtesy of Defense Evaluation Research Agency, Farnborough, U.K.)*

Systems Engineering

Now that we've described stress, strain, and vibration, we can return to the systems engineering problem of putting the spacecraft structure together. As we saw in Figure 13-58 the challenge is to create a structure that satisfies all of the mission, system, and subsystem-level requirements and constraints. It must have the right mechanical behavior to keep from breaking, or even deforming during launch. And it needs a smart mechanical configuration that accommodates all of the payloads and subsystems within the mass and volume constraints.

One of the first jobs the structural engineers must do is to translate all of the general requirements and constraints into specific values for

- Strength—minimum necessary to ensure the structure won't break
- Stiffness—minimum necessary to ensure the structure remains stable and won't bump into the side of the fairing during launch
- Fundamental frequency—avoid resonance with launch vehicle vibration modes
- Mass properties—compute mass, volume, center of gravity, moments of inertia
- Mechanical interfaces—what needs to attach where and how

They then use these values as the starting point for the structural design process. Realize that all of these requirements potentially conflict with one another. For example, analysis may indicate the structure needs some additional strength in one area to handle launch loads. This could dislocate a sensor that would like to be in that location to do its job.

As we've shown, we must design the spacecraft structure with the necessary mechanical behavior to meet the demands of the launch, and function throughout the entire life of the mission. By far, the most demanding requirements typically come from the launch environment. Early in the space systems engineering process, selecting a launch vehicle defines specific mission requirements and constraints that directly flow to the structure. We'll briefly review how the launch environment creates these severe demands on the mechanical behavior of the spacecraft's structure. Then we'll look at the subtler demands on the mechanical configuration imposed by the rest of the subsystems and payloads.

The Launch Environment

During launch, the spacecraft is subjected to a violent loading environment, filled with accelerations, shocks, noise, and vibrations. Our first concern, of course, is to ensure the primary structure doesn't break during this exciting ride into space. From this standpoint, we're most concerned about designing a structure to withstand the static load caused by launch acceleration. These accelerations, or "g-loads," produce a compressive force on the vehicle and spacecraft through its longitudinal axis (long axis of the launch vehicle). A load of one-g is normal for the launch vehicle at Earth's surface. You feel a one-g load pushing you up in your chair as you read this chapter.

Figure 13-67 shows a typical launch acceleration profile. Notice the load builds rapidly to 4 g's in the first few minutes, then falls to near zero before building again. A four g load is a force equal to four times the weight of the spacecraft. For example, it's the force you'd feel, if your car could accelerate at nearly 40 m/s^2 (130 ft./s^2). That's 0 to 100 km/hr (60 m.p.h.) in 0.7 seconds! The points on the acceleration profile where the acceleration drops to near zero are during staging.

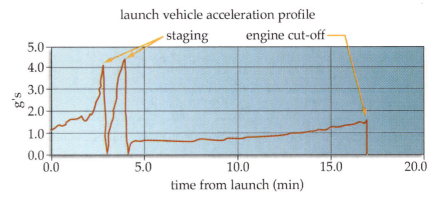

Figure 13-67. Launch Loads. Loads change drastically during a typical ride into space. Loads build to a peak at staging then drop off to near zero before building again.

Launch vehicles use stages to get to orbit more efficiently. Launch vehicles consist of a series of sub-vehicles called *stages*, each with large rockets and propellant tanks that do their jobs and then drop off to save weight. We'll discuss more about why launch vehicles use stages in Chapter 14. When the first stage burns out, the vehicle's acceleration drops to zero before the second stage ignites and the acceleration builds again. The launch-load plot shows a 3-stage launch profile. At the end of the third stage, the acceleration drops to zero and stays there because the launch vehicle reached orbit. After this, spacecraft rockets may fire for final orbital maneuvering, however, the acceleration during these later phases of the flight are much less than the 4 g's or more experienced during launch.

As the acceleration loads suddenly drop to zero during staging, the entire structure recoils similar to a spring that has a load released. Structural engineers statistically combine these low frequency transient loads with the predicted acceleration loads to produce a *quasi-static load* for the spacecraft structural design. For example, based on analysis for a new launch vehicle, engineers may compute a quasi-static load of 6.5 g's in all axes. Thus, the design team must ensure the spacecraft has sufficient strength that it will not buckle or fracture when attached to the launch vehicle and subjected to a 6.5 g load in all axes.

As with all structural design problems, engineers like to build in some extra margin of comfort, or safety factor. A *safety factor* is a multiplier used to reduce (but never eliminate!) the chance of failure. For example, structural analysis may indicate a particular strut needs a minimum yield stress of 100×10^6 N/m^2. However, to give it an extra safety margin, engineers may use a 1.3 safety factor and specify a yield stress of 130×10^6 N/m^2 when selecting the material to use.

In addition to strength, the structure must also have sufficient stiffness. Recall there are two aspects of stiffness

- How much a structure flexes when subjected to a load

- The structure's fundamental frequency

As you can imagine, any structure, no matter how securely we bolt it down, will flex, back and forth, when subjected to loads and vibrations. The structure (including solar panels, antennas, and anything else hanging from it) must be stiff enough that it will not flex so much that it goes outside the prescribed volume envelope allocated within the launch-vehicle fairing. Figure 13-68 shows the allowable spacecraft envelope under the fairing in a typical launch vehicle.

However, the second aspect of stiffness—fundamental frequency—can be far more troublesome for spacecraft structural design. This trouble comes from launch vibrations. A ride into space is very bumpy. If not properly designed, the spacecraft could literally shake apart before it gets to orbit. During launch, launch vehicles generate random vibrations over a wide spectrum of frequencies. These vibrations come from the rocket engine combustion that transfers through the vehicle in the form of shocks and noise. Engine vibrations generated during normal operation conduct through the launch vehicle structure, directly to the spacecraft. These vibrations are present during the entire ride into space.

Shocks occur during stage separation and other points in the flight, when pyrotechnic actuators fire to release attachment bolts and other mechanisms. They usually last only a small fraction of a second but can produce a load of over a 1000 g's.

And, launches are noisy! If you ever witnessed a launch, even from miles away, you can actually feel the sound vibrating through your body.

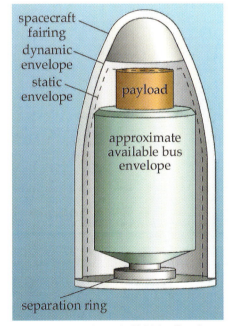

Figure 13-68. Launch Vehicle Envelope. We must carefully design a spacecraft to fit within a specific launch envelope inside the launch vehicle fairing.

These acoustic loads can exceed 140 decibels (dB). In comparison, the loudest rock music ever recorded was only about 120 dB. A *decibel* is a logarithmic ratio of sound pressure to a reference. Thus, a 6-dB increase in sound level represents a 2-fold increase in pressure. So a ride into space can be more than six times louder than the loudest rock concert! All this noise is mainly a problem for big spacecraft with large surface areas, especially solar panels. This noise can actually cause vibration loading on large spacecraft during lift-off, and as the vehicle goes through the sound barrier, that is worse than the acceleration loads.

As we discussed earlier, our primary concern with vibrations is reducing the effects of resonance. Recall that a structure resonates when it vibrates at or near its fundamental frequency. Resonance causes the vibration amplitude to increase rapidly, leading to high stress levels and possibly, failure. To reduce the effects of this resonance problem (we can never completely eliminate it!), engineers design the primary spacecraft structure to have a fundamental frequency that is well above or well below the major forcing frequency of the launch vehicle. Fortunately, we know the launch vehicle resonance frequency before launch, even for a new vehicle. For example, the launch vehicle providers may know that, of all the random vibrations during launch, the most severe ones that transmit to the spacecraft are about 25 Hz. Thus, the spacecraft designers must ensure the fundamental frequency of the primary structure is well above or below this value.

That kind of vibration analysis prevents primary structural damage, but what about the rest of the spacecraft? Unfortunately, because launch vibrations cover a wide spectrum of frequencies, some parts of the secondary structure will resonate. It's unavoidable. To prevent problems, engineers secure bolts and fasteners with wire locks and other techniques, so they don't shake loose during launch. Manufacturers coat sensitive electronic components, such as circuit boards, with a layer of epoxy or other resin that holds solder joints and other pieces in place and dampens vibrations.

Mechanical Configuration

Designing a generic structure to survive the well-defined launch environment is a fairly straightforward engineering exercise. Designing a spacecraft's mechanical configuration that survives while still satisfying the competing demands of the payload and subsystems is far more complicated. The most important requirements for the mechanical configuration are mass properties and mechanical interfaces.

Mass properties include the mass and volume budgets imposed on the entire mission by the launch vehicle lift capacity and payload fairing volume. These system-level requirements then flow down to subsystem budgets. For example, the mission may require an on-orbit ΔV of 1000 m/s. To provide this velocity change, the orbital control subsystem would need a minimum of 1000 kg of propellant with a volume of 750 liters. That much mass and volume may compete with other subsystems for extra spacecraft budget flexibility.

Another important set of requirements that comes from the mass properties is the center of gravity and moments of inertia. Optimum values for both of these are critical for attitude and orbit control. If we don't know the center of gravity well, then rocket thrusters may produce unwanted torques when they fire. Without the proper moment of inertia, the spacecraft won't be stable while spinning.

In addition to mass properties, important mechanical interface requirements drive the configuration. Generally speaking, these requirements tell us what components need to go where and how to attach them. Specifically, this process can dictate

- Field of view
- Thermal properties
- Assembly, integration, and testing (AIT) needs

Antennas, sensors, solar panels, radiators, and payloads may all compete for valuable external "real estate" to view space, the Sun, or Earth. These field of view requirements may conflict with each other, if, for example, an antenna shades a solar panel, or a radiator can't face into the cold of space to emit heat.

As we saw in Section 13.2, passive thermal control can depend heavily on using the structure to conduct heat. Certain subsystems, such as the batteries, may need to be in one part of the spacecraft to maintain their desired temperature.

Most important, we must design the mechanical configuration with assembly, integration, and testing in mind. To begin with, all of the spacecraft components must interconnect by a complex *wiring harness*, to carry power and data, as shown in Figure 13-69. The propulsion subsystem needs pipes to carry propellant between the storage tanks and engines. These harnesses and pipes need brackets to hold them securely during launch. Mechanisms must be carefully integrated with structures, such as a boom and the solar panels they must deploy or actuate. While all of this designing may sound trivial, without careful consideration to harness, pipe placement, or mechanism integration, the spacecraft may be impossible to assemble.

Throughout the integration and testing process, we may assemble and disassemble the spacecraft many times (especially if we find problems during testing). A structure that we design from the start for easy disassembly and re-assembly can save hundreds of hours in the process. We must also consider handling the spacecraft and its subsystems during AIT. Especially for expensive spacecraft, weighing thousands of kilograms, they need strong, convenient attachment points for handling equipment and to safely move them around, as shown in Figure 13-70. Finally, we must consider the ease of access for last minute items done only immediately prior to flight, such as installing pyrotechnic actuators and loading propellant. We don't want to take the whole spacecraft apart on the pad just to fill it with gas!

Figure 13-69. Wiring Harness. Here we show the wiring harness for a microsatellite. Even on very small satellites, the problem of connecting all subsystems together with power and data can be a major design driver. *(Courtesy of the U.S. Air Force Academy)*

Figure 13-70. Spacecraft Assembly. During the assembly process, handling jigs, like the one shown here, safely move the spacecraft to allow engineers access to every part of the structure. *(Courtesy of Defense Evaluation Research Agency, Farnborough, U.K.)*

Design and Analysis Tools

As you may imagine, the trade-offs between mechanical configuration and mechanical behavior require a number of iterations before engineers finish the structural design. To make these trade-offs, engineers have a number of analysis and design tools in their toolbox. Until recently, the tools they used to design the structure of a spacecraft weren't much different than the tools used to design the Pyramids—pencil, paper, and ruler. There are two important aspects of this process that are still used in modern computer aided design (CAD)—analysis and drawings.

The basic principles we reviewed earlier form the basis for structural analysis. Whether we do it on the back of a papyrus or with finite element analysis on a computer, the approach is basically the same—given specifications for loads, materials, and dimensions; calculate stress, strain, mass properties, and other design parameters.

By far, the most important design tools are mechanical drawings, and they can be as simple as a sketch on the back of an envelope or as complex as a 3-D rendering of the entire system. Either way, drawings communicate the design to the entire team. A picture may be worth a thousand words, but a good engineering drawing is worth volumes. From the simplest component to the entire system, drawings tell the analyst what assumptions to make, tell the machinist how to make the part, and tell the subsystem engineers where to put their components.

Figure 13-71 shows the manufacturing drawing for a simple box used to house the FireSat magnetometer. We must specify all the dimensions and material, and show all the views. Throughout the design process, the configuration for even this simple box could change many times as we trade requirements. Thus, another very important function of drawings is to maintain *configuration control* by documenting these changes. Notice that in the lower right-hand corner of the figure, specific details are given about the project, such as the designer and other information. In contrast, Figure 13-72 shows an exploded assembly drawing for the entire magnetometer. Here, details are less important than communicating the overall configuration for each major component in the sensor. In either case, mechanical drawings carry out a maxim of design engineers, older than Stonehenge: "Draw it the way you want to build it, and build it the way it's drawn." Figure 13-73 shows the final assembly drawing for the entire FireSat structure.

Figure 13-73. FireSat Comes Together. This assembly drawing gives an "exploded" view of FireSat showing how all the major structural components go together. *(Courtesy of Surrey Satellite Technology, Ltd., U.K.)*

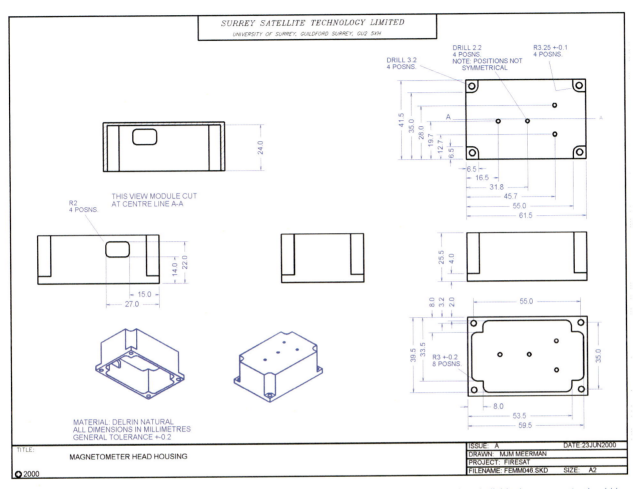

Figure 13-71.   Manufacturing Drawing of a Simple Box. Manufacturing drawings communicate how individual components should be fabricated. We must carefully draw and specify details of even a simple box used to house the FireSat magnetometer, such as the one shown here, to ensure proper manufacturing. *(Courtesy of Maarten Meerman, Surrey Satellite Technology, Ltd., U.K.)*

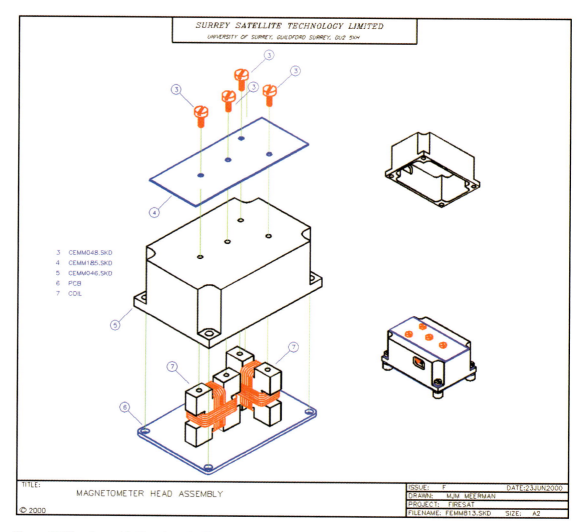

SURREY SATELLITE TECHNOLOGY LIMITED
UNIVERSITY OF SURREY, GUILDFORD SURREY, GU2 5XH

3   CEMM048.SKD
4   CEMM185.SKD
5   CEMM046.SKD
6   PCB
7   COIL

TITLE:
MAGNETOMETER HEAD ASSEMBLY

© 2000

ISSUE:	F	DATE:23JUN2000
DRAWN:	MJM MEERMAN	
PROJECT:	FIRESAT	
FILENAME: FEMM813.SKD		SIZE: A2

Figure 13-72.   Assembly Drawing of a Magnetometer. Assembly drawings communicate how individual components fit together to make a subsystem. This assembly drawing illustrates how all the individual pieces of the FireSat magnetometer come together. *(Courtesy of Maarten Meerman, Surrey Satellite Technology, Ltd., U.K.)*

Testing

During the design phase, engineers perform rigorous analysis of the structure to estimate its strength, stiffness, and natural frequency. Once we build it, we must qualify the entire structure prior to launch. Structural *qualification* is a mandatory set of tests imposed by the launch-vehicle provider to validate whether the design meets mission (specifically launch) requirements and constraints.

As part of the qualification test, we put the spacecraft on a large shaker table, as shown in Figure 13-74, that applies a pre-determined series of loads and vibrations. The results of these tests tell engineers the mechanical behavior of the structure (strength, stiffness and fundamental frequency) and its mass properties. We must compare this information to our analyses to see if actual values match predictions. We may subject the entire spacecraft, or large pieces of it, to simulated launch noise by putting them in front of massive speakers in an acoustic chamber and blasting them with sound to see how they vibrate. We give the analyses to the launch provider to verify the structure will survive its trip into space.

Figure 13-74.   Shaking It Up. Before launch, we test spacecraft on shaker tables to ensure they can endure launch vibrations. *(Courtesy of Naval Research Laboratory)*

In addition to qualifying the structure as a whole "spaceworthy" for launch, test engineers must test individual mechanisms sufficiently, under flight conditions, to satisfy mission designers that they will work when they need to. This testing can be far more difficult than it sounds. For example, if pyrotechnic actuators are necessary, it's impossible to test the one that must work for the mission, because the test destroys the actuator. By design, these are "one shot" items. The best we can do is select one at random from the same factory lot and test it. If it works during testing, we can assume (and hope) an identical one from the same lot, installed for launch, will work as well. (Of course, actuator manufacturers thoroughly test numerous samples and report the success rate.) We'll examine some of the economic implications of this system reliability in Chapter 16. Fortunately, we can test, and then reset, most high-cycle mechanisms prior to launch. But even then, it is often difficult to simulate free fall conditions on Earth. For example, a mechanism would need only a tiny force to rotate a massive solar array in free fall. However, it may be difficult or impossible to simulate these same conditions under 1-g on Earth. Designers must use a combination of ingenuity and analysis (and good fortune) to certify that the mechanisms are ready for launch.

Section Review

Key Terms

axial loads
bending moment, M
brittle
coefficient of thermal
    expansion, α
composites
composite materials
compression
compressive load
configuration control
damping
decibel
dry weight
ductile
dynamic loads
elastically
elastic region
fatigue
fundamental frequency
homogeneous
lateral loads
load
mechanical behavior
mechanical configuration
mechanisms
modulus of elasticity, E
natural frequency
out-gassing
primary structure
proportional limit
qualification
quasi-static load
resonance
residual strain
safety factor
secondary structure
shear
shear stress
spring constant, k
stages
static loads
stiff
stiffness

Key Concepts

➤ The spacecraft structure accounts for 10%–20% of the dry weight of the entire spacecraft

   • The primary structure carries all the major loads. The most stressful of these are the vibrations and other loads associated with launch.

   • The secondary structure holds all the subsystems together and attaches them to the primary structure

   • Low-cycle mechanisms operate only once or a few times to deploy booms and solar panels or open (or close) propellant valves

   • High-cycle mechanisms operate repeatedly during a mission and include solar array pointing devices, propellant valves, reaction wheels, and other moving parts

➤ The structural design process balances requirements and constraints on mechanical configuration with those on mechanical behavior

➤ The loads on a structure are either constant (static) or changing with time (dynamic). Static loads can be either

   • Pushing (compression) or pulling (tension), which causes stress and strain

   • Bending—can lead to more stress

   • Changes in temperature—heating causes expansion; cooling causes contraction. Both cause stress in the structure

➤ Dynamic loads primarily cause vibrations

   • Every structure has a natural frequency of vibration, depending on its mass and inherent stiffness. Designers must ensure the natural frequency of the spacecraft structure is different from the natural frequency of the booster; otherwise, dangerous resonance can occur, which could tear the structure apart.

➤ Designers choose material for spacecraft structures based on various qualities

   • Mechanical properties: strength and stiffness

   • Physical properties—density and reaction to the space environment

   • Ease of fabrication

   • Cost

➤ During the space systems engineering process, we must define structural requirements and constraints for

   • Strength—minimum necessary to ensure the structure won't deform or break

Continued on next page

Section Review (Continued)

Key Terms (Continued)

strain, ε
strength
stress, σ
structural design
tensile load
tension
torsional loads
ultimate failure point
ultimate strength, F_u
vibrations
wiring harness
yield point
yield strength
Young's modulus

Key Equations

$$\sigma = \frac{F}{A}$$

$$\varepsilon = \frac{\Delta L}{L}$$

$$\Delta L = \alpha(\Delta T)L$$

$$f_{fund} = 2\pi\sqrt{\frac{k}{m}}$$

$$\sigma = E\varepsilon$$

Key Concepts (Continued)

- Stiffness—minimum necessary to ensure the structure remains stable and won't flex into the side of the fairing during launch
- Fundamental frequency—avoid resonance with the launch vehicle vibration modes
- Mass properties—find the mass, volume, center of gravity, and moments of inertia
- Mechanical interfaces—what needs to attach where and how

➤ The launch environment is the biggest driver of mechanical behavior requirements

- Launch acceleration ("g-loads"), shocks, noise, and vibrations combine to produce a total quasi-static load the structure must withstand
- The primary structure's fundamental frequency must be different from the major forcing frequency of the launch vehicle to prevent resonance
- Some resonance in the secondary structure is inevitable. We must deal with it by locking bolts, coating electronics, and other measures.

➤ Mechanical configuration comes from requirements for

- Field of view
- Thermal properties
- Assembly, integration, and testing (AIT)

➤ One of the most important design and analysis tools for structural design are mechanical drawings. We use them to communicate and document configuration, fabrication, integration, and other vital information.

➤ During qualification testing, we subject the structure to loads and vibrations to determine its strength, stiffness, fundamental frequency, and mass properties. We also test mechanisms to ensure they'll function under flight conditions.

Example 13-4

Problem Statement

Your mechanical engineering team for the FireSat project has developed a structural configuration for the spacecraft, as shown in Figure 13-72. To accommodate the subsystems and payload within the available 0.30 × 0.30 × 0.30 m volume, the FireSat primary structure will be composed of 0.3 m × 0.3 m, 4-mm-thick aluminum honeycomb panels, connected by a 0.3 m long, 0.15 m outside diameter cylindrical "thrust tube" that will carry the primary launch loads. The payload sensor, sun sensors, and magnetometers will attach to the top panel. The launch vehicle attach fitting will connect to the base panel along with the gravity gradient boom and rocket thruster. Secondary, non-load bearing structures will include the subsystem electronics, which will be contained in four module boxes placed around the thrust tube attached to the base panel and the four solar arrays mounted on thin honeycomb panels. Four 316 Stainless steel bolts (F_{cy} = 290 MPa) will hold the spacecraft to the launch vehicle and will be cut for deployment by pyrotechnic devices. The Falcon launch vehicle will impart a peak acceleration of 20 g's lateral load during ascent. Assume the spacecraft will have a total mass at the maximum value of 15 kg, and during peak lateral acceleration, 2 of the 4 attachment bolts will experience the total tensile load. Designers are proposing to use 10-mm diameter bolts for the job. Determine if these bolts will be sufficient to provide a safety factor of 10.

Problem Summary

Given: acceleration = 20 g
spacecraft mass = 15 kg
no. bolts = 2
F_{cy} = 240 × 10^6 Pa
safety factor = 10.0
Find: Minimum diameter for the attachment bolts

Conceptual Solution

1) Find the maximum launch load on the spacecraft
launch load = (acceleration) (spacecraft mass)

2) Find the design load

$$\text{design load} = \frac{(\text{launch load})(\text{safety factor})}{\text{no. bolts}}$$

3) Find the minimum cross sectional area for the bolts

$$A_{min} = \frac{\text{design load}}{F_{cy}}$$

4) Solve for the minimum bolt diameter

$$A_{min} = \pi(\text{bolt radius})^2$$

$$\text{bolt radius} = \sqrt{\frac{A_{min}}{\pi}}$$

bolt diameter = 2 (bolt radius)

Analytic Solution

1) Find the maximum launch load on the spacecraft
launch load = (acceleration) (spacecraft mass)

$$= (20 \text{ g}) (15 \text{ kg}) \left(\frac{9.798 \frac{m}{s^2}}{1g} \right)$$

launch load = 2.939 × 10^3 N

2) Find the design load

$$\text{design load} = \frac{(\text{launch load})(\text{safety factor})}{\text{no. bolts}}$$

$$= \frac{(2.939 \times 10^3 N)(10.0)}{2}$$

design load = 1.47 × 10^4 N

3) Find the minimum cross sectional area for the bolts

$$A_{min} = \frac{\text{design load}}{F_{cy}} = \frac{1.47 \times 10^4 N}{240 \times 10^6 Pa}$$

$$A_{min} = 61.238 \text{ mm}^2$$

4) Solve for the minimum bolt diameter

$$A_{min} = \pi(\text{bolt radius})^2$$

$$\text{bolt radius} = \sqrt{\frac{A_{min}}{\pi}} = \sqrt{\frac{61.238 mm^2}{\pi}}$$

bold radius = 4.415 mm
bolt diameter = 2 (bolt radius)
bolt diameter = 8.83 mm

Interpreting the Results

During launch, two of the four attachment bolts must hold the spacecraft down against 20 g's lateral acceleration. For the proposed bolt material of 316 grade stainless steel, the minimum diameter to provide a safety factor of 10 is 8.83 mm. Therefore, the proposed use of 10-mm bolts is more than sufficient.

Example 13-5

t a temperature of 25° C, a beam forming part of a pace truss structure is 10.1 m long. As the beam enters ll sunlight, it reaches a temperature of 100° C. If the pefficient of thermal expansion is $3 \times 10^{-5}/°$ C, how uch strain occurs in the beam?

roblem Summary

iven: $L_{initial} = 10.1$ m
$T_{initial} = 25°$ C
$\alpha = 3 \times 10^{-5}/°$ C
$T_{final} = 100°$ C

ind: ε

Conceptual Solution

) Solve for expansion, δ and ΔL

$\delta = \alpha\,(\Delta T)\,L_{initial}$
$\Delta L = \delta$

) Solve for strain, ε

$\varepsilon = \dfrac{\Delta L}{L_{initial}}$

Analytical Solution

1) Solve for expansion, δ

$\delta = \alpha\,(\Delta T)\,L_{initial}$
$\delta = (3 \times 10^{-5}/°\ C)\,(100°\ C - 25°\ C)\,(10.1\ m)$
$\delta = 0.0227$ m
$\Delta L = 0.0227$ m

2) Solve for strain, ε

$$\varepsilon = \frac{\Delta L}{L_{initial}} = \frac{0.0227\ m}{10.1\ m} = 2.2 \times 10^{-3}\frac{m}{m}$$

Interpreting the Result

After entering the sunlight the beam will expand by 0.0227 m (almost a full inch). This is a strain of 2.248×10^{-3} m/m.

References

Asimov, Isaac. *Asimov's Biographical Encyclopedia of Science and Technology.* Garden City, NJ: Doubleday and Company, Inc., 1972.

Beer, Ferdinand P. and Russel E. Johnson, Jr. *Statics and Mechanics of Materials.* New York, NY: McGraw-Hill Inc., 1992.

Chang, Prof. I. Dee (Stanford University), Dr. John Billingham (NASA Ames), Dr. Alan Hargen (NASA Ames). "Colloquium on Life in Space." Spring, 1990.

Chetty, P.R.K. *Satellite Power Systems: Energy Conversion, Energy Storage, and Electronic Power Processing.* George Washington University Short Course 1507. October 1991.

Chetty, P.R.K. *Satellite Technology and Its Applications.* New York, NY: THB Professional and Reference Books, McGraw-Hill, Inc., 1991.

Doherty, Paul. "Catch a Wave." *Exploring.* Vol. 16, No. 4, (Winter, 1992): 18–22.

Gere, James M. and Stephen P. Timoshenko. *Mechanics of Materials.* Boston, MA: PWS Publishers, 1984.

Gonick, Larry and Art Huffman. *The Cartoon Guide to Physics.* New York, NY: Harper Perennial, 1991.

Gordon, J.E. *Structures: Why Things Don't Fall Down,* New York, NY: Da Capp Press, Inc., 1978.

Gunston, Bill. *Jane's Aerospace Dictionary.* New Edition. London, U.K.: Jane's Publishing Co., Ltd., 1986.

Holman, J.P. *Thermodynamics.* New York, NY: McGraw-Hill Book Company, 1980.

MacElroy, Robert D. Course Notes from AA 129, Life in Space, Stanford University, 1990.

Nicogossian, Arnauld E., Huntoon, Carolyn Leach, Sam L. Pool. *Space Physiology and Medicine.* 2nd Edition, Philadelphia, PA: Lea & Febiger, 1989.

Pitts, Donald R. and Leighton E. Sissom. *Heat Transfer.* Schaum's Outline Series. New York, NY: McGraw-Hill, Inc., 1977.

Poole, Lon. "Inside the Processor," *MacWorld.* October 1992, pp. 136–143.

Sarafin, Thomas P. *Spacecraft Structures and Mechanisms.* Dordrecht, the Netherlands: Kluwer Academy Publishers, 1995.

Wertz, James R. and Wiley J. Larson. *Space Mission Analysis and Design.* Third edition. Dordrecht, Netherlands: Kluwer Academic Publishers, 1999.

Mission Problems

13.1 Communication and Data-handling Subsystem (CDHS)

1 Describe the inputs, outputs, and basic processes within the spacecraft communication data-handling subsystem (CDHS).

2 Explain the difference between data and information. Give examples of each.

3 We can build a rudimentary communication system using two tin cans connected by a string. What is the carrier signal for this system? What is the modulator? The demodulator?

4 What is the difference between AM and FM? Draw a simple graph to illustrate what each would look like.

5 Draw the input/output diagram for a simple data-handling subsystem and define all of its components.

6 Define CPU, RAM, and ROM, and give examples of each in a common personal computer.

7   Define bit, byte, and processing speed.

8   What are the two types of software in a data-handling subsystem? What does each do? Give examples in a desktop personal computer.

9   Explain the difference between analog and digital data. Give examples of each.

10 Chamber pressure in a rocket engine is varying at a frequency of 100 Hz. What is the minimum digital sample rate needed to capture this analog data?

11 What five factors determine the requirements for the communication and data-handling subsystem (CDHS) design? What determines each of these?

12 What two budgets are critical to the CDHS design? Define each.

13 A new high resolution remote-sensing mission will use advanced 2048 × 2048 pixel CCD arrays. Mission users would like to capture images every 15 seconds along its 90 min polar orbit and downlink all of them for ground analysis through a single ground station. If 10 bits per pixel are needed to encode the spectral data, and the pass time at the ground station is 12 min, what is the required data rate to return one orbit's worth of data?

14 Designers are looking at the data-handling subsystems for two different spacecraft. One will orbit Pluto to take high-resolution photographs of this cold, mysterious planet and transmit them at a low data rate to the operations center on Earth. The other will be in low-Earth orbit to detect forest fires and transmit their location to ground stations. Discuss the trade-offs in complexity for the CDHS for each of these spacecraft.

13.2   Electrical Power Subsystem (EPS)

15 What are the basic inputs, outputs, and processes within a spacecraft EPS?

16 Define charge, voltage, current, resistance, and power.

17 What is Coulomb's Law?

18 What is Ohm's Law?

19 A 30-amp circuit has a 10 Ω resistor in it. What is the voltage drop across the resistor?

20 A spacecraft designer wants to supply 500 W of power at 28 V. What will be the current?

21   What are the three basic energy sources available to spacecraft?

22   What is the basic function of a photovoltaic cell? Describe the significance of the IV curve.

23   A spacecraft in Earth orbit has its solar panels at a 15° angle to the Sun. If the efficiency of the solar cells is 22%, what is the power output per square meter? (Assume the average incident solar power density for Earth is 1358 W/m^2).

24   What three environmental factors can degrade solar cell's performance?

25   A spacecraft is in a low-Earth orbit at an altitude of 350 km. What is the maximum time of eclipse for this orbit?

26   What types of chemical-power systems are used onboard spacecraft?

27   What is the difference between primary and secondary batteries?

28   Define depth of discharge. How does it affect battery life?

29   A spacecraft with a requirement for 500 W of continuous power is in an orbit with a maximum eclipse time of 32 minutes. If the maximum depth of discharge (DOD) for its batteries is 30%, what battery capacity does it need? (Give your answer in $W \cdot hr$).

30   Give an example of a mission which would need an Radioisotope Thermoelectric Generator. Discuss some of the issues associated with their use.

31   Within the EPS, what are the two stages that prepare the electrical power for the payload and bus?

32   Explain the difference between direct energy transfer and peak power tracking.

33   Explain the difference between centralized versus decentralized power conditioning.

34   What EPS concerns do mission planners have when they determine the end-of-life power requirements?

35   What is the difference between peak power and orbit average power? Explain the trade-offs between designing to the one versus the other.

36 As a follow-on to the successful Magellan mission, scientists are thinking about launching another SAR spacecraft to map the surface of Venus (R = 6051 km). The spacecraft will orbit at an altitude of 500 km. The total planned bus power is 120 W. The SAR payload will need an additional 500 W but will only operate 20% of each orbit while in sunlight. Compute the orbit average power, minimum battery capacity at 30% DOD, and eclipse time. Use $\mu_{Venus} = 3.249 \times 10^5 (km^3/s^2)$

13.3 Environmental Control and Life-support Subsystem (ECLSS)

37 What two primary functions does a spacecraft's environmental control and life-support subsystem perform?

38 Define thermal equilibrium.

39 Describe the potential sources of heat for a spacecraft.

40 Describe the three mechanisms of heat transfer. Give examples of each from everyday life.

41 Describe the difference between reflectivity, transmissivity, absorptivity, and emissivity.

42 You're holding a 0.1 m long metal rod that is 0.01 m in diameter in a pot of boiling water (100° C). If the thermal conductivity of the rod is 100 W/K m, and your hand is at normal body temperature (37° C), what is the rate of energy transfer along the rod?

43 A section of Space Shuttle tile is exposed to 1200 K on re-entry. If the tile is 0.1 m thick with an area of 0.01 m², what is the rate of heat transfer through the tile? The thermal conductivity of shuttle tile is about 0.108 W/K m.

44 A 1 m² spacecraft radiator with an emissivity of 0.85 will eject 100 joules of heat per second. What temperature must it be?

45 A cube 1 m on a side is in interplanetary space, the same distance from the Sun as Earth is from the Sun. If the absorptivity, α, of the surface is 0.3 and the emissivity, ε, is 0.7, what's the thermal-equilibrium temperature of the cube? Assume no internal heat.

46 What is the difference between active and passive thermal control?

47 List the various ways a spacecraft can transfer heat internally.

48 Discuss ways a spacecraft can eject heat.

49 Cars have "radiators," but do they really radiate? Explain.

50 How does multi-layer insulation protect a spacecraft from external heat sources?

51 Define latent heat of fusion and describe how we can use this principle onboard a spacecraft.

52 List the inputs and outputs of the human "system."

53 Why do Shuttle crews reduce the cabin pressure twelve hours before an Extra Vehicular Activity?

54 What variables go into the life-support budget for a manned space mission?

55 Discuss analysis and testing issues for spacecraft thermal control subsystems.

56 Mission analysts are thinking about adding one-half of a solar panel to one of the facets of the FireSat spacecraft described in Example 13-3 currently covered in multi-layer insulation (MLI). Assuming all other data and assumptions remain the same, how will this change affect the equilibrium temperature in sunlight and eclipse?

13.4   Structures and Mechanisms

57 Approximately what percentage of a typical spacecraft's mass is structure?

58 What is the difference between primary and secondary structures?

59 What two types of mechanisms can we use on spacecraft? Give examples of each.

60 List the loads that a spacecraft structure may be subjected to. Give examples of each.

61 Suppose we subject a cylindrical rod 10 m long with a diameter of 0.1 m to a 1000 N axial tensile load. What is the stress in the rod?

62 If the rod in Problem 61 deforms by 0.001 m, what is the strain?

63 If we bolt the rod in Problem 61 to the side of the spacecraft on one end and subject it to a 200 N lateral force on the other end, what is the bending moment?

64 The rod in Problem 61 has a coefficient of thermal expansion of 0.01/° C. If the 10 m length was measured at room temperature (21° C), what will the strain be if the temperature is 41° C? If the temperature is 1° C?

65 What is natural frequency? Why must engineers be concerned with it?

66 What is the natural frequency, in Hz, of a 10 kg spring with a spring constant of 0.1 N/m?

67 An Earth-observation satellite will launch on a vehicle with high-amplitude ascent vibrations at a frequency of 25 Hz. If the mass of the spacecraft is 1000 kg, what spring constant must the structure avoid to prevent resonance?

68 What factors do engineers consider in choosing a material for a structure?

69 A 5-m beam is subjected to a 100 N/m^2 compressive stress. If the Young's modulus of the material is 50×10^9 N/m^2, what is the strain in the beam?

70 Why are composite materials so popular for certain structural members?

71 What significant loads affect a spacecraft during launch?

72 How does a spacecraft's structural design affect its assembly, integration, and testing?

73 Describe the importance of engineering drawings to communicate design issues in a project.

Mission Profile—HST

Earth's atmosphere has always been a problem for astronomers—it distorts and attenuates incoming light. The Hubble Space Telescope (HST) was designed to provide a remarkable new view of the galaxy and beyond. Placed above Earth's atmosphere, Hubble can detect objects 25 times fainter than any visible from Earth's surface. Thus, astronomers see a universe almost 250 times larger than what is visible on Earth.

Mission Overview

Hubble's mission is to provide an orbiting platform for space-based astronomy, avoiding the interference of Earth's atmosphere. HST was designed with three main abilities

✓ High angular resolution to provide fine image detail

✓ Ultraviolet performance to provide ultraviolet images and spectra

✓ High sensitivity to detect very faint objects

Mission Data

✓ Circular orbit at 607 km (377 mi.) altitude and 28.5° inclination

✓ Pointing accuracy of 0.00000278° (0.01 ± 0.007 arcsec) for up to 24 hours. This means it could focus on a penny at a distance of more than 200 km.

✓ Large aperture mirror built using honeycomb-sandwich, reducing weight by a factor of four over solid glass

✓ Internal structure holds optical components aligned within 2.54×10^{-4} cm (1/10,000 in.) through extreme temperature changes

Mission Impact

At the start of this 15-year NASA mission, Hubble's thousands of observations have lent substantial credibility to its creators' claims. It has made many outstanding new discoveries and will continue to make observations that place the HST program at the forefront of astronomy. With future servicing missions Hubble's usefulness will continue to improve. It will remain on the cutting edge of science and technology for years to come.

Hubble's faint-object camera captured this image of a rare cosmic sight—gravitational lens G2237 + 0305—sometimes called the "Einstein Cross." This photograph shows four false images around a single quasar approximately eight billion light years away. The multiple images are caused by an aspect of the theory of relativity. The mass of the galaxies between us and the quasar (the fuzzy image in the middle) distorts space, causing the light to refract as if it passed through a big lens. *(Courtesy of the Association of Universities for Research in Astronomy, Inc./Space Telescope Science Institute)*

For Discussion

• After it was launched, users noticed HST's mirror had a flaw reducing its effectiveness (which was eventually fixed during a Shuttle mission). What lesson does this teach us about the design and management of space projects?

• How do we justify the cost of purely scientific missions like HST?

Contributors

Mari D. Brenneman and John D. Slezak, the U.S. Air Force Academy

References

National Aeronautics and Space Administration. *Hubble Space Telescope Update: 18 Months in Orbit.* Washington: Government Printing Office, 1992.

National Aeronautics and Space Administration. *Hubble Space Telescope, Media Reference Guide.* Sunnyvale, CA: Lockheed Missiles & Space Co., Inc.

The Space Shuttle Discovery rockets into the sky, powered by its two mighty, solid-rocket boosters and three main engines. *(Courtesy of NASA/Johnson Space Center)*

Rockets and Launch Vehicles

14

Outline

14.1 Rocket Science
Thrust
The Rocket Equation
Rockets

14.2 Propulsion Systems
Propellant Management
Thermodynamic Rockets
Electrodynamic Rockets
System Selection and Testing
Exotic Propulsion Methods

14.3 Launch Vehicles
Launch-vehicle Subsystems
Staging

In This Chapter You'll Learn to...

☛ Explain some of the basic principles of rocket science

☛ Discuss the various types of rocket systems and their operating principles

☛ Describe launch-vehicle subsystems and their key design issues

☛ Discuss the principles of rocket staging and how to determine the velocity change from a staged launch vehicle

You Should Already Know...

❏ Newton's Laws of Motion, the conservation of linear momentum, and the definition of kinetic energy (Chapter 4)

❏ The functions of spacecraft subsystems (Chapter 11)

❏ The definitions of charge and electric field (Chapter 13)

Mankind will not remain on Earth forever, but in its quest for light and space will at first timidly penetrate beyond the confines of the atmosphere, and later will conquer for itself all the space near the Sun.

Konstantin E. Tsiolkovsky
father of Russian cosmonautics

Space Mission Architecture. This chapter deals with the Launch Vehicles segment of the Space Mission Architecture, introduced in Figure 1-20.

Rockets take spacecraft where they need to go in space. Rockets form the core of the propulsion subsystems found on everything from fireworks to Space Shuttles to the Star Ship *Enterprise*. Propulsion subsystems

- Get spacecraft into space
- Move them around after they get there
- Change their attitude (the direction they're pointing)

Figure 14-1 characterizes all of these propulsion-system functions. In Chapter 9 we saw how much velocity change, ΔV, a launch vehicle needs to get from Earth's surface into orbit. Launch vehicles rely on their propulsion subsystems to produce this huge velocity change. After a spacecraft gets into space, its propulsion subsystem provides the necessary ΔV to take it to its final mission orbit, and then provides orbital corrections and other maneuvers throughout the mission lifetime.

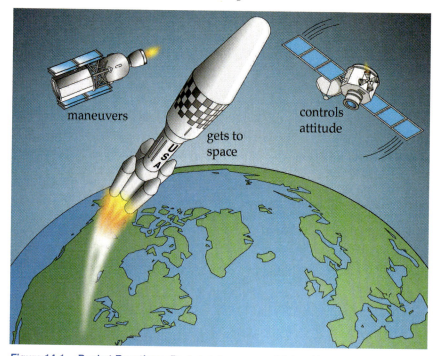

Figure 14-1.   Rocket Functions. Rockets take spacecraft into orbit, move them around in space, and help control their attitude.

Propulsion is also essential for attitude control. In Chapter 12 we learned how spacecraft use thrusters as attitude actuators. These thrusters are either the sole method of attitude control (similar to the Shuttle's reaction-control system), or they complement the primary system for large slewing maneuvers or to provide a means for momentum dumping. In this chapter we peel back the mysteries of rocket science to see how rockets work and how rocket scientists put together propulsion subsystems for spacecraft and launch vehicles.

14.1 Rocket Science

In This Section You'll Learn to...

- Explain the basic operating principles of rockets from a systems perspective
- Define and determine important parameters describing rocket performance—thrust, specific impulse, density specific impulse, and velocity change
- Explain how rockets convert stored energy into thrust
- Explain basic trade-offs in rocket design

You can't be a real rocket scientist until you can explain how a rocket works. In this section, we'll dissect rockets to see how all that noise, smoke, and fire can hurtle a spacecraft into space. Let's start with the big picture. A rocket is basically a system that takes mass plus energy and converts them into a force to move a vehicle. The input mass for a rocket is generally called *propellant*. The force produced by a rocket we call *thrust*. Figure 14-2 shows the block diagram for this simplified version of a rocket system.

Our examination of rocket systems begins by looking at the output—thrust. This requires us to dust off Newton's Laws to see how high speed exhaust going in one direction, pushes a vehicle in another. Next we'll see how this thrust, over time, produces a velocity change for the vehicle, and—most important for mission planning—how to calculate this effect and ensure we have enough propellant to get our vehicle where we want it to go. We'll then turn our attention to the process at the heart of a rocket: how it converts stored energy plus some mass into the high-speed exhaust. We'll tie all these concepts together by looking at the simplest example of a rocket—cold-gas thrusters—to see how varying some of the inputs and design variables changes the thrust and the overall system efficiency.

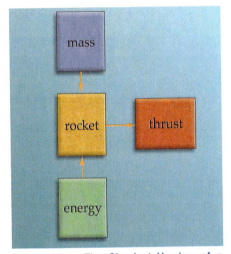

Figure 14-2. The Simplest Version of a Rocket System. The rocket's basic function is to take mass, add energy, and convert them into thrust, a force large enough to move a vehicle.

Thrust

A rocket ejects mass at high speed in one direction so a vehicle can go in the other. The simplest example of this is a balloon. All of us have blown up a toy balloon and let go of the stem to watch it fly wildly around the room, as shown in Figure 14-3. What makes the balloon go? Recall from Chapter 4 where we introduced Newton's Third Law

For every action there is an equal but opposite reaction.

When you blow into a balloon, you force air into it, making the rubber skin tighten, increasing the internal air pressure, and storing mechanical

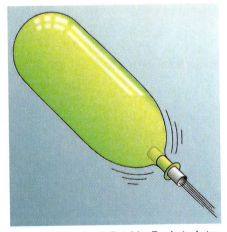

Figure 14-3. An Inflatable Rocket. A toy balloon is a simple example of a rocket. When we let go of the stem, "rocket propulsion" causes it to fly wildly around the room.

energy like a spring. When you let go of the stem, the air pressure has an escape route, so the skin releases, forcing the air out under pressure. Following Newton's Third Law, as the air, which has mass, is forced out in one direction (the action), an equal force pushes the balloon in the opposite direction (the reaction).

Let's look at this action/reaction situation in a bit more detail to see where the force comes from. Consider a rocket scientist perched in a wagon armed with a load of rocks, as shown in Figure 14-4. If he's initially at rest and begins to throw the rocks in one direction, because of Newton's Third Law, an equal but opposite force will move him (and the wagon load of rocks) in the opposite direction.

To throw the rocks, the scientist has to apply a force to them. This force is identical in magnitude, but opposite in direction, to the force applied to the scientist and thus, the wagon. However, remember the concept of conservation of linear momentum we discussed in Chapter 4. It tells us the change in speed of the rock (because it has less mass) will be greater than the change in speed of the wagon.

The rock's mass leaves at a rate we call the *mass flow rate*, $\Delta m / \Delta t = \dot{m}$ (kg/s). Recall from Chapter 4 that linear momentum is mass times velocity. If the speed of the rocks is V_{exit}, the ejected mass has a change in momentum of $\dot{m} V_{exit}$. But remember, momentum is always conserved! So as the momentum of the ejected mass (rocks) goes in one direction, the momentum of the rocket (or wagon in this case) goes in the other direction, as shown in Figure 14-4. This is the basic principle that produces rocket thrust. A rocket expends energy to eject mass out one end at high velocity, pushing it (and the attached vehicle) in the opposite direction.

Dropping the vector notation to look at magnitudes only, we can say

$$\dot{P}_{rocket} = \dot{P}_{exhaust} \quad Pp$$
$$\dot{P}_{rocket} = \dot{m} V_{exit}$$

(14-1)

where

$\dot{P}_{rocket}$ = time rate of change of the rocket's momentum (N)

$\dot{P}_{exhaust}$ = time rate of change of the exhausted mass' momentum (N)

$\dot{m}$ = mass flow rate of the exhaust products (kg/s)

V_{exit} = exit velocity of the exhaust (m/s)

Notice this momentum change has the same units as force. This is the force on the rocket we defined to be the thrust. As we'll see later, depending on the type of rocket used, the effective thrust delivered may be slightly different than $\dot{m} V_{exit}$. For this reason, we define a more comprehensive term called *effective exhaust velocity, C,* so we can express rocket thrust as

$$\boxed{F_{thrust} = \dot{m} C}$$

(14-2)

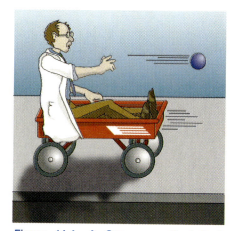

Figure 14-4. A One-person Rocket. A person throwing rocks out the back of a wagon illustrates the basic principles of a rocket. Muscles apply force to the rock, accelerating it in one direction, causing an equal but opposite force on the person and the wagon, pushing them in the opposite direction.

where

F_{thrust} = rocket's total thrust (N)

C   = effective exhaust velocity (m/s)

$\dot{m}$ = mass flow rate (kg/s)

This relationship should make sense from our wagon example. The scientist can increase the thrust on the wagon by either increasing the rate at which he throws the rocks (higher $\dot{m}$) or by throwing the rocks faster (higher C). Or he could do both. For example, if he threw bowling balls, he could produce a high $\dot{m}$ but with lower velocity than if he were throwing small pebbles.

Of course, exhaust velocities for typical rockets are much, much higher than anyone can achieve by throwing rocks. For typical chemical rockets similar to the Space Shuttle's, the exhaust velocity can be as high as 3 km/s. Because these high velocities are hard to visualize, it's useful to think about the raw power involved in a rocket engine. Recall from Chapter 4 that kinetic energy is

$$KE = \frac{1}{2} mV^2 \qquad (14\text{-}3)$$

We define *power* as energy expended per unit time. Thus, the power in the jet exhaust of a rocket is

$$P_J = \frac{1}{2} \dot{m} \, C^2 \qquad (14\text{-}4)$$

where

P_J = Jet power in a rocket (J/s = W)

At lift-off, the Space Shuttle's three main engines plus its solid-rocket boosters produce 26.6 billion watts of power. That is equivalent to 13 Hoover Dams! We'll see the effect of all that power next.

The Rocket Equation

To better understand how we use the thrust produced by rockets to get a vehicle where we want it to go, we first need to introduce a new concept—impulse. Impulse will help us understand the total velocity change rockets deliver.

Impulse

So a rocket produces thrust that pushes on a vehicle. Then what happens? If you push on a door, it opens. If you hit a ball with a bat, it flies out to left field. Returning to our scientist in the wagon, realize that to give the rocks their velocity, he has to apply a force to them over some length of time. Force applied to an object over time produces an *impulse*. We explain impulse using Newton's Second Law, also introduced in Chapter 4. Dropping vector notation, we express this law as

$$F = \frac{\Delta p}{\Delta t}$$

If we multiply both sides by Δt, we get

$$F \Delta t = \Delta p$$

The left side of this equation represents a force, F, such as the force we would generate if we hit a baseball with a bat over some time, Δt. When our bat hits that fast ball speeding over home plate, it seems like the impact is instantaneous, but the bat actually contacts the ball for a fraction of a second, applying its force to the ball during that time. We see the result of that force acting over time on the right side of the equation, where Δp represents the resulting change in momentum.

From this relationship we realize that to change momentum, we can apply a large force acting over a short time (like a bat hitting a ball) or a smaller force acting over a longer time (like an ant moving a bread crumb). We define *total impulse, I,* to be force (assumed constant here) times time, or change in momentum

$$\boxed{I \equiv F\Delta t = \Delta p} \qquad (14\text{-}5)$$

where

I   = total impulse (N s)

F   = force (N)

Δt = time (s)

Δp = momentum change (N s)

Impulse works the same way for rockets as it does for baseballs. We want to change the velocity and hence the momentum of our rocket, so we must apply some impulse. This impulse comes from the thrust acting over a time interval. But as we showed, we can produce the same impulse for a rocket by applying a small thrust over a long time or a large thrust over a short time.

Although total impulse is useful for telling us the total effect of rocket thrust, it doesn't give us much insight into the rocket's efficiency. To compare the performance of different types of rockets, we need a new parameter we call specific impulse. *Specific impulse, I_{sp},* is the ratio of the total impulse to the propellant's mass required to produce that impulse (how much "bang for the buck").

$$\boxed{I_{sp} \equiv \frac{I}{\Delta m_{propellant}\, g_o}} \qquad (14\text{-}6)$$

where

I_{sp} = specific impulse (s)

I   = total impulse (N s)

Δm = change in the propellant's mass (kg)

g_o = gravitational acceleration constant = 9.81 m/s² (We use this value for g_o so we can compare one rocket's performance with another's.)

Substituting for total impulse

$$I_{sp} = \frac{F_{thrust} \Delta t}{\Delta m_{propellant}\ g_o}$$

$$\boxed{I_{sp} = \frac{F_{thrust}}{\dot{m}\ g_o}} \qquad (14\text{-}7)$$

where

I_{sp} = specific impulse (s)

F_{thrust} = force of thrust (N)

$\dot{m}$ = propellant's mass flow rate (kg/s)

g_o = gravitational acceleration constant = 9.81 m/s²

I_{sp} represents rocket efficiency, the ratio of what we get (momentum change) to what we spend (propellant). So the higher the I_{sp}, the more efficient the rocket.

Earlier, we found the force of thrust in terms of the mass flow rate and the effective exhaust velocity. By substituting Equation (14-2) into Equation (14-7), we get another useful expression for I_{sp}.

$$\boxed{I_{sp} = \frac{C}{g_o}} \qquad (14\text{-}8)$$

where

C = effective exhaust velocity (m/s)

Notice g_o is a constant value representing the acceleration due to gravity at sea level, which we use to calibrate the equation. This means *no matter where we go in the universe, we humans will use the same value of g_o to measure rocket performance.*

As a measure of rocket performance, I_{sp} is like the miles per gallon (m.p.g.) rating given for cars. The higher the I_{sp} is for a rocket, the more ΔV it will deliver for a given mass of propellant. Another way to think about I_{sp} is that the faster a rocket can expel propellant, the more efficient it is.

Realize that we express I_{sp} in terms of propellant mass. For some space missions, especially those involving small satellites, such as the UoSAT-12 spacecraft shown in Figure 14-5, conserving volume can be just as important, or even more important, than conserving mass. Therefore, comparing only the I_{sp} of two systems may not tell us the whole story. For this reason, we define another useful term called *density specific impulse, I_{dsp},* which we find by multiplying the rocket's I_{sp} times the specific gravity of the propellants at nominal storage conditions, δ_{av}.

Figure 14-5. Small Satellites and Rocket Efficiency. Small satellites, such as UoSAT-12 shown here, can be as much volume constrained as mass constrained. Therefore, density specific impulse, I_{dsp}, becomes another important measure of rocket performance. *(Courtesy of Surrey Satellite Technology, Ltd., U.K.)*

$$I_{dsp} = \delta_{av} I_{sp} \tag{14-9}$$

where

I_{dsp} = density specific impulse (s)

δ_{av} = average specific gravity of propellants at nominal storage conditions

[*Note: Specific gravity* is the ratio of the density of a substance to water (e.g., specific gravity of water = 1.0, specific gravity of kerosene = 0.8)]

By comparing the mass and volume between different system options, mission planners can do more realistic trade-offs. We'll see the trade-off between I_{sp} and I_{dsp} in Example 14-1. In Section 14.2, we'll review other important factors to consider when selecting a rocket for a given mission.

Velocity Change

When we take a long trip in our car, we have to make sure we'll have enough gas in the tank to get there. This concern is even more important for a trip into space where there are no gas stations along the way. But how do we determine how much "gas," or propellant, we need for a given mission?

Naturally, some rockets are more efficient than others. For example, one rocket may need 100 kg of propellant to change velocity by 100 m/s while another needs only 50 kg. To figure how much propellant we need for a given trip, we must have a relationship between the velocity change and the amount of propellant used. We find this relationship by setting the thrust equal to the momentum change.

$$F_{thrust} = \dot{m}C = \frac{\Delta p_{rocket}}{\Delta t}$$

where

F_{thrust} = effective thrust from the rocket (N)

$\dot{m}$ = propellant's mass flow rate (kg/s)

C = effective exhaust velocity (m/s)

$\frac{\Delta p_{rocket}}{\Delta t}$ = rocket's time rate of change of momentum (N)

From this relationship we can derive the *ideal rocket equation*. (See Appendix C.9 for the complete derivation.) It tells us how much ΔV we get for a certain amount of propellant used.

$$\Delta V = C \ln\left(\frac{m_{initial}}{m_{final}}\right) \tag{14-10}$$

where

ΔV = velocity change (m/s)

C = effective exhaust velocity (m/s)

ln = natural logarithm of the quantity in the parentheses

$m_{initial}$ = vehicle's initial mass, before firing the rocket (kg)

m_{final} = vehicle's final mass, after firing the rocket (kg)

Equation (14-10) is one of the most useful relationships of rocket propulsion. Armed with this equation, we can determine how much propellant we need to do almost anything, from stopping the spin of a spacecraft in orbit, to launching a satellite to another solar system. Notice that we're taking the natural logarithm of the ratio of initial to final mass. *The difference between initial and final mass represents the amount of propellant used.* ΔV is also a function of the effective exhaust velocity. This relationship should make sense because, as the propellant moves out of the nozzle faster, momentum changes more, and the rocket goes faster.

We can substitute the definition of I_{sp} into the rocket Equation (14-10) to compute the ΔV for a rocket, if we know the I_{sp}, as well as the initial and final rocket mass.

$$\Delta V = I_{sp}g_o \ln\left(\frac{m_{initial}}{m_{final}}\right) \qquad (14\text{-}11)$$

where

ΔV = vehicle's velocity change (m/s)

I_{sp} = propellant's specific impulse (s)

g_o = gravitational acceleration at sea level (9.81 m/s^2)

ln   = natural logarithm of the quantity in the parentheses

$m_{initial}$ = vehicle's initial mass, before firing the rocket (kg)

m_{final} = vehicle's final mass, after firing the rocket (kg)

In Example 14-1 we'll see how useful this equation can be for space mission planning.

Rockets

Now that we've seen what rockets do—expel high speed exhaust in one direction so a space vehicle can go in the other—let's take a closer look at how they do it. Figure 14-6 illustrates this simplified view of a rocket system. For purposes of discussion, we can break this process into two steps. First, energy must be *transferred* to the propellant in some form. Second, the energized propellant must be *converted* into high speed exhaust. Figure 14-7 shows this expanded view of a rocket system.

There are only two basic types of rockets currently in use. Their classification depends on the form of energy that is transferred to the propellant and converted to high speed exhaust. These are

- Thermodynamic rockets—rely on thermodynamic energy (heat and pressure)
- Electrodynamic rockets—rely on electrodynamic energy (electric charge and electric and magnetic fields)

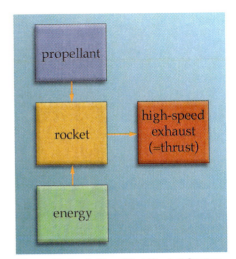

Figure 14-6.   A Simplified Rocket System. Rockets take in propellant and energy to produce a high speed exhaust. Conservation of momentum between the exhaust and the rocket produces thrust.

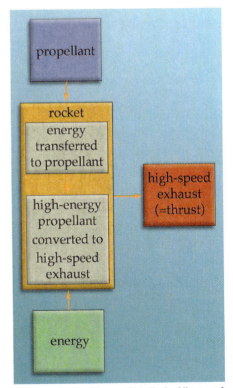

Figure 14-7. More Detailed View of Rockets. Energy is first transferred to the incoming mass. This high-energy mass is then converted to high-speed mass, producing thrust.

Thermodynamic energy is in the form of heat and pressure. This is something we're all familiar with. A covered pot of water on the stove reaches high temperature and produces high pressure steam. Most of us have seen how the thermodynamic energy in steam drives trains or produces electricity in power plants. In a *thermodynamic rocket*, thermodynamic energy transfers to the propellant in the form of heat and pressure. A propellant can produce heat through a chemical reaction or from external sources such as electricity, solar, or nuclear energy. Gaseous or liquid propellants are delivered to the rocket under pressure, supplying additional thermodynamic energy. However, for now, the result is the most important thing. Once energy transfers to the propellant, we have a high temperature, high pressure gas, or in other words, a gas with lots of thermodynamic energy. Air in a toy balloon or high-pressure gases from burning liquid hydrogen and liquid oxygen inside the Shuttle main engines are two extreme examples.

Later in this section, we'll look at the simplest type of rocket, a cold-gas thruster, that relies on gas under pressure alone as its source of thermodynamic energy. In Section 14.2 we'll look at other, more complex and efficient types of thermodynamic rockets.

Electrodynamic rockets rely on electrodynamic energy. *Electrodynamic energy* relates to the energy available from charged particles moving in electric and magnetic fields. This is the energy that makes our hair stand on end when we get a shock and makes magnets stick to some metals. To understand this, we need to go back to the concept of charge introduced in Chapter 13.

Recall, *charge* is a fundamental property of matter, like mass, and can be either positive or negative. Like charges repel each other and opposite charges attract. Typically, a molecule of propellant has the same number of protons and electrons making it electrically neutral. However, if one or more electrons can be "stripped off," the resulting molecule will have a net positive charge, making it an *ion*, as illustrated in Figure 14-8. To create the ion, the electrical-power subsystem (EPS) must supply some electrodynamic energy. Unlike a thermodynamic rocket, where the inherent energy of the energized propellant is quite high, the inherent energy of an ion is relatively low. However, once a particle is charged, additional electrodynamic energy can easily accelerate it to very high velocities. Later in this section, we'll look at how we do this inside an electrodynamic rocket.

The form of energy transferred to the propellant determines how it can be converted to high speed exhaust. In the rest of this section, we'll look at the two ways of doing this

- Thermodynamic expansion—using nozzles
- Electrodynamic acceleration—using electric and magnetic fields

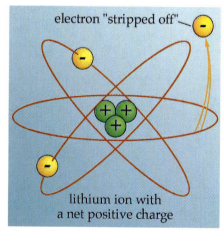

Figure 14-8.   An Ion. We create an ion when we "strip off" outer shell electrons from a neutral atom or molecule, leaving a net positive charge. Electric or magnetic fields can then accelerate this ion.

Thermodynamic Expansion—Nozzles

By far, the most commonly used types of rockets rely on nozzles. *Nozzles* convert the thermal energy produced by chemical, nuclear, or electrical sources into kinetic energy through thermodynamic expansion. In Figure 14-9, we show the huge nozzles used by the Saturn V F-1 engines that propelled astronauts to the Moon. To understand how nozzles do this, we must first understand a bit about fluid mechanics. After we understand how fluids behave, we can look at nozzles to see how they convert low-speed, high-temperature gasses into high-velocity exhaust. We'll then look at how we can predict and measure the performance of thermodynamic rockets and look at a simple example of a cold-gas rocket to see how all these principles fit together.

Fluid Mechanics. Fluid mechanics is the science of fluid (gasses or liquid) behavior. Let's start by looking at one of the simplest examples, the air in a balloon, as shown in Figure 14-10. Assuming the air in the balloon behaves as a "perfect gas," we can relate the pressure, density, and temperature of the air using the perfect-gas law

$$P = \rho RT \tag{14-12}$$

where

P = pressure (N/m^2)

ρ = density (kg/m^3)

T = temperature (K)

R = specific gas constant (J/kgK) = Ru/M

    Ru = universal gas constant = 8314.41 (J/kmole K)

    M = molecular mass of the gas (kg/kmole)

As we did in Chapter 4, we must review some basic assumptions that will make our discussion valid. We assume that it behaves as a perfect gas and has the following properties

- No heat transfer into or out of the fluid—this is known as an *adiabatic flow*

- The flow is reversible—meaning total energy is conserved. (A flow that is adiabatic and reversible is called an *isentropic flow.*)

- Flow in one dimension, because changes in the other two dimensions is negligible

- "Frozen flow"—meaning all chemical reactions occur inside the combustion chamber

- Steady flow—meaning mass flow rate is constant. (Remember that energy and momentum are conserved.)

These same assumptions will apply throughout this chapter in our discussion of thermodynamic rockets. Because these refer to ideal conditions, we often say the results apply to an "ideal rocket."

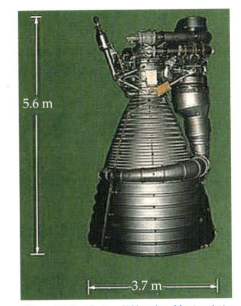

Figure 14-9.   Saturn V Nozzles. Most rockets rely on nozzles to convert thermal energy into kinetic energy through thermodynamic expansion. We show the huge nozzles for the Saturn V F-1 engines here. *(Courtesy of NASA/ Marshall Space Flight Center)*

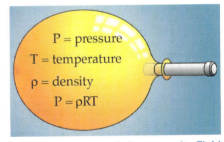

Figure   14-10. A   Balloon   and   Fluid Mechanics. Basic principles of fluid mechanics are at work in a simple balloon. The air inside obeys the perfect gas law, so its pressure, temperature, and volume are all related, based on the properties of the gas inside.

Earlier we discussed mechanisms for transferring energy to the propellant. In most cases, the propellant arrives in the combustion chamber under pressure so it already has some mechanical energy. The most common means of adding energy is through heat, such as that produced in chemical reactions (combustion). To track the total energy contained in the exhaust products, we need to introduce a new concept. We describe the total energy in fluid systems in terms of their *specific enthalpy, h.*

$$h = u + Pv \qquad (14\text{-}13)$$

where

h = specific enthalpy (J/kg)
u = internal energy (J/kg)
P = pressure (N/m^2)
v = specific volume (m^3/kg)

Computing enthalpy allows us to separate the energy in the exhaust due to the internal energy (heat) from the mechanical energy (pressure). To understand the usefulness of this concept, we need to put some fluid into motion. Let's connect our balloon to a pipe and watch the gas flow, as shown in Figure 14-11. We can compute the mass flow rate by multiplying the density of the exhaust by the flow velocity and the cross-sectional area of the pipe

$$\dot{m} = \rho VA \qquad (14\text{-}14)$$

where

$\dot{m}$ = fluid's mass flow rate (kg/s)
ρ = fluid's density (kg/m^3)
V = fluid's velocity (m/s)
A = pipe's cross-sectional area (m^2)

Now what happens when we vary the pipe's cross-sectional area? If we reduce the pipe's area, as shown in Figure 14-12, the flow velocity must *increase* to maintain a constant mass flow rate. We can imagine this concept by considering a garden hose. When we hold our thumb over the outlet, decreasing the area, the flow rate increases. This increase in flow velocity to maintain constant mass flow rate due to a constriction in area is called the *venturi effect* (Figure 14-12). Looking at it the other way, if we increase the pipe's cross-sectional area, the flow velocity decreases to maintain the same mass flow rate.

These effects are common ones, so they're fairly intuitive. However, it only works for *low-speed* flows. For very *high-speed* flows, the opposite effect takes place. As the area increases, the flow speeds up! How can this happen? For the case of steady, incompressible flow, with no heat transfer to or work done by the system (called an *isentropic* process) the sum of the specific enthalpy, h, and one half the square of the flow velocity, $1/2V^2$, is constant, due to conservation of energy.

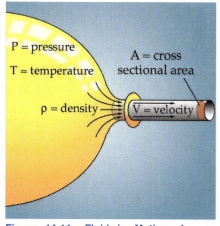

Figure 14-11. Fluid in Motion. As gas escapes a balloon and flows through a pipe, the mass flow rate depends on the fluid's velocity and density, and the pipe's cross-sectional area.

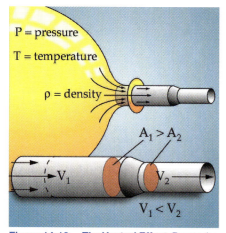

Figure 14-12.   The Venturi Effect. Due to the venturi effect, when a fluid hits a constriction in area, the flow velocity increases to maintain constant mass flow rate. Likewise, if the area increases, the flow rate decreases.

$$h + \frac{1}{2}V^2 = \text{constant} \qquad\qquad (14\text{-}15)$$

where

h = u + Pv, specific enthalpy from Equation (14-13)

V = flow's velocity (m/s)

Note the second term is the same as the specific kinetic energy (independent of the mass) of the flow. Thus, the internal energy and gas pressure can be traded for kinetic energy. In other words, the *velocity of the flow can increase at the expense of enthalpy and vice versa.* This relationship is the *Bernoulli Principle,* named after its discoverer, Italian mathematician, Daniel Bernoulli (1700–1782). This is one of the most important concepts in science. It helps us explain the dynamics of weather and how birds and planes fly.

But this still doesn't immediately explain how the flow speed increases when the cross-sectional area expands for very high-speed flow. To understand this, we need to delve a little deeper into the behavior of high speed gasses by looking at the speed of sound.

In 1947, Chuck Yeager piloted the Bell X-1 rocket plane and became the first person to "break the sound barrier" by traveling faster than the speed of sound. The speed of sound represents the velocity at which a pressure disturbance moves through a medium. In other words, if this book falls on the floor, it'll create a pressure disturbance (sound) that will travel out from the source at a specific speed that depends on characteristics of the air in the room. You'll hear the sound a fraction of a second before someone on the other side of the room. For short distances like these, it seems almost instantaneous. But if you've ever witnessed a thunderstorm, you've seen flashes of lightning in the distance and heard the thunder a few seconds later. By counting the seconds between seeing the lightning and hearing the thunder, and by knowing the speed of sound, you can get a good estimate of the distance to the storm. We can find the speed of sound, a_o, from

$$a_o = \sqrt{\gamma\, R\, T} \qquad\qquad (14\text{-}16)$$

where

a_o = speed of sound (m/s)

γ = ratio of specific heats (dimensionless)

R   = specific gas constant (J/kgK)

T   = temperature (K)

Notice in Equation (14-16) we introduce the *ratio of specific heats, γ.* This parameter is constant for a particular gas (or gas mixture) at a given temperature and pressure and depends on the molecular make-up of the gas. We can compute it using various gas modeling techniques that are beyond the scope of our discussion here, or we can measure it experimentally. As we'll see later, γ is an extremely useful parameter for calculating

$M_a < 1$
(subsonic)

converging nozzle (dA < 1)
velocity increases (dV > 1)

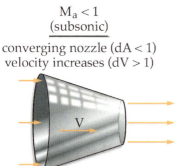

diverging nozzle (dA > 1)
velocity decreases (dV < 1)

$M_a > 1$
(supersonic)

converging nozzle (dA < 1)
velocity decreases (dV < 1)

diverging nozzle (dA > 1)
velocity increases (dV > 1)

Figure 14-13. Changing Mach Number Versus Changing Area. For subsonic flow ($M_a < 1$) the flow velocity increases when area decreases along the nozzle. For supersonic flow ($M_a > 1$) the flow velocity increases when the area increases.

rocket efficiency and other characteristics. Applying Equation (14-16) to air at 20° C, we get a speed of sound of 346 m/s. Therefore, if we count 3 seconds between a flash of lightning and hearing its thunder we know the storm is only 1.0 km away (head for cover!).

The ratio of the velocity of the flow (or the velocity of the *Bell* X-1) to the speed of sound is the *Mach Number, M_a*.

$$M_a = \frac{V}{a_o} \qquad (14\text{-}17)$$

Using Equations 14-16 and 14-17 and applying some additional thermodynamic analysis, we can develop a relationship between Mach Number, M_a, the change in area, dA, and change in flow velocity, dV.

$$\frac{dA}{A} = (M_a^2 - 1)\frac{dV}{V} \qquad (14\text{-}18)$$

where

dA = infinitesimal area change (m^2)

A   = pipe's cross-sectional area (m^2)

M_a = Mach Number

dV = infinitesimal velocity change (m/s)

V   = flow's velocity (m/s)

Looking closely at this relationship, we can see that for subsonic flow ($M_a < 1$) a positive change in area (expansion) leads to a negative change in velocity (decrease). However, for supersonic flow ($M_a > 1$) the opposite must occur, a positive area change (expansion) leads to a positive velocity change (increase). Figure 14-13 illustrates these relationships. This flip-flop in relationships between area and velocity above and below the speed of sound is due to conservation of energy (the trade-off between enthalpy and velocity) and conservation of mass.

Without this effect, rockets as we know them wouldn't be possible. We take advantage of this principle to convert the enthalpy of gasses, in the form of heat and pressure in the combustion chamber, into kinetic energy using nozzles. Just as with our balloon example, the process begins in the combustion chamber where high pressure (and usually high tempera-ture) exhaust products are created with little or no velocity ($M_a \ll 1$). From the combustion chamber, the products flow first into a converging section of the nozzle where we now know the velocity of the flow increases. The narrowest portion of the nozzle is the throat. By design, in the throat, the flow velocity reaches the speed of sound ($M_a = 1$) and we say the flow is "choked." As the nozzle expands beyond the throat, the velocity continues to increase to supersonic speeds before being exhausted at the nozzle exit. Figure 14-14 shows a simplified combustion chamber and nozzle diagram.

Nozzles. The more we expand the exhaust through the nozzle, the higher the exit velocity. But there are practical limits. Equation (14-15) showed us that the increase in velocity comes at the expense of the enthalpy of the gasses. Recall, enthalpy is a measure of the gasses' internal energy (heat) plus mechanical energy (pressure). Therefore, as the gasses gain velocity through expansion in the nozzle, they lose enthalpy, both temperature and pressure. Theoretically, we would need an infinitely long nozzle to expand the exhaust to zero exit pressure (vacuum) under ideal conditions, as we'll see later. In practice, of course, this isn't possible. Instead, rocket scientists design nozzles that are long enough for the conditions in which they operate. As we'll see next, the most important condition to consider is the outside air pressure.

Earlier, we used our rocket scientist in the wagon example to show that rocket thrust equals the mass flow rate, $\dot{m}$, times the effective exhaust velocity, C.

$$F = \dot{m}\,C$$

However, this represents only the thrust produced from momentum change, or *momentum thrust*. But for rockets using nozzles to convert thermal energy into kinetic energy, the thrust due to the momentum of the exhaust is only part of the story.

As we learned, unless a nozzle is infinitely long, the exhaust will have some exit pressure, P_{exit}. This pressure also contributes to the rocket's thrust. To see this, consider an imaginary "control volume" drawn around a rocket, as shown in Figure 14-15. Acting on the boundaries of this volume, we've drawn the atmospheric pressure $P_{atmosphere}$ on all sides except at the nozzle exit. At the nozzle exit, the pressure is P_{exit}, drawn inward for consistency. Notice that due to symmetry, $P_{atmosphere}$ cancels everywhere except in the direction parallel to momentum thrust over an area equal to the nozzle exit area, A_{exit}. The net force exerted on the rocket from this pressure differential is the *pressure thrust*. It equals the difference between exit pressure, P_{exit}, and atmospheric pressure, $P_{atmosphere}$, times the exit area, A_{exit}. It's magnitude is

$$F_{pressure\ thrust} = A_{exit}\,(P_{exit} - P_{atmosphere}) \qquad (14\text{-}19)$$

where

$F_{pressure\ thrust}$ = pressure thrust (N)
A_{exit} = nozzle's exit area (m^2)
P_{exit} = exit pressure (N/m^2)
$P_{atmosphere}$ = atmospheric pressure (N/m^2)

By adding together momentum thrust and pressure thrust, we express the magnitude of the total thrust on a rocket, using the total rocket thrust equation

$$F_{thrust} = \dot{m}\,V_{exit} + A_{exit}\,(P_{exit} - P_{atmosphere}) \qquad (14\text{-}20)$$

where

F_{thrust} = rocket 's total thrust (N)
$\dot{m}$ = mass flow rate (kg/s)

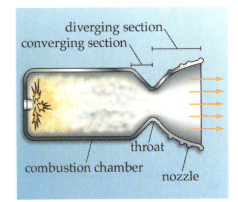

Figure 14-14. Standard Combustion Chamber and Nozzle Configuration. A standard thermodynamic rocket has two main parts—a combustion chamber (where energy transfers to a propellant) and the nozzle (where high energy combustion products convert to high-velocity exhaust).

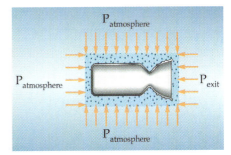

Figure 14-15. Pressure Thrust. Pressure thrust on a rocket results from the difference between the exit pressure and atmospheric pressure at the nozzle exit. Here we show a standard rocket in a "control volume" to illustrate this pressure difference.

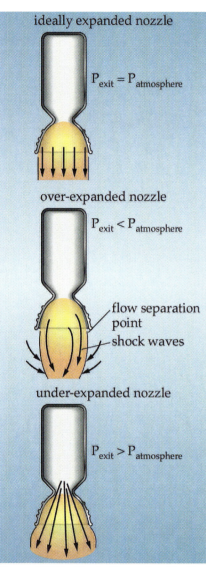

Figure 14-16.   Nozzle Expansion. To effectively convert all the enthalpy available in the combustion products to high-velocity flow, we need the nozzle exit pressure (P_{exit}) to equal the outside atmospheric pressure ($P_{atmosphere}$). When $P_{exit} < P_{atmosphere}$, the flow is overexpanded, causing shock waves that decrease flow velocity. When $P_{exit} > P_{atmosphere}$, the flow is underexpanded meaning not all available enthalpy converts to velocity. Here, we show all three expansion cases. In practice, we need an infinitely long nozzle to achieve perfect expansion in a vacuum.

V_{exit} = exit velocity of exhaust products (m/s)

To further simplify this equation, we can now fully define the effective exhaust velocity, C, to be

$$C = V_{exit} + \frac{A_{exit}}{\dot{m}}(P_{exit} - P_{atmosphere})$$ (14-21)

where

C = effective exhaust velocity (m/s)

So we're back to where we started with

$$F = \dot{m}\,C$$

Only now we know a lot more about where C comes from and how nozzle expansion affects it.

A casual glance at the total rocket thrust in Equation (14-21) may lead us to conclude we'd want to make $P_{exit} \gg P_{atmosphere}$ to maximize total thrust. Although this would appear to increase the amount of thrust generated, a big loss in overall efficiency would actually reduce the total effective thrust. Recall that for supersonic flow, as the gasses expand they *increase* in velocity while, due to the Bernoulli Principle, they *decrease* in pressure. Thus, the higher the V_{exit}, the lower the P_{exit}. For the ideal case, the pressure thrust should be zero ($P_{exit} - P_{atmosphere} = 0$), which means the exit pressure exactly equals the atmospheric pressure ($P_{exit} = P_{atmosphere}$). In this case, we maximize the exit velocity and thus the momentum thrust.

But what happens when $P_{exit} \neq P_{atmosphere}$? When this happens, we have a rocket that's not as efficient as it could be. We can consider two possible situations

- *Over-expansion:* $P_{exit} < P_{atmosphere}$. This is often the case for a rocket at lift-off. Because many launch pads are near sea level, the atmospheric pressure is at a maximum. This atmospheric pressure can cause shock waves to form at the nozzle's lip. These shock waves represent areas where kinetic energy turns back into enthalpy (heat and pressure). In other words, they rob kinetic energy from the flow, lowering the exhaust velocity and thus decreasing the overall thrust.

- *Under-expansion:* $P_{exit} > P_{atmosphere}$. In this case, the exhaust gasses have not expanded as much as they could have within the nozzle and thus, there's a "loss" in the sense that we've not converted all the enthalpy we could have into velocity. This is the normal case for a rocket operating in a vacuum, because P_{exit} is always higher than $P_{atmosphere}$ ($P_{atmosphere} = 0$ in vacuum). Unfortunately, we'd need an infinitely long nozzle to expand the flow to zero pressure, so in practice we must accept some loss in efficiency.

Figure 14-16 illustrates all cases of expansion. In Section 14.3, we'll see how we deal with this problem for launch-vehicle rocket engines.

The total expansion in the nozzle depends, of course, on its design. We define the nozzle's *expansion ratio, ε*, as the ratio between the nozzle's exit area, A_e, and the throat area, A_t

$$\varepsilon = \frac{A_e}{A_t}$$

(14–22)

where

ε = nozzle's expansion ratio (unitless)

A_e = nozzle's exit area (m^2)

A_t = engine's throat area (m^2)

Later in this section, we'll see how varying expansion ratio can also affect engine performance.

Characteristic Exhaust Velocity. We know rocket thrust depends on effective exhaust velocity, C, the rocket's output. But how do we measure C? Unfortunately, we can't just stick a velocity sniffer into superheated rocket exhaust. Therefore, when we're doing rocket experiments, we need to have some other, more measurable parameters available. As we'd expect, we can vary rocket output (thrust) by changing the inputs: $\dot{m}$, of the propellant going into the combustion chamber, and the resulting pressure in the combustion chamber, P_c. In addition to these dynamic parameters, there are also important physical dimensions of the rocket that we can vary, such as the area of the nozzle throat, A_t, and the expansion ratio, ε. To understand the relationship between these design variables, we define another important rocket performance parameter called *characteristic exhaust velocity, C** (c-star), in terms of chamber pressure, P_c, mass flow rate, $\dot{m}$, and throat area, A_t.

$$C^* = \frac{P_c A_t}{\dot{m}}$$

(14-23)

where

C^* = characteristic exhaust velocity (m/s)

P_c = chamber pressure (N/m^2)

A_t = engine's throat area (m^2)

$\dot{m}$ = exhaust's mass flow rate (kg/s)

The nice thing about C^* is that not only can we easily measure it experimentally, we can also compute it by modeling combustion characteristics. Rocket scientists use a variety of computer codes to predict the characteristic exhaust velocity. These techniques are beyond the scope of the discussion here, but for a complete description see *Space Propulsion Analysis and Design* [Humble et. al., 1995]. One of the most important parameters to find using these modeling techniques is the ratio of specific heats, γ, for the products in the combustion chamber. Knowing this value, we can compute C^* using

$$C^* = \frac{a_o}{\Gamma} \qquad (14\text{-}24)$$

where

a_o = speed of sound in the fluid (m/s)

$$\Gamma = \gamma\left(\frac{2}{(\gamma + 1)}\right)^{\left[\frac{(\gamma + 1)}{(2\gamma - 2)}\right]} \quad \text{(unitless)}$$

By comparing C^* from experiments with C^* from our predictions, we can determine how efficiently the rocket transfers the available energy to the propellants relative to an ideal value. This is especially useful as it allows us to measure the performance of the combustion chamber (energy transfer) independent from the nozzle (energy conversion). Typically, no rocket (or rocket scientist for that matter) is perfect. However, if the measured C^* is over 95% of predicted ideal C^*, we consider it good performance.

As we learned earlier, absolute rocket performance is defined in terms of I_{sp} and I_{dsp}, which are functions of effective exhaust velocity, C. While it is fairly straight forward to determine delivered I_{sp} experimentally by measuring delivered thrust and the total mass of propellant used, it is also important to be able to predict ideal I_{sp} for a given type of rocket. Fortunately, we can compute it directly from C^* and the ratio of specific heats for a given reaction.

$$I_{sp_{measured}} = \frac{F_{measured}}{g_o \dot{m}_{measured}} \qquad (14\text{-}25)$$

where

$I_{sp_{measured}}$ = measured specific impulse from experiments (s)
$F_{measured}$ = measured thrust from experiments (N)

$$I_{sp_{ideal}} = \frac{C^*}{g_o}\gamma\left[\left(\frac{2}{\gamma - 1}\right)\left(\frac{2}{\gamma + 1}\right)^{\frac{\gamma + 1}{\gamma - 1}}\right]^{\frac{1}{2}} \qquad (14\text{-}26)$$

where

$I_{sp_{ideal}}$ = theoretical ideal specific impulse from prediction (s)
C^* = characteristic exhaust velocity (m/s)
γ = ratio of specific heats for the gas (dimensionless)
[*Note:* This equation assumes infinite expansion of the exhaust, which is, of course, impossible to do.]

As we'd expect, one of the primary goals of rocket design is to maximize performance. We express rocket performance most often in terms of mass efficiency (I_{sp}). How do we maximize I_{sp}? We've just seen that effective exhaust velocity, and hence specific and density specific impulse are all functions of C^*. From Equation (14-24), we know C^* depends on the speed of sound in the combustion chamber, a_o, which depends on the gas constant, R, and the combustion temperature, T. Looking at Equation (14-16), we can see that the higher the temperature, the higher the speed of sound and thus, from Equations (14-24) and (14-26), the higher the I_{sp}. What about R? Remember from Equation (14-12), R is *inversely* proportional to the propellant's molecular mass. *Molecular mass* is a measure of the mass per molecule of propellant. Thus, to improve I_{sp} for thermodynamic rockets, we try to maximize the combustion temperature while minimizing the molecular mass of the propellant. We can express this relationship more compactly as

$$I_{sp} \propto \sqrt{\frac{T_{combustion}}{M}} \qquad (14\text{-}27)$$

where

I_{sp} = specific impulse (s)
$T_{combustion}$ = combustion temperature (K)
M   = molecular mass (kg/mole)

[*Note:* The symbol "$\propto$" means proportional to]

As a result, the most efficient thermodynamic systems operate at the highest temperature with the propellants having the lowest molecular mass. For this reason, hydrogen is often the fuel of choice because it has the lowest possible molecular mass and achieves high temperatures in combustion. Unfortunately, the low molecular mass also means low density. Thus, while hydrogen systems achieve high I_{sp}, they often don't do well in I_{dsp}. Designers must trade mass versus volume efficiency depending on the mission requirements.

Finally, we like to know what total thrust our rocket produces. After all, that's why we have a rocket in the first place! We can relate characteristic exhaust velocity, C*, to effective exhaust velocity, C, through yet another parameter called the *thrust coefficient, C_F*.

$$C_F = \frac{C}{C^*} \qquad (14\text{-}28)$$

where

C_F = thrust coefficient (unitless)
C   = effective exhaust velocity (m/s)
C* = characteristic exhaust velocity (m/s)

We can also relate the thrust coefficient to the thrust, F, chamber pressure, P_c, and throat area, A_t through

$$C_F = \frac{F}{(P_c A_t)} \qquad (14\text{-}29)$$

where

F   = thrust (N)
P_c = chamber pressure (N/m^2)
A_t = throat area (m^2)

Using this parameter, we can compare the measured rocket thrust to the ideal from theoretical modeling, to determine how well a nozzle converts enthalpy into kinetic energy. Like characteristic exhaust velocity, the thrust coefficient gives us a way to determine the performance of a nozzle, independent of the combustion chamber. Again, no nozzle is perfect. However, a well-designed nozzle, for the correct expansion conditions, should achieve at least 95% of predicted performance.

Summary. Let's review what we've discussed about thermodynamic rockets. Figure 14-17 further expands our systems view of a thermodynamic rocket and summarizes important performance parameters. Recall, there are two important steps to the rocket propulsion process—energy transfer and acceleration. These two steps take place in the combustion chamber and nozzle, respectively. The most important output is the thrust that moves a vehicle from point A to point B.

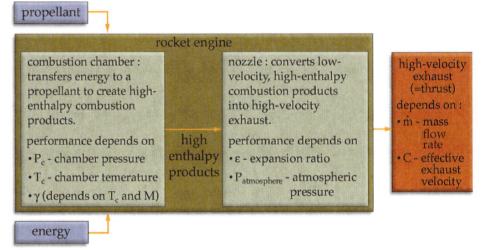
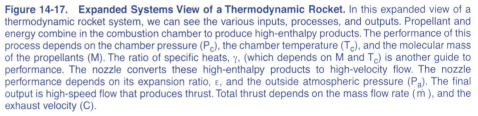

Figure 14-17.   Expanded Systems View of a Thermodynamic Rocket. In this expanded view of a thermodynamic rocket system, we can see the various inputs, processes, and outputs. Propellant and energy combine in the combustion chamber to produce high-enthalpy products. The performance of this process depends on the chamber pressure (P_c), the chamber temperature (T_c), and the molecular mass of the propellants (M). The ratio of specific heats, γ, (which depends on M and T_c) is another guide to performance. The nozzle converts these high-enthalpy products to high-velocity flow. The nozzle performance depends on its expansion ratio, ε, and the outside atmospheric pressure (P_a). The final output is high-speed flow that produces thrust. Total thrust depends on the mass flow rate ($\dot{m}$), and the exhaust velocity (C).

Now that we've filled your head with the behavior of gasses (or blown a lot of hot air, depending on how you look at it), let's put all these principles together by looking at one specific example—the simplest type of thermodynamic rocket in use, a cold-gas rocket.

Cold-gas Rockets. A *cold-gas rocket* uses primarily mechanical energy in the form of pressurized propellant as its energy source, similar to the toy balloon example we talked about at the beginning of the chapter. While spacecraft designers don't send balloons into orbit, the basic principles of cold-gas rockets aren't that different. A coiled spring stores mechanical energy that can be converted to work, such as running an old-fashioned, wind-up watch. Similarly, any fluid under pressure has stored mechanical energy that can be used to do work. Any rocket system containing fluids under pressure (and virtually all do) uses this mechanical energy in some way. As we'll see, usually this energy is a minor contribution to the overall energy of the propellant. However, for cold-gas rockets, this is the primary energy the propellant has.

Table 14-1 summarizes basic principles and propellants used by cold-gas rockets and Figure 14-18 shows a diagram of a simple cold-gas system.

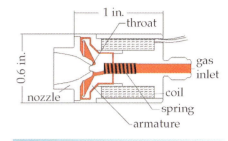

Table 14-1. Summary of Cold-gas Rockets.

Operating Principle	Uses the mechanical energy contained in a compressed gas and thermodynamically expands the gas through a nozzle producing high-velocity exhaust
Propellants	Helium (He), Nitrogen (N_2), Carbon dioxide (CO_2), or virtually any compressed gas
Advantages	• Extremely simple • Reliable • Safe, low-temperature operation • Short impulse bit (thrust pulses)
Disadvantages	Low I_{sp} and I_{dsp} compared to other types of rockets
Example	UoSAT-12 Cold-gas thrusters Propellant = N_2, P_c = 4 bar, Thrust = 0.1 N, I_{sp} = 65 s

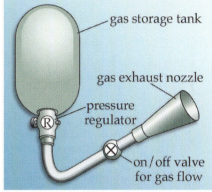

Figure 14-18. A Cold-gas Thruster. A cold-gas thruster is perhaps the simplest example of a rocket. In a typical thruster, shown in the cross-sectional drawing (upper), a gas enters from the right and stays behind the solenoid seal until it opens on command releasing the gas through the nozzle. *(Courtesy of Polyflex Aerospace, Ltd., U.K.)*

Cold-gas rockets are very reliable and can be turned on and off repeatedly, producing very small, finely controlled thrust pulses (also called *impulse bits*)—a desirable characteristic for attitude control. A good example of them is on the manned maneuvering unit (MMU) that was used by Shuttle astronauts. The MMU, shown in Figure 14-19, uses compressed nitrogen and numerous small thrusters to give astronauts complete freedom to maneuver.

Unfortunately, due to their relatively low thrust and I_{sp}, we typically use cold-gas systems only for attitude control or limited orbital maneuvering on small spacecraft. Even so, they can serve as a good example of trading some of the basic rocket parameters we've talked about in this section.

Figure 14-20 presents the results of analyzing five variations of the same basic cold-gas rocket. From the baseline design using nitrogen as the propellant at room temperature (298 K) and 5 bar (72.5 p.s.i.) chamber pressure, the figure shows the effects of increasing P_c or T_c, and reducing ε, or lowering the molecular mass (and γ) of the propellant by switching to helium. From this analysis, we can draw some general conclusions about basic trade-offs in thermodynamic rocket design

- Increasing chamber pressure increases thrust (with little or no effect on specific impulse or density specific impulse)

- Increasing chamber temperature increases specific impulse and density specific impulse with a slight decrease in thrust

Figure 14-19. Manned Maneuvering Unit (MMU). The MMU relies on small nitrogen cold-gas rockets to move astronauts around in space. *(Courtesy of NASA/Johnson Space Center)*

- Decreasing expansion ratio (under-expanded condition) decreases specific impulse, density specific impulse and thrust
- Decreasing propellant molecular mass increases specific impulse at the expense of decreasing density specific impulse and thrust

We'll apply these results to the specific problem of picking a thruster for the FireSat mission in Example 14-1.

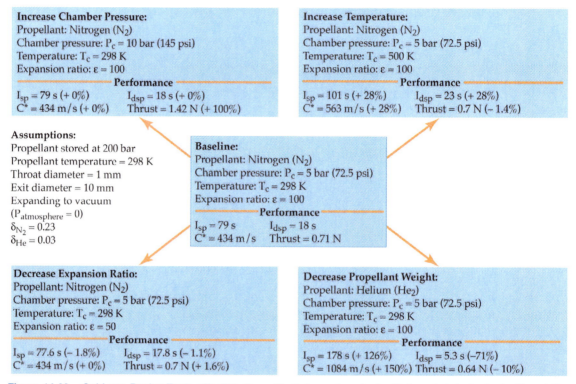

Increase Chamber Pressure:
Propellant: Nitrogen (N_2)
Chamber pressure: $P_c = 10$ bar (145 psi)
Temperature: $T_c = 298$ K
Expansion ratio: $\varepsilon = 100$
Performance
$I_{sp} = 79$ s (+ 0%) $I_{dsp} = 18$ s (+ 0%)
$C^* = 434$ m/s (+ 0%)   Thrust = 1.42 N (+ 100%)

Increase Temperature:
Propellant: Nitrogen (N_2)
Chamber pressure: $P_c = 5$ bar (72.5 psi)
Temperature: $T_c = 500$ K
Expansion ratio: $\varepsilon = 100$
Performance
$I_{sp} = 101$ s (+ 28%) $I_{dsp} = 23$ s (+ 28%)
$C^* = 563$ m/s (+ 28%)   Thrust = 0.7 N (– 1.4%)

Assumptions:
Propellant stored at 200 bar
Propellant temperature = 298 K
Throat diameter = 1 mm
Exit diameter = 10 mm
Expanding to vacuum
($P_{atmosphere} = 0$)
$\delta_{N_2} = 0.23$
$\delta_{He} = 0.03$

Baseline:
Propellant: Nitrogen (N_2)
Chamber pressure: $P_c = 5$ bar (72.5 psi)
Temperature: $T_c = 298$ K
Expansion ratio: $\varepsilon = 100$
Performance
$I_{sp} = 79$ s $I_{dsp} = 18$ s
$C^* = 434$ m/s   Thrust = 0.71 N

Decrease Expansion Ratio:
Propellant: Nitrogen (N_2)
Chamber pressure: $P_c = 5$ bar (72.5 psi)
Temperature: $T_c = 298$ K
Expansion ratio: $\varepsilon = 50$
Performance
$I_{sp} = 77.6$ s (– 1.8%) $I_{dsp} = 17.8$ s (– 1.1%)
$C^* = 434$ m/s (+ 0%)   Thrust = 0.7 N (+ 1.6%)

Decrease Propellant Weight:
Propellant: Helium (He_2)
Chamber pressure: $P_c = 5$ bar (72.5 psi)
Temperature: $T_c = 298$ K
Expansion ratio: $\varepsilon = 100$
Performance
$I_{sp} = 178$ s (+ 126%) $I_{dsp} = 5.3$ s (–71%)
$C^* = 1084$ m/s (+ 150%) Thrust = 0.64 N (– 10%)

Figure 14-20.   Cold-gas Rocket Trade-offs. This figure illustrates various trade-offs in rocket design by looking at the simplest type of rocket—a cold-gas thruster. From the baseline case using nitrogen as the propellant, the specific impulse, density specific impulse, characteristic exhaust velocity, C*, and thrust are affected by changing chamber pressure, chamber temperature, expansion ratio, and the molecular mass of the propellant. *(Analysis courtesy of the Johnson Rocket Company)*

Electromagnetic Acceleration

We've spent considerable time in this section discussing thermodynamic expansion and acceleration of exhaust using nozzles to convert propellant with thermodynamic energy into high-speed flow. But there is a second method for propellant acceleration currently gaining wider use on spacecraft—electrodynamic acceleration. To take advantage of this method, we must start with a charged propellant. In Chapter 13 we presented the force of attraction (or repulsion) between charges. Recall, this force depends on the strength of the charges involved and the distance between them. We expressed this as Coulomb's Law

$$\vec{F} = K\frac{Qq}{R^2}\hat{R}$$ (14-30)

where
$\vec{F}$ = electrostatic force vector on charge 1 (N)
K = constant (9×10^9) (N m^2/C^2)
Q = value of charge 1 (coulombs, C)
q = value of charge 2 (coulombs, C)
R = distance between charges (m)
$\hat{R}$ = unit vector in the direction of the force (unitless)

An *electric field* exists when there is a difference in charge between two points. That is, there is a large imbalance between positive and negative charges in a confined region. We call the energy an electric field can transmit to a unit charge the *electrical potential*, described in terms of volts/m. The resulting force on a unit charge is called an *electrostatic force*.

If you've ever rubbed a balloon through your hair and stuck it to a wall, then you've seen a simple example of electrostatic force in action. As you rubbed the balloon through your hair it picked up a net positive charge. When placed against the wall (initially neutral) the positive charges on the surface are pushed away leaving a net negative charge. The opposite charges attract each other creating a force strong enough to keep the balloon in place, despite the pull of gravity. The strength of the electrostatic force is a function of the charge and the electric field.

$$\vec{F}_i = m_i\vec{a}_i = q_i\vec{E}$$ (14-31)

where
$\vec{F}_i$ = electrostatic force vector (N)
m_i = mass of the charged particle (kg)
$\vec{a}_i$ = acceleration vector of the charged particle (m/s^2)
q_i = charge on the particle (coulombs, C)
$\vec{E}$ = electric field vector (electrical potential) (V/m)

Figure 14-21 illustrates this principle. Notice, the direction of the force is parallel to the electric field.

Electrodynamic rockets take advantage of this principle to create thrust. In the simplest application, they only need some charged propellant and an electric field. As with any rocket, the two key performance parameters are thrust, F, and specific impulse, I_{sp}. From Equation (14-2), we know thrust depends on the mass flow rate, $\dot{m}$, and the effective exhaust velocity, C

$$F = \dot{m}\,C$$

From Equation (14-8) we know specific impulse, I_{sp}, directly relates to C by

$$C = I_{sp}g_o$$

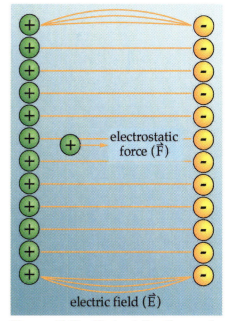

Figure 14-21. Electrostatic Force. An electric field exists when there is an imbalance between positive and negative charges in a confined region. This field will impart an electrostatic force on a charged particle within the field, accelerating it.

In an electrodynamic rocket, we achieve high $\dot{m}$ by having a high density of charged propellant. High exhaust velocity comes from having a strong electric field and/or applying the electrostatic force for a longer time. We can summarize these effects on performance as follows

- Higher charge density $\rightarrow$ higher $\dot{m}$ $\rightarrow$ higher thrust

- Stronger electric field $\rightarrow$ stronger electrostatic force on the propellant $\rightarrow$ higher acceleration $\rightarrow$ higher exhaust velocity $\rightarrow$ higher I_{sp}

Thus, by varying the charge density and the applied field, we can create a wide range of thruster designs. Naturally, there are practical design issues that limit how high we can increase each of these parameters. Let's start with charge density.

Charge density is limited by the nature of the propellant and how it is charged. Earlier, we defined an ion as a positively charged propellant molecule that has had one or more electrons "stripped off." Ions are handy in that they are simple to accelerate in an electric field. Unfortunately, when we try to pack lots of positive ions into a small, confined space, they tend to repel each other. This creates a practical limit to the achievable charge density.

One way around this density limit is to create a plasma with the propellant. A *plasma* is an electrically neutral mixture of ions and free electrons. Common florescent lamps or neon lights create a plasma when turned on. When a gas, such as neon, is in a strong electric field, the electrons become only weakly bound to the molecules creating a "soup" of ions and free electrons. The glow results from electrons jumping back and forth between energy states within the molecule. Because it is electrically neutral, a plasma can contain a much higher charge density than a collection of ions alone.

So far we've only considered the acceleration effect from an applied electric field. However, whenever we apply an electric field to a plasma, it creates (induces) a magnetic field. Charged particles also accelerate by magnetic fields but at right angles to the field, instead of parallel to it. To determine the combined effect of electric and magnetic fields on a charged particle, we must look at the cross product of their interaction.

$$\vec{F}_{em} = m_i \vec{a}_{em} = q_i(\vec{E} + \vec{V}_i \times \vec{B})$$ (14-32)

where

$\vec{F}_{em}$ = electromagnetic force on a charged particle (N)

m_i = charged particle's mass (kg)

a_{em} = electromagnetic acceleration (m/s^2)

$\vec{E}$ = electric field vector (V/m)

$\vec{V}_i$ = particle's velocity vector (m/s)

B = magnetic field vector (tesla)

Some types of electrodynamic rockets rely on this combined effect to produce thrust. However, for most cases, the electrostatic force dominates and we can ignore the effect of the magnetic field for simple analysis of performance.

When we consider the second parameter that limits thruster performance, the strength of the electric field, we can focus mainly on the practical limits of applied power. From Equation (14-4) the jet power of a rocket is

$$P_J = \frac{1}{2}\dot{m}\,C^2 \qquad (14\text{-}33)$$

Thus, for a given charge density, the exhaust velocity increases with the square root of the power. That is, if we apply 4 times the power, the exhaust velocity will only double. As we'd expect, there are practical limits to the amount of power available in any spacecraft, thus limiting the ultimate performance of electrodynamic rockets.

While exhaust velocity (and I_{sp}) goes up with power, there is a trade-off between thrust and exhaust velocity, as illustrated by the following relationship

$$\boxed{F \propto \frac{2P}{C}} \qquad (14\text{-}34)$$

where

F = thrust (N)

P = power (W)

C = effective exhaust velocity (m/s)

[*Note:* The symbol "$\propto$" means proportional to]

Therefore, when designing an electrodynamic thruster, for the same input power you can have high exhaust velocity or high thrust, but not both at the same time. In Section 14.2 we'll look at some specific examples of electrodynamic thrusters and compare their performance. It is important to note that this relationship is derived from conservation of energy and applies to electrodynamic rockets as well as thermodynamic rockets that rely on an external energy source (such as resistojets and arcjets as well, see in Section 14.2).

Section Review

Key Terms

adiabatic flow
Bernoulli Principle
characteristic exhaust velocity, C*
charge
cold-gas rocket
density specific impulse, I_{dsp}
effective exhaust velocity, C
electric field, $\vec{E}$
electrical potential
electrodynamic energy
electrodynamic rocket
electrostatic force
expansion ratio, ε
ideal rocket equation
impulse
impulse bits
ion
isentropic
isentropic flow
Mach Number, M_a
mass flow rate
molecular mass
momentum thrust
nozzles
plasma
power
pressure thrust
propellant
ratio of specific heats, γ
specific enthalpy, h
specific impulse, I_{sp}
thermodynamic energy
thermodynamic rocket
thrust
thrust coefficient, C_F
total impulse, I
venturi effect

Key Equations

$$F_{thrust} = \dot{m}C$$

$$I \equiv F\Delta t = \Delta p$$

$$I_{sp} \equiv \frac{I}{\Delta m_{propellant}\, g_o}$$

Key Equations (Continued)

$$I_{sp} = \frac{F_{thrust}}{\dot{m}\, g_o}$$

$$I_{sp} = \frac{C}{g_o}$$

$$\Delta V = I_{sp}g_o\ln\left(\frac{m_{initial}}{m_{final}}\right)$$

$$P = \rho RT$$

$$h = u + Pv$$

$$\dot{m} = \rho VA$$

$$h + \frac{1}{2}V^2 = constant$$

$$a_o = \sqrt{\gamma\, R\, T}$$

$$M_a = \frac{V}{a_o}$$

$$\frac{dA}{A} = (M_a^2 - 1)\frac{dV}{V}$$

$$C = V_{exit} + \frac{A_{exit}}{\dot{m}}(P_{exit} - P_{atmosphere})$$

$$\varepsilon = \frac{A_e}{A_t}$$

$$C^* = \frac{P_c A_t}{\dot{m}}$$

$$I_{sp_{ideal}} = \frac{C^*}{g_o}\gamma\left[\left(\frac{2}{\gamma-1}\right)\left(\frac{2}{\gamma+1}\right)^{\frac{\gamma+1}{\gamma-1}}\right]^{\frac{1}{2}}$$

$$I_{sp} \propto \sqrt{\frac{T_{combustion}}{M}}$$

$$C_F = \frac{C}{C^*}$$

$$\vec{F} = K\frac{Qq}{R^2}\hat{R}$$

$$\vec{F}_i = m_i\vec{a}_i = q_i\vec{E}$$

$$\vec{F}_{em} = m_i\vec{a}_{em} = q_i(\vec{E} + \vec{V}_i \times \vec{B})$$

$$F \propto \frac{2P}{C}$$

Key Concepts

➤ As a system, a rocket takes in mass and energy and converts them into thrust

- Rocket thrust is a result of Newton's Third Law *"For every action, there is an equal but opposite reaction."* Rockets eject high-velocity mass in one direction causing the rocket to go in the other direction.

- Total thrust delivered depends on the velocity of the mass ejected (effective exhaust velocity, C) and how much mass is ejected in a given time (mass flow rate, $\dot{m}$)

- You can find the amount of velocity change, ΔV, a rocket delivers for a given amount of propellant using the rocket equation

- Specific impulse, I_{sp}, measures a rocket's efficiency in terms of propellant mass. The higher the I_{sp}, the less propellant mass needed to deliver the same total impulse. I_{sp} is a function of a rocket's exhaust velocity.

- Density specific impulse, I_{dsp}, describes a rocket's efficiency in terms of propellant volume. The higher the I_{dsp}, the less propellant volume needed to deliver the same total impulse.

Continued on next page

▒ Section Review (Continued)

Key Concepts (Continued)

➤ Within a rocket system, there are two main processes at work

- First, energy must be transferred to the propellant (in the form of heat, pressure, or charge)
- Second, the energized propellant must be converted to high-velocity exhaust

➤ We classify rockets based on the form of energy they use

- Thermodynamic rockets—rely on thermodynamic energy (heat and pressure)
- Electrodynamic rockets—rely on electrodynamic energy from charged particles moving in electric and magnetic fields

➤ Thermodynamic expansion within a rocket nozzle depends on fluid mechanics

- Hot gasses within a rocket's combustion chamber obey (more or less) the perfect gas law
- Specific enthalpy, h, describes the total energy in the combustion products
- Rockets trade specific enthalpy, h, for kinetic energy (velocity)
- The Bernoulli Principle relates how the Mach Number of the flow changes depending on whether the area is contracting or expanding
  - If the flow is initially subsonic ($M_a < 1$), the flow will increase in velocity as it enters the converging section of the nozzle
  - By design, at the nozzle throat, $M_a = 1$
  - As the flow enters the diverging section of the nozzle, its velocity continues to increase to $M_a \gg 1$

➤ Nozzle performance depends on the total expansion and the external atmospheric pressure

- Effective exhaust velocity is a function of momentum thrust plus pressure thrust. Momentum thrust results from the action/reaction of expelled mass. Pressure thrust results from the difference in pressure between the exhaust gases and the atmospheric pressure. An ideally expanded nozzle produces no pressure thrust.

➤ Experimental rocket development depends on predicting and measuring several important parameters

- Characteristic exhaust velocity, C*, relates the chamber pressure, nozzle throat area, and mass flow rate. We can predict an ideal value for C* for a given set of propellants using thermodynamic models.
- Ideal specific impulse, I_{sp}, is a function of the combustion temperature and the molecular mass of the propellants. High I_{sp} results from the highest temperature and lowest molecular mass (e.g., hydrogen).

➤ Electrodynamic rockets use electric and magnetic fields to accelerate charged particles in a propellant. Charges can be either positive or negative. Like charges repel and opposite charges attract.

- An electric field applies an electrostatic force to charged particles. The force of acceleration, hence the thrust and exhaust velocity, depends on the strength of the field and the charge on the particle.
- Higher Charge Density → higher $\dot{m}$ → higher thrust. We produce ions when we strip electrons from neutral molecules, leaving a net positive charge. Plasmas can achieve a higher charge density because they are an electrically neutral mixture of ions and electrons.
- Stronger electric field → stronger electrostatic force on the propellant → higher acceleration → higher velocity → higher exhaust velocity → higher I_{sp}. The available power limits the strength of the electric field.

Example 14-1

Problem Statement

The propellant tank for the FireSat spacecraft must fit within a cylinder 0.15-m in diameter and 0.15-m long. Designers have selected a cold-gas rocket for the mission to provide station keeping. However, there is some disagreement over the type of propellant to use. Helium gas will provide an effective exhaust velocity of 1766 m/s, while nitrogen will provide only 787 m/s. Assuming either gas would be stored at 200 bar and at a temperature of 298 K, which propellant option will provide the most mission ΔV? Recall that the deployed mass of FireSat will be 15 kg.

Problem Summary

Given: $m_{spacecraft} = 15$ kg
$P_{storage} = 200$ bar
$T_{storage} = 298$ K
bar $= 1.0 \times 10^5$ Pa
$C_{N_2} = 787$ m/s
$C_{He} = 1766$ m/s
tank diameter $= 0.15$ m
tank length $= 0.15$ m
$m_{He} = 4.00$ kg/1000 mole
$m_{N_2} = 28.01$ kg/1000 mole
$R = 8314.41$ J/1000 mole K

Find: total ΔV for each propellant

Conceptual Solution

1) Determine total volume of propellant available

$$\text{tank volume} = \pi \left(\frac{\text{tank diameter}}{2} \right)^2 \text{tank length}$$

2) Solve Equation (14-12) for the storage density of each propellant

$$R_{N_2} = \frac{R}{M_{N_2}}$$

$$\rho_{N_2} = \frac{P_{storage}}{R_{N_2} T_{storage}}$$

$$R_{He} = \frac{R}{M_{He}}$$

$$\rho_{He} = \frac{P_{storage}}{R_{He} T_{storage}}$$

3) Determine the mass of each propellant available

$m_{N_2} = \text{tank volume } \rho_{N2}$

$m_{He} = \text{tank volume } \rho_{He}$

4) Determine total ΔV available from each propellant

$$m_{final\ N_2} = m_{spacecraft} - m_{N_2}$$

$$\Delta V_{N_2} = C_{N_2} \ln \left(\frac{m_{spacecraft}}{m_{final\ N_2}} \right)$$

$$m_{final\ He} = m_{spacecraft} - m_{He}$$

$$\Delta V_{He} = C_{He} \ln \left(\frac{m_{spacecraft}}{m_{final\ He}} \right)$$

Analytical Solution

1) Determine total volume of propellant available

$$\text{tank volume} = \pi \left(\frac{\text{tank diameter}}{2} \right)^2 \text{tank length}$$

$$= 3.1416 \left(\frac{0.15\ \text{m}}{2} \right)^2 (0.15)$$

$$\text{tank volume} = 2.651 \times 10^{-3}\ \text{m}^3 = 2.651\ \text{liters}$$

2) Solve Equation (14-12) for the storage density of each propellant

$$R_{N_2} = \frac{R}{M_{N_2}} = \frac{8314.41\ \text{J}/1000\ \text{mole K}}{28.01\ \text{kg}/1000\ \text{mole}} = 296.8 \frac{\text{J}}{\text{kg}}$$

$$\rho_{N_2} = \frac{P_{storage}}{R_{N_2} T_{storage}} = \frac{200\ \text{bar}}{\left(296.8 \frac{\text{J}}{\text{kgK}} \right) (298\ \text{K})}$$

$$= \frac{200 \times 10^5\ \text{Pa}}{8.446 \times 10^4 \frac{\text{J}}{\text{kg}}}$$

$$\rho_{N_2} = 226.1 \frac{\text{kg}}{\text{m}^3}$$

Example 14-1 (Continued)

$$R_{He} = \frac{R}{M_{He}} = \frac{8314.41 \text{ J}/1000 \text{ mole K}}{4.00 \text{ kg}/1000 \text{ mole}}$$

$$= 2078.6 \frac{\text{J}}{\text{kgK}}$$

$$\rho_{He} = \frac{P_{storage}}{R_{He}T_{storage}} = \frac{200 \text{ bar}}{\left(2078.6 \frac{\text{J}}{\text{kgK}}\right)(298 \text{ K})}$$

$$= \frac{200 \times 10^5 \text{Pa}}{6.1942 \times 10^5 \frac{\text{J}}{\text{kg}}}$$

$$\rho_{He} = 32.29 \frac{\text{kg}}{\text{m}^3}$$

3) Determine the mass of each propellant available

3) Determine the mass of each propellant available

$$m_{N_2} = \text{tank volume } (\rho_{N_2}) = 2.651 \text{ liter} \left(226.1\frac{\text{kg}}{\text{m}^3}\right)$$

$$m_{N_2} = 0.5994 \text{ kg}$$

$$m_{He} = \text{tank volume } \rho_{He} = 2.651 \text{ liter} \left(32.29\frac{\text{kg}}{\text{m}^3}\right)$$

$$m_{He} = 0.0856 \text{ kg}$$

4) Determine total ΔV available from each propellant

$$m_{\text{final } N_2} = m_{spacecraft} - m_{N_2} = 15 \text{ kg} - 0.5994 \text{ kg}$$
$$= 14.40 \text{ kg}$$

$$\Delta V_{N_2} = C_{N_2} \ln\left(\frac{m_{spacecraft}}{m_{\text{final } N_2}}\right) = \left(787\frac{\text{m}}{\text{s}}\right) \ln\left(\frac{15 \text{ kg}}{14.40 \text{ kg}}\right)$$

$$\Delta V_{N_2} = 32.09 \text{ m/s}$$

$$m_{\text{final He}} = m_{spacecraft} - m_{He} = 15 \text{ kg} - 0.0856 \text{ kg}$$
$$= 14.91 \text{ kg}$$

$$\Delta V_{He} = C_{He} \ln\left(\frac{m_{spacecraft}}{m_{\text{final He}}}\right) = \left(1766\frac{\text{m}}{\text{s}}\right) \ln\left(\frac{15 \text{ kg}}{14.91 \text{ kg}}\right)$$
$$= \Delta V_{He} = 10.107 \text{ m/s}$$

Interpreting the Results

Even though helium provides more than twice the effective exhaust velocity of nitrogen gas, for a volume-constrained mission such as FireSat, nitrogen provides three times the total ΔV. So, we'll pick nitrogen because we get more ΔV for the same volume of gas. Thus, for many small satellite missions, density specific impulse, I_{dsp}, is more important than specific impulse, I_{sp}.

14.2 Propulsion Systems

▬▬ In This Section You'll Learn to...

☞ Describe the key components of propulsion subsystems

☞ Explain the basic operating principles for the different types of rockets currently in use and compare their relative advantages and disadvantages

☞ Discuss future concepts for exotic propulsion subsystems that produce thrust without mass

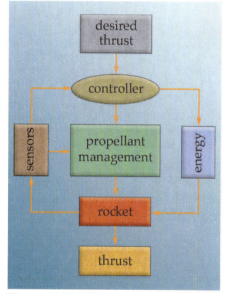

Figure 14-22.   Block Diagram of a Complete Propulsion Subsystem. A propulsion subsystem uses the desired end state (specific thrust at a specific time), plus inputs from sensors, to determine commands for the propellant management and the energy control systems to produce the system output—thrust.

Section 14.1 gave us an exhaustive look at rockets as a system. We saw how they take two inputs, propellant plus energy, and convert them into thrust. But rockets, as important as they are, comprise only one part of an entire propulsion subsystem. In this section, we'll concentrate less on rocket theory and more on propulsion-system technology to learn what essential components we need, and how they're put together.

Figure 14-22 shows a block diagram for an entire propulsion system. To design a specific system, we start with the desired thrust, usually at some very specific time. The propulsion-system controller manages these inputs and formulates commands to send to the propellant management actuators to turn the flow of propellant on or off. For some systems, the controller also manages the energy input to the rocket. For example, in an electrodynamic rocket, the system has to interface with the spacecraft's electrical-power subsystem (EPS) to ensure it provides the required power. The controller uses sensors extensively to monitor the temperature and pressure of the propellant throughout the system.

One of the two key inputs to a rocket is propellant. In this section, we'll start by looking at propellant management, how to store liquid or gaseous propellants, and supply them to the rocket as needed. We'll then review in detail most of the thermodynamic and electrodynamic rocket technologies currently in use or on the drawing boards. Following this discussion, we'll look briefly at important factors for selecting and testing propulsion subsystems. Finally, rocket scientists are always striving to improve propulsion subsystem performance, so we'll look at what exotic concepts may someday take us to the stars.

Propellant Management

All rockets need propellant. The job of storing propellant and getting it where it needs to go at the right time is called *propellant management*. The propellant management portion of a propulsion subsystem has four main tasks

- Propellant storage
- Pressure control
- Temperature control
- Flow control

Let's look briefly at the requirements and hardware for each of these tasks.

Just as your car has a gas tank to store gasoline, propulsion subsystems need tanks to store propellant. We normally store gaseous propellants, such as nitrogen for cold-gas rockets, in tanks under high pressure to minimize their volume. Typical gas storage pressures are 200 bar (2900 p.s.i.) or more. Unfortunately, we can't make a liquid propellant denser by storing it under pressure. However, depending on how we pressurize the liquid propellant for delivery to the combustion chamber, we may need to design the storage tanks to take high pressure as well. In any case, propellant tanks are typically made from aluminum, steel, or titanium and designed to withstand whatever pressure the delivery system requires.

As we presented in Section 14.1, combustion chamber pressure is an important factor in determining rocket thrust. This pressure depends on the delivery pressure of the propellants. Pressurizing the flow correctly is another function of propellant management. There are two approaches to achieving high-pressure flow: pressure-fed systems and pump-fed systems.

As Figure 14-23 shows, a *pressure-fed propellant system* relies on either a gaseous propellant stored under pressure, or a separate tank attached to the main tank, filled with an inert, pressurized gas, such as nitrogen or helium, to pressurize and expel a liquid propellant. The high-pressure gas "squeezes" the liquid propellant out of the storage tank at the same pressure as the gas, like blowing water out of a straw.

To minimize volume, the storage pressure of the gas is typically much higher than the pressure needed in the combustion chamber. To reduce, or regulate the high pressure in the storage tank to the lower pressure for propellant delivery, we typically use mechanical regulators. As high pressure gas flows into a *regulator*, the gas pushes against a carefully designed diaphragm. The resulting balance of forces maintains a constant flow rate but at a greatly reduced output pressure. For example, a gas stored at 200 bar may pass through a regulator that reduces it to 20 bar before it goes into a liquid propellant tank. Pressure regulators are common devices, found in most rocket plumbing systems. Scuba tanks use regulators to reduce high pressure air, stored in the tank, to a safe, lower pressure for breathing.

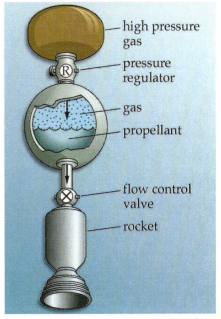

Figure 14-23.   Pressure-fed Propellant System. In a pressure-fed propulsion subsystem, high pressure gas forces the liquid propellant into the combustion chamber under pressure, much like blowing liquid through a straw.

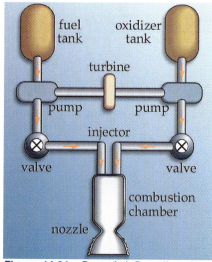

Figure 14-24. Pump-fed Propellant Management. In a pump-fed system, turbine-driven pumps use mechanical energy to increase the pressure of the propellants for delivery to the combustion chamber.

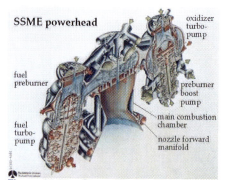

Figure 14-25. Space Shuttle Main Engine (SSME). The SSMEs use turbine pumps to feed the liquid hydrogen and oxygen to the combustion chamber. *(Courtesy of NASA/ Marshall Space Flight Center)*

Figure 14-26. Keeping Things Cool. Cryogenic propellants, such as liquid oxygen and liquid hydrogen on the Shuttle, must remain hundreds of degrees below zero. The cap on the top of the main tank helps control propellant boil off. *(Courtesy of NASA/Johnson Space Center)*

The main drawback of pressure-fed systems is that the amount of liquid propellant in the tank (or tanks) relates directly to the amount of pressurizing gas needed. For very large propulsion subsystems, such as on the Space Shuttle, the propellant-management subsystem must deliver enormous quantities of high-pressure propellant to the combustion chamber each second. To do this using a pressure-fed system would require additional large, high-pressure gas tanks, making the entire launch vehicle larger and heavier. Instead, most launch vehicles use pump-fed delivery systems.

Pump-fed delivery systems rely on pumps to take low pressure liquid and move it toward the combustion chamber at high pressure, as shown in Figure 14-24. Pumps impart kinetic energy to the propellant flow, increasing its pressure. Modern cars use electrical power to turn a small pump that moves gasoline from the tank to the engine under pressure. On the Space Shuttle, massive turbo pumps burn a relatively small amount of H_2 and O_2 to produce mechanical energy. This energy takes the liquid propellants normally stored at a few bars and boosts the feed pressure to over 480 bar (7000 p.s.i.) at a flow rate of 2.45×10^5 liters/s (6.5×10^4 gal./s). Spinning at over 30,000 r.p.m., the Shuttle propellant pumps could empty an average-sized swimming pool in only 1.5 seconds! (Figure 14-25)

Regardless of the propellant-delivery system, the pressure of propellants and pressurizing gasses must be constantly monitored. *Pressure transducers* are small electromechanical devices used to measure the pressure at various points throughout the system. This information is fed back to the automatic propellant controller and sent to ground controllers via telemetry channels.

Temperature control for propellant and pressurant gases is another important propellant-management function. The ideal gas law tells us that a higher gas temperature causes a higher pressure and vice versa. The propellant-management subsystem must work with the spacecraft's environmental control and life support subsystem (ECLSS) to maintain gases at the right temperature and to prevent liquid propellants from freezing or boiling. In the deep cold of outer space, there is a danger of propellants freezing. For instance, hydrazine, a common spacecraft propellant, freezes at 0° C. Usually, the spacecraft ECLSS maintains the spacecraft well above this temperature, but in some cases exposed propellant lines and tanks may need heaters to keep them warm.

On launch vehicles, propellant thermal management often has the opposite problem. It must maintain liquid oxygen (LOX) and liquid hydrogen (LH$_2$) at temperatures hundreds of degrees below zero, centigrade. Using insulation helps control the temperature, however, some boil off of propellants prior to launch is inevitable and must be planned for, as shown in Figure 14-26.

Finally, the propellant-management subsystem must control the flow of gases and liquids. It does this using valves. Valves come in all shapes and sizes to handle different propellants, pressures, and flow rates. Technicians use fill and drain valves to fill (and sometimes, drain) the tanks prior to launch. Tiny, electrically controlled, low-pressure valves pulse cold-gas thrusters on and off to deliver precise, micro-amounts of thrust. Large pyrotechnic valves mounted below liquid-propellant tanks keep them sealed until ignition. When the command arrives, a pyrotechnic charge fires, literally blowing the valve open, allowing the propellant to flow. Of course, these types of valves are good for only one use. Propellant control valves and pyrotechnic isolation valves are examples of high-cycle and low-cycle mechanisms we discussed in Chapter 13.

To protect against over pressure anywhere in the system, *pressure-relief valves* automatically release gas if the pressure rises above a preset value. *Check valves* allow liquid to flow in only one direction, preventing back-flow in the wrong direction. Other valves throughout the system ensure propellant flows where it needs to when the system controller sends the command. Some of these other valves lead to redundant lines that ensure the propellant flows even when a main valve malfunctions.

Let's briefly review the components needed for propellant management. Propellants and pressurant gas are stored in tanks. Below the tanks, valves control the flow throughout the system and regulators reduce the pressure where needed. Transducers and other sensors measure pressure and temperature at various points in the system. Figure 14-27 shows a possible schematic for the FireSat propulsion subsystem, based on using a single, cold-gas thruster. Now that we've shown how propellant gets to the rocket, let's look at various types, shapes, and sizes of rockets.

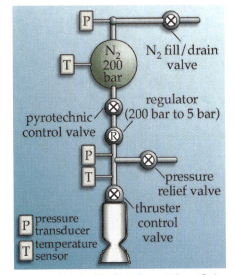

Figure 14-27. FireSat Propulsion Subsystem. Even a relatively simple system such as a cold-gas thruster for a small satellite like FireSat requires tanks, valves, and sensors to measure and control the propellant flow.

Astro Fun Fact
Lost in Space

(Courtesy of NASA/Goddard Space Flight Center)

Mars Observer was a NASA mission to study Mars, including its surface, atmosphere, interior, and magnetic field from Martian orbit. The mission was designed to operate for one full Martian year (687 Earth days) to permit observations of the planet through its four seasons. The spacecraft also carried a radio relay package designed to receive information from the planned Mars Balloon Experiment carried on the planned Russian Mars '94 mission for retransmission to Earth. Observer was launched successfully on September 25, 1992, by a Titan IIIe-TOS from Cape Canaveral, Florida, and cruised well until arriving at Mars. However, for reasons unknown at the time, contact was lost with Observer just as it arrived on August 21, 1993, when operators sent a command to ignite its thrusters to enter Mars orbit. Speculations were that it may have blown up during ignition, been destroyed by a meteorite, or may have simply frozen after having lost orientation. Later investigation indicated that the problem was probably due to a propulsion subsystem explosion caused by propellant leaking past a faulty check valve. The loss of Mars Observer underscores the costs and risks of interplanetary exploration.

Wade, Mark. "The Mars Observer." Encyclopedia Astronautica. 10 October 1999.

Thermodynamic Rockets

As we described in Section 14.1, thermodynamic rockets transfer thermodynamic energy (heat and pressure) to a propellant and then convert the energized propellant into high-speed exhaust using nozzles. There are a wide variety of thermodynamic rockets currently available or being considered. We can classify these based on their source of energy

- Cold gas—use mechanical energy of a gas stored under pressure
- Chemical—rely on chemical energy (from catalytic decomposition or combustion of propellants) to produce heat
- Solar thermal—use concentrated solar energy to produce heat
- Thermoelectric—use the heat produced from electrical resistance
- Nuclear thermal—use the heat from a nuclear reaction

Because we examined simple cold-gas rockets in detail in the last section, here we'll review the other four types and compare their relative performances.

Chemical Rockets

The vast majority of rockets in use today rely on chemical energy. When we strike a match, the match head ignites the wood and the flame results from a combustion process. The fuel—the wood in the match—is chemically combining with the oxygen in the air to form various chemical by-products (CO, CO_2, water, etc.) and, most importantly, heat. In *chemical rockets*, the propellants release energy from their chemical bonds during combustion. The Space Shuttle relies on chemical rockets, as shown in Figure 14-28. In the Shuttle main engines, liquid hydrogen (H_2) and liquid oxygen (O_2) combine in the most basic of chemical reactions

$$2H_2 + O_2 \rightarrow 2H_2O + \text{Heat} \tag{14-35}$$

All combustion reactions must have a *fuel* (such as hydrogen) plus an *oxidizer* (such as oxygen). These two combine, liberating a vast amount of heat and creating by-products that form the exhaust. The heat transfers to the combustion products, raising their temperatures. This chemical reaction and energy transfer takes place in the *combustion chamber*. Although the propellants arrive in the combustion chamber under pressure, delivered by the propellant-management subsystem, this mechanical energy is small compared to the thermal energy released by the chemical reaction.

Chemical rockets generally fall into one of three categories

- Liquid
- Solid
- Hybrid

Let's briefly review the operating principles and performance parameters of each.

Figure 14-28.   Chemical Rockets. Chemical rockets use the energy stored in the propellants. The Space Shuttle main engines and solid-rocket boosters are two examples of chemical rockets. *(Courtesy of NASA/Johnson Space Center)*

Liquid-chemical Rockets. Liquid-chemical rockets are usually one of two types: bipropellant or monopropellant. As the name implies, *bipropellant rockets* use two liquid propellants. One is a fuel, such as liquid hydrogen (LH_2) and the other, an oxidizer, such as liquid oxygen (LOX) (Figure 14-29). Brought together under pressure in the combustion chamber by the propellant-management subsystem, the two compounds chemically react (combust), releasing vast quantities of heat and producing combustion products (these vary depending on the propellants). To insure complete, efficient combustion, the oxidizer and fuel must mix in the correct proportions. The *oxidizer/fuel ratio (O/F)* is the proportion, by mass, of oxidizer to fuel. The optimum O/F is called the *stoichiometric* (chemically balanced) combination.

Some propellant combinations, such as hydrogen and oxygen, won't spontaneously combust on contact. They need an igniter, just as your car needs a spark plug, to get started. This need, of course, increases the complexity of the system somewhat. So propellant chemists strive to find combinations that react on contact. We call these propellants *hypergolic* because they don't need a separate means of ignition. The combination of hydrazine (N_2H_4) plus nitrogen tetroxide (N_2O_4) is an example of hypergolic propellants.

Another important feature in selecting a propellant is its storability. Although the liquid hydrogen and liquid oxygen combination in the Space Shuttle main engines offers high performance (specific impulse around 455 s), they require supercooling to hundreds of degrees below zero, centigrade. Because of their low storage temperature, we call these propellants *cryogenic*. Unfortunately, it is difficult to maintain these extremely low temperatures for long periods (days or months). When the mission calls for long-term storage, designers turn to *storable propellants* such as hydrazine and nitrogen tetroxide that can remain stable at room temperature for a very long time (months or even years).

The Titan, an early intercontinental ballistic missile (ICBM), used hypergolic, storable propellants, because the missiles stayed deep in underground silos for many years. The Shuttle uses these propellants in it's orbital-maneuvering engines and reaction-control thrusters. The majority of spacecraft use storable, hypergolic liquid rockets for maneuvering. The penalty paid for the extra convenience of spontaneous combustion and long-term storage is a much lower performance (I_{sp} ~300 s) than the cryogenic option. In addition, current hypergolic combinations are extremely toxic and require special handling procedures to prevent propellant release. Table 14-2 summarizes key points about bipropellant rockets. Figure 14-30 shows a a photograph of the LEROS hypergolic, bipropellant engine that has been used on a variety of missions for final orbit insertion.

As the name implies, *monopropellant* chemical rockets use only a single propellant. These propellants are relatively unstable and easily decompose through contact with a suitable catalyst.

Hydrogen peroxide (H_2O_2) is one example of a monopropellant. People use a low-concentration (3%), drug-store variety of this compound to disinfect a bad scrape, or to bleach hair. Rocket-grade hydrogen peroxide,

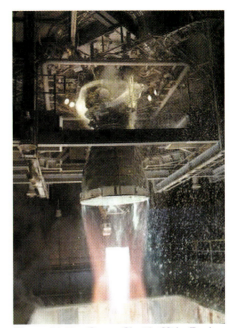

Figure 14-29.   Space Shuttle Main Engine (SSME). This SSME uses liquid hydrogen and liquid oxygen in a test at the Stennis Research Center. *(Courtesy of NASA/Stennis Research Center)*

Figure 14-30.   LEROS Bipropellant Engine. The LEROS engine, produced by the ARC Royal Ordnance in the United Kingdom, is just one example of a bipropellant rocket. It uses nitrogen tetroxide (N_2O_4) with hydrazine (N_2H_4) to deliver a total thrust of 400 N at 317 s specific impulse. This engine has been reliably used on a Mars mission and other deep space and near-Earth missions for orbital insertion. *(Courtesy of British Aerospace Royal Ordnance)*

Figure 14-31.   Monopropellant Rocket. This 22 N (5 lbf) hydrazine monopropellant engine built by Kaiser-Marquardt delivers 235 s specific impulse. At only 20 cm (8 in.) long by 4 cm (1.5 in.) wide, it can be easily integrated into a variety of spacecraft for attitude control and small orbital maneuvers. *(Courtesy of Kaiser-Marquardt)*

Table 14-2.   Bipropellant Rockets.

Operating Principle

A liquid oxidizer and a liquid fuel react in a combustion process, liberating heat and creating exhaust products that thermodynamically expand through a nozzle.

Typical Propellants

Oxidizers: Liquid oxygen (LOX), HTP = high-test hydrogen peroxide (>85% H_2O_2), nitrogen tetroxide (N_2O_4)

Fuels: Liquid hydrogen (LH_2), kerosene (RP-1: "rocket propellant-1" C_4H_8), hydrazine (N_2H_4)

Advantages	Disadvantages
• High I_{sp}	• Must manage two propellants
• Can be throttled	• Intense combustion heat creates thermal control
• Can be re-started	problems for chamber and nozzle

also called high-test peroxide (HTP), has a concentration of 85% or more. It is relatively safe to handle at room temperatures, but when passed through an appropriate catalyst (such as silver), it readily decomposes into steam (H_2O) and oxygen, releasing significant heat in the process. Typical HTP reactions exceed 630° C. This relatively high temperature, combined with the molecular mass of the reaction products, gives HTP monopropellant rockets an I_{sp} of about 180 s. The X-15 rocket plane and Scout launch vehicle used these types of thrusters successfully.

By far, the most widely used monopropellant today is hydrazine (N_2H_4). It readily decomposes when exposed to a suitable catalyst, such as iridium, producing an I_{sp} of about 230 s. The main disadvantage of hydrazine is its high toxicity. This problem means technicians need specialized handling procedures and equipment during all testing and launch operations.

The biggest advantage of monopropellant over bipropellant systems is simplicity. The propellant-management subsystem maintains only one set of tanks, lines, and valves. Unfortunately, there is a significant penalty in performance for this added simplicity (2/3 the I_{sp} of a comparable bipropellant system or less). However, for certain mission applications, especially station keeping and attitude control on large communication satellites, this trade-off is well worth it. The benefit grows when we use hydrazine as the fuel with nitrogen tetroxide in a large bipropellant rocket for initial orbital insertion, and then by itself in a smaller, monopropellant rocket for station keeping. Such "dual-mode" systems take advantage of the flexibility offered by hydrazine to maximize overall system performance and simplicity. Table 14-3 summarizes key points about monopropellant rockets. Figure 14-31 shows a typical monopropellant engine.

Solid-chemical Rockets. The fireworks we watch on the 4th of July, are a good example of solid rockets at work. Solid rockets date back thousands of years to the Chinese, who used them to confuse and frighten their enemies on the battlefield. In modern times, these rockets create thrust for intercontinental ballistic missiles, as well as space-launch vehicles.

Table 14-3. Monopropellant Rockets.

Operating Principle
A single propellant decomposes using a catalyst, releasing heat and creating by-products that thermodynamically expand through a nozzle.

Typical Propellants
Hydrazine (N_2H_4), HTP = high-test hydrogen peroxide (>85% H_2O_2)

Advantages	Disadvantages
• Simple, reliable	• Lower I_{sp} than bipropellant
• One propellant to manage	
• Lower temperature reactions means fewer thermal problems in the chamber and nozzle	

Just as a liquid bipropellant rocket combines fuel and oxidizer to create combustion, a *solid rocket* contains a mixture of fuel, oxidizer, along with a binder, blended in the correct proportion and solidified into a single package called a *motor*. A typical composite solid-rocket fuel is powdered aluminum. The most commonly used solid-rocket motor oxidizer is ammonium perchlorate (AP). Together, the fuel and oxidizer comprise about 85%–90% of the rocket motor's mass, with an oxidizer/fuel ratio of about 8:1. The remaining mass of the motor consists of a binder that holds the other ingredients together and provides overall structural integrity. Binders are usually a hard, rubber-like compound, such as hydroxyl-terminated polybutadiene (HTPB). During combustion, the binder also acts as additional fuel.

As we learned in Section 14.1, rocket thrust depends on mass flow rate. In a solid-rocket motor, this rate depends on the propellant's burn rate (kg/s) and the burning surface area (m^2). The faster the propellant burns and the greater the burning surface area, the higher the mass flow rate and the higher the resulting thrust. The propellant's burn rate depends on the type of fuel and oxidizer, their mixture ratio, and the binder material. The total burning area depends primarily on the inside shape of the solid propellant. During casting, designers can shape the hollow inner core (grain design) of the solid propellant to adjust the surface area available for burning, so they can control the burning rate and thrust (Figure 14-32). The Space Shuttle's solid-rocket motors, for example, have a star-shaped core, shown in Figure 14-33, specifically tailored so the thrust decreases 55 seconds into the flight to reduce acceleration and the effects of aerodynamic forces.

Because solid-rocket-motor combustion depends on the exposed propellant's surface area, manufacturers must carefully mold the propellant mixture to prevent cracks. Burning occurs on any exposed surface, even along undetected cracks in the propellant grain. Investigators linked the Space Shuttle *Challenger* accident to an improperly sealed joint between solid-motor segments. This open seal exposed the motor case to hot gases, burning it through and causing the accident.

The *Challenger* disaster highlighted another drawback of solid motors—once they start, they are very difficult to stop. With a liquid rocket, we can command valves to close, turning off the flow of propellant and shutting

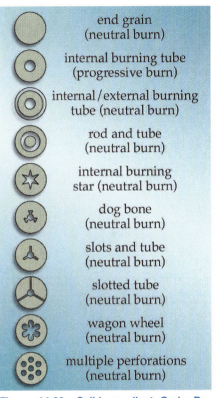

Figure 14-32. Solid-propellant Grain Designs. By altering the grain design, engineers cause progressive or neutral burn rates. Shaded areas indicate propellant and blank areas indicate empty space. *(Courtesy of Space Propulsion Analysis and Design by Humble, et. al.)*

end grain (neutral burn)

internal burning tube (progressive burn)

internal/external burning tube (neutral burn)

rod and tube (neutral burn)

internal burning star (neutral burn)

dog bone (neutral burn)

slots and tube (neutral burn)

slotted tube (neutral burn)

wagon wheel (neutral burn)

multiple perforations (neutral burn)

Figure 14-33. Solid-propellant Shape. The "star" shape of the Space Shuttle SRB controls the burning rate, hence the thrust profile, of the motor. *(Courtesy of NASA/Kennedy Space Center)*

Figure 14-34.   Solid-rocket Boosters. Many launch vehicles, such as the Delta II shown here, rely on solid-rocket motors to get them off the ground. *(Courtesy of NASA/Marshall Space Flight Center)*

off the engine. Solid motors burn until all the propellant is gone. To stop one prior to that requires blowing off the top or splitting it open along its side, releasing internal pressure and thus, stopping combustion. These are not very practical solutions on the way to orbit!

Despite their drawbacks, solid motors are used on a variety of missions, because they offer good, cost-effective performance in a simple, self-contained package that doesn't require a separate propellant-management subsystem. One important use of solid motors is to augment liquid engines on launch vehicles. For instance, without the solid-rocket boosters, the Space Shuttle couldn't get off the ground. Several expendable launch vehicles use various combinations of strap-on solid motors to give users a choice in payload-lifting capacity, without the need to redesign the entire vehicle. For example, three, six, or nine solid motors can be added to the Delta II launch vehicle, shown in Figure 14-34, depending on the payload mass. Solid motors also provide thrust for strap-on upperstages for spacecraft needing a well-defined velocity change (ΔV) to go from a parking orbit into a transfer orbit.

A solid-rocket motor's specific impulse depends on the fuel and oxidizer used. After mixing the propellants and casting the motor, manufacturers can't change the I_{sp} or thrust. Specific impulse for typical solid motors currently in use range from 200–300 seconds, somewhat more than a liquid monopropellant rocket but slightly less than a typical, liquid bipropellant engine. Their big performance advantage is in terms of I_{dsp}. For example, the Shuttle's solid-rocket boosters (SRBs), have a I_{dsp} 6% less than the I_{dsp} of the liquid main engines (SSMEs), even though the I_{sp} for the SSMEs is almost 70% higher. This makes solid motors ideal for volume-constrained missions needing a single, large ΔV. Table 14-4 summarizes key points about solid-rocket motors.

Table 14-4.   Solid Rockets.

Operating Principle
An oxidizer and fuel are blended together in a single, solid grain along with a binder. Combustion takes place along any exposed surface producing heat and by-products that are thermodynamically expanded through a nozzle.

Typical Propellants
Fuel: Aluminum; oxidizer: Ammonium perchlorate (AP); Binder: Hydroxyl-terminated polybutadiene (HTPB)

Advantages	Disadvantages
• Simple, reliable	• Susceptible to cracks in the grain
• No propellant management needed	• Can't restart
• High I_{dsp} compared to bipropellant	• Difficult to stop
• No combustion chamber cooling issues	• Modest I_{sp}

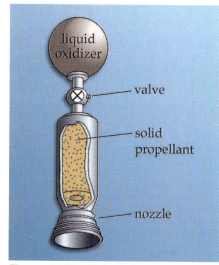

Figure 14-35.   Hybrid Rocket Motor. A hybrid rocket uses a solid fuel with a liquid oxidizer. This offers the flexibility of a liquid system with the simplicity and density of a solid motor. Falling in between liquids and solids in performance, hybrids have yet to see applications on launch vehicles or spacecraft.

Hybrid-chemical Rockets. *Hybrid-propulsion systems* combine aspects of liquid and solid systems. A typical hybrid rocket uses a liquid oxidizer and a solid fuel. The molded fuel grain forms the combustion chamber and the oxidizer is injected into it, as illustrated in Figure 14-35. A separate sparking system or a superheated oxidizer initiates combustion.

The hybrid-combustion process is similar to burning a log in the fireplace. Oxygen from the air combines with the log (fuel) in a fast oxidation process and burns. If we take away the air (turn off the flow of the oxidizer), the fire goes out. If we use a bellows or blow on the fire, we increase the flow of air, and the fire grows.

A properly designed hybrid rocket offers the flexibility of a liquid system with the simplicity and density of a solid motor. Hybrids are safe to handle and store, similar to a solid, but can be throttled and restarted, similar to a liquid engine. Their efficiencies and thrust levels are comparable to solids. For example, one interesting hybrid configuration uses high-test peroxide (HTP) oxidizer with HTPB (rubber, the same used as a binder for solid motors) or with polyethylene (plastic) fuel. At an O/F of 8:1 this system offers an I_{sp} of around 290 s and I_{dsp} of around 3.8×10^5 kg/m^3 s. It has the added advantage that the HTP can be used alone as a monopropellant making it a "dual-mode" system. Unfortunately, at this time hybrid-rocket research and applications lag far behind liquid and solid systems and have yet to see operational use on launch vehicles or spacecraft. Their most dramatic, recent application has been on the world 2-wheeled speed record attempt vehicle shown in Figure 14-36. Table 14-5 summarizes key points about hybrid rockets.

Figure 14-36. Maximum Impulse! The Gillette Mach 3 Challenger used HTP/HTPB hybrid motors producing over 10,000 N (2248 lb.) thrust to reach a peak speed of 365 m.p.h. in an effort to set the world two-wheeled speed record. *(Courtesy of Richard Brown, Project Machinery)*

Table 14-5.   Hybrid Rockets.

Operating Principle
Hybrid rockets typically use a liquid oxidizer with a solid fuel. The oxidizer is injected into a hollow port (or ports) within the fuel grain where combustion takes place along the boundary with the surface.

Typical Propellants
Oxidizers: Liquid oxygen (LOX), nitrous oxide (N_2O), high-test hydrogen peroxide (>85% H_2O_2)
Fuels: HTPB = hydroxyl-terminated polybutadiene (rubber), PE = polyethylene (plastic)

Advantages	Disadvantages
• Simpler than a bipropellant system with similar performance	• Limited heritage
• Safer, more flexible than solids	• Modest I_{sp}
• No combustion-chamber cooling issues	

Chemical-rocket Summary. Table 14-6 compares the I_{sp} and I_{dsp} of the thermodynamic rockets we've discussed in this section and compares their performance and key features.

Solar-thermal Rockets

In chemical rockets, the heat is a by-product of a chemical reaction. But rockets can produce heat in other ways, then transfer it directly to the propellant using conduction and/or convection. We discussed these heat-transfer mechanisms in Chapter 13. One convenient source of heat is the Sun. If you've ever played with a magnifying glass on a sunny day you've seen the power of solar energy to produce heat. By concentrating solar energy using mirrors or lenses, a rocket can create extremely high temperatures (up to 2400 K) on a focused point. By passing a propellant, such as hydrogen, through this point, it can directly absorb the heat, reaching

Table 14-6.   Thermodynamic Rocket Comparison. LH$_2$ = liquid hydrogen; kerosene, RP-1 = "rocket propellant-1;" N$_2$O$_4$ = nitrogen tetroxide; N$_2$H$_4$ = hydrazine; HTPB = hydroxyl-terminated polybutadiene (rubber); PE = polyethylene (plastic); HTP = high-test hydrogen peroxide (>85% H$_2$O$_2$).

Type	Propellant Combinations (O/F) [Specific Gravity]	I_{sp} (s)	I_{dsp} (s)	Advantages	Disadvantages
Liquid	--	--	--	• High I_{sp} • Can be throttled • Can be re-started	• Must manage two propellants • Intense combustion heat creates thermal-control problems for chamber and nozzle
Bipropellant	LO$_2$/LH$_2$ (5 : 1) [1.15 : 0.07]	477	462	• High I_{sp} • Environmentally friendly propellants	• Cryogenic fuel and oxidizer difficult to store
	LO$_2$/Kerosene (RP-1) (2.25:1) [1.15 : 0.8]	370	385	• Storable fuel • Good I_{dsp}	• Cryogenic oxidizer
	N$_2$O$_4$/N$_2$H$_4$ (1.9 : 1) [1.43 : 1.00]	334	429	• Storable propellants • Good I_{sp}	• Toxic propellants
Mono-propellant	--	--	--	• Simple, reliable • One propellant to manage • Lower temperature reactions means fewer thermal problems in chamber and nozzle	• Lower I_{sp} than bipropellant
	N$_2$H$_4$ (hydrazine) (N/A) [1.00]	245	246	• Large flight heritage	• Toxic
	H$_2$O$_2$ (90% hydrogen peroxide) (N/A) [1.37]	181	247	• Environmentally friendly propellant	• Little flight heritage
Solid	NH$_4$ClO$_4$ (AP)/Al (includes a binder e.g. HTPB) (3.5 : 1) [1.95 : 1.26]	300	539	• Simple, reliable • No propellant management needed • High I_{dsp} compared to bipropellant • No combustion-chamber cooling issues	• Susceptible to cracks in the propellant grain • Difficult to stop • Can't re-start • Modest I_{sp}
Hybrid	H$_2$O$_2$ (90%)/PE (8 : 1) [1.37 : 0.90]	333	437	• Simpler than a bipropellant system with similar performance • Safer, more flexible than solids • No combustion-chamber cooling issues • Restart	• Limited heritage • Modest I_{sp}

very high temperature before expanding through a nozzle to achieve high exhaust velocity. In this way, *solar-thermal rockets* use the limitless power of the sun to produce relatively high thrust with high I_{sp}.

The natural advantage of a solar-thermal rocket is the abundant source of solar energy, eliminating the need to produce the energy on the spot or carry it along as chemical energy. It can use virtually any propellant. The best I_{sp}, of course, comes from using hydrogen. Theoretical and experimental results indicate a liquid-hydrogen, solar-thermal rocket could achieve a specific impulse of more than 800 s. Basic engineering problems limit thrust levels, due to inefficiencies in transferring heat between the thermal mass that absorbs solar energy and the propellant. However, thrusts in the several-newton range should be achievable.

Another important operational challenge for solar-thermal rockets is deploying and steering large mirrors to collect and focus solar energy. Naturally, they would not be effective in eclipse or for interplanetary missions far from the Sun. Several concepts for solar-thermal rockets have been proposed, such as the one shown in Figure 14-37. However, up to now, none have been tested in orbit. Table 14-7 summarizes key features of solar-thermal rockets.

Table 14-7.   Solar-thermal Rockets.

Operating Principle
Lenses or mirrors concentrate solar energy onto a heat-transfer chamber. A propellant, such as liquid hydrogen, flows through the chamber, absorbs heat, and then expands through a nozzle.

Typical Propellants
Virtually any propellant can be used, but hydrogen produces the best I_{sp}

Advantages	Disadvantages
• Limitless energy supply, can be refueled and re-used	• Needs intense, direct sunlight
• Potentially very high I_{sp} (~800 s with H_2)	• Must carefully point a large mirror or lens
	• No flight heritage

Figure 14-37.   Solar-thermal Rocket. This solar-thermal rocket concentrates solar energy on a thermal mass that reaches very high temperature (up to 2400 K). In this concept, liquid hydrogen flows through the thermal mass, absorbing the energy and then expanding through a nozzle, producing a thrust of over 6 N (1.6 lbf) at an I_{sp} of 750 s. *(Courtesy of NASA/ Marshall Space Flight Center)*

Thermoelectric Rockets

Of course, solar energy is only available when the Sun is shining. A spacecraft in eclipse, or far from the Sun, needs another heat source. On Earth, we commonly use electrical energy to produce heat—to heat our homes or toast our bread. This heat comes from electrical resistance (friction) of the current flowing through a wire. If you hold your hand next to a conventional light bulb, you'll feel the heat produced by the resistance of the filament in the bulb. For space applications, the energy source is the electrical energy provided by the spacecraft's electrical-power subsystem (EPS). By running electricity through a simple resistor, or by creating an arc discharge, similar to a spark plug, we can create heat. *Thermoelectric rockets* transfer this heat to the propellant by conduction and convection.

One of the simplest examples of a thermoelectric rocket is a *resistojet*. This type works much like an electric tea kettle. As we show in Figure 14-38, electrical current flows through a metal-heating element inside a combustion chamber. The resistance (or electrical friction) in the metal causes it to heat up. As propellant flows around the heating element, heat transfers to it via convection, increasing its temperature before it expands through a nozzle.

This simple principle can be applied to virtually any propellant (NASA even investigated using urine on the Space Station as a propellant!). The resistojet concept can significantly increase the specific impulse of a conventional cold-gas rocket, making it, in effect, a hot-gas rocket with increased I_{sp} (recall our cold-gas rocket trade-off example in Section 14.1 where we increased I_{sp} from 79 s to 101 s by increasing temperature from 298 to 500 K). Resistojets also improve the performance of conventional

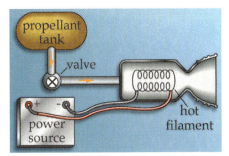

Figure 14-38.   Resistojet. A resistojet uses electrical resistance to produce heat inside a thrust chamber. This heat transfers to the propellant via convection to the propellant flowing through the chamber which then expands through a nozzle.

Figure 14-39. Resistojets at Work. The International Space Station uses resistojets, such as this one, to help maintain its orbit and attitude. *(Courtesy of NASA/Johnson Space Center)*

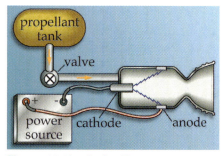

Figure 14-40. Arcjet Thruster. An arcjet thruster works by passing a propellant through an electric arc, rapidly increasing its temperature before expanding it out a nozzle.

Figure 14-41. Arcjets at Work. The Argos spacecraft, shown here, tested a powerful 26-kW ammonia (NH_3) arcjet, setting a record for the most powerful electric-propulsion system ever tested in orbit. Its I_{sp} was 800 s, and its thrust was 2 N. *(Courtesy of the U.S. Air Force)*

hydrazine monopropellant rockets by heating the exhaust products, thus boosting their I_{sp} by about 50% (from 200 s to over 300 s). The direct benefit of a resistojet rocket comes from adding heat to the propellant, so the hotter it gets, the higher its I_{sp} and I_{dsp}. Hydrazine resistojets are gaining wide use, as mission designers become increasingly able to trade extra electrical power for a savings in propellant mass. Astronauts on the International Space Station, for example, rely on hydrazine resistojets, shown in Figure 14-39, to maintain the ISS's final mission orbit and attitude.

Another method for converting electrical energy into thermal energy is by using a spark or electric arc. To form an arc, we create a gap in an electrical circuit and charge it with a large amount of electricity. When the electrical potential between the two points gets high enough, an arc forms (during a thunderstorm we see this as dazzling displays of lightning). An *arcjet rocket* passes propellant through a sustained arc, increasing its temperature. Arcjet systems can achieve relatively high I_{sp} (up to 1000 s) with small but significant thrust levels (up to 1 newton). Like resistojets, arcjets can use almost any propellant. Current versions use hydrazine, liquid hydrogen, or ammonia. We show a schematic for an arcjet system in Figure 14-40. The ARGOS spacecraft, shown in Figure 14-41, was launched in 1999 to test a 26-kW ammonia arcjet, producing a thrust of about 2 N with a specific impulse over 800 s.

As expected, the primary limitation on thermoelectric-rocket thrust and efficiency is the amount of power available. Recall from Section 14.1, where we saw a simple relationship between input power, thrust, and effective exhaust velocity or specific impulse in Equation (14-34)

$$F \propto \frac{2P}{C} = \frac{2P}{I_{sp}g_o} \qquad (14\text{-}36)$$

where

F   = thrust (N)

P   = power (W)

C   = effective exhaust velocity (m/s)

I_{sp} = specific impulse (s)

g_o = gravitational acceleration = 9.81 m/s^2

[*Note:* The symbol "$\propto$" means proportional to]

Using this simple equation, we can fine tune the design of a thermoelectric thruster, trading thrust versus power versus specific impulse. For example, if we double the power input we can increase thrust by a factor of 4 for the same specific impulse. Table 14-8 summarizes key features of thermoelectric rockets.

Nuclear-thermal Rockets

Another potentially useful heat source in space is nuclear energy. On Earth, nuclear reactors harness the heat released by the fission of uranium to produce electricity. Water absorbs this heat, making steam that turns turbine generators. In much the same way, a *nuclear-thermal rocket* uses its

Table 14-8. **Thermoelectric Rockets.**

Operating Principle
Heat comes from an electric resistance or a spark discharge inside a heat transfer chamber. A propellant flows through the chamber, absorbs heat, and then expands through a nozzle.

Typical Propellants
Hydrazine, water, ammonia, or virtually any propellant

Advantages	**Disadvantages**
• Simple, reliable	• Requires large amounts of
• Can be used as an "add on" to conventional monopropellant rocket to boost I_{sp} ~50%	onboard electrical power
• High-power arcjets offer very high I_{sp} (>800 s with NH_3)	• Relatively low thrust (<1 N)

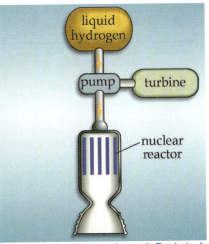

Figure 14-42. Nuclear-thermal Rocket. A nuclear-thermal rocket uses a nuclear reactor to heat a propellant, such as liquid hydrogen. The superheated propellant then expands through a nozzle.

propellant, such as liquid hydrogen, to flow around the nuclear core, absorbing thermal energy. As we show in Figure 14-42, propellant enters the reaction chamber where it absorbs the intense heat from the nuclear reaction. From there, thermodynamic expansion through a nozzle produces high thrust (up to 10^6 N) and high I_{sp} (up to 1000 s using hydrogen).

Because of their relatively high thrust and better efficiencies, nuclear-thermal rockets offer a distinct advantage over chemical systems, especially for manned planetary missions. These missions must minimize transit time to decrease the detrimental effects of free fall, as well as exposure to solar and cosmic radiation on the human body, as discussed in Chapter 3. Ironically, future astronauts may escape the danger of space radiation by using the energy from a nuclear reactor to propel them to their destination faster. Extensive research into nuclear-thermal rockets was done in the U.S. in the 1960s as part of the NERVA program, as shown in Figure 14-43. Additional work was done in the 1980s when great theoretical advances occurred in heat transfer. Unfortunately, environmental and political concerns about safe ground testing of nuclear-thermal rockets (let alone the potential political problems of trying to launch a fully fueled nuclear reactor) has severely reduced research into this promising technology. Table 14-9 summarizes key features of nuclear-thermal rockets.

Table 14-9. **Nuclear-thermal Rockets.**

Operating Principle
Heat comes from nuclear fission inside a reactor. A propellant, such as liquid hydrogen, flows through the reactor, absorbs heat, and then expands through a nozzle.

Typical Propellants
Virtually any propellant can be used, but hydrogen produces the best I_{sp}

Advantages	**Disadvantages**
• Long-term energy supply, can be refueled and re-used	• Environmental and political problems with testing and launching nuclear
• Potentially very high I_{sp} (~1000 s with H_2)	reactors
• High thrust (~10^6 N)	• No flight heritage

Figure 14-43. Nuclear Engine for Rocket Vehicle Applications (NERVA) Rocket. The NERVA program tested nuclear-thermal rockets from 1947 until 1972. Future missions may depend on their impressive performance to take humans to Mars. *(Courtesy of the Report of the Synthesis Group on America's Space Exploration Initiative)*

Electrodynamic Rockets

While thermodynamic rockets offer relatively high thrust over a very wide range (10^{-1} to 10^6 newtons) basic problems in heat transfer pose practical limits on the maximum specific impulse (up to 1000 s or so, for nuclear rockets). To achieve the higher efficiencies demanded by future, more challenging interplanetary and commercial missions, we need to take a different approach—electrodynamic rockets.

As we discussed in Section 14.1, electrodynamic rockets rely on electric and/or magnetic fields to accelerate a charged propellant to very high velocities (more than 10 times the exhaust velocity and I_{sp} of the Shuttle main engines). However, this high I_{sp} comes with a price tag—high power requirement and low thrust. Recall from Equation (14-34), the relationships among power, exhaust velocity, and thrust for electrodynamic rockets (as well as thermodynamic rockets using an external power source)

$$F \propto \frac{2P}{C} \qquad (14\text{-}37)$$

Power, of course, is always a limited commodity on a spacecraft. No matter how much there is, it's always nice to have more, especially for an electrodynamic thruster. However, given a finite amount of power, P, Equation (14-37) tells us we can have higher exhaust velocity, C, only at the expense of reduced thrust, F. As a result, practical limits on power availability make electrodynamic thrusters unsuitable for launch vehicles or when a spacecraft needs a quick, large impulse, such as when it brakes to enter a capture orbit. Even so, because electrodynamic rockets offer very high I_{sp}, mission planners are increasingly willing to sacrifice power and thrust (and the extra time it will take to get a spacecraft where it needs to go) in order to save large amounts of propellant mass.

As we indicated in Section 14.1, there are several ways to use electric and/or magnetic fields to accelerate a charged propellant. Here we'll focus on the two primary types of electrodynamic rockets currently in use operationally

- Ion (or electrostatic) thrusters—use electric fields to accelerate ions
- Plasma thrusters—use electric and magnetic fields to accelerate a plasma

Ion Thrusters

An *ion thruster* (also called an *electrostatic thruster*) uses an applied electric field to accelerate an ionized propellant. Figure 14-44 illustrates the basic configuration of an ion thruster. To operate, the thruster first ionizes a propellant by stripping off the outer shell of electrons, making positive ions. It then accelerates these ions by applying a strong electric field. If the engine ejected the positive ions without neutralizing them, the spacecraft would eventually accumulate a negative charge due to the leftover electrons. To prevent this, as Figure 14-44 illustrates, it uses a neutralizer source at the exit plane to eject electrons into the exhaust, making it neutrally charged.

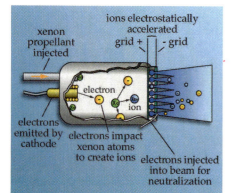

Figure 14-44. Simple Ion Thruster. An ion thruster is the simplest example of an electrodynamic rocket. A strong electric field accelerates an ionized propellant to high velocity (>30 km/s). To prevent charging of the spacecraft, negative ions are injected into the exhaust, neutralizing it.

The ideal ion-thruster propellant is easy to ionize, store and handle. Early ion-thruster research used mercury and cesium, because these metals are easy to ionize. Unfortunately, they are also toxic, making them difficult to store and handle. Currently, the most popular propellant for ion thrusters is xenon. Xenon is a safe, inert gas that stores as a dense gas (1.1052 kg/l) under a moderate pressure of 58.4 bar at room temperature. This high density propellant also gives ion thrusters excellent density specific impulse, I_{dsp}.

Ion thrusters offer an electrically efficient propulsion option with very high specific impulse (as high as 10,000 s). About 90% of the power goes to accelerate the propellant. Because of their efficiency, ion thrusters have been used on a variety of space missions. Perhaps their most exciting application is on interplanetary missions. NASA's Deep Space 1 mission, shown in Figure 14-45, was the first to rely on an ion rocket for the primary propulsion subsystem beyond Earth orbit.

Plasma Thrusters

As we discussed in Section 14.1, there is a practical limit to the number of ions that we can pack into a small volume inside a thruster. However, a neutral plasma can have a much higher charge density. *Plasma thrusters* can take advantage of this fact to offer slightly higher thrust than ion thrusters for the same power input at the expense of somewhat lower I_{sp} and electrical efficiency (we don't get something for nothing when it comes to rockets). Plasma thrusters use the combined effect of electric and magnetic fields to accelerate the positive ions within a plasma.

There are two types of plasma thrusters that have been used in space

- Hall-effect thrusters (HET)
- Pulsed-plasma thrusters (PPT)

Currently, the most widely used type of plasma thruster is the *Hall-effect thruster (HET)*. HETs take advantage of a unique effect called a "Hall current" that occurs when we apply a radial magnetic field to a conducting plasma. The interaction of the magnetic field with the resulting electric field creates a force that accelerates the positive ions in the plasma, as illustrated in Figure 14-46. Figure 14-47 shows a photograph of an operating HET. Note the circular-shaped plume that results from using the radial magnetic field. Russian scientists pioneered many of the modern advances in HETs, having run them for several years for orbital station-keeping applications. Because the propellant requirements for plasma thrusters are the same as for ion thrusters, xenon is also the most widely used propellant.

A second type of plasma thruster is called a pulsed-plasma thruster (PPT). Unlike all other types of rockets that operate continuously, *pulsed-plasma thrusters (PPT)* operate in a noncontinuous, pulsed mode. Unlike ion and plasma thrusters, PPTs use a solid propellant, usually Teflon (PTFE). A high voltage arc pulses over the exposed surface of the propellant, vaporizing it and creating an instant plasma. The resulting induced

Figure 14-45. Ion Engines in Space. NASA's Deep Space 1 mission used an ion thruster for its primary propulsion. The engine operated at 2.3-kw producing a thrust of 0.09 N with a specific impulse of 3100 s. *(Courtesy of NASA/Jet Propulsion Laboratory)*

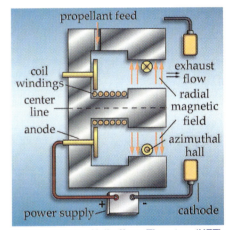

Figure 14-46. Hall-effect Thruster (HET) Diagram. In a HET, the interaction of an applied magnetic field with the resulting electric field creates the force that accelerates the positive ions within the plasma. This diagram shows a crossaction of the radial chamber.

Figure 14-47. Hall-effect Thruster (HET) in Operation. Notice the circular shape of the plasma. HETs take advantage of the unique properties of a radial magnetic field to accelerate a propellant, such as Xenon that has been heated to create a plasma. *(Courtesy of Primex Aerospace Company)*

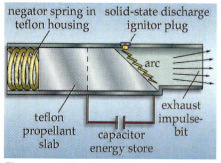

Figure 14-48. Pulsed-plasma Thruster (PPT). PPTs create a plasma by pulsing an electric arc over the surface of a solid propellant, such as Teflon (PTFE). The induced magnetic field accelerates the plasma. As the teflon slab shrinks, the negator spring gradually feeds it into the arc.

magnetic field accelerates the plasma. Figure 14-48 shows a schematic for a simple PPT. A number of missions have used PPTs for spacecraft station keeping. Their advantage is their precisely controlled, low-thrust levels. Because they operate in a pulsed mode, they don't need continuous high power. Instead, they can gradually store electrical energy in a capacitor for release in high power bursts (the same technique used in a camera flash). This low-power, pulsed operating mode makes them suitable for many small satellite applications.

Compared to ion and stationary plasma thrusters, PPTs are relatively low in energy-conversion efficiency (20%). However, they provide respectable I_{sp} (700 s to 1500 s) but with low thrust (10^{-3} to 10^{-5} N). Their biggest potential advantage is in ease of integration. Because they don't require any additional propellant management, they can be built as simple, self-contained units that, in principle, we can easily bolt onto a spacecraft. Table 14-10 summarizes key information about the electrodynamic rockets we've discussed.

Table 14-10. Electrodynamic Rockets. *(Adapted from Space Propulsion Analysis and Design).*

Type	Propellant	Operating Principle	Electrical Efficiency	Thrust (N)	I_{sp} (s)
Ion (or electrostatic) thruster	Xenon	Applied electric field accelerates an ionized propellant	90%	0.1 – 1.0	2000 – 10,000
Hall-effect thruster (HET)	Xenon	Combined electric and magnetic fields produce a "Hall effect" that accelerates ions within a plasma	60%	0.1 – 1.0	~2000
Pulsed-plasma thruster (PPT)	Teflon (PTFE)	An electric arc pulses over a solid propellant, vaporizing it and creating a plasma. Interaction between the applied electric field and resulting magnetic field accelerates the plasma.	20%	10^{-5} – 10^{-3}	~1500

System Selection and Testing

So far, we've looked at all the pieces that make up propulsion subsystems and many of the various rocket-technology options available. In Chapter 11, we presented the orbital-control budget, which tells us the total ΔV needed throughout the mission, as one important driver of the propulsion subsystem design. Using the tools from this chapter, such as the rocket equation, we can translate this into a propellant mass and volume requirements for a given rocket technology. But many questions about propulsion-subsystem design and applications remain.

- How do mission planners select the best-technology rocket from this large menu of available systems? How do researchers decide which is the best technology to pursue for future applications?

- How are new or improved systems tested and declared fit for flight?

Let's start with the problem of technology selection. As with most technology decisions, there is rarely one, best answer for any given application. Sometimes, as with the case of our FireSat example, the severe constraints on volume, power, and mass, coupled with the modest ΔV requirements, leaves only a few realistic options—cold-gas thrusters, or possibly, a monopropellant system. Even when we narrow the field, the choice of the right propulsion subsystem for a given mission depends on a number of factors that we must weigh together.

One way to trade various rocket options is to select one with the lowest total cost. But here, cost represents much more than simply the engine's price tag. The total cost of a propulsion system includes at least eight other factors, in addition to the bottom-line price tag, that we must consider before making a final selection [Sellers, 1998]. These factors include

- Mass performance—measured by I_{sp}

- Volume performance—measured by I_{dsp}

- Time performance—how fast it completes the needed ΔV, measured by total thrust

- Power requirements—how much total power the EPS must deliver

- Safety costs—how safe the system (including its propellant) is and how difficult it is to protect people working with the system

- Logistics requirements—how difficult it is to transport the system and propellant to the launch site and service it for flight

- Integration cost—how difficult the system is to integrate and operate with other spacecraft subsystems and the mission operations concept

- Technical risk—what flight experience does it have or how did it perform in testing

Different missions (and mission planners) naturally place a higher value on some of these factors than on others. Example 14.1 showed that for the FireSat mission, a helium cold-gas system had lower mass cost, but its volume cost was prohibitive. Other missions, such as a complex-commercial mission, may place high priority on reducing technical risk. For them, a new type of plasma rocket, even if it offers lower mass cost, may be too risky when they consider all other factors. When asking what's the best option for a given mission, "it depends" is usually the best answer!

After selecting a system, engineers must conduct a rigorous testing and qualifying process to declare it safe for use. New rocket development usually progresses from relatively crude, engineering-model testing under atmospheric conditions, to more elaborate testing of flight models under high-altitude or vacuum conditions. Of course, for specialized systems such as electrodynamic thrusters (e.g. ion thrusters or HETs), engineers can only do tests under vacuum conditions, using highly accurate thrust stands to measure micronewtons (10^{-3} N) of thrust. During experimental testing, rocket scientists carefully measure mass flow rates, chamber pressures, temperatures, and other parameters, and compare them to predicted values based on thermochemical and other models.

Figure 14-49. Propellant Loading Operations. Transferring toxic propellants from storage containers to rocket tanks is dangerous and requires safety suits with breathing apparatus and special handling equipment. *(Courtesy of British Aerospace Royal Ordinance)*

Figure 14-50. Rocket Testing. From initial development through flight testing, rockets and propulsion systems undergo rigorous testing to measure performance and ensure safe, reliable operations. This photograph shows the NSTAR ion thruster used for the NASA Deep Space 1 mission in its test-stand configuration. *(Courtesy of NASA/Jet Propulsion Laboratory)*

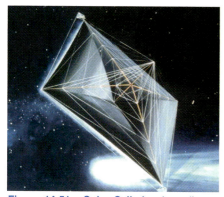

Figure 14-51. Solar Sail. A solar sail captures the minute pressure exerted by solar radiation. Even a very large surface area (1 km²) would only generate about 5 N of thrust. However, this thrust is essentially "free" as no mass is expended. Thus, the solar sail is free to sail around the inner solar system (where solar radiation is most intense). *(Courtesy of NASA/Jet Propulsion Laboratory)*

Because rockets typically involve high pressures, high temperatures, high voltages, and hazardous chemicals, safety issues are a primary concern. These concerns carry through from initial development of new rockets to servicing of proven systems while preparing them for flight. In the case of launch-vehicle propulsion, human lives may depend on safe, reliable operation. As discussed earlier, special loading procedures and equipment insure safe handling of hazardous propellants. Figure 14-49 shows skilled technicians performing propellant-loading operations.

Ensuring system reliability involves a complex series of ground tests that measure performance over many conditions. These conditions can range from relatively simple tests, to ensure the system doesn't leak at flight pressure, to complicated tests that require widely varying O/F ratios and expansion conditions. In addition to performance, all the typical space-environment testing done for other subsystems, such as thermal and vacuum testing discussed in Chapter 13, must also be accomplished for the propulsion subsystem. Figure 14-50 shows an ion-thruster setup for testing in a vacuum chamber.

Exotic Propulsion Methods

Chemical rockets have given us access to space and taken spacecraft beyond the solar system. Electrodynamic rockets offer a vast increase in mass efficiency, making exciting new missions possible. However, to really open space to colonization, and allow humans to challenge the stars, we need bold, new approaches. *Exotic propulsion systems* are those "far out" ideas still on the drawing boards. While there are many exotic variations to the rockets we've already discussed (such as using high energy density or meta-stable chemicals, nuclear fusion, or antimatter to create super-heated products), here we focus on even more unconventional types of propulsion—ones that produce thrust *without* ejecting mass

- Solar sails
- Tethers

We'll first look at how these far-out concepts may one day give us even greater access to the solar system. Then we'll go beyond that to look at the unique challenges of interstellar flight.

Solar Sails

In Chapter 12, we discussed solar pressure as one source of disturbance torque on a spacecraft. Light, when thought of as photons, imparts a tiny force to any surface it strikes. Just as a conventional sail harnesses the force of the wind to move a ship, a *very* large *solar sail* can harness the force of solar pressure to propel a spaceship without ejecting mass! Of course, the farther it goes from the sun the less solar pressure it can collect, so a solar sail would work best inside Mars' orbit. Figure 14-51 shows an artist's concept of a solar sail.

How large would a sail need to be? We can determine this force from

$$F = \frac{F_s}{c}A_s(1 + \rho)\cos I \qquad (14\text{-}38)$$

where

F = force on the sail (N)
F_s = solar constant = 1358 W/m^2 at Earth's orbit around the Sun
c = speed of light = 3×10^8 m/s
A_s = surface area (m^2)
ρ = surface reflectance (where $\rho = 1$ for a perfect reflector)
I = incidence angle to the Sun (deg)

To produce just five newtons (about one pound) of thrust near Earth, we'd need one square kilometer (0.62 mi. on a side) of sail! To achieve escape velocity from a low-Earth orbit (assuming a total spacecraft mass of only 10 kg), this force would have to be applied for more than 17 years! Of course, a solar sail uses no propellant, so the thrust is "free." As long as travellers aren't in a hurry, a solar sail offers a cheap way to get around. Some visionaries propose that solar sails can be used to maneuver mineral-rich asteroids closer to Earth to allow for orbital mining operations.

Tethers

Another imaginative means of propulsion that doesn't need propellant, *tethers*, uses very long cables. Recall in Chapter 12 that we discussed the use of gravity-gradient booms to vertically stabilize spacecraft. Typically, these booms are only a few meters long. By using a small mass at the end of a very long tether, tens or even hundreds of kilometers long, we produce the same stabilizing effect. But even more interesting effects become possible as well.

Picture a large spacecraft, such as the Shuttle, in a circular orbit. Now, imagine a small payload deployed upward (away from Earth), from the Shuttle at the end of a very long tether, as shown in Figure 14-52. Recall from Chapter 4, that when we compute orbital velocities, we assume we are dealing with point masses, affected only by gravity. From an orbital-mechanics standpoint, this point-mass assumption is valid only at the center of mass of the Shuttle/payload system. If the payload mass is small compared to the Shuttle's mass, the system's center of mass will not move significantly when it deploys. Thus, the orbital velocity of the system will stay about the same. What does this mean for the payload? Secured by the tether, it is pulled along in orbit at the Shuttle's orbital velocity. But the payload is well above the Shuttle. In Chapter 4, we said that orbital velocity depends on the distance from Earth's center. Therefore, because the payload is higher than the Shuttle, its proper circular, orbital velocity should be somewhat *slower* than the velocity it maintains due to the tether. Or, said another way, the tether forces the payload to travel *faster* than orbital mechanics would dictate for its altitude.

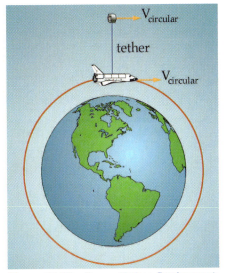

Figure 14-52. Space Tether Deployment. This diagram illustrates a payload deployed upward, away from Earth's center, at the end of a long tether.

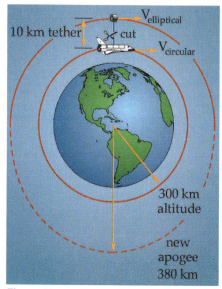

Figure 14-53.   Tether Orbit Boost. A payload deployed 10km higher than the Space Shuttle in a 300 km orbit will be boosted to a 310 × 380 km orbit when the tether is cut.

Figure 14-54. Tether Experiment. This artist's concept shows a small mass deployed downward on a long tether. *(Courtesy of NASA/ Marshall Space Flight Center)*

Now, what happens if we suddenly cut the tether? Orbital mechanics would take over and the payload would suddenly find itself at a velocity too fast for a circular orbit at its altitude. The situation would be as if its velocity were suddenly increased by firing a rocket. It would enter an elliptical orbit with a higher apogee, one-half orbit later. Analysis indicates this new apogee altitude would be higher than the original circular orbit by 7 times the length of the tether [Humble et. al., 1995]. In other words, if the payload's original altitude were 310 km and the tether's length were 10 km, the payload's new elliptical orbit would be 310 × 380 km, as illustrated in Figure 14-53.

If the payload were deployed downward instead of upward, the opposite would happen. Its orbit would shrink, so that half an orbit after the tether releases, the payload would reach perigee. This technique was used by the Small Expendable-tether Deployment System (SEDS) mission in 1993 to successfully deorbit a small payload [Humble et. al., 1995].

Of course, tether propulsion isn't completely "free." We still need to add the mass of the tether and its deployment motors and gears to a spacecraft. And we need extra electrical power to operate the tether-deployment mechanisms. However, once we put these systems in place, we could conceivably use the tether system repeatedly to boost or de-orbit payloads.

Space Shuttle astronauts have performed a number of experiments to investigate the exciting possibilities of tethers. So far, these experiments have focused on the practical problems of deploying, controlling, and reeling in a small payload at the end of a long tether. Figure 14-54 shows an artist's concept of a tether experiment. Future applications for tethers are truly unlimited. A series of rotating tether stations could be used to "sling-shot" payloads, passing them from one to the other, all the way from low-Earth orbit to the Moon. Another exiting use of tethers is for power generation. A conducting tether passing through Earth's magnetic field could generate large amounts of electrical power. [Forward, SciAm, Feb. 99].

Interstellar Travel

The ultimate dream of space exploration is to someday travel to other star systems, as depicted in TV shows like *Star Trek* and *Babylon 5*. Actually, the first human-built "star ships" are already on their way out of the solar system. Launched in 1972 and 1973, NASA's Pioneer 10, shown in Figure 14-55, and Pioneer 11 probes became the first spacecraft to leave our local planetary neighborhood and begin their long journey to the stars. Unfortunately, at their present velocities, they're not expected to pass near another stellar body in over 2 million and 4 million years, respectively!

Obviously, these travel times are far too long to be useful for scientists who want to be around to review the results from the mission. Hollywood's version of rocket science can take advantage of hyperspace and warp drive to allow round-trip times to nearby stars during a single episode. Unfortunately, real-world rocket science is far from using these amazing means of propulsion.

Assuming we could develop efficient, onboard energy sources, such as fusion or antimatter, and rely on ion or other extremely efficient types of rockets to achieve very high specific impulse, there is still the limit imposed by the speed of light. If a rocket could thrust continuously for several years, even at a very low-thrust level, it would eventually reach a very high velocity. However, as its speed approached a significant fraction of the speed of light, interesting things would begin to happen.

One aspect of Albert Einstein's theory of relativity says that, as an object's velocity approaches light speed, its perception of time begins to change relative to a fixed observer. This time adjustment leads to the so-called "twin paradox," illustrated in Figure 14-56. To visualize this concept, imagine a set of twins, two sisters. If one sister leaves her twin and departs on a space mission that travels near the speed of light, when she returns, she'll find her twin much older than she is! In other words, while the mission may seem to have lasted only a few years for her, tens or even hundreds of years may have passed for her twin on Earth.

We express this *time dilation* effect, sometimes called a tau (τ) factor, using the Lorentz transformation

$$\tau = \frac{t_{starship}}{t_{Earth}} = \sqrt{1 - \frac{V^2}{c^2}} \qquad (14\text{-}39)$$

where

$t_{starship}$ = time measured on a starship (s)

t_{Earth} = time measured on Earth (s)

V = starship's velocity (km/s)

c = speed of light = 300,000 km/s

The *tau factor, τ*, tells us the ratio of time aboard a speeding starship compared to Earth time, as demonstrated in Example 14-3. As the spacecraft's velocity approaches light speed, τ gets very small, meaning that time on the ship passes much more slowly than it does on Earth. While this may seem convenient for readers thinking about a weekend journey to the star Alpha Centauri (4.3 light years away), Einstein's theory also places a severe "speed limit" on would-be space travelers. As a spacecraft's velocity increases, its effective mass also increases. Thus, as the ship's velocity approaches light speed, it needs more thrust than it did at lower speeds to get the same velocity change. To attain light speed, it'd need an infinite amount of thrust to accelerate it's infinite mass. For this reason alone, travel at or near the speed of light is well beyond current technology.

But who knows what the future holds? For years, scientists and engineers said travel beyond the speed of sound, the so-called "sound barrier," was impossible. But in October 1947, Chuck Yeager proved them all wrong while piloting the Bell X-1. Today, jet planes routinely travel at speeds two and three times the speed of sound. Perhaps by the 23rd century some future Chuck Yeager will break another speed barrier and take a spacecraft beyond the speed of light.

Figure 14-55. Pioneer 10 Boldly Goes. NASA's Pioneer 10 spacecraft became the first interstellar probe when it left the solar system to start a million-year journey to the stars. *(Courtesy of NASA/Jet Propulsion Laboratory)*

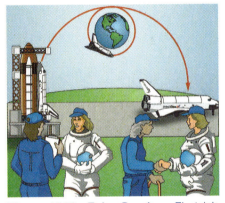

Figure 14-56. Twin Paradox. Einstein's theory of relativity tells us that if one twin leaves Earth and travels at speeds near the speed of light, when she returns she'll find her twin will have aged more than she.

▪ Section Review

Key Terms

arcjet rocket
bipropellant rockets
check valves
chemical rockets
combustion chamber
cryogenic
electrostatic thruster
exotic propulsion systems
fuel
Hall-effect thruster (HET)
hybrid-propulsion systems
hypergolic
ion thruster
monopropellant
motor
nuclear-thermal rocket
oxidizer
oxidizer/fuel ratio (O/F)
plasma thrusters
pressure-fed propellant system
pressure-relief valves
pressure transducers
propellant management
pulsed-plasma thrusters (PPT)
pump-fed delivery system
regulator
resistojet
solar sail
solar-thermal rockets
solid rocket
stoichiometric
storable propellants
tau factor, τ
tethers
thermoelectric rockets
time dilation

Key Equations

$$\tau = \frac{t_{starship}}{t_{Earth}} = \sqrt{1 - \frac{V^2}{c^2}}$$

Key Concepts

➤ All propulsion subsystems have the same basic elements

• Controller—to control and manage all the other elements

• Energy source—either "built-in" to the propellant (in cold-gas or chemical systems), supplied by the electrical-power subsystem (in the case of electrothermal, electrostatic, or electromagnetic thrusters), or supplied from other sources, such as solar or nuclear energy

• Propellant-management subsystem—to regulate and control the propellant flow

• Sensors—to monitor temperature, pressure, and other important parameters

• Rocket—to produce thrust

➤ Chemical rockets are the most common rockets currently in use. Three basic types are

• Liquid

• Solid

• Hybrid

➤ Table 14-1 compares the various types of chemical systems

➤ Solar-thermal rockets use concentrated solar energy to heat a propellant to high temperature

➤ Nuclear-thermal rockets use the heat produced by a nuclear reaction to produce high-temperature propellant

➤ Thermoelectric rockets use heat produced by electrical resistance to create high-temperature propellant These include

• Resistojets

• Arcjets

➤ Electrodynamic systems include

• Ion (also called electrostatic) thrusters— applied electric field accelerates an ionized propellant, such as xenon

Continued on next page

▆▆ Section Review (Continued)

Key Concepts (Continued)

- Plasma thrusters
  - Hall-effect thrusters (HETs)—combine electric and magnetic fields to produce a "Hall-effect" current that accelerates ions within a plasma (such as xenon)
  - Pulsed-plasma thrusters (PPTs)—a pulsed electric arc discharges over a solid propellant, vaporizing it and creating a plasma. The interaction between the applied electric field and the resulting magnetic field accelerates the plasma.

➤ Exotic propulsion methods allow for ΔV without expending propellant. These ideas include

- Solar sails—capture the minute pressure of solar photons to produce thrust just as conventional sails capture the force of the wind
- Tethers—take advantage of gravity-gradient differences to raise and lower spacecraft orbits

➤ Exotic systems may one day propel spacecraft at near light speed. When this happens, we'll have to worry about time dilation and other problems predicted by Einstein's theory of relativity.

Example 14-2

Problem Statement

Thermochemical modeling for a new hybrid rocket motor indicates the theoretical characteristic exhaust velocity, C^*, is 1260 m/s with ratio of specific heats of the combustion products, γ of 1.19. During experimental testing, the total mass flow rate was computed to be 0.15 kg/s, the chamber pressure was measured at 22 bar, and the thrust was measured at 400 N. The test rocket nozzle had a throat diameter of 1.0 cm. Analyze the combustion efficiency for the test motor. Compare the measured versus theoretical specific impulse for this motor.

Problem Summary

Given: $C^*_{theoretical} = 1260$ m/s
$\dot{m} = 0.15$ kg/s
$g_o = 9.81$ m/s^2
$\gamma = 1.19$
$D_t = 1.0$ cm
$P_c = 22$ bar
$F = 400$ N

Find: Theoretical versus measured C^*, theoretical versus measured I_{sp}

Conceptual Solution

1) Compute the nozzle throat area

$$A_t = \pi \left(\frac{D_t}{2}\right)^2$$

2) Compute the measured characteristic exhaust velocity, C^*, using Equation (14-23) and compare to theoretical value

$$C^*_{measured} = \frac{P_c A_t}{\dot{m}}$$

3) Compute measured specific impulse

$$I_{sp\,measured} = \frac{F}{g_o \dot{m}}$$

4) Compute theoretical specific impulse from Equation (14-26)

$$I_{sp\,theoretical} = \frac{C^*_{theoretical}}{g_o} \gamma \left[\left(\frac{2}{\gamma-1}\right)\left(\frac{2}{\gamma+1}\right)^{\frac{\gamma+1}{\gamma-1}}\right]^{\frac{1}{2}}$$

5) Compare measured versus theoretical I_{sp}

$$\frac{I_{sp\,measured}}{I_{sp\,theoretical}}$$

Analytical Solution

1) Compute the nozzle throat area

$$A_t = \pi \left(\frac{D_t}{2}\right)^2 = \pi \left(\frac{1.0\ cm}{2}\right)^2$$

$$A_t = 7.854 \times 10^{-5}\ m^2$$

2) Compute the measured characteristic exhaust velocity, C^*, using Equation (14-23) and compare to theoretical value

$$C^*_{measured} = \frac{P_c A_t}{\dot{m}}$$

$$= \frac{(22\ bar)\left(101 \times 10^3 \frac{N/m^2}{bar}\right)(7.854 \times 10^{-5} m^2)}{0.15\ kg/s}$$

$$C^*_{measured} = 1.152 \times 10^3\ m/s$$

$$\frac{C^*_{measured}}{C^*_{theoretical}} = \frac{1152\ m/s}{1260\ m/s} = 0.914$$

3) Compute measured specific impulse

$$I_{sp\,measured} = \frac{F}{g_o \dot{m}} = \frac{400\ N}{(9.81\ m/s^2)(0.15\ kg/s)}$$

$$I_{sp\,measured} = 271.831\ s$$

Example 14-2 (Continued)

) Compute theoretical specific impulse from Equation (14-26)

$$I_{sp \, theoretical} = \frac{C^*_{theoretical}}{g_o} \gamma \left[\left(\frac{2}{\gamma - 1} \right) \left(\frac{2}{\gamma + 1} \right)^{\frac{\gamma + 1}{\gamma - 1}} \right]^{\frac{1}{2}}$$

$$= \frac{(1260 \text{ m/s})}{9.81 \text{ m/s}^2} (1.19) \left[\left(\frac{2}{1.19 - 1} \right) \cdot \left(\frac{2}{1.19 + 1} \right)^{\frac{1.19 + 1}{1.19 - 1}} \right]^{\frac{1}{2}}$$

$$I_{sp \, theoretical} = 293.925 \text{ s}$$

) Compare measured versus theoretical I_{sp}

$$\frac{I_{sp \, measured}}{I_{sp \, theoretical}} = \frac{271.831 \text{ s}}{293.925 \text{ s}} = 0.925$$

Interpreting the Results

Comparing theoretical analysis to measured performance, the experimental hybrid-rocket motor is achieving only about 92% of predicted characteristic exhaust velocity, C^*, and a similar proportion of theoretical I_{sp}. This represents good but probably not the best achievable performance. While 100% is usually unobtainable, 96% or better should be possible. Rocket scientists on the project should continue to look at ways to improve combustion efficiency.

14.3 Launch Vehicles

In This Section You'll Learn to...

☞ Discuss the various subsystems that make up a launch vehicle

☞ Discuss the advantages of launch-vehicle staging

☞ Determine the velocity change staging can achieve

Now that we've seen the types of rockets available, and how propulsion subsystems fit together, let's see how they're used to solve perhaps the most important problem of astronautics—getting into space. Launch vehicles come in many different shapes and sizes, from the mighty Space Shuttle to the tiny Pegasus, as shown in Figure 14-57. In this section, we start by examining the common elements of modern launch vehicles. Looking at launch vehicles as systems, we'll review the various subsystems that work together to deliver a payload into orbit and focus on the unique requirements for the massive propulsion subsystems needed to do the job. Finally, we'll look at staging to see why launch vehicles come in sections that are used and discarded on the way to orbit.

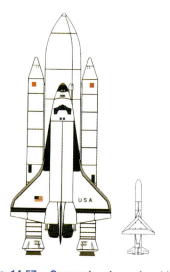

Figure 14-57. Comparing Launch-vehicle Sizes. Launch vehicles come in all shapes and sizes, from the massive Space Shuttle with a total lift-off mass of 2,040,000 kg to the tiny Pegasus (XL) at a mere 24,000 kg. *(Courtesy of NASA/Johnson Space Center and Orbital Sciences Corporation)*

Launch-vehicle Subsystems

In Chapter 11 we introduced the subsystems that comprise the spacecraft bus. A launch vehicle needs most of the same subsystems to deliver a payload (the spacecraft) from the ground into orbit. The two biggest differences between a launch vehicle and a spacecraft are the total operation time (about 10 minutes versus 10+ years) and the total velocity change needed (>8 km/s versus 0–1 km/s). Let's start by looking at the challenges of launch-vehicle propulsion to see how we must adapt the technologies discussed earlier in this chapter to the challenging launch environment. Then we'll briefly review the other subsystems needed to support these large rockets to safely deliver spacecraft (and people) into space.

Propulsion Subsystem

The launch-vehicle propulsion subsystem presents several unique challenges that sets it apart from the same subsystem on a spacecraft. These include

- Thrust-to-weight ratio—must be greater than 1.0 to get off the ground

- Throttling and thrust-vector control—may need to vary the amount and direction of thrust to decrease launch loads and to steer

- Nozzle design—nozzles face varying expansion conditions from the ground to space

Let's go through each of these challenges in more detail.

Thrust-to-weight ratio. To get a rocket off the ground, the total thrust produced must be greater than the vehicle's weight. We refer to the ratio of the thrust to the vehicle's weight as the *thrust-to-weight ratio*. Thus, a launch-vehicle's propulsion system must produce a thrust-to-weight ratio greater than 1.0. For example, the thrust-to-weight ratio at lift-off for the Atlas launch vehicle is about 1.2, and for the Space Shuttle it's about 1.6.

Even though chemical rockets aren't as efficient as some rocket types discussed in the last section, they offer very high thrust and, more importantly, very high thrust-to-weight ratios. For this important reason, current launch vehicles only use chemical rockets, specifically cryogenic ($LH_2 + LO_2$), storable (hydrazine + N_2O_4) bipropellant, or solid rockets.

Throttling and thrust-vector control. For virtually all spacecraft applications, rocket engines are either on or off. There is rarely a need to vary their thrust by *throttling* the engines. However, launch vehicles often need throttling, greatly adding to the complexity (and cost!) of their propulsion subsystems.

One reason for throttling has to do with the high aerodynamic forces on a vehicle as it flies through the atmosphere. Within the first minute or so of launch, the vehicle's velocity increases rapidly while it is still relatively low in altitude, where the atmosphere is still fairly dense. Passing through this dense atmosphere at high velocity produces *dynamic pressure* on the vehicle. Without careful attention to design and analysis, these launch loads could literally rip the vehicle apart. During design, engineers assume some maximum value, based on their extensive analysis of expected launch conditions, that the vehicle can't exceed without risking structural failure. Prior to each launch, they must carefully measure and analyze the winds and other atmospheric conditions over the launch site to insure the vehicle won't exceed it's design tolerances. In many cases, they must design in a specifically tailored thrust profile for the vehicle, which decreases, or "throttles down" during the peak dynamic pressure. The Space Shuttle, for example, reduces the main engines' thrust from 104% to 65%, during this phase of flight, and the burn profile of the solid-rocket boosters' propellant grain is specifically tailored to reduce thrust a similar amount to keep dynamic pressure below a predetermined, safe level.

Another reason for throttling is to keep total acceleration below a certain level. Astronauts strapped to the top of a launch vehicle feel the thrust of lift-off as an acceleration or *g-load* that pushes them back into their seats. From Newton's laws in Chapter 4, we know the total acceleration depends on the force (thrust) and the total mass of the vehicle. If the engine thrust is constant, the acceleration will gradually increase as the vehicle gets lighter due to expended propellant. This means the acceleration tends to increase over time. To keep the overall g-load on the Space Shuttle under 3 g's, the main engines throttle back about six minutes into the launch to decrease thrust to match the burned propellant.

Some vehicles also need throttling for landing. The decent stage engine in the Lunar Excursion Module (LEM), shown in Figure 14-58, used during the Apollo missions, allowed astronauts to throttle the engine over a range of 10%–100%, so they could make a soft touch down on the lunar surface.

Figure 14-58.   Throttle Back for Landing. The Lunar Excursion Module (LEM) used a throttleable bipropellant engine, allowing the astronauts to control their descent to the lunar surface. *(Courtesy of NASA/Johnson Space Center)*

Finally, launch-vehicle rockets often have the unique requirement to vary their thrust direction for steering. This *thrust-vector control (TVC)* can gimbal the entire engine to point the thrust in the desired direction. The Space Shuttle, for example, can vary the thrust direction for each main engine by ±10 degrees. Of course, the mechanical gears and hydraulic actuators needed to move massively thrusting rocket engines can be quite complicated. Earlier rockets used simpler methods of thrust-vector control. The V-2 rocket, for example, used large, movable ablative vanes stuck directly into the exhaust stream to change the vehicle's direction. Other launch vehicles use separate steering rockets or direct injection of gasses into the exhaust flow to change the thrust direction.

Nozzle design. In Section 14.1, we discussed the importance of the external atmospheric pressure and the nozzle expansion ratio to overall engine performance. We prefer not to have a rocket nozzle either over-expanded or under-expanded, but instead designed for ideal expansion. In comparison, spacecraft rocket engines always work within a vacuum, so designers simply use the greatest expansion ratio practical for the best performance. For launch-vehicle rocket engines, the choice of expansion ratio isn't so simple.

During launch, the external pressure on the first stage engines goes from sea level (1 bar or 14.5 p.s.i.) to near zero (vacuum) in just a few minutes. Ideally, we want the nozzle to increase its expansion ratio throughout the trajectory to change the exit pressure as atmospheric pressure decreases. Unfortunately, with current technology, the hardware to do this weighs too much. Instead, the nozzle is typically designed to reach ideal expansion at some design altitude about 2/3 of the way from the altitude of engine ignition to the altitude of engine cutoff.

For example, if we design a rocket to go from sea level to 60,000 meters, a reasonable choice for the design exit pressure would be the atmospheric pressure at about 40,000 meters altitude. As a result, our rocket would (by design) be over-expanded below 40,000 meters and under-expanded above 40,000 meters. As we see in Figure 14-59, a nozzle designed in this way offers better overall performance than one designed to be ideally expanded only at sea level.

One ingenious way around this nozzle design problem is to use a completely different kind of nozzle. During our discussion of nozzle expansion issues in Section 14.1, we focussed on conventional "bell-shaped" nozzles for simplicity. However, another, more versatile design may one day be used on launch vehicles—an aerospike. An old idea, the aerospike nozzle has only recently become the focus of large-scale development to support NASA's X-33 program. Unlike a conventional bell nozzle, where all exhaust-gas expansion takes place inside the nozzle, the aerospike allows expansion on the outside. In the linear aerospike design being developed by Boeing/Rocketdyne, shown in Figure 14-60, the throat is at the edge of a sloping "ramp" that forms the nozzle. The total expansion is determined by the atmospheric pressure, as well as the shape and length of the ramp. The big advantage of this design for a launch vehicle is that it offers near ideal expansion from lift-off to orbit, adding up

Figure 14-60.   An Inside-out Nozzle. Unlike a conventional bell-shaped nozzle, the linear aerospike nozzle, being developed by Boeing/ Rocketdyne for NASA's X-33 program allows expansion to take place outside. This offers the advantage of near ideal expansion from lift-off to orbit adding up to 10% efficiency, crucial for the goal of a single-stage-to-orbit rocket. *(Courtesy of NASA/Marshall Space Flight Center)*

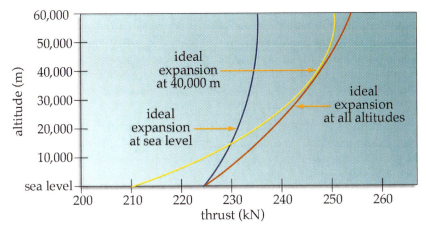

Figure 14-59. Thrust Versus Altitude for Different Nozzle Designs. Because we can't build an ideally expanded nozzle for all altitudes, we typically design them for ideal expansion 2/3 of the way up. In this example, we design a nozzle that must work from sea level to 60,000 m altitude for ideal expansion at 40,000 m. This design offers better overall performance than a nozzle designed for ideal expansion at either sea level or 60,000 m.

to 10% efficiency—crucial for the goal of single-stage-to-orbit vehicles. We'll see what makes single-stage-to-orbit so challenging later in this section.

Navigation, Guidance, and Control Subsystem

In Chapter 12, we discussed the control problems handled by the spacecraft's attitude and orbit control subsystem (AOCS). A launch vehicle must deal with these same problems in a much more dynamic environment. The navigation, guidance, and control (NGC) subsystem keeps the launch vehicle aligned along the thrust vector to prevent dangerous side loads, keeps the thrust vector pointed according to the flight profile, and ensures the vehicle reaches the correct position and velocity for the desired orbit.

As with all control systems, the NGC subsystem has actuators and sensors. The primary launch-vehicle actuators are the main engines, that make use of TVC and throttling to get the rocket where it needs to go. NGC sensors typically include accelerometers and gyroscopes to measure acceleration and attitude changes. Even though the accuracy of these sensors drifts over time, they are usually sufficiently accurate for the few minutes needed to reach orbit. New launch vehicles are starting to rely on the Global Positioning System (GPS) for additional position, velocity, and attitude information.

Communication and Data Handling

Throughout launch, the vehicle must stay in contact with the Launch Control Center. There, flight controllers constantly monitor telemetry from the launch-vehicle subsystems to ensure they're functioning nominally. To do this, the vehicle needs a communication and data-handling subsystem to process onboard data and deliver telemetry to the control center. Data-handling equipment for launch vehicles is very similar to the equipment used on spacecraft as we discussed in Chapter 13. Computers process sensor information and compute commands for actuators, as well as monitor other onboard processes. On expendable vehicles, these subsystems can be relatively simple because they need to work for only a few minutes during launch and won't be exposed to long periods of space radiation. However, the vibration and acoustic environments require these systems to be very rugged.

Communication equipment is also very similar in concept to those found on spacecraft. However, for safety reasons, operators need an independent means of tracking a launch vehicle's location on the way to orbit. In the Launch Control Center, Range Safety Officers monitor a launch vehicle's trajectory using separate tracking radar, ready to send a self-destruct command if it strays beyond the planned flight path to endanger people or property.

Electrical Power

Electrical power requirements for launch vehicles are typically quite modest compared to a spacecraft's. Launch vehicles need only enough power to run the communication and data-handling subsystems, as well as sensors and actuators. Because of their limited lifetimes, expendable launch vehicles typically rely on relatively simple batteries for primary power during launch. The Space Shuttle uses fuel cells powered by hydrogen and oxygen, as explained in Chapter 13.

Structure and Mechanisms

Finally, we must design the launch vehicle's structures and mechanisms to withstand severe loads and perform the numerous mechanical actuations and separations that must happen with split-second timing. A typical launch vehicle can have 10's or even 100's of thousands of individual nuts, bolts, panels, and load-bearing structures that hold the subsystems in place and take the loads and vibrations imposed by the engines' thrust and the atmosphere's dynamic pressure. In Chapter 13, we discussed some of the unique challenges of designing a spacecraft to survive the launch environment. The challenges for a launch vehicle are even more severe.

Because the majority of a launch vehicle's volume contains propellant tanks, these tend to dominate the overall structural design. Often, the tanks become part of the primary load-bearing structure. For example, the Atlas launch vehicle, shown in Figure 14-61, uses a thin-shelled tank that literally inflates with a small positive pressure to create the necessary structural rigidity.

In addition to the problem of launch loads and vibrations, 100's of individual mechanisms must separate stages and perform other dynamic actions throughout the flight. These mechanisms are usually larger than similar mechanisms on spacecraft. During staging, large sections of the vehicle's structure must literally break apart, usually by explosive bolts.

Figure 14-61.   Atlas Inflated for Launch. To provide structural rigidity on the Atlas launch vehicle, the thin aluminum skin of the propellant tanks were inflated like balloons. *(Courtesy of the U.S. Air Force)*

Gimbaling the massive engines to change their thrust direction requires large hinges, hydraulic arms, and supporting structure.

Launch-vehicle designers have the challenge of carefully integrating all of these structures and mechanisms with the engines, tanks, and other subsystems, to create a compact, streamlined vehicle. Sadly, for expendable vehicles, all the painstaking design and expensive construction and testing to build a reliable launch vehicle burns up or drops in the ocean within 10 minutes after launch! Figure 14-62 shows a cut-away view of the Ariane V launch vehicle. As you can see, the majority of the structure is propellant tanks and engines. All the other subsystems squeeze into small boxes, tucked into the secondary structure. Notice there are several sets of engines on this, and all launch vehicles currently in use. Each set comprises a separate stage. Next, we'll see why all these stages are needed to get a spacecraft to orbit.

Staging

Getting a payload into orbit is not easy. As we showed in Section 14.2, the state-of-the-art in chemical rockets (the only type currently available with a thrust-to-weight ratio greater than 1.0) can deliver a maximum I_{sp} of about 470 s. Given the velocity change, ΔV, needed to get into orbit, as we determined in Chapter 9, and the hard realities of the rocket equation, designers must create a launch vehicle that is mostly propellant. In fact, over 80% of a typical launch vehicle's lift-off mass is propellant. Large propellant tanks, that also add mass, contain all of this propellant. Of course, the larger mass of propellant tanks, and other subsystem, the less mass is available for payload. One way of reducing the vehicle's mass on the way to orbit is to get rid of stuff that's no longer needed. After all, why carry all that extra tank mass along when the rocket engines empty the tanks steadily during launch? Instead, why not split the propellant into smaller tanks and then drop them as they empty? Fighter planes, flying long distances, use this idea in the form of "drop tanks." These tanks provide extra fuel for long flights and can be dropped when they are empty, to lighten and streamline the plane. This is the basic concept of staging.

Stages consist of propellant tanks, rocket engines, and other supporting subsystems that are discarded to lighten the launch vehicle on the way to orbit. As the propellant in each stage is used up, the stage drops off, and the engines of the next stage ignite (hopefully) to continue the vehicle's flight into space. As each stage drops off, the vehicle's mass decreases, meaning a smaller engine can keep the vehicle on track into orbit. Figure 14-63 shows an artist's concept of the Saturn I vehicle staging on the way to orbit.

Table 14-11 gives an example of how staging can increase the amount of payload delivered to orbit. For this simple example, notice the two-stage vehicle can deliver more than twice the payload to orbit as a similar-sized, single-staged vehicle with the same total propellant mass—even after adding 10% to the structure's overall mass to account for the extra engines and plumbing needed for staging. This added payload-to-orbit capability is why all launch vehicles currently rely on staging.

Figure 14-62.   Ariane V Cut-away. Most of the mass and volume of this giant booster consists of propellant tanks. *(Courtesy of Arianespace/European Space Agency/Centre National D'Etudes Spatiales)*

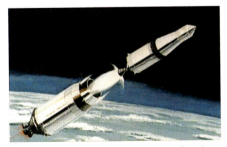

Figure 14-63.   Saturn I during Staging. When a launch vehicle, such as the Saturn I shown here in an artist's concept, stages, it shuts off the lower-stage rocket, separates it, and ignites the rocket on the next stage to continue into orbit. *(Courtesy of NASA/Kennedy Space Center)*

In Table 14-11, for both cases, the mass of the payload delivered to orbit compared to the mass of the entire launch vehicle is pretty small—5% or less. About 80% of a typical vehicle is propellant. The other 15% or so includes structure, tanks, plumbing, and other subsystems. Obviously, we could get more payload into space if the engines were more efficient. However, with engines operating at or near the state-of-the-art, the only other option, as the examples show, is to shed empty stages on the way into orbit.

Table 14-11. Comparing Single-stage and Two-stage Launch Vehicles.

Launch Vehicle	Parameters	Payload to Orbit
Single Stage $m_{payload}$ = 84 kg $m_{structure}$ = 250 kg $m_{propellant}$ = 1500 kg engine I_{sp} = 480 s	ΔV_{design} = 8000 m/s I_{sp} = 480 s $m_{structure}$ = 250 kg $m_{propellant}$ = 1500 kg	$m_{payload}$ = 84 kg
Two Stage $m_{payload}$ = 175 kg $m_{propellant}$ = 750 kg $m_{structure}$ = 140 kg engine I_{sp} = 480 s $m_{propellant}$ = 750 kg $m_{structure}$ = 140 kg engine I_{sp} = 480 s	ΔV_{design} = 8000 m/s **Stage 2** I_{sp} = 480 s $m_{structure}$ = 140 kg $m_{propellant}$ = 750 kg **Stage 1** I_{sp} = 480 s $m_{structure}$ = 140 kg $m_{propellant}$ = 750 kg	$m_{payload}$ = 175 kg

Now let's see how we use the rocket equation to analyze the total ΔV we get from a staged vehicle. We start with

$$\Delta V = I_{sp} g_o \ln\left(\frac{m_{initial}}{m_{final}}\right)$$

Recognize that, for a staged vehicle, each stage has an initial and a final mass. Also, the I_{sp} may be different for the engine(s) in different stages. To get the total ΔV of the staged vehicle, we must add the ΔV for each stage.

This gives us the following relationship for the ΔV of a staged vehicle with n stages.

$$\Delta V_{total} = \Delta V_{stage\ 1} + \Delta V_{stage\ 2} + \ldots + \Delta V_{stage\ n} \qquad (14\text{-}40)$$

$$\Delta V_{total} = I_{sp\ stage\ 1} g_o \ln\left(\frac{m_{initial\ stage\ 1}}{m_{final\ stage\ 1}}\right)$$

$$+ I_{sp\ stage\ 2} g_o \ln\left(\frac{m_{initial\ stage\ 2}}{m_{final\ stage\ 2}}\right) \qquad (14\text{-}41)$$

$$\vdots$$

$$+ I_{sp\ stage\ n} g_o \ln\left(\frac{m_{initial\ stage\ n}}{m_{final\ stage\ n}}\right)$$

where

ΔV_{total}	= total ΔV from all stages (m/s)
$I_{sp\ stage\ n}$	= specific impulse of stage n (s)
g_o	= gravitational acceleration at sea level (9.81 m/s^2)
$m_{initial\ stage\ n}$	= initial mass of stage n (kg)
$m_{final\ stage\ n}$	= final mass of stage n (kg)

What is the initial and final mass of stage 1? The initial mass is easy; it's just the mass of the entire vehicle at lift-off. But what about the final mass of stage 1? Here we have to go to our definition of final mass when we developed the rocket equation. Final mass of any stage is the initial mass of that stage (including the mass of subsequent stages) less the propellant mass burned in that stage. So for stage 1

$$m_{final\ stage\ 1} = m_{initial\ vehicle} - m_{propellant\ stage\ 1}$$

Similarly, we can develop a relationship for the initial and final mass of stage 2, stage 3, and so on.

$$m_{initial\ stage\ 2} = m_{final\ stage\ 1} - m_{structure\ stage\ 1}$$

$$m_{final\ stage\ 2} = m_{initial\ stage\ 2} - m_{propellant\ stage\ 2}$$

Example 14-4 shows how to compute the total ΔV for a staged vehicle. Overall, staging has several unique advantages over a one-stage vehicle. It

- Reduces the vehicle's total mass for a given payload and ΔV requirement
- Increases the total payload mass delivered to space for the same-sized vehicle
- Increases the total velocity achieved for the same-sized vehicle
- Decreases the engine efficiency (I_{sp}) required to deliver a same-sized payload to orbit

But, as the old saying goes, "There ain't no such thing as a free lunch" (or launch)! In other words, all of these staging advantages come with a few drawbacks. These include

- Increased complexity because of the extra sets of engines and their plumbing
- Decreased reliability because we add extra sets of engines and the plumbing
- Increased total cost because more complex vehicles cost more to build and launch

Another interesting limitation of staging has to do with the law of diminishing returns. So far, you may be ready to conclude that if two stages are good, four stages must be twice as good. But this isn't necessarily the case. Although a second stage significantly improves performance, each additional stage enhances it less. By the time we add a fourth or fifth stage, the increased complexity and reduced reliability offsets the small performance gain. That's why most launch vehicles currently in use have only three or four stages.

As we'll see in Chapter 16 in more detail, getting into space is expensive. In some cases, the price per kilogram to orbit is more than the price per kilogram of gold! In an ongoing effort to reduce the cost of access to space, researchers are looking for ways to make launch vehicles less expensive. One of the most promising ways is to make the entire vehicle reusable. One company, Kistler Aerospace, is attempting to do this with a two-stage vehicle design (see the Mission Profile at the end of this chapter).

However, the ultimate goal would be a single-stage-to-orbit vehicle that could take off and land as a single piece, offering airline-like operations. However, the technical challenges in propulsion and materials to overcome the limitations of a single stage are formidable. The goal of NASA's X-33 program, shown in Figure 14-64, is to push the state of the art in rocket engines (the aerospike design described earlier), materials, computer-aided design and fabrications, and operations. One day, the successors to this pioneering program may give all of us the ability to live and work in space routinely.

Figure 14-64.   Single-stage-to-orbit (SSTO). The X-33 is a prototype SSTO vehicle that promises to revolutionize access to space. *(Courtesy of NASA/Marshall Space Flight Center)*

Astro Fun Fact
It's a Rocket! It's a Plane!
No! It's a Combined Cycle!

Launch vehicles using conventional rockets must carry along fuel and oxidizer to reach orbit. But during the early part of the trajectory, they're blasting through the atmosphere, which is about 30% oxygen already. Carrying along their own oxidizer through the atmosphere is like taking their own sand to the beach! Of course, jet engines already take advantage of this fact. Jets are called "air breathing" because they burn oxygen from the atmosphere with their onboard fuel. Unfortunately, conventional jet engines can only operate at speeds up to about three times the speed of sound (Mach 3). Supersonic-combustion ram jets (or "scramjets") promise to push this envelope to Mach 7 or beyond, but still not fast enough to reach orbital speed (Mach 25). However, by combining the best of both worlds—jets and rockets—we may be able to significantly increase the overall propulsion subsystem efficiency and make single-stage-to-orbit a reality. Engineers and rocket scientists are researching combined-cycle propulsion subsystems that use jets, scramjets, and rockets to deliver launch vehicles into orbit. In practice, these systems would use jets and/or scramjets while low and relatively slow in the atmosphere, then gradually transition to conventional rocket engines to accelerate to orbital velocity. Such an approach promises to deliver an overall system-specific impulse of 600 s or greater and may one day pave the way for cheap access to space.

Czysz, Paul A. "Combined Cycle Propulsion—Is it the Key to Achieving Low Payload to orbit Costs." ISABE paper No. 99–7183, 14th International Symposium on Air Breathing Engines, Florence, Italy, Sept. 5–10, 1999.

▰ Section Review

Key Terms

dynamic pressure
g-load
stages
throttling
thrust-to-weight ratio
thrust-vector control, TVC

Key Equations

$$\Delta V_{total} = I_{sp \; stage \; 1} g_o \ln\left(\frac{m_{initial \; stage \; 1}}{m_{final \; stage \; 1}}\right)$$

$$+ I_{sp \; stage \; 2} g_o \ln\left(\frac{m_{initial \; stage \; 2}}{m_{final \; stage \; 2}}\right)$$

$$\vdots$$

$$+ I_{sp \; stage \; n} g_o \ln\left(\frac{m_{initial \; stage \; n}}{m_{final \; stage \; n}}\right)$$

Key Concepts

➤ Launch-vehicle subsystems are similar in many ways to spacecraft subsystems, discussed in Chapter 13. The primary differences include

- Total lifetime (minutes rather than years)
- Propulsion subsystem requirements (see the next bullet)

➤ Launch-vehicle propulsion subsystems must be designed for

- Thrust-to-weight ratio greater than 1.0
- Throttling and thrust-vector control
- Optimum nozzle expansion ratio

➤ By staging launch vehicles, we can

- Reduce the total vehicle mass for a given payload and ΔV requirement
- Increase the total mass of the payload delivered to space for the same-sized vehicle
- Increase the total velocity achieved for the same-sized vehicle
- Decrease the engine efficiency (I_{sp}) required to deliver a same-sized payload to orbit

➤ But staging also has several disadvantages

- Increased complexity, because the vehicle needs extra engines and plumbing
- Decreased reliability, because we add extra sets of engines and the plumbing for the upper stages
- Increased total cost, because a more complex vehicle costs more to build and launch

Example 14-3

Problem Statement

Imagine you are preparing the new Falcon launch vehicle for its first mission from Kennedy Space Center. The vehicle must deliver a total ΔV (ΔV_{design}) of 10,000 m/s. The total mass of the second stage, including structure and propellant, is 12,000 kg, 9000 kg of which is propellant. The payload mass is 2000 kg. The I_{sp} of the first stage is 350 seconds and of the second stage is 400 seconds. The structural mass of the first stage is 8000 kg. What mass of propellant must be loaded on the first stage to achieve the required ΔV_{design}? What is the vehicle's total mass at lift-off?

Problem Summary

Given: 2 stages

$m_{payload} = 2000$ kg

$m_{structure-2} + m_{propellant-2} = 12,000$ kg

$m_{propellant-2} = 9000$ kg

$m_{structure-1} = 8000$ kg

$I_{sp-1} = 350$ s

$I_{sp-2} = 400$ s

$\Delta V_{design} = 10,000$ m/s

Find: $m_{propellant-1}$

$m_{initial}$

Conceptual Solution

1) Determine the $\Delta V_{stage\ 2}$

$$\Delta V_{stage\ 2} = I_{sp\ 2}g_0 \times$$

$$\ln\left(\frac{m_{structure\ 2} + m_{propellant\ 2} + m_{payload}}{m_{structure\ 2} + m_{payload}}\right)$$

2) Determine the required ΔV of stage 1

$$\Delta V_{stage\ 1} = \Delta V_{design} - \Delta V_{stage\ 2}$$

3) Determine the initial mass of stage 1

$$\Delta V_{stage\ 1} = I_{sp\ 1}g_0 \times$$

$$\ln\left(\frac{m_{initial}}{m_{structure\ 2} + m_{propellant\ 2} + m_{payload} + m_{structure\ 1}}\right)$$

4) Determine the mass propellant in stage 1

$$m_{propellant\ 1} = m_{initial} -$$

$$(m_{stucture\ 1} + m_{structure\ 2} + m_{propellant\ 2} + m_{payload})$$

Analytical Solution

1) Determine

$$\Delta V_{stage\ 2} = I_{sp\ 2}g_0 \times$$

$$\ln\left(\frac{m_{structure\ 2} + m_{propellant\ 2} + m_{payload}}{m_{stucture\ 2} + m_{payload}}\right)$$

$$= (400\ s)(9.81 m/s^2)\ \ln\left(\frac{12,000 kg + 2000 kg}{3000 kg + 2000 kg}\right)$$

$$\Delta V_{stage\ 2} = 4040\ m/s$$

2) Determine the required ΔV of the first stage

$$\Delta V_{stage\ 1} = \Delta V_{design} - \Delta V_{stage\ 2}$$

$$= 10,000\ m/s - 4040\ m/s$$

$$\Delta V_{stage\ 1} = 5960\ m/s$$

3) Determine the initial mass of stage 1

$$\Delta V_{stage\ 1} = I_{sp\ 1}g_0 \times$$

$$\ln\left(\frac{m_{initial}}{m_{structure\ 1} + m_{structure\ 2} + m_{propellant\ 2} + m_{payload}}\right)$$

$$m_{initital} = (8000\ kg + 3000\ kg + 9000\ kg + 2000\ kg)$$

$$e^{\left[\frac{5960 m/s}{(350 s)(9.81 m/s^2)}\right]}$$

$$m_{initial} = 124,821\ kg$$

4) Determine mass of propellant in stage 1

$$m_{propellant\ 1} = m_{initial} -$$

$$(m_{structure\ 1} + m_{structure\ 2} + m_{propellant\ 2} + m_{payload})$$

$$= 124,821 - (8000\ kg + 3000\ kg + 9000\ kg + 2000 kg)$$

$$m_{propellant-1} = 102,821\ kg$$

Interpreting the Results

The total mass of this launch vehicle at lift-off is 124,821 kg (113 tons). About 82% of this mass is propellant in the first stage alone (102,821 kg/124,821 kg). Less than 2% of the total lift-off mass is payload (2000 kg/124,821 kg).

References

Einstein, Albert. *Relativity: The Special and the General Theory.* New York, NY: Bonanza Books, 1961. Distributed by Crown Publishers, Inc. for the estate of Albert Einstein.

Forward, Robert L. and Robert P. Hoyt. "Space Tethers." *Scientific American.* pp. 86–87. February, 1999.

Humble, Ronald., Gary N. Henry, and Wiley J. Larson. *Space Propulsion Analysis and Design.* New York, NY: McGraw-Hill, Inc., 1995.

Isakowitz, Steven J. *International Reference Guide to Space Launch Systems.* Washington, D.C.: American Institute of Aeronautics and Astronautics (AIAA), 1999.

Sellers, Jerry J. et. al., "Investigation into Cost-Effective Propulsion System Options for Small Satellites." *Journal of Reducing Space Mission Cost.* Vol. 1, No. 1, 1998.

Wertz, James R. and Wiley J. Larson. *Space Mission Analysis and Design.* Third edition. Dordrecht, Netherlands: Kluwer Academic Publishers, 1999.

Mission Problems

14.1 Rocket Science

1   What three things are rockets used for on launch vehicles and spacecraft?

2   Describe the inputs and outputs of the simplest version of a rocket system.

3   Define rocket thrust and explain where it comes from in terms of Newton's Third Law of Motion.

4   Describe the relationship between rocket thrust, propellant mass flow rate and exhaust velocity.

5   A small experimental rocket engine delivers an effective exhaust velocity of 1800 m/s with a mass flow rate of 800 gm/s. What is the effective jet power of the exhaust?

6   A major league batter can swing his bat with force of 1000 N. If the bat is in contact with a ball for 0.1 s, what is the change in momentum of the ball? What is the total impulse imparted?

7   Define specific impulse? What are its units?

8   During mission analysis, engineers determine a communications spacecraft will need a total impulse of 10,000 N-s during its lifetime for orbit maintenance. How much total thrust time would be needed, if the spacecraft uses a single 10 N thrust engine? If it uses a 2- or a 1-N thrust engine?

9   Imagine a 100-kg astronaut is stranded in space with only a large bag (50 kg) of Moon rocks (0.5 kg each). To get back to the space shuttle, she'll have to throw the rocks in one direction to move in the other. If she can throw 1 rock per second at a velocity of 30 m/s, what is the specific impulse of her makeshift rocket? What is her thrust? What total ΔV can she generate?

10   What is the ideal rocket equation? Define each term.

11   While on its way into orbit, the Space Shuttle, with an initial mass of 100,000 kg, burns 1000 kg of propellant through its orbital maneuvering system's engines (C = 2000 m/s). How much ΔV does it achieve?

12 A remote-sensing spacecraft needs to correct its orbit by 10 m/s. If the effective exhaust velocity of the orbital-maneuvering thruster is 1000 m/s, and the spacecraft's initial mass including propellant is 1000 kg, how much propellant will the maneuver require?

13 Describe the two separate processes that take place within a rocket?

14 What are the two main categories of rockets in current use and how are they classified?

15 What are the two forms of thermodynamic energy? What forms do electrodynamic energy take?

16 What is charge? What is an ion?

17 What basic function does a nozzle serve for a thermodynamic rocket?

18 Nitrogen gas (molecular mass 28 kg/kmole) for a cold-gas thruster is being stored in a tank at 100 bar at temperature of 290 K. What is the density of the propellant?

19 What two conditions must be satisfied for isentropic flow? What is "frozen flow?"

20 Define specific enthalpy.

21 Water (1 gm/cc) is flowing through a 1-cm diameter hose at a rate of 3 kg/min. What is the flow velocity? If a kink in the hose reduces the effective diameter to 0.5 cm, to what velocity will the flow increase to maintain the same mass flow rate?

22 The Mach-1 Challenger rocket bike is attempting to break the sound barrier on the Utah salt flats. If the ratio of specific heats for air is 1.4, what will be the speed of sound at noon with the air temperature at 36° C? What velocity will the rocket bike need to reach Mach 1.3? Use $M_{air} = 28.97$ kg/kmole.

23 What is the venturi effect? Give an every day example. Why is it important for thermodynamic rockets?

24 During deployment, the rocket nozzle for an upperstage was damaged, causing the effective exit area to decrease. What effect will this have on the effective exhaust velocity? On the pressure thrust?

25 Describe the difference between momentum thrust and pressure thrust. What is the ideal pressure?

26 Describe the difference between over-expanded, under-expanded, and ideally expanded rocket nozzles. What is the exit pressure for an ideally expanded nozzle?

27 An upperstage for an interplanetary spacecraft has a nozzle with an exit diameter of 0.25 m. If the expansion ratio is 200:1, what is the nozzle throat diameter?

28 An improved upperstage solid-rocket motor has a nozzle throat diameter of 0.5 cm. If the theoretical characteristic exhaust velocity, C^*, for the new propellant combination is 1400 m/s, and the operational chamber pressure will be 20 bar, how long will it take to use the 10 kg of propellant?

29 Thermochemical analysis for a new rocket engine design using LOX and RP-1 predicts the combustion temperature will be 3415 K with a ratio of specific heats, γ, of 1.225. If the molecular mass of the combustion products is 21.79 kg/kmole, what is the theoretical characteristic exhaust velocity, C^*, and specific impulse, I_{sp}, for this new engine?

30 What two qualities of a thermodynamic rocket engine affect the I_{sp} it can produce? How do we choose these qualities to produce the highest I_{sp}?

31 Describe the basic process that takes place within a rocket combustion chamber. What does the performance of this process depend on? Describe the process that takes place within the nozzle. What does its performance depend on?

32 Explain the basic operating principle of a cold-gas rocket. Give an example of their application.

33 Engineers are analyzing the effect of replacing the regulators used in the astronaut's MMU that take the tank storage pressure from 200 bar down to the chamber pressure of 5 bar. If a new regulator will reduce the pressure only to 10 bar, what will be the expected impact on thrust and system specific impulse? (*Hint:* Refer to the test results in Figure 14-20)

34 Discuss the effect of replacing nitrogen (M = 28 kg/kmole) in a cold-gas system with carbon dioxide (M = 44 kg/kmole).

35 Describe how we can use an electric field to accelerate ions.

36 Describe the relationship between charge density and thrust. Between electric-field strength and specific impulse.

37 Define plasma and give an every-day example of one. What is the main advantage of using plasma for electrodynamic propulsion?

38 Mission designers are hoping to double the power available for an electrodynamic rocket on a deep-space probe. Assuming thrust for the engine remains constant, what will be the effect on specific impulse?

39 Engineers are evaluating two different thrusters for a new communication satellite. System 1 has an I_{sp} of 100 s and system 2 has an I_{sp} of 150 s. If the total ΔV the system must deliver over the life of the spacecraft is 500 m/s, how much propellant will they save by using system 2 instead of system 1? Assume the initial mass in both cases is 1000 kg.

14.2 Propulsion Systems

40 List the key elements of any propulsion subsystem and describe how they relate to each other?

41 What are the four main tasks of propulsion-system propellant management?

42 What are the primary differences between pressure-fed and pump-fed propulsions systems? What are some of the advantages and disadvantages of each?

43 Using the proposed propulsion subsystem for the FireSat mission presented in Figure 14-26, go through each component and describe its function.

44 List the five basic types of thermodynamic rockets and describe their basic operating principles?

45 Describe the functional difference between bipropellant and monopropellant liquid-chemical rockets. Compare their relative advantages and disadvantages.

46 Give an example of a hypergolic propellant combination. What is the advantage of hypergolics over non-hypergolics?

47 A sub-orbital launch vehicle powered by a hybrid-rocket motor has a stoichiometric oxidizer-to-fuel ratio, O/F of 8:1. If the rocket equation predicts a total propellant mass of 900 kg will be needed, what mass of this will be oxidizer?

48 Describe the basic operating principle for solid-rocket motors and list some of their advantages and disadvantages. Give examples of their applications.

49 Explain how the shape of the propellant grain can affect the thrust profile for a solid-rocket motor. Give an example of how this can be used to operational advantage.

50 Describe the basic operating principle of a hybrid rocket. Compare their advantages and disadvantages to chemical bipropellant systems.

51 Describe the basic operating principle for solar-thermal rockets. Compare their advantages and disadvantages to chemical bipropellant systems.

52 List the two main types of thermoelectric rocket thrusters, describe their basic operating principles, and compare their performances. Describe the fundamental design trade-offs engineers have among power, thrust, and specific impulse.

53 Explain the operating principle for nuclear-thermal rockets and discuss the technical and political issues associated with their future applications.

54 What are the two main types of electrodynamic rockets currently in use? What is the primary difference between the two?

55 List the two primary types of plasma thrusters and describe their basic operating principles.

56 Discuss future applications of electrodynamic rockets on interplanetary, communication, and remote-sensing missions. What are advantages and disadvantages of these systems versus chemical-propulsion options?

57 List the various factors that we must consider when determining the total "cost" of a particular propulsion subsystem option. Describe how the relative importance of these factors would differ between an experimental-science mission conducted by university students and a communication mission conducted by a commercial aerospace company.

58 What is the basic operating principle of a solar sail? Where does its thrust come from?

59 Ambitious asteroid miners plan to use a large solar sail to maneuver an asteroid closer to the orbital processing facility. What thrust will a sail 10 km × 10 km generate at 1 AU? What acceleration will this impart to a 10^6 kg asteroid? Assume the reflectivity, ρ, is 0.9.

60 Describe how we could use a tether to de-orbit a spacecraft upperstage. Is the ΔV provided by the tether completely "free?"

61 A small satellite is being deployed from the International Space Station (currently in a 300 km circular orbit) upwards at the end of a 20-km tether. What will be the approximate new apogee of the satellite after it releases the tether?

62 The starship Endeavour travels at 80% of the speed of light for one year (relative to the crew.) How much time will pass relative to those of us on Earth?

14.3   Launch Vehicles

63 What are the two biggest differences affecting the design of launch vehicles versus spacecraft?

64 What unique challenges are presented by launch-vehicle propulsion subsystems versus spacecraft propulsion subsystems?

65 Give examples for why a launch vehicle may need to throttle its rocket engines.

66 The first stage for a launch vehicle is designed to burn out at an altitude of 50,000 m. What would be an optimum design altitude for the rocket nozzle?

67 Describe the differences and similarities between launch vehicles and spacecraft in these subsystems: NGC, communication and data-handling, electrical power, and structure and mechanisms.

68 What are the advantages and disadvantages of staging?

69 Rocket scientists are testing a new three-stage rocket for delivering small payloads to low-Earth orbit. It has these characteristics

- I_{sp} stage 1 = 300 s
- I_{sp} stage 2 = 350 s
- I_{sp} stage 3 = 400 s
- Payload mass = 1500 kg
- Structure mass stage 1 = 10,000 kg
- Structure mass stages 2 and 3 = 7500 kg each
- Propellant mass stage 1 = 50,000 kg
- Propellant mass stage 2 = 40,000 kg
- Propellant mass stage 3 = 35,000 kg

a) What is the initial mass of the entire vehicle?

b) What is the final mass of stage 1?

c) What is the ΔV of stage 1?

d) What is the initial mass of stage 2?

e) What is the final mass of stage 2?

f) What is the ΔV of stage 2?

g) What is the initial mass of stage 3?

h) What is the final mass of stage 3?

i) What is the ΔV of stage 3?

j) What is the total ΔV of the booster?

70 A two-stage rocket with the following character-istics must produce a total ΔV of 8000 m/s.

- I_{sp} stage 1 = 400 s

- I_{sp} stage 2 = 450 s

- Payload mass = 100 kg

- Structure mass stage 1 = 8000 kg

- Structure mass stage 2 = 6000 kg

If the ΔV for stage 2 is 3000 m/s, what is the vehicle's total mass at lift-off?

71 NASA is working on a new two-stage rocket. The I_{sp} of stage 1 is 300 s. The I_{sp} of stage 2 is 400 s. The structural mass of stage 1 is 1000 kg and of stage 2 is 800 kg. Propellant loading is 75,000 kg for stage 1 and 50,000 kg for stage 2. Engineers want to place a 1200 kg payload into a circular orbit at an altitude of 200 km. The proposed launch site is Kennedy Space Center (L_o = 28.5°) with an inclination of 28.5°. Assume ΔV_{losses} are 800 m/s. Can this rocket do the job? [*Hint:* You may need to review Example 14-3 and the ΔV_{design} calculations from Chapter 9.]

For Discussion

72 What would be the best rocket technologies, or set of technologies, to use for a manned Mars mission?

73 Given the current state-of-the-art in rocket technology, what other launch-vehicle design options could offer lower cost access to space?

Kistler Aerospace Corporation, based in Kirkland, Washington, is developing the K-1 reusable aerospace vehicle. Kistler engineers and operators will conduct flight tests and commercial operations from Woomera, South Australia. Kistler is developing the K-1 entirely with private capital and plans to build a fleet of five vehicles operating from sites in Woomera and Nevada. A team of contractors, including Lockheed Martin Michoud Space Systems, Northrop Grumman Corporation, GenCorp Aerojet, Draper Laboratory, Allied Signal Aerospace, Irvin Aerospace, and Oceaneering Thermal Systems is manufacturing the vehicle.

Mission Overview

The two-stage K-1 launches vertically. After stage separation, the first stage, or Launch Assist Platform (LAP), reignites its center engine for return to the launch site, landing with parachutes and airbags. The second stage, or Orbital Vehicle, continues into orbit and deploys its payload. It then enters a phasing orbit for 22 hours before deorbiting and returning to the launch site using parachutes and airbags.

Mission Data

- The overall K-1 vehicle is 36.9 m (121 ft.) long and weighs 382,300 kg (841,000 lbm.) at liftoff.
- The LAP is 18.3 m (60 ft.) long, 6.7 m (22 ft.) in diameter, and weighs 250,500 kg (551,000 lbm.) at launch.
- The OV is 18.6 m (61 ft.) long, has a cylindrical diameter of 4.3 m (14 ft.), and weighs 131,800 kg (290,000 lbm.) fully fueled.
- Both stages use liquid oxygen (LOX) and kerosene propellants
  - The LAP is powered by two AJ26-58 and one AJ26-59 engine
    I_{sp} = 331 s (vacuum)
    Thrust = 4540 kN (1,020,000 lbf.)
  - The OV is powered by an AJ26-60 engine
    I_{sp} = 346 s (vacuum)
    Thrust = 1760 kN (395,000 lbf.)
- The K-1 will initially service inclinations between 45° and 60°, and between 84° and 99°
  - 4600 kg (10,140 lbm.) into a 200 km (110 nmi.) circular orbit at a 45° inclination
  - 2800 kg (6,170 lbm.) into a 200 km (110 nmi.) circular orbit at a 98° inclination

Mission Impact

The K-1 is designed to significantly reduce the cost of delivering payloads to low-Earth orbit (LEO) by reusing each vehicle up to 100 times. The K-1 is also designed to provide launch-on-demand capability and schedule flexibility, with a nine-day turnaround between consecutive launches of the same vehicle. Kistler is targeting the market for LEO commercial communication satellites as well as government customers.

Artist's concept of the K-1 during stage separation at 43.3 km or 142,000 ft. *(Courtesy of Kistler Aerospace Corporation)*

For Discussion

- What other advantages or disadvantages will fully reusable systems like the K-1 have over expendable launch vehicles?
- How can mission designers benefit from the launch-on-demand capability resulting from reusable launch vehicles?
- Should the government continue to fund development of completely expendable launch vehicles, or should it move towards more use of commercially available, reusable launch services?
- How should the government license reusable launch vehicles?

Contributor

Debra Facktor Lepore, Manager, Payload Systems. Kistler Aerospace Corporation.

References

Kistler Aerospace Corporation. *K-1 Payload User's Guide.* May 1999.

Space operations experts work "behind the scenes" at the Mission Control Center of NASA's Johnson Space Center, which supports the Space Shuttle while it's in orbit. *(Courtesy of NASA/Johnson Space Center)*

Space Operations

In This Chapter You'll Learn to...

- Describe the major functions of space operations systems
- Identify the main parts of a space mission's communication network
- Explain basic communication principles and determine key parameters of system design
- Describe key tasks performed by teams throughout the mission lifetime
- Explain the use of basic tools for effective team and project management

You Should Already Know...

- Elements of a space mission architecture (Chapter 1)
- The space systems engineering process (Chapter 11)
- Electromagnetic spectrum and black-body radiation (Chapter 11)

Outline

15.1 Mission Operations Systems
Spacecraft Manufacturing
Operations
Communication
Satellite Control Networks

15.2 Mission Management and Operations
Mission Teams
Mission Management
Spacecraft Autonomy

Astronaut Swigert: "Okay Houston, we've had a problem here."
Mission Control Capcom: "This is Houston, say again please."

*Fateful words that began
the dramatic Apollo 13 ordeal*

Space Mission Architecture. This chapter deals with the Mission Operations Systems segment of the Space Mission Architecture, introduced in Figure 1-20.

Figure 15-1. Mission Control Center (MCC). NASA engineers and operators needed all of the MCC capabilities to rescue the stranded Apollo 13 astronauts. Here, flight controllers see and talk to one of the astronauts. *(Courtesy of NASA/Johnson Space Center)*

Throughout most of the space age, whenever a U.S. manned spacecraft reaches orbit, world attention switches from the launch site to Houston, Texas, the hub of U.S. manned space programs. On TV we see a bunch of people seated at consoles, staring at computer monitors. But what are all these people doing? Why does it seem to take so many of them to support just one mission? Here we'll explore how a mission really happens. Throughout the book, we've referred to the operations concept for a mission—the unifying principles that describe the relationships among hardware systems, mission data and services, and their ultimate users—people. Here, we'll turn our attention to what this concept involves, who does it, what they do, and how. We'll explore the entire breadth of space mission operations, from mission design, to launch, to collecting data on orbit.

Space operations is how we really get a mission off the drawing board and into space. These activities include everything, from "cradle to grave," for a given mission. Remember the Space Mission Architecture icon we introduced in Chapter 1. In this chapter, we'll look at the remaining elements of this architecture to see how the final pieces of the mission puzzle fit together. As we'll see, mission operations systems support the entire mission and serve as the "glue" that ties the other elements together. Space Mission Management and Operations wraps all the various elements—the mission, orbits, spacecraft, launch vehicles, and operations systems—into a single, well-integrated package.

To organize, design, assemble, integrate, test, launch, and operate a mission, we need lots of tools. Everything from simple personal computers to massive launch complexes go into getting missions off the ground—and keeping them there. We'll start by looking at some of the tools that make up these critical operations systems. We'll focus on the "big ticket" items including assembly, integration, and test (AIT) facilities, launch sites, and the far-flung communication networks that bind managers, operators, and astronauts into an efficient team. Then we'll turn our attention to the true stars of the show—people.

People are the most important part of any space mission. While most of our attention in this book has been on systems and orbits, it is people who plan, support, direct, and even risk their lives to complete a mission. With the seven simple words quoted at the beginning of this chapter, the crew of Apollo 13 set in motion a heroic effort by the ground-control team at the Mission Control Center in Houston, Texas. Hundreds of engineers and operators worked around the clock to devise a plan to save three astronauts facing death 384,400 km (238,862 mi.) from home (Figure 15-1). The resourcefulness of these dedicated men and women and the cool reactions of the crew ensured their safe return to Earth. We'll see how mission managers and operators organize the space systems engineering process to produce the hardware and software a mission depends on.

15.1 Mission Operations Systems

- ☛ Identify important operations systems needed during spacecraft manufacturing, launch, and operations

- ☛ Explain basic principles of communication and identify the elements of a communication architecture

- ☛ Apply basic principles of radio communications to understand link design

- ☛ Describe components of major NASA and DoD satellite-control networks

Lucky for us, when we decide to build a house, we can take for granted that all the tools we need are on hand or readily available. Otherwise, our first step in the process would be to design a mill to cut the lumber to build the house. And we'd need a shop to make the saws, hammers, and other tools, to create the mill to make the lumber to build the house. And before that we'd...well, you get the idea.

When the first space missions began, operators faced a situation almost this frustrating. They had to invent nearly everything to make the mission possible—launch vehicles, computers, space suits, and even Velcro™. Fortunately, today's missions build on more than 40 years of space heritage and can leverage many of the existing tools.

In this section, we'll look at some of these tools and see how important they are for making space missions possible. We'll use the term *mission operations systems* to include any facilities or infrastructure needed to design, assemble, integrate, test, launch, or operate a space mission (Figure 15-2). For the most part, operations systems stay in the background of any mission, quietly doing their jobs. But without them, space missions couldn't generate the products the designers had in mind. We don't want to take for granted the unsung operator heroes of the space program. However, we can't present all of the thousands of separate operations systems that support even a simple mission, such as FireSat, in a single chapter. Instead, we'll look at some of the critical operations systems that function during the three basic phases of a spacecraft's life—building, launching, and operating. We'll focus on

- Spacecraft Manufacturing—the systems that support design, assembly, integration, and testing

- Launch—the systems that bring the spacecraft and launch vehicle together and get them safely off the pad

- Operations—mainly communication systems, such as the web of radio links that track and relay data to and from the spacecraft

Figure 15-2.   Schriever Air Force Base. At this satellite control center, operators send commands and receive telemetry from most of the Defense Department satellites. The white "golf-ball" building on the right houses a dish antenna for transmitting and receiving signals. *(Courtesy of the U.S. Air Force)*

Spacecraft Manufacturing

In Chapters 11–14 we presented the space systems engineering process and applied it to the basic design challenges for individual spacecraft subsystems. Now we can turn our attention to the operations systems needed to support this process from a blank sheet of paper through final testing.

Throughout the systems-engineering process, spacecraft design teams rely on a wide variety of design and analysis tools. Some of these, such as off-the-shelf orbital simulation software, we described in Chapter 11. Increasingly, in an effort to cut mission costs, smart mission managers look for such versatile hardware or software tools that they can re-use throughout the mission lifetime, and for subsequent missions. For example, the same program used during preliminary spacecraft design to estimate propellant consumption during a critical maneuver could also be used to perform the final calculations prior to the real maneuver on orbit.

After the spacecraft design has been scrutinized during several detailed technical reviews, assembly and integration begins. Technicians fabricate structural components using conventional machine tools, such as lathes and drill presses, or computer-aided manufacturing equipment, such as the ones shown in Figure 15-3. These computer-driven tools allow the technicians to turn electronic drawings directly into finished pieces. Other specially trained technicians assemble electronic components by hand on electronic work benches using soldering guns and other conventional tools from the electronics industry.

To insure the highest possible quality, most spacecraft components are assembled and integrated in dedicated clean rooms. A *clean room*, such as the one shown in Figure 15-4, is a specially designed space with a carefully controlled level of particulate in the air. How clean is clean depends on the rating for a given room. For example, a Class 1000 clean room has fewer than 1000 particles that are less than 0.01 mm in size per cubic meter of volume (in comparison, a typical home may have billions

Figure 15-3.   Computer-aided Manufacturing. Computer-aided manufacturing equipment, such as the 3-axis Knee Mill shown here, allow designers to turn electronic drawings directly into hardware. *(Courtesy of CNC Automation, Inc. at www.cncauto.com)*

Figure 15-4.   Clean Room. Clean rooms provide a dust free environment for assembling sensitive spacecraft components and help to enforce a careful, rigid discipline during the assembly process to ensure the highest quality of workmanship. Workers shown here are wearing standard clean room uniform items—coats, hats, etc. Queen Elizabeth retains her prerogative in headgear. *(Courtesy of Surrey Satellite Technology, Ltd., U.K.)*

or more particles of this size or bigger per m^3—so start dusting!). Clean rooms serve two primary purposes. The first, and most obvious, is that they limit the exposure of sensitive components, such as sensor lenses, to particulate that could damage them or reduce their performance in space.

The second purpose of a clean room is psychological. Arguably, some spacecraft components could be safely assembled in a garage and probably function quite well in space (and, for some missions, have). However, the discipline imposed by working in a clean room creates a carefully-controlled work environment that helps to prevent carelessness and mistakes. When technicians wear crisp, new white smocks with disposable booties, a hair net, and rubber gloves they tend to be far more conscious of the importance of the task at hand and more likely to make careful, deliberate moves near expensive equipment and spacecraft parts.

Within the clean room, other specialized *ground support equipment (GSE)* helps during subsystem and system assembly, integration, and testing. Custom-built mechanical-handling equipment, or "jigs," hold, rotate, and move the spacecraft structure and individual components. This type of machinery is especially important when we're dealing with very large structures. Figure 15-5 shows the Space Shuttle orbiter being moved into position over the external tank prior to mating, using a massive overhead crane. Other dedicated GSE provides power, communications, and other support to individual subsystems during AIT. For example, in Chapter 14 we discussed the need for dedicated propellant handling equipment for loading fuel and oxidizer prior to flight. This type and other GSE accompany the spacecraft during final checkout prior to launch.

Figure 15-5.   Specialized Ground Support Equipment (GSE). The photograph shows a specially designed crane used to move the 100,000 kg Space Shuttle Orbiter into place over the external tank. Spacecraft of all sizes need specialized ground support equipment for AIT and during launch preparation. *(Courtesy of NASA/Johnson Space Center)*

After we assemble and integrate the spacecraft, integrated testing can begin. Prior to this time, individual components and subsystems typically undergo their own testing. These tests can be as simple as screening and functional checkout or include most of the same tests done during integrated testing. We discussed most of these tests in Chapter 13. *Functional testing*, as the name implies, determines how well the subsystems, as well as the complete integrated spacecraft, work as required under a range of operational scenarios. *Environmental testing* ensures the spacecraft can survive the heat, cold, vacuum, radiation, vibration, and g-loading it will experience throughout the mission. Typically, functional and environmental tests go hand in hand. We naturally want to ensure our spacecraft works before we subject it expensive thermal/vacuum cycling. After the thermal/vacuum tests, we must repeat the functional tests to determine if anything broke. If something did, we must fix it and then test again.

In addition to thermal/vacuum facilities, other operations systems used during the testing phase include

- Shaker table—subjects the spacecraft structure to the dynamic loading environment it will experience during launch

- Acoustic chamber—subjects the spacecraft structure to the high acoustic, launch loads (noise)

Figure 15-6. Anechoic Chamber. The unusual chamber walls absorb radio frequency energy to prevent stray reflections from interfering with the sensitive instruments that record the transmitted energy levels. Here the AFRISTAR satellite, built by Worldspace, Inc., undergoes antenna testing in the Mistral test range in Toulouse, France. *(Courtesy of Matra Marconi Space)*

- Anechoic chamber—tests onboard radio equipment to ensure antennas are functioning and produce the correct signal strength (Figure 15-6)

- Solar simulation chamber—simulates the solar radiation input to test solar cell output, as well as thermal-control-system design

Even for a very small-scale mission, such as a our FireSat example, we need to have access to most of the AIT operations systems we've discussed. We need clean rooms for assembly, integration, and testing of subsystems, especially the sensitive lenses of the payload sensors. We subject subsystems, as well as the entire spacecraft, to thermal and thermal/vacuum cycling to ensure components will function in the harsh space environment. We may need an anechoic chamber to test and verify the transmitter antenna patterns. Finally, the launch provider would require us to subject the entire structure to simulated launch loads, vibrations, and g forces, to ensure it won't fall apart on the way to orbit. This last test is most important from the standpoint of the launch vehicle provider, especially when a spacecraft travels as a secondary payload at the behest of the main payload. As a "guest" passenger, a spacecraft can't do anything that may jeopardize the main mission. (No one, other than us, may care if FireSat actually works after it gets into orbit, but the people paying for the main mission don't want FireSat to break their expensive spacecraft on the way!)

Launch

The launch can sometimes account for nearly 30% of a mission's cost. Not only is the launch vehicle expensive, but we also pay for the complex operations systems that provide the infrastructure to get it and our spacecraft safely off the ground and into space. These systems include

- The launch site and its associated range

- The launch pad

- Payload and vehicle processing facilities

- Launch operations centers

Figure 15-7. Kennedy Space Center (KSC). NASA officials chose this site along the eastern Florida coast for launch safety and the added velocity from Earth's rotation rate. *(Courtesy of NASA/Kennedy Space Center)*

Launch sites are typically chosen based on geography and safety. As we discussed in Chapter 9, launch site latitude determines the minimum orbital inclination available from a given site. For example, Space Shuttle launches from the Kennedy Space Center (Figure 15-7), located at 28.5 deg latitude, can reach a minimum inclination of 28.5 deg. Therefore, from the standpoint of geography, the ideal location for a launch site is on the equator. The Kourou Launch Site, used by the Ariane launch vehicles, is located at 4 deg north latitude. This latitude gives Ariane a distinct advantage over launch sites with a higher north or south latitude for launching large communication spacecraft bound for geosynchronous orbit, while still allowing for launches into polar orbits.

With any site is an associated range. The *range* refers to the large area around the site that extends underneath the launch vehicle's trajectory. All of this area does not have to actually be on the site, but it must be clear of population centers and under the control of the launch site authority. For example, the Vandenberg Air Force Base launch site in California includes a large area along the coast, north of Santa Barbara. For safety reasons, launches from Vandenberg must go toward the south and the Western Range is responsible for monitoring all launches from Vandenberg and keeping the area downrange of the launch site clear and safe. Figure 15-8 shows the available launch inclinations for the Vandenberg and Kennedy launch sites. Notice that due to range safety concerns, not all inclinations that are physically possible are allowed operationally.

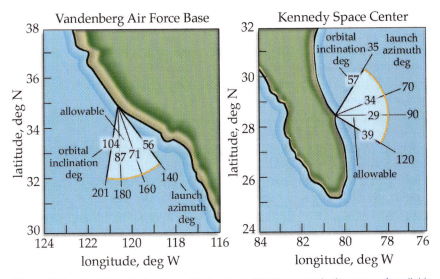

Figure 15-8.   Available Inclinations. While physical limits constrain the range of available inclinations (see Chapter 9), politics and safety also play a part. Here we see the range of available inclinations and corresponding launch azimuths for launches from Kennedy Space Center, Florida, and the U.S. Air Force's launch facility at Vandenberg AFB, California.

One way around the problem of a fixed launch site is to build a mobile one. The Sea Launch platform, shown in Figure 15-9, designed and built by an international consortium led by Boeing in the U.S. and Krounechev in the Ukraine, is a large, converted oil rig that they can tow to any location for launch. This mobility allows the mission planners to place their booster and payload right on the equator to take maximum advantage of Earth's rotation and launch directly into an equatorial orbit, saving important mass for on-orbit operations at geostationary altitude.

For the FireSat mission, recall that we assumed that our spacecraft would travel as a secondary payload on the hypothetical Falcon Launch Vehicle. As a secondary payload, it takes the role of a "hitchhiker," dependent on the primary payload to set the specific requirements for launch and launch-vehicle interface. Essentially, it gets to go along for the ride, taking advantage of most of the launch infrastructure already in place for the primary payload.

Figure 15-9.   Sea Launch. The sea launch complex, built by a consortium of Boeing Aerospace, U.S.A., and Krounechev Aerospace, Ukraine, gets around the inherent limitations of a fixed launch site by loading all the necessary operations systems onto a single, large floating platform. They can then move this mobile platform to a favorable location, such as directly over the equator for launches into geostationary orbits. *(Courtesy of The Boeing Company)*

Operations

The final phase of a mission is operations—the time when a spacecraft finally gets down to business. Another set of operations systems support this phase, perhaps the most important of which is the spacecraft tracking and communication network.

Even before the launch vehicle lifts off the pad, this complex communication network sends health and status data to eager controllers in the mission operations center. This network also tracks the trajectory of the vehicle into orbit. Throughout the mission lifetime, teams of people on the ground depend on these vital radio communication links to keep them in contact with the spacecraft and to deliver the all-important mission data (Figure 15-10).

Recall that for the FireSat example, we assumed we'd rely on a single ground station located in Colorado Springs, Colorado. FireSat would communicate directly with this station whenever it came into range. The mission success depends on this single link. Fortunately, we can distribute mission data to users through the Internet, a complex communication system already in place. Other, more complex, missions depend on other, more complex, operations networks. All of these networks depend on the process of radio communication.

We looked at the spacecraft side of communication, performed by the communication and data-handling subsystem (CDHS) in Chapter 13. Now let's step back to look at the overall problem of linking the ground system to spacecraft. In general, communication is the exchange of messages and information. For space missions, *communication* is the exchange of commands and engineering data between the spacecraft and ground controllers, as well as the processing and transmitting of payload data to users. In this section we'll focus on the different ways spacecraft communicate—the communication architecture. Figure 15-11 shows an example of a communication architecture. The *communication architecture* is the configuration of satellites and ground stations in a space system and the network that links them together. It has four elements

- *Spacecraft*—the spaceborne elements of the system
- *Ground stations*—Earth-based antennas, transmitters, and receivers that talk to the spacecraft
- *Control center*—the command authority that controls the spacecraft and all other elements in the network
- *Relay satellites*—additional spacecraft that link the primary spacecraft with ground stations

Information moves between these elements on various links

- *Downlink*—data sent from the primary spacecraft to a ground station
- *Uplink*—data sent from a ground station to the primary spacecraft
- *Forward link*—data sent from a ground station to the primary spacecraft through a relay satellite

Figure 15-10. Defense Support Program (DSP) Control Site. All space missions require some type of ground site with antennas to maintain radio contact with orbiting spacecraft. This site links the DSP facility at Buckley Air National Guard Base, Colorado, with its geostationary satellites. Large, directional antennas operate under the "golf-balls" that protect them from the elements. *(Courtesy of the U.S. Air Force)*

- *Return link*—data sent from the primary spacecraft to a ground station through a relay satellite
- *Crosslink*—data sent on either the forward link or return link

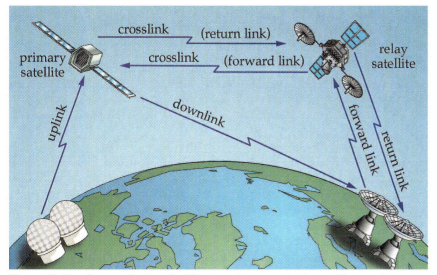

Figure 15-11.   Communication Architecture. A communication architecture for space missions consists of ground- and space-based elements tied together through different communication paths or links.

Data received via the downlink is also known as telemetry. *Telemetry* literally means "far measurement" and includes information on the health and status of the spacecraft and its payload, as well as the communication network that transmits the telemetry to the control center. After launch, telemetry monitoring by operators moves into high gear. The telemetry system takes thousands of separate measurements—rocket engine temperature, tank pressure, battery voltage, attitude, and many, many more.

Operators view all of this data on displays in the operations center and stand poised to react if something goes wrong in real time. *Real time* refers to reacting second-by-second to events as they happen during the mission. Operators use this term to distinguish real-time operations from simulated or post-flight operations. For example, on deep space missions to Jupiter, the time delay between an event happening and receiving the telemetry from the event in the control center could be hours. In this case, operators don't have the luxury of reacting in real-time and must carefully plan ahead and build in capability for the spacecraft to function on its own, without human intervention.

To receive mission data, antennas on the ground must point at the spacecraft whenever it's within range. This means ground stations must know where the spacecraft is at all times. One of the ways they do this is to rely on world-wide tracking networks (Figure 15-12). In Chapter 8 we discussed the problem of tracking and predicting a spacecraft trajectory. This process relies on ground- or space-based parts of the communication network to take measurements in one of three ways

Figure 15-12.   Iridium Tracking Site. This ground station in Yellowknife, Canada, is one of several that tracks the Iridium constellation of 66 satellites. *(Courtesy of Motorola, Inc.)*

- Receiving and analyzing a spacecraft's telemetry signal
- Using radar to measure the range, azimuth, and elevation similar to air traffic control radars that track airplanes
- Sending the spacecraft a signal that an onboard transponder receives and retransmits, so the ground system can measure the spacecraft's range, azimuth, and elevation

The operations team uses this tracking data to determine position and velocity vectors for the spacecraft. From there, they can derive the spacecraft's classic orbital elements, as we explained in Chapter 5, and predict its path as we showed in Chapter 8. (So that's what all that was for!)

Throughout every phase of the launch and on orbit, operators send commands to the launch vehicle or spacecraft via the uplink. *Commands* are instruction sets telling the onboard computers to take some specific action or update some critical part of the software. Commands can tell the spacecraft to charge batteries, fire rockets, or reorient a sensor to point at a new target.

Commands can either be real-time or stored. The computer implements *real-time commands* immediately upon receipt. Operators send *stored commands* with a time tag so the computer will carry it out some time in the future or after some future mission event, such as an antenna deployment. Most important commands are sent in two stages. This means the command travels to the spacecraft through an uplink and the communication subsystem echoes it back to the operators through the downlink. After operators confirm that the command received by the spacecraft is the one sent, they send a second command to enable it in the onboard software.

Two-stage commands ensure that important information doesn't get garbled during transmission and provide a measure of security from outside interference (we don't want someone else taking control of our spacecraft!). Because the spacecraft communication links are so important to mission performance, most spacecraft have built-in commands to follow if the communication link is lost. A simple timing switch, or "watch-dog timer" onboard keeps track of the time a spacecraft goes without communication. If this exceeds a certain preset amount, the spacecraft will put itself into a "safe mode" to ensure the communication link can be more easily restored and the mission resumed.

Because this communication process is so important, in this section we'll examine the basic principles of radio communication in greater detail. In Chapter 13 we examined the communication and data-handling subsystem within the spacecraft. There we described the process of data manipulation needed to send and receive radio signals. Here we'll look at the radio-frequency signals to better understand how we stay in contact with our spacecraft. We'll start by looking at the communication process, then examine the basic principles of communication links.

Communication

Each of us communicates every day. We talk to our friends in the hallway. We talk to our family over the phone. Let's take a moment to dissect the communication process, so we can better understand the problems faced in building communication systems for spacecraft.

Imagine that you go to a friend's house to talk about your French homework. What conditions are necessary for you and your friend to communicate? We need to consider

- Distance
- Language
- Speed
- Environment

First of all, you should be within hearing distance so your words reach your friend. The farther away you are from your friend, the louder you must talk to be heard. If you're too far away to hear each other, you can't communicate. Another issue to consider is language. You and your friend must be able to understand and speak the same language. If she knows French much better than you do, it will be difficult to communicate. Next, there is the speed at which you talk. Have you ever listened to someone who speaks very fast? If you have, you know that it's hard to catch every word. If they speak too rapidly, you can't process the words fast enough to understand their message.

So, if you and your friend are within range and speaking the same language at a reasonable rate, you should communicate, right? Wrong! You also need to consider the environment. Imagine that, as you and your friend are talking, her little brothers and sisters run into the room screaming loudly. Their screaming represents noise. If the screaming is too loud, you'll need to raise your voice so your friend can still hear you above the noise. In other words, your *signal*—the volume and content of your message—must be louder than the noise. That is, to be heard, your *signal-to-noise ratio* must be greater than 1.0. The important quantity for communication is the ratio of the volume of your speech to the volume of their noise.

Now let's see what all this has to do with spacecraft communications. To communicate effectively from one spacecraft to another or to a ground station, we must consider the distance or range between the speaker—called the *transmitter*—and the listener—called the *receiver*. We must also have a transmitter and receiver that understand the language or code that each other uses. Recall, in Chapter 13 we discussed techniques, such as amplitude modulation (AM) or frequency modulation (FM) for overlaying a message onto a carrier signal. The transmitter and receiver must be using the same language or modulation scheme. Furthermore, the receiver must handle the transmitter's message speed or *data rate*. Finally, the volume or *signal strength* at the receiver must be higher than the overall noise in the system. To see all these concepts in practice, we can now focus on some basics of spacecraft communication.

Communication Links

As we described in our discussion of the communication and data-handling subsystem (CDHS) in Chapter 13, spacecraft communications use the radio frequencies of electromagnetic radiation. Your car stereo illustrates the basic principles of radio, as shown in Figure 15-13. When you turn on your radio, you receive signals from the radio station in the form of electromagnetic (EM) radiation. Remember, EM radiation comes from an accelerating charge. As charges accelerate in the radio station's transmitter antenna, an electric field forms and induces a magnetic field, which induces an electric field, and so on. James Maxwell (1831–1879) first developed this concept. The frequency at which this charge accelerates determines the frequency of the EM radiation. The faster the charge accelerates, the higher the frequency.

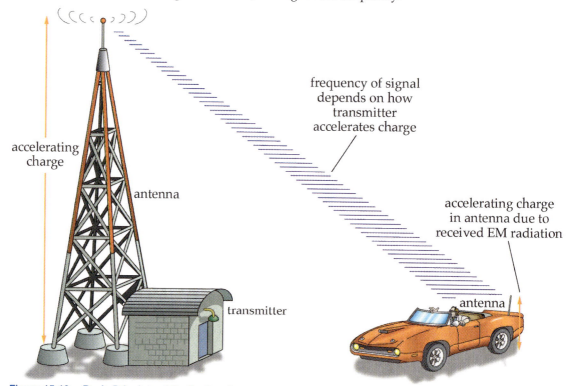

Figure 15-13.   Basic Principle of Radio. A radio station produces a signal by accelerating a charge in the antenna. The signal then travels out as EM radiation until it's received by your car antenna, where charges again accelerate to produce the music you hear.

The station broadcasts a *carrier signal* at a specified frequency, regulated and licensed by the Federal Communication Commission (FCC) in the U.S. The transmitter then super-imposes the message being sent—music, news, or mission data—on top of the carrier signal, using some type of modulation scheme. The schemes we're most familiar with are amplitude or frequency modulation (AM and FM, see Figure 13-9). Spacecraft applications use other schemes as well. This signal travels outward from the station's antenna and hits your radio antenna. There, more charges accelerate. Your receiver detects this charge movement in the antenna and re-translates it to the original signal. The receiver demodulates the AM or FM signal to separate the message from the carrier signal and, suddenly, you're listening to tunes while cruising down the road.

Now we want to take a closer look at communication systems to understand some of the basic principles and limitations. Let's use a light bulb to demonstrate some of these key principles. Similar to a radio transmitter, a light bulb emits EM radiation, but at a different frequency— visible light. If we put a light bulb in the center of a room, as shown in Figure 15-14, light radiates outward in all directions (assuming it's a perfect bulb with no light blockage). The intensity or brightness of the light at some distance from the bulb is called the *power-flux density, F*. Of course, the farther we get from the light bulb, the dimmer it appears. In other words, the power-flux density, which we perceive as brightness, decreases as we move farther away. From test measurements, we know the brightness actually decreases with the square of the distance, because all the output is distributed over the surface of a sphere surrounding the source. We express this as

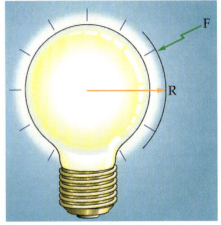

Figure 15-14.   Power-flux Density. An ideal light bulb radiates equally in all directions. The brightness, or power-flux density, F, at any given distance, R, depends on the bulb's output, P.

$$F = \frac{Power}{Surface\ area\ of\ a\ sphere} = \frac{P}{4\pi R^2} \qquad (15\text{-}1)$$

where

F = power-flux density (W/m^2)

P = power rating of the light bulb (W)

R = distance from the bulb (m)

We know that visible light, like that of a light bulb, is simply electromagnetic radiation. Radiation, moving equally in all directions, similar to our light bulb example from Figure 15-14, is called *omnidirectional* or *isotropic*. Now, what if we wanted to increase the brightness or power-flux density in only one direction using the same bulb? As Figure 15-15 shows, that's just what a flashlight does. This time we're still using our ideal light bulb, but we've put a parabolic-shaped mirror on one side of it. Thus, most of the light in one direction reflects off the mirror and heads in the opposite direction, and we have a directed beam of light—a spotlight— rather than an omnidirectional source. Doing this, we effectively concentrated most of the light energy into a smaller area. As a result, we get a brightness in that one direction that is much, much greater than it was when the bulb emitted light isotropically. We've "gained" extra power density by using the parabolic mirror.

The flashlight example illustrates the basic principle of an antenna. Instead of broadcasting in all directions, wasting all that energy, specially designed "dish" antennas allow us to focus the energy on a particular point of interest, such as a receiving antenna. Spacecraft often rely on directional antennas that point toward the receiver at the ground station, making more efficient use of their transmitter's power. Ground stations usually employ another directional (dish) antenna to better receive the

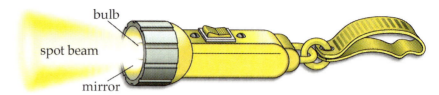

Figure 15-15.   Directed Output from a Light Bulb. A parabolic mirror can direct the bulb output to give us an effective spot light. The mirror allows us to focus the bulb's energy in one direction, thus increasing the gain.

signal, as well as transmit commands back to the spacecraft. Similar to our flashlight's mirror, these dish antennas are often parabolic-shaped to transmit and receive the radio energy efficiently.

An important antenna parameter is its gain. The *gain* of an antenna is the ratio of the energy it transmits in its primary direction to the energy that would be available from an omnidirectional source. In other words, the gain for an omnidirectional antenna is 1, whereas the gain for a directed antenna is greater than 1. The general expression for gain is

$$G = \frac{\text{Energy on target with a directed antenna}}{\text{Energy on target with an omnidirectional antenna}}$$

We can relate these two values for energy to an antenna's area, its efficiency, and the wavelength of the energy it's transmitting by

$$G = \frac{4\pi\,A\eta}{\lambda^2} = \frac{4\pi\,A_e}{\lambda^2} \qquad (15\text{-}2)$$

where

G = gain (unitless)
A = physical area of the antenna (m^2)
η = antenna's efficiency (0.55–0.75 for parabolic antennas)
A_e = antenna's effective area (= $A\eta$, m^2)
λ = signal's wavelength (m)

This relationship tells us that if we want to increase the gain of an antenna (and transmit our message more efficiently), we can either increase its effective area or decrease our signal's wavelength. We use the same expression for the gain of transmitting and receiving antennas.

If we multiply the transmitter's power output by its antenna gain, we get an expression that represents the amount of power an isotropic transmitter would have to emit to get the same amount of power on a target. We call this the *effective isotropic radiated power (EIRP)*.

$$\text{EIRP} = P_t\,G_t \qquad (15\text{-}3)$$

where

EIRP = effective isotropic radiated power (W)
P_t = transmitter's power output (W)
G_t = transmitter's antenna gain (unitless)

How much of the transmitter's power does the receiver collect? Think about collecting rainfall in a bucket. The amount of rain water collected depends on how hard it's raining—the rain's density—and the bucket's size or cross-sectional area. Similarly, the signal strength at a receiver is a function of the power-flux density at the receiver and the area of the receiver's antenna. The resulting expression for the signal gathered by the receiving antenna is then

$$S = \left(\frac{P_t G_t}{4\pi R^2}\right) A_{e_{receiver}} \tag{15-4}$$

where

S $\quad$ = received signal strength (W)

$\left(\dfrac{P_t G_t}{4\pi R^2}\right)$ = transmitter's effective power spread over a sphere of radius, R (W)

$A_{e_{receiver}}$ = receiving antenna's effective area (m^2)

Solving the right-hand expression in Equation (15-2) for $A_{e_{receiver}}$ and substituting into Equation (15-4) results in

$$S = P_t G_t \left(\frac{\lambda}{4\pi R}\right)^2 G_r \tag{15-5}$$

where

S $\quad$ = received signal strength (W)

P_t $\quad$ = transmitter's power output (W)

G_t $\quad$ = transmitter's antenna gain (unitless)

$\left(\dfrac{\lambda}{4\pi R}\right)^2$ = space loss term (0 < space loss < 1.0) (unitless)

G_r $\quad$ = receiver's antenna gain (computed the same way as the transmitter's antenna gain) (unitless)

Notice we have a term representing space loss. *Space loss* is not a loss in the sense of power being absorbed in the atmosphere; rather, it accounts for the way energy spreads out as an electromagnetic wave travels away from a transmitting source. As distance increases, this term becomes smaller, which means space losses get worse. This situation makes sense. The greater the distance between a transmitter and receiver, the greater the total space losses (smaller space loss term). When this term is multiplied by the transmitter's power, and the receiver's and transmitter's antenna gains, the total signal strength, S, gets smaller for longer distances.

So we now have several ways to increase the received signal

- Increase the transmitter's power—P_t

- Increase the transmitter's antenna gain, concentrating the focus of the energy—G_t

- Increase the receiver's gain so it collects more of the signal—G_r

- Decrease the distance between the transmitter and receiver—R

A few pages back we discussed the concept of signal-to-noise (S/N) ratio in communication systems. So far in this discussion we've talked about the received signal, S. Earlier, when we discussed communicating across a room, noise came from some rambunctious kids. But where does noise come from for a radio signal? One important source of radio noise is heat. Recall from our discussion of black-body radiation in Chapter 11 that any object having a temperature greater than absolute 0 K emits EM radiation. While a receiver is running, just like your TV set, it gets hot and produces EM radiation as noise. The amount of noise power is given by

$$N = kTB \qquad (15\text{-}6)$$

where

N = noise power (W)
k = Boltzmann's constant = 1.381×10^{-23} joules/K
T = receiver system's temperature (K)
B = receiving system's bandwidth (Hz)

Bandwidth is the range of frequencies the receiver is designed to receive. For example, the range of human eyesight, or the bandwidth of our eyes, is about 3.90×10^{14} Hz to 8.13×10^{14} Hz, which is a bandwidth of 4.23×10^{14} Hz. This represents the small portion of the EM spectrum we can see—visible light. Note that the noise in the receiver increases as the bandwidth increases. This should make sense, because the more information a receiver attempts to receive, the more likely it'll pick up noise. Ideally, we try to reduce the receiver temperature as much as possible and restrict the bandwidth of interest to minimize the noise.

Combining Equation (15-5) and Equation (15-6), we get the signal-to-noise ratio for a radio signal

$$\frac{S}{N} = \left(\frac{P_t G_t}{kB}\right)\left(\frac{\lambda}{4\pi R}\right)^2\left(\frac{G_r}{T}\right) \qquad (15\text{-}7)$$

where

S/N = signal-to-noise ratio (unitless)
P_t = transmitter's power (W)
G_t = transmitter's gain (unitless)
k   = Boltzmann's constant = 1.381×10^{-23} joules/K
B   = receiving system's bandwidth (Hz)
λ = signal's wavelength (m)
R   = range between the transmitter and receiver (m)
G_r = receiver's gain (unitless)
T   = receiver system's temperature (K)

Remember, for effective communication, the signal-to-noise ratio must be greater than or equal to 1.0. (The voice you hear must be louder than the background noise in the room.) To improve the S/N we can

- Increase the strength of the signal using the methods outlined above

- Reduce the signal's bandwidth—B

- Reduce the receiver's temperature—T

So far we haven't said much about changing the signal's frequency or wavelength. What effect does this have? Looking at Equation (15-7), we'd expect that increasing the wavelength would improve the S/N ratio, but remember the relationship for gain, given in Equation (15-2). The transmitter and receiver gains are inversely related to wavelength. That is, as wavelength increases (lower frequency), gain decreases. This means the net effect of increasing wavelength (decreasing frequency) is to decrease the antenna gains and thus reduce the S/N ratio. In other words, all other system parameters being equal, higher frequency gives us improved S/N. We show all these relationships in action in Examples 15-2 applied to our FireSat scenario.

Satellite Control Networks

Now that we've looked at the theoretical aspects of communication networks, let's look at some examples of control networks in place to support NASA and the DoD space missions. NASA has two networks for tracking and receiving data from space. The Spaceflight Tracking and Data Network (STDN) mostly tracks and relays data for the Space Shuttle and other near-Earth missions. The STDN includes ground-based antennas at White Sands, New Mexico (Figure 15-16), as well as space-based portions using the Tracking and Data Relay Satellites (TDRS) in geostationary orbits. The deep-space tracking network (DSN) includes very large antennas (more than 70 m in diameter), used for tracking and receiving data from interplanetary space missions. These antennas are located in Madrid, Spain; Canberra, Australia (Figure 15-17); and Goldstone, California.

Figure 15-16. Tracking and Data Relay Satellite's (TDRS) Second Terminal. This ground station controls NASA's TDRS constellation and receives telemetry and mission data from many satellites, including the Space Shuttle. *(Courtesy of NASA/White Sands)*

The U.S. Space Command is responsible for tracking and controlling all DoD spacecraft. To do this, the DoD has two networks. The Space Surveillance Network (SSN) is a world-wide network of high-power radars (Figure 15-18) that track approximately 8100 objects in Earth orbit. These objects range from the Space Shuttle orbiter to Astronaut Ed White's glove, lost during a Gemini mission. The radar data passes to the Space Surveillance Center in Cheyenne Mountain AFB, Colorado, where orbital analysts maintain the space catalog. This catalog contains the current classic orbital elements of all the stuff in orbit large enough to track (anything bigger than about 10 cm [3.9 in.] long).

The Air Force uses other networks to control spacecraft, the largest being the Air Force Satellite Control Network (AFSCN). It consists of spacecraft communication sites in such interesting locations as Guam, Diego Garcia, and Hawaii. These stations connect with control centers at Schriever Air Force Base, Colorado, and Onizuka Air Force Base, Sunnyvale, California, where Air Force engineers and space operators command and control almost all Department of Defense satellites. We

Figure 15-17.   Deep Space Network (DSN). This complex of giant antennas at Canberra, Australia, keeps a constant watch for radio signals from NASA's interplanetary satellites, such as Galileo and Cassini. *(Courtesy of NASA/Goddard Space Flight Center)*

Figure 15-18.   Space Surveillance Network (SSN) Radar. This phased array radar at Eglin Air Force Base, Florida, tracks thousands of orbiting objects that make up the space catalog. *(Courtesy of the U.S. Air Force)*

talked about the missions of these spacecraft in earlier chapters; each type (communication, navigation, etc.) requires a team of specialists. Some missions, such as early warning, require so much ground support that they have their own dedicated control stations. Control of these Defense Support Program (DSP) satellites, which provide early warning of enemy missile launches, requires special ground stations in the U.S.

Fortunately, the operational requirements for our FireSat mission are simple enough that we don't have to rely too heavily on the AFSCN or the NASA STDN. It may be necessary to compare the position and velocity information computed onboard using GPS with the independent values determined by the SSN. Recall, the U.S. Space Command uses the SSN to track and catalog the whereabouts of thousands of satellites and pieces of space junk in orbit. However, we still need a dedicated ground station for our spacecraft operations.

Recall that we plan to use a single primary ground station located in Colorado Springs, Colorado. This simple station will need only a single transmit/receive antenna and the necessary communication gear to operate them and communicate with the spacecraft. Our operators can do other operations, such as monitoring subsystem performance, generating commands and collecting mission data, and distributing it to users, using off-the-shelf personal computers running some specialized software. Figure 15-19 shows a similar, simple ground station used to operate microsatellites at the U.S. Air Force Academy. In the next section we'll delve into the specific responsibilities of the dedicated operations team.

Figure 15-19.   A Simple Ground Station. A simple ground station to control small satellites can be assembled from personal computers and off-the-shelf communications gear. This photograph shows the ground station used to control the FalconSAT spacecraft at the U.S. Air Force Academy. *(Courtesy of the U.S. Air Force Academy)*

624

Section Review

Key Terms

bandwidth
carrier signal
clean room
commands
communication
communication
    architecture
control center
crosslink
data rate
downlink
effective isotropic radiated
    power (EIRP)
environmental testing
forward link
functional testing
gain
ground stations
ground support
    equipment (GSE)
isotropic
mission operations
    systems
omnidirectional
power-flux density, F
range
real time
real-time commands
receiver
relay satellites
return link
signal
signal strength
signal-to-noise ratio
space loss
spacecraft
stored commands
telemetry
transmitter
uplink

Key Concepts

➤ Mission operations systems includes the facilities and infrastructure to design, assemble, integrate, test, launch, and operate a space mission

➤ Spacecraft manufacturing—systems used to support design and AIT

• Design and analysis software tools

• Clean rooms

• Ground support equipment

• Test facilities, such as thermal/vacuum chambers, shaker tables, acoustic chambers, anechoic chambers, and solar simulation chambers

➤ Launch—systems needed to bring the spacecraft and launch vehicle together and get them safely off the ground, such as

• The launch site and its associated range

• The launch pad

• Payload and vehicle processing facilities

• Launch operations centers

➤ Operations—the communication architecture, the web of radio links that track and relay data to and from a spacecraft, and ground-based operators

➤ Communication architecture is the configuration of satellite and ground stations in a space system and the network that links them together. It has four major elements—spacecraft, ground stations, a control center, and relay satellites.

• Information moves between elements of the communication architecture on various links—uplink, downlink, forward link, return link, and cross link

• Commands are instructions sent by ground controllers to the spacecraft telling it what to do and when to do it. Real-time commands execute immediately. The spacecraft's data-handling subsystem stores other commands in memory for later execution

➤ To communicate effectively, whether talking to a friend across a noisy room or to a spacecraft at the edge of the solar system, we must meet four requirements

• Transmitter and receiver must be close enough to one another

• The language or code used for the message must be common to the transmitter and receiver

• The speed or data rate of the message must be slow enough for the receiver to interpret

• The volume or strength of the signal must be greater than any noise. That is, the signal-to-noise ratio must be greater than 1.

Continued on next page

▦ Section Review (Continued)

Key Equations

$$G = \frac{4\pi \, A\eta}{\lambda^2} = \frac{4\pi \, A_e}{\lambda^2}$$

$$S = P_t G_t \left(\frac{\lambda}{4\pi R}\right)^2 G_r$$

$$\frac{S}{N} = \left(\frac{P_t G_t}{kB}\right)\left(\frac{\lambda}{4\pi R}\right)^2\left(\frac{G_r}{T}\right)$$

Key Concepts (Continued)

➤ The basic principle of radio involves accelerating charges in a transmitter's antenna to generate electromagnetic radiation. The receiver's antenna detects this radiation, which accelerates charges there. The receiver's antenna passes the received signal into the receiver for demodulation and preparation for use.

➤ Electromagnetic energy from any source, represented by the power-flux density, decreases in strength with the square of the distance

➤ An antenna can focus electromagnetic energy in one direction. The increase in power-flux density achieved using an antenna is the antenna gain.

➤ To increase the received signal strength, we can

• Increase the transmitter's power

• Increase the transmitter or receiver antenna's gain

• Decrease the distance between the transmitter and receiver

➤ Noise in a radio signal can come from the black-body radiation, which the receiver temperature emits; it's a function of the receiver bandwidth (range of frequencies)

➤ To increase the signal-to-noise ratio we can

• Increase the signal's strength

• Reduce the signal's bandwidth

• Reduce the receiver's temperature

• Increase the carrier frequency

➤ NASA and the DoD depend on worldwide satellite-control networks

• NASA uses geostationary Tracking and Data Relay Satellites (TDRS) along with ground based Spacecraft Tracking and Data Network (STDN). For planetary missions they use the Deep Space Tracking Network (DSN).

• The U.S. Space Command is responsible for tracking and controlling all DoD spacecraft. They rely on the Air Force Satellite Control Network (AFSCN).

Example 15-1

Problem Statement

We can think of the Sun as a "perfect light bulb" radiating isotropically. If the Sun's power-flux density, F, on the Earth is 1358 W/m², what is the Sun's power output? Distance to the Sun is about 1.496×10^{11} m.

Conceptual Solution

1) Solve the power-flux density relationship for power output

$$F = \frac{\text{Power}}{\text{Surface area of a sphere}} = \frac{P}{4\pi R^2}$$

$$P = (F)\, 4\pi R^2$$

Problem Summary

Given: $F = 1358$ W/m², $R = 1.496 \times 10^{11}$ m
Find: P output

Problem Diagram

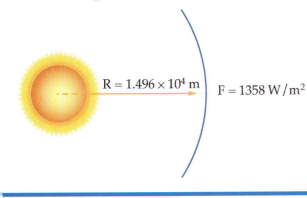

$R = 1.496 \times 10^4$ m $F = 1358$ W/m²

Analytical Solution

1) Solve the power-flux density relationship for power output

$$P = (F)\, 4\pi R^2$$

$$P = (1358 \text{ W/m}^2)\, (4\pi)\, (1.496 \times 10^{11} \text{ m})^2$$

$$= 3.819 \times 10^{26} \text{ W}$$

Interpreting the Results

The Sun puts out a lot of power—millions of times greater than the outputs of all the power plants on Earth. Even at a distance of almost 150 million km (93 million mi.), its intensity is still about 1358 W per square meter.

Example 15-2

Problem Statement

Engineers are designing the communication system for FireSat to ensure vital Telemetry, Tracking, and Commanding data gets through to our operators. The communication frequency they've chosen is 2 GHz (2×10^9 Hz, which is in the S-band) with a bandwidth of 2000 Hz. The spacecraft transmit and receive antenna will be an omnidirectional dipole with a gain of 1.0. Transmitter power output is 2 W. The link will operate over a maximum distance of 4000 km. If the receiver temperature is 800 K, with the receiver antenna efficiency of 0.75, what receiver-antenna diameter do we need to have a signal-to-noise ratio of 10? What is the corresponding ground transmitter power needed to have the same S/N at FireSat for uplink?

Problem Summary

Given: $P_{FireSat} = 2$ W
$S/N = 10$
$f = 2 \times 10^9$ Hz
$R = 4000$ km
$T = 800$ K
$B = 2 \times 10^3$ Hz
$k = 1.381 \times 10^{-23}$ J/K
$G_{FireSat} = 1.0$
$\eta_{ground} = 0.75$

Find: Diameter of the receive antenna, D_{ground} and the ground transmitter's power, P_{ground}

Conceptual Solution

1) Find the communication wavelength

$$\lambda = \frac{c}{f}$$

2) Solve Equation (15-7) for the required ground receiver's gain, G_{ground}

$$G_{ground} = S/N \left(\frac{kB}{P_{FireSat} G_{FireSat}} \right) \left(\frac{4\pi R}{\lambda} \right)^2 T$$

3) Solve for the required receiver-antenna area and diameter

$$A_{ground} = \frac{G_{ground} \lambda^2}{4\pi \, \eta_{ground}}$$

$$D_{ground} = 2\sqrt{\frac{A_r}{\pi}}$$

4) Solve Equation (15-7) for the ground transmitter power

$$P_{ground} = S/N \left(\frac{kB}{G_{ground}} \right) \left(\frac{\lambda}{4\pi R} \right)^{-2} \left(\frac{T}{G_{FireSat}} \right)$$

Analytical Solution

1) Find the communication wavelength

$$\lambda = \frac{c}{f} = \frac{3 \times 10^8 \, m/s}{2 \times 10^9 \, Hz}$$

$$\lambda = 0.15 \, m$$

2) Solve Equation (15-7) for the required ground receiver's gain, G_{ground}

$$G_{ground} = S/N \left(\frac{kB}{P_{FireSat} G_{FireSat}} \right) \left(\frac{4\pi R}{\lambda} \right)^2 T$$

$$= 10 \left(\frac{1.381 \times 10^{-23} \, J/K \quad 2 \times 10^3 \, Hz}{(2W)(1.0)} \right)$$

$$\left(\frac{4\pi \; 4 \times 10^6 \, m}{0.15 \, m} \right)^2 (800 \, K)$$

$$G_{ground} = 12.397 \, (\text{unitless})$$

3) Solve for the required receiver-antenna area and diameter

$$A_{ground} = \frac{G_{ground} \lambda^2}{4\pi \, \eta_{ground}} = \frac{12.397(0.15 \, m)^2}{4\pi(0.75)}$$

$$A_{ground} = 0.03 \, m^2$$

$$D_{ground} = 2\sqrt{\frac{A_{ground}}{\pi}} = 2\sqrt{\left(\frac{0.03 \, m^2}{\pi} \right)}$$

$$D_{ground} = 0.194 \, m$$

Example 15-2 (Continued)

4) Solve Equation (15-7) for the ground transmitter power

$$P_{ground} = S/N \left(\frac{kB}{G_{ground}} \right) \left(\frac{\lambda}{4\pi R} \right)^{-2} \left(\frac{T}{G_{FireSat}} \right)$$

$$= 10 \left(\frac{1.381 \times 10^{-23} \text{J/K } 2000 \text{ Hz}}{12.397} \right)$$

$$\left(\frac{0.15 \text{ m}}{4\pi \, 4 \times 10^6 \text{m}} \right)^{-2} \left(\frac{800 \text{ K}}{1.0} \right)$$

$$P_{ground} = 2 \text{ W}$$

Interpreting the Results

Even with a relatively small transmit power and low-gain omnidirectional antennas, the FireSat ground station will only need a 20-cm diameter dish antenna to receive with a S/N of 10. The ground transmitter will need the same transmit power, 2 W, to achieve the same S/N for the uplink.

15.2 Mission Management and Operations

(with Julie Chesley, the U.S. Air Force Academy)

▣ In This Section You'll Learn to...

☛ Discuss important responsibilities of various management and operations teams during each phase of a mission

☛ Describe basic principles of team management and some useful management tools

☛ Discuss advantages of spacecraft autonomy

Recall from Chapter 1, the "sections" of the space mission architecture wheel include orbits and trajectories, spacecraft, launch vehicles, and mission operations systems. So far, our focus has been on these technologies. It's easy to glamorize a billion-dollar spacecraft and streamlined rockets thundering into the sky. But behind all this hardware are ordinary people like you and me: people who design the spacecraft, order the parts, assemble the components, keep track of the budgets, and ensure all the mission computers run. People make space missions happen.

In this section, we'll turn our attention to the people-centered activities of running a space mission. While technologies form the sections of the mission-architecture wheel, with the mission at the hub, mission management and operations pulls the circle together. It's where the "rubber meets the road" and the day-to-day work of the mission really gets done.

Collectively, *mission management and operations* involves all activities, from "cradle to grave," needed to take a mission from just an idea in someone's head to useful mission data on the owner's desk. This boils down to the formidable task of spending minimum resources (labor, money, and time) to achieve maximum result (data returned, company profit, or other success metric). One of the biggest challenges of mission management and operations is to control the space systems engineering process. Recall from Chapter 11, this process, shown again in Figure 15-20, takes us from a basic mission need, through definition of system and subsystem requirements and constraints, to eventually arriving at a completely designed, assembled, and tested space system. Along the way, mission managers must

- Define mission objective(s)
- Define mission requirements and constraints
- Define and derive system requirements and constraints
- Actively manage the "requirements loop" to trade mission and system requirements and constraints

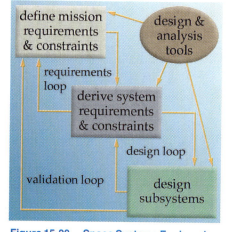

Figure 15-20.   Space Systems Engineering Process. We introduced the space systems engineering process in Chapter 11. One of the primary purposes of Mission Management is to "drive" this process to achieve the highest possible performance with the minimum risk and time.

- Create analysis and design tools needed to effectively trade among options

- Establish any necessary operations systems to support the mission

- Actively manage the "design loop" to trade requirements and constraints to develop a final design

- Oversee the "validation loop" to make sure we pay for what we want and we get what we pay for

All this must be done while juggling the often-competing demands of mission sponsors (bosses or customers) and mission implementers (spacecraft engineers, operators, project scientists, technicians, and support personnel). Mission managers and operators must carefully spend scarce resources—time, money, and people—while monitoring the eternal trade-offs among cost, schedule, and performance. How can they manage all this? By being skilled at managing large, sometimes geographically dispersed teams, while keeping track of a thousand and one technical and administrative details, all critical to the final product.

We'll start by looking at the various teams that make up a typical space project to see the challenges and responsibilities they face during each phase of a mission. Then we'll look at the challenge of managing the most precious mission resource—people. Finally, we'll return to the problem of managing these far-flung teams and some of the tools that are available for keeping large projects on track.

Mission Teams

If it's difficult to list all the operations systems that go into supporting a mission, then it's impossible to list all the teams of people who use these systems to get the mission off the ground. Some teams or individuals, such as the project manager and his or her team, may be intimately involved with nearly every detail of the mission from the initial concept through the end-of-life operations. While other teams or individuals, such as the launch team, may only show up to do their critical part for a short period, and then go off to support another mission.

Table 15-1 lists the major types of tasks that teams must perform throughout the mission lifetime and gives some examples. As you can imagine, during the various mission phases, some of these tasks require more attention than others. For example, during mission design and manufacturing, the focus is on systems engineering and system assembly, integration, and testing (AIT), with less emphasis on flight control (because the system isn't built yet). On the other hand, during the operations phase, AIT tasks are complete and the emphasis is on flight control. Other tasks, such as mission management, span the life of the mission.

As we did with operations systems in the previous section, here we'll divide the important personalities and responsibilities that make a mission possible into three basic teams—manufacturing, launch, and operations—and go through some of the important characteristics and functions for each.

Table 15-1. Major Mission Tasks and Examples. This table lists major tasks that mission personnel perform throughout the life of a mission and gives some examples of each. During various phases of a mission, some of these tasks receive more attention than others.

Mission Tasks	Examples
Mission Management	• Tracking and controlling a project's cost, schedule, and performance • Juggling money, time, facilities, people, and other resources • Managing teams and project morale
Mission Planning and Analysis	• Planning mission timelines and sequencing events • Analyzing trade-offs between competing technical options • Defining flight rules to govern actions during nominal and off-nominal flight conditions
Systems Engineering	• Defining and validating system and subsystem-level requirements • Applying analysis and design tools to define system architectures • Designing subsystems and constituent components
System Assembly, Integration, and Testing (AIT)	• Screening components for form, fit, and function • Assembling components to build subsystems and integrating subsystems to build systems • Testing subsystems and systems to ensure they perform under flight conditions
Simulations and Training	• Developing computer software to simulate major mission events • Practicing operational procedures using simulations
Flight Control	• Monitoring and interpreting telemetry to determine a spacecraft's health and status • Tracking a spacecraft's or launch vehicle's position and velocity • Sending commands to spacecraft to change operating conditions or fix problems
System Maintenance and Support	• Performing routine maintenance to clean rooms, thermal/vacuum chambers, and other operations systems • Updating ground software to enhance performance or correct problems
Data Processing and Handling	• Distributing mission data to users • Analyzing and archiving spacecraft engineering data

Mission Design and Manufacturing Teams

Broadly speaking, design and manufacturing teams get the mission started. They take it from simply stated user needs to a gleaming new spacecraft, with all the necessary operations systems, ready for launch. From the initial concept through flight readiness, the focus of the Manufacturing Team is on

- Systems engineering—Defining system and subsystem requirements and constraints and taking them through the complete design of the spacecraft and other operations systems

- Mission planning and analysis—Planning mission timelines and analyzing engineering performance data to determine operational scenarios and training requirements

- System AIT—Assembling subsystems from individual components, integrating the entire spacecraft, and performing environmental and functional testing

In Chapters 11–14 we presented the space systems engineering process and described its application to designing a spacecraft. As you can imagine, a large team of engineers, scientists, technicians, and support personnel must work together to translate basic user requirements into a fully designed spacecraft. The design team must work closely with the manufacturing team to ensure that their design can actually be built. *Design-for-manufacture* techniques focus on reducing overall mission costs by making hardware easier to fabricate and assemble. These efforts can support an overall program of *design-to-cost*, on which budget-strapped programs increasingly must focus.

In parallel with this spacecraft design effort, other teams must apply the same systems-engineering process to create critical mission operations systems, as discussed in Section 15.1. As we described in Chapter 11, the mission-operations concept drives the entire system design. This concept describes how users will get their data or use the mission services—what they need and when. To be most effective, the design and manufacturing teams must work hand in hand with the launch and operations teams, as we'll discuss next, to bring a real-world perspective to this process. For example, throughout the FireSat systems-engineering process, the whole team must work to refine how the spacecraft, ground controllers, and the various Forest Services will work together to conduct the mission. Many of their suggestions will be critical to the design of all the systems, from the spacecraft bus and payload, to the communication network, and the launch-site's ground-support equipment.

The second task that design and manufacturing teams focus on is mission planning and analysis. Part of this planning involves constructing a mission timeline. The *mission timeline* is a detailed script clearly defining all major events that occur during the mission, and when they must occur. From the rollout of the launch vehicle to the mission's end, the sequence of events must be carefully organized, to ensure that one action doesn't get ahead of another and cause trouble. For example, we'd like to keep the spacecraft running on ground power until the last possible minute before launch to save onboard batteries. The mission timeline would lay out exactly when in the countdown sequence to cut off external power.

Another important focus of planning and analysis is understanding personnel training requirements. Operators on the launch- and orbital-operations teams spend lots of time training. When it's time for launch they need to be intimately familiar with every aspect of the mission and what to do when something goes wrong. The best way they can do that is the same way a musician gets to Carnegie Hall—practice, practice, practice. For operators, practice involves complex dress rehearsals, known as *simulations*. During mission planning and analysis, the whole team must decide how they will train for the actual mission and define simulator requirements to support that training. Figure 15-21 shows Space Shuttle astronauts rehearsing procedures in one of the Shuttle simulators in Houston, Texas.

Figure 15-21. Shuttle Simulation. Years before their first flight, astronauts work through every phase of a mission, using simulators such as this motion-based simulator housed at the NASA Johnson Space Center in Houston, Texas. *(Courtesy of NASA/Johnson Space Center)*

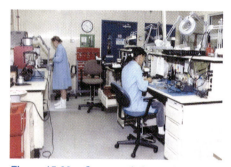

Figure 15-22.   Spacecraft Assembly. During assembly, skilled technicians, such as the ones shown here, perform delicate touch labor to manufacture circuit boards and other critical components. *(Courtesy of Surrey Satellite Technology, Ltd., U.K.)*

Figure 15-23.   Spacecraft Integration. Pulling together components into completed systems is part of the spacecraft assembly process. Here, an engineer performs final integration of the FalconSAT spacecraft. *(Courtesy of the U.S. Air Force Academy)*

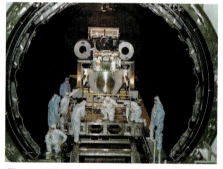

Figure 15-24.   Spacecraft Testing. From components to subsystems to entire spacecraft, testing is vital to ensure that everything will function in the harsh space environment. In this photograph engineers are preparing the Advanced X-ray Astrophysics Facility spacecraft for thermal/vacuum testing. *(Courtesy of NASA/Marshall Space Flight Center)*

While these training requirements for a small mission, such as FireSat, are relatively modest, for complex missions, such as a Space Shuttle Mission to service the Hubble Space Telescope, hundreds of operators must carefully rehearse with the astronauts practicing their job to ensure everyone can react to any contingency. Operators working with dedicated engineering teams must define hardware and software requirements to conduct simulations and plan coordinated training programs. For completely new missions, the process of designing, building, and testing complex simulators can take months or even years to complete.

Finally, after the mission team sets requirements and designs the systems, it's time to start "bending metal" and build something. In Chapters 11–14 we touched on some individual subsystem requirements for assembly, integration, and testing (AIT). AIT is a coordinated team effort by assembly technicians, system integrators, test conductors, and appropriate engineers and scientists, backed by administrative support personnel.

During assembly, skilled technicians, such as the ones shown in Figure 15-22, perform delicate touch labor to populate circuit boards and weave long wire harnesses. Throughout integration, they connect together individual components and subsystems to form the finished system. Figure 15-23 shows an engineering team performing final integration of the FalconSAT microsatellite.

Testing takes place throughout AIT, not just at the end. Technicians test individual components for functionality at each stage of the AIT process. After the entire spacecraft is integrated, highly trained test conductors take the spacecraft through dedicated test campaigns in vibration and thermal/vacuum facilities. Figure 15-24 shows test conductors preparing a spacecraft for thermal/vacuum testing. After AIT is complete, the mission design and analysis team performs one final, detailed review of the spacecraft status. During this formal process, sometimes called a flight-readiness review, senior engineers and managers go through technical data and test reports with a fine-toothed comb to ensure nothing has been over looked, all procedures have been followed (all i's dotted and t's crossed), and the spacecraft is finally ready for flight. With that review completed, the spacecraft gets shipped to the launch site, where the launch teams take over to prepare it for its ride into space.

Launch Teams

The launch teams' job starts long before the spacecraft arrives at the pad. Composed of hundreds or even thousands of people, they focus on getting the launch vehicle and its precious payload into orbit safely. While the tasks of the launch teams span the whole range of those listed in Table 15-1, their primary focus is on two major tasks

- System AIT—integrating the spacecraft to the launch vehicle and performing a final check-out
- Flight Control—monitoring the launch vehicle telemetry and trajectory and sending commands to make corrections as needed to deliver the payload to the promised orbit

Even before the spacecraft arrives at the launch site, the first of the launch teams is busy preparing clean rooms for the final spacecraft checkout, and assembling the launch vehicle and its payload interface. After the spacecraft arrives, a dedicated launch campaign swings into action. Usually consisting of lead engineers who've lived with the spacecraft from its conception through AIT, along with launch-vehicle experts, the launch-campaign team integrates the spacecraft with the launch vehicle, servicing it for flight (charging the batteries and filling the propellant tanks), and performing final tests and checkouts. After the launch teams finish these tasks and before lift-off, they begin loading millions of gallons of propellant into the launch vehicle.

Finally, everything is in place. The launch vehicle, with the spacecraft safely tucked inside the nose fairing, waits on the pad, and all that remains is to give the "go" for launch. At this point, the true business of flight control takes over. The flight-control team for launch (also called the launch-control team), composed of operators who monitor the launch vehicle and spacecraft systems, follow the lead of their *launch director*. In addition, dozens or even hundreds of support personnel run the tracking stations worldwide, ensure the range is clear of stray airplanes and ships, and keep a critical eye on the weather.

This last task can stop a launch more easily than a hardware problem or software glitch. Weather forecasters at the launch site carefully watch the skies for thick clouds, rain, or lightning, and they send weather balloons high above the launch area to measure upper-level winds. If these winds are too high, the launch vehicle may not survive the excessive dynamic pressure as it accelerates into orbit.

When everything and everybody is in place, the launch vehicle and spacecraft are happy and healthy, and the weather is cooperating, launch controllers take part in the final *launch-readiness review*. Similar to the flight-readiness review that happens before the spacecraft leaves the manufacturing facility, the launch-readiness review is a formal process that includes spacecraft manufacturers, users, and mission managers, who go through technical data, telemetry, test procedures, weather, and day-of-launch analysis before deciding to launch. Only when everyone on the team is completely satisfied that all risks have been minimized (they can never be completely eliminated), do mission managers give the final GO for launch.

3...2...1...Ignition...Lift-off! (Figure 15-25) While crowds of people applaud the thunderous launch, flight controllers are busy watching telemetry and tracking data to take action if something goes wrong.

One of the most attentive people monitoring this tracking data is someone with a thankless job—the *range-safety officer (RSO)*. The RSO sits with one finger poised over a button that can send a destruct signal to the launch vehicle as soon as it leaves the pad. Why would anyone do such a thing? Recall in Chapter 9 that we discussed the inclinations physically attainable from a given launch site. In Section 15.1 we discussed other constraints that further limit the orbits we can use from a certain site. Chief among these constraints is safety. Because most launch vehicles drop stages

Figure 15-25.   Titan IV Liftoff. At Vandenberg Air Force Base, California, a Titan IV launch vehicle roars off the launch pad, headed for space. *(Courtesy of the U.S. Air Force)*

on their way into orbit, it isn't safe to launch them over populated areas. That's why most launch sites are located along a coast, so the launch trajectories go over the water and avoid population centers. The RSO must carefully monitor a launch vehicle's trajectory into orbit and send commands to destroy it if it veers off course or threatens life or property.

If all goes well, the launch vehicle follows its planned trajectory, and the RSO doesn't have to push any buttons. But other flight controllers on the launch team keep busy during the entire launch. During Space Shuttle launches, the Booster Officer ("Booster") monitors the solid-rocket motors' performance through burnout and separation, as well as the main engines from ignition to cutoff, 8.5 minutes into the flight. Booster has to be ready to override a faulty engine sensor or other problem at a moments notice to avert a premature engine shutdown or catastrophic failure. The Flight Dynamics Officer ("FIDO") monitors the trajectory as carefully as the RSO and prepares to recommend abort options to return to the launch site or land at a downrange landing site in Africa, if performance is less than predicted. The Guidance and Procedures Officer, ("GUIDO") tracks the health of the onboard navigation and guidance system to ensure the Shuttle steers correctly into orbit and stands ready to assist the crew with off-nominal procedures. In addition to these flight controllers, dozens more in the control center monitor every possible subsystem and stand ready to recommend courses of action in case of problems.

After the launch vehicle is safely in orbit, the job of the launch team is done for that mission. Months and years of training must meet the test in a few minutes from lift-off to orbit. Then it's up to the operations team to take over and run the mission until it ends.

Operations Teams

Finally, after years, or even decades, of planning, designing, building, running simulations, and enduring the dramatic events of launch, the spacecraft is ready to do its job. Operating the spacecraft is the responsibility of the *mission-operations team*. Members of this team are also called *flight controllers*, the *flight-control team*, or simply, *operators*.

At the head of the flight-control team is a team leader, called the *operations director* (or *flight director* for Space Shuttle missions), who coordinates the input from other team members. The operations director sits in the "hot seat." He or she must make the final decisions on what to do throughout the mission.

Under the operations director, team members hold positions that follow the spacecraft's functional lines. *Subsystem specialists* are experts on individual parts of the spacecraft. For example, one person may monitor the electrical-power subsystem, while another watches the propulsion subsystem. The final members of the team are the *payload specialists*. They're responsible for the payload—its health, status, and operation. It's up to them to point cameras or antennas to collect valuable mission data. They process and deliver this data to users quickly and efficiently. Figure 15-26 shows some Shuttle mission operators at their consoles.

Figure 15-26. Mission Operators. Mission operators, shown here at the Space Shuttle's Mission Control Center in Houston, Texas, send commands to their spacecraft and receive health and status telemetry, as well as mission data. *(Courtesy of NASA/Johnson Space Center)*

Because the job of the operations team may last months or even years, at some point they get involved with nearly every major mission task described in Table 15-1. However, their main focus is on four key responsibilities

- Simulation and training—preparing for launch and on-orbit operations, as well as contingency procedures

- Flight control—monitoring the spacecraft telemetry and trajectory and sending commands to make corrections or other adjustments to deliver the payload data to mission users

- Data processing and handling—receiving, analyzing, storing, and distributing mission data to engineers and users

- System maintenance and support—maintaining and supporting all the hardware and software operations systems to keep the mission flying

Typically, the flight-control teams assemble months or even years before a flight. Until the launch, the team focuses on rehearsing for the mission by taking part in detailed simulations. During a simulation (or "sim" for short), devious trainers feed simulated mission data and anomaly scenarios to operators at their consoles. Throughout this training, the operators see almost every problem that could conceivably occur during the mission. By learning to deal calmly and efficiently with "worst on worst" cases, operators develop the skills and confidence to deal with routine anomalies that inevitably occur. Figure 15-27 shows an astronaut practicing in the Neutral Buoyancy Laboratory for an upcoming flight.

And training never ends until the mission ends. Especially for long-running mission operations, team members transfer in and out. As operators move on to other jobs, they must train their replacements and document how well their procedures worked. The simulations used before the mission started help in training new operators and rehearsing contingencies.

Operations teams really earn their wages performing flight operations. For Space Shuttle missions, the flight-control team headquartered in NASA's Mission Control Center in Houston, Texas, officially takes over after the Shuttle clears the launch tower. For other missions, the operations team may not officially assume responsibility until the launch vehicle places the spacecraft into the desired parking orbit. After the spacecraft arrives in its parking orbit, the operators must monitor the transfer to the final mission orbit, if required.

With the spacecraft safely in its mission orbit, the operators must verify that all subsystems are working normally during a period of on-orbit checkout, often called *commissioning*. During commissioning, operators carefully check each subsystem to determine if it can support the payload. Years before, operators and mission engineers worked out detailed procedures, called *flight rules*, telling them how each subsystem should work and what to do during any contingency to save the mission.

Figure 15-27.   Astronaut Practices. Shuttle astronaut, Tamara Jernigan, practices her on-orbit procedures floating in a deep pool of water to simulate free-fall conditions on orbit, while mission operators practice their duties in support. *(Courtesy of NASA/Johnson Space Center)*

Figure 15-28.   Telemetry Monitoring. For relatively routine missions, such as the Global Positioning System (GPS) spacecraft shown here, the operations team only needs to take a daily "snapshot" of telemetry to make sure everything is working correctly.

Figure 15-29.   Satellite Pour l'Observation de la Terre (SPOT). This small constellation takes Earth observations and sends them through numerous downlink sites to the operations center at Toulouse, France. *(Courtesy of French Space Agency CNES)*

After the spacecraft is fully commissioned and ready to start performing its mission, the operations team moves into normal operations. Normal operations is as close as a space mission ever gets to "routine."

For Space Shuttle missions, "routine" operations involve working around the clock to ensure mission performance and safety. Commands go directly to the Space Shuttle's computers, or sometimes, in the form of the Shuttle's teletype instructions, to the astronauts. For less complex and more mature missions, such as the Global Positioning System (GPS) shown in Figure 15-28, controllers may take only a "snap-shot" of telemetry from most of the satellites in the constellation and send commands once a day or so.

The third, and perhaps the most important task performed by the operations team (at least from the standpoint of mission users), is collecting, processing, and distributing payload data. (After all, the spacecraft was built and launched to satisfy a user's need.) To do this task, operators may need to oversee antenna pointing and the data traffic through communication satellites, or they may need to process large amounts of data sent from remote-sensing spacecraft, such as the Satellite Pour l'Observation de la Terre (SPOT), shown in Figure 15-29. For the first example, operator intervention may be fairly routine and requires very little work. For the SPOT example, operators may spend hundreds of hours processing volumes of data that rival the amount stored in the entire Library of Congress!

Finally, in addition to their training, commanding, and data-handling duties, the operations team must also maintain the operations systems that support them, such as the complex-communication networks that keep them in touch with their spacecraft. This effort involves routine maintenance at remote-tracking sites, upgrades to control-center hardware and software, and even new relay-satellite links. One vital member of the operations team that oversees all these systems on a day-to-day basis throughout the mission is the ground-systems specialist (called the Integrated Communications Officer or "INCO" on Shuttle missions). The *ground-system specialist* links the operations team and the spacecraft. This team member maintains the computers in the control center and ensures the complex communication network, which links the ground systems to the spacecraft, is connected and running. He or she also gathers data from the spacecraft and tracking sites, delivers it to the control center, and relays commands from operators to the spacecraft.

Mission Management

Notice in our discussion of design and manufacturing, launch, and operations we didn't specifically mention mission management. However, this task receives attention throughout the mission lifetime. Because of the vital importance of coordinating the efforts of hundreds or even thousands of individuals scattered around the world on various teams, while juggling billion-dollar budgets and national resources, we'll focus on two problems of team and project management.

Team Management

Although the technical aspects of designing and building space systems are extremely challenging, managing the people on the teams that do it creates other unique challenges. Project team leaders must ensure the team clearly understands the mission objectives, their individual roles in the success of the mission and the interrelationships among the various organizational tasks. Early on, team leaders must establish an effective communication and decision-making process that works in the specific project environment. Since before the Pyramids, many projects have failed, not because they didn't have the necessary resources, but because of poor communication or conflict between team members. Team leaders can address many of the potential problem areas by paying attention to the norms set for the team, the level of team cohesion, and the method of conflict resolution.

Team norms are standards of conduct that guide team member behavior. They typically aren't written down, but are understood throughout the team as accepted means of behaving. They can help a team by letting members know what is okay to do, and what is considered wrong. For example, on a design team, a norm could be that the group respects everyone's ideas, regardless of their title or expertise. This idea can be helpful to the team, because good ideas often come from people who have less experience on a project. The norm allows them to speak out and have their ideas considered.

The level of cohesiveness on a team can also enhance team effectiveness, because team members of highly cohesive teams are typically very committed to team success. Team leaders can enhance cohesiveness by establishing common goals and by increasing team interaction. Research has shown that if a team spends a lot of time together, and can agree on the goal they are working towards, typically the team becomes cohesive and committed to mission success.

Finally, the methods a team uses to manage conflict can also impact it's effectiveness. While a lot has been written about methods of conflict resolution, the important thing for a team leader to consider is that one style is not appropriate in all situations. Many times, the appropriate response is to compromise or accommodate others to reach an agreement. At other times, competition or collaboration is the preferred method. It's the team leader's job to help team members examine the issues and choose the appropriate resolution techniques.

Management Tools

As we've described, managing a team takes a combination of leadership, management, and psychology skills. But even the best leader or, most skilled manager, can be easily overwhelmed by the number of details generated by even a moderately sized project. For multibillion dollar international space programs, such as the International Space Station (ISS), shown in Figure 15-30 the number of details to track is staggering. Fortunately, astute project managers have a number of useful tools available in their kit to help them keep things on schedule and within budget. In this section we'll look at just of few of the most commonly used project-management techniques and the principles behind them.

Figure 15-30.   Big Project Management. For large multibillion dollar projects, such as the International Space station (ISS), shown here in an artist's concept, the number of details project managers must keep track of is staggering. To do this, they use a number of helpful management tools. *(Courtesy of NASA/Johnson Space Center)*

A large part of project management reduces to one thing—planning. Abraham Lincoln once said that if he had only six hours to chop down a tree, he'd spend five hours sharpening the ax. Detailed, thorough planning, done early, saves time, effort, and frustration later in the project. There are a variety of tools available to systematically plan nearly every aspect of a project. We'll start by looking at the work breakdown structure (WBS), then see how we can use the WBS for more detailed timeline planning using PERT and/or Ghant Charts.

Work Breakdown Structures. After we define the overall project objective—build FireSat within three years for two million dollars—we must specify the particular tasks that need to be accomplished. One tool we can use is the work breakdown structure. Simply put, a *work breakdown structure (WBS)* separates a project into manageable pieces for estimating what we need to do to complete the whole project. We can also use it to determine the resources required (people and dollars), as well as the time needed for each piece.

For example, at the project level, we can break the tasks for building FireSat into the following four major areas

1) Project Management

2) Systems Engineering

3) Subsystem Design and Fabrication

4) Subsystem and System-level Testing

We can refine each of these areas into sub-tasks and sub-sub-tasks until we've completely defined the project. Figure 15-31 shows an example of how we can look at one of the major project areas, AOCS, one of its sub-tasks, fabrication, and five of its sub-sub-tasks. To fully define the complete WBS for even a relatively small project, such as FireSat, would take considerable time and fill many pages. For very large projects, such as the multibillion dollar, DoD Milstar satellite system shown in Figure 15-32, the WBS alone fills volumes of documentation!

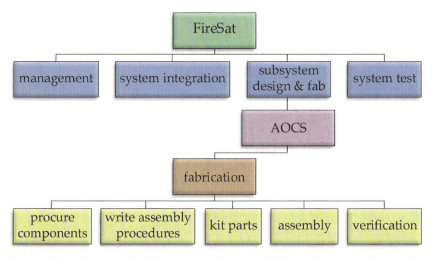

Figure 15-32. Milstar Satellite. For major space projects, such as the DoD Milstar satellite shown here, the work breakdown structure (WBS) becomes amazingly complex and fills volumes of documentation. But this level of detail is essential for efficient project management. *(Courtesy of the U.S. Air Force)*

Figure 15-31.   Example Work Breakdown Structure (WBS) for the FireSat Project. The WBS allows us to systematically divide an entire project into a set of major tasks, sub-tasks and sub-sub-tasks. For example, this figure illustrates how we can divide fabricating the attitude and orbit control subsystem (AOCS) into five sub-tasks. We could separate each of these sub-tasks even further until we've defined the entire project in detail.

After breaking down all the activities that comprise the project, our next step is to determine how long each activity will take. While there are a number of ways to estimate task duration, prior experience is our best gauge. If we have done a task before, we have a much better idea of how long it will take in the future. When historical data is not available, estimating techniques are available to help us with task-duration estimates.

Network Modeling. After we have decided on the activities and have a reasonable estimate of how long each activity will take, we can begin scheduling the tasks. Although there are a number of popular methods for network scheduling, the *Program Evaluation and Review Technique (PERT)* and the *Critical-path Method (CPM)* have received widespread use in project management. Despite conceptual differences between these two methods (based primarily on the method used to estimate activity durations), for practical purposes, most managers use them interchangeably. The principal benefits of these methods are that they

- Allow us to visually examine the interrelationships among the activities

- Show us which activities we can do simultaneously

- Help us focus on the tasks that are most critical to completing the project on time

To begin the CPM scheduling process, we need to look at each of our activities (developed in the WBS) and determine logical relationships among them to determine precedence requirements by asking ourself these questions

- Which activities should we finish *immediately before* we start this activity?

- Which activities can we do *at the same time* as this activity?

- Which activities *can't we start* until after finishing this activity?

Table 15-2 gives an example of simplified precedence requirements for the fabrication step in developing the FireSat attitude and orbit-control subsystem, as well as the approximate duration of each activity.

Table 15-2.   FireSat Project Data. As part of the Critical-path Method of project management we need to list each activity in the WBS, determine which ones come before it, and estimate the task duration.

Activity	Predecessors	Duration (Months)
A – Procure components	None	6
B – Kit parts	A	4
C – Assemble sun-sensor components	B	6
D – Assemble GPS components	B	9
E – Verification	C, D	3
Total		28

Table 15-2 indicates that Activity E (Verification) will take three months to complete and cannot start until activities C and D are completed. This analysis would seem to tell us that it will take 28 months to complete this part of the project. But realize this analysis assumes that each activity is done one at a time. How many total months will the project take to complete if we can do some of the activities at the same time? Also, how do we determine what tasks we should give priority to, so the entire project won't be delayed?

Drawing a network diagram can help answer these more complex questions, by illustrating project data graphically. Figure 15-33 shows a network diagram for these FireSat tasks.

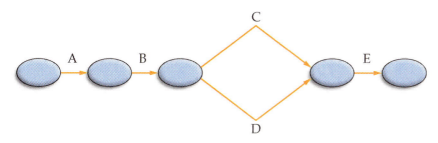

Figure 15-33.   FireSat Network Diagram. This figure shows the sequencing of activities for the FireSat attitude and orbit-control subsystem.

To draw a network diagram, we imagine the project as a "roadmap" of predecessor and successor activities. The A and B roads (known as arcs) each begin at a common starting point (called a "node," which denotes the beginning or ending of an activity). Activity C must begin at the same node where activity B ends, because it is a successor to B. Activity E must begin at the node where C and D end, because C and D must be complete prior to starting E. More complex programs require many more diagrams, but we draw them similarly, taking great care to insure the network reflects the information in the project's data table.

Drawing the project activities in a network diagram simplifies the remaining analysis considerably and helps provide us with valuable information such as: when we can schedule each activity, how much we can delay an activity (without affecting the overall completion time), and the earliest-start and finish times for each activity. We can compute these times using the CPM rules for the "forward pass" and "backward pass."

The forward pass finds the earliest-start (ES) and earliest-finish (EF) times for each activity by moving from left to right on a project diagram. All activities that do not have a predecessor (in our example, activity A) can start immediately, so they have an earliest-start time of zero. Then we add the project duration times to the project-start time and that sum is the earliest-finish time for that activity. In our example, the earliest-finish times for A and B are 6 and 10 months respectively.

The earliest-start time (ES) for subsequent activities should use the longest, earliest-finish (EF) time coming into the node. For example, the earliest-start time for activity E is 19 months, because activities C and D must be complete before E can start (the EF for C is 16, the EF for D is 19). By completing the forward pass, we find that the project duration is only 22 months and not the sum of the durations of all the activities (28 months).

We now know the earliest we can expect to start and finish each of the activities, however, we may also need to know what is the *latest* time we can start and finish each activity without delaying the overall project. We obtain such information using the backward pass. The backward pass is the opposite of the forward pass—it works from right to left, as we find the latest-start (LS) and latest-finish (LF) time for each activity. To begin the backward pass, we start at the last node on the network diagram that identifies the project duration of 22 months. This time is the latest we can finish the final activities and not delay the project. So in our example, the latest-finish for activity E is 22 months. To find the latest start, we simply subtract the activity duration (3 months for E) from the latest-finish time (22 months). Therefore, the latest-start for activity E is at 19 months (22-3). Figure 15-34 shows the earliest-start, earliest-finish, latest-start, and latest-finish times of all activities for the FireSat attitude and orbit-control subsystem example.

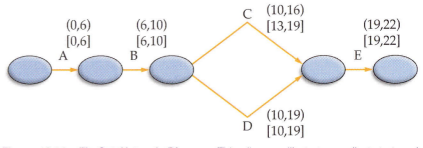

Figure 15-34. FireSat Network Diagram. This diagram illustrates earliest-start and earliest-finish times (in parentheses) and latest-start and latest-finish times [in brackets] found by using the forward-pass and backward-pass approaches. For this example of the FireSat attitude determination and control subsystem, all tasks proceed in series, with the exception of C and D which proceed in parallel. Of these, only task C can slip as much as six months without delaying this entire part of the project.

Now that we know the earliest-start, earliest-finish, latest-start, and latest-finish times for each activity, as well as the project duration, the next question is what activities are the most important to monitor to prevent project delays? The answer to that question is we should monitor the activities on the *critical path*, activities that we can't delay without delaying the project. Any activity on the critical path will have zero slack. *Slack* is the amount of time we can delay an activity without delaying the overall project. By identifying the activities that have zero slack (A, B, D, E), we have identified the critical path for this part of the project. We may delay all other activities (in this case, only activity C) by the amount of slack that they have, without serious effect on the overall project.

In this section, we have illustrated some of the uses of network modeling. One of the more powerful aspects of CPM is the ability to accurately control a project by shifting resources as needed. If one activity is taking longer than expected, we can find another activity that has slack and move resources from the one activity to the other to keep it on schedule. Although this example is simple, this modeling technique is especially useful on large, complex projects. For example, the purchase, construction, and launch of DoD satellites involve literally thousands of activities—all of which program managers must track to prevent costly delays.

Project Control. One of the more challenging questions we can ask a mission manager in the midst of a complex project is, "how is it going?" Without some previously developed performance indicators, the best answer we may get is a gut feeling, but "gut feeling" doesn't go over well with worried stock holders or mission sponsors. To address this, the mission manager must develop a control plan, one that highlights problem areas and gives the team a sense of their progress. Table 15-3 lists key characteristics for effective program performance indicators.

Table 15-3.   Characteristics of Effective Program Performance Indicators.

Aligned with goals	Performance indicators need to tie to project objectives so that our measures are only measuring mission-critical areas
Simple and easy to understand	We can't act on information if we don't understand what a measure is telling us about the project
Actionable	We must do something about the measure, if there is a deviation, otherwise, the information is of little use
Timely	The indicator must provide information in enough time for us to make changes, if there is a problem
Flexible	Measures should be changeable, if necessary, depending on external circumstances

While there are several indicators that we can use to control a project, one indicator that is gaining wide attention in government and industry projects is called earned value. Basically, *earned value* tells us the value of the work that is completed at any time. For example, let's assume we budgeted $500,000 and 24 months for developing the first FireSat spacecraft. If, after 12 months (half the time allocated for the task), we find that we've spent $250,000 using only schedule and budget data, we can assume we are tracking perfectly—half the time spent should equal half the budget spent. However, what if we are only 25% complete with the spacecraft? Schedule and budget numbers alone won't tell us that. Somehow we need to compare the resources spent to the progress made. Earned value lets us see whether the work done is keeping up with the spending rate. Computing earned value involves converting the percent complete on each activity to a dollar amount by multiplying the total budgeted cost for the activity with the percent complete. This measure now gives us a better idea of our progress, in addition to tracking costs and schedules.

Spacecraft Autonomy

This overview of mission management and operations shows that many people work to get a spacecraft off the drawing board and into space. Supporting this army of people, along with the intricate communication network that ties them together, is expensive. It's easy to see why operations alone often account for as much as one-third of a large program's cost. For this reason, engineers and operators are constantly looking for new ways to streamline operations and cut costs.

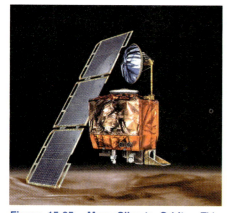

Figure 15-35.   Mars Climate Orbiter. This interplanetary mission ended abruptly when operators sent an erroneous command to fire the orbital-insertion rocket engine too long. The spacecraft was never heard from again and presumably burned up in Mars' atmosphere or crashed into the surface. *(Courtesy of NASA/Jet Propulsion Laboratory)*

One way to do this is to place more functions onboard the spacecraft, reducing the need for costly operations-team members. Mission *autonomy* refers to the ability of a spacecraft to perform some or all of its functions without human intervention. Commercial missions, ever focused on the bottom-line budget, turn to autonomy as a way to cut costs by decreasing the number of control centers and operators.

For certain missions, some degree of autonomy is essential. Interplanetary missions, for example, must operate with long time delays because of the extremely long distances. Their onboard software needs to deal with any contingency without waiting for advice from Earth-bound operators. Finally, there is another strong argument for increased spacecraft autonomy. Human errors have been a source of mission failure. The Russian Phobos mission and the NASA Mars Climate Orbiter (Figure 15-35) are two prominent examples of human error leading to a mission's end.

Increased onboard computer power, coupled with the desire to drive down costs while enhancing mission reliability, all point toward greater autonomy on spacecraft and even on launch vehicles. Someday, as one NASA engineer suggested, we may send probes on their merry way to the planets with simple instructions to phone home if they find anything interesting!

▦ Section Review

Key Terms

autonomy
commissioning
critical path
Critical-path Method (CPM)
design-for-manufacture
design-to-cost
earned value
flight-control team
flight controllers
flight director
flight rules
ground-system specialist
launch director
launch-readiness review
mission management and
    operations
mission-operations team
mission timeline
operations director
operators
payload specialists
Program Evaluation and Review
    Technique (PERT)
range-safety officer (RSO)
simulations
slack
subsystem specialists
team norms
work breakdown structure (WBS)

Key Concepts

➤ Mission Management and Operations involves all activities from "cradle to grave" needed to take a mission from concept, through launch, through mission end. Major mission tasks throughout the mission life are listed in Table 15-1 with examples. Different teams focus on different tasks during the mission

- Mission-design and manufacturing teams focus on system engineering, mission planning and analysis, and system assembly, integration, and testing (AIT)

- Launch teams focus on system AIT (integrating the launch vehicle and spacecraft) and flight control (during countdown and launch)

- Operations teams consist of an operations director (called a flight director for the Space Shuttle) supported by subsystems, payload, and ground-system specialists. Their focus is on simulation and training, flight control, data processing and handling, and system maintenance and support.

➤ Mission management covers the personnel and project control tools needed to get a mission started and keep it moving on schedule and within budget

- Team management includes

    – Establishing effective team communication and decision-making processes

    – Developing the norms for the team, the level of team cohesion, and the methods for conflict resolution

- Management tools include

    – Work breakdown structures—systematic hierarchy of system elements

    – Network modeling—determining the relationship between elements in the work breakdown structure. This includes the Critical-path Method (CPM) which helps managers see which tasks are most critical for keeping a project on schedule.

    – Project control involves tracking effective performance indicators, described in Table 15-3

➤ Spacecraft autonomy refers to having routine and off-nominal decisions made by spacecraft using onboard software, instead of by ground-based operators, to decrease cost, reduce human error, and provide for more rapid response

References

Boden, Daryl G. and Wiley J. Larson. *Cost-Effective Space Mission Operations*. New York, NY: McGraw-Hill, Inc., 1996.

Feldman, Daniel C. and Hugh J. Arnold. *Managing Individual and Group Behavior in Organizations*. New York, NY: McGraw-Hill, Inc., 1983.

Hackman, J. Richard. (ed. M. Dunnette) "Group Influences on Individuals." *Handbook of Industrial and Organizational Psychology*. New York, NY: John Wiley & Sons, 1976.

Morgan, Walter L., Gary D. Gordon. *Communications Satellite Handbook*. New York, NY: John Wiley & Sons, 1989.

Pratt, Timothy and Charles W. Bostian. *Satellite Communications*. New York, NY: John Wiley & Sons, 1986.

Rockwell International Space Systems Group, Space Shuttle System Summary, 1980.

Shaw, M.E. *Group Dynamics*. 3rd ed. New York, NY: McGraw-Hill, Inc., 1981.

Thomas, Kenneth. (ed. M.D. Dunnette) "Conflict and Conflict Management." *Handbook of Industrial and Organizational Behavior*. New York, NY: John Wiley & Sons, 1976.

Mission Problems

15.1   Mission Operations Systems

1 Describe what goes into the mission operations systems. Why do we say mission operations is often in the background for a space mission?

2 What's the purpose of a clean room during spacecraft manufacturing? What does it mean to have a Class 10,000 clean room? Is that cleaner than a Class 1000 clean room?

3 List and describe four types of test facilities used for spacecraft integration and tests.

4 What four pieces make up a launch complex that prepares and boosts a payload into orbit?

5 For a space mission, what communications must take place? Describe the four pieces of a Space Communication Architecture.

6 For space missions, what are real-time operations? If real-time communication isn't possible, what are the alternative types of operations?

7 What is a spacecraft "safe" mode and how can it save a mission?

8 What four conditions must be compatible for two people or two spacecraft to communicate? What does it mean to have a signal-to-noise ratio greater than 1.0?

9 Describe the steps a radio signal must take to get from a radio station to your car speaker in a form you can enjoy?

10 Compute the average power-flux density from the Sun at Mercury's orbital radius, 57.9×10^9 m. The total radiance from the Sun is 3.826×10^{26} W.

11 How does a directional antenna increase the radiated power over an omnidirectional antenna? Define the gain of an antenna. How efficient is a parabolic antenna?

12 What does an antenna's effective isotropic radiated power depend on?

13 Satellite A and B have the same communication equipment (transmitter, antenna, etc.) onboard to communicate with the same ground station. If satellite A is in an orbit 1000 km higher than satellite B, which will have a poorer signal-to-noise ratio? Why"

14 Engineers plan to use a 2×10^9 Hz link with a 300 Hz bandwidth to communicate with a remote-sensing spacecraft in an orbit at 400 km altitude. The transmitter-antenna gain is 300 and the receiver-antenna diameter is 30 m. Transmitter power will be 13 W. If the receiver temperature is 300 K, compute the signal-to-noise ratio for the uplink signal. Will this be an effective link?

15 List four ways to increase the amount of transmitted-signal strength that a receiver collects.

16 Where does noise come from in a communication signal?

17 List three ways to increase the signal-to-noise ratio in a communication signal.

18 Why would it be a bad idea to use the U.S. Air Force's Satellite Control Network to uplink commands to our FireSat satellite? Could we ask the Space Surveillance Network to track our 0.3 m × 0.3 m × 0.3 m spacecraft?

15.2   Mission Management and Operations

19 To create a completely designed, assembled, and tested space system, what eight things must mission managers do?

20 List eight operations tasks that must be performed during the life of a space mission.

21 What is design for manufacture? How does design-to-cost help mission managers plan a space mission?

22 Describe a mission timeline and how managers use it to script a space mission.

23 What tests lead to the flight-readiness review? What happens at this review?

24 Describe the roles of the launch director and range-safety officer in getting a launch vehicle with its spacecraft safely into orbit.

25 What positions comprise an operations team? Which one takes charge of the team for the Space Shuttle?

26 Describe the job of an operations-team leader. What tools are available to assist them in their management tasks?

27 List three benefits to network scheduling methods, such as the Program Evaluation and Review Technique and the Critical-path Method.

28 What five characteristics of effective program control help the program manager gain confidence with owners and users that a successful mission is underway?

29 For the FireSat Network Diagram, Figure 15-33, why did we need the earliest-start and earliest-finish times? How and why did we compute the latest-start and latest-finish times?

30 How does spacecraft autonomy save mission costs? Why aren't all spacecraft totally autonomous?

For Discussion

31 Why are mission operations so expensive? Suggest some ways to decrease their costs.

32 What type of communication network would we need to support the human exploration of Mars? Describe the number and type of elements needed and the various links.

Mission Profile—Apollo 13

Apollo 13 was the thirteenth in the Apollo series missions beginning with Apollo 1 in January, 1967, and ending with Apollo 17, December, 1972. Previous Apollo missions successfully accomplished many "firsts," including the first lunar orbit, and first manned lunar landing. Apollo 13 was planned as the third lunar landing attempt.

Mission Overview

This mission was planned with the primary objectives of exploring the Moon, surveying and sampling the Imbrium Basin, deploying and activating the Apollo Lunar Surface Experiments Package (ALSEP), further developing human's capability to work in the lunar environment, and photographing candidate exploration sites. The mission was abandoned due to a rupture in the service module's oxygen tank.

Mission Data

✓ Launch: Saturday, April 11, 1970, at 13:13 CST

✓ Crew: James A. Lovell, Jr., John L. Swigert, Jr., and Fred W. Haise, Jr.

✓ Milestones: 46 hours 43 minutes Joe Kerwin, the CapCom on duty, said, "The spacecraft is in real good shape as far as we are concerned. We're bored to tears down here."

✓ 55 hours 55 minutes Oxygen tank No. 2 exploded, causing No. 1 tank also to fail. The Apollo 13 command module's normal supply of electricity, light, and water was lost.

✓ Astronauts used the lunar module as a lifeboat in space

✓ Astronauts and ground support manually navigated Apollo 13 to a safe re-entry

✓ Landing: 17 April, 1970

✓ Mission Duration: 142 hrs 54 mins 41 s

The Accident Review Board concluded that wires which had been damaged during pre-flight testing in oxygen tank No. 2 shorted and the teflon insulation caught fire, causing the explosion. With the oxygen stores depleted, the command module was unusable, so the mission had to be aborted. The crew transferred to the lunar module (named Aquarius) and powered down the command module (named Odyssey) until it was time to use the re-entry capsule for landing.

Mission Impact

Although none of the primary mission objectives was accomplished, the Apollo 13 mission can be called a "successful failure." It was the first in the Apollo Program requiring an emergency abort. The excellent performance of the lunar-module system in a back-up capacity and the training of the flight crew and ground-support personnel resulted in the safe and efficient return of the crew.

Damaged Apollo 13 Service Module. This photo shows the side panel blown away from the Service Module. *(Courtesy of NASA/ Johnson Space Center)*

Contributor

Nathan Kartchner, the U.S. Air Force Academy

References

NASA/Goddard Space Flight Center. National Space Science Data Center website. http://nssdc.gsfc. nasa.gov/cgi-bin/database/www-nmc?70-029A.

NASA/Johnson Space Center. Apollo 13 website. http://cass.jsc.nasa.gov/pub/expmoon/ Apollo13/Apollo13.html.

NASA/Johnson Space Center. Images website. http:/ /images.jsc.nasa.gov/iams/html/pao/as13.htm

NASA/Kennedy Space Center. Apollo 13 website. http://www.ksc.nasa.gov/history/apollo/apollo-13/apollo-13.html.

NASA/Langley Research Center. Abstracts website. http://lava.larc.nasa.gov/ABSTRACTS/LV-1998-00042.html.

Retrieving and repairing broken satellites for later reuse is one of many potential space businesses of the future. *(Courtesy of NASA/Johnson Space Center)*

Using Space

16

Wiley J. Larson
the U.S. Air Force Academy

▤ In This Chapter You'll Learn to...

- ☛ Appreciate the balance between the political, economic, and technical dimensions of space missions
- ☛ Explain current trends in government and commercial space activities
- ☛ Discuss the political reasons that nations pursue space activities, and the legal and regulatory environment for these missions
- ☛ Discuss the economic factors that drive space missions and affect their cost from beginning to end

▤ You Should Already Know...

- ❏ Nothing. You need no specific tools to understand this chapter, but it will make more sense if you're familiar with the rest of the book.

▤ Outline

16.1 The Space Industry
Globalization
Commercialization
Capital Market Acceptance
Emergence of New Industry
    Leaders

16.2 Space Politics
Political Motives
Laws, Regulations, and Policies

16.3 Space Economics
Life-cycle Costs
Cost Estimating
Return on Investment
The FireSat Mission

There is just one thing I can promise you about the outer space program—your tax dollar will go farther.

Werner VonBraun

So far, our primary focus in this book has been on the technical aspects of space. We've looked at each element of the space mission architecture and how they're designed and integrated into a cohesive whole. But technology, no matter how advanced, is only part of the story. In the real world, two other factors—politics and economics—can be just as important (and often more important) in getting a mission off the ground. If you imagine the space mission architecture, as shown, to be the "wheel" around which a mission turns, economics provides the power to turn that wheel, and politics determines the direction it will go. Technology, economics, and politics form three legs of a space mission triad, illustrated in Figure 16-1. Like the legs of a three-legged stool, any mission must have each of these, of equal strength, to have a firm base to build upon.

In this chapter, we'll take a step back to look at the big picture of space missions from the perspective of government and industry. Our goal is to understand some of the key forces that drive the space industry and how they're related to current trends. For NASA and DoD missions, politics, including international politics drives much of what is done. But more and more, economics, or cost-effectiveness, determines what missions governments can do and how they're done. Technology facilitates these missions, and often, new technology helps make missions possible or more cost-effective. Unlike government, industry is in business to make a profit. Industry will use whatever technology it can to achieve that goal, but industry is subject to political and economic pressures, as well, that must be carefully factored into any sound business plan.

We'll start by analyzing important trends in space and how they're shaping the face of the industry. Then we'll turn our attention to space politics to see why countries pursue space activities and how they control them through laws, regulations, and policies. Finally, we'll look at the bottom line of space—economics. We'll review important factors that drive up the cost of doing business in space and the issues that current and future space entrepreneurs must consider.

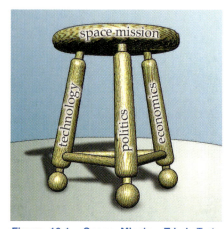

Figure 16-1.   Space Mission Triad. Technology, economics, and politics form the three legs of the space mission triad. Any mission must have strong support on all three legs to stand.

16.1 The Space Industry

In This Section You'll Learn to...

☞ List and describe emerging trends in the space industry today

☞ Explain important markets for commercial space activities

Space has become an integral part of all our lives. Space assets form an invisible infrastructure that most people take for granted. Recent surveys show that the average person in the U.S. uses space assets about nine times per day, usually without realizing it! Space provides television, radio, weather information, location or navigation capabilities, information and pictures for newspapers, internet and web sites, and telephony to name a just a few.

The way we perceive these space activities has changed dramatically over the last 40 years. Before 1957, space travel was something that only a few people, science fiction writers, and technologists, even dreamed about. The launch of the tiny Sputnik spacecraft, which did nothing but broadcast electronic "beeps," stunned the world and changed our perception of space forever. At that time, going into space was an experimental journey into the great unknown. It was truly a mysterious entity and we had a lot to learn about its environment and the technology needed to get us there.

Looking at a snapshot of 1969, when the manned space program was at its climax, reveals a totally different picture after only 12 years of effort. Primarily two nations, the former Soviet Union and United States, were in a struggle to see who would be preeminent in space. Space enjoyed special status, and funding for space projects was plentiful. The result was the rapid development of propulsion, electronics, materials, and other technologies that made grandiose missions such as Apollo and the Mariner planetary probes possible. But spacecraft were hand built, one at a time using technologies not widely available, thus space missions were still an expensive undertaking and remained the exclusive domain of the super powers.

Thirty years later, at the dawn of a new millennium, many people think space is a routine endeavor. In addition to many hand-built systems, constellations of mass produced spacecraft, using off-the-shelf hardware, exist. Computer processing capability and memory is readily available for space systems. Many countries have space-related capabilities, and there are a wide variety of launch options available. International cooperation is a necessity, and competition and profit drive many of the activities. Many people dream of careers in space-related areas. Over the span of about 40 years, space activities have blossomed.

As of the year 2000, there were about 500 functional spacecraft orbiting Earth. But the demand for space services continues to grow and plans for many new systems are in various stages of development. By some predictions [*Via Satellite*], this number could double or triple by 2010!

What will all these missions do? Who will build them? Who will pay for them? According to projections, the majority of these new spacecraft will be for telecommunications, and the U.S. industry and government will sponsor about 60% of them. But what about the rest? To understand these we need to look deeper into the state of the global space industry.

In his draft manuscript, *Space Power* [Sullivan, 1998], Dr. Brian Sullivan identifies several recent trends that provide insights to our next decade in space. These trends include

- Globalization
- Commercialization
- Capital market acceptance
- Emergence of new industry leaders

In this section, we'll examine each of these trends in turn, to see how they will shape the face of the 21st century space industry.

Globalization

Space has always provided a global perspective of Earth. But during the early days of the Space Age, that perspective was only available to a few super-power nations. Today, space is truly a global activity. This proliferation of space activities to many countries around the world is called *globalization*. More than 20 countries now have active national programs related to developing space infrastructure, with the United States, Europe, Russia, China, and Japan leading the way.

As of 2000, the U.S. Government or industry owned about 50% of the existing spacecraft in orbit. Ten years from now, when the total number of spacecraft has potentially tripled, that fraction is expected to decline to 35% or 45%, due to the growth of international space activities.

A number of factors have fueled this globalization of space. Increased accessibility of space technology, caused by the explosion in the microelectronics industry, has made the task of designing and building a spacecraft far cheaper and simpler than it was 30 years ago. This accessibility, coupled with the wide availability of relatively low-cost launch opportunities, has made it possible for almost any country to undertake their own dedicated space program for a relatively small investment. Figure 16-2 shows FASat-A, the first national spacecraft of Chile.

In addition to their own indigenous programs, many developing nations have become significant purchasers of space-related products and services such as space-based telecommunications systems and remote-sensing data. Emerging markets in Central Europe, Russia, Africa, South America, and the Pacific Rim represent significant opportunities for the space-based telecommunication industry. These opportunities have led to a number of firms expanding internationally through mergers, acquisitions, and strategic-partner arrangements.

Figure 16-2.   Chile in Space. The increased globalization of the space industry has made it possible for developing countries to undertake their own dedicated space programs. The FASat-A micro satellite, shown here with its project team, is the first national spacecraft of Chile. *(Courtesy of Surrey Satellite Technology, Ltd., U.K.)*

Commercialization

Twenty years ago space was primarily a government's endeavor. Today, the trend slants more toward commercial ventures—*commercialization*. By the end of 1997, a total of about $100 billion had been spent on commercial space activities since their inception. But this number is growing rapidly. Experts predict that private industry will invest an estimated $125–$150 billion more over the next three to five years. [*Via Satellite*, July, 1999] Between the years 2000 and 2010, they expect continuing high levels of profit to bring an estimated $650–$800 billion in revenue to the space industry worldwide. By 2010, cumulative U.S. corporate investments in space alone could reach $500–$600 billion. That same year, revenues from global commercial space activities are estimated at $500–$600 billion. By 2020, the U.S. space industry could be producing 10%–15% of U.S. Gross Domestic Product. We'll have to wait and see.

In comparison, in 1996, global government (civil and military) space expenditures totaled about $50 billion, about 70% of that was U.S. Government, civil and military, space-funding profiles are interesting to note. NASA's funding has been relatively stable and is projected to remain at $13–$14 billion in present dollars, per year, over the next five to ten years. From Congressional estimates for U.S. defense funding, the U.S. military is expected to spend about $35–$40 billion per year on space. Wow! This means that the total funding by the U.S. Government could be as much as $54 billion per year on space activities in the 2010 timeframe, while, the U.S. space industry is projecting revenues of 10 times that amount during the same period!

The key point here is that commercial spending on space is beginning to overshadow government spending. Industry will lead the way and the government will become a follower in many applications. Obviously, the government will still be the key player in national security-related technologies and systems, but much of the infrastructure for space will become commercially driven.

Part of the reason for this trend is government policies toward industry and space activities. The global trend toward *deregulation* of the telecommunications industry has created a large number of new competitors, services, and markets for the space industry. Additional space-related opportunities come from commercializing many traditional government-run space activities. For example, Europe has established private marketing organizations for launch vehicles (Arianespace) and remote-sensing spacecraft data (Spot Image). In the U.S., much of the public-domain remote-sensing data is virtually free for the taking, while Europeans establish clearinghouses for data that they sell to the public. In the U.S., government-owned national launch ranges are now licensed to private concerns, and many suppliers of defense-related space infrastructure, who formerly sold exclusively to the government, are now permitted to compete commercially. Mission operations for the Space Shuttle and many unmanned spacecraft missions are gradually being handed over to industry on a commercial basis.

Another reason for the increased commercial space activities is that space activities for their own sake are becoming profitable. The *global space industry* can be divided into four primary market areas

- Space-related services (the real money maker)
- Spacecraft design and manufacturing
- Space operations systems design and manufacturing
- Launch services

Of these, space-related services is by far the most lucrative—and growing—area. In 1997 alone, space services generated $19.4 billion in revenue, compared to $13.5 billion, $11.0 billion, and $7.5 billion for the other three areas respectively. Let's look at three of the most important areas of space-related services, telecommunications, navigation, and remote sensing.

Telecommunications currently is the largest portion of commercial space activities. This area was the first and continues to be the most significant money maker for the space industry. Hundreds of government and commercial organizations worldwide own, operate, and use space systems to support a variety of critical communication needs. Spacecraft, as an integral part of the world's telecommunications infrastructure, provide long-distance data transmission, television broadcasting, telephony, and other services. In the developing world, spacecraft are delivering basic telephone service to millions of people for the first time. These countries are using this communication boon to support rapid economic growth. In the U.S. and Europe, space-based assets are enabling exciting new services, such as personal communications systems, distance learning, tele-medicine, and private networks.

Figure 16-3.   Commercial Communication Missions. The Iridium satellite, shown here, is part of a constellation of commercial space-craft providing worldwide telephone services. *(Courtesy of Motorola, Inc.)*

Competition is fierce and the risks are high in the telecommunications industry—vast fortunes are at stake. Corporations that invested in ground-based infrastructure compete fiercely with other companies, trying to provide basically the same services using space-based systems. This competition is evident in the mobile-telephone market and in the distribution of television and radio programming. Figure 16-3 shows an Iridium satellite that forms part of a constellation of small satellites competing to provide worldwide telephone services.

Navigation is the next most profitable space-related service area. This entire market niche has been a direct spin-off of the *Global Positioning System (GPS)*. The GPS allows a user with a small handset to receive highly accurate location, velocity, and time information, virtually anywhere on or above Earth. Industry is finding ways to use this basic GPS information combined with other value-added capabilities to increase profit in traditional areas and to generate revenue from new markets. For example, shipping companies are using GPS to locate inventory being transported by ship, plane, and truck worldwide, streamlining their tracking systems. Automobile manufacturers are now offering GPS systems to provide trip tracking and emergency response assistance as an additional option, along with stereos and sport trim, as shown in Figure 16-4.

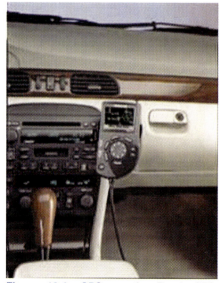

Figure 16-4.   GPS on the Road. GPS receivers are becoming integral parts of our lives. Here we see a GPS system incorporated into a car to keep the driver from getting lost. *(Courtesy of Magellan Corporation)*

The concept of "value added" is especially important in the space-related service area. Imagine that you are on the golf course and someone hands you a GPS receiver. You instantly know your exact location. This, by itself, wouldn't help your game, but what if you could program in the locations of all 18 holes? Then you could get an exact measurement of the distance and direction to the next hole—making it easier to select the right golf club for the shot. Going further, if the GPS receiver incorporated a wind indicator providing speed and direction, you could better estimate how to adjust the direction to hit the ball. Value-added features are a very powerful money maker that can be added to basic services, such as GPS, making them more accessible and useful to Earth-bound users.

The commercial exploitation of remote-sensing data by adding value to public domain, governmental data has also proven to be useful and profitable. Obviously, military remote-sensing missions launched for national security reasons do not have to be commercially viable. Other remote-sensing missions, such as NASA's Earth Science Enterprise, generate incredible amounts of environmental data, paid for by U.S. taxpayers. Companies have created profitable markets by manipulating this public domain data in various ways to provide a variety of useful products and services for customers around the world.

Over the last 20 years, data from Earth remote-sensing spacecraft have become important in helping to predict the weather, improve public safety, map Earth's features and infrastructure, manage natural resources, and study environmental change. In the future, the U.S. and other countries are likely to increase reliance on these systems to gather useful data about Earth. Turning remotely sensed data into useful information requires adequate storage and computer systems capable of managing, organizing, sorting, distributing, and manipulating the data at exceptional speeds.

Earth remote sensing is now a broad-based international activity. This fact has transformed the ground rules for intergovernmental cooperation and offers new opportunities to reduce the costs and improve the effectiveness of overlapping, national, remote-sensing programs. We're already seeing companies generate profits by processing easily accessed information. Private companies have taken the lead in linking data sources to data users by turning raw data into productive information. In addition, several corporations have begun to market raw data from privately financed, remote-sensing systems. The OrbView system, shown in Figure 16-5, is a privately-financed mission, aimed at the commercial, remote-sensing market.

Geographic Information Systems (GIS) using space-based resources provide detailed terrestrial data needed by a variety of markets (Figure 16-6). For example, farmers use GIS tools to analyze and manage their crops, thereby improving crop yields and enhancing competitiveness in an increasingly global marketplace. Insurance companies use GIS data to assess claims following a flood or fire disaster. Timber companies, government agencies, and environmental groups use GIS data to monitor forests. Even the fast-food chain, McDonald's, uses remote-sensing data to estimate population growth patterns in and near cities to determine where best to locate new stores.

Figure 16-5. Commercial Remote Sensing. The OrbView spacecraft, developed by Orbital Sciences Corporation, is a commercial remote-sensing system, aimed at making money from satellite imagery. *(Courtesy of OrbImage)*

Figure 16-6. The U.S. Air Force Academy as Mapped Using a Geographic Information System (GIS). A GIS is a computer-based mapping tool, tying geographic location of features (where things are) to descriptive information (what are things like). This descriptive information allows us to better model a complex "real world." *(Courtesy of Chris Benson, the U.S. Air Force Academy)*

Capital Market Acceptance

Where is all the money coming from to support these exciting new commercial missions? Fortunately, there is another trend within capital markets making investments in space ventures more common place. Financial communities around the world are increasingly aware of the explosive growth potential of space-related products and services, and are more willing to invest in these relatively high-risk ventures. When the Iridium company set out to provide worldwide space-based telephony services from small handsets, they had to raise billions of dollars from investors to get their 66-spacecraft constellation off the ground. Unfortunately, the company declared bankruptcy soon after deployment in 1999, leaving investors, users, and competitors waiting to see their fate.

While the problems with Iridium have made the investment community more cautious about which space ventures to invest in, they don't seem to have stopped the availability of capital for new missions. Successful financial performance in other more traditional, geostationary telecommunications missions should continue to attract investors, thereby firmly establishing the space industry in the capital markets. While *capital market acceptance* is still not as widespread as it is for information technology and Internet ventures, the financial community has begun to recognize that many ventures with a space component are not as risky as previously thought and offer a real opportunity for moneymaking.

Driving this acceptance by capital markets has been the continued convergence of traditional terrestrial technologies with space technologies. Telecommunications and information (Internet and web TV) technologies are continuing to merge and fuel commercial growth for advanced information and communication products and services for a very mobile, worldwide community. The inherent "look-down" advantages of space-based capabilities continues to be an effective means for delivering services and gathering information on a national, regional, and global basis.

Emergence of New Industry Leaders

The final recent trend has been the rapidly changing face of companies within the space industry. There have been so many changes, that you almost need a program to recognize all the players.

The small-to-medium-sized firms have generally been on the forefront of much of the commercial innovation within the space industry. They often possess the low-cost structures and commercially oriented market behavior necessary to capitalize quickly on emerging opportunities and to compete effectively. Given the substantial size of the worldwide space industry and the emergence of numerous commercially viable niches, many of these companies can experience ample growth without inviting significant competition. Orbital Sciences Corporation, makers of the Pegasus launch vehicle, as shown in Figure 16-7, has built a successful business by focusing on the launch and operations of small satellites.

On the other hand, we are seeing many mergers and acquisitions occurring within the U.S. aerospace industry—Lockheed and Martin Marietta, RCA and Martin Marietta, Ford Aerospace and Lockheed Martin, Boeing and Rockwell, and Boeing and McDonnell Douglas, to name a few. Similar actions are taking place within the European community. The result of all these mergers is that there are now fewer, larger companies competing for the same government space contracts.

In the next section we'll look at the political environment, in which all these emerging trends are taking place. Following that, we'll look at the third leg of the space mission triad to understand the role of economics in governmental and commercial missions.

Figure 16-7. Market Niches. Orbital Sciences Corporation has built a successful business by focusing on the launch and operations of small satellites. Here we show a test flight of their Pegasus launch vehicle, with the mother plane at the top, a chase plane in the middle, and the Pegasus launch vehicle at the bottom. *(Courtesy of NASA/Dryden Flight Research Center)*

Section Review

Key Terms

capital market
    acceptance
commercialization
deregulation
Geographic Information
    Systems (GIS)
Global Positioning
    System (GPS)
global space industry
globalization

Key Concepts

➤ Several recent trends give us insight into space mission in the next decade

- Globalization—increasingly, smaller, emerging nations are joining the traditional space superpowers to participate in the high frontier

- Commercialization—commercial missions are beginning to dominate the space industry over traditional military and government space activities

- Capital market acceptance—the growth in commercial space missions has been helped by capital markets, which recognize that space offers a good area for investment with the potential for significant returns at relatively high, but understandable risk. This growth has been further fueled by the convergence of terrestrial and space technologies, especially in the area of telecommunication.

- Emergence of new market leaders—new, small companies have emerged to take advantage of market niches of space services and technology, while larger, traditional aerospace companies continue to merge

16.2 Space Politics

▓ In This Section You'll Learn to...

- ☞ Examine the political motivations for space activities
- ☞ List and discuss the seven key principles that guide international space law
- ☞ Describe the functions of the International Telecommunications Union
- ☞ Discuss the impact of national policies on the conduct of space missions

Politics has dominated space missions from the very beginning. The "space race" between the U.S. and U.S.S.R., described in Chapter 2, pitted the two superpowers against each other in a political and technical contest to see which side could upstage the other in the new high ground. In fact, until the early 1960's, the primary motivation for space missions was political. It wasn't until the first communication spacecraft demonstrated the value of geosynchronous missions to relay telephone and television around the world that the commercial space industry took off.

In this section, we'll look at the political dimension of the space triad that we introduced at the start of the chapter, to see why and how politics continues to shape the direction of space programs. We'll start by looking at the some of the political motivations that drive government-sponsored space missions, then see how governmental laws, regulations, and policies reach into the market place to affect the environment for commercial missions as well.

Political Motives

Governments set policies and spend resources to achieve national or international objectives. Space programs are just one more area of activity, along with agriculture, education, and economic programs that governments can pursue (or not pursue) depending on their political will. Space industry expert Jim Oberg [Oberg, 1999] notes that governments typically pursue space activities for one (or a combination of) the following reasons

- Promote national image and foreign policy objectives
- Enhance national and regional security
- Advance science and technology
- Support national industries

Let's briefly examine each of these political motivations for space missions.

Promote National Image and Foreign Policy Objectives

A strong space program can enhance a country's *national image* and help to promote its foreign policy objectives. This was evident during the space race, as the U.S. and U.S.S.R. tried to outdo the other to demonstrate the superiority of their political system. The demonstration of U.S. technical prowess in its space program, for example, advertises its overall national competence in systems and technologies related to military capabilities.

Space programs can also be useful tools for cementing relationships among friendly nations to further foreign policy objectives. The International Space Station, as shown in Figure 16-8, is serving to enhance foreign relations between the U.S., Russia, the European Community, Japan, Canada, and other partners in the program.

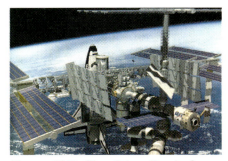

Figure 16-8. The International Space Station (ISS). As its name implies, the ISS is a tool to promote international relations and foreign policy objectives, while furthering the goals of science. *(Courtesy of NASA/Ames Research Center)*

Enhance National and Regional Security

Space has become an integral part of national security. Spy satellites track enemy troop movements and identify targets. Early warning satellites watch for missile launches. GPS satellites help to steer planes, ships, and even bombs to their destinations. Communication spacecraft provide the infrastructure to link commanders with their far-flung forces around the globe. Space assets have become a vital "force multiplier" allowing the modern military to do more with less. The Gulf War and the Kosovo conflicts brought home to military leaders their near total dependence on space assets to accomplish their missions.

The U.S. Department of Defense (DoD) alone spends $26 billion per year on space missions for national security. While spending in many other areas within the military has shrunk in the years following the end of the Cold War, space spending has increased. If anything, the need to pursue space missions to promote *national* and *regional security* will continue to increase as space grows in importance to the international economy.

The growing importance of space for national security has also raised the issue of "space control." In future military conflicts, should countries look up helplessly while their enemies use space assets to their advantage, or should the battleground move to the new high ground, as adversaries attack each other's space assets? As we'll see, these issues come in direct conflict with international law.

Advance Science and Technology

Another reason for nations to support space activities is to promote their level of competence in science and technology, projects vital to their economic and industrial base. Funding for advanced space technologies goes to universities, national laboratories, and commercial industries. At the university level, this support attracts students into technical fields and works to increase the pool of qualified specialists. At national laboratories, basic and applied research in space technologies can often transfer to other domestic needs, such as energy and transportation.

Support National Industries

The final reason that nations pursue space activities is to enhance the competitive advantage of their own national industries. In the modern global economy, nations need vigorous, competitive national industries to generate exports to offset imports—or risk the stability of their national economies. "Buy American" summarizes the U.S. policy of only buying products and services from local, homegrown industries. This policy creates profits for those companies, which they can use to invest in research and development, manufacturing, and other assets to increase their ability to compete with similar companies in other countries in the world marketplace.

The goal, of course, is to create marketable space-related products and services—information and information flow, data products, spacecraft, launch vehicles, ground systems, and associated hardware and software. But, to remain competitive, national industries must vigorously pursue space applications for business and profit, and fund their own in-house basic and applied research to maintain a competitive edge in the designing, manufacturing, deploying, and operating of space systems. This includes the innovation of modern and efficient facilities for producing many spacecraft rapidly at low cost, the ability to operate space systems economically but safely, and the strategy to leverage other technologies into space-related applications.

Critics of space programs often complain about "all that taxpayer money being wasted on space." However, by establishing and funding national space programs, governments indirectly or directly subsidize their national industries and help to grow the overall economy. This creates positive feedback as a strong economy makes it easier to fund a strong space program.

As we've seen, nations pursue space activities for a variety of reasons. It's important to note that a broadly supported space program will support all of these objectives. However, in some cases these objectives can conflict. For example, the goal of enhancing national security can directly conflict with the goal of supporting national industries. For national security reasons, there is an advantage to keeping technological know-how within a nation's borders. Yet, to be competitive, national industries need to be able to sell this capability on the world market place. Attempting to balance these sometimes-competing objectives falls into the area of laws and public policy, as we'll see next.

Laws, Regulations, and Policies

When they're not pursuing space activities for their own ends as described earlier, governments work to shape the context in which all space missions must operate. They do this by developing rules of international law that all space-faring nations, and multinational companies, must follow, and by setting up national and international bodies to oversee the implementations of those laws and policies. Let's start by looking at some principles of international space law.

Principles of Space Law

We can summarize international space law with seven main principles that guide our daily use of space [Houston, 1999]. These seven principles come from five space treaties to which various countries subscribe. They include the: Outer Space Treaty, Registration Convention, Liability Convention, Rescue Agreement, and Moon Agreement [UN, 1972].

For the most part these seven principles represent tradition, which, to some extent, has been overcome by rapidly changing technology, terminology, and concepts. Nonetheless, these seven principles are useful for gaining insights into our present situation in space.

- International law applies to outer space—everyone should conduct activities in space in accordance with international law. Space is not a "lawless" arena. Just because you're on the far side of the Moon, doesn't mean you can escape the long arm of international law.

- Obligation to use space for peaceful purposes—no nuclear or other weapons of mass destruction in space. The moon and other celestial bodies will be used exclusively for peaceful purposes. As space operations become more and more important to military missions, this principle will become more contentious.

- Right to use outer space, but not to appropriate—it's possible for all to use, but not possible for anyone to own. This principle may one day soon come in conflict with "space prospectors" who want to stake their own private claim on outer space real estate.

- Register space objects—owners must provide a general indication of functions, orbit, frequency use, and date of launch for any object put in space. Every object launched must be registered with the United Nations, which tracks who owns what.

- State responsibility and supervision for private activities—states (or nations) should oversee the activities of industry and academia in space to ensure compliance with treaties. It is up to each nation to police their own people and industries to ensure they comply with international law in space.

- Retention of jurisdiction and control—every nation that registers a space object is responsible for it even if it stops functioning. If you launched it, it's your responsibility to keep track of it.

- Liability for damage—registered owners of space hardware are responsible for damage caused in space, on Earth, and in the atmosphere. If you launch it, and it comes crashing down on someone's head—before it reaches orbit or after 50 years—its your responsibility.

These seven concepts tend to guide the use of space. But keep in mind that they're typical practices, based on treaties and international understandings that are subject to change by a nation if it so deems. One of the ways the international community works to implement these laws is through international agencies that oversee and regulate space activities. One of the most important of these is the International Telecommunications Union.

The International Telecommunications Union

The *International Telecommunications Union (ITU)* is a specialized agency of the United Nations that manages, among other issues, the geostationary orbital-slot assignments, as well as frequency allocation for its international member states. The ITU is the recognized body that provides

- Interference-free and equitable access to orbital positions and radio frequencies

- A first-come, first-served approach to the positions and frequencies, assuming that the members use the resources allocated to them responsibly

ITU member states receive the right to use and not to own the resources, perpetually. They have a limited right to sell or barter the resources. Member states have the right to replace failed spacecraft in orbital locations that they have been allocated. Some countries that have permission to use orbital slots have tried to sell or trade the slots for other resources. The ITU frowns on that practice.

National Policies

In the U.S., as in many nations, there is a government agency that manages frequency use within its boundaries. The *Federal Communications Commission (FCC)* manages frequency allocation for operations over the U.S. territory. Large, global communication companies must work very hard and long to obtain permission from the ITU and FCC, as well as other national agencies in the countries they plan to operate. As you can imagine, trying to provide global communication coverage may involve hundreds of countries and their individual agencies and associated politics.

Governments may have a number of different (sometimes competing) agencies with oversight responsibility for various aspects of space missions. The U.S. *Department of Transportation (DOT)*, for example, has responsibility for cars, railroads, ships, and airplanes, plus it also works with NASA, DoD, the Federal Aviation Administration (FAA), and other agencies to regulate launch vehicles and launch sites. The proposed new Kodiak Island launch site in Alaska, as shown in Figure 16-9, will undergo a variety of reviews by DOT and other agencies before and during operations.

Administering governmental laws and regulations by these agencies can have wide-ranging impact on commercial space activities. For example, the U.S. State Department must approve exports of all space technology. Preparing the necessary documentation can add cost and create delays on the sale of spacecraft, launch, and other services to foreign customers. These policies may restrict or even prohibit the sale of certain technologies in the interests of national security. Imagery from commercial remote sensing spacecraft, for example, undergoes careful scrutiny.

Figure 16-9. Kodiak Island Launch Site. U.S. Government agencies, such as the Department of Transportation, have oversight into diverse areas that impact space operations, such as the proposed new commercial launch site, shown here on Kodiak Island, Alaska. *(Courtesy of Alaska Aerospace Development, Corp.)*

In some cases, governmental policies to protect and support one segment of the national space industry may have detrimental effects on others. For example, to protect national launch capability, the US government imposes restrictions on the use of foreign launchers by government and commercial missions. While this ensures U.S. launch vehicles will be used, in some cases this means the price of the launch will be higher than it could have been, increasing the cost of the entire mission.

In the next section, we'll turn our attention to cost and other economic considerations for space missions. We'll then revisit some of these political issues to see how they may affect our FireSat mission.

Section Review

Key Terms

Department of Transportation
   (DOT)
Federal Communications
   Commission (FCC)
International Telecommunications
   Union (ITU)
national and regional security
national image

Key Concepts

➤ Governments pursue space activities for a variety of political motives
  - Promote national image and foreign policy objectives
  - Enhance national and regional security
  - Advance science and technology
  - Support national industries

➤ International space law derives from traditions and several space-related treaties. We can summarize these as seven basic principles
  - International law applies to outer space
  - Obligation to use space for peaceful purposes
  - Right to use outer space, but not to appropriate
  - Register space objects
  - State responsibility for and supervision of private activities
  - Retention of jurisdiction and control
  - Liability for damage

➤ The International Telecommunication Union (ITU), along with related national agencies, regulates the scarce frequency allocations to government and commercial space activities

16.3 Space Economics

In This Section You'll Learn to...

- ☞ Discuss the factors that contribute to the total life-cycle cost for a space mission
- ☞ Describe the importance of cost estimating for mission planning
- ☞ Explain the concept of internal rate of return and how it effects the investment climate for commercial space missions
- ☞ Discuss some of the political and economic issues for the FireSat missions

The final leg of our space mission triad, and the bottom line for any space endeavor, is economics. No matter how advanced the technology, or committed the political will, if the organization can't afford it, the mission won't get off the ground. In this section we'll explore some important aspects of space economics, starting with cost (How much do missions cost? Why are they so expensive?). Then we'll review, briefly, ways of estimating mission costs and look at the challenge of turning a profit with new space missions by earning a solid return on investment. Finally, we'll return to our FireSat mission example to see how politics and economics would shape the course of that mission.

Life-cycle Costs

Imagine that instead of going to buy a car from a dealer, you decide to create your own car. That's right. You decide to design and build your own new car, just the way you want it. So you step your way through the systems engineering process that we introduced in Chapter 11 and have applied in the last four chapters. You start by defining your requirements and constraints. Then you create a system design that includes various drawings and specifications for the car—engine (maybe a little bigger than normal), transmission (a smooth five speed), drive train, suspension (using hydraulics so you can raise and lower the car), body features (very aerodynamic), interior (plush), instrumentation (maybe all digital), safety features, and more.

You then move to the subsystem specifications, to identify all of the hardware and software you'll need to build the car, including what materials to use, and which bolts and lights—all the details, including where you'll buy all the components.

Next you build the car in a facility—maybe your garage. Along the way, you test individual subsystems—engines, drive train, power seats—making sure they meet your original requirements. Finally, the car passes all the tests and you're ready to hit the road!

Of course, to operate your car, an infrastructure must provide roads, traffic management, and road maintenance, as well as many other necessities. You must provide a place to park your car, insurance, maintenance, and fuel. These items represent the operations cost of your car. After many years of use, say ten years, you realize that your car is worn out and you need to haul it to the junkyard. Now you can start all over again to build a new car!

During this process you've gone through the entire cycle of your car's life—conceptual design, development, production, delivery or deployment, operations, and retirement. Together, these phases are called a product's *life cycle*. All of the costs associated with each phase equal its *life-cycle cost*. If you add all of the costs we just described for your custom-made car, the total may be hundreds of thousands, if not, millions of dollars. Ouch! Obviously, it's much cheaper to simply buy a ready-made car from a dealer. Why? Because the automobile manufacturers have done the design, development, production and testing work for you. By spreading, or amortizing, these costs over millions of customers for the same model, the cost for each individual car is far cheaper than you could ever make it yourself. Now let's see how these basic economic principles apply to space systems.

Life-cycle costs for a space mission includes the cost of four key activities

- Design
- Manufacture
- Launch
- Operations

So what do space missions cost? It depends! Spacecraft costs vary dramatically. There is a broad range of cost options available today, ranging from ultra expensive to almost "free." For example, the total cost of the Apollo Program—tens of flights, landing on the moon, etc.—inflated to today's dollars (fiscal year 2000), is about $150 billion. Today, that's the total NASA budget for the next 11–12 years!

In contrast, the DoD and NASA designed the Clementine spacecraft to fly to the Moon and take images of the lunar surface. The life-cycle cost of the mission was $85 million (in fiscal year 1994). The spacecraft bus and payload costs were about $57 million; launch was $21 million on a Titan II; and ground and mission operations for about 10-12 weeks cost about $7 million. Clementine was an example of a limited-cost mission (Figure 16-10). In contrast, let's look at the PoSat-1 spacecraft, by Surrey Satellite Technology Limited, in the United Kingdom, for the Portuguese government. The PoSat-1 mission costs were as follows: spacecraft and payload cost about $1.6 million (fiscal year 1994); launch was about $0.27 million; and, mission and ground operations cost about $0.3 million for the first year. Total life-cycle costs for one year of operations for PoSat-1 was about $2.1 million (Figure 16-11). Now, let's take a closer look at where all this money gets spent.

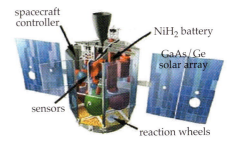

Figure 16-10.   Clementine Mission. NASA's Clementine mission to take images of the Moon, held costs down, yet produced great results. *(Courtesy of the Naval Research Laboratory)*

Figure 16-11. PoSat-1 Mission. This low-cost mission successfully imaged Earth's surface and monitored the space-radiation environment. *(Courtesy of Surrey Satellite Technology, Ltd., U.K.*

Astro Fun Fact
Automobile and Spacecraft Transportation Costs

We all have a healthy respect for the cost of transportation. The U.S. Government allows business owners to deduct 31 cents per mile for business travel in an automobile. It costs 12 to 15 cents per mile to fly coast to coast (in economy, triple that cost in first class). Let's analyze the cost of operating a car. Assume a new car costs $20,000 and we intend to use the car for ten years with no residual value. Over the ten years of operation we spend about $7500 for fuel, $4000 for maintenance, $5000 for insurance, and $1000 for license plates—a total life-cycle cost of $37,500. We drive about 150,000 miles over the ten years—about 25 cents per mile. Not bad. Now let's look at the same issues for a low-Earth orbiting spacecraft that costs about $200 million. Launch costs are about $200 million and operations and maintenance costs are about $100 million. The total life-cycle cost is $500 million—a lot more than our automobile. The spacecraft circles Earth about 15 times per day, resulting in a total distance traveled during ten years of 2.4 x 10⁹ million miles. The cost per mile of our spacecraft is a mere 20 cents per mile. What a good deal!

Contributed by Wiley Larson, the U.S. Air Force Academy

Figure 16-12.   Redundancy in Action. The need to create highly reliable spacecraft often leads designers to build redundancy into critical systems. The folded solar panels shown on the PanAmSat-5 spacecraft here must be deployed for mission success, so we use redundant techniques to insure they deploy. *(Courtesy of Hughes Space and Communications Company)*

Design Costs

Let's start by looking at how economic issues affect the design. First, realize that spacecraft are very complex beasts—a traditional spacecraft is electronically equivalent of about 200 color television sets. That's a lot of components. A failure in many of them could spell disaster.

To ensure the overall system is reliable, the first step is to make sure we design in reliability and use individual components that are as reliable as possible. Spacecraft manufacturers usually employ only space-qualified components. *Space qualification* is an exhaustive program that checks every step in making the component, from the raw materials to packing and storage. Sounds expensive? It is! A space component typically must meet a military standard (MIL STD) or NASA standard, which makes it cost up to ten times more than the commercial version.

The next step in achieving system reliability is often to design in redundancy. Spacecraft sometimes use duplicate or triplicate components, so that, wherever possible, a single failure can't cause a whole system to fail. If the spacecraft carries a crew, reliability is even more important and redundancy in all critical systems is built in, as we find on the Space Shuttle that has 5 identical main computers and 3 identical subsets of most avionics and control subsystems!

Let's illustrate this with a simple example. Suppose we want to unfold a solar panel from the side of our spacecraft when it's in orbit, such as the one shown in Figure 16-12. While one spring would be enough, the panel could be hinged with at least two springs to make sure it fully deploys.

A strap with an exploding bolt at each end could be used to hold it in place during launch. To deploy the panel, we send a signal to both bolts to make them explode. If either works, the strap is free, and the panel unfolds.

Exploding bolts are an example of an interesting type of component—the sort we can't actually test before installing on the flight vehicle. Instead, through space qualification, we can tightly control the manufacture of the bolts, so we can safely assume that, if a few bolts drawn at random from a batch, all go "bang" on command, then they all should. Exploding bolts tend to be very reliable, but for some missions, we may still want to duplicate critical systems, in case one fails.

We do all of this in a quest for reliability. *Reliability* is defined as the probability that a given component will do a certain job under certain conditions for a specified amount of time. This enthusiasm for reliability can lead to over-specification and hence higher system costs. One example is specifying the ability of electronic devices to resist the effects of the space environment. As we saw in Chapter 3, radiation and charged particles can harm spacecraft components. Government and industry have developed special computers to survive the space environment, but because so few are needed, they are very expensive and can lag about ten years behind the Earth-bound state-of-the-art. Result—in some cases spacecraft costing hundreds of millions of dollars with computers no better than you find in a washing machine!

In many cases, spacecraft specifications are unnecessarily demanding. They represent a combination of "worst cases" that rarely occur for most operational spacecraft in Earth orbit. To cut costs, and increase performance, commercial missions, more and more, are choosing to use commercial off-the-shelf (COTS) computers (such as the ones in your personal computer and car) and other components, instead of the far more expensive, "space qualified" versions. This approach has proven to be cost-effective, especially for low-cost, small satellite missions, such as UoSAT-2 shown in Figure 16-13.

The main reason for most of the stringent specifications is the quest for "zero defects" or nearly perfect performance, to protect people and ensure mission success. For instance, the public furor over the Challenger accident underscores the widely held belief that any failure is unacceptable. This demand for "perfect" systems inevitably drives up cost. In general, cost and reliability have an exponential relationship, as shown in Figure 16-14. For example, going from 80%–90% reliability costs significantly more than going from 50%–60%. As reliabilities move to 95% or 98%, the design costs can get excessively high.

Figure 16-13. Commercial Off-the-shelf (COTS) Components in Space. The UoSAT-2 micro satellite, built completely with COTS components, was launched in 1983 to perform store-and-forward communications and space-environment research. It was still functioning in 2000, 17 years later, demonstrating the robustness of COTS hardware in space for these applications. *(Courtesy of Surrey Satellite Technology, Ltd., U.K.)*

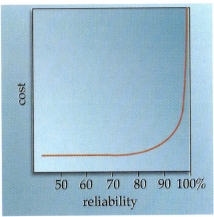

Figure 16-14. Cost Versus Reliability. Reliability is expensive. As the graph shows, increasing the reliability for a system, whether it's a toaster or the Space Shuttle, drives the costs up exponentially. For example, it may double the cost to go from 80%–90% reliability. But the costs may quadruple again to go from 90%–95%. No system can ever be 100% reliable.

Manufacturing Costs

Other spacecraft cost drivers occur during manufacturing. Even though we may want to launch only one spacecraft, we may need to build four different models! These include

- An engineering model
- A test model
- A flight model
- A flight spare

Usually cost constraints don't allow us to build this many models. We can afford to build one, or maybe two. *Engineering models* help us to ensure all the subsystems work together and are used in non-destructive tests. Figure 16-15 shows the engineering model of the Apollo Lunar Excursion Module. We subject the *test models* to the sort of testing that might cause permanent damage, such as vibration and shock, or exposure to radiation, vacuum, and heat. In Chapters 13 and 15 we discussed most of the tests spacecraft can undergo. Ideally, designers would like to subject their spacecraft to as many tests as possible; but when money is tight, designers may have to rely on analysis alone and hope that their assumptions and models are correct.

Early problems with the Hubble Space Telescope underscores the importance of testing. Due to budget constraints, a key end-to-end test of the Hubble optical system was cut from the program. Soon after launch, operators learned that Hubble's vision was slightly blurry due to a manufacturing defect. While this problem was later corrected during an astronaut repair mission, this Hubble experience illustrates the importance of end-to-end testing of the entire spacecraft, to uncover problems that may not appear during individual subsystem tests. Of course, this complete test is easier said than done, especially if the system's behavior depends on some feature of the space environment we can't reproduce on Earth, such as free-fall.

The *flight model* is the one that launches. We may build an additional *flight spare* in case something goes wrong with the flight model just before launch, and to have a duplicate on the ground for testing and trouble-shooting. In practice it's sometimes possible to refurbish the engineering or test models as a flight spare.

As you'd expect, manufacturing and testing procedures are another important cost driver. Not only are the test articles expensive, and the procedures labor-intensive, but we must factor in the cost of the facilities. In Chapter 15, we discussed all the operations systems used during spacecraft assembly, integration, and testing (AIT), such as clean rooms, handling jigs, and thermal-vacuum chambers. The cost of establishing this entire infrastructure for a single mission would normally be prohibitive. Instead, companies or governments try to amortize these costs over many missions to drive down the price tag for individual spacecraft.

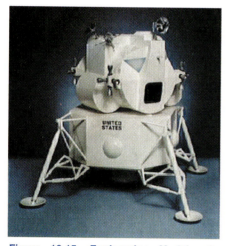

Figure 16-15. Engineering Models. As many as four different models are built for some space programs. The engineering model is typically used to ensure all subsystems function together correctly. This is the engineering model for the Apollo Program's Lunar Excursion Module. *(Courtesy of NASA/ Johnson Space Center)*

Launch Costs

What happens to the result of all this tender loving care? We bolt a spacecraft on the top of a massive rocket, containing hundreds of tons of explosive propellant, and fire it into space, as Figure 16-16 illustrates. In fact, we do a bit more than that. Launch, and the spacecraft's operation and control on orbit, take the same care and attention to detail as design and manufacturing and, therefore, are also expensive and complicated.

The cost of launching a spacecraft varies greatly, from around $5000 per kg for a proposed advanced launch system, up to $30,000 per kg for a small, commercial rocket. As with other economic activities, the price per kilogram appears to be less as we launch more kilograms. But this may be a false bargain. The cost of assembling several tons of payloads to take advantage of a large, cheap launch vehicle may exceed the apparent savings.

In addition to the cost of the launch vehicle, we must consider all the incidental costs of operations systems to integrate the spacecraft and launch vehicle, run the launch site, and monitor telemetry before, during, and immediately after launch. Someone must also pay the capital cost of the launch site. A site, such as the European Space Agency's launch center in Kourou, French Guiana, may cost on the order of $1 billion to set up. If it's used for ten launches per year, the interest on that money alone is about $10 million per launch!

Figure 16-16.   As Good as Gold? Launch is one of the most costly phases of a mission. Launch costs, at $5000–$30,000 per kg, exceed the price of gold! *(Courtesy of NASA/Kennedy Space Center)*

Operations Costs

After successfully launching the satellite, someone must pay to operate it. The cost for mission operations can vary dramatically. Part of the cost of operations is related to the infrastructure necessary to connect the operators with their spacecraft and give them the tools that they need to do their job. But even if the facilities already exist, another large part of the cost of operations is related to the number of people required. So you say, "How many people does it take to operate a spacecraft?" As usual, it depends!

We conducted a survey of several spacecraft operations organizations and counted all the people involved in operating telecommunications spacecraft in geostationary orbit. They took into account full- and part-time people, including people in the control room, people doing maintenance, administrators, and secretaries. Then we estimated the number of full-time equivalent people. The results were as follows: 27 for DoD, 24 for NASA, 21 for the European Space Operations Center, and 12 for a commercial organization. If we assume the total annual cost (includes salary, benefits, office space, taxes) for each of these people is about $120,000, we get a range in cost of operations for one geostationary space-craft for one year of between $1.5 million and $3.3 million. For a ten-year mission this equates to between $15 million and $33 million.

The famous pictures of "Mission Control, Houston" depicted in Figure 16-17, shows a team of nearly 100 people managing one spacecraft. A center like that costs about $100 million per year to run, so we have to

Figure 16-17. Mission Operations Costs. Maintaining a large, highly trained team of mission operators, such as the ones shown here in the Space Shuttle Operation Center in Houston, Texas, greatly adds to the life-cycle cost of a mission. *(Courtesy of NASA/Johnson Space Center)*

seek more efficient approaches to make missions more affordable. As discussed in Chapter 15, the more the spacecraft can do for itself, the more autonomously it can operate, and the less expensive an operations infrastructure it needs on the ground for support.

Space mission teams go to all that effort and expense to analyze their spacecraft's design, build it with the best components, test it exhaustively, launch it, and assemble a well-trained operations team to run it. What if it still fails? Who pays? For government-sponsored missions, we do! Government missions are "self-insured," or, in other words, the taxpayer foots the bill. The Challenger disaster not only caused the irreplaceable loss of seven brave astronauts, it also destroyed a multi-billion dollar national asset. The cost of the replacement orbiter, Endeavor, was paid for out of the national budget.

For commercial missions, the cost of failure impacts their bottom line. Just as we buy car insurance to cover the cost of repairs and legal liability, in the case of accidents, companies often buy "space insurance" to protect themselves from known mission risks. For example, a corporation may purchase launch insurance for its spacecraft, so that if there is a launch failure, they will receive enough money to build and launch another one.

Insurance companies cover failure of the rocket or the spacecraft, typically paying for another launch and a new satellite, as they did in the case of the FASat-A spacecraft, as shown in Figure 16-18. Of course, this adds to the overall mission cost. Insurance companies charge a premium of 15%–30% of the total costs for the satellite and launch. For a $100 million spacecraft that's a $15–$30 million premium. Wow! On orbit, after everything checks out, the premium may get down to between 2% and 3% per year—only $2–$3 million per year, still a major drain on mission revenues. Coverage for lost revenue if the satellite doesn't enter service may be available in addition, but is often prohibitively expensive.

Why is insurance so expensive? Typically one in ten launches fails in some way; that is, the payload doesn't arrive in the correct orbit or work after it gets there. The premium charged reflects the insurance company's confidence in the particular spacecraft and launch vehicle. Just as a 16-year old driving a new sports car will pay more for insurance than a 40-year old driving a 10-year old sedan, insurance for an entirely new spacecraft on a new launch vehicle (or one with a recent history of failure) could be far more expensive than for a proven design on a well-proven launch vehicle.

Given the high cost of insurance, some companies choose to "self-insure," similar to the government, and pass the cost of failure on to the investors. Either way, it's a risky business strategy. If a company pays for insurance, and the mission is successful, the insurance "unnecessarily" adds 15%–30% to the cost of the mission, right off the top. But if it doesn't buy insurance, and there's a failure, it'll have to pay the entire cost of the spacecraft and launch all over again!

Figure 16-18. Insurance Claim. The FASat-A spacecraft failed to separate from the primary payload soon after launch in 1997. Fortunately, the mission sponsors bought insurance that paid for building and launching FASat-B several years later. *(Courtesy of Surrey Satellite Technology, Ltd., U.K.)*

Astro Fun Fact
Space Shuttle to the Rescue!

Normally, it's not practical to send a repair team to fix a broken spacecraft. But in 1992 NASA did just that! They used the Shuttle to rescue a faulty communications satellite belonging to Intelsat. Although the cost of the Shuttle flight was approximately $500 million, and the cost of the satellite saved—approximately $50 million. Although on the surface this may appear to be a poor trade-off, the lifetime value of the spacecraft was much greater when the potential, cumulative revenues for the corporation and the insurance cost were considered.

Costs in Perspective

For most of us on hamburger budgets, discussions of mission costs in the millions and billions of dollars simply make our eyes glaze over. Let's try to put those costs into perspective.

We can compare spacecraft costs to other recognizable items, such as gold. Today, gold sells for about $14,000 per kg. In comparison, total spacecraft costs for fairly large, traditional missions range from about $30,000–$160,000 per kg, 2 to 10 times more costly than gold! The cost to launch a spacecraft into orbit ranges from about $5000 per kg to about $30,000 per kg—again, more precious than gold.

So, space is expensive, but many are working to make space more cost-effective and affordable. Politically, NASA and DoD must reduce the life-cycle cost of doing their missions, so Congress will continue to support them. "Faster, better, cheaper" has become the rallying cry to lead that effort. Economically, the life-cycle cost must decrease so that companies doing business in space can remain in business and make a profit. Technically, all players are trying to find technologies that will allow them to reduce the overall cost of space missions.

Cost Estimating

With space missions so expensive, companies don't want to get too far into a program unless they know they have enough money in their piggy bank to pay for it. To protect themselves, and give assurance to program sponsors and investors, mission managers try to estimate spacecraft and total mission costs, as early and as often as possible. If you're estimating the cost to build your custom car that we discussed earlier, at least you can check the price of most of the parts in a catalog. And you can make a reasonable estimate of the cost of facilities and labor by asking experienced mechanics and comparing it to other industries.

But to accurately predict the total cost for a completely new mission, such as to put humans on Mars, like the one shown in Figure 16-19, is extremely hard to do. There are no on-line catalogs for Mars landing

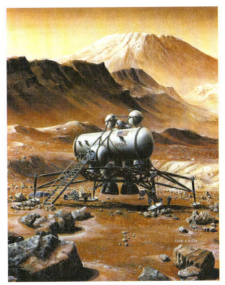

Figure 16-19. Red Planet Budgets. This artist's concept shows future astronauts on the surface of Mars. But before the mission gets onto the drawing boards, sponsors must try to estimate total cost using various cost estimating relationships (CERs). *(Courtesy of NASA/Ames Research Center)*

shuttles and no similar missions to compare operations costs! Instead, mission planners attempt to determine the total mission cost by breaking a large mission into small pieces and trying to estimate the cost of each. Over the years, the aerospace industry has developed a variety of *cost estimating relationships (CERs)* for individual subsystems and entire missions. Mission sponsors use these CERs as planning tools prior to committing large funds to a project.

Most CERs use parametric data to compare apples with apples, as closely as possible. For example, based on historical data, a spacecraft propulsion subsystem using conventional chemical rockets may cost $1000 per 1 m/s ΔV. Thus, if a mission calls for 1.2 km/s ΔV, for planning purposes the manager could assume the cost for a new propulsion subsystem will be about $1.2M. By applying similar relationships for other subsystems based on mass, data rate, and other parameters, we can gradually add up the cost for a mission.

At best, mission cost estimating is an inexact science. Because the techniques are based on historical data, they often can't reflect the effects of new technologies or cost reduction techniques. However, at least they bound the problems and give a starting point for mission planning. They also give investors some idea of how to estimate the financial risks for a mission. We'll examine these investment issues in more detail next.

— *Astro Fun Fact* —
It's the Launch Rate,
Not the Technology!

While NASA and industry focuses on newer and better technologies, such as the X-33 single-stage-to-orbit prototype, to reduce the cost of access to space from the current average of $10,000/kilogram, economic analysis of the problem indicates they may be barking up the wrong tree. Analysts looked at the total cost of operating a fleet of Boeing 747 aircraft in the same way as the Space Shuttle fleet. If you assume you had 4 conventional 747s, flying only 8–10 missions per year, and adjusting for payload, you find the cost per pound (or per kilogram) is nearly the same as flying the Space Shuttle! This indicates that the problem isn't technology, its launch rate. The more you fly, the cheaper it gets. To reduce the cost per kilogram to orbit to $1000, you would need a fleet of about 50 vehicles flying over 130 times per year. Of course, this creates a chicken/egg dilemma for potential entrepreneurs looking at building such a fleet. You can't justify the investment to build such a fleet until you have the demand for that many flights. And you won't have demand until the cost comes down. Perhaps one day we'll break this cycle and cheap access to space will be a reality for us all.

Czysz, P.A. "Combined Cycle Propulsion—Is It the Key to Achieving Low Payload to orbit Costs?" 14th International Symposium on Airbreathing Engines. Florence, Italy: 5–10 September, 1999.

Return on Investment

Let's compare and contrast government-funded and commercially financed missions. Often in government-funded space missions, the goal is to obtain as much science or other information as possible for the funds spent. A big challenge for governments is matching the funds available with the mission requirements and corresponding costs incurred in developing, launching, and operating a particular mission. The goal has been to accomplish the mission, and so sponsors go back to congress each year asking for money until they achieve the goal.

On the other hand, commercial, space-related businesses usually provide a product or service with the goal of making a profit for their investors. For example, they may provide telephone or television services to their customers, who (hopefully!) are willing to pay a price for the service. The goal is to have more revenue than expenses when it's all over, producing profit to be divided among the shareholders.

In a commercial venture, investors provide financing up front to pay for the design, manufacture, launch, and operation of the system, with the expectation of earning a good return on their investment. Earlier in the chapter we discussed the fact that financial markets are becoming more accepting of space-related projects, where previously they perceived space as too risky of an investment. The mentality of these investors is much like ours. For example, we'd probably feel fairly comfortable putting $100 in a bank where we are 99% sure that in one year we'll get back $105 (a simple 5% interest rate compounded yearly). However, if we had the option of investing that same $100 for one year with a 50% chance of getting back $140 (a simple 40% interest rate compounded yearly), we would probably be much more cautious. Right? Investors in space-related businesses are equally cautious. They may be willing to invest millions of dollars for a period of time, if the chances that they will get their money back are acceptable, and if the return on their investment would be high enough. These are both big "ifs!"

One method used to attract these high-roller investors is to offer a high *internal rate of return (IRR)*. The IRR is the simple interest rate that equates to the present value of the expected revenues compared to the initial investment. More simply, it's the interest rate that they'd make if they added all the revenues and expenses they expect to have over a time period, taking into account inflation, and then compared that number to the investment they originally made. Since space-related business is still considered by many to be somewhat risky, venture capital investors like to see an internal rate of return of 40% or more, otherwise they'd rather invest in a new fast-food chain or Internet company. The approach used by many venture-capital firms is to invest in, say, ten risky ventures with potentially high rates of return, hoping that two or three of them will pan out.

The IRR calculation depends highly on the amount and timing of revenues and expenses, as well as the cost of money (loans) to keep the business flowing. Because of this fact, a commercial mission's design and development depends largely on the company's business plan. To be credible, and attract investors, the plan must carefully analyze the potential market for the proposed space product or services, assess the competition, and carefully estimate expected costs and revenues. We'll look at some of these issues for our FireSat mission example next.

The FireSat Mission

In Chapter 11 we described our hypothetical FireSat mission and provided the following mission requirements and constraints

- Mission Objective—detect and localize forest fires (>4 hectares) worldwide and provide information to users within 24 hrs

- Users—National Forest Services in the U.S. and worldwide

- Operations Concept—assume the mission will need a six-satellite constellation to provide 24-hour notification to users. All operations will take place from an existing ground station in Colorado Springs, Colorado. Spacecraft will collect and store mission data onboard and relay it to the ground station when they pass overhead. The ground station will notify users via the Internet.

- Mission Constraints

  - Life-cycle cost—<$10M

  - Schedule—first three spacecraft ready for launch in two years

  - Performance—minimum to detect fires >4 hectares from 500-km circular, polar orbit, and relay data to the ground station

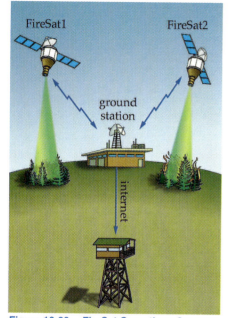

Figure 16-20.   FireSat Operations Concept. This operations concept uses two FireSat satellites in low-Earth orbit, one ground station, and an Internet link that passes forest fire information to users.

Throughout Chapters 11–15, we've focused exclusively on the technical aspects of the FireSat mission, gradually completing conceptual designs for the spacecraft's payload and subsystems. We show the resulting space mission architecture in Figure 16-20. Now let's consider some of the potential political and economic aspects of such a mission.

Mission Politics

The political concerns for even a small mission, such as FireSat, would, of course, vary greatly depending on whether we mean it to be government-sponsored or purely commercial. As a government program, we would have to view it in the context of the objectives for national space programs, as discussed in Section 16.2. For example, by offering to share its data with international partners, our government could use the system as a tool to persuade equatorial countries to increase efforts to preserve rain forests, which are often illegally burned to clear them for farmlands. Government spending on the program could also help to support university or national laboratory research, and create jobs within industry.

A purely commercial FireSat mission would still be subject to international, as well as domestic laws and regulations. The company leading the mission would need approval from the FCC and ITU for uplink and downlink frequencies. The sharing of mission data and other technical details may also be restricted by national policies concerning technology transfer or laws involving trade with "rogue" nations. After these political hurdles are overcome, mission sponsors then need to consider how to pay for it all.

Mission Economics

Economic issues would again depend on whether the mission were government or privately sponsored. Notice within the mission constraints, we capped the total mission cost at $10M. This means we have a design-to-cost mission, whereby designers will trade performance and other requirements to reach the target cost. For a government mission, as long as political support remains (and there is money in the budget), the mission will go forward. Even if the costs exceed the cap, political support may allow it to proceed anyway. (Such cost over runs on government space programs, especially military, used to be quite common when political support was strong. However, budget tightening has forced all government programs to hold the line on costs.)

Another economic difference between a government and commercial FireSat mission is the approach taken with respect to the cost cap. For a government program, there is typically little incentive to reduce costs below some minimum threshold to insure the mission will maintain support. However, for commercial missions, every dollar saved translates directly into increased profit for investors (with greater bonuses and stock options for mission engineers and operators!).

As discussed earlier in this section, life-cycle costs for FireSat will come from design, manufacture, the launch, and the operations. Even though this is a "bare bones" mission, with the launch opportunities donated, there are other, hidden costs to consider. "There ain't no such thing as a free launch," still applies. A "free" launch may be offered because the Falcon launch vehicle is untested. This will translate into higher insurance premiums. Or the launch vehicle providers may expect a financial stake in the business venture in return.

To build a sound commercial business plan, mission entrepreneurs need to estimate the total mission life-cycle costs based on past experience with other missions or by using standard industry cost estimating relationships. For example, using the small satellite cost model from Wertz, 1999, we expect the spacecraft to cost about $1.5 million. Budget planners would have to factor in the availability and cost for critical operations systems, such as clean rooms and test facilities, as well as the need for hardware and software upgrades at the existing operations center in Colorado.

They would then need to compare these costs to the expected revenue from the mission. They could sell subscriptions to their forest-fire warning service to government and private agencies on an annual-fee basis, for example, and the trade-off between price and the number of customers carefully analyzed. Only then could they approach potential investors with hard data on their expected internal rate of return.

We've covered a lot of ground (well, space, actually) in this book. The bottom line is that we need to keep our eyes on the big picture, while we design new missions or simply try to understand why missions and systems look the way they do. If we consider, not only the technical aspects of a mission, but the politics and economics as well, we will better understand our mission. More and more our space missions are constrained, not by technology, but by what we choose to do and what is economically feasible.

▰ Section Review

Key Terms

cost estimating
    relationships (CERs)
engineering models
flight model
flight spare
internal rate of return
    (IRR)
life-cycle cost
life cycle
reliability
space qualification
test models

Key Concepts

➤ Life-cycle costs include costs incurred during all phases of a space mission: design, manufacture, launch, and operations

- Design costs are influenced by the redundancy and associated complexity of systems

- Manufacturing costs are driven by the type and number of models needed (engineering, test, flight, and spare), the total testing and associated infrastructure required

- Currently, launch costs exceed the cost of gold per kilogram

- Operations costs vary greatly for government and commercial missions. Increased use of onboard autonomy can help to reduce these costs. Insurance costs are another important factor contributing to operations costs.

➤ Mission planners use cost estimating relationships to provide a starting point for mission design to determine if their budgets match their requirements and estimate the possible return on investments

➤ The FireSat mission illustrates the differences in approaches between government and commercially sponsored missions

References

Oberg, James. *Space Power Theory.* Government Printing Office, 676-460, 1999.

Houston, Alice, et al., editor. *Keys to Space.* (Chapter 12.1 by P. Tuinder). New York, NY: McGraw-Hill Companies, Inc., 1999.

Sullivan, Brian. Initial manuscript of *Space Power.* 1998.

Via Satellite, Vol. XIV, Number 1, Phillips Business Information, Inc., July, 1999.

Wertz, James R. and Wiley J. Larson. *Space Mission Analysis and Design.* Third edition. Dordrecht, Netherlands: Kluwer Academic Publishers, 1999.

Mission Problems

16.1   The Space Industry

1   List and describe emerging trends in the space industry.

2   List the four primary market areas for commercial space activities. Give examples of each.

3   What are the three most important areas for space-related services? Give examples of each.

4   Describe what we mean by a value-added service.

5   Give examples of how we might use space capabilities to exploit the deregulation of electrical power utilities in the world.

16.2   Space Politics

6   According to industry expert Jim Oberg, what are the four primary reasons nations pursue space activities? Give specific historic examples of each.

7   List and discuss the seven key principles that fashion international space activities.

8   Compare and contrast the Federal Communications Commission in the U.S. and the International Telecommunications Union.

16.3   Space Economics

9   Define the life-cycle cost of a space mission. What four mission activities is it composed of?

10   Explain why space-qualified hardware is more expensive than hardware tested on Earth.

11   Define reliability. What is the relationship between it and cost?

12 Describe the difference between an engineering model, test model, flight model, and flight spare.

13 Describe your thought process in deciding to invest in a fairly risky space venture. What risks would you take and what return on investment would be appropriate?

14 Discuss the concept of space mission insurance. For what types of mission would it be most appropriate? Least appropriate?

15 Describe the effects on return of investment of a launch failure, and subsequent mission delay, on a business venture in space.

16 Describe two key differences between government and commercially sponsored space missions.

For Discussion

17 Research information on the NASA budget on the internet to try to answer the following questions.

What fraction goes to missions versus administration?

What national priorities emerge from a look at the budget?

18 Study some of the everyday machines you encounter (automobile, photocopier, toaster, washing machine, etc.) How would the design and manufacture be different if no one could service the machine after it was built?

19 Can you think of any other products, like contact lenses, which are so valuable per kg that it would be economical to manufacture them in space?

20 How would the planning and execution of a space project change as you vary the internal rate of return required by investors?

21 Can you think of "users" who could determine the needs of programs that space agencies are planning?

22 Air freight costs around a few dollars per kg. Space launch costs several thousand dollars per kg. How might we use space if the costs of launch fell to approach those of air freight? Since the very first satellite, Sputnik, weighing in at 84 kg (185 lb.), satellites have grown in size, weight, and corresponding cost. Billion-dollar satellites taking ten years to complete are common. Beginning in 1979, researchers at the University of Surrey, U.K., inspired by advances in computer technology, decided to buck this trend. They've built and launched a series of "micro-satellites" (all under 100 kg [220 lb.]), proving that spacecraft don't have to be big and expensive.

Mission Profile—Iridium

The Iridium constellation was the first large commercial, orbiting, mobile communication system in the world. From the outset, Motorola's vision and mission statements indicated a very ambitious project—to provide global, mobile communication services for a respectable return on investment.

Mission Overview

Iridium's mission statement is very straightforward

Global Personal Communications
Anyone..Anywhere..Anytime

Iridium's business and mission objectives are clear as well

- Build a global, handheld, mobile, voice communication system
- Maintain Motorola's premier position in mobile communications
- Ensure a profitable project reflecting the risk
- Increase market share for telephones and pagers around the globe
- Create a funding approach reflecting the global reach
- Establish a partnership for production, based upon high quality
- Establish a realizable path for frequency allocation

Notice that the objectives for a commercial enterprise are key to a successful business. These objectives provide a clear statement of what the owners want to accomplish, as well as some guidance as to how they will conduct the business and mission.

We show the resulting mission concept and mission architecture in the next figure. We emphasize the strong tie between the customer resources (pagers, phone booths, mobile telephones), spacecraft (connected to each other via crosslinks), and the terrestrial infrastructure (shown as gateways from space into the ground

system and the system control centers). Notice that the spacecraft communicate with each other using a very specific K-band frequency, and with the customers using very specific, L-band frequency. To use these absolutely critical frequencies, Iridium System team members had to negotiate with hundreds of organizations globally to get concurrence—demonstrating successful negotiations in politics, regulations, and economics.

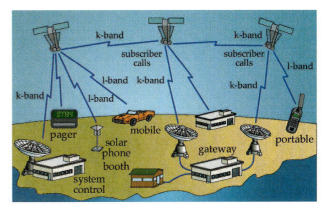

Mission Data

✓ The Iridium LLC faced many challenges in the financial and regulatory arenas. For example, the company needed to acquire $3.4 B in financing to meet the financial needs of the program. They did this by attracting about 20 corporations located worldwide and managing the large international consortium carefully to keep them in the program.

✓ Obtaining regulatory approval globally was one of their biggest challenges. Working with the U.S. Federal Communications Commission (FCC), they obtained the L-band frequency license necessary for launch and operations in the U.S. They also obtained agreement on their approach from the Worldwide Administrative Radio Conference of 1992. By involving each country in the overall network, and sharing in the financial pie, they were successful in obtaining authority to use frequencies controlled by over 100 countries—an incredible feat!

✓ On the technical side, Iridium System engineers designed and implemented a system that virtually connects any point on the globe to any other point—almost instantly. They created a digitally switched telephone network that needed 66 operational spacecraft in LEO, in six near-polar orbital planes to provide continuous, real-time coverage of the entire Earth. Further, this space-based system incorporated sophisticated digital onboard processing and cross linking between spacecraft. The cross-linked network is the key to the system and is the primary difference between it and traditional transponder or "bent-pipe" systems (usually in geostationary orbits).

✓ Manufacturing is another arena where the Iridium LLC shined. With their partners and unique, commercially oriented approach, they manufactured 72 spacecraft in only 12 months. Obviously, they changed the traditional approach in many ways, taking cues from existing spacecraft manufacturers and high-volume manufacturers in the automotive, appliance, and cellular-telephone industries. The result, that will have significant impact on the space-manufacturing industry, is a spacecraft production rate of one completed every 4.3 days!

✓ The launch campaign for the Iridium LLC is as impressive as the manufacturing aspect. They had to launch the 66-spacecraft constellation with associated spares quickly because they needed to produce a revenue stream from their customers. So, they launched 72 spacecraft in 13 months, using three international launch systems with no launch failures!

✓ The Iridium LLC also established a standard in cost-effective mission operations as well. Typically, it takes between 10 and 30 people to operate a single commercial communication spacecraft. The Iridium System requires between 2 and 5 operators for each spacecraft. Granted, because they operate 70-plus spacecraft, they have a lot of people, but they conduct some very efficient operations.

Mission Impact

The Iridium LLC has established a standard for the industry in spacecraft manufacturing, streamlined launch operations and austere mission operations. Their experience in the marketplace provides invaluable insights for other companies to heed.

For Discussion

- How do you think the spacecraft manufacturing approach changes if you're building 70 spacecraft instead of just one?
- After researching the Iridium System on the Internet, describe how they conduct mission management and operations.
- After researching the Iridium LLC business case, why do you think they went bankrupt?

Contributor

Wiley J. Larson, the U.S. Air Force Academy

References

Swan, Peter A. The Iridium Case Study. Teaching Science and Technology, Inc., 1999.

Math Review

A.1 Trigonometry

Trigonometric Functions

Trigonometric functions allow us to compute the sizes of angles and the lengths of sides in geometric figures. If we have a right triangle ABC, as in Figure A-1, with hypotenuse (longest side) of length C and angle θ, we can find the lengths of the other sides by using the *sine* and *cosine* functions. Basic trigonometric functions define the relationships among sides and angles in a right triangle

- cosine θ = cos θ = B / C

- sine θ = sin θ = A / C

- tangent θ = tan θ = A / B

Outline

A.1 Trigonometry
  Trigonometric Functions
  Angle Measurements
  Spherical Trigonometry

A.2 Vector Math
  Definitions
  Vector Components
  Vector Operations
  Transforming Vector
    Coordinates

A.3 Calculus
  Definitions

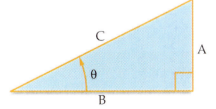

Figure A-1. Right Triangle. We can determine the angles and lengths of the sides of a right triangle using trigonometric functions.

Thus,

- B = C cos θ

- A = C sin θ

One way we remember this relationship is that sine is opposite side over hypotenuse, cosine is adjacent side over hypotenuse, and tangent is opposite side over adjacent side. That is, SOH-CAH-TOA.

Example

Find the length of side A if C = 2 and θ = 15°

$$A = C \sin \theta = 2 \sin (15°) = (2)(0.2588) = 0.5176$$

Now if we take a different perspective and look at a vector in an $\hat{I}\hat{J}$ coordinate system, as shown in Figure A-2, we can see that $A_I = A\cos\theta$ and $A_J = A\sin\theta$.

We can also see that as θ approaches 0°, sin θ goes to 0 and cos θ goes to 1. Also, as θ approaches 90°, sin θ goes to 1 and cos θ goes to 0. Summarizing,

$$\cos 0° = \sin 90° = 1$$
$$\cos 90° = \sin 0° = 0$$
$$\cos 180° = \sin 270° = -1$$
$$\cos 45° = \sin 45° = 0.707$$

A graph of the cosine and sine functions is in Figure A-3.

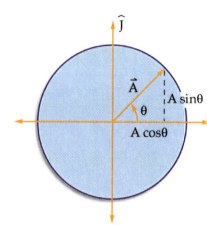

Figure A-2. Vector Components as Function of Angle. The vector $\vec{A}$ h components in the $\hat{I}$ direction (Acosθ) and direction (Asinθ). As θ increases, $\vec{A}$ mov around the circle, similar to the second hand a clock in reverse.

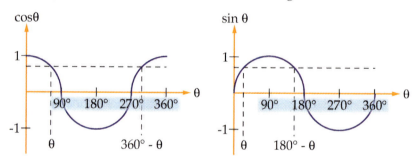

Figure A-3.   Trigonometric Functions. The cosine and sine functions repeat periodically, as shown.

Also notice the following important relationships

$$\cos \theta = \cos (360° - \theta) \qquad \text{(A-1)}$$
$$\sin \theta = \sin (180° - \theta) \qquad \text{(A-2)}$$

This means when we take the inverse of a trigonometric function we get two possible answers! That is

$$\text{arc} \cos x = \cos^{-1} x = \theta \text{ and } (360° - \theta)$$
$$\text{arc} \sin y = \sin^{-1} y = \theta \text{ and } (180° - \theta)$$

Example

Find the angle(s) whose sine is 0.65.

$$\sin^{-1} (0.65) = 40.54° \text{ and } (180° - 40.54° = 139.46°)$$

Find the angle(s) whose cosine is 0.65.

$$\cos^{-1} (0.65) = 49.46° \text{ and } (360° - 49.46° = 310.54°)$$

Angle Measurements

Engineers measure angles in one of two units: degrees or radians. Of course there are 360° in a circle. The measure of radians came about by looking at the relationship between the diameter and the circumference of a circle. The constant number (π) is the ratio of a circle's circumference (C) to its diameter (D).

$$\pi = C/D = 3.14159...$$

So, a circle contains 2π radians.

$$2\pi \text{ rads} = 360°$$

Spherical Trigonometry

The preceding discussion develops from angles and sides measured on a plane. However, when measuring angles and sides on the surface of a sphere such as Earth, things are different. For example, the sum of the angles in a spherical triangle can be greater than 180°, as shown in Figure A-4.

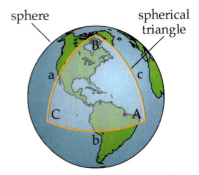

Figure A-4. Spherical Triangles. To find the sides and angles of a triangle drawn on the surface of a sphere, we must use spherical trigonometry.

Oblique Spherical Triangles. (Adapted from *Space Mission Analysis and Design* [1999].)

The following rules apply for any spherical triangle

The Law of Sines

$$\frac{\sin a}{\sin A} = \frac{\sin b}{\sin B} = \frac{\sin c}{\sin C}$$

The Law of Cosines for Sides

$$\cos a = \cos b \cos c + \sin b \sin c \cos A$$
$$\cos b = \cos c \cos a + \sin c \sin a \cos B$$
$$\cos c = \cos a \cos b + \sin a \sin b \cos C$$

The Law of Cosines for Angles

$$\cos A = -\cos B \cos C + \sin B \sin C \cos a$$
$$\cos B = -\cos C \cos A + \sin C \sin A \cos b$$
$$\cos C = -\cos A \cos B + \sin A \sin B \cos c$$

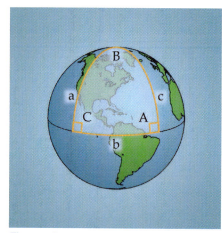

Figure A-5. Right Spherical Triangles. When at least one angle of a spherical triangle is 90°, the formulas for computing the sides and angles are simplified.

Right Spherical Triangles. (Adapted from *Space Mission Analysis and Design* [1999].)

When one angle of a spherical triangle is 90°, we call the triangle a right spherical triangle. The previous formulas simplify for a right spherical triangle

The Law of Sines

$$\frac{\sin a}{\sin A} = \frac{\sin b}{\sin B} = \sin c$$

The Law of Cosines for Sides

$$\cos a = \cos b \cos c + \sin b \sin c \cos A$$

$$\cos b = \cos c \cos a + \sin c \sin a \cos B$$

$$\cos c = \cos a \cos b$$

The Law of Cosines for Angles

$$\cos A = \sin B \cos a$$

$$\cos B = \sin A \cos b$$

$$0 = -\cos A \cos B + \sin A \sin B \cos c$$

A.2 Vector Math

Definitions

Scalar. A *scalar* is a quantity that has magnitude only. Speed, energy, and temperature are examples of scalars. None of these quantities has a unique meaning in a particular direction. A single letter, such as E for total mechanical energy, denotes a scalar quantity.

Vector. A *vector* is a quantity that has magnitude and direction. For example, if we ask you where you went on your hike, you could say, "I went north." But this wouldn't tell us much. "How far?" If instead you answered, "I went five miles," we still wouldn't know much. "Five miles in what direction?" If, however, you answered "I went five miles north," we would know both the distance you hiked (magnitude) and the direction (north). Position, velocity, and angular momentum are examples of vector quantities. A letter with an arrow over it, such as $\vec{R}$ for the position vector, denotes a vector quantity.

Unit Vector. A *unit vector* is a vector having a magnitude of one and we use it to determine direction only. For example, when we define a three-dimensional coordinate system, we do so using three orthogonal (mutually perpendicular) unit vectors such as $\hat{X}\hat{Y}\hat{Z}$ or $\hat{I}\hat{J}\hat{K}$. A letter with a caret or hat over it, such as $\hat{I}$ for the I unit vector, denotes a unit vector.

Vector Components

We can resolve a vector into components along each of three directions in an orthogonal coordinate system. If we have the vector $\vec{A}$ in the $\hat{I}\hat{J}\hat{K}$ coordinate system, as shown in Figure A-6, we can resolve it into its three components as

$$\vec{A} = A_I\hat{I} + A_J\hat{J} + A_K\hat{K}$$

Vector Operations

Magnitude of a Vector. Magnitude is the scalar part of a vector. We find it by taking the square root of the sum of the squares of each of its components. That is,

$$\left|\vec{A}\right| = A = \sqrt{A_I^2 + A_J^2 + A_K^2} \qquad (A-3)$$

Note: We use $\left|\vec{A}\right|$ and A to denote the magnitude of the vector $\vec{A}$.

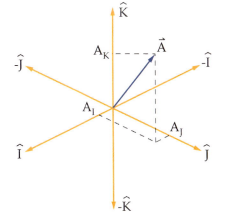

Figure A-6.   Components of a Vector. Here we show the three components of a vector.

Example

Find the magnitude of the vector $\vec{B} = 3\hat{I} + 1\hat{J} - 2\hat{K}$

$$B = \sqrt{3^2 + 1^2 + (-2)^2} = \sqrt{9 + 1 + 4} = \sqrt{14} = 3.74$$

To create a unit vector (of magnitude one), we divide each component by the magnitude of the original vector.

Example

Find a unit vector in the direction of $\hat{B}$

$$\hat{B} = \frac{3}{3.74}\hat{I} + \frac{1}{3.74}\hat{J} - \frac{2}{3.74}\hat{K}$$

$$\hat{B} = 0.8\hat{I} + 0.27\hat{J} - 0.53\hat{K}$$

$$\left|\hat{B}\right| = \sqrt{(0.8)^2 + (0.27)^2 + (-0.53)^2} = 1.0$$

Vector Addition. *The result of vector addition is a vector.* To add or subtract vectors we must add or subtract individual like components. That is, for two vectors, $\vec{A}$ and $\vec{B}$,

$$\vec{A} + \vec{B} = (A_I + B_I)\hat{I} + (A_J + B_J)\hat{J} + (A_K + B_K)\hat{K} \qquad \text{(A-4)}$$

Example

Find the sum of two vectors $\vec{A}$ and $\vec{B}$ where

$$\vec{A} = 4\hat{I} - 3\hat{J} + 1\hat{K} \text{ and } \vec{B} = 3\hat{I} + 1\hat{J} - 2\hat{K}$$

$$\vec{A} + \vec{B} = (4 + 3)\hat{I} + (-3 + 1)\hat{J} + (1 - 2)\hat{K} = 7\hat{I} - 2\hat{J} - 1\hat{K}$$

Vector Multiplication. There are two ways to multiply vectors. The first way results in a scalar and we call it the scalar or *dot product*. The second way results in a vector and is the vector or *cross product*.

Scalar or Dot Product. *The result of a scalar or dot product is a scalar quantity.* The dot product of two vectors $\vec{A}$ and $\vec{B}$ multiplies the amount of $\vec{A}$ that is in the direction of $\vec{B}$ by the magnitude of $\vec{B}$. To do this when we have two vectors $\vec{A}$ and $\vec{B}$, we use trigonometry to find the amount of $\vec{A}$ that is in the direction of $\vec{B}$, as shown in Figure A-7.

$$\vec{A} \text{ which is in the direction of } \vec{B} = A\cos\theta$$

When we multiply this value by the amount of $\vec{B}$ in the direction of $\vec{B}$ (which is simply the magnitude of $\vec{B}$), we get the value for the dot product

$$\boxed{\vec{A} \cdot \vec{B} = AB\cos\theta} \qquad \text{(A-5)}$$

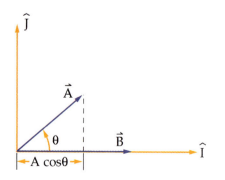

Figure A-7. The Dot Product. The dot product of two vectors, $\vec{A}$ and $\vec{B}$, represents the amount of A in the direction of B. This is also the "projection" of $\vec{A}$ onto $\vec{B}$.

We can also find this value by multiplying the individual like components of the two vectors and adding the result, as follows

$$\vec{A} \cdot \vec{B} = (A_I B_I) + (A_J B_J) + (A_K B_K) \qquad \text{(A-6)}$$

By combining these two approaches for the dot product, we can determine the angle between two vectors. If we know the components of the two vectors, we can use Equation (A-6) to find their dot product. Then, by rearranging Equation (A-5), we can get a relationship for the angle between them.

$$\theta = \cos^{-1} \frac{\vec{A} \cdot \vec{B}}{AB} \qquad \text{(A-7)}$$

Example

Find the angle between the two vectors $\vec{A}$ and $\vec{B}$ where

$$\vec{A} = 4\hat{I} - 3\hat{J} + 1\hat{K} \text{ and } \vec{B} = 3\hat{I} + 1\hat{J} - 2\hat{K}$$

First we find the dot product of the two vectors using Equation (A-4)

$$\vec{A} \cdot \vec{B} = (4)(3) + (-3)(1) + (1)(-2) = 12 - 3 - 2 = 7$$

Next we need the magnitudes of $\vec{A}$ and $\vec{B}$ which we find using Equation (A-1)

$$A = \sqrt{A_I^2 + A_J^2 + A_K^2} = \sqrt{4^2 + (-3)^2 + 1^2} = \sqrt{16 + 9 + 1} = \sqrt{26} = 5.1$$

$$B = \sqrt{3^2 + 1^2 + (-2)^2} = \sqrt{9 + 1 + 4} = \sqrt{14} = 3.74$$

Now we can use Equation (A-5) to solve for the angle between the two vectors.

$$\theta = \cos^{-1} \frac{\vec{A} \cdot \vec{B}}{AB} = \cos^{-1} \frac{7}{(5.1)(3.74)} = \cos^{-1}(0.37) = 68.5° \text{ (or } 291.5°)$$

(Note the inverse cosine produces two possible results.)

Properties of the Dot Product

- The dot product is commutative: $\vec{A} \cdot \vec{B} = \vec{B} \cdot \vec{A}$

- The dot product of like vectors equals the square of the magnitude of that vector: $\vec{A} \cdot \vec{A} = A^2$. This implies

$$\hat{I} \cdot \hat{I} = \hat{J} \cdot \hat{J} = \hat{K} \cdot \hat{K} = 1 \text{ (recall } \cos 0° = 1)$$

- The dot product of perpendicular vectors is zero. This means

$$\hat{I} \cdot \hat{J} = \hat{J} \cdot \hat{K} = \hat{I} \cdot \hat{K} = 0 \text{ (recall } \cos 90° = 0)$$

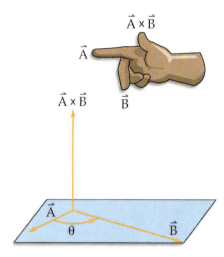

$\vec{A} \times \vec{B}$

$\vec{A}$

$\vec{A} \times \vec{B}$ $\vec{B}$

$\vec{A}$ θ $\vec{B}$

Figure A-8.   The Cross Product. The cross product of two vectors $\vec{A}$ and $\vec{B}$ (both in the plane of the page) results in a new vector that is perpendicular to both A and B and thus, comes out of the page.

Vector or Cross Product. The vector or cross product of two vectors results in a vector quantity. Let's assume two vectors lie in the same plane. The cross product of these two vectors forms a third vector, which is perpendicular to this plane. We use the "right-hand rule" to find the direction of this third vector. If we point our index finger along the first vector and our middle finger along the second vector, our thumb points in the direction of the resultant vector, as shown in Figure A-8.

We find the cross product by solving the determinant

$$\vec{A} \times \vec{B} = \begin{vmatrix} \hat{I} & \hat{J} & \hat{K} \\ A_I & A_J & A_K \\ B_I & B_J & B_K \end{vmatrix} \tag{A-8}$$

Then, we can evaluate the solution, component by component, as follows

- The $\hat{I}$ component is

$$\begin{vmatrix} \hat{I} & & \\ & A_J & A_K \\ & B_J & B_K \end{vmatrix} = [(A_J)(B_K) - (B_J)(A_K)]\hat{I}$$

- The $\hat{J}$ component is

$$\begin{vmatrix} & \hat{J} & \\ A_I & & A_K \\ B_I & & B_K \end{vmatrix} = -[(A_I)(B_K) - (B_I)(A_K)]\hat{J}$$

[Note the minus sign in front!]

- The $\hat{K}$ component is

$$\begin{vmatrix} & & \hat{K} \\ A_I & A_J & \\ B_I & B_J & \end{vmatrix} = [(A_I)(B_J) - (B_I)(A_J)]\hat{K}$$

The result is

$$\vec{A} \times \vec{B} = [(A_J)(B_K) - (B_J)(A_K)]\hat{I}$$
$$-[(A_I)(B_K) - (B_I)(A_K)]\hat{J} + [(A_I)(B_J) - (B_I)(A_J)]\hat{K} \tag{A-9}$$

If we know the angle between the two vectors, θ, we can also find the magnitude of the cross-product vector. That is

$$\left| \vec{A} \times \vec{B} \right| = AB \sin \theta \tag{A-10}$$

Example

Find the cross product of two vectors $\vec{A}$ and $\vec{B}$ where

$$\vec{A} = 4\hat{I} - 3\hat{J} + 1\hat{K} \text{ and } \vec{B} = 3\hat{I} + 1\hat{J} - 2\hat{K}$$

First we set up the determinant per Equation (A-8).

$$\vec{A} \times \vec{B} = \begin{vmatrix} \hat{I} & \hat{J} & \hat{K} \\ 4 & -3 & 1 \\ 3 & 1 & -2 \end{vmatrix}$$

Evaluating, as in Equation (A-9), we get

$$\vec{A} \times \vec{B} = [(-3)(-2) - (1)(1)]\hat{I} - [(4)(-2) - (3)(1)]\hat{J}$$

$$+ [(4)(1) - (3)(-3)]\hat{K}$$

$$\vec{A} \times \vec{B} = [(6) - (1)]\hat{I} - [(-8) - (3)]\hat{J} + [(4) - (-9)]\hat{K}$$

$$\vec{A} \times \vec{B} = 5\,\hat{I} + 11\,\hat{J} + 13\,\hat{K}$$

This allows us to rewrite the cross product as

$$\vec{A} \times \vec{B} = AB\sin\theta\,\hat{n}$$

where $\hat{n}$ is a unit vector formed by the right-hand rule in the direction of $\vec{A} \times \vec{B}$.

- Properties of the cross product

  - Cross product is *not* commutative: $\vec{A} \times \vec{B} \neq \vec{B} \times \vec{A}$

    Note: $(\vec{A} \times \vec{B}) = -(\vec{B} \times \vec{A})$

  - Distributive Law: $\vec{A} \times (\vec{B} + \vec{C}) = \vec{A} \times \vec{B} + \vec{A} \times \vec{C}$

  - Associative Law: $c(\vec{A} \times \vec{B}) = (c\vec{A}) \times \vec{B} = \vec{A} \times (c\vec{B})$ (where c is a scalar)

  - Because the angle between parallel vectors is zero,

    $$\hat{I} \times \hat{I} = \hat{J} \times \hat{J} = \hat{K} \times \hat{K} = 0 \quad \sin(0°) = 0$$

  - Because the angle between perpendicular vectors is 90°,

$\hat{I} \times \hat{J} = \hat{K}$	$\hat{J} \times \hat{I} = -\hat{K}$	note that $\sin 90° = 1$
$\hat{J} \times \hat{K} = \hat{I}$	$\hat{K} \times \hat{J} = -\hat{I}$	
$\hat{K} \times \hat{I} = \hat{J}$	$\hat{I} \times \hat{K} = -\hat{J}$	

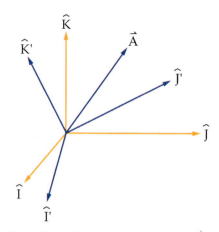

Figure A-9. We can express the vector $\vec{A}$ in either the $\hat{I}$ $\hat{J}$ $\hat{K}$ system or the $\hat{I}'\hat{J}'\hat{K}'$ system. In either case, the vector $\vec{A}$ is the same.

Transforming Vector Coordinates

We can write a vector in different coordinate frames to make writing equations of motion easier. For example, as shown in Figure A-9, we can write $\vec{A}$ as

$$\vec{A} = x\hat{I} + y\hat{J} + z\hat{K} = a\hat{I}' + b\hat{J}' + c\hat{K}'$$

These two descriptions represent the same vector. Therefore, we need a method of rotating or transforming the descriptions from one frame to another (without changing the vector's magnitude or direction). To do so, we use

• Positive rotation about the $\hat{I}$ axis through an angle α (ROT1)

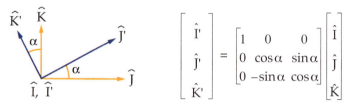

$$\begin{bmatrix} \hat{I}' \\ \hat{J}' \\ \hat{K}' \end{bmatrix} = \begin{bmatrix} 1 & 0 & 0 \\ 0 & \cos\alpha & \sin\alpha \\ 0 & -\sin\alpha & \cos\alpha \end{bmatrix} \begin{bmatrix} \hat{I} \\ \hat{J} \\ \hat{K} \end{bmatrix}$$

• Positive rotation about the $\hat{J}$ axis through an angle β (ROT2)

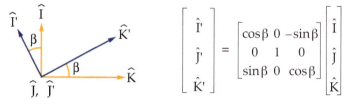

$$\begin{bmatrix} \hat{I}' \\ \hat{J}' \\ \hat{K}' \end{bmatrix} = \begin{bmatrix} \cos\beta & 0 & -\sin\beta \\ 0 & 1 & 0 \\ \sin\beta & 0 & \cos\beta \end{bmatrix} \begin{bmatrix} \hat{I} \\ \hat{J} \\ \hat{K} \end{bmatrix}$$

• Positive rotation about the $\hat{K}$ axis through an angle γ (ROT3)

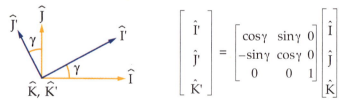

$$\begin{bmatrix} \hat{I}' \\ \hat{J}' \\ \hat{K}' \end{bmatrix} = \begin{bmatrix} \cos\gamma & \sin\gamma & 0 \\ -\sin\gamma & \cos\gamma & 0 \\ 0 & 0 & 1 \end{bmatrix} \begin{bmatrix} \hat{I} \\ \hat{J} \\ \hat{K} \end{bmatrix}$$

• Multiple rotations

For multiple rotations about two or more axes, we find the total transformation matrix by multiplying the transformation matrices for each axis. For example, transforming with a ROT1 followed by a ROT2, the vector from the $\hat{I}$ $\hat{J}$ $\hat{K}$ coordinates to the $\hat{I}''\hat{J}''\hat{K}''$ coordinates, would be

$$[ROT\,2][ROT\,1]\begin{bmatrix} \hat{I} \\ \hat{J} \\ \hat{K} \end{bmatrix} = \begin{bmatrix} \hat{I}'' \\ \hat{J}'' \\ \hat{K}'' \end{bmatrix}$$

A.3 Calculus

Definitions

Derivative. The *derivative* represents the rate of change of one parameter with respect to another. Calculus was developed to analyze changing parameters. For example, if we're traveling north in our car, our position vector is changing over time. The rate at which our position changes over time is our velocity. Thus, if we go 25 miles north (in a straight line) in 30 minutes, our velocity is simply

$$\text{velocity} = \vec{V} = \frac{\text{change in position}}{\text{change in time}} = \frac{\Delta \vec{R}}{\Delta t} = \frac{25 \text{ miles north}}{30 \text{ minutes}}$$

$$= 0.83\left(\frac{\text{mi.}}{\text{min}}\right)\text{north} = 50 \text{ m.p.h. north}$$

Note that in mathematics and engineering, we use the Greek letter Δ (delta) to represent a change in something. So, we define velocity as the derivative of position with respect to time. Similarly, the rate of change of velocity with respect to time (for instance, when you step on the gas) is the derivative of velocity, which is acceleration.

The derivative is really an *instantaneous* rate of change rather than a change over time. For this reason, we use the letter d to represent a change over a short time interval.

$$\text{acceleration} = \vec{a} = \frac{\text{change in velocity}}{\text{change in time}}$$

$$= \lim_{\Delta t \to 0} \frac{\Delta \vec{V}}{\Delta t}$$

$$= \frac{d\vec{V}}{dt} = \dot{\vec{V}} \text{ or } \text{``}\vec{V} \text{ dot''}$$

Note: In this text we use the "dot" notation to denote a derivative with respect to time. That is

$$dy/dt = \dot{y} \text{ for a first derivative with respect to time and}$$
$$d^2y/dt^2 = \ddot{y} \text{ for a second derivative with respect to time, etc.}$$

Because acceleration is the derivative of velocity and velocity is the derivative of position, we can say that acceleration is the *second* derivative of position.

$$\vec{a} = \frac{d\vec{V}}{dt} = \frac{d}{dt}\left(\frac{d\vec{R}}{dt}\right) = \frac{d^2\vec{R}}{dt^2} = \ddot{\vec{R}}$$

To better illustrate the meaning of the derivative, consider the function that describes the distance an object travels at constant acceleration

$$x = 1/2 \, a \, t^2 \qquad \text{(A-11)}$$

where

x = position at time t

a = acceleration

t = time

If we let a = 9.798 m/s², we can graph the position with respect to time, as shown in Figure A-10. If we want to know the velocity after five seconds, we would take the derivative of the function, which is

$$\dot{x} = a \, t = (9.798 \text{ m/s}^2) t \qquad \text{(A-12)}$$

and substitute t = 5 to get

$$\dot{x} \, (t = 5) = 48.99 \text{ m/s} \qquad \text{(A-13)}$$

But the derivative is also the slope of the curve at this point, which is the instantaneous change in position per change in time, as we show in Figure A-10.

Integral. The *integral* represents the cumulative effect of one parameter changing with respect to another. On a graph of one parameter vs. another, the integral is the area under the curve. For example, if we're traveling in a car at some velocity for some amount of time, our change in position will be the integral of velocity over the period. That is, we're adding up all the position changes over time to get the total distance. The integral is essentially the reverse of the derivative. Because acceleration is the derivative of velocity over time, we say that velocity is the integral of acceleration over time.

Velocity = sum of accelerations over time

$$\vec{V} = \int \vec{a} \, dt = \int \frac{d\vec{V}}{dt} dt = \int d\vec{V}$$

If we want to know our velocity after constantly accelerating over some time, we simply integrate the acceleration from the initial time (0) to the final time (t) to get the familiar relationship

$$\vec{V} = \int_0^t \vec{a} \, dt = \vec{a}(t - 0) + \vec{V}_0 = \vec{a}t + \vec{V}_0$$

Note that the V_0 representing the initial velocity appears because a constant of integration must always be added to the result of the integration. To get our position after this acceleration, we simply integrate our result from above one more time (position is the second integral of acceleration, because acceleration is the second derivative of position).

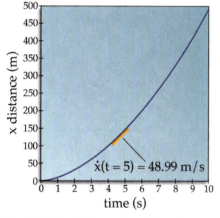

Figure A-10.   Distance Over Time. Here we see a plot of the function x = 1/2 (9.798) t². The derivative of the function at t = 5 represents the velocity at this time and we see it as the slope of the curve at that point.

$$\vec{R} = \int_0^t (\vec{a}t + V_0)dt = \frac{1}{2}\vec{a}t^2 + \vec{V}_0t + \vec{R}_0$$

One way to look at the integral is as if we were adding together tiny little area slices under a curve. If we can approximate each little slice as a rectangle, we can easily find the area of each slice and then add them together to get the entire area under the curve.

Returning to our simple example of a falling object's velocity from Equation (A-12), we can plot this function, as shown in Figure A-11. We can then approximate the distance traveled after five seconds using one-second-wide rectangles, as shown under the curve in Figure A-11, to get

$$x \cong (1)(0) + (1)(9.798) + (1)(19.596) + (1)(29.394) + (1)(39.192)$$

$$\cong 97.980 \text{ m}$$

To get the exact distance traveled, we take the integral of Equation (A-12) and solve at t = 5 to get

$$x = \int_0^5 \dot{x}\, dt = \int_0^5 (9.798 \text{ m/s}^2)t\, dt = 1/2(9.798 \text{ m/s}^2)t^2\Big|_0^5$$

$$= 4.899\,(5)^2 - 4.899\,(0)$$

$$= 122.475 \text{ m}$$

The difference between 97.980 m and 122.475 m is a result of the little bits of uncounted areas under the curve but above the rectangles, in Figure A-11. The integral is exact, so the 122.475 m is correct. Summing rectangles under a curve approximates the value for the integral, but to maintain accuracy, we must keep the rectangles' widths as small as possible.

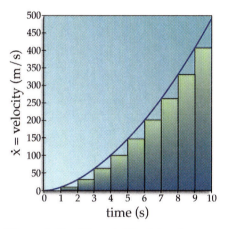

Figure A-11. Integral. The integral represents the cumulative effect of something over time. Here the integral represents the area under the curve found by adding together each little rectangle.

Units and Constants

B.1 Canonical Units

When analyzing spacecraft motion across literally astronomical distances, using traditional measures (meters and kilometers or feet and miles) becomes cumbersome. The gravitational parameter of the central body, μ, is also an unwieldy number when expressed in conventional units (3.986×10^5 km^3/s^2 for example). Therefore, when solving astronautics problems it's sometimes convenient to use a system of normalized units called *canonical units*.

We'll define canonical units for the Earth and the Sun, but we could easily do so for any other central body.

Canonical Units for Earth

1 Distance Unit (DU) = Earth's Radius

$$1 \text{ DU} = 6378.137 \text{ km} = 3963.191 \text{ mi.}$$

1 Earth Time Unit (TU_{Earth} or $\text{TU}_{\oplus}$) = time for a satellite in a circular orbit of 1 DU radius (just skimming Earth's surface) to travel 1 DU (1 radian of arc)

$$1 \text{ TU}_{\oplus} = 13.44685116 \text{ min} = 806.8110953 \text{ s}$$

1 DU/$\text{TU}_{\oplus}$ (Earth speed unit) = speed of a satellite in a circular orbit

$$1 \text{ DU}/\text{TU}_{\oplus} = 7.90536599 \text{ km/s} = 25936.24015 \text{ ft./s}$$

Earth's Gravitational Parameter

$$\mu_{\text{Earth}} \equiv 1 \frac{\text{DU}^3}{\text{TU}_{\oplus}^2} = 3.986005 \times 10^5 \frac{\text{km}^3}{\text{s}^2} = 1.4076444 \times 10^{16} \frac{\text{ft.}^3}{\text{s}^2}$$

Outline

B.1 **Canonical Units**
Canonical Units for Earth
Solar Canonical Units

B.2 **Unit ConversionsConstants**

B.3 **Constants**

B.4 **Greek Alphabet**

All data adapted from
Space Mission Analysis and Design
and *Fundamentals of Astrodynamics*

Solar Canonical Units

1 Astronomical Unit (AU) = Mean distance from Earth to the Sun (radius of Earth's orbit)

$$1 \text{ AU} = 149.59787 \times 10^6 \text{ km} = 4.908125 \times 10^{11} \text{ ft.}$$

1 Solar Time Unit (TU_{Sun}) = time for Earth in its (nearly) circular orbit around the Sun at 1 AU radius to travel 1 AU (1 radian of arc)

$$1 \text{ TU}_{Sun} = 58.132441 \text{ days} = 5.0226429 \times 10^6 \text{ s}$$

1 AU/TU_{Sun} (Solar speed unit) = Earth's speed in its 1 AU-radius circular orbit

$$1 \text{ AU}/\text{TU}_{Sun} = 29.784692 \text{ km}/\text{s} = 9.7719329 \times 10^4 \text{ ft.}/\text{s}$$

Sun's Gravitational Parameter

$$\mu_{Sun} \equiv 1 \frac{AU^3}{TU_{Sun}^2} = 1.327124 \times 10^{11} \frac{km^3}{s^2} = 4.6868016 \times 10^{21} \frac{ft.^3}{s^2}$$

B.2 Unit Conversions

In this book, we use the metric system of units, officially known as the *International System of Units*, or *SI*, except that we sometimes express angular measurements in degrees rather than the SI unit of radians. By international agreement, the fundamental SI units of length, mass, and time are as follows (see National Bureau of Standards Special Publication 330, 1986):

- The *meter* is the length of the path traveled by light in a vacuum over 1/299,792,458 of a second
- The *kilogram* is the mass of the international prototype of the kilogram
- The *second* is the duration of 9,192,631,770 periods of the radiation corresponding to the transition between two hyperfine levels of the ground state of the cesium-133 atom

Additional base units in the SI system are the *ampere* for electric current, the *kelvin* for thermodynamic temperature, the *mole* for amount of substance, and the *candela* for intensity of light. Mechtly [1977] neatly summarizes SI units for science and technology.

The names of multiples and submultiples of SI units take the following prefixes

Factor by which unit is multiplied	Prefix	Symbol
10^{18}	exa	E
10^{15}	peta	P
10^{12}	tera	T
10^{9}	giga	G
10^{6}	mega	M
10^{3}	kilo	k
10^{2}	hecto	h
10	deka	da
10^{-1}	deci	d
10^{-2}	centi	c
10^{-3}	milli	m
10^{-6}	micro	μ
10^{-9}	nano	n
10^{-12}	pico	p
10^{-15}	femto	f
10^{-18}	atto	a

For each quantity listed below, the SI unit and its abbreviation are in brackets. For convenient use in computers, we've listed conversion factors with the greatest available accuracy. Note that some conversions are exact definitions and some (speed of light, astronomical unit) depend on the value of physical constants. All notes are on the last page of the list.

To convert from	To	Multiply by	Notes
Acceleration [meters/second^2, m/s^2]			
Foot/second^2	m/s^2	3.048×10^{-1}	E
Free fall (standard), g	m/s^2	9.80665	E
Angular Acceleration [radians/second^2, rad/s^2]			
Degrees/second^2	rad/s^2	$\pi/180$ $\approx 0.01745329251994329577$	E
Revolutions/second^2, rev/s^2	rad/s^2	2π $\approx 6.283185307179586477$	E
Revolutions/minute^2	rad/s^2	$\pi/1800$ $\approx 1.745329251994329577 \times 10^{-3}$	E
Revolutions/minute^2	deg/s^2	0.1	E
Radians/second^2, rad/s^2	deg/s^2	$180/\pi$ $\approx 57.29577951308232088$	E
Revolutions/second^2, rev/s^2	deg/s^2	360	E

Angular Measure [radian, rad]. This book uses degree (abbreviated deg) as the basic unit.

To convert from	To	Multiply by	Notes
Degree	rad	$\pi/180$ $\approx 0.01745329251994329577$	E
Minute (of arc)	rad	$\pi/10,800$ $\approx 2.908882086657216 \times 10^{-4}$	E
Radian	deg	$= 180/\pi$ $\approx 57.29577951308232088$	E
Second (of arc)	rad	$\pi/648,000$ $\approx 4.8481368110954 \times 10^{-6}$	E

To convert from	To	Multiply by	Notes
Angular Momentum [kilogram · meter^2/second, kg · m^2/s]			
Gram · cm^2/second	kg · m^2/s	1.0×10^{-7}	E
Pound mass · inch^2/second	kg · m^2/s	2.926397×10^{-4}	
Slug · inch^2/second	kg · m^2/s	9.415402×10^{-3}	
Pound mass · foot^2/second	kg · m^2/s	4.214011×10^{-2}	
Inch · pound force · second	kg · m^2/s	1.129848×10^{-1}	
Slug · foot^2/second =			
Foot · pound force · second	kg · m^2/s	1.355818	
Angular Velocity [radian/second, rad/s]			
Degrees/second	rad/s	$\pi/180$ $\approx 0.01745329251994329577$	E
Radians/second	deg/s	$180/\pi$ $\approx 57.29577951308232088$	E
Revolutions/minute, r.p.m.	rad/s	$\pi/30$ $\approx 0.01047197551196597746$	E
Revolutions/second	rad/s	2π $\approx 6.283185307179586477$	E
Revolutions/minute, r.p.m.	deg/s	6.0	E
Revolutions/second	deg/s	360	E

To convert from	To	Multiply by	Notes
Area [meter^2, m^2]			
Acre	m^2	4.046856422×10^3	E
Foot^2, ft^2	m^2	0.09290304	E
Hectare	m^2	1.0×10^4	E
Inch^2, in^2	m^2	6.4516×10^{-4}	E
Mile^2 (U.S. statute)	m^2	2.58998811×10^6	M
Yard^2, yd^2	m^2	0.083612736	E
(Nautical mile)^2	m^2	3.429904×10^6	E
Density [kilogram/meter^3, kg/m^3]			
Gram/centimeter^3	kg/m^3	1.00×10^3	E
Pound mass/inch^3	kg/m^3	$2.767990471020 \times 10^4$	M
Pound mass/foot^3	kg/m^3	16.01846337396	M
Slug/foot^3	kg/m^3	$5.153788183932 \times 10^2$	M

Energy or Torque [joule ≡ newton · meter ≡ kilogram · meter^2/s^2,
J ≡ N · m ≡ kg · m^2/s^2]

To convert from	To	Multiply by	Notes
British thermal unit, Btu (mean)	J	$1.05505585262 \times 10^3$	M
Calorie (mean), cal	J	4.1868	M
Kilocalorie (mean), kcal	J	4.1868×10^3	M
Electron volt, eV	J	$1.60217733 \times 10^{-19}$	W
Erg ≡ gram · cm^2/s^2 = pole · cm · oersted	J	1.0×10^{-7}	E
Foot poundal	J	$4.214011009380 \times 10^{-2}$	M
Foot lbf = slug · foot^2/s^2	J	1.3558179483314	M
Kilowatt hour, kW · hr	J	3.60×10^6	E
Ton equivalent of TNT	J	4.184×10^9	E

Force [newton ≡ kilogram · meter/second^2, N ≡ kg · m /s^2]

To convert from	To	Multiply by	Notes
Dyne	N	1.0×10^{-5}	E
Pound force (avoirdupois), lbf ≡ slug · foot/s^2)	N	4.4482216152605	E

Length [meter, m]

To convert from	To	Multiply by	Notes
Angstrom, Å	m	1.0×10^{-10}	E
Astronomical unit, (IAU)	m	$1.4959787066 \times 10^{11}$	AA
Astronomical unit, (radio)	m	1.4959789×10^{11}	W
Earth's equatorial radius	m	6.37813649×10^6	W
Foot	m	0.3048	E
Inch	m	2.54×10^{-2}	E
Light year	m	$9.46073047258 \times 10^{15}$	W
Micron (μm)	m	1.0×10^{-6}	E
Mil (10^{-3} inch)	m	2.54×10^{-5}	E
Mile (U.S. statute), mi.	m	1.609344×10^3	E
Nautical mile (U.S.), NM	m	1.852×10^3	E
Parsec (IAU)	m	$3.08567759749 \times 10^{16}$	W
Yard	m	0.9144	E

To convert from	To	Multiply by	Notes
Mass [kilogram, kg]			
Atomic unit (electron)	kg	$9.1093877 \times 10^{-31}$	W
Metric ton	kg	1.0×10^3	E
Pound mass, lbm (avoirdupois)	kg	0.45359237	E
Slug	kg	14.59390293721	W
Short ton (2000 lbm)	kg	9.0718474×10^2	E
Solar mass	kg	1.9891×10^{30}	W
Moment of Inertia [kilogram $\cdot$ meter^2, kg $\cdot$ m^2]			
Gram $\cdot$ centimeter^2	kg $\cdot$ m^2	1.0×10^{-7}	E
Pound mass $\cdot$ inch^2	kg $\cdot$ m^2	$2.9263965343 \times 10^{-4}$	
Pound mass $\cdot$ foot^2	kg $\cdot$ m^2	$4.21401100938 \times 10^{-2}$	W
Slug $\cdot$ inch^2	kg $\cdot$ m^2	$9.415402418968 \times 10^{-3}$	W
Inch $\cdot$ pound force $\cdot$ s^2	kg $\cdot$ m^2	0.1129848290276	W
Slug $\cdot$ foot^2 = ft $\cdot$ lbf $\cdot$ s^2	kg $\cdot$ m^2	1.3558179483314	E
Power [watt $\equiv$ joule/second $\equiv$ kilogram $\cdot$ meter^2/second^3, W $\equiv$ J/s $\equiv$ kg $\cdot$ m^2/s^3]			
Foot pound force/second	W	1.355817948331	W
Horsepower (550 ft lbf/s)	W	$7.456998715823 \times 10^2$	W
Horsepower (electrical)	W	7.46×10^2	E
Solar luminosity	W	3.845×10^{26}	W
Pressure or Stress [pascal $\equiv$ newton/meter^2 $\equiv$ kilogram/(meter $\cdot$ second^2), Pa $\equiv$ N/m^2 $\equiv$ kg/m s^2]			
Atmosphere	Pa	1.01325×10^5	E
Bar	Pa	1.0×10^5	E
Centimeter of mercury (0° C)	Pa	$1.333223874145 \times 10^3$	W
Dyne/centimeter^2	Pa	0.10	E
Inch of mercury (0° C)	Pa	$3.386388640341 \times 10^3$	W
Pound force/foot^2	Pa	47.88025898034	W
Pound force/inch^2, p.s.i.	Pa	$6.894757293168 \times 10^3$	W
Torr (0° C)	Pa	$1.33322368421052 \times 10^2$	W
Solid Angle [steradian, sr]			
Degree^2, deg^2	sr	$(\pi/180)^2$ $\approx 3.046174197867085993 \times 10^{-4}$	E
Steradian, sr	deg^2	$(180/\pi)^2$ $\approx 3.282806350011743794 \times 10^3$	E
Stress (see Pressure)			
Temperature [kelvin, K]			
Celsius, °C	K	$t_K = t_C + 273.15$	E
Fahrenheit, °F	K	$t_K = (5/9)\,(t_F + 459.67)$	E
Fahrenheit, °F	C	$t_C = (5/9)\,(t_F - 32.0)$	E

To convert from	To	Multiply by	Notes
Time			
Sidereal day	s	8.6164100352×10^4	W
Solar day, average	s	8.64×10^4	W
Julian century	d	36,525	E
Gregorian calendar century	d	36,524.25	E
Torque (see Energy)			
Velocity [meter/second, m/s]			
Foot/minute	m/s	5.08×10^{-3}	E
Kilometer/hour	m/s	$(3.6)^{-1} = 0.277777...$	E
Foot/second	m/s	0.3048	E
Miles/hour	m/s	0.44704	E
Knot (international)	m/s	0.5144444444...	E
Miles/minute	m/s	26.8224	E
Miles/second	m/s	1.609344×10^3	E
Astronomical unit/ sidereal year	m/s	4.740388554×10^3	W
Velocity of light, c	m/s	2.99792458×10^8	E
Volume [meter^3, m^3]			
Foot^3	m^3	$2.8316846592 \times 10^{-2}$	E
Gallon (U.S. liquid)	m^3	$3.785411784 \times 10^{-3}$	E
Liter	m^3	1.0×10^{-3}	E

NOTES

E   (*Exact*) indicates that the conversion is exact by definition of the non-SI unit or that it came from other exact conversions
M   Values from Mechtly
W   Values from Wertz
AA Values from *Astronomical Almanac*
H   Values from Weast

B.3 Constants

Table B-1.   Fundamental Physical Constants.

Quantity	Symbol	Value	Units	Relative Uncertainty (1 σ, ppm)
Speed of light in a vacuum	c	299,792,458	m/s	(exact)
Universal Gravitational Constant	G	6.67259×10^{-11}	$N \cdot m^2/kg^2$	128
Planck constant	h	$6.6260755 \times 10^{-34}$	J·s	0.60
Elementary charge	e	$1.60217733 \times 10^{-19}$	C	0.30
Electron mass	m_e	$9.1093897 \times 10^{-31}$	kg	0.59
Proton mass	m_p	$1.6726231 \times 10^{-27}$	kg	0.59
Proton-electron mass ratio	m_p/m_e	1836.152701	--	0.02
Neutron mass	m_n	$1.6749286 \times 10^{-27}$	kg	0.59
Boltzmann constant, R/NA	k	1.380658×10^{-23}	J/K	8.5
Stefan-Boltzmann Constant	σ	5.67051×10^{-8}	$W/m^2 \cdot K^4$	34
Electron volt	eV	$1.60217733 \times 10^{-19}$	J	0.30
Atomic mass unit	u	$1.6605402 \times 10^{-27}$	kg	0.59

Table B-2. Spaceflight Constants.

Quantity	Value	Units
$\mu_{Earth} = G\, m_{Earth}$	3.986004418×10^5	km^3/s^2
$\mu_{Sun} = G\, m_{Sun}$	1.327124×10^{11}	km^3/s^2
$\mu_{Moon} = G\, m_{Moon}$	4.902798882×10^3	km^3/s^2
$\mu_{Earth\ and\ Moon}$ $= G\, m_{Earth\ and\ Moon}$	4.035031135×10^5	km^3/s^2
Obliquity of the ecliptic at Epoch 2000	23.43928111	deg
Precession of the equinox	1.396971278	deg/century
Flattening factor for Earth	1/298.25642	
Earth's equatorial radius	6378.13649	km
1 AU	$1.49597870691 \times 10^{11}$	m
Mean lunar distance	3.84401×10^8	m
Solar constant	1358	W/m^2 at 1 AU
Acceleration due to gravity at Earth's equatorial radius	9.798	m/s^2
Acceleration due to gravity at standard sea level	9.80665	m/s^2
1 solar day	1.00273790935	sidereal days
Earth's rotation rate	15 0.25 0.25068447733746215	deg/sidereal hr deg/sidereal min deg/solar min

B.4 Greek Alphabet

Table B-3.   Greek Alphabet. [Adapted from The American Heritage Dictionary]

Symbol	Name	Symbol	Name
A α	alpha	N ν	nu
B β	beta	Ξ ξ	xi
Γ γ	gamma	O o	omicron
Δ δ	delta	Π π	pi
E ε	epsilon	P ρ	rho
Z ζ	zeta	Σ σ	sigma
H η	eta	T τ	tau
Θ θ	theta	Υ υ	upsilon
I ι	iota	Φ φ	phi
K κ	kappa	X χ	chi or khi
Λ λ	lambda	Ψ ψ	psi
M μ	mu	Ω ω	omega

References

American Heritage Dictionary. Boston, MA: Houghton Mifflin Company, 1985.

Cohen, E. Richard and Taylor, B.N. *CODATA Bulletin No. 63*, Pergamon Press, Nov. 1986.

Hagen, James B. and Boksenberg, A., eds. *Astronomical Almanac.* Washington, D.C.: U.S. Government Printing Office. 1991.

Mechtly, E. A. *The International System of Units.* Champaign, IL: Stipes Publishing Company, 1977.

Weast, R. C., ed. *CRC Handbook of Chemistry and Physics.* Boca Raton, FL: CRC Press, 1985.

Wertz, James R., ed. *Spacecraft Attitude Determination and Control.* Holland: D. Reidel Publishing Company, 1978.

Wertz, James R. and Wiley J. Larson. *Space Mission Analysis and Design.* Third edition. Dordrecht, Netherlands: Kluwer Academic Publishers, 1999.

Derivations

C.1 Restricted Two-body Equation of Motion

In Chapter 4 we developed the restricted two-body equation of motion

$$\ddot{\vec{R}} + \frac{\mu}{R^2}\hat{R} = 0 \qquad (C\text{-}1)$$

Knowing that $\hat{R} = \dfrac{\vec{R}}{R}$, we can write it as

$$\ddot{\vec{R}} + \frac{\mu}{R^3}\vec{R} = 0 \qquad (C\text{-}2)$$

Several of the following derivations will use this fundamental relationship.

Outline

C.1 Restricted Two-body Equation of Motion

C.2 Constants of Motion
Proving Specific Mechanical Energy is Constant
Proving Specific Angular Momentum is Constant

C.3 Solving the Two-body Equation of Motion

C.4 Relating the Energy Equation to the Semimajor Axis

C.5 The Eccentricity Vector

C.6 Deriving the Period Equation for an Elliptical Orbit

C.7 Finding Position and Velocity Vectors from COEs

C.8 $V_{burnout}$ in SEZ Coordinates

C.9 Deriving the Rocket Equation

C.10 Deriving the Potential Energy Equation and Discovering the Potential Energy Well

C.2 Constants of Motion

Proving Specific Mechanical Energy is Constant

We can prove that the specific mechanical energy, ε, of an orbit is constant by beginning with the restricted two-body equation of motion in Equation (C-2).

$$\ddot{\vec{R}} + \frac{\mu}{R^3}\vec{R} = 0$$

We then take the dot product of both sides with $\dot{\vec{R}}$.

$$\dot{\vec{R}} \cdot \left(\ddot{\vec{R}} + \frac{\mu}{R^3}\vec{R}\right) = \dot{\vec{R}} \cdot 0$$

or

$$\dot{\vec{R}} \cdot \ddot{\vec{R}} + \frac{\mu}{R^3}\vec{R} \cdot \dot{\vec{R}} = 0 \cdot \dot{\vec{R}} = 0$$

Note: $\dot{\vec{R}} = \vec{V}$ and $\ddot{\vec{R}} = \dot{\vec{V}}$ so

$$\vec{V} \cdot \dot{\vec{V}} + \frac{\mu}{R^3}\vec{R} \cdot \dot{\vec{R}} = 0 \qquad (C\text{-}3)$$

From the definition of the dot product, we know for any two vectors $\vec{a}$ and $\vec{b}$

$$\vec{a} \cdot \vec{b} = a\, b\, \cos\theta$$

where

θ = angle between the two vectors

Thus, to see how to proceed, we use a specific example of a dot product

$$\vec{a} \cdot \vec{a} = a\, a\, \cos\theta$$

If we recognize that $\vec{a}$ is parallel to itself, then we know the angle between them is zero and $\cos\theta = 1$. Thus, substituting for $\cos\theta$, we get

$$\vec{a} \cdot \vec{a} = a^2 \qquad (C\text{-}4)$$

Proceeding with the proof, we must take the derivative of both sides of Equation (C-4)

$$\frac{d}{dt}(\vec{a} \cdot \vec{a}) = \frac{d}{dt}(a^2)$$

Applying the chain rule from calculus, the derivatives become

$$\vec{a} \cdot \dot{\vec{a}} + \dot{\vec{a}} \cdot \vec{a} = 2a\,\dot{a}$$

$$2(\vec{a} \cdot \dot{\vec{a}}) = 2a\,\dot{a}$$

thus

$$\vec{a} \cdot \dot{\vec{a}} = a\,\dot{a} \tag{C-5}$$

[*Note:* $\left|\dot{\vec{a}}\right| \neq \dot{a}$]

Now, if we recognize that this result is the same as the two expressions on the left side of Equation (C-3), we can rewrite it as

$$V\dot{V} + \frac{\mu}{R^3}R\dot{R} = 0$$

or

$$V\dot{V} + \frac{\mu}{R^2}\dot{R} = 0 \tag{C-6}$$

To recognize this equation as usable, we suppose there are two variables x and y, such that

$$x = \frac{V^2}{2} \tag{C-7}$$

and

$$y = \frac{-\mu}{R} \tag{C-8}$$

To get where we need to go, we take their derivatives

$$\frac{dx}{dt} = V\dot{V}$$

and

$$\frac{dy}{dt} = \frac{\mu}{R^2}\dot{R}$$

Now, we substitute these expressions for the corresponding quantities in Equation (C-6) and get

$$\frac{dx}{dt} + \frac{dy}{dt} = \frac{d}{dt}(x + y) = 0 \tag{C-9}$$

Now, we substitute our expressions for x and y, to get

$$\frac{d}{dt}\left(\frac{V^2}{2} - \frac{\mu}{R}\right) = 0 \tag{C-10}$$

From Chapter 4, we know the term in parenthesis is the specific mechanical energy, ε

$$\varepsilon = \frac{V^2}{2} - \frac{\mu}{R} \qquad \text{(C-11)}$$

Finally, if $\frac{d}{dt}(\varepsilon) = 0$, which we showed, then ε = constant, because, when we integrate this differential equation, we get a constant of integration on the right side.

Proving Specific Angular Momentum is Constant

We can prove the specific angular momentum, $\vec{h}$, of an orbit is constant by taking the cross product of the two-body equation of motion, Equation (C-2), with the position vector, $\vec{R}$

$$\vec{R} \times \left(\ddot{\vec{R}} + \frac{\mu}{R^3}\vec{R} \right) = \vec{R} \times 0$$

$$\vec{R} \times \ddot{\vec{R}} + \frac{\mu}{R^3}(\vec{R} \times \vec{R}) = 0$$

Because the cross product of parallel vectors is zero, the second term goes to zero, and we're left with

$$\vec{R} \times \ddot{\vec{R}} = 0 \qquad \text{(C-12)}$$

Now realize that

$$\frac{d}{dt}(\vec{R} \times \dot{\vec{R}}) = (\dot{\vec{R}} \times \dot{\vec{R}}) + (\vec{R} \times \ddot{\vec{R}})$$

where $\dot{\vec{R}} \times \dot{\vec{R}} = 0$, (parallel vectors, again), leaving only

$$\frac{d}{dt}(\vec{R} \times \dot{\vec{R}}) = \vec{R} \times \ddot{\vec{R}}$$

Substituting this quantity into Equation (C-12), we get

$$\frac{d}{dt}(\vec{R} \times \dot{\vec{R}}) = 0$$

but $\dot{\vec{R}} = \vec{V}$, so we get

$$\frac{d}{dt}(\vec{R} \times \vec{V}) = 0$$

Recall from Chapter 4 that the specific angular momentum is

$$\vec{h} = \vec{R} \times \vec{V}$$

Thus,

$$\frac{d}{dt}(\vec{h}) = 0$$

When we integrate both sides of this equation, we get

$$\vec{h} = \text{constant}$$

which proves that specific angular momentum is constant.

C.3 Solving the Two-body Equation of Motion

To find a solution that describes satellite motion, we begin with the two-body equation of motion from Equation (C-2)

$$\ddot{\vec{R}} + \frac{\mu}{R^3}\vec{R} = 0 \qquad\qquad \text{(C-2)}$$

We can't solve for $\vec{R}$ as a function of time in closed form, but we can find an exact solution using variable substitution. We cross both sides of the equation with the specific angular momentum vector, $\vec{h}$

$$\ddot{\vec{R}} \times \vec{h} + \frac{\mu}{R^3}(\vec{R} \times \vec{h}) = 0 \times \vec{h} = 0$$

using the cross-product identity $\vec{a} \times \vec{b} = -(\vec{b} \times \vec{a})$, we get

$$\ddot{\vec{R}} \times \vec{h} = \frac{\mu}{R^3}(\vec{h} \times \vec{R}) \qquad\qquad \text{(C-13)}$$

Beginning with the left-hand side of Equation (C-13), we take the derivative using the chain rule to get

$$\frac{d}{dt}(\dot{\vec{R}} \times \vec{h}) = \ddot{\vec{R}} \times \vec{h} + \dot{\vec{R}} \times \dot{\vec{h}}$$

But $\vec{h} = \text{constant}$ so $\dot{\vec{h}} = 0$, thus

$$\frac{d}{dt}(\dot{\vec{R}} \times \vec{h}) = \ddot{\vec{R}} \times \vec{h} \qquad\qquad \text{(C-14)}$$

Now we turn to the right-hand side of Equation (C-13). From the vector identity

$$(\vec{a} \times \vec{b}) \times \vec{c} = \vec{b}(\vec{a} \cdot \vec{c}) - \vec{a}(\vec{b} \cdot \vec{c})$$

we can say

$$\vec{h} \times \vec{R} = (\vec{R} \times \vec{V}) \times \vec{R} = \vec{V}(\vec{R} \cdot \vec{R}) - \vec{R}(\vec{V} \cdot \vec{R})$$

From the definition of the dot product and Equation (C-4), we know $\vec{R} \cdot \vec{R} = R^2$, so

$$\vec{h} \times \vec{R} = \vec{V}R^2 - \vec{R}(\dot{\vec{R}} \cdot \vec{R})$$

From Equation (C-5), we know $\vec{\dot{R}} \cdot \vec{R} = R\dot{R}$, so we get

$$\vec{h} \times \vec{R} = \vec{V}R^2 - \vec{R} R \dot{R}$$

Multiplying both sides by μ/R^3, we get

$$\frac{\mu}{R^3}(\vec{h} \times \vec{R}) = \frac{\mu}{R^3}(\vec{V}R^2 - \vec{R} R \dot{R}) \qquad (C-15)$$

Now, from the derivative of a quotient, we realize

$$\mu\frac{d}{dt}\left(\frac{\vec{R}}{R}\right) = \mu\frac{d}{dt}(\vec{R} R^{-1})$$

$$= \mu(\vec{\dot{R}} R^{-1} - \vec{R} R^{-2}\dot{R})$$

We factor out $1/R^3$ from the right side, to get

$$\mu\frac{d}{dt}\left(\frac{\vec{R}}{R}\right) = \frac{\mu}{R^3}(\vec{V}R^2 - \vec{R} R \dot{R}) \qquad (C-16)$$

Equating Equation (C-15) and Equation (C-16), we get

$$\frac{\mu}{R^3}(\vec{h} \times \vec{R}) = \mu\frac{d}{dt}\left(\frac{\vec{R}}{R}\right) \qquad (C-17)$$

Substituting Equation (C-14) and Equation (C-17) into Equation (C-13), we end up with

$$\frac{d}{dt}(\vec{\dot{R}} \times \vec{h}) = \mu\frac{d}{dt}\left(\frac{\vec{R}}{R}\right)$$

Integrating both sides, we get

$$\vec{\dot{R}} \times \vec{h} = \mu\frac{\vec{R}}{R} + \vec{B} \qquad (C-18)$$

where

$\vec{B}$ = constant vector of integration

Now we dot both sides of Equation (C-18) with $\vec{R}$

$$\vec{R} \cdot (\vec{\dot{R}} \times \vec{h}) = \vec{R} \cdot \left(\mu\frac{\vec{R}}{R} + \vec{B}\right) \qquad (C-19)$$

From the vector identity

$$\vec{a} \cdot (\vec{b} \times \vec{c}) = (\vec{a} \times \vec{b}) \cdot \vec{c}$$

715

we have

$$\vec{R} \cdot (\dot{\vec{R}} \times \hat{h}) = (\vec{R} \times \dot{\vec{R}}) \cdot \hat{h}$$

$$= (\vec{R} \times \vec{V}) \cdot \hat{h}$$

$$= \vec{h} \cdot \hat{h}$$

$$= h^2$$

so, when we substitute into Equation (C-19), it becomes

$$h^2 = \vec{R} \cdot \left(\mu \frac{\vec{R}}{R} + \vec{B} \right) \qquad \text{(C-20)}$$

Expanding the right-hand side of Equation (C-20), we have

$$\vec{R} \cdot \left(\mu \frac{\vec{R}}{R} + \vec{B} \right) = \frac{\mu}{R}(\vec{R} \cdot \vec{R}) + (\vec{R} \cdot \vec{B})$$

$$= \frac{\mu}{R}(R^2) + \vec{R} \cdot \vec{B}$$

$$= \mu R + R B \cos \nu$$

where

ν = angle between $\vec{R}$ and $\vec{B}$

Thus, we end up with

$$h^2 = \mu R + R B \cos \nu$$

Solving for the magnitude of the position vector, R, we get

$$R = \frac{h^2 / \mu}{1 + (B/\mu) \cos \nu}$$

Now, let $h^2/\mu = k_1$ and $B/\mu = k_2$. Substituting for these terms, we have

$$R = \frac{k_1}{1 + k_2 \cos \nu} \qquad \text{(C-21)}$$

which is the solution of the restricted two-body equation of motion in terms of two constants k_1, k_2, and angle ν. This solution derives purely from the dynamics of the problem. From geometry, we know the polar form of the equation for a conic section is also

$$R = \frac{k_1}{1 + k_2 \cos \nu}$$

where

R $\;=\;$ magnitude of the position vector (km)

$k_1 = p =$ semilatus rectum shown in Figure C-1 (km)

$k_2 = e =$ eccentricity (unitless)

v $\;=\;$ true anomaly (deg or rad)

so,

$$R = \frac{p}{1 + e\cos v}$$

If we use the expression for p,

$$p = a\,(1 - e^2)$$

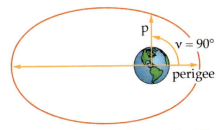

Figure C-1.   Semilatus Rectum. We define the semilatus rectum, p, as the distance from the center of Earth to the orbit where true anomaly, v, is 90°.

where

p $\;=\;$ semilatus rectum or semiparameter (km)

a $\;=\;$ semimajor axis (km)

e $\;=\;$ eccentricity (unitless)

we get the familiar solution to the two-body equation of motion which relates dynamics to geometry and shows that all objects moving under the influence of gravity travel along conic sections.

$$R = \frac{a(1 - e^2)}{1 + e\cos v} \qquad\qquad \text{(C-22)}$$

C.4 Relating the Energy Equation to the Semimajor Axis

We start with the equation for specific angular momentum

$$\vec{h} = \vec{R} \times \vec{V} = RV\cos\phi$$

When we apply it at perigee, where the flight-path angle $\phi = 0$, we get

$$h = R_{perigee}V_{perigee}$$

Now, we recall the relationship for specific mechanical energy

$$\varepsilon = \frac{V^2}{2} - \frac{\mu}{R}$$

Because ε is constant everywhere along an orbit, we can examine it at perigee. We realize from above that

$$V_{perigee}^2 = \frac{h^2}{R_{perigee}^2}$$

Substituting this squared velocity term into the specific mechanical energy equation, we have

$$\varepsilon = \frac{h^2}{2R_{perigee}^2} - \frac{\mu}{R_{perigee}} \tag{C-23}$$

From the polar equation of a conic, Equation (C-22)

$$R = \frac{a(1-e^2)}{1+e\cos\nu}$$

at perigee $\nu = 0$, so

$$R_{perigee} = \frac{a(1-e^2)}{1+e} = \frac{a(1-e)(1+e)}{1+e}$$

$$R_{perigee} = a(1-e) \tag{C-24}$$

Note that $p = a(1-e^2) = h^2/\mu$. Thus,

$$h^2 = \mu a(1-e^2)$$

Substituting for h^2 in Equation (C-23), gives us

$$\varepsilon = \frac{\mu a(1-e^2)}{2R_{perigee}^2} - \frac{\mu}{R_{perigee}}$$

and substituting for $R_{perigee}$ in Equation (C-24), we get

$$\varepsilon = \frac{\mu a(1-e^2)}{2a^2(1-e)^2} - \frac{\mu}{a(1-e)}$$

Next, we form a common denominator and add the fractions

$$= \frac{\mu(1-e^2)}{2a(1-e)^2} - \frac{2\mu(1-e)}{2a(1-e)(1-e)}$$

$$= \frac{\mu(1-e^2) - 2\mu(1-e)}{2a(1-e)^2}$$

Then we simplify the expression to get

$$\varepsilon = \frac{-\mu[2(1-e) - (1-e^2)]}{2a(1-e)^2}$$

$$= \frac{-\mu}{2a}\left[\frac{2 - 2e - 1 + e^2}{1 - 2e + e^2}\right]$$

$$= \frac{-\mu}{2a}\left[\frac{1 - 2e + e^2}{1 - 2e + e^2}\right]$$

$$\boxed{\therefore \varepsilon = \frac{-\mu}{2a}} \qquad \text{(C-25)}$$

This relationship says that specific mechanical energy is inversely proportional to the orbit's semimajor axis (its size), a. It's valid for all conic sections.

C.5 The Eccentricity Vector

In deriving the solution to the two-body equation of motion, we developed a constant vector, $\vec{B}$. We know the magnitude of this vector relates to eccentricity, by $B = \mu e$. We can also define an eccentricity vector, $\vec{e}$, in the same direction as $\vec{B}$

$$\vec{e} = \vec{B}/\mu \qquad\qquad\qquad (C\text{-}26)$$

To develop a more useful relationship for $\vec{e}$, we begin with the relationship we developed in solving the two-body problem

$$\dot{\vec{R}} \times \vec{h} = \mu\frac{\vec{R}}{R} + \vec{B} \qquad\qquad (C\text{-}18)$$

solving for $\vec{B}$, we get

$$\vec{B} = (\dot{\vec{R}} \times \vec{h}) - \mu\frac{\vec{R}}{R}$$

Substituting this expression for $\vec{B}$ into Equation (C-26), we get

$$\vec{e} = \frac{\left[(\dot{\vec{R}} \times \vec{h}) - \mu\dfrac{\vec{R}}{R}\right]}{\mu}$$

$$\vec{e} = \frac{\dot{\vec{R}} \times \vec{h}}{\mu} - \frac{\vec{R}}{R} \qquad\qquad (C\text{-}27)$$

Substituting for $\vec{h} = \vec{R} \times \vec{V}$ and $\dot{\vec{R}} = \vec{V}$

$$\vec{e} = \frac{\vec{V} \times (\vec{R} \times \vec{V})}{\mu} - \frac{\vec{R}}{R}$$

Applying the triple cross product identify, $\vec{a} \times (\vec{b} \times \vec{c}) = \vec{b}(\vec{a} \cdot \vec{c}) - \vec{c}(\vec{a} \cdot \vec{b})$, to the numerator of the first fraction, we have

$$\vec{e} = \frac{\vec{R}(\vec{V} \cdot \vec{V}) - \vec{V}(\vec{V} \cdot \vec{R})}{\mu} - \frac{\vec{R}}{R}$$

When we dot a vector by itself, we get $\vec{V} \cdot \vec{V} = V^2$, so we use this relationship to get

$$\vec{e} = \frac{\vec{R}V^2 - \vec{V}(\vec{V} \cdot \vec{R})}{\mu} - \frac{\vec{R}}{R}$$

Now we multiply both sides by μ to get

$$\vec{e}\mu = \vec{R}V^2 - \vec{V}(\vec{V} \cdot \vec{R}) - \mu\frac{\vec{R}}{R}$$

We arrange terms to get

$$\vec{e}\mu = \vec{R}\left(V^2 - \frac{\mu}{R}\right) - \vec{V}(\vec{V} \cdot \vec{R})$$

Finally, we divide by μ to arrive at the vector equation for eccentricity

$$\vec{e} = \frac{1}{\mu}\left[\left(V^2 - \frac{\mu}{R}\right)\vec{R} - (\vec{R} \cdot \vec{V})\vec{V}\right] \qquad \text{(C-28)}$$

How do we know what direction $\vec{e}$ points? To determine that direction, we begin with the relationship for $\vec{e}$ in Equation (C-27)

$$\vec{e} = \frac{\dot{\vec{R}} \times \vec{h}}{\mu} - \frac{\vec{R}}{R} \qquad \text{(C-29)}$$

We can express $\vec{R}$ and $\vec{V}$ in the perifocal coordinate system, $\hat{P}\hat{Q}\hat{W}$, shown in Figure C-2, where the

- Origin is Earth's center
- Fundamental plane is the orbital plane
- Principal direction is toward perigee

$\vec{R}$ and $\vec{V}$ then become

$$\vec{R} = R\cos v\hat{P} + R\sin v\hat{Q}$$

$$\vec{V} = \sqrt{\frac{\mu}{P}}\left[-\sin v\hat{P} + (e + \cos v)\hat{Q}\right]$$

In terms of $\vec{R} \times \vec{V}$, we start by solving for $\vec{h}$

$$\vec{h} = \vec{R} \times \vec{V} = \begin{vmatrix} \hat{P} & \hat{Q} & \hat{W} \\ R\cos v & R\sin v & 0 \\ -\sqrt{\frac{\mu}{p}}\sin v & \sqrt{\frac{\mu}{p}}(e + \cos v) & 0 \end{vmatrix}$$

$$= \sqrt{\frac{\mu}{p}}[R\cos v((e + \cos v) + R\sin^2 v)]\hat{W}$$

$$= \sqrt{\frac{\mu}{p}}[eR\cos v + R\cos^2 v + R\sin^2 v]\hat{W}$$

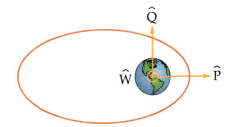

Figure C-2. $\hat{P}\hat{Q}\hat{W}$ System. The $\hat{P}\hat{Q}\hat{W}$ system has its origin at Earth's center. $\hat{P}$ points in the direction of the eccentricity vector, $\hat{W}$ is perpendicular to the orbital plane in the direction of the angular momentum vector, and $\hat{Q}$ completes the right-hand rule.

Applying the trigonometric identity $\cos^2\theta + \sin^2\theta = 1$, we get

$$\hat{h} = \sqrt{\frac{\mu}{p}}[eR\cos v + R]\hat{W}$$

Then, factoring out R, gives us

$$\hat{h} = \sqrt{\frac{\mu}{p}}[R(1 + e\cos v)]\hat{W}$$

Now we look at the first cross product in Equation (C-27)

$$\dot{\vec{R}} \times \hat{h} = \vec{V} \times \hat{h} = \begin{vmatrix} \hat{P} & \hat{Q} & \hat{W} \\ \sqrt{\frac{\mu}{p}}(-\sin v) & \sqrt{\frac{\mu}{p}}(e + \cos v) & 0 \\ 0 & 0 & \sqrt{\frac{\mu}{p}}[R(1 + e\cos v)] \end{vmatrix}$$

$$= \left(\sqrt{\frac{\mu}{p}}\right)^2 \left\{ [R(e + \cos v)(1 + e\cos v)]\hat{P} - [(-\sin v)(R)(1 + e\cos v)]\hat{Q} \right\}$$

From the solution to the two-body equation of motion in Equation (C-22) and the definition of semiparameter, p

$$p = R (1 + e\cos v)$$

Substituting p for $R(1 + e\cos v)$, we get

$$\vec{V} \times \hat{h} = \frac{\mu}{p}\left\{ [(e + \cos v)p]\hat{P} + [p\sin v]\hat{Q} \right\}$$

$$\vec{V} \times \hat{h} = \mu(e + \cos v)\hat{P} + \mu\sin v\hat{Q}$$

Substituting into Equation (C-27), gives us

$$\hat{e} = \frac{\mu(e + \cos v)\hat{P} + \mu\sin v\hat{Q}}{\mu} - \frac{R\cos v\hat{P} + R\sin v\hat{Q}}{R}$$

$$= (e + \cos v)\hat{P} + \sin v\hat{Q} - \cos v\hat{P} - \sin v\hat{Q}$$

$$\hat{e} = e\hat{P} + \cos v\hat{P} + \sin v\hat{Q} - \cos v\hat{P} - \sin v\hat{Q}$$

Simplifying this expression yields the result

$$\hat{e} = e\hat{P}$$

Therefore, $\hat{e}$ points at perigee.

C.6 Deriving the Period Equation for an Elliptical Orbit

The orbital geometry in Figure C-3 shows that the horizontal component of velocity is

$$V \cos \phi, \text{ or simply, } R\dot{v}$$

Knowing $|\vec{h}| = |\vec{R} \times \vec{V}| = RV \cos\phi$, we can express the specific angular momentum, h, as

$$h = \frac{R^2 dv}{dt} \Rightarrow dt = \frac{R^2}{h} dv$$

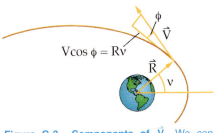

Figure C-3. Components of $\hat{V}$. We can break out the components of the velocity vector, $\hat{V}$, as shown.

From elementary calculus we know that the differential element of area, dA, as shown in Figure C-4, swept out by the radius vector as it moves through an angle, dv, is

$$dA = \frac{1}{2}R(Rdv) = \frac{1}{2}R^2 dv$$

So, we can rewrite the above equation as

$$dt = \frac{2}{h} dA$$

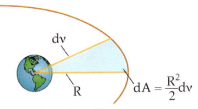

Figure C-4. A satellite sweeps out a small area, dA, per unit time.

which proves Kepler's Second Law that "equal areas are swept out by the radius vector in equal time intervals" because h is constant for an orbit. Integrating this equation through one period yields the following

$$P = \frac{2\pi ab}{h}$$

where

P   = period (s)

a   = semimajor axis (km)

b   = semiminor axis (km)

πab = total area of an ellipse

From the geometry of an ellipse, we use $b = \sqrt{a^2 - c^2} = \sqrt{a^2(1 - e^2)} = \sqrt{ap}$ and because we have $h = \sqrt{\mu p}$, from the definition of specific angular momentum, we substitute for b and h to get

$$P = \frac{2\pi a \sqrt{ap}}{\sqrt{\mu p}} = \frac{2\pi}{\sqrt{\mu}} a^{3/2} = 2\pi \sqrt{\frac{a^3}{\mu}}$$

C.7 Finding Position and Velocity Vectors from COEs

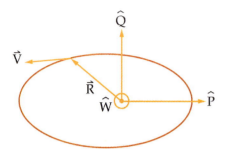

Figure C-5. The Perifocal Coordinate System. $\hat{P}$ points at perigee. $\hat{Q}$ is 90° from $\hat{P}$, in the direction of the spacecraft's motion, $\hat{W}$ is perpendicular to the orbital plane (out of the page).

We start by defining a new coordinate system, the *Perifocal System (PQW)*, as shown in Figure C-5, the

- Origin is Earth's center
- Fundamental plane is the orbital plane
- Principal direction is toward perigee

Writing $\vec{R}$ and $\vec{V}$ in terms of $\hat{P}$ $\hat{Q}$ $\hat{W}$

$$\vec{R} = R\cos\nu\hat{P} + R\sin\nu\hat{Q}$$

$$\vec{V} = \frac{d\vec{R}}{dt} = (\dot{R}\cos\nu - R\dot{\nu}\sin\nu)\hat{P} + (\dot{R}\sin\nu + R\dot{\nu}\cos\nu)\hat{Q}$$

What about $\dot{\hat{P}}$ and $\dot{\hat{Q}}$? Answer: PQW is an inertial reference frame, so the derivatives equal zero. Although we know R and ν, we don't know $\dot{R}$ and $\dot{\nu}$. So, we use the solution to the two-body equation of motion

$$R = \frac{p}{1 + e\cos\nu} = p(1 + e\cos\nu)^{-1}$$

To find $\dot{R}$, we take the time derivative of the expression, recognizing that ν is the only quantity to vary.

$$\dot{R} = -p(-e\dot{\nu}\sin\nu)(1 + e\cos\nu)^{-2} = \frac{pe\dot{\nu}\sin\nu}{(1 + e\cos\nu)^2}$$

To find $\dot{\nu}$, we must look at orbital geometry

$$|\vec{h}| = |\vec{R} \times \vec{V}| = RV\sin\theta = RV\cos\phi$$

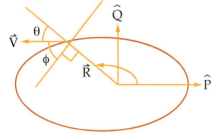

Figure C-6. $\vec{R}$ and $\vec{V}$ in the Perifocal System.

The component of $\vec{V}$ normal to $\vec{R}$ is $R \cdot R\dot{\nu}$ (tangential velocity, from Section C.6),

$$R\dot{\nu} = V\cos\phi$$

So, substituting for h gives us

$$h = R^2\dot{\nu}$$

Then, rearranging this equation, we get

$$\dot{\nu} = \frac{h}{R^2}$$

Substituting this expression for $\dot{v}$ into the equation for $\dot{R}$, gives us

$$\dot{R} = \frac{p\,e\,h\sin v}{R^2(1+e\cos v)^2} \quad \text{but } R = \frac{p}{1+e\cos v}$$

so, we substitute for R

$$\dot{R} = \frac{pe\,\sin v(1+e\cos v)^2}{p^2(1+e\cos v)^2} = \frac{e\sqrt{\mu p}\sin v}{p}$$

$$\dot{R} = \sqrt{\frac{\mu}{p}}e\sin v$$

Now, using the polar equation of a conic for R and the expression we just derived for $\dot{v}$, we multiply them and have

$$R\dot{v} = \frac{p}{1+e\cos v}\frac{h}{R^2} = \frac{ph(1+e\cos v)^2}{p^2(1+e\cos v)} = \frac{\sqrt{\mu p}(1+e\cos v)}{p}$$

$$R\dot{v} = \sqrt{\frac{\mu}{p}}(1+e\cos v)$$

Going back to the equation for $\vec{V}$, and substituting for $\dot{R}$ and $\dot{v}$, we get

$$\vec{V} = \left[\sqrt{\frac{\mu}{p}}e\sin v\cos v - \sqrt{\frac{\mu}{p}}(1+e\cos v)\sin v\right]\hat{P}$$

$$+ \left[\sqrt{\frac{\mu}{p}}e\sin v\sin v + \sqrt{\frac{\mu}{p}}(1+e\cos v)\cos v\right]\hat{Q}$$

When we distribute the quantities in parentheses, we get

$$\vec{V} = \sqrt{\frac{\mu}{p}}(e\sin v\cos v - \sin v - e\cos v\sin v)\hat{P}$$

$$+ \sqrt{\frac{\mu}{p}}(e\sin^2 v + \cos v + e\cos^2 v)\hat{Q}$$

Simplifying this equation by subtracting like terms and applying the trigonometric identity, $\sin^2\theta + \cos^2\theta = 1$, we get

$$\vec{V} = \sqrt{\frac{\mu}{p}}(-\sin v)\hat{P} + \sqrt{\frac{\mu}{p}}(e+\cos v)\hat{Q}$$

Finally, we can write $\vec{R}, \vec{V}$ in the perifocal coordinate frame as

$$\vec{R}_{PQW} = \frac{a(1-e^2)}{1+e\cos v}\left[\cos v\hat{P} + \sin v\hat{Q}\right]$$

$$\vec{V}_{PQW} = \sqrt{\frac{\mu}{p}}\left[-\sin v\hat{P} + (e+\cos v)\hat{Q}\right]$$

which are the position and velocity vectors entirely in terms of the Classic Orbital Elements (COEs).

The next step in this problem is to transform the coordinates of the $\vec{R}$ and $\vec{V}$ vectors from the $\hat{P}\hat{Q}\hat{W}$ system to the $\hat{I}\hat{J}\hat{K}$ system. This step requires three separate transformation matrices using the remaining COEs, i, ω, and Ω. (For basic vector transformations, see Appendix A.2, "Transforming Vector Coordinates.") To get a vector from the $\hat{P}\hat{Q}\hat{W}$ system into the $\hat{I}\hat{J}\hat{K}$ frame, we begin with a rotation about the $\hat{W}$ axis (a rotation about the third axis, or ROT3) through a negative argument of perigee angle, $-\omega$, to bring $\hat{P}$ into the equatorial plane, as shown in Figure C-7. The matrix for this operation is

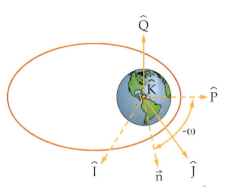

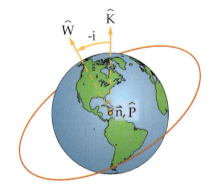

Figure C-7.   Rotation 3 of $-\omega$ about $\hat{W}$.
(*Note:* W is out of the page.)

$$ROT3(-\omega) = \begin{bmatrix} \cos\omega & -\sin\omega & 0 \\ \sin\omega & \cos\omega & 0 \\ 0 & 0 & 1 \end{bmatrix}$$

This rotation aligns $\hat{P}$ with the ascending node vector, $\hat{n}$. Next, we rotate the system about this new $\hat{P}/\hat{n}$ axis through an angle of minus inclination, $-i$, to bring $\hat{Q}$ to the equatorial plane, as shown in Figure C-8. This step takes a ROT1 matrix

$$ROT1(-i) = \begin{bmatrix} 1 & 0 & 0 \\ 0 & \cos i & -\sin i \\ 0 & \sin i & \cos i \end{bmatrix}$$

Figure C-8.   Rotation 1 of $-i$ about the new $\hat{P}/\hat{n}$ axis.

This rotation aligns $\hat{W}$ with $\hat{K}$. Finally, we rotate the system about the $\hat{W}/\hat{K}$ axis through a negative right ascension of the ascending node angle, $-\Omega$, to align $\hat{P}$ with $\hat{I}$ and $\hat{Q}$ with $\hat{J}$, as shown in Figure C-9. The ROT3 matrix is

$$ROT3(-\Omega) = \begin{bmatrix} \cos\Omega & -\sin\Omega & 0 \\ \sin\Omega & \cos\Omega & 0 \\ 0 & 0 & 1 \end{bmatrix}$$

Putting it all together for this specific transformation, we have

$$\vec{R}_{IJK} = [A]\vec{R}_{PQW}$$

$$\vec{V}_{IJK} = [A]\vec{V}_{PQW}$$

where

$\vec{R}_{IJK}$, $\vec{R}_{PQW}$ = position vectors in the $\hat{I}\hat{J}\hat{K}$ and $\hat{P}\hat{Q}\hat{W}$ systems

$\vec{V}_{IJK}$, $\vec{V}_{PQW}$ = velocity vectors in the $\hat{I}\hat{J}\hat{K}$ and $\hat{P}\hat{Q}\hat{W}$ systems

[A]   = combined transformation matrix from $\hat{P}\hat{Q}\hat{W}$ to $\hat{I}\hat{J}\hat{K}$

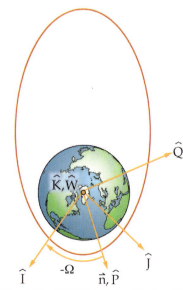

Figure C-9.   Rotation 3 of $-\Omega$ about the new $\hat{W}/\hat{K}$ axis.

$$[A] = [ROT3(-\Omega)][ROT1(-i)][ROT3(-\omega)] = \begin{bmatrix} A_{11} & A_{12} & A_{13} \\ A_{21} & A_{22} & A_{23} \\ A_{31} & A_{32} & A_{33} \end{bmatrix}$$

where

$A_{11} = \cos\Omega \, \cos\omega - \sin\Omega \, \sin\omega \, \cos i$

$A_{12} = -\cos\Omega \, \sin\omega - \sin\Omega \, \cos\omega \, \cos i$

$A_{13} = \sin\Omega \, \sin i$

$A_{21} = \sin\Omega \, \cos\omega + \cos\Omega \, \sin\omega \, \cos i$

$A_{22} = -\sin\Omega \, \sin\omega + \cos\Omega \, \cos\omega \, \cos i$

$A_{23} = -\cos\Omega \, \sin i$

$A_{31} = \sin\omega \, \sin i$

$A_{32} = \cos\omega \, \sin i$

$A_{33} = \cos i$

We can rotate any vector in the $\hat{P}\hat{Q}\hat{W}$ coordinate system to the $\hat{I}\hat{J}\hat{K}$ system using this [A] matrix, as long as we know the COEs.

C.8 $V_{burnout}$ in SEZ Coordinates

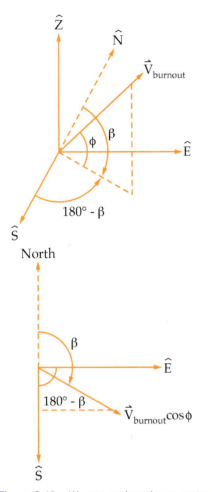

Figure C-10. We can analyze the geometry of the launch site with respect to known burnout conditions, such as flight-path angle, ϕ, and launch azimuth, β, to determine what the burnout velocity vector should be.

Using the flight-path angle, ϕ, and the launch azimuth angle, β, we can derive the components for the burnout velocity, $\vec{V}_{burnout}$, as shown in Figure C-10. We measure ϕ from the horizon to the velocity vector and β clockwise from due north to the projection of the velocity vector on the horizontal plane. For the zenith component

$$V_{burnout_{zenith}} = V_{burnout} \sin\phi$$

To get the south and east components, we must project the magnitude of the burnout velocity onto the horizontal plane, again using the flight-path angle, ϕ

$$V_{burnout_{projection}} = V_{burnout} \cos\phi$$

Then, using this result and the azimuth angle, β, we get the east component

$$V_{burnout_{east}} = V_{burnout} \cos\phi \sin(180° - \beta)$$

We can simplify this equation by using the trigonometric identity

$$\sin(180° - \beta) = \sin\beta$$

$$V_{burnout_{east}} = V_{burnout} \cos\phi \sin\beta$$

And for the south component, we get

$$V_{burnout_{south}} = V_{burnout} \cos\phi \cos(180° - \beta)$$

Then simplifying with the trigonometric identity

$$\cos(180° - \beta) = -\cos\beta$$

$$V_{burnout_{south}} = -V_{burnout} \cos\phi \cos\beta$$

C.9 Deriving the Rocket Equation

Newton's Second Law states that the sum of the external forces on an object equals its change in momentum, or, in equation form

$$\sum F_{external} = d\frac{(mV)}{dt} = d\frac{(p)}{dt}$$

where

F = forces on an object (N)
m = mass of the object (kg)
V = object's velocity (m/s)
p = object's linear momentum (kg m/s)
t   = time (s)
d = differential operator (unitless)

Consider a rocket expelling mass at a constant velocity, V_{exit}, relative to the vehicle. If we ignore gravity, drag, and air pressure, no external forces act on the body.

$$\therefore \sum F_{external} = 0 = d\frac{(p)}{dt}$$

But the real question is, "what is the momentum, p, of the rocket?"

$$p(t) = \underbrace{m(t)V(t)}_{(1)} + \underbrace{\int_{o}^{t} dm(t)(V(t) - V_{exit})}_{(2)} \qquad (C\text{-}30)$$

where

(1) = current momentum of the rocket
(2) = sum of the momentum associated with all of the mass ejected as combustion products from time zero to the present

If we assume the rocket, shown in Figure C-11, burns fuel at some rate, $\dot{m}$, the time rate of change of the vehicle mass is $\dot{m} < 0$ (negative because the vehicle is losing mass). But dm (t) = $-\dot{m}$ dt > 0 (the mass of the fuel expelled by the rocket is increasing with time). Now, we substitute $\dot{m}$ dt for dm in Equation (C-30)

$$\therefore p(t) = m(t)V(t) - \int_{0}^{t} \dot{m}(V(t) - V_{exit})dt \qquad (C\text{-}31)$$

Differentiating

$$\frac{dp(t)}{dt} = \cancel{\dot{m}(t)V(t)} + m(t)\dot{V}(t) - \cancel{\dot{m}(t)V(t)} + \dot{m}(t)V_{exit} = 0$$

Notice $\dot{m}(t)V(t)$ terms cancel, so we have

$$\therefore m(t)\dot{V}(t) = -\dot{m}(t)V_{exit}$$

Figure   C-11.   Absolute   Velocity. The absolute velocity of the ejected mass, dm (t) is $V(t) - V_{exit}$.

We divide both sides by m(t) to get

$$\dot{V}(t) = -\frac{\dot{m}(t)}{m(t)}V_{exit} = -\frac{\dot{m}}{m}V_{exit}$$

Integrating both sides from initial conditions to final conditions gives us

$$\int_{V_o}^{V_f} dV = \int_{m_o}^{m_f} \frac{-dm}{m} V_{exit}$$

$$\therefore (V_f - V_o) = \Delta V = (-V_{exit} \ln m]_{m_o}^{m_f})$$

We evaluate the natural logarithm of m at the initial and find masses to get

$$\Delta V = V_{exit} \ln \frac{m_o}{m_f}$$

From Chapter 14, $V_{exit} = C$, the effective exhaust velocity, so we have

$$\boxed{\Delta V = C \ln\left(\frac{m_o}{m_f}\right)} \tag{C-32}$$

This is the rocket equation, which gives the velocity change due to burning fuel $(m_o - m_f)$ and exhausting the combustion products at a speed of C, the effective exhaust velocity.

C.10 Deriving the Potential Energy Equation and Discovering the Potential Energy Well

We start with a basic physics principle that in a conservative energy field (such as gravity), the amount of work done raising or lowering an object equals the integral of the force vector, $\vec{F}$, dotted with the differential distance vector, $d\vec{s}$. We also know that the work done raising or lowering an object equals its change in potential energy, ΔPE. So, in equation form, we say

$$W = \int \vec{F} \cdot d\vec{s} = -\Delta PE \qquad \text{(C-33)}$$

where

W   = work done in a conservative field (Nm)
$\vec{F}$ = applied force vector (N)
$d\vec{s}$ = object's differential displacement vector (m)
ΔPE = object's change in potential energy (Nm)

If we limit the object's motion to one dimension (up and down), we get

$$W_z = \int F_z dz = -\Delta PE_z$$

Thus in a gravity field, using our two-body assumptions from Chapter 4, we have

$$W = \int \vec{F}_g d\vec{R} \qquad \text{(C-34)}$$

where

$\vec{R}$ = object's position vector from Earth's center (km)

Now, we need to use Newton's Law of Universal Gravitation to get an expression for $\vec{F}_g$

$$\vec{F}_g = \frac{-Gm_1\,m_2}{R^2}\hat{R}$$

Next, we arbitrarily choose up as positive, so $\vec{F}_g$ points down, which is negative.

Near Earth, we combine Gm_1 and call it μ_{Earth}, so we have

$$\vec{F}_g = \frac{-\mu m_2}{R^2}\hat{R}$$

Now, we substitute our simplified expression for $\vec{F}_g$ into Equation (C-34), and get

$$W = \int \frac{-\mu m_2}{R^2} dR$$

which acts along the $\hat{R}$ direction.

When we integrate, we get

$$W = \frac{\mu m_2}{R} + C$$

And the potential energy change is

$$\Delta PE = -W = \frac{-\mu m_2}{R} - C \qquad \text{(C-35)}$$

where

C = constant of integration (Nm)

If we ever want to compute a value for ΔPE, we must establish a reference point, so we can set a value for C.

In astronautics, we choose to set potential energy equal to zero at $R = \infty$. When we substitute these values into Equation (C-35), we get

$$0 = \left. \frac{-\mu m_2}{R} \right|_{\lim R \to \infty} - C$$

When we take the limit of $\frac{-\mu m_2}{R}$ as R approaches infinity, that term is zero. So, $C = 0$ for our chosen reference condition. Thus, the potential energy in Equation (C-35) becomes

$$PE = \frac{-\mu m_2}{R}$$

So, starting at $R = \infty$, where PE = 0, for every other finite radius, the potential energy is negative, growing more negative toward Earth's center, where the expression does not yield a finite answer.

Solar and Planetary Data

D.1 Physical Properties of the Sun

Table D-1. Physical Properties of the Sun. [Larson & Wertz, *Space Mission Analysis and Design*, 1999.]

Quantity	Value
Radius of the photosphere	6.95508×10^5 km
Angular diameter of the photosphere at 1 AU	0.53313 deg
Mass	1.9891×10^{30} kg
Mean density	1.410 g/cm^3
Total radiation emitted	3.826×10^{26} J/s
Total radiation per unit area at 1 AU	1358 W/m^2
Apparent visual magnitude at 1 AU	−26.75
Absolute visual magnitude (magnitude at distance of 10 parsecs)	+4.82
Color index, B-V	+0.65
Spectral type	G2 V
Effective temperature	5777 K
Inclination of the equator to the ecliptic	7.25 deg
Adopted period of sidereal rotation ($\phi = 17°$)	25.38 days
Corresponding synodic rotation period (relative to Earth)	27.345 days
Mean sunspot period	11.04 years
Dates of former maxima	1968.9, 1980.0, 1989.6
Mean time from maximum to subsequent minimum	6.2 years

Outline

D.1 Physical Properties of the Sun

D.2 Physical Properties of the Earth

D.3 Physical Properties of the Moon

D.4 Planetary Data

D.5 Spheres of Influence for the Planets

D.2 Physical Properties of the Earth

Table D-2. Physical Properties of the Earth. [Larson & Wertz, *Space Mission Analysis and Design*, 1999.]

Quantity	Value
Equatorial radius, a	6378.13649 km
Flattening factor (ellipticity), $f \equiv (a - c)/a$	$1/298.25642 \approx 0.00335282$
Polar radius,[*] c	6356.7517 km
Mean radius,[*] $(a^2 c)^{1/3}$	6371.00 km
Eccentricity,[*] $(a^2 - c^2)^{1/2}/a$	0.081819301
Surface area	5.100657×10^8 km^2
Volume	1.083207×10^{12} km^3
Ellipticity of the equator $(a_{max} - a_{min})/a_{mean}$	$\sim 1.6 \times 10^{-5}$
Longitude of the maxima	14.805° W, 165.105° E
Ratio of the mass of the Sun to the mass of the Earth	332945.9
Gravitational parameter, $Gm_{Earth} \equiv \mu_{Earth}$	$3.986004418 \times 10^{14}$ m^3/s^2
Mass of Earth	5.9737×10^{24} kg
Mean density	5.5548 g/cm^3
Gravitational field constants	$J2 = +1.08263 \times 10^{-3}$ $J3 = -2.53231 \times 10^{-6}$ $J4 = -1.62043 \times 10^{-6}$
Mean distance of Earth center from Earth-Moon barycenter	4671 km
Average lengthening of the day	0.0015 s/century

[*] Based on adopted values of f and a.

D.3 Physical Properties of the Moon

Table D-3. Physical Properties of the Moon. [Adapted from *The Astronomical Almanac*, Nautical Almanac Office, U.S. Naval Observatory, Government printing office, 2000, except where noted.]

Quantity	Value
Equatorial radius	1737.4 km
Surface area	37.9×10^6 km^2 [*]
Ratio of the mass of the Moon to the mass of the Earth	0.0123
Mass of the Moon	7.3483×10^{22} kg
Mean density	3.34 g/cm^3
Gravitational field parameters	$J2 = +0.2027 \times 10^{-3}$
Semimajor axis of lunar orbit	384,400 km
Gravitational parameter, $Gm_{Moon} \equiv \mu_{Moon}$	4.902794×10^3 km^3/s^2 [**]
Sidereal orbit period	27.321661 solar days
Sidereal rotation period	27.321661 solar days
Orbital eccentricity	0.054900489
Orbital inclination with respect to Earth's equator	$18.28° - 28.58°$

[*] Heiken, Grant H., David T. Vaniman, Bevan M. French. *Lunar Sourcebook*. Cambridge, U.K.: Cambridge University Press, 1991.

[**] Wertz, James R. and Wiley J. Larson (ed.). *Space Mission Analysis and Design*. Dordrecht, Netherlands: Kluwer Academic Publishers, 1999.

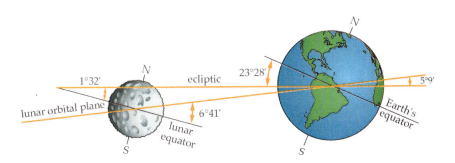

Figure D-1. Relationship Between Inclinations of the Earth and Moon.

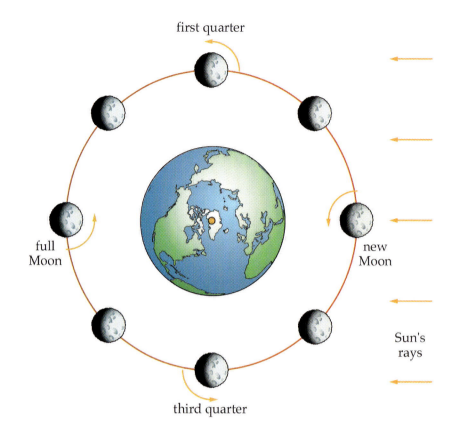

Figure D-2.   Revolution of the Moon Around the Earth.

D.4 Planetary Data

Table D-4.   Planetary Data. [Adapted from *The Astronomical Almanac*, Nautical Almanac Office, U.S. Naval Observatory, Government printing office, 2000, except where noted.]

Planet	Mean Equatorial Radius (km)	Mass (kg)	Distance from Sun (AU)	Orbital Period (Years)	Orbital Eccentricity	Orbital Inclination (deg)	Atmosphere[*]	Surface Gravity[*] Earth=1g	Gravitational Parameter μ[*] (km^3/s^2)
Mercury	2439.7	3.3022×10^{23}	0.387	0.241	0.206	7.0	None	0.352	2.094×10^4
Venus	6051.8	4.869×10^{24}	0.723	0.615	0.007	3.39	CO_2	0.8874	3.249×10^5
Earth	6378.14	5.9742×10^{24}	1.0	1.0	0.017	0	$N_2 + O_2$	1.0	3.986×10^5
Mars	3397.2	6.4191×10^{23}	1.524	1.881	0.093	1.85	CO_2	0.37	4.269×10^4
Jupiter	71,492	1.8988×10^{27}	5.204	11.862	0.049	1.30	$H_2 + H_e$	N/A	1.267×10^8
Saturn	60,268	5.685×10^{26}	9.582	29.458	0.056	2.49	$CH_4 + NH_3$	N/A	3.7967×10^7
Uranus	25,559	8.6625×10^{25}	19.20	84.014	0.046	0.77	$H_2 + H_e$	N/A	5.7918×10^6
Neptune	24,764	1.0278×10^{26}	30.05	164.79	0.011	1.77	$H_2 + H_e$	N/A	6.806×10^6
Pluto	1195	1.5×10^{22}	39.24	247.7	0.244	17.2	thin CH_4	0.0603	798.04

[*] Values from McGraw-Hill Encyclopedia of Science and Technology. 7th ed., Vol. 13. New York, NY: McGraw-Hill, Inc., 1992.
Note: 1 AU = 149.6×10^6 km.

D.5 Spheres of Influence for the Planets

Table D-5. **Sphere of Influence Radii for the Planets.** (Computed using method from Chapter 7.)

Planet	Distance from Sun ($\times 10^6$ km)	Mass of Planet ($\times 10^{24}$ kg)	Radius of SOI ($\times 10^3$ km)
Mercury	57.9	0.33022	112
Venus	108.2	4.869	616
Earth	149.6	5.9742	925
Mars	228.0	0.64191	577
Jupiter	778.5	1898.8	48,216
Saturn	1433.4	568.5	54,802
Uranus	2872.3	86.625	51,740
Neptune	4495.4	102.78	86,710
Pluto	5870.2	0.015	3308

Table D-6. **Synodic Periods for Missions from Earth to the Planets.**

Planet	Synodic Period (Years)
Mercury	0.32
Venus	1.60
Mars	2.13
Jupiter	1.09
Saturn	1.04
Uranus	1.01
Neptune	1.01
Pluto	1.00

References

Allen, C. W. *Astrophysical Quantities*. Third Edition. London, England: The Athlene Press, 1973.

American Ephemeris and Nautical Almanac. London, England: Her Majesty's Stationery Office, 1961.

H. M. Nautical Almanac Office. *Explanatory Supplement to the Astronomical Ephemeris*. London, England: Her Majesty's Stationery Office, 1961.

Larsen, Dennis G. and Richard Holdaway, eds. *The Astronomical Almanac, 2000*. Nautical Almanac Office, U.S. Naval Observatory, and H. M. Nautical Almanac Office. Washington, D.C.: U.S. Government Printing Office, 2000.

Hartman, William K. *Moon and Planets*. Belmont, CA: Wadsworth, Inc., 1983.

Hedgley, David R., Jr. *An Exact Transformation from Geocentric to Geodetic Coordinates for Nonzero Altitudes*. NASA TRR-458, Flight Research Center, 1976.

Hedman, Edward L., Jr. A High Accuracy Relationship Between Geocentric Cartesian coordinates and Geodetic Latitude and Altitude. *J. Spacecraft*. 7: 993-995, 1970.

Heiken, Grant, David Vaniman, and Bevan M. French. *Lunar Sourcebook*. Cambridge, U.K.: Cambridge University Press, 1991.

Muller, Edith A. and Jappel Arndst, eds. *International Astronomical Union Proceedings of the Sixteenth General Assembly, Grenoble, 1976*. Dordrecht, Holland: D. Reidel Publishing Co., 1977.

Wertz, James R. and Wiley J. Larson. *Space Mission Analysis and Design*. Third edition. Dordrecht, Netherlands: Kluwer Academic Publishers, 1999.

Motion of Ballistic Vehicles

E.1 Equation of Motion

Ballistic trajectories are the paths followed by nonthrusting objects, such as baseballs or intercontinental ballistic missiles (ICBMs), moving under the influence of gravity. We can use the geocentric-equatorial coordinate system to describe this motion. The equation of motion is

$$\ddot{\vec{R}} + \frac{\mu}{R^2}\hat{R} = 0 \tag{E-1}$$

using three assumptions

- Most of the trajectory is outside Earth's atmosphere: $\vec{F}_{drag} = 0$

- Start time at burnout: $\vec{F}_{thrust} = 0$

- Other forces are negligible compared to gravity: $\vec{F}_{other} = 0$

A ballistic trajectory, like an orbit, is defined by six initial conditions (ICs)

- Radius at burnout, $R_{burnout}$
- Velocity at burnout, $V_{burnout}$
- Flight-path angle at burnout, $\phi_{burnout}$
- Azimuth angle at burnout, $\beta_{burnout}$
- Latitude at burnout, $L_{burnout}$
- Longitude at burnout, $l_{burnout}$

The shape of a ballistic trajectory is an ellipse which intersects Earth's surface at two points—launch and impact, as shown in Figure E-1.

Outline

E.1 **Equation of Motion**
Ground-track Geometry
Trajectory Geometry
Maximum Range
Time of Flight
Rotating-Earth Correction
Error Analysis

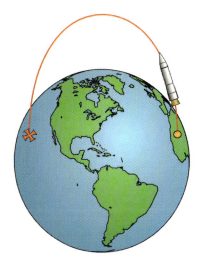

Figure E-1. Basic Trajectory for an ICBM.

Ground-track Geometry

The ground track of a ballistic trajectory is the arc of a great circle. To determine this range angle, Λ, from the launcher to the target, we start with L_o and l_o, the latitude and longitude of the launcher, and L_t and l_t, the latitude and longitude of the target. Then, using spherical trigonometry, as shown in Figure E-2, we get

$$\cos\Lambda = \sin L_o \sin L_t + \cos L_o \cos L_t \cos\Delta l \qquad \text{(E-2)}$$

where

$$\Delta l = l_t - l_o$$

Note that this setup assumes that Earth isn't rotating. This equation gives two values for the range angle, Λ. The smaller one, Λ, is the short way around the Earth; the larger value, $360° - \Lambda$, is for the long way. To convert from a range angle in degrees to range in kilometers, multiply by 10,000 km/90°. We show the range-angle geometry in Figure E-2.

Figure E-2.   Ballistic Trajectories. To visualize ballistic trajectories, it's helpful to slice earth open like an apple, revealing a launch site (launch site latitude, L_o, target latitude, L_t, and the range angle, Λ. The range traces over Earth's surface.

One of the initial conditions to locate the trajectory is the burnout azimuth angle, β. We measure this angle clockwise from true north to the trajectory, as shown in Figure E-3.

$$\cos\beta = \frac{\sin L_t - \sin L_o \cos\Lambda}{\cos L_o \sin\Lambda} \qquad \text{(E-3)}$$

A polar plot of the trajectory is useful to determine the correct quadrant for the azimuth angle, as shown in Figure E-4.

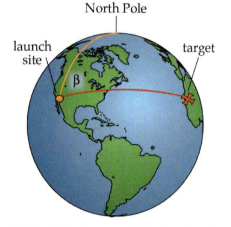

Figure E-3. Launch Azimuth. Launch azimuth, β, is the angle measured from true north at the launch site, clockwise to the launch direction.

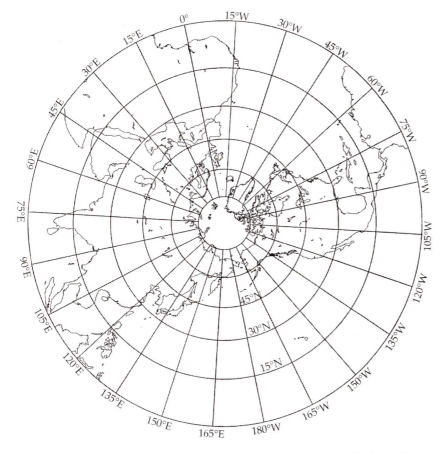

Figure E-4.   Polar Plot. Polar plots help us visualize the ground tracks of ballistic objects.

Trajectory Geometry

Define a trajectory parameter

$$Q_{burnout} = \frac{V_{burnout}^2}{V_{circular}^2} = \frac{V_{burnout}^2 R_{burnout}}{\mu}$$ (E-4)

- $Q_{burnout} < 1.0$: This restricts the booster to go only the short way to a target. Because most ballistic rockets use the short way to a target, they need a $Q_{burnout}$ less than one.

- $Q_{burnout} \geq 1.0$: This implies $V_{burnout} \geq V_{circular}$, which means the rocket can place a payload into orbit at a radius, $R_{burnout}$. This also means the booster can reach any point on Earth using either the short or long way.

- $Q_{burnout} = 2.0$: This implies $V_{burnout} = \sqrt{\dfrac{2\mu}{R_{burnout}}}$. This is Earth's escape velocity. A booster with $Q_{burnout} \geq 2.0$ would leave on an escape trajectory (parabolic or hyperbolic) and would therefore be useless for getting from one point on Earth to another.

Another of the angles describing the trajectory is the flight-path angle, ϕ, defined as the angle between the local horizontal and the velocity vector. Based on the value of the trajectory parameter, $Q_{burnout}$, the ballistic vehicle can follow either a high or low path, as Figure E-5 shows. The following figure shows the various paths available

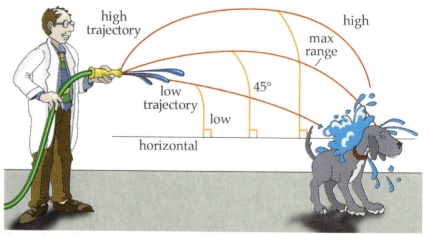

Figure E-5. Flight-path Angle and Trajectory. Whether you're squirting a hose or launching a missile, the effect of flight-path angle is the same. Maximum range is achieved with a flight-path angle of 45° (for very short trajectories). Two other angles will get you to the same spot—a low, direct trajectory or a high, arcing trajectory.

To solve for the flight-path angle

$$\phi_{burnout_{low}} = \frac{1}{2}\left\{ \sin^{-1}\left[\left(\frac{2 - Q_{burnout}}{Q_{burnout}}\right) \sin\frac{\Lambda}{2} \right] - \frac{\Lambda}{2} \right\}$$ (E-5)

This angle results in a low, direct trajectory.

$$\phi_{burnout_{high}} = \frac{1}{2}\left\{180° - \sin^{-1}\left[\left(\frac{2-Q_{burnout}}{Q_{burnout}}\right)\sin\frac{\Lambda}{2}\right] - \frac{\Lambda}{2}\right\} \quad (E-6)$$

This angle gives a high, arcing trajectory.

Maximum Range

For a specified value of the trajectory parameter, we can determine the maximum range achievable for that ballistic vehicle. Given the trajectory parameter, $Q_{burnout}$

$$\Lambda_{max} = 2\sin^{-1}\left(\frac{Q_{burnout}}{2-Q_{burnout}}\right) \quad (E-7)$$

To find the flight-path angle for launch to achieve the maximum range angle, Λ_{max}

$$(\phi_{burnout})_{max\ range} = 45° - \frac{\Lambda_{max}}{4} \quad (E-8)$$

Similarly, to avoid over-designing a missile, we can solve for the minimum value of $Q_{burnout}$ needed to reach some range angle, Λ

$$Q_{burnout_{min}} = \frac{2\sin\frac{\Lambda}{2}}{1+\sin\frac{\Lambda}{2}} \quad (E-9)$$

Time of Flight

Time of flight can be determined in two ways. The first way was previously discussed in Chapter 8. This involves using the definition of the ballistic trajectory as an elliptical path and solving Kepler's equation. The second method uses two charts based on these equations.

Figure E-6 shows a chart that relates the ratio of time of flight (TOF) to the period of a circular orbit and to the total range angle, Λ. The graph also contains lines for the trajectory parameter, $Q_{burnout}$, and for the flight-path angle, ϕ.

To find the TOF for the trajectory, you must first compute the range angle, Λ, and have the value for the radius at burnout, $R_{burnout}$. In Figure E-6, the vertical axis is a ratio, $TOF/P_{circular}$. We earlier defined $Q_{burnout}$ as the ratio of the square of $V_{burnout}$ to the square of $V_{circular}$. Similarly, we set up a ratio of the TOF of a trajectory to the period of a circular orbit at that radius of burnout. Let's step through how we use this chart to find TOF

- Find the value of Λ on the horizontal axis. Move vertically until you intersect the given value of $Q_{burnout}$ for the problem. The values for $Q_{burnout}$ are in increments of 0.05, so if your value is between curves, you must estimate.

- Find the intersection with the appropriate $Q_{burnout}$ curve to get two possibilities—a high and low trajectory. The one above the max range line is the high path, and the one below is the low path.

- Estimate the value for flight-path angle from the lines for ϕ. These lines are in 10° increments, so you may need to interpolate.

- Move left to the vertical axis to find the value of the ratio TOF/$P_{circular}$.

- Find the appropriate value of the circular orbit period at $R_{burnout}$ by using the equation for period

$$P_{circular} = 2\pi\sqrt{\frac{R_{burnout}^3}{\mu}}$$

Be careful of units. The value of $P_{circular}$ needs to be in minutes.

- Multiply the ratio by the circular orbit period to find

$$TOF = \left(\frac{TOF}{P_{circular}}\right)P_{circular}$$

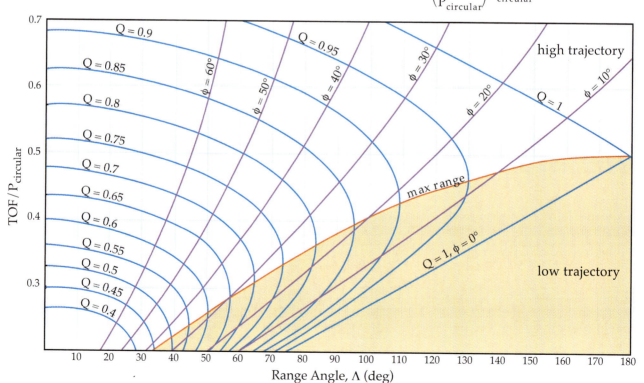

Figure E-6.   Range Angle in Degrees Versus TOF/P$_{circular}$.

Recall that all trajectories with $Q_{burnout} < 1.0$ have two options: a high trajectory and a low trajectory. We must solve each case separately for the time of flight.

Rotating-Earth Correction

Earth's rotation at $15°/hr$ has the following effect on trajectories

- Eastward launches
  - Target moves away from the launcher
  - Range angle, Λ, *increases* (from nonrotating solution)
  - Flight-path angle, ϕ, must *increase* on the *low* trajectory and *decrease* on the *high* trajectory
- Westward launches
  - Target moves toward the launcher
  - Range angle, Λ, *decreases* (from nonrotating solution)
  - Flight-path angle, ϕ, must *decrease* on the *low* trajectory and *increase* on the *high* trajectory

To account for the rotation we adjust the range-angle equation to

$$\cos \Lambda = \sin L_o \sin L_t + \cos L_o \cos L_t \cos(\Delta l + \omega_{Earth} TOF) \quad \text{(E-10)}$$

We can't solve directly for the range angle, Λ, so we must start with a "guess" and then iterate until we reach a solution. We're given L_o and l_o, as well as L_t and l_t. To find Λ, β, ϕ, we must go through the following algorithm

- Solve for the nonrotating range angle
$$\cos \Lambda = \sin L_o \sin L_t + \cos L_o \cos L_t \cos \Delta l$$
- Find the time of flight, TOF, with $\Lambda_{nonrotating}$ by using the chart in Figure E-6 for high trajectory
- Plug this time into the rotating range-angle equation and get a value for $\Lambda_{rotating}$
$$\cos \Lambda = \sin L_o \sin L_t + \cos L_o \cos L_t \cos(\Delta l + \omega_{Earth} TOF)$$
- Compute a new TOF using the new value of $\Lambda_{rotating}$
- Repeat the last two steps until the difference between successive values of range angle is small enough (usually $0.5°$ to $1°$)
- Solve for β and ϕ for the chosen trajectory
- Repeat all of the above for low trajectory

Error Analysis

Errors in any of the six initial conditions for a ballistic trajectory will cause it to miss the target. We categorize how we miss the target in terms

Figure E-7.   Conventions for Error Analysis.

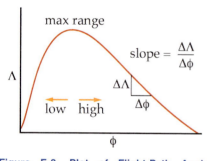

Figure E-8. Plot of Flight-Path Angle Versus Range Angle.

of downrange errors (in the direction of motion) and crossrange errors (perpendicular to the direction of motion). We use the following sign convention, as shown in Figure E-7.

- Landing long or to the right of the target is a positive error
- Landing short or to the left of the target is a negative error

Downrange error ($\Delta\Lambda$) has three causes

- Burning out at higher or lower altitude

$$\Delta\Lambda = \left(\frac{\partial\Lambda}{\partial R_{burnout}}\right)\Delta R_{burnout}$$

$$\frac{\partial\Lambda}{\partial R_{burnout}} = \frac{4\mu}{V_{burnout}^2 R_{burnout}^2}\frac{\sin^2\frac{\Lambda}{2}}{\sin 2\phi}\frac{180°}{\pi}\frac{deg}{km} \qquad (E\text{-}11)$$

- Burning out at higher or lower velocity

$$\Delta\Lambda = \left(\frac{\partial\Lambda}{\partial V_{burnout}}\right)\Delta V_{burnout}$$

$$\frac{\partial\Lambda}{\partial V_{burnout}} = \frac{8\mu}{(V_{burnout})^3(R_{burnout})}\frac{\sin^2\frac{\Lambda}{2}}{\sin 2\phi}\frac{180°}{\pi}\frac{deg}{m/s} \qquad (E\text{-}12)$$

- Burning out at a different flight-path angle. This is also shown in Figure E-8

$$\Delta\Lambda = \left(\frac{\partial\Lambda}{\partial\phi_{burnout}}\right)\Delta\phi_{burnout}$$

$$\frac{\partial\Lambda}{\partial\phi_{burnout}} = \frac{2\sin(2\phi_{burnout}+\Lambda)}{\sin(2\phi_{burnout})}-2\frac{deg}{deg} \qquad (E\text{-}13)$$

These three ratios showing the change in range angle due to a change in some initial condition are called influence coefficients. Because they are rather complicated to compute, we can use estimates of the influence coefficients called rule-of-thumb values.

Crossrange errors (ΔC) have two causes

- Displacing the launch site left or right of the trajectory

$$\Delta C = (\Delta y_{burnout})\cos\Lambda \qquad (E\text{-}14)$$

- Burning out with a larger or a smaller azimuth angle

$$\Delta C = \Delta\beta\left(111.1\frac{km}{deg}\right)\sin\Lambda \qquad (E\text{-}15)$$

References

Bate, Roger R., Donald D. Mueller, and Jerry E. White. *Fundamentals of Astrodynamics.* New York, N.Y.: Dover Publications, Inc., 1971.

Answers to Numerical Mission Problems

Chapter 1
Space in Our Lives

None

Chapter 2
Exploring Space

None

Chapter 3
The Space Environment

None

Chapter 4
Understanding Orbits

8) $H = 0.25 \text{ kg} \cdot \text{m}^2/\text{s}$

9) $H = 0.006283 \text{ kg} \cdot \text{m}^2/\text{s}$

10) $F_g = 0.05336 \text{ N}$

11) $g = 9.722 \text{ m}/\text{s}^2$

12) $V = 76.669 \text{ m}/\text{s}$

   $t = 7.825 \text{ s}$

23) a) $a = 7565.5 \text{ km}$

   b) $e = 0.1074$

   c) $R = 8374.926 \text{ km}$

      $\text{Alt} = 1996.926 \text{ km}$

   d) Negative

25) b) $\vec{h} = -49{,}441 \ \hat{K} \ \text{km}^2/\text{s}$

   d) $\varepsilon = -31.999 \ \text{km}^2/\text{s}^2$

26) $KE_{truck} = 0.844 \text{ kg} \cdot \text{km}^2/\text{s}^2$

   $V_{space} = 7.473 \text{ km}/\text{s}$

   $KE_{space} = 279{,}249 \text{ kg} \cdot \text{km}^2/\text{s}^2$

27) $\text{Alt} = 35{,}863.08 \text{ km}$

30) a) $V_{circ} = 3.993 \text{ km}/\text{s}$

   b) $\Delta V_{esc} = 1.654 \text{ km}/\text{s}$

   c) $\Delta KE = 7.972 \text{ km}^2/\text{s}^2$

Chapter 5
Describing Orbits

4) $\varepsilon = -4.73 \dfrac{\text{km}^2}{\text{s}^2}$

16) a) $\vec{h} = -40{,}307.38\hat{I} + 50{,}036.88\hat{J} - 6920.24\hat{K}\dfrac{\text{km}^2}{\text{s}}$

   b) $i = 96.15°$

   c) $\bar{n} = -50{,}036.88\hat{I} - 40{,}307.38\hat{J} \ (\text{km}^2/\text{s})$

   d) $\Omega = 218.85°$

17) a) $\bar{e} = 0.135\hat{I} + 0.092\hat{J} - 0.120\hat{K}$

   b) $\omega = 216.7°$

   c) $\nu = 319.3°$

22) a) $\Delta N = 240°$

   b) $\text{Lat}_{max \text{ and } min} = 25°$

23) a) $i \approx 50°$

   b) $P = 3 \text{ hrs}$

Chapter 6
Maneuvering in Space

5) a) $\varepsilon_{transfer} = -29.73 \text{ km}^2/\text{s}^2$

   b) $\Delta V_1 = 0.1 \text{ km}/\text{s}$

   c) $\Delta V_2 = 0.11 \text{ km}/\text{s}$

   d) $\text{TOF} = 45.5 \text{ min}$

10) $\Delta V_{simple} = 4.45 \text{ km}/\text{s}$

11) $\Delta V_{simple} = 4.71 \text{ km}/\text{s}$

12) a) $\varepsilon_{transfer} = -12.11 \text{ km}^2/\text{s}^2$

b) $\Delta V_1 = 2.08 \text{ km}/\text{s}$

c) $\Delta V_{combined} = 1.70 \text{ km}/\text{s}$

14) a) TOF = 2642.6 s

b) $\omega_{interceptor} = 0.001205 \text{ rad}/\text{s}$

$\omega_{target} = 0.001173 \text{ rad}/\text{s}$

c) $\alpha_{lead} = 3.10 \text{ rad}$

d) $\phi_{final} = 0.0436 \text{ rad}$

e) wait time = 19.8 hrs

15) a) TOF = 5877 s

b) a = 7038.8 km

c) $\Delta V_1 = 0.23 \text{ km}/\text{s}$

Chapter 7
Interplanetary Travel

8) $\Delta V = 2.32 \text{ km}/\text{s}$

9) a) $a_{transfer} = 1.289 \times 10^8 \text{ km}$

$\varepsilon_{transfer} = -514.7 \text{ km}^2/\text{s}^2$

b) $V_{\infty \text{ Earth}} = 2.49 \text{ km}/\text{s}$

c) $V_{\infty \text{ Venus}} = 2.72 \text{ km}/\text{s}$

d) $\Delta V_{boost} = 3.5 \text{ km}/\text{s}$

e) $\Delta V_{retro} = 3.32 \text{ km}/\text{s}$

f) $\Delta V_{mission} = 6.82 \text{ km}/\text{s}$

10) TOF = 146.1 days

11) Radius of SOI = 66,183 km

12) $\phi_{final} = 106°$

Chapter 8
Predicting Orbits

2) TOF = 1390.09 s = 23.17 min

3) TOF = 21.96 min

4) u = 155.6°

5) $\nu_{future} = 157.39°$

9) Time = 816.67 days

10) $\dot{\Omega} \cong -8.1 \text{ deg}/\text{day}$, westward

11) $\nu_{future} = 233.1°$

$a_{future} = 6876 \text{ km}$

$e_{future} = 0.0288$

$i_{future} = 75°$

$\Omega_{future} = 0°$

$\omega_{future} = 0°$

Chapter 9
Getting to Orbit

7) LST = 0300 hrs

8) LST = 0320 hrs

12) 1) One

2) Two

14) a) LWST = 135°

b) LWST = 0900 hrs

c) Wait 21 hrs

17) e) $\alpha = 30°$

f) $\gamma = 60.24°$

g) $\delta = 6.91°$

h) $LWST_{AN} = 0928$ hrs

j) $\beta_{AN} = 60.24°$

k) $LWST_{AN} = 2032$ hrs

l) 6 hrs and 2 min

m) $\beta_{DN} = 119.76°$

18) Launch at 2036 $\beta_{AN} = 46.17°$

19) a) 13°

b) 15 hrs, 21 min, and 36 s

21) 408.7 m/s

24) a) $\left| \vec{V}_{burnout} \right| = 7451.8$ m/s

b) $\vec{V}_{launch\ site} = 408.7$ m/s $\hat{E}$

c) $\vec{V}_{needed} = \begin{bmatrix} 0 \\ 7043.1 \\ 3732.4 \end{bmatrix}$ m/s

d) $\left| \vec{V}_{needed} \right| = 7971$ m/s

e) $\Delta V_{design} = 8971$ m/s

25) $\Delta V_{design} = 8558.2$ m/s

Chapter 10
Returning From Space: Re-entry

11) 24.813 g's, alt = 28,251 m

12) 20,348 m

18) 118,391 W/m^2

21) 0.34 g's

Chapter 11
Space Systems Engineering

11) 6×10^{14} Hz

16) $\lambda_m = 3.574$ μm

$E = 2.452 \times 10^4$ W/m^2

23) $1.22e^{-6}$ rad $= 6.99e^{-5}$ deg

24) h = 8.1967×10^3 km

Chapter 12
Space Vehicle Control Systems

9) Torque = 2.0 Nm

$\Delta H = 2.0$ kg m^2/s^2

12 1.273×10^{-4} Nm

13 1.811×10^{-4} N

14 0.077 N

30) $a_g = -9.6765$ m/s^2

34) $\vec{V}_{to\ go} = \begin{bmatrix} 35.11 \\ 7439 \\ 14.4 \end{bmatrix}_{SEZ}$ m/s

Chapter 13
Spacecraft Subsystems

13) 2.1×10^7 bits/s

19) $V_R = 300$ V

20) $i = 17.86$ A

23) $P_{OUT} = 288.6$ W/m^2

25) $TOE_{max} = 36.3$ min

29) 888.9 W hr

36) orbit average power = 220 W

    TE = 36.5 min

    battery capacity = 243.4 W hr

42) $q = -4.94$ W

43) $q = -9.69$ W

44) $T = 213.43$ K = $-59.57°$ C

45) $T = 203.37$ K = $-69.6°$ C

61) $\sigma = 127{,}389$ N/m^2

62) $\varepsilon = 0.0001$

63) $M = 2000$ N $\cdot$ m

64) $\varepsilon = 0.2$ ($41°$ C)

    -0.2 ($1°$ C)

66) $f_{natural} = 0.0159$ Hz

67) $K = 24.67 \times 10^6$ N/m

69) $\varepsilon = 2 \times 10^{-9}$

Chapter 14
Rockets and Launch Vehicles

5) $P_J = 1.296$ MW

6) $\Delta p = 100$ Ns

8) $\Delta t = 1000$ s   10 N engine

    = 5000 s   2 N engine

    = 10,000 s   1 N engine

9) $F = 15$ N

    $I_{sp} = 3.06$ s

    $\Delta V = 12.16$ m/s

11) $\Delta V = 20.1$ m/s

12) $m_{propellant} = 9.95$ kg

18) $\rho = 116.1$ kg/m^3

21) $V = 0.637$ m/s

    $V_{kink} = 2.546$ m/s

22) $a_o = 352.4$ m/s

    Mach $1.3 = 458.1$ m/s

27) $D_t = 0.0177$ m

28) $\Delta t = 356.5$ s

29) $I_{sp_{ideal}} = 383.9$ s

    $a_o = 1263$ m/s

    $C^* = 1747$ m/s

33) From Figure 14-20: I_{sp} no change; Thrust increases

34) I_{sp}, F, and C^* decrease

39) Savings of 111.2 kg

47) $m_{oxidizer} = 800$ kg

59) F = 860.1 N

$a = 8.60 \times 10^{-4} \text{ m/s}^2$

61) $R_{apogee} = 460$ km

62) 1.667 years on Earth

66) $Alt_{optimum} = 33{,}333$ m

69) a) 151,500 kg

b) 101,500 kg

c) 1178.75 m/s

d) 91,500 kg

e) 51,500 kg

f) 1973.43 m/s

g) 44,000 kg

h) 9000 kg

i) 6227.25 m/s

j) 9379.43 m/s

70) $m_{i_{tot}} = 71{,}644.7$ kg

71) Yes, because $\Delta V_{tot} > \Delta V_{design}$

Chapter 15
Space Operations

10) $F = 9081 \text{ W/m}^2$

14) $\dfrac{S}{N} = 6.07 \times 10^{11}$ ($\eta = 55\%$), yes

8.27×10^{11} ($\eta = 75\%$)

Chapter 16
Using Space

None

Index

—A—

ablation   346
ablative cooling   346
absorbed energy   385
absorptivity   484
accelerating charge   618
acceleration   118, 510
accelerometer   435, 439
acceptable operating ranges, ECLSS   373
accuracy, re-entry   341
acoustic chamber   611
acoustic load   512
action and reaction   115
active actuators   422
active sensor   391
active thermal control   485
actuator   404, 437
acute dosages, radiation   94
Adams, John Couch   41
adiabatic flow   541
advance science and technology   663
aerobraking   350, 352
aerodynamic drag   416
Air Force Satellite Control Network (AFSCN) 623
albedo   482
Aldrin, Buzz   49
Alfonsine Tables   34
algebra   34
Almagest   34
alternating current   465
altitude   309
altitude of maximum deceleration   341
altitude of maximum heating rate   341
amortize   672
amperes   463
amplitude   451, 618
amplitude modulation   451, 456
analog data   455
analog to digital conversion   456
Andromeda   76
anechoic chamber   457, 612
angle of attack   350
angle of incidence   466
angular momentum   112, 123, 140, 412, 426
angular momentum vector   419

angular resolution   389
angular velocity   112, 211, 412, 426
anode   469
aperture   389
apogee   135, 141, 156
Apollo program   49
    Apollo 1 mission   489
    Apollo 13 mission   608, 651
    Apollo Applications Program (AAP)   287
    Apollo capsule   324
    Apollo mission   151
    Apollo-Soyuz mission   50
apparent solar day   293
application software   454
arcjet rocket   572
argument of latitude   164, 262
argument of perigee   156, 159, 161, 171, 274
    quadrant for   172
Ariane IV   60
Ariane Structure for Auxiliary Payloads
    (ASAP)   60
Ariane V   17, 591
Arianespace   60, 657
Aristotle   33
Armstrong, Neil   49
arrival at the target planet   227
ascending node   158, 162, 181
ascending node vector   169
ascending-node launch opportunity   300
assembly drawing   516
assembly, integration, and test (AIT)   513, 608, 634
asteroids   84
astrodynamics   80
Astrolabe   34
astrolabe   34
astrology   33, 266
astronautics   11, 13
astronauts   491
astronomy   33, 266
Atlas   590
atmosphere   73, 81
atmospheric
    density   81
    drag   273, 413
    pressure   81

re-entry 324
   windows 385
atomic oxygen 82
attitude accuracy 409
attitude and orbit control 370
attitude and orbit control subsystem
   (AOCS) 371, 406, 589
attitude control 370
attitude determination and control
   subsystem (ADCS) 407, 430
attitude dynamics 412
attitude-control budget 371
Aurora Borealis 88
autonomy 646
average mean motion 281
axial load 501
azimuth 261

—B—

backward pass 643
ballistic coefficient (BC) 330, 342, 345
ballistic missile 45
bandwidth 622
barbecue mode 485
batteries
   nickel-cadmium 469
   nickel-hydrogen 469
   primary 469
   secondary 469
Becquerel, Henri 42
beginning-of-life 467
bending moment 502
Bernoulli Principle 543
Bernoulli, Daniel 543
biased momentum systems 427
bicycle wheels in space 426
bioregenerative life-support subsystems 493
bipropellant rockets 565
bit 453
bitflip 88
black body 387
black-body radiation 622
block diagrams 403
body frame 408
body-mounted solar array 466
Boeing Sea Launch 310
Boltzmann's constant 622

Booster Officer 636
bow shock 87
Brahe, Tycho 36
bremsstrahlung radiation 95
brittle 507
Bruno, Giordano 36
Buckley Air National Guard Base 614
burnout velocity 309, 313
bytes 453

—C—

calculus 114
Callisto 55
cameras 389
capillary wick system 487
capital market acceptance 656, 660
carbon dioxide 489
carrier signal 451, 618
Cartwheel Galaxy 109
Cassini mission 27, 56
cathode 469
Centaurus A 56
center of gravity 513
central processing unit (CPU) 453
CERISE mission 84
Challenger 62
chamber pressure 551
chamber temperature 551
Chandra X-ray Observatory 3, 56
characteristic exhaust velocity 547
charge 462
charge-coupled device (CCD) 390
charged particles 74, 75, 554
charging 88
Charon 57
check valves 563
chemical energy 468
chemical rockets 564
chemical-energy systems 468
chlorofluorocarbons 387
choked flow 544
circle 106, 134
circuits 462, 464
circular equatorial orbit 165
circular orbit 164
circular trajectory 106
Clarke, Arthur C. 352

classic orbital elements (COEs) 155, 161, 261, 272, 616
   alternates 165
clean room 610, 672
Clementine mission 58
Clementine spacecraft 669
closed-loop control system 405, 408
co-apsidal orbits 194
coefficient of drag 329
coefficient of thermal expansion 504
cold welding 82, 83
cold-gas rocket 550
   trade-offs 552
Columbia space vehicle 290
combined plane change 203, 205
combined-cycle propulsion systems 595
combustion chamber 545, 564
comet 36, 84
commands, spacecraft 616
commercial off-the-shelf (COTS) 671
commercial spending 657
commercialization of space 59, 657
commissioning 637
communicate 2
communication 6, 614, 617
communication and data handling 589
communication and data-handling subsystem (CDHS) 371, 449, 618
communication architecture 614
communication components 452
communication networks 623
compass 87
composite material 508
composite structure 509
composites 508
compression load 501
compressive load 501
computational fluid dynamics (CFD) 342
computer aided design (CAD) 514
computer-aided manufacturing equipment 610
condenser 487
conduction 83, 337, 482
conductive heat transfer 337
conductor 464, 483
configuration control 514
conflict resolution 639
Congreve, William 44
conic section 106, 134, 225

conservation of angular momentum 420
conservation of linear momentum 534
conservation of momentum 123
conservative field 124, 140
constellation 16
constraints 632
contact forces 80
control 434
control center 614
control systems 403
controller 404, 429, 438
control-moment gyroscope (CMG) 427, 429
convection 83, 337, 483
convective heat transfer 337
co-orbital rendezvous 213
coordinate system 107, 131
   fundamental plane 131
   origin 131
   principal direction 131
Copernicus, Nicolaus 32, 35
coplaner orbits 194
corridor width 341
cosmic year 76
cosmonauts 490
cost 631
cost estimating 675
cost estimating relationships (CERs) 676
cost versus reliability 671
coulomb 462
Coulomb's Law 462
critical path 644
Critical-path Method (CPM) 641
cross product 169
crosslink 615
cryogenic 565
cryogenic coolers 488
cyrogenic propellants 562

—D—

damper 424
damping 506
data 450
   budget 457
   handling 453
   rate 457
   sampling 456
data budgets 372

data processing and handling — 632, 637
data rate — 617
debris, space — 84
deceleration profiles — 336, 343
decibel — 512
decision-making process — 639
decreased hydrostatic gradient — 91
Deep Space Network antennas — 257
deep-space tracking network (DSN) — 623
Defense Satellite Communication System — 6
Defense Support Program
  (DSP) mission — 488, 614, 624
degree — 36
Delta II — 568
delta-V — 194
  boost — 240
  mission — 245
  retro — 244
demodulator — 452
density specific impulse — 537
density, atmospheric — 81
Department of Transportation (DOT) — 666
depth-of-discharge (DOD) — 469
deregulation — 657
  telecommunications — 657
derivative control — 430
descending node — 158
descending-node launch opportunity — 300
design and manufacturing teams — 632
design velocity — 314
design-for-manufacture — 375, 633
design-to-cost — 365, 633, 679
Dewar flasks — 488
digital data — 455
dinosaurs — 331
dipole — 416
direct current — 465
direct energy transfer — 472
direct orbit — 157, 183
direct solar input — 481
directional antenna — 619
disturbance torques — 413
Dnepr — 60
Dobson Unit (DU) — 11
docking — 208, 219
dot product — 168
downlink — 614
downrange angle — 309

drag
  atmospheric — 81, 132, 273, 329
  coefficient — 417
  torque — 417
dry weight — 498
dual-spin spacecraft — 424
dual-spin systems — 423
ductile — 507
dust, cosmic — 84
duty cycle — 475
dynamic envelope — 511
dynamic loads — 501
dynamic pressure — 329, 587

—E—

earliest-finish (EF) time — 643
earliest-start (ES) time — 643
early-warning satellites — 7
earned value — 645
Earth
  angular radius — 467
  curvature — 106
  departure — 227
  escaping — 238
  infrared — 481
  magnetic field — 413
  rotation rate — 293
  sensors — 418
  shine — 481
Earth Science Enterprise mission — 659
Earth shine — 386
eccentric anomaly — 264
eccentricity — 37, 136, 157, 161
eccentricity vector — 168
Echo I — 6, 48
eclipse time — 468
ecliptic plane — 223
economics — 654, 668
edema — 92
Edgeworth-Kuiper Disk — 57
effective exhaust velocity — 534, 537, 553
effective isotropic radiated power (EIRP) — 620
effective program performance indicators — 645
Einstein, Albert — 42
elastic region — 506
elastically — 506

electric 75
    current 463
electric field 463, 553, 618
electrical
    charge 540
    potential 463, 553
    power 590
electrical power subsystem (EPS) 372, 448, 461
    example 474
    solar powered 472
electricity 462
electrodes 468
electrodynamic 539
    acceleration 540
    energy 540
    rockets 540, 574, 576
electrolyte 468
electromagnetic
    energy 382
    radiation 74, 93, 384, 386, 618
    spectrum 3, 74, 382
electrons 75
electrostatic force 462, 553
electrostatic thruster 574
elevation angle 261
ellipse 106, 134
emerging markets 656
emission 346
emissivity 346, 485
Endeavour 290
end-of-life (EOL) 467, 474
energy 124, 384
energy sources 465
engineering models 672
engineering performance data 632
environmental control and life support 372
environmental control and life-support
    subsystem (ECLSS) 373, 448, 480
      acceptable operating ranges 373
environmental testing 611
epicycles 34
equants 34
equation of motion 108, 134
equatorial orbit 157
error analysis 108
error signal 430
escape trajectory 137
estimating mission 668

Europa 55
European Space Agency (ESA) 55
evaporator 487
Evolved Expendable Launch Vehicle (EELV) 60
exotic propulsion systems 578
expansion ratio 547, 552
exploration 6
Explorer 1 mission 47, 88
extravehicular activity (EVA) 219, 489

—F—

Far Infrared and Submillimetre Telescope
    (FIRST) 57
FASat-A spacecraft 656, 674
faster, better, cheaper 57
Federal Communication Commission
    (FCC) 618, 666, 683
feedback control system 405
field-of-view (FOV) 16, 388
final frontier 2, 3
final phase angle 212
FireSat 364
    drawing 514
    mission 382
    operations concept 365
    propulsion system 563
First Point of Aries 132
flares, solar 86
flash evaporator 486
flight
    control 632, 634, 637
    controllers 636
    director 21, 636
    model 672
    rules 637
    spare 672
Flight Dynamics Officer 636
flight-control team 18, 635, 636
flight-path angle 135, 136, 309, 313
flight-readiness review 635
fluid shift 91
focal length 389
foci 135
focus 38, 135
food in space 491
forest fires from space 382
forward link 614

forward pass   643
Foucault Pendulum   420
Fourier, J. B. J.   483
Fourier's Law   483
free fall   4, 40, 80, 91, 488
frequency   384, 623
frequency allocation   666
frequency modulation   618
frozen flow   541
fuel   564
fuel cells   470
functional testing   611
fundamental frequency   505, 509
fundamental plane   131
future missions   10

—G—

Gagarin, Yuri A.   48
gain   620
galactic cosmic rays (GCRs)   86, 93
Galilei, Galileo   32, 39, 118
Galileo mission   27, 53, 55
Galileo spacecraft   254
Gamma Ray Observatory   3
gamma rays   85
Gemini mission   193, 208, 219
geocentric   33
    model   34
geocentric-equatorial coordinate system   131, 157
Geographic Information Systems (GIS)   659
geometrical parameters   135
geostatic   33
geostationary orbit   164, 182
geosynchronous orbit   164, 182
gimbal   420
Glenn, John   48, 53
g-load   587
global cellular phone service   59
global perspective   3
Global Positioning System (GPS)   9, 16, 43, 59, 61,
                          368, 421, 445, 589, 658
global space industry   658
Global Surveyor mission, Mars   57
globalization   656
GLONASS   9
Goddard, Robert H.   45
government   654

spending   657
gravitational acceleration   105, 325
gravitational constant   117
gravitational parameter   118, 156
gravity   110, 132
    assist   27, 222, 252, 254
    force of   117
    gradient   413, 414
    turn   309
    well   313
Gravity Probe B   436
gravity-gradient stabilization   422
gravity-gradient torque   414
great circle   179
greenhouse effect   387
Greenwich Mean Time (GMT)   294
Greenwich Meridian   132
Greenwich, England   294
ground resolution versus angular resolution   390
ground stations   614
ground support equipment (GSE)   611
ground tracks   154, 179
ground-system specialist   638
guidance   434, 438
Guidance and Procedures Officer   636
gyroscopes   419
gyroscopic stiffness   413

—H—

half-life   470
Hall-effect thruster (HET)   575
hardened   86
heat pipes   487
heat sink   345
heat sources   481
heat transfer   82, 482
heating rate   338, 344
heliocentric   35, 223
heliocentric transfer orbit   238
heliocentric-ecliptic coordinate system   223
Herschel, William   41
Hohmann Transfer   194, 222, 230–234
Hohmann, Walter   194
home-heating system   404
homogeneous   508
HST Optical Systems Test platform   21

Hubble Space Telescope (HST)   3, 10, 52, 56,
                                  192, 529
    thermal-induced vibrations   504
Hubble, Edwin   42
Huggins, William   41
human system   489
Huygens mission   56
hybrid rocket   568
hybrid-propulsion systems   568
hydrazine   565
hydrogen peroxide   565
hydrostatic gradient   91
hydroxyl-terminated polybutadiene   567
hyperbola   106, 134
hyperbolic
    Earth departure   238
    excess velocity   238
    planetary arrival   243
    trajectory   238
hyperbolic-departure trajectory   238
hypergolic   565

—I—

Ida   55
ideal rocket equation   538
IKONOS   189
impulse   535
impulse bits   551
impulsive burn   195
inclination   156–158, 161, 168, 182, 203, 298
inclination auxiliary angle   300, 303
indirect orbit   157
industry   654
inertia   110
inertial navigation system   437
inertial reference frame   130
information   450
initial conditions   108
input/output (I/O) devices   453
integral control   430
Integrated Communications Officer (INCO) 638
integrated testing   457
intelligence, surveillance, and reconnaissance
    (ISR)   61
Intercontinental Ballistic Missile (ICBM) 326
    re-entry   335
internal rate of return (IRR)   677, 679

internal thermal control   486
International Astronomical Union (IAU)   235
International Celestial Reference Frame (ICRF) 132
International Extreme Ultraviolet Hitchhiker
    Experiment (IEUHE) mission   21
international space law   665
International Space Station   11
International Space Station (ISS)   54, 360, 663
International Space Station (ISS) mission 10
International Telecommunication Union (ITU) 452
International Telecommunications Union (ITU) 666
interplanetary rendezvous   249
interplanetary transfer   222
interstellar travel   580
intimate-contact devices   492
Io   27, 55
ion   540
    engines   575
    thruster   574
Iridium System   660
    business   683
    mission objectives   683
    mission   16, 59, 366
    mission statement   683
    space manufacturing   684
isentropic   542
    flow   541
isotropic   619
iteration   265

—J—

J2 effect   274
jet power   535
Jupiter   10, 27, 55

—K—

K-1 launch vehicle   60
Kapton™   486
Kennedy launch site   613
Kennedy Space Center (KSC)   203, 298, 311
Kennedy, John F.   48, 151
Kepler, Johannes   32, 37, 155, 266
Kepler's Equation   265
Kepler's Laws
    Kepler's First Law   38
    Kepler's Second Law   38
    Kepler's Third Law   38

Kepler's Problem   261, 280
kinetic energy   124, 126, 141, 314, 326, 535
Kistler Aerospace   60
Kodiak Island launch site   666
Korolev, Sergei P.   47
Kourou, French Guyana   310

—L—

Lagrange Libration Points (LLP)   241
Landsat mission   7, 51, 399
latent heat of fusion   346, 487
latent heat of vaporization   487
lateral load   501
latest-finish (LF) time   644
latest-start (LS) time   644
launch   609, 612, 635
    ascending-node opportunity   300
    azimuth   302
    costs   673
    descending-node opportunity   300
    director   635
    environment   510
    geometry   303
    loads   510
    opportunity   299
    services   658
    site   292
    site latitude   303
    team   631
    time   292
    vehicle   17, 308
    vehicle ascent, phases of   308
    vehicle envelope   511
    vehicles   586
        Ariane   17, 60, 591
        Atlas   590
        Delta II   568
        K-1   605
        Pegasus   60, 661
        Proton   54, 60
        Titan   56, 60
        Tsyklon   60
        Vanguard   46
    window   291, 298, 308
launch-control team   635
launch-direction auxiliary angle   301, 303
launch-readiness review   635

launch-site latitude   301
launch-site longitude   295
launch-vehicle subsystems   586
launch-window location angle   301, 303
launch-window sidereal time (LWST)   299, 301
launch-window situations   304
laws   662
lead angle   211
level of cohesiveness   639
Leverrier, Urbain   41
Liability Convention   665
life cycle   669
life-cycle cost   669
    design   669
    launch   669
    manufacture   669
    operations   669
life-support budget   493
light year   42, 76
line of nodes   158
linear aerospike nozzle   588
linear momentum   111
lines of latitude   179
lines of longitude   179
link budget   372, 457
liquid-chemical rockets   565
lithium hydroxide canisters   492
loads, types of   501
local sidereal time (LST)   294, 299
logical relationships   642
Long Duration Exposure Facility (LDEF) mission   82
longitude of perigee   165
low-cost launch opportunities   656
Lunar Excursion Module (LEM)   587
Lunar Prospector mission   57
Lunar-orbit rendezvous   208

—M—

Mach Number   544
Magellan mission   10, 55
    mapper   257
    Space Probe   257
magnetic field   75, 87, 618
magnetic field strength   416
magnetic torque   416
magnetic torquer   425

magnetometer 421
    assembly drawing 516
magnetopause 87
magnetosphere 87, 93
magnetotail 87
magnification 389
major axis 135
manned maneuvering unit (MMU) 551
manufacturing
    costs 672
    drawing 515
    team 633
map makers 8
Mariner mission 10, 51
Mars 10, 37
Mars Climate Orbiter mission 646
Mars Global Surveyor mission 353
Mars Observer mission 57, 563
Mars Pathfinder mission 53
Mars Polar Lander mission 58
mass 110
mass flow rate 534
mass moment of inertia 410, 426
mass properties 509, 512
Mathematical Principles of Natural
    Philosophy 41
maximum deceleration 336
    altitude of 341
maximum heating rate 338
    altitude of 341
Maxwell, James 618
mean anomaly 263
mean motion 262
mean solar day 293
mean solar time 293
mechanical
    behavior 500
    configuration 500
    energy 140
    interfaces 509
mechanisms 498
Mercator projection 179
Mercury 10
Mercury program 48
Meteor spacecraft 483
meteoroids 84
meteors 2, 331
micrometeoroid 84

Milky Way 2, 42, 76
Milstar mission 6
minor axis 135
minute of arc 36
Mir Space Station 50, 54
Miranda 27
mission 13
    analysis 376
    constraints 678
    director 21
    management 632, 638
    management and operations 19, 630
    managers 631
    objective 363, 639, 678
    operations center 614
    operations system 18, 609
    operations team 20
    planning and analysis 632
    politics 678
    statement 13, 363
    timeline 633
Mission Control Center (MCC), Houston 19
mission-operations team 636
model
    engineering 672
    flight 672
    test 672
MODEM 452
modulation 451
modulator 452
modulus of elasticity 507
molecular mass 549
Molniya orbit 164, 276
moment arm 113
moment of inertia 112, 123, 415, 513
momentum 111, 123
    dumping 429
    thrust 545
    wheels 427
momentum-control devices 426
monopropellant rocket 565
Moon 117
Moon Agreement 665
Morse code 451
Motion Analysis Process (MAP) 107, 130, 223, 328
motor 567
multi-layer insulation (MLI) 486
Mylar™ 486

—N—

nadir	34
national	663
national image	663
national security	663
natural frequency	504
navigation	6, 9, 658
sensor	435
navigation, guidance, and control (NGC)	589
navigation, guidance, and control (NGC) subsystem	434, 438
NAVSTAR	445
near infrared camera and multi-object spectrometer (NICMOS)	499
Near-Earth Asteroid Rendezvous (NEAR) mission	58
needed velocity	438
Neptune	10, 27, 32
network scheduling	641
neutral charge	462
neutrons	75
new industry leaders	656, 661
Newton, Isaac	40, 109
Newton's Laws	
Newton's First Law of Motion	111
Newton's Law of Universal Gravitation	116, 133, 228, 414, 463
Newton's Second Law of Motion	114, 224, 328, 411, 535
Newton's Third Law of Motion	115, 533
Newtonian flow	342
Next Generation Space Telescope (NGST)	56
nickel-cadmium battery	469
nickel-hydrogen battery	469
Nimbus 4	11
nitrogen	489
nitrogen tetroxide	565
nodal displacement	180
nodal regression rate	274
noise	510
non-spherical shape	260
North American Aerospace Defense Command (NORAD)	84
North American Aviation's X-15	46
North Pole	87
Northern Lights	88

nozzle	545
ideally expanded	546
over-expanded	546
under-expanded	546
nozzle design	586, 588
nozzle expansion ratio	588
nuclear energy	470
Nuclear Engine for Rocket Vehicle Applications (NERVA)	573
nuclear fusion	74
nuclear-thermal rocket	572, 573
Nyquist criteria	456

—O—

Oberth, Hermann J.	45
objectives	
advance science and technology	662
foreign policy	662
national image	662
national industries support	662
regional security	662
space mission	13, 662
oblateness	273
ocean tides	117
ohm	464
Ohm's Law	464
omnidirectional antenna	619
Onizuka Air Station	20
open-loop control system	404
operating-system software	454
operational scenarios	632
operations	609, 614
concept	13, 364, 678
costs	673
director	21, 636
operators	631, 636
orbit average power	475
orbit cranking	253
orbit pumping	253
orbital	
control	370, 434
maneuvers	192
motion	104
period	142
perturbations	272
plane	143, 292, 295
orbital maneuvering system (OMS)	335, 356

Orbital Sciences Corporation 661
orbital-control budget 371
orbital-prediction problem 260
orbital-slot assignments 666
orbits 15, 104
    circular 164
    circular equatorial 165
    direct 157
    equatorial 157
    geostationary 164, 182
    geosynchronous 164, 182
    indirect 157
    Molniya 164
    polar 157
    prograde 157
    retrograde 157
    semi-synchronous 164
    Sun-synchronous 164
OrbView 659
origin, coordinate system 131
orthostatic intolerance 92
Outer Space Treaty 665
out-gassing 83, 508
overshoot boundary 327, 344
oxidation 82
oxidizer 564
oxidizer/fuel ratio (O/F) 565
oxygen 489
ozone 11, 82, 93

—P—

Palapa A and B satellites 6
parabola 106, 134
parabolic trajectory 238
paraffin 487
parallax 35
parking orbit 17, 192
partial pressure 489
passive actuators 422
passive sensors 391
passive thermal control 485
patched-conic approximation 224, 227, 230
Pathfinder mission 10, 57
payload 15, 366, 368
    data 638
    design process 393
    sensors 388
    specialists 636

peak power 475
peak power tracking 472
Peenemuende 69
Pegasus launch vehicle 60, 661
performance 631
perigee 135, 141, 156
perigee rotation rate 275
period 38, 142, 181
peripheral vision 389
personal computer 453
perturbation 41, 261, 272
phase angle 212
phase-change material 487
phasing of the planets 248
phasing orbit 213
Philolaus 37
photons 86, 384, 387, 415
photovoltaic 465
photovoltaic cells 85
Pioneer mission 10, 581
pitch 408
pitch over 309
Planck, Max 57
Planck's black-body radiation curve 387
plane changes 203
plane-change angle 204
plant 403
plant model 405, 434
plasma 75, 554
plasma thrusters 575
Pluto 57, 235
pointing accuracy 409
polar equation of a conic 136
polar orbit 157
policies 662
politics 654, 662
PoSat-1 spacecraft 669
position 155
position vector 135
post-flight operations 615
potential 463
potential energy 124, 141, 314
power 535
    beginning-of-life 467
    budget 474
    conditioning and distribution 473
    end-of-life 467
    supply 472–474

power-flux density 619
precedence requirements 642
precession 274, 413
predicting orbits 280
President Kennedy's address to Congress 151
pressure
    atmospheric 81
    regulator 561
    thrust 545
    transducers 562
pressure-fed propellant system 561
pressure-relief valves 563
primary batteries 469
primary focus 135
primary payload 613
primary structure 499
Prime Meridian 294
principal direction 131
prograde orbit 157
Program Evaluation and Review Technique
    (PERT) 641
Project Phoenix 101
project-management techniques 640
propagate 262
propellant 533
    budget 374
    cryogenic 565
    hypergolic 565
    loading 578
    management 561
    storable 565
proportional control 430
proportional limit 506
propulsion subsystem 373
    FireSat 563
propulsion system block diagram 560
propulsion-system functions 532
Proton launch vehicle 54, 60
protons 75
Proxima Centauri 76
Ptolemy 32, 34
pulsed-plasma thrusters (PPT) 575
pump-fed delivery systems 562
pyrotechnic actuators 511

—Q—

quadrant check for true anomaly 173
qualification test 517
quasi-static load 511

—R—

radar 392
radiation 84, 337, 384, 484
radiative cooling 346, 347
radiative heat transfer 337
radiators 486
radio frequency 452
radioisotope thermoelectric generators
    (RTGs) 470–472
radius of apoapsis 135
radius of periapsis 135
RADs 93
random access memory (RAM) 453
range 261, 613
Ranger 7-9 missions 51
range-safety officer (RSO) 635
ratio of specific heats 543
reaction wheels 427
    operation 428
read only memory (ROM) 453
real time 615
real-time commands 616
real-time operations 615
received signal strength 621
receiver 617
Redstone rocket 47
redundancy 670
re-entry
    accuracy 325, 341
    coordinate system 328
    corridor 326, 340, 344
    deceleration 325
    design 341
    flight-path angle 328, 333
    heating 325
    profile 351
    velocity 333
reference frame 130
reflected energy 385
reflection 389
reflectivity 484
refraction 389

regional security	663
Registration Convention	665
regulations	662
relative biological effectiveness (RBE)	93
relative velocity	229
relativity	40
relay satellites	614
reliability	671
remote sensing	6, 382, 658
data	659
mission	383, 659
satellites	7
rendezvous	208, 210, 248
Lunar orbit	208
requirements	632
derived	367
trading	367
Rescue Agreement	665
residual strain	506
resistance	464
resistojet	571
resistor	464
resolution	389
resonance	505
restricted two-body equation of motion	134, 272
restricted two-body problem	130
retrograde orbit	157, 183, 311
return link	615
return on investment	677
right ascension of the ascending node	156, 158, 161, 170, 203, 274
quadrant for	171
right-hand rule	112, 131
ring-laser gyroscope	421
rocket	17, 437, 532
equation	535, 539
nozzle	541
propulsion	2
scientist	533
staging	591
testing	578
roentgen equivalent man (REM)	94

—S—

safe mode	616
safety factor	511
Sagan, Carl	77
Salyut space stations	321, 490
Satellite Pour l'Observation de la Terre (SPOT)	7, 638
satellites	39
Saturn	10, 27, 56
Saturn I	591
Saturn V	361
scan rate	389
schedule	631
Schriever Air Force Base	20, 62, 623
science and exploration	10
scintillation	3
Sea Launch platform	613
search for extraterrestrial intelligence (SETI)	101
secondary battery	469
secondary payload	613
secondary structure	499
semiconductor crystal	465
semimajor axis	136, 142, 156, 161, 167
semiminor axis	135, 136
semi-synchronous orbit	164
sensor	404
separation ring	511
shaker table	611
Shapley, Harlow	42
shear	503
shear stress	503
shock front	87
shock wave	337, 344
attached	337
detached	337
shocks	510, 511
Shoemaker-Levy 9	55
sidereal day	294
sidereal time	294, 296
signal	403, 617
signal strength	617
signal-to-noise ratio	617, 622
simple plane change	203
simplifying assumptions	108
simulation and training	632, 637
simulations	633
single event phenomenon (SEP)	88
single event upset (SEU)	88, 455
single-stage-to-orbit (SSTO)	594
Skylab mission	50, 82, 287
slack	644
slew rate	409

Sojourner mission   10, 57
solar
    cell efficiency   465
    cells   85, 465
    constant   415
    cycle, 11-year   273
    energy   3, 465
    flares   75
    particle events   75
    pressure   86
    radiation pressure   277
    sail   578
    simulation chamber   612
    system   10, 74
    time   296
    wind   75, 86
Solar and Heliospheric Observatory
   (SOHO) mission   56
solar-powered electrical power subsystem   472
solar-radiation pressure   413, 415
solar-thermal rockets   569, 570
solid-chemical rockets   566
solid-propellant grain designs   567
solid-rocket boosters   568
Somnium   39
South Atlantic Anomaly   95, 455
South Pole   87
south-east-zenith (SEZ) coordinate system   310
Soviet Salyut   321
space   73
    catalog   623
    debris   84
    food   491
    heat sources   481
    insurance   674
    junk   84
    loss   621
    manufacturing   80
    operations systems design and
        manufacturing   658
    qualification   670
    systems engineering process   377
    technology   2
    vehicle control   402
    vehicle dynamics   434
    walk   219
    water   491

space mission
    architecture   13, 364
    operations   608
    subject   367
    users   363
space missions
    Apollo   151
    Apollo-Soyuz   50
    Cassini   56
    CERISE   84
    Clementine   58
    Explorer 1   47, 88
    Galileo   27, 53
    Global Surveyor, Mars   57
    Hubble Space Telescope (HST)   3, 10
    Huygens   56
    International Extreme Ultraviolet
        Hitchhiker Experiment   21
    International Space Station (ISS)   10
    Iridium   16, 59
    Landsat   51
    Long Duration Exposure Facility (LDEF)   82
    Lunar Prospector   57
    Magellan   55
    Mariner   51
    Mars Observer   57
    Mars Pathfinder   53
    Mars Polar Lander   58
    Near-Earth Asteroid Rendezvous
        (NEAR)   58
    Pathfinder   10, 57
    Pioneer   10
    Rangers 7-9   51
    Skylab   50, 82
    Sojourner   10, 57
    Solar and Heliospheric Observatory
        (SOHO)   56
    Stardust   57
    Surveyor   51
    Transition Region and Coronal Explorer
        (TRACE)   56
    Ulysses   55
    Viking   10, 52
    Voyager   10, 27, 52, 76
    Wild-2   57
Space Mission Management and Operations   608

Space Shuttle   50, 54, 290, 325, 326, 356
    ascent   356
    Challenger   567
    computers   454
    de-orbit burn   357
    external tank (ET)   356
    main engine (SSME)   356, 562, 565
    orbiter   356
    radiators   486
    reaction control system (RCS)   356
    re-entry   357
    solid-rocket boosters (SRBs)   356
    Space Transportation System (STS)   356
    toilet   492
Space Surveillance Network (SSN)   623
space systems engineering process   448, 630
Space Transportation System (STS)   356
Space Warfare Center   62
space-based telescope   389
spacecraft   2, 13, 614
    assembly   513
    assembly, integration, and testing (AIT)   672
    attitude actuators   422
    attitude sensors   417
    bus   15, 366, 369
    charging   88
    Clementine   669
    commands   616
    control   370
    design and manufacturing   658
    design process   375
    dipole   416
    FASat-A   656, 674
    forces acting upon   133, 224
    Galileo   254
    manufacturing   610
    Meteor   483
    motion   104
    Pioneer 10   581
    pitch   408
    PoSat-1   669
    roll   408
    Sputnik   46
    subsystems   370
    testing   517
    thermal analysis techniques   494
    thermal control   485
    yaw   408

Spaceflight Tracking and Data Network (STDN)   623
space-funding profiles   657
space-qualified components   670
space-related products   660
space-related services   658, 660
Spartan 201 Solar Observer   21
spatial resolution   390
specific angular momentum   143, 157
specific angular momentum vector   170
specific enthalpy   542
specific gravity   538
specific impulse   536
specific mechanical energy   140, 142, 156,
                                     167, 195, 197, 231
    equation   230
spectroscopy   41
speed of light   75
sphere of influence (SOI)   228, 252
spherical triangle   300
spin rate   123
spin stabilization   423
Spot Image   657
spring constant   505
Sputnik II spacecraft   46
sputtering   88
stages   17, 510
standard atmospheric pressure   83
star sensor   419
Stardust mission   57
Starry Messenger   39
static envelope   511
static loads   501
steady flow   541
Stefan-Boltzmann's constant   484
Stefan-Boltzmann's equation   388
Stefan-Boltzmann's Law   484
stiffness   505, 509, 511
stoichiometric combination   565
storable propellants   565
stored commands   616
strain   503
strength   506, 509
stress   503
stress-strain curves   507
structural
    design   500
    fatigue   504
    load   499

structures and mechanisms	373, 498, 590
subject, space mission	15
sublunar realm	33
subsonic flow	544
subsystem specialists	636
Sun	10, 74
Sun sensor	419
Sun-centered transfer	227
sun-synchronous orbit	164, 275
sun-tracking array	466
superlunar realm	33
supernova	36
supersonic flow	544
support national industries	664
data	664
ground systems	664
hardware and software	664
information	664
launch vehicles	664
spacecraft	664
surface reflectance	415
Surrey Satellite Technology, Ltd.	60, 669
Surveyor mission	51
swath width	16, 389
synodic period	249
synthetic aperture radar (SAR)	257, 392
system	403
system assembly, integration, and testing (AIT)	632, 634
system block diagram	403
system maintenance and support	632, 637
systems engineering	361, 457, 632
systems engineering process	362, 610

—T—

tangential burn	195
tangential velocity	194, 309
tau factor	581
team leaders	639
team norms	639
technologies	
information	660
space	660
telecommunications	660
terrestrial	660
technology	654
telecommunications	658

telemetry	450, 615
tensile load	501
test models	672
testing the model	108
tether	580
deployment	579
experiment	580
orbital boost	580
theory of relativity	43
thermal	
conductivity	483
control	481
cycling	494
equilibrium	346, 480
node	494
vacuum facility	494
thermal-protection systems (TPS)	342, 345
thermal-vacuum chamber	83, 672
thermocouples	470
thermodynamic	539
energy	540
expansion	541
rocket	540, 550, 564
rocket comparison	570
thermoelectric rocket	571, 573
third body	132
third-body gravitational effects	277
throttling	587
thrust	132, 437, 533
thrust coefficient	549
thruster	18, 425
thrust-to-weight ratio	587
thrust-vector control (TVC)	586, 588
time dilation	581
time of flight (TOF)	198, 211, 213, 248, 262
Titan launch vehicle	56, 60
Toledan tables	34
topocentric-horizon frame	310
torque	411
torque applied to a spinning disk	413
torsional loads	501
total angular momentum	428
total energy	124, 199
total heat load	339
total impulse	536
total mechanical energy	124, 126, 325
Total Ozone Mapping Spectrometer (TOMS)	10

total rocket thrust 545
total system power budget 474
total velocity change 196
Tracking and Data Relay Satellites (TDRS) 623
tracking data 261
trajectory 15
transcendental equation 265, 282
transfer ellipse 231
transfer orbit 17, 195, 197
transfer time of flight 248
Transition Region and Coronal Explorer
  (TRACE) mission 56
transmissivity 484
transmitted energy 384
transmitter 617
Treaty of Versailles 69
trigonometry 34
true anomaly 135, 136, 156, 160, 172
true longitude 165
Tsiolkovsky, Konstantin E. 44
Tsyklon launch vehicle 60
twin paradox 581
two-body equation of motion 134
two-body problem 225

—U—

U.S. Air Force Academy 624
U.S. Space Command 623
ultimate failure point 506
ultimate strength 507
ultraviolet radiation 86, 385
Ulysses mission 55
undershoot boundary 327
unexpected thrusting 277
Universe 2, 10, 33
uplink 614
Upper Atmospheric Research Satellite
  (UARS) 486
upperstage 17
Uranus 10, 27, 32, 41
users 678
  space product 13

—V—

V infinity at Earth 232
V-2 46, 69

vacant focus 135
vacuum 82
vacuum phase 309
value added 659
Van Allen radiation belts 86, 88
Vandenberg Air Force Base launch site 613
Vanguard launch vehicle 46
vector 112
VEEGA (Venus, Earth, Earth Gravity Assist) 253, 254
vehicle shape
  accuracy, effect on 344
  deceleration, effect on 343
  heating rate, effect on 344
  re-entry corridor, effect on 344
velocity 155, 309
  change 196, 298, 538
  Earth 231
  needed 313
  transfer orbit 232
  vector 135
velocity change
  combined plane change 205
  simple plane change 204
velocity needed 313
Venera 1 51
venturi effect 542
Venus 10, 55, 392
vernal equinox direction 131, 223, 296
vertical ascent 308
vestibular functions 92
vibrations 504, 510
Viking mission 10, 52
voltage 463
volts 463
Vomit Comet 92
von Braun, Wernher 45
Voyager mission 10, 27, 52, 76, 77

—W—

wait time 212, 249
waste management 491
watch-dog timer 616
water in space 491
watt 464
wavelength 74, 384, 623
weather forecasts 8
weight 109

Wien's Displacement Law 387
Wild-2 mission 57
wiring harness 513
word 453
work breakdown structure (WBS) 640
World Administrative Radio Conference
    (WARC) 452, 683

—X—

X-33 61, 594
X-rays 85

—Y—

yaw 408
yield point 506
yield strength 507
Young's modulus 507
Yuri Gagarin 10

—Z—

Zarya module 54
Zenit launch vehicle 60
zenith 34
zero gravity 79
zero-bias system 427